Encyclopaedia of Mathematical Sciences
Volume 122

Operator Algebras and Non-Commutative Geometry III

Subseries Editors:
Joachim Cuntz Vaughan F.R. Jones

B. Blackadar

Operator Algebras

Theory of C*-Algebras and von Neumann Algebras

Bruce Blackadar
Department of Mathematics
University of Nevada, Reno
Reno, NV 89557
USA
e-mail: bruceb@math.unr.edu

Founding editor of the Encyclopedia of Mathematical Sciences:
R. V. Gamkrelidze

Library of Congress Control Number: 2005934456

Mathematics Subject Classification (2000):
46L05, 46L06, 46L07, 46L08, 46L09, 46L10, 46L30, 46L35, 46L40, 46L45, 46L51,
46L55, 46L80, 46L85, 46L87, 46L89, 19K14, 19K33

ISSN 0938-0396
ISBN-10   3-540-28486-9 Springer Berlin Heidelberg New York
ISBN-13   978-3-540-28486-4 Springer Berlin Heidelberg New York

This work is subject to copyright. All rights are reserved, whether the whole or part of the material is concerned, specifically the rights of translation, reprinting, reuse of illustrations, recitation, broadcasting, reproduction on microfilm or in any other way, and storage in data banks. Duplication of this publication or parts thereof is permitted only under the provisions of the German Copyright Law of September 9, 1965, in its current version, and permission for use must always be obtained from Springer. Violations are liable for prosecution under the German Copyright Law.

Springer is a part of Springer Science+Business Media
springer.com
© Springer-Verlag Berlin Heidelberg 2006
Printed in Germany

The use of general descriptive names, registered names, trademarks, etc. in this publication does not imply, even in the absence of a specific statement, that such names are exempt from the relevant protective laws and regulations and therefore free for general use.

Typesetting: by the author and TechBooks using a Springer LaTeX macro package

Cover design: *design & production* GmbH, Heidelberg

Printed on acid-free paper   SPIN: 11543510   46/TechBooks   5 4 3 2 1 0

to my parents
and
in memory of Gert K. Pedersen

Preface to the Encyclopaedia Subseries on Operator Algebras and Non-Commutative Geometry

The theory of von Neumann algebras was initiated in a series of papers by Murray and von Neumann in the 1930's and 1940's. A von Neumann algebra is a self-adjoint unital subalgebra M of the algebra of bounded operators of a Hilbert space which is closed in the weak operator topology. According to von Neumann's bicommutant theorem, M is closed in the weak operator topology if and only if it is equal to the commutant of its commutant. A *factor* is a von Neumann algebra with trivial centre and the work of Murray and von Neumann contained a reduction of all von Neumann algebras to factors and a classification of factors into types I, II and III.

C^*-algebras are self-adjoint operator algebras on Hilbert space which are closed in the norm topology. Their study was begun in the work of Gelfand and Naimark who showed that such algebras can be characterized abstractly as involutive Banach algebras, satisfying an algebraic relation connecting the norm and the involution. They also obtained the fundamental result that a commutative unital C^*-algebra is isomorphic to the algebra of complex valued continuous functions on a compact space – its spectrum.

Since then the subject of operator algebras has evolved into a huge mathematical endeavour interacting with almost every branch of mathematics and several areas of theoretical physics.

Up into the sixties much of the work on C^*-algebras was centered around representation theory and the study of C^*-algebras of type I (these algebras are characterized by the fact that they have a well behaved representation theory). Finite dimensional C^*-algebras are easily seen to be just direct sums of matrix algebras. However, by taking algebras which are closures in norm of finite dimensional algebras one obtains already a rich class of C^*-algebras – the so-called AF-algebras – which are not of type I. The idea of taking the closure of an inductive limit of finite-dimensional algebras had already appeared in the work of Murray-von Neumann who used it to construct a fundamental example of a factor of type II – the "hyperfinite" (nowadays also called approximately finite dimensional) factor.

One key to an understanding of the class of AF-algebras turned out to be K-theory. The techniques of K-theory, along with its dual, Ext-theory, also found immediate applications in the study of many new examples of C^*-algebras that arose in the end of the seventies. These examples include for instance "the noncommutative tori" or other crossed products of abelian C^*-algebras by groups of homeomorphisms and abstract C^*-algebras generated by isometries with certain relations, now known as the algebras $\mathcal{O}_n$. At the same time, examples of algebras were increasingly studied that codify data from differential geometry or from topological dynamical systems.

On the other hand, a little earlier in the seventies, the theory of von Neumann algebras underwent a vigorous growth after the discovery of a natural infinite family of pairwise nonisomorphic factors of type III and the advent of Tomita-Takesaki theory. This development culminated in Connes' great classification theorems for approximately finite dimensional ("injective") von Neumann algebras.

Perhaps the most significant area in which operator algebras have been used is mathematical physics, especially in quantum statistical mechanics and in the foundations of quantum field theory. Von Neumann explicitly mentioned quantum theory as one of his motivations for developing the theory of rings of operators and his foresight was confirmed in the algebraic quantum field theory proposed by Haag and Kastler. In this theory a von Neumann algebra is associated with each region of space-time, obeying certain axioms. The inductive limit of these von Neumann algebras is a C^*-algebra which contains a lot of information on the quantum field theory in question. This point of view was particularly successful in the analysis of superselection sectors.

In 1980 the subject of operator algebras was entirely covered in a single big three weeks meeting in Kingston Ontario. This meeting served as a review of the classification theorems for von Neumann algebras and the success of K-theory as a tool in C^*-algebras. But the meeting also contained a preview of what was to be an explosive growth in the field. The study of the von Neumann algebra of a foliation was being developed in the far more precise C^*-framework which would lead to index theorems for foliations incorporating techniques and ideas from many branches of mathematics hitherto unconnected with operator algebras.

Many of the new developments began in the decade following the Kingston meeting. On the C^*-side was Kasparov's KK-theory – the bivariant form of K-theory for which operator algebraic methods are absolutely essential. Cyclic cohomology was discovered through an analysis of the fine structure of extensions of C^*-algebras These ideas and many others were integrated into Connes' vast *Noncommutative Geometry* program. In cyclic theory and in connection with many other aspects of noncommutative geometry, the need for going beyond the class of C^*-algebras became apparent. Thanks to recent progress, both on the cyclic homology side as well as on the K-theory side, there is now a well developed bivariant K-theory and cyclic theory for a natural class of topological algebras as well as a bivariant character taking K-theory to cyclic theory. The 1990's also saw huge progress in the classification theory of nuclear C^*-algebras in terms of K-theoretic invariants, based on new insight into the structure of exact C^*-algebras.

On the von Neumann algebra side, the study of subfactors began in 1982 with the definition of the *index* of a subfactor in terms of the Murray-von Neumann theory and a result showing that the index was surprisingly restricted in its possible values. A rich theory was developed refining and clarifying the index. Surprising connections with knot theory, statistical mechanics and quantum field theory have been found. The superselection theory mentioned above turned out to have fascinating links to subfactor theory. The subfactors themselves were constructed in the representation theory of loop groups.

Beginning in the early 1980's Voiculescu initiated the theory of free probability and showed how to understand the free group von Neumann algebras in terms of random matrices, leading to the extraordinary result that the von Neumann algebra M of the free group on infinitelymany generators has full fundamental group, i.e. pMp is isomorphic to M for every non-zero projection $p \in M$. The subsequent introduction of free entropy led to the solution of more old problems in von Neumann algebras such as the lack of a Cartan subalgebra in the free group von Neumann algebras.

Many of the topics mentioned in the (obviously incomplete) list above have become large industries in their own right. So it is clear that a conference like the one in Kingston is no longer possible. Nevertheless the subject does retain a certain unity and sense of identity so we felt it appropriate and useful to create a series of encylopaedia volumes documenting the fundamentals of the theory and defining the current state of the subject.

In particular, our series will include volumes treating the essential technical results of C^*-algebra theory and von Neumann algebra theory including sections on noncommutative dynamical systems, entropy and derivations. It will include an account of K-theory and bivariant K-theory with applications and in particular the index theorem for foliations. Another volume will be devoted to cyclic homology and bivariant K-theory for topological algebras with applications to index theorems. On the von Neumann algebra side, we plan volumes on the structure of subfactors and on free probability and free entropy. Another volume shall be dedicated to the connections between operator algebras and quantum field theory.

October 2001

subseries editors:
Joachim Cuntz
Vaughan Jones

Preface

This volume attempts to give a comprehensive discussion of the theory of operator algebras (C*-algebras and von Neumann algebras.) The volume is intended to serve two purposes: to record the standard theory in the Encyclopedia of Mathematics, and to serve as an introduction and standard reference for the specialized volumes in the series on current research topics in the subject.

Since there are already numerous excellent treatises on various aspects of the subject, how does this volume make a significant addition to the literature, and how does it differ from the other books in the subject? In short, why another book on operator algebras?

The answer lies partly in the first paragraph above. More importantly, no other single reference covers all or even almost all of the material in this volume. I have tried to cover all of the main aspects of "standard" or "classical" operator algebra theory; the goal has been to be, well, encyclopedic. Of course, in a subject as vast as this one, authors must make highly subjective judgments as to what to include and what to omit, as well as what level of detail to include, and I have been guided as much by my own interests and prejudices as by the needs of the authors of the more specialized volumes.

A treatment of such a large body of material cannot be done at the detail level of a textbook in a reasonably-sized work, and this volume would not be suitable as a text and certainly does not replace the more detailed treatments of the subject. But neither is this volume simply a survey of the subject (a fine survey-level book is already available [Fil96].) My philosophy has been to not only state what is true, but explain why: while many proofs are merely outlined or even omitted, I have attempted to include enough detail and explanation to at least make all results plausible and to give the reader a sense of what material and level of difficulty is involved in each result. Where an argument can be given or summarized in just a few lines, it is usually included; longer arguments are usually omitted or only outlined. More detail has been included where results are particularly important or frequently used in the sequel, where the results or proofs are not found in standard references, and in the few cases where new arguments have been found. Nonetheless, throughout the volume the reader should expect to have to fill out compactly written arguments, or consult references giving expanded expositions.

I have concentrated on trying to give a clean and efficient exposition of the details of the theory, and have for the most part avoided general discussions of the nature of the subject, its importance, and its connections and applications in other parts of mathematics (and physics); these matters have been amply treated in the introductory article to this series. See the introduction to [Con94] for another excellent overview of the subject of operator algebras.

There is very little in this volume that is truly new, mainly some simplified proofs. I have tried to combine the best features of existing expositions and arguments, with a few modifications of my own here and there. In preparing this volume, I have had the pleasure of repeatedly reflecting on the outstanding talents of the many mathematicians who have brought this subject to its present advanced state, and the theory presented here is a monument to their collective skills and efforts.

Besides the unwitting assistance of the numerous authors who originally developed the theory and gave previous expositions, I have benefited from comments, suggestions, and technical assistance from a number of other specialists, notably G. Pedersen, N. C. Phillips, and S. Echterhoff, my colleagues B.-J. Kahng, V. Deaconu, and A. Kumjian, and many others who set me straight on various points either in person or by email. I am especially grateful to Marc Rieffel, who, in addition to giving me detailed comments on the manuscript, wrote an entire draft of Section II.10 on group C*-algebras and crossed products. Although I heavily modified his draft to bring it in line with the rest of the manuscript stylistically, elements of Marc's vision and refreshing writing style still show through. Of course, any errors or misstatements in the final version of this (and every other) section are entirely my responsibility.

Speaking of errors and misstatements, as far-fetched as the possibility may seem that any still remain after all the ones I have already fixed, I would appreciate hearing about them from readers, and I plan to post whatever I find out about on my website.

No book can start from scratch, and this book presupposes a level of knowledge roughly equivalent to a standard graduate course in functional analysis (plus its usual prerequisites in analysis, topology, and algebra.) In particular, the reader is assumed to know such standard theorems as the Hahn-Banach Theorem, the Krein-Milman Theorem, and the Riesz Representation Theorem (the Open Mapping Theorem, the Closed Graph Theorem, and the Uniform Boundedness Principle also fall into this category but are explicitly stated in the text sinced they are more directly connected with operator theory and operator algebras.) Most of the likely readers will have this background, or far more, and indeed it would be difficult to understand and appreciate the material without this much knowledge. Beginning with a quick treatment of the basics of Hilbert space and operator theory would seem to be the proper point of departure for a book of this sort, and the early sections will be useful even to specialists to set the stage for the work and establish notation and terminology.

September 2005 *Bruce Blackadar*

Contents

I	**Operators on Hilbert Space**		1
	I.1	Hilbert Space	1
		I.1.1 Inner Products	1
		I.1.2 Orthogonality	2
		I.1.3 Dual Spaces and Weak Topology	3
		I.1.4 Standard Constructions	4
		I.1.5 Real Hilbert Spaces	5
	I.2	Bounded Operators	5
		I.2.1 Bounded Operators on Normed Spaces	5
		I.2.2 Sesquilinear Forms	6
		I.2.3 Adjoint	7
		I.2.4 Self-Adjoint, Unitary, and Normal Operators	8
		I.2.5 Amplifications and Commutants	9
		I.2.6 Invertibility and Spectrum	10
	I.3	Other Topologies on $\mathcal{L}(\mathcal{H})$	13
		I.3.1 Strong and Weak Topologies	13
		I.3.2 Properties of the Topologies	14
	I.4	Functional Calculus	17
		I.4.1 Functional Calculus for Continuous Functions	18
		I.4.2 Square Roots of Positive Operators	19
		I.4.3 Functional Calculus for Borel Functions	19
	I.5	Projections	19
		I.5.1 Definitions and Basic Properties	20
		I.5.2 Support Projections and Polar Decomposition	21
	I.6	The Spectral Theorem	23
		I.6.1 Spectral Theorem for Bounded Self-Adjoint Operators	23
		I.6.2 Spectral Theorem for Normal Operators	25
	I.7	Unbounded Operators	27
		I.7.1 Densely Defined Operators	27
		I.7.2 Closed Operators and Adjoints	29

		I.7.3	Self-Adjoint Operators	30
		I.7.4	The Spectral Theorem and Functional Calculus for Unbounded Self-Adjoint Operators	32
	I.8	Compact Operators		36
		I.8.1	Definitions and Basic Properties	36
		I.8.2	The Calkin Algebra	37
		I.8.3	Fredholm Theory	37
		I.8.4	Spectral Properties of Compact Operators	40
		I.8.5	Trace-Class and Hilbert-Schmidt Operators	41
		I.8.6	Duals and Preduals, σ-Topologies	43
		I.8.7	Ideals of $\mathcal{L}(\mathcal{H})$	44
	I.9	Algebras of Operators		47
		I.9.1	Commutant and Bicommutant	47
		I.9.2	Other Properties	48
II	C*-Algebras			51
	II.1	Definitions and Elementary Facts		51
		II.1.1	Basic Definitions	51
		II.1.2	Unitization	53
		II.1.3	Power series, Inverses, and Holomorphic Functions	54
		II.1.4	Spectrum	54
		II.1.5	Holomorphic Functional Calculus	55
		II.1.6	Norm and Spectrum	57
	II.2	Commutative C*-Algebras and Continuous Functional Calculus		59
		II.2.1	Spectrum of a Commutative Banach Algebra	59
		II.2.2	Gelfand Transform	60
		II.2.3	Continuous Functional Calculus	61
	II.3	Positivity, Order, and Comparison Theory		63
		II.3.1	Positive Elements	63
		II.3.2	Polar Decomposition	67
		II.3.3	Comparison Theory for Projections	72
		II.3.4	Hereditary C*-Subalgebras and General Comparison Theory	75
	II.4	Approximate Units		79
		II.4.1	General Approximate Units	79
		II.4.2	Strictly Positive Elements and σ-Unital C*-Algebras	81
		II.4.3	Quasicentral Approximate Units	82
	II.5	Ideals, Quotients, and Homomorphisms		82
		II.5.1	Closed Ideals	83
		II.5.2	Nonclosed Ideals	85
		II.5.3	Left Ideals and Hereditary Subalgebras	89
		II.5.4	Prime and Simple C*-Algebras	93
		II.5.5	Homomorphisms and Automorphisms	95

II.6	States and Representations 100	
	II.6.1	Representations 101
	II.6.2	Positive Linear Functionals and States 103
	II.6.3	Extension and Existence of States 106
	II.6.4	The GNS Construction 107
	II.6.5	Primitive Ideal Space and Spectrum 111
	II.6.6	Matrix Algebras and Stable Algebras 116
	II.6.7	Weights 118
	II.6.8	Traces and Dimension Functions 121
	II.6.9	Completely Positive Maps 124
	II.6.10	Conditional Expectations 132
II.7	Hilbert Modules, Multiplier Algebras, and Morita Equivalence 137	
	II.7.1	Hilbert Modules 137
	II.7.2	Operators 141
	II.7.3	Multiplier Algebras 144
	II.7.4	Tensor Products of Hilbert Modules 147
	II.7.5	The Generalized Stinespring Theorem 149
	II.7.6	Morita Equivalence 150
II.8	Examples and Constructions 154	
	II.8.1	Direct Sums, Products, and Ultraproducts 154
	II.8.2	Inductive Limits 156
	II.8.3	Universal C*-Algebras and Free Products 158
	II.8.4	Extensions and Pullbacks 167
	II.8.5	C*-Algebras with Prescribed Properties 176
II.9	Tensor Products and Nuclearity 179	
	II.9.1	Algebraic and Spatial Tensor Products 180
	II.9.2	The Maximal Tensor Product 180
	II.9.3	States on Tensor Products 182
	II.9.4	Nuclear C*-Algebras 184
	II.9.5	Minimality of the Spatial Norm 186
	II.9.6	Homomorphisms and Ideals 187
	II.9.7	Tensor Products of Completely Positive Maps . . . 190
	II.9.8	Infinite Tensor Products 191
II.10	Group C*-Algebras and Crossed Products 192	
	II.10.1	Locally Compact Groups 193
	II.10.2	Group C*-Algebras 197
	II.10.3	Crossed products 199
	II.10.4	Transformation Group C*-Algebras 205
	II.10.5	Takai Duality 211
	II.10.6	Structure of Crossed Products 212
	II.10.7	Generalizations of Crossed Product Algebras 212
	II.10.8	Duality and Quantum Groups 214

III Von Neumann Algebras 221
III.1 Projections and Type Classification 222
- III.1.1 Projections and Equivalence 222
- III.1.2 Cyclic and Countably Decomposable Projections .. 225
- III.1.3 Finite, Infinite, and Abelian Projections 227
- III.1.4 Type Classification 231
- III.1.5 Tensor Products and Type I von Neumann Algebras 232
- III.1.6 Direct Integral Decompositions 237
- III.1.7 Dimension Functions and Comparison Theory ... 240
- III.1.8 Algebraic Versions 243

III.2 Normal Linear Functionals and Spatial Theory 244
- III.2.1 Normal and Completely Additive States 245
- III.2.2 Normal Maps and Isomorphisms of von Neumann Algebras 248
- III.2.3 Polar Decomposition for Normal Linear Functionals and the Radon-Nikodym Theorem ... 257
- III.2.4 Uniqueness of the Predual and Characterizations of W*-Algebras 259
- III.2.5 Traces on von Neumann Algebras 260
- III.2.6 Spatial Isomorphisms and Standard Forms 269

III.3 Examples and Constructions of Factors 275
- III.3.1 Infinite Tensor Products 275
- III.3.2 Crossed Products and the Group Measure Space Construction 280
- III.3.3 Regular Representations of Discrete Groups 288
- III.3.4 Uniqueness of the Hyperfinite II_1 Factor 291

III.4 Modular Theory 293
- III.4.1 Notation and Basic Constructions 293
- III.4.2 Approach using Bounded Operators 295
- III.4.3 The Main Theorem 295
- III.4.4 Left Hilbert Algebras 296
- III.4.5 Corollaries of the Main Theorems 299
- III.4.6 The Canonical Group of Outer Automorphisms and Connes' Invariants 302
- III.4.7 The KMS Condition and the Radon-Nikodym Theorem for Weights 306
- III.4.8 The Continuous and Discrete Decompositions of a von Neumann Algebra 310
  - III.4.8.1 The Flow of Weights. 312

III.5 Applications to Representation Theory of C*-Algebras 313
- III.5.1 Decomposition Theory for Representations 313
- III.5.2 The Universal Representation and Second Dual .. 318

IV Further Structure ... 323
IV.1 Type I C*-Algebras ... 323
- IV.1.1 First Definitions ... 323
- IV.1.2 Elementary C*-Algebras ... 326
- IV.1.3 Liminal and Postliminal C*-Algebras ... 327
- IV.1.4 Continuous Trace, Homogeneous, and Subhomogeneous C*-Algebras ... 329
- IV.1.5 Characterization of Type I C*-Algebras ... 337
- IV.1.6 Continuous Fields of C*-Algebras ... 340
- IV.1.7 Structure of Continuous Trace C*-Algebras ... 344

IV.2 Classification of Injective Factors ... 350
- IV.2.1 Injective C*-Algebras ... 352
- IV.2.2 Injective von Neumann Algebras ... 353
- IV.2.3 Normal Cross Norms ... 360
- IV.2.4 Semidiscrete Factors ... 362
- IV.2.5 Amenable von Neumann Algebras ... 365
- IV.2.6 Approximate Finite Dimensionality ... 367
- IV.2.7 Invariants and the Classification of Injective Factors ... 367

IV.3 Nuclear and Exact C*-Algebras ... 368
- IV.3.1 Nuclear C*-Algebras ... 368
- IV.3.2 Completely Positive Liftings ... 374
- IV.3.3 Amenability for C*-Algebras ... 378
- IV.3.4 Exactness and Subnuclearity ... 383
- IV.3.5 Group C*-Algebras and Crossed Products ... 391

V K-Theory and Finiteness ... 395
V.1 K-Theory for C*-Algebras ... 395
- V.1.1 K_0-Theory ... 396
- V.1.2 K_1-Theory and Exact Sequences ... 402
- V.1.3 Further Topics ... 408
- V.1.4 Bivariant Theories ... 411
- V.1.5 Axiomatic K-Theory and the Universal Coefficient Theorem ... 413

V.2 Finiteness ... 418
- V.2.1 Finite and Properly Infinite Unital C*-Algebras ... 418
- V.2.2 Nonunital C*-Algebras ... 423
- V.2.3 Finiteness in Simple C*-Algebras ... 430
- V.2.4 Ordered K-Theory ... 434

V.3 Stable Rank and Real Rank ... 444
- V.3.1 Stable Rank ... 445
- V.3.2 Real Rank ... 452

V.4 Quasidiagonality ... 457
- V.4.1 Quasidiagonal Sets of Operators ... 457
- V.4.2 Quasidiagonal C*-Algebras ... 460

        V.4.3   Generalized Inductive Limits 464

References 479

Index 505

I

Operators on Hilbert Space

I.1 Hilbert Space

We briefly review the most important and relevant structure facts about Hilbert space.

I.1.1 Inner Products

I.1.1.1 A *pre-inner product* on a complex vector space X is a positive semidefinite hermitian sesquilinear form $\langle \cdot, \cdot \rangle$ from X to $\mathbb{C}$, i.e. for all $\xi, \eta, \zeta \in X$ and $\alpha \in \mathbb{C}$ we have $\langle \xi+\eta, \zeta \rangle = \langle \xi, \zeta \rangle + \langle \eta, \zeta \rangle$, $\langle \alpha\xi, \eta \rangle = \alpha \langle \xi, \eta \rangle$, $\langle \eta, \xi \rangle = \overline{\langle \xi, \eta \rangle}$, and $\langle \xi, \xi \rangle \geq 0$. An *inner product* is a positive definite pre-inner product, i.e. one for which $\langle \xi, \xi \rangle > 0$ for $\xi \neq 0$.

We have made the convention that inner products are linear in the first variable and conjugate-linear in the second, the usual convention in mathematics, although the opposite convention is common in mathematical physics. The difference will have no effect on results about operators, and in the few places where the convention appears in arguments, a trivial notational change will convert to the opposite convention. When dealing with Hilbert modules (II.7.1), it is natural to use inner products which are conjugate-linear in the first variable, with scalar multiplication written on the right.

I.1.1.2 A pre-inner product on a vector space induces a "seminorm" by $\|\xi\|^2 = \langle \xi, \xi \rangle$. The pre-inner product and "seminorm" satisfy the following relations for any ξ, η:

$|\langle \xi, \eta \rangle| \leq \|\xi\| \|\eta\|$ (CBS inequality)
$\|\xi + \eta\| \leq \|\xi\| + \|\eta\|$ (Triangle inequality)
$\|\xi + \eta\|^2 + \|\xi - \eta\|^2 = 2(\|\xi\|^2 + \|\eta\|^2)$ (Parallelogram law)
$\langle \xi, \eta \rangle = \frac{1}{4}(\|\xi + \eta\|^2 - \|\xi - \eta\|^2 + i\|\xi + i\eta\|^2 - i\|\xi - i\eta\|^2)$ (Polarization)

All these are proved by simple calculations. (For the CBS inequality, use the fact that $\langle \xi + \alpha\eta, \xi + \alpha\eta \rangle \geq 0$ for all $\alpha \in \mathbb{C}$, and in particular for $\alpha =$

$-\langle\eta,\xi\rangle/\langle\eta,\eta\rangle$ if $\langle\eta,\eta\rangle \neq 0$.) By the triangle inequality, the "seminorm" is really a seminorm. From the CBS inequality, we have that $\|\xi\| = \max_{\|\zeta\|=1} |\langle\xi,\zeta\rangle|$ for every ξ. If $\langle\cdot,\cdot\rangle$ is an inner product, it follows that if $\langle\xi,\zeta\rangle = 0$ for all ζ, then $\xi = 0$, and that $\|\cdot\|$ is a norm.

The CBS inequality is attributed, in different forms and contexts, to A. Cauchy, V. Buniakovsky, and H. Schwarz, and is commonly referred to by various subsets of these names.

I.1.1.3 An inner product space which is complete with respect to the induced norm is a *Hilbert space*. The completion of an inner product space is a Hilbert space in an obvious way. A finite-dimensional inner product space is automatically a Hilbert space.

The standard example of a Hilbert space is $L^2(X,\mu)$, the space of square-integrable functions on a measure space (X,μ) (or, more precisely, of equivalence classes of square-integrable functions, with functions agreeing almost everywhere identified), with inner product $\langle f,g\rangle = \int_X f\bar{g}\,d\mu$. If S is a set, let μ be counting measure on S, and denote $L^2(S,\mu)$ by $l^2(S)$. Denote $l^2(\mathbb{N})$ by l^2; this is the space of square-summable sequences of complex numbers.

I.1.1.4 The definition and basic properties of l^2 were given in 1906 by D. Hilbert, who was the first to describe something approaching the modern notion of a "Hilbert space," and developed by E. Schmidt, F. Riesz, and others in immediately succeeding years. The L^2 spaces were also studied during this period, and the Riesz-Fischer theorem (isomorphism of $L^2(\mathbb{T})$ and $l^2(\mathbb{Z})$ via Fourier transform) proved. The definition of an abstract Hilbert space was, however, not given until 1928 by J. von Neumann [vN30a].

I.1.2 Orthogonality

I.1.2.1 Hilbert spaces have a geometric structure similar to that of Euclidean space. The most important notion is *orthogonality*: ξ and η are orthogonal if $\langle\xi,\eta\rangle = 0$, written $\xi \perp \eta$. If S is a subset of a Hilbert space $\mathcal{H}$, let

$$S^\perp = \{\xi \in \mathcal{H} : \langle\xi,\eta\rangle = 0 \text{ for all } \eta \in S\}.$$

$S^\perp$ is a closed subspace of $\mathcal{H}$, called the *orthogonal complement* of S.

I.1.2.2 If S is a closed convex set in a Hilbert space $\mathcal{H}$, and $\xi \in \mathcal{H}$, and $\eta_1, \eta_2 \in S$ satisfy $\|\xi - \eta_1\|^2$, $\|\xi - \eta_2\|^2 < dist(\xi, S)^2 + \epsilon$, where $dist(\xi, S) = \inf_{\zeta \in S} \|\xi - \zeta\|$, then by the parallelogram law

$$\|\eta_1 - \eta_2\|^2 = 2(\|\xi - \eta_1\|^2 + \|\xi - \eta_2\|^2) - 2\|\xi - \tfrac{1}{2}(\eta_1 + \eta_2)\|^2 < 2\epsilon$$

so there is a unique "closest vector" $\eta \in S$ satisfying $\|\xi - \eta\| = dist(\xi, S)$. If S is a closed subspace, then the closest approximant η to ξ in S is the "orthogonal projection" of ξ onto S, i.e. $\xi - \eta \perp S$. It follows that every $\xi \in \mathcal{H}$ can be uniquely written $\xi = \eta + \zeta$, where $\eta \in S$ and $\zeta \in S^\perp$. Thus $(S^\perp)^\perp = S$. Note that completeness is essential for the results of this paragraph.

I.1.2.3 A set $\{\xi_i\}$ of vectors in an inner product space is *orthonormal* if $\langle \xi_i, \xi_j \rangle = \delta_{ij}$, i.e. it is a mutually orthogonal set of unit vectors. A maximal orthonormal set in a Hilbert space is an *orthonormal basis*. If $\{\xi_i\}$ is an orthonormal basis for $\mathcal{H}$, then every vector $\eta \in \mathcal{H}$ can be uniquely written as $\sum_i \alpha_i \xi_i$ for $\alpha_i = \langle \eta, \xi_i \rangle \in \mathbb{C}$; and $\|\eta\|^2 = \sum_i |\alpha_i|^2$. Every Hilbert space has an orthonormal basis, and the cardinality of all orthonormal bases for a given $\mathcal{H}$ is the same, called the *(orthogonal) dimension* of $\mathcal{H}$. Two Hilbert spaces of the same dimension are isometrically isomorphic, so the dimension is the only structural invariant for a Hilbert space. A Hilbert space is separable if and only if its dimension is countable. In this case, an orthonormal basis can be generated from any countable total subset by the Gram-Schmidt process; in particular, any dense subspace of a separable Hilbert space contains an orthonormal basis (this is useful in applications where a concrete Hilbert space such as a space of square-integrable functions contains a dense subspace of "nice" elements such as polynomials or continuous or smooth functions.) Every separable infinite-dimensional Hilbert space is isometrically isomorphic to l^2.

I.1.3 Dual Spaces and Weak Topology

Bounded linear functionals on a Hilbert space are easy to describe:

I.1.3.1 THEOREM. [RIESZ-FRÈCHET] Let $\mathcal{H}$ be a Hilbert space, and ϕ a bounded linear functional on $\mathcal{H}$. Then there is a (unique) vector $\xi \in \mathcal{H}$ with $\phi(\eta) = \langle \eta, \xi \rangle$ for all $\eta \in \mathcal{H}$; and $\|\xi\| = \|\phi\|$.

So the dual space $\mathcal{H}^*$ of $\mathcal{H}$ may be identified with $\mathcal{H}$ (the identification is conjugate-linear), and the weak (= weak-*) topology is given by the inner product: $\xi_i \to \xi$ weakly if $\langle \xi_i, \eta \rangle \to \langle \xi, \eta \rangle$ for all η. In particular, a Hilbert space is a reflexive Banach space and the unit ball is therefore weakly compact. A useful way of stating this is:

I.1.3.2 COROLLARY. Let $\mathcal{H}$ be a Hilbert space. If (ξ_i) is a bounded weak Cauchy net in $\mathcal{H}$ (i.e. $(\langle \xi_i, \eta \rangle)$ is Cauchy for all $\eta \in \mathcal{H}$), then there is a (unique) vector $\xi \in \mathcal{H}$ such that $\xi_i \to \xi$ weakly, i.e. $\langle \xi_i, \eta \rangle \to \langle \xi, \eta \rangle$ for all $\eta \in \mathcal{H}$.

The weak topology is, of course, distinct from the norm topology (strictly weaker) if $\mathcal{H}$ is infinite-dimensional: for example, an orthonormal sequence of vectors converges weakly to 0. The next proposition gives a connection.

I.1.3.3 PROPOSITION. Let $\mathcal{H}$ be a Hilbert space, $\xi_i, \xi \in \mathcal{H}$, with $\xi_i \to \xi$ weakly. Then $\|\xi\| \leq \liminf \|\xi_i\|$, and $\xi_i \to \xi$ in norm if and only if $\|\xi_i\| \to \|\xi\|$.

PROOF: $\|\xi\|^2 = \lim \langle \xi_i, \xi \rangle \leq \|\xi\| \liminf \|\xi_i\|$ by the CBS inequality.

$$\|\xi_i - \xi\|^2 = \langle \xi_i, \xi_i \rangle - \langle \xi_i, \xi \rangle - \langle \xi, \xi_i \rangle + \langle \xi, \xi \rangle$$

which goes to zero if and only if $\|\xi_i\| \to \|\xi\|$.

I.1.3.4 If $\mathcal{H}$ is infinite-dimensional, then the weak topology on $\mathcal{H}$ is not first countable, and a weakly convergent net need not be (norm-)bounded. It is an easy consequence of Uniform Boundedness (I.2.1.3) that a weakly convergent sequence is bounded. For example, if $\{\xi_n\}$ is an orthonormal sequence in $\mathcal{H}$, then 0 is in the weak closure of $\{\sqrt{n}\xi_n\}$, but no subsequence converges weakly to 0 ([Hal67, Problem 28]; cf. [vN30b]). If $\mathcal{H}$ is separable, then the restriction of the weak topology to the unit ball of $\mathcal{H}$ is metrizable [Hal67, Problem 24].

I.1.4 Standard Constructions

There are two standard constructions on Hilbert spaces which are used repeatedly, direct sum and tensor product.

I.1.4.1 If $\{\mathcal{H}_i : i \in \Omega\}$ is a set of Hilbert spaces, we can form the Hilbert space direct sum, denoted $\bigoplus_\Omega \mathcal{H}_i$, as the set of "sequences" (indexed by Ω) $(\cdots \xi_i \cdots)$, where $\xi_i \in \mathcal{H}_i$ and $\sum_i \|\xi_i\|^2 < \infty$ (so $\xi_i \neq 0$ for only countably many i). The inner product is given by

$$\langle (\cdots \xi_i \cdots), (\cdots \eta_i \cdots) \rangle = \sum_i \langle \xi_i, \eta_i \rangle.$$

This is the completion of the algebraic direct sum with respect to this inner product. If all $\mathcal{H}_i$ are the same $\mathcal{H}$, the direct sum is called the *amplification* of $\mathcal{H}$ by $card(\Omega)$. The amplification of $\mathcal{H}$ by n (the direct sum of n copies of $\mathcal{H}$) is denoted by $\mathcal{H}^n$; the direct sum of countably many copies of $\mathcal{H}$ is denoted $\mathcal{H}^\infty$ (or sometimes $l^2(\mathcal{H})$).

I.1.4.2 If $\mathcal{H}_1, \mathcal{H}_2$ are Hilbert spaces, let $\mathcal{H}_1 \odot \mathcal{H}_2$ be the algebraic tensor product over $\mathbb{C}$. Put an inner product on $\mathcal{H}_1 \odot \mathcal{H}_2$ by

$$\langle \xi_1 \otimes \xi_2, \eta_1 \otimes \eta_2 \rangle = \langle \xi_1, \eta_1 \rangle_1 \langle \xi_2, \eta_2 \rangle_2$$

extended by linearity, where $\langle \cdot, \cdot \rangle_i$ is the inner product on $\mathcal{H}_i$. (It is easily checked that this is an inner product, in particular that it is positive definite.) Then the Hilbert space tensor product $\mathcal{H}_1 \otimes \mathcal{H}_2$ is the completion of $\mathcal{H}_1 \odot \mathcal{H}_2$. Tensor products of finitely many Hilbert spaces can be defined similarly. (Infinite tensor products are trickier and will be discussed in III.3.1.1.)

If $\mathcal{H}$ and $\mathcal{H}'$ are Hilbert spaces and $\{\eta_i : i \in \Omega\}$ is an orthonormal basis for $\mathcal{H}'$, then there is an isometric isomorphism from $\mathcal{H} \otimes \mathcal{H}'$ to the amplification of $\mathcal{H}$ by $card(\Omega)$, given by $\sum_i \xi_i \otimes \eta_i \mapsto (\cdots \xi_i \cdots)$. In particular, $\mathcal{H}^n$ is naturally isomorphic to $\mathcal{H} \otimes \mathbb{C}^n$ and $\mathcal{H}^\infty \cong \mathcal{H} \otimes l^2$.

If $(\mathcal{X}, \mu)$ is a measure space and $\mathcal{H}$ is a separable Hilbert space, then the Hilbert space $L^2(X, \mu) \otimes \mathcal{H}$ is naturally isomorphic to $L^2(X, \mu, \mathcal{H})$, the set of weakly measurable functions $f : X \to \mathcal{H}$ (i.e. $x \mapsto \langle f(x), \eta \rangle$ is a complex-valued measurable function for all $\eta \in \mathcal{H}$) such that $\int_X \|f(x)\|^2 \, d\mu(x) < \infty$, with inner product

$$\langle f, g \rangle = \int_X \langle f(x), g(x) \rangle \, d\mu(x).$$

The isomorphism sends $f \otimes \eta$ to $(x \mapsto f(x)\eta)$.

Unless otherwise qualified, a "direct sum" or "tensor product" of Hilbert spaces will always mean the Hilbert space direct sum or tensor product.

I.1.5 Real Hilbert Spaces

One can also consider *real inner product spaces*, real vector spaces with a (bilinear, real-valued) inner product $(\cdot, \cdot)$ satisfying the properties of I.1.1.1 for $\alpha \in \mathbb{R}$. A *real Hilbert space* is a complete real inner product space. A real Hilbert space $\mathcal{H}_{\mathbb{R}}$ can be complexified to $\mathbb{C} \otimes_{\mathbb{R}} \mathcal{H}_{\mathbb{R}}$; conversely, a (complex) Hilbert space can be regarded as a real Hilbert space by restricting scalar multiplication and using the real inner product $(\xi, \eta) = Re\langle \xi, \eta \rangle$. (Note, however, that these two processes are not quite inverse to each other.) All of the results of this section, and many (but by no means all) results about operators, hold verbatim or have obvious exact analogs in the case of real Hilbert spaces.

Throughout this volume, the term "Hilbert space," unless qualified with "real," will always denote a *complex* Hilbert space; and "linear" will mean "complex-linear."

I.2 Bounded Operators

I.2.1 Bounded Operators on Normed Spaces

I.2.1.1 An operator (linear transformation) T between normed vector spaces $\mathcal{X}$ and $\mathcal{Y}$ (by convention, this means the domain of T is all of $\mathcal{X}$) is *bounded* if there is a $K \geq 0$ such that $\|T\xi\| \leq K\|\xi\|$ for all $\xi \in \mathcal{X}$. The smallest such K is the *(operator) norm* of T, denoted $\|T\|$, i.e.

$$\|T\| = \sup_{\|\xi\|=1} \|T\xi\|.$$

An operator is continuous if and only if it is bounded. The set of bounded operators from $\mathcal{X}$ to $\mathcal{Y}$ is denoted $\mathcal{L}(\mathcal{X}, \mathcal{Y})$ (the notation $\mathcal{B}(\mathcal{X}, \mathcal{Y})$ is also frequently used). $\mathcal{L}(\mathcal{X}, \mathcal{X})$ is usually denoted $\mathcal{L}(\mathcal{X})$ (or $\mathcal{B}(\mathcal{X})$). $\mathcal{L}(\mathcal{X}, \mathcal{Y})$ is closed under addition and scalar multiplication, and the norm is indeed a norm, i.e. we have $\|S + T\| \leq \|S\| + \|T\|$ and $\|\alpha T\| = |\alpha| \|T\|$ for all $S, T \in \mathcal{L}(\mathcal{X}, \mathcal{Y})$, $\alpha \in \mathbb{C}$. The space $\mathcal{L}(\mathcal{X}, \mathcal{Y})$ is complete if $\mathcal{Y}$ is complete. The norm also satisfies $\|ST\| \leq \|S\| \|T\|$ for all $T \in \mathcal{L}(\mathcal{X}, \mathcal{Y})$, $S \in \mathcal{L}(\mathcal{Y}, \mathcal{Z})$.

$\mathcal{L}(\mathcal{X})$ is thus a Banach algebra if $\mathcal{X}$ is a Banach space. The identity operator on $\mathcal{X}$, denoted I, is a unit for $\mathcal{L}(\mathcal{X})$.

I.2.1.2 It is easy to see (most easily using the adjoint) that an element of $\mathcal{L}(\mathcal{X}, \mathcal{Y})$ is weakly continuous, i.e. continuous when $\mathcal{X}$ and $\mathcal{Y}$ are given their weak topologies. Thus if $\mathcal{X}$ is a Hilbert space (or, more generally, a reflexive Banach space) with closed unit ball B, and $T \in \mathcal{L}(\mathcal{X}, \mathcal{Y})$, then $T(B)$ is weakly compact and hence weakly (and therefore norm-) closed since the weak topology on $\mathcal{Y}$ is Hausdorff.

We recall three fundamental results about bounded operators on Banach spaces.

I.2.1.3 THEOREM. [UNIFORM BOUNDEDNESS] Let $\mathcal{X}$ and $\mathcal{Y}$ be Banach spaces, $\mathcal{T}$ a subset of $\mathcal{L}(\mathcal{X}, \mathcal{Y})$. If $\{\|T\xi\| : T \in \mathcal{T}\}$ is bounded for each $\xi \in \mathcal{X}$, then $\{\|T\| : T \in \mathcal{T}\}$ is bounded.

I.2.1.4 THEOREM. [OPEN MAPPING] Let $\mathcal{X}$ and $\mathcal{Y}$ be Banach spaces, $T \in \mathcal{L}(\mathcal{X}, \mathcal{Y})$. If T maps $\mathcal{X}$ onto $\mathcal{Y}$, then T is an open mapping.

I.2.1.5 THEOREM. [CLOSED GRAPH] Let $\mathcal{X}$ and $\mathcal{Y}$ be Banach spaces, and $T : \mathcal{X} \to \mathcal{Y}$ an operator (recall this means that the domain of T is all of $\mathcal{X}$). If the graph of T is closed in $\mathcal{X} \times \mathcal{Y}$, then T is bounded.

All are consequences of the Baire Category Theorem. On Hilbert spaces, there are direct proofs not using the Baire Category Theorem; cf. [Hal67]. Even the first proofs of I.2.1.3 on general Banach spaces used a "gliding hump" argument in place of the Baire Category Theorem. The proof using Baire Category was given in [BS27]. I.2.1.4 and I.2.1.5 were first proved in [Ban32]. Nowadays, I.2.1.3 is usually obtained as a corollary of I.2.1.5, and I.2.1.5 as a corollary of I.2.1.4.

Good general references for various aspects of operator theory are [DS88a], [DS88b], [Hal67], [HP74], [RSN55], [Ped89a].

I.2.2 Sesquilinear Forms

I.2.2.1 If $\mathcal{X}$ and $\mathcal{Y}$ are Hilbert spaces, a *sesquilinear form* on $(\mathcal{X}, \mathcal{Y})$ is a function $(\cdot, \cdot) : \mathcal{X} \times \mathcal{Y} \to \mathbb{C}$ which is linear in the first variable and conjugate-linear in the second. The form is *bounded* if there is a K such that

$$|(\xi, \eta)| \leq K\|\xi\|\|\eta\|$$

for all $\xi \in \mathcal{X}$, $\eta \in \mathcal{Y}$. The smallest such K is called the *norm* of the form.

I.2.2.2 PROPOSITION. Let $(\cdot, \cdot)$ be a bounded sesquilinear form on $(\mathcal{X}, \mathcal{Y})$, with norm K. Then there is a (unique) bounded operator T from $\mathcal{X}$ to $\mathcal{Y}$ such that $(\xi, \eta) = \langle T\xi, \eta \rangle$ for all $\xi \in \mathcal{X}$, $\eta \in \mathcal{Y}$. This operator has $\|T\| = K$.

The proof is an easy application of the Riesz-Frèchet Theorem (I.1.3.1).

I.2.2.3 Conversely, of course, any $T \in \mathcal{L}(\mathcal{X}, \mathcal{Y})$ defines a bounded sesquilinear form $(\xi, \eta) = \langle T\xi, \eta \rangle$. There is thus a one-one correspondence between bounded operators and bounded sesquilinear forms. Each point of view has its advantages. Some early authors such as Hilbert used the form approach exclusively instead of working with operators, but some parts of the theory become almost impossibly difficult and cumbersome (e.g. composition of operators) from this point of view. I. Fredholm and F. Riesz emphasized the operator point of view, which is superior for most purposes.

I.2.2.4 The results of this subsection hold equally for operators on real Hilbert spaces.

I.2.3 Adjoint

I.2.3.1 If $\mathcal{H}_1$ and $\mathcal{H}_2$ are Hilbert spaces, then each $T \in \mathcal{L}(\mathcal{H}_1, \mathcal{H}_2)$ has an *adjoint*, denoted T^*, in $\mathcal{L}(\mathcal{H}_2, \mathcal{H}_1)$, defined by the property

$$\langle T^*\eta, \xi \rangle_1 = \langle \eta, T\xi \rangle_2$$

for all $\xi \in \mathcal{H}_1$, $\eta \in \mathcal{H}_2$, where $\langle \cdot, \cdot \rangle_i$ is the inner product on $\mathcal{H}_i$. (The existence of T^* follows immediately from I.2.2.2; conversely, an operator with an adjoint must be bounded by the Closed Graph Theorem.) Adjoints have the following properties for all $S, T \in \mathcal{L}(\mathcal{H}_1, \mathcal{H}_2)$, $\alpha \in \mathbb{C}$:

(i) $(S+T)^* = S^* + T^*$, $(\alpha T)^* = \bar{\alpha} T^*$, and $(T^*)^* = T$;
(ii) $\|T^*\| = \|T\|$ and $\|T^*T\| = \|T\|^2$.

Properties (i) are obvious, and the ones in (ii) follow from the fact that $\|T\| = \sup_{\|\xi\|=\|\eta\|=1} |\langle T\xi, \eta \rangle| \geq \sup_{\|\xi\|=1} |\langle T\xi, \xi \rangle|$.

If $\mathcal{H} = \mathbb{C}^n$, then $\mathcal{L}(\mathcal{H})$ can be naturally identified with $\mathbb{M}_n = M_n(\mathbb{C})$, and the adjoint is the usual conjugate transpose.

The range $\mathcal{R}(T)$ and the null space $\mathcal{N}(T)$ of a bounded operator T are subspaces; $\mathcal{N}(T)$ is closed, but $\mathcal{R}(T)$ is not necessarily closed. The fundamental relationship is:

I.2.3.2 PROPOSITION. *If $\mathcal{X}$ and $\mathcal{Y}$ are Hilbert spaces and $T \in \mathcal{L}(\mathcal{X}, \mathcal{Y})$, then $\mathcal{R}(T)^\perp = \mathcal{N}(T^*)$.*

I.2.3.3 An operator on a real Hilbert space has an adjoint in the same manner. We will primarily have occasion to consider conjugate-linear operators on (complex) Hilbert spaces. A bounded conjugate-linear operator T from $\mathcal{H}_1$ to $\mathcal{H}_2$ has a (unique) conjugate-linear adjoint $T^* : \mathcal{H}_2 \to \mathcal{H}_1$ with

$$\langle T^*\eta, \xi \rangle_1 = \langle T\xi, \eta \rangle_2$$

for all $\xi \in \mathcal{H}_1$, $\eta \in \mathcal{H}_2$ (note the reversal of order in the inner product). The uniqueness of T^* is clear; existence can be proved by regarding T as an operator between the real Hilbert spaces $\mathcal{H}_1$ and $\mathcal{H}_2$ with inner product $(\xi, \eta) = Re\langle \xi, \eta \rangle$, or directly using involutions.

An *involution* on a (complex) Hilbert space $\mathcal{H}$ is a conjugate-linear isometry J from $\mathcal{H}$ onto $\mathcal{H}$ with $J^2 = I$. Involutions exist in abundance on any Hilbert space: if (ξ_i) is an orthonormal basis for $\mathcal{H}$, then $\sum \alpha_i \xi_i \mapsto \sum \overline{\alpha}_i \xi_i$ is an involution (and every involution is of this form). If $\xi, \eta \in \mathcal{H}$, then $\langle J\eta, J\xi \rangle = \langle \xi, \eta \rangle$.

If $T : \mathcal{H}_1 \to \mathcal{H}_2$ is a bounded conjugate-linear operator, and J is an involution on $\mathcal{H}_1$, then TJ is a bounded linear operator from $\mathcal{H}_1$ to $\mathcal{H}_2$; $T^* = J(TJ)^*$ is the adjoint of T (the proof uses the fact that $\langle J\xi, \eta \rangle = \langle J\eta, \xi \rangle$ for all $\xi, \eta \in \mathcal{H}_1$, i.e. J is "self-adjoint.")

I.2.4 Self-Adjoint, Unitary, and Normal Operators

I.2.4.1 DEFINITION. Let $\mathcal{H}$ be a Hilbert space, and $T \in \mathcal{L}(\mathcal{H})$. Then

T is *self-adjoint* if $T = T^*$.
T is a *projection* if $T = T^* = T^2$.
T is an *isometry* if $T^*T = I$.
T is *unitary* if $T^*T = TT^* = I$.
T is *normal* if $T^*T = TT^*$.

Self-adjoint and unitary operators are obviously normal. The usual definition of an isometry is an operator T such that $\|T\xi\| = \|\xi\|$ for all ξ. For such a T, it follows from polarization that $\langle T\xi, T\eta \rangle = \langle \xi, \eta \rangle$ for all ξ, η, and hence that $T^*T = I$. Conversely, if $T^*T = I$, then it is obvious that T is an isometry in this sense, so the definitions are equivalent. If T is an isometry, then TT^* is a projection. Projections will be described in I.5.

If T is unitary, then T and T^* are isometries, and conversely; an invertible or normal isometry is unitary.

Isometries and unitary operators between different Hilbert spaces can be defined analogously.

I.2.4.2 If T is a normal operator on $\mathcal{H}$, and $\xi \in \mathcal{H}$, then

$$\|T\xi\|^2 = \langle T\xi, T\xi \rangle = \langle T^*T\xi, \xi \rangle = \langle TT^*\xi, \xi \rangle = \langle T^*\xi, T^*\xi \rangle = \|T^*\xi\|^2.$$

(Conversely, if $\|T\xi\| = \|T^*\xi\|$ for all ξ, then by polarization $\langle T\xi, T\eta \rangle = \langle T^*\xi, T^*\eta \rangle$ for all ξ, η, so T is normal.) Thus, if T is normal, $\mathcal{N}(T) = \mathcal{N}(T^*) = \mathcal{R}(T)^\perp$.

I.2.4.3 EXAMPLES.

(i) Let (X, μ) be a measure space, and $\mathcal{H} = L^2(X, \mu)$. For every bounded measurable function $f : X \to \mathbb{C}$ there is a *multiplication operator* $M_f \in \mathcal{L}(\mathcal{H})$ defined by $(M_f \xi)(x) = f(x)\xi(x)$ (of course, this only makes sense almost everywhere). M_f is a bounded operator, and $\|M_f\|$ is the essential supremum of $|f|$. $M_f M_g = M_{fg}$ for all f, g, and $M_f^* = M_{\bar{f}}$, where $\bar{f}$ is the complex conjugate of f. Thus all M_f are normal. In fact, the Spectral Theorem implies that every normal operator looks very much like a multiplication operator. M_f is self-adjoint if and only if $f(x) \in \mathbb{R}$ a.e., and M_f is unitary if and only if $|f(x)| = 1$ a.e.

(ii) Perhaps the most important single operator is the *unilateral shift* S, defined on l^2 by

$$S(\alpha_1, \alpha_2, \ldots) = (0, \alpha_1, \alpha_2, \ldots).$$

S is an isometry, and S^* is the backwards shift defined by

$$S^*(\alpha_1, \alpha_2, \ldots) = (\alpha_2, \alpha_3, \ldots).$$

$S^*S = I$, but $SS^* \neq I$ (it is the projection onto the space of sequences with first coordinate 0). S can also be defined as multiplication by z on the Hardy space H^2 of analytic functions on the unit disk with L^2 boundary values on the circle $\mathbb{T}$ (if $L^2(\mathbb{T})$ is identified with $l^2(\mathbb{Z})$ via Fourier transform, H^2 is the subspace of functions whose negative Fourier coefficients vanish).

I.2.5 Amplifications and Commutants

I.2.5.1 If $\mathcal{H}_1$ and $\mathcal{H}_2$ are Hilbert spaces, there are natural extensions of operators in $\mathcal{L}(\mathcal{H}_i)$ to operators on $\mathcal{H}_1 \otimes \mathcal{H}_2$: if $T \in \mathcal{L}(\mathcal{H}_1)$, define $T \otimes I \in \mathcal{L}(\mathcal{H}_1 \otimes \mathcal{H}_2)$ by

$$(T \otimes I)(\xi \otimes \eta) = T\xi \otimes \eta.$$

Similarly, for $S \in \mathcal{L}(\mathcal{H}_2)$, $I \otimes S$ is defined by $(I \otimes S)(\xi \otimes \eta) = \xi \otimes S\eta$. $T \mapsto T \otimes I$ and $S \mapsto I \otimes S$ are isometries. The images of $\mathcal{L}(\mathcal{H}_1)$ and $\mathcal{L}(\mathcal{H}_2)$ are denoted $\mathcal{L}(\mathcal{H}_1) \otimes I$ and $I \otimes \mathcal{L}(\mathcal{H}_2)$ respectively, called *amplifications*.

I.2.5.2 More generally, if $S_1, T_1 \in \mathcal{L}(\mathcal{H}_1)$, $S_2, T_2 \in \mathcal{L}(\mathcal{H}_2)$, we can define $S_1 \otimes S_2 \in \mathcal{L}(\mathcal{H}_1 \otimes \mathcal{H}_2)$ by

$$(S_1 \otimes S_2)(\xi \otimes \eta) = S_1\xi \otimes S_2\eta.$$

Then
$$S_1 \otimes S_2 = (S_1 \otimes I)(I \otimes S_2) = (I \otimes S_2)(S_1 \otimes I)$$
$$(S_1 + T_1) \otimes S_2 = (S_1 \otimes S_2) + (T_1 \otimes S_2)$$
$$S_1 \otimes (S_2 + T_2) = (S_1 \otimes S_2) + (S_1 \otimes T_2)$$
$$(S_1 \otimes S_2)(T_1 \otimes T_2) = S_1 T_1 \otimes S_2 T_2$$
$$(S_1 \otimes S_2)^* = S_1^* \otimes S_2^*$$
$$\|S_1 \otimes S_2\| = \|S_1\| \|S_2\|.$$

I.2.5.3 If $\mathcal{S} \subseteq \mathcal{L}(\mathcal{H})$, then the *commutant* of $\mathcal{S}$ is

$$\mathcal{S}' = \{T \in \mathcal{L}(\mathcal{H}) : ST = TS \text{ for all } S \in \mathcal{S}\}.$$

$\mathcal{S}'$ is a closed subalgebra of $\mathcal{L}(\mathcal{H})$ containing I. We denote the *bicommutant* $(\mathcal{S}')'$ by $\mathcal{S}''$. We obviously have $\mathcal{S} \subseteq \mathcal{S}''$, and $\mathcal{S}_1' \supseteq \mathcal{S}_2'$ if $\mathcal{S}_1 \subseteq \mathcal{S}_2$. It follows that $\mathcal{S}' = (\mathcal{S}'')'$ for any set $\mathcal{S}$.

I.2.5.4 PROPOSITION. If $\mathcal{H}_1, \mathcal{H}_2$ are Hilbert spaces and $\mathcal{H} = \mathcal{H}_1 \otimes \mathcal{H}_2$, then we have $(\mathcal{L}(\mathcal{H}_1) \otimes I)' = I \otimes \mathcal{L}(\mathcal{H}_2)$ (and, similarly, $(I \otimes \mathcal{L}(\mathcal{H}_2))' = \mathcal{L}(\mathcal{H}_1) \otimes I$). Thus, if $\mathcal{S} \subseteq \mathcal{L}(\mathcal{H}_1)$, then $(\mathcal{S} \otimes I)'' = \mathcal{S}'' \otimes I$.

The proof is a straightforward calculation: if $\{\xi_i\}$ is an orthonormal basis for $\mathcal{H}_1$, use the operators $T_{ij} \otimes I \in \mathcal{L}(\mathcal{H}_1) \otimes I$, where $T_{ij}(\eta) = \langle \eta, \xi_i \rangle \xi_j$.

A similar calculation (cf. [KR97a, 5.5.4]) shows:

I.2.5.5 PROPOSITION. Let $\mathcal{H}$ and $\mathcal{H}'$ be Hilbert spaces, $S_1, \ldots, S_n \in \mathcal{L}(\mathcal{H})$, $T_1, \ldots, T_n \in \mathcal{L}(\mathcal{H}')$. Then $\sum S_i \otimes T_i = 0$ in $\mathcal{L}(\mathcal{H} \otimes \mathcal{H}')$ if and only if it is 0 in $\mathcal{L}(\mathcal{H}) \odot \mathcal{L}(\mathcal{H}')$, i.e. the natural map from $\mathcal{L}(\mathcal{H}) \odot \mathcal{L}(\mathcal{H}')$ to $\mathcal{L}(\mathcal{H} \otimes \mathcal{H}')$ is injective.

I.2.6 Invertibility and Spectrum

In this section, we will essentially follow the first comprehensive development and exposition of this theory, given by F. Riesz [Rie13] in 1913, abstracted to Banach spaces and Hilbert spaces (which did not exist in 1913!) and with some modern technical simplifications.

I.2.6.1 If $T \in \mathcal{L}(\mathcal{X}, \mathcal{Y})$ ($\mathcal{X}, \mathcal{Y}$ Banach spaces) is invertible in the algebraic sense, i.e. T is one-to-one and onto, then T^{-1} is automatically bounded by the Open Mapping Theorem. So there are two ways a bounded operator T can fail to be invertible:

T is not bounded below (recall that T is *bounded below* if $\exists \epsilon > 0$ such that $\|T\xi\| \geq \epsilon \|\xi\|$ for all ξ)
$\mathcal{R}(T)$ is not dense.

These possibilities are not mutually exclusive. For example, if T is a normal operator on a Hilbert space, then $\mathcal{N}(T) = \mathcal{R}(T)^\perp$, so if $\mathcal{R}(T)$ is not dense, then T is not one-to-one and hence not bounded below. Thus a normal operator on a Hilbert space is invertible if and only if it is bounded below.

I.2.6.2 DEFINITION. Let $\mathcal{X}$ be a normed vector space, $T \in \mathcal{L}(\mathcal{X})$. The *spectrum* of T is

$$\sigma(T) = \{\lambda \in \mathbb{C} : T - \lambda I \text{ is not invertible}\}$$

The spectrum of T can be thought of as the set of "generalized eigenvalues" of T (note that every actual eigenvalue of T is in the spectrum). The first definition (not the same as the one here – essentially the reciprocal) and name "spectrum" were given by Hilbert; the modern definition is due to F. Riesz.

I.2.6.3 If $|\lambda| > \|T\|$, then the series $-\lambda \sum_{n=0}^{\infty} (\lambda^{-1} T)^n$ converges absolutely to an inverse for $T - \lambda I$, so $\sigma(T) \subseteq \{\lambda : |\lambda| \leq \|T\|\}$.

It follows from general theory for (complex) Banach algebras (cf. II.1.4) that $\sigma(T)$ is nonempty and compact. Thus the *spectral radius*

$$r(T) = \max\{|\lambda| : \lambda \in \sigma(T)\}$$

makes sense and satisfies $r(T) \leq \|T\|$.

I.2.6.4 The *numerical range* of an operator T on a Hilbert space $\mathcal{H}$ is

$$W(T) = \{\langle T\xi, \xi\rangle : \|\xi\| = 1\}.$$

Set

$$w(T) = \sup\{|\lambda| : \lambda \in W(T)\}.$$

It is obvious that $W(T^*) = \{\bar{\lambda} : \lambda \in W(T)\}$, so $w(T^*) = w(T)$; if $T = T^*$, then $W(T) \subseteq \mathbb{R}$. (Conversely, if $W(T) \subseteq \mathbb{R}$, then $T = T^*$: for $\xi, \eta \in \mathcal{H}$, expand $\langle T(\xi + \eta), \xi + \eta\rangle \in \mathbb{R}$ and $\langle T(\xi + i\eta), \xi + i\eta\rangle \in \mathbb{R}$.)

I.2.6.5 PROPOSITION. If $T \in \mathcal{L}(\mathcal{H})$, then $\sigma(T) \subseteq \overline{W(T)}$ and $r(T) \leq w(T) \leq \|T\|$.

To see that $\sigma(T) \subseteq \overline{W(T)}$, note that if $\lambda \in \sigma(T)$, then either $T - \lambda I$ or $(T - \lambda I)^*$ is not bounded below, so there is a sequence (ξ_n) of unit vectors such that either $(T - \lambda I)\xi_n \to 0$ or $(T - \lambda I)^* \xi_n \to 0$. In either case, we obtain that $\langle (T - \lambda I)\xi_n, \xi_n\rangle \to 0$.

I.2.6.6 Note that $r(T) = w(T) = \|T\|$ does not hold for general operators: for example, let T be the (nilpotent) operator on $\mathbb{C}^2$ given by the matrix $\begin{bmatrix} 0 & 0 \\ 1 & 0 \end{bmatrix}$. Then $\sigma(T) = \{0\}$, so $r(T) = 0$; but $\|T\| = 1$. It can be shown by an easy calculation that $W(T) = \{\lambda : |\lambda| \leq 1/2\}$, so $w(T) = 1/2$. (There are also operators on infinite-dimensional Hilbert spaces whose numerical range is not closed.)

I.2.6.7 DEFINITION. A (necessarily self-adjoint) operator T on a Hilbert space $\mathcal{H}$ is *positive* if $\langle T\xi, \xi\rangle \geq 0$ for all ξ. Write $T \geq 0$ if T is positive, and write $\mathcal{L}(\mathcal{H})_+$ for the set of positive operators on $\mathcal{H}$.

If $T \in \mathcal{L}(\mathcal{H})$, then $T^*T \geq 0$ (this is also true if T is conjugate-linear); if $T = T^*$, then $\lambda I - T \geq 0$ for $\lambda \geq \|T\|$. Sums and nonnegative scalar multiples of positive operators are positive.

If S, T are self-adjoint operators, write $S \leq T$ if $T - S \geq 0$, i.e. if $\langle S\xi, \xi\rangle \leq \langle T\xi, \xi\rangle$ for all ξ.

I.2.6.8 Let S be a positive operator, and let m and M be the inf and sup of $W(S)$ respectively. The form $(\xi, \eta) = \langle S\xi, \eta \rangle$ is positive semidefinite, and thus satisfies the CBS inequality, yielding, for all ξ, η,

$$|\langle S\xi, \eta\rangle|^2 \leq \langle S\xi, \xi\rangle \langle S\eta, \eta\rangle \leq \langle S\xi, \xi\rangle M\|\eta\|^2 \leq M^2 \|\xi\|^2 \|\eta\|^2.$$

Setting $\eta = S\xi$ and dividing through by $\|S\xi\|^2$, we obtain, for all ξ,

$$\langle S^2\xi, \xi\rangle = \langle S\xi, S\xi\rangle = \|S\xi\|^2 \leq M\langle S\xi, \xi\rangle \leq M^2\|\xi\|^2.$$

So we conclude:

I.2.6.9 PROPOSITION.
(i) $\|S\| = M$ and $S^2 \leq MS$.
(ii) If S is bounded below, then $m > 0$.

I.2.6.10 PROPOSITION. If T is a self-adjoint operator on a Hilbert space H, and m and M are the inf and sup of $W(T)$ respectively, then $\sigma(T) \subseteq [m, M] \subseteq \mathbb{R}$ and $m, M \in \sigma(T)$. Thus $r(T) = w(T)$.
PROOF: $\sigma(T) \subseteq [m, M]$ by I.2.6.5. Applying I.2.6.9(ii) to $T - mI$ and $MI - T$, we obtain $m, M \in \sigma(T)$.

I.2.6.11 PROPOSITION. Let T be a self-adjoint element of $\mathcal{L}(\mathcal{H})$. Then $r(T) = \|T\|(= w(T))$.

There are several simple proofs known. Perhaps the best proceeds by applying I.2.6.9 to $S = (T+I)/2$ if $-I \leq T \leq I$ (so $0 \leq S \leq I$) to conclude that $S^2 \leq S$, so $I - T^2 = 4(S - S^2) \geq 0$, $0 \leq T^2 \leq I$, $\|T^2\| = \|T\|^2 \leq 1$ by I.2.6.9. In particular, if $T \geq 0$, then $\sigma(T) \subseteq [0, \|T\|]$ and $\|T\| \in \sigma(T)$.

The statement $r(T) = w(T) = \|T\|$ is true for an arbitrary normal operator T (II.1.6.3).

Here is a useful generalization of I.2.6.8 (cf. [KR97b, 10.5.7]).

I.2.6.12 PROPOSITION. Let $S_1, \ldots, S_n \in \mathcal{L}(\mathcal{H})_+$ with $\|S_1 + \cdots + S_n\| \leq M$, and $\xi_1, \ldots, \xi_n \in \mathcal{H}$. Then $\sum_{k=1}^n \|S_k\xi_k\|^2 \leq M \sum_{k=1}^n \langle S_k\xi_k, \xi_k\rangle$.
PROOF: Define a pre-inner product on $\mathcal{H}^n$ by $(\bigoplus \eta_k, \bigoplus \zeta_k) = \sum_{k=1}^n \langle S_k\eta_k, \zeta_k\rangle$. Set $\eta_k = \xi_k$ and $\zeta_k = \sum_{j=1}^n S_j\xi_j$ for all k, and apply the CBS inequality.

I.2.6.13 COROLLARY. Let $S_1, \ldots, S_n \in \mathcal{L}(\mathcal{H})_+$ with $\|S_1 + \cdots + S_n\| \leq M$. If $\lambda_1, \ldots, \lambda_n \in \mathbb{C}$ and $T = \sum_{k=1}^n \lambda_k S_k$, then $T^*T \leq M \sum_{k=1}^n |\lambda_k|^2 S_k$. In particular, $\|T\| \leq M \max |\lambda_k|$.

Another application of the CBS inequality gives:

I.2.6.14 PROPOSITION. Let T be a positive operator on a Hilbert space $\mathcal{H}$, and $\xi, \eta \in \mathcal{H}$. Then $\langle T(\xi+\eta), \xi+\eta \rangle \leq [\langle T\xi, \xi \rangle^{1/2} + \langle T\eta, \eta \rangle^{1/2}]^2$.

PROOF: We have $\langle T(\xi+\eta), \xi+\eta \rangle = \langle T\xi, \xi \rangle + 2Re\langle T\xi, \eta \rangle + \langle T\eta, \eta \rangle$, and by the CBS inequality $Re\langle T\xi, \eta \rangle \leq [\langle T\xi, \xi \rangle \langle T\eta, \eta \rangle]^{1/2}$.

I.2.6.15 COROLLARY. Let T be a positive operator on $\mathcal{H}$, and $\{\xi_i\}$ a total set of vectors in $\mathcal{H}$. If $\langle T\xi_i, \xi_i \rangle = 0$ for all i, then $T = 0$.

An important special case is worth noting:

I.2.6.16 COROLLARY. Let $\mathcal{H}_1$ and $\mathcal{H}_2$ be Hilbert spaces, and T a positive operator on $\mathcal{H}_1 \otimes \mathcal{H}_2$. If $\langle T(\xi_1 \otimes \xi_2), \xi_1 \otimes \xi_2 \rangle = 0$ for all $\xi_1 \in \mathcal{H}_1$, $\xi_2 \in \mathcal{H}_2$, then $T = 0$. (Actually ξ_1 and ξ_2 need only range over total subsets of $\mathcal{H}_1$ and $\mathcal{H}_2$ respectively.)

I.3 Other Topologies on $\mathcal{L}(\mathcal{H})$

$\mathcal{L}(\mathcal{H})$ has several other useful topologies in addition to the norm topology.

I.3.1 Strong and Weak Topologies

I.3.1.1 DEFINITION.
The *strong operator topology* on $\mathcal{L}(\mathcal{H})$ is the topology of pointwise norm-convergence, i.e. $T_i \to T$ strongly if $T_i\xi \to T\xi$ for all $\xi \in \mathcal{H}$.
The *weak operator topology* on $\mathcal{L}(\mathcal{H})$ is the topology of pointwise weak convergence, i.e. $T_i \to T$ weakly if $\langle T_i\xi, \eta \rangle \to \langle T\xi, \eta \rangle$ for all $\xi, \eta \in \mathcal{H}$.

The word "operator" is often omitted in the names of these topologies; this usually causes no confusion, although technically there is a different topology on $\mathcal{L}(\mathcal{H})$ (more generally, on any Banach space) called the "weak topology," the topology of pointwise convergence for linear functionals in $\mathcal{L}(\mathcal{H})^*$. This topology, stronger than the weak operator topology, is rarely used on $\mathcal{L}(\mathcal{H})$. In this volume, "weak topology" will mean "weak operator topology," and "strong topology" will mean "strong operator topology." (Caution: in some references, "strong topology" or "strong convergence" refers to the norm topology.)

I.3.1.2 If (T_i) is a (norm-)bounded net in $\mathcal{L}(\mathcal{H})$, then $T_i \to T$ in the strong [*resp.* weak] operator topology if and only if $T_i\xi \to T\xi$ in norm [*resp.* weakly] for a dense (or just total) set of ξ. Boundedness is essential here; a strongly convergent net need not be bounded. (A weakly convergent *sequence* must be bounded by the Uniform Boundedness Theorem.)

I.3.1.3 There are variants of these topologies, called the σ-*strong (or ultrastrong) operator topology* and the σ-*weak (or ultraweak) operator topology*. A (rather unnatural) definition of these topologies is to first form the amplification $\mathcal{H}^\infty = \mathcal{H} \otimes l^2$; then the restriction of the weak [*resp.* strong] topology on $\mathcal{L}(\mathcal{H}^\infty)$ to the image $\mathcal{L}(\mathcal{H}) \otimes I$ is the σ-weak [*resp.* σ-strong] topology on $\mathcal{L}(\mathcal{H})$. An explicit description of the topologies in terms of sets of vectors in $\mathcal{H}$ can be obtained from this. A more natural definition will be given in I.8.6; the σ-strong and σ-weak topologies are actually more intrinsic to the Banach space structure of $\mathcal{L}(\mathcal{H})$ than the weak or strong topologies.

I.3.1.4 As the names suggest, the weak operator topology is weaker than the strong operator topology, and the σ-weak is weaker than the σ-strong. The weak is weaker than the σ-weak, and the strong is weaker than the σ-strong. The weak and σ-weak topologies are not much different: they coincide on all (norm-)bounded sets, as follows easily from I.3.1.2. Similarly, the strong and σ-strong coincide on bounded sets. All are locally convex Hausdorff vector space topologies (the strong operator topology is generated by the family of seminorms $\|\cdot\|_\xi$, where $\|T\|_\xi = \|T\xi\|$; the others are generated by similar families of seminorms.) All of these topologies are weaker than the norm topology, and none are metrizable (cf. III.2.2.28) except when $\mathcal{H}$ is finite-dimensional, in which case they coincide with the norm topology. (The unit ball of $\mathcal{L}(\mathcal{H})$ is metrizable in both the strong and weak topologies if $\mathcal{H}$ is separable.)

I.3.1.5 EXAMPLES. Let S be the unilateral shift (I.2.4.3). Then $S^n \to 0$ weakly, but not strongly. $(S^*)^n \to 0$ strongly, but not in norm.

I.3.1.6 This example shows that the adjoint operation is not strongly continuous. (It is obviously weakly continuous.) One can define another, stronger topology, the *strong-* operator topology*, by the seminorms $T \mapsto \|T\xi\|$ and $T \mapsto \|T^*\xi\|$ for $\xi \in \mathcal{H}$, in which the adjoint is continuous. There is similarly a σ-strong-* operator topology.

Even though the adjoint is not strongly continuous, the set of self-adjoint operators is strongly closed (because it is weakly closed).

I.3.2 Properties of the Topologies

I.3.2.1 Addition and scalar multiplication are jointly continuous in all the topologies. Multiplication is separately continuous in all the topologies, and is jointly strongly and strong-* continuous on *bounded* sets. However, multiplication is not jointly strongly (or strong-*) continuous. Example I.3.1.5 shows that multiplication is not jointly weakly continuous even on bounded sets: S^n and $(S^*)^n$ both converge to 0 weakly, but $(S^*)^n S^n = I$ for all n. The separate weak continuity of multiplication implies that $\mathcal{S}'$ is weakly closed for any $\mathcal{S} \subseteq \mathcal{L}(\mathcal{H})$.

I.3.2.2 The (closed) unit ball of $\mathcal{L}(\mathcal{H})$ is complete in all these topologies [although $\mathcal{L}(\mathcal{H})$ itself is not weakly or strongly complete if $\mathcal{H}$ is infinite-dimensional: if T is an unbounded everywhere-defined operator on $\mathcal{H}$, for each finite-dimensional subspace F of $\mathcal{H}$ let $T_F = T \circ P_F$, where P_F is the orthogonal projection onto F. Then (T_F) is a strong Cauchy net in $\mathcal{L}(\mathcal{H})$ which does not converge weakly in $\mathcal{L}(\mathcal{H})$. However, $\mathcal{L}(\mathcal{H})$ *is* σ-strongly and σ-strong-* complete (II.7.3.7, II.7.3.11)!] In fact, since the σ-weak operator topology coincides with the weak-* topology on $\mathcal{L}(\mathcal{H})$ when regarded as the dual of the space of the trace-class operators (I.8.6), the unit ball of $\mathcal{L}(\mathcal{H})$ is compact in the weak operator topology. This can be proved directly using Banach limits, which are also useful in other contexts:

I.3.2.3 Let Ω be a fixed directed set. Let $C_b(\Omega)$ be the set of all bounded functions from Ω to $\mathbb{C}$. With pointwise operations and supremum norm, $C_b(\Omega)$ is a Banach space (in fact, it is a commutative C*-algebra). For $i \in \Omega$, let $\pi_i : C_b(\Omega) \to \mathbb{C}$ be evaluation at i; then π_i is a homomorphism and a linear functional of norm 1 (in fact, it is a pure state (II.6.2.9)). A *pure Banach limit* on Ω is a weak-* limit point ω of the net (π_i) in the unit ball of $C_b(\Omega)^*$ (which is weak-* compact). We usually write $\lim_\omega (\lambda_i)$ for $\omega((\lambda_i))$. A pure Banach limit has the following properties:

(i) $\lim_\omega (\lambda_i + \mu_i) = (\lim_\omega \lambda_i) + (\lim_\omega \mu_i)$.
(ii) $\lim_\omega (\lambda \mu_i) = \lambda (\lim_\omega \mu_i)$.
(iii) $\lim_\omega (\lambda_i \mu_i) = (\lim_\omega \lambda_i)(\lim_\omega \mu_i)$.
(iv) $\lim_\omega \overline{\lambda_i}$ is the complex conjugate of $\lim_\omega \lambda_i$.
(v) $\liminf |\lambda_i| \leq |\lim_\omega \lambda_i| \leq \limsup |\lambda_i|$.
(vi) If $\lambda_i \to \lambda$, then $\lambda = \lim_\omega \lambda_i$.

A *Banach limit* on Ω is an element of the closed convex hull of the pure Banach limits on Ω, i.e. an element of $\cap_{i \in \Omega} \overline{co}\{\pi_j | j \geq i\}$. A Banach limit has the same properties except for (iii) in general.

I.3.2.4 Banach limits can also be defined in $\mathcal{L}(\mathcal{H})$. Again fix Ω and a Banach limit ω on Ω. If (T_i) is a bounded net in $\mathcal{L}(\mathcal{H})$, for $\xi, \eta \in \mathcal{H}$ define $(\xi, \eta) = \lim_\omega \langle T_i \xi, \eta \rangle$. Then $(\cdot, \cdot)$ is a bounded sesquilinear form on $\mathcal{H}$, so, by I.2.2.2, there is a $T \in \mathcal{L}(\mathcal{H})$ with $(\xi, \eta) = \langle T\xi, \eta \rangle$ for all ξ, η. Write $T = \lim_\omega T_i$. This Banach limit has the following properties:

(i) $\lim_\omega (S_i + T_i) = (\lim_\omega S_i) + (\lim_\omega T_i)$.
(ii) $\lim_\omega (\lambda T_i) = \lambda (\lim_\omega T_i)$.
(iii) $\|\lim_\omega T_i\| \leq \limsup \|T_i\|$.
(iv) $\lim_\omega (T_i^*) = (\lim_\omega T_i)^*$.
(v) $\lim_\omega T_i$ is a weak limit point of (T_i).
(vi) If $T_i \to T$ weakly, then $T = \lim_\omega T_i$.

Property (v) implies the weak compactness of the unit ball of $\mathcal{L}(\mathcal{H})$.

The most useful version of completeness for the strong operator topology is the following:

I.3.2.5 PROPOSITION. Let (T_i) be a bounded increasing net of positive operators on a Hilbert space $\mathcal{H}$ (i.e. $T_i \leq T_j$ for $i \leq j$, and $\exists K$ with $\|T_i\| \leq K$ for all i). Then there is a positive operator T on $\mathcal{H}$ with $T_i \to T$ strongly, and $\|T\| = \sup \|T_i\|$. T is the least upper bound for $\{T_i\}$ in $\mathcal{L}(\mathcal{H})_+$.

For the proof, an application of the CBS inequality yields that, for any $\xi \in \mathcal{H}$, $(T_i\xi)$ is a Cauchy net in $\mathcal{H}$, which converges to a vector we call $T\xi$. It is easy to check that T thus defined is linear, bounded, and positive, and $T_i \to T$ strongly by construction.

It follows that every decreasing net of positive operators converges strongly to a positive operator.

I.3.2.6 COROLLARY. Let (T_i) be a monotone increasing (or decreasing) net of positive operators converging weakly to an operator T. Then $T_i \to T$ strongly, and T is the least upper bound (or greatest lower bound) of $\{T_i\}$.

Indeed, if (T_i) is increasing and weakly convergent to T, then $0 \leq T_i \leq T$ for all i, so $\|T_i\| \leq \|T\|$.

Here is a simple, but useful, fact which is closely related (in fact, combined with I.8.6 gives an alternate proof of I.3.2.5):

I.3.2.7 PROPOSITION. Let (T_i) be a net in $\mathcal{L}(\mathcal{H})$, and suppose $T_i^*T_i \to 0$ weakly. Then

(i) $T_i \to 0$ strongly.
(ii) If $\{T_i\}$ is bounded, then $T_i^*T_i \to 0$ strongly.

PROOF: (i): If $\langle T_i^*T_i\xi, \eta\rangle \to 0$ for all ξ, η, then for all ξ,

$$\|T_i\xi\|^2 = \langle T_i\xi, T_i\xi\rangle = \langle T_i^*T_i\xi, \xi\rangle \to 0.$$

(ii) If $\|T_i\| \leq K$ for all i, then for any ξ, $\|T_i^*T_i\xi\| \leq K\|T_i\xi\| \to 0$.

I.3.2.8 COROLLARY. Let (S_i) be a bounded net of positive operators converging weakly to 0. Then $S_i \to 0$ strongly.

PROOF: Set $T_i = S_i^{1/2}$ (I.4.2).

Caution: I.3.2.6 and I.3.2.8 do not mean that the strong and weak topologies coincide on the positive part of the unit ball of $\mathcal{L}(\mathcal{H})$!

They do, however, coincide on the set of unitaries:

I.3.2.9 PROPOSITION. The strong, weak, σ-strong, σ-weak, strong-*, and σ-strong-* topologies coincide on the group $\mathcal{U}(\mathcal{H})$ of unitary operators on $\mathcal{H}$, and make $\mathcal{U}(\mathcal{H})$ into a topological group.

The agreement of the strong and weak topologies on $\mathcal{U}(\mathcal{H})$ follows immediately from I.1.3.3. The adjoint (inverse) is thus strongly continuous on $\mathcal{U}(\mathcal{H})$, so the strong and strong-* topologies coincide. The σ-topologies agree with the others on bounded sets. Multiplication is jointly strongly continuous on bounded sets.

Note, however, that the closure of $\mathcal{U}(\mathcal{H})$ in $\mathcal{L}(\mathcal{H})$ is not the same in the strong, weak, and strong-* topologies. The strong closure of $\mathcal{U}(\mathcal{H})$ consists of isometries (in fact, all isometries; cf. [Hal67, Problem 225]), and thus $\mathcal{U}(\mathcal{H})$ is strong-* closed; whereas the weak closure includes coisometries (adjoints of isometries) also. Thus the unilateral shift S is in the strong closure, but not the strong-* closure, and S^* is in the weak closure but not the strong closure. In fact, the weak closure of $\mathcal{U}(\mathcal{H})$ is the entire closed unit ball of $\mathcal{L}(\mathcal{H})$ [Hal67, Problem 224].

I.3.2.10 THEOREM. [DD63] Let $\mathcal{H}$ be an infinite-dimensional Hilbert space. Then $\mathcal{U}(\mathcal{H})$ is contractible in the weak/strong/strong-* topologies, and the unit sphere $\{\xi \in \mathcal{H} : \|\xi\| = 1\}$ of $\mathcal{H}$ is contractible in the norm topology.

PROOF: First suppose $\mathcal{H}$ is separable, and hence $\mathcal{H}$ can be identified with $L^2[0,1]$. For $0 < t \le 1$, let $V_t \in \mathcal{L}(\mathcal{H})$ be defined by $[V_t(f)](s) = t^{-1/2}f(s/t)$ (setting $f(s) = 0$ for $s > 1$). Then V_t is an isometry, $V_1 = I$, and $t \to V_t$ and $t \to V_t^*$ are strongly continuous. If $P_t = V_t V_t^*$, then P_t is a projection (it is the projection onto the subspace $L^2[0,t]$) and $t \mapsto P_t$ is strongly continuous. In fact, P_t is multiplication by $\chi_{[0,t]}$. Then, as $t \to 0$, $V_t \to 0$ weakly and $P_t \to 0$ strongly, and the map from $\mathcal{U}(\mathcal{H}) \times [0,1]$ to $\mathcal{U}(\mathcal{H})$ defined by

$$(U,t) \mapsto V_t U V_t^* + (I - P_t)$$

for $t > 0$ and $(U,0) \mapsto I$ is continuous and gives a contraction of $\mathcal{U}(\mathcal{H})$. Similarly, if $f \in L^2[0,1]$, $\|f\| = 1$, set

$$H(f,t) = t^{1/2} V_t f + (1-t)^{1/2} \chi_{[t,1]};$$

this gives a contraction of the unit sphere of $\mathcal{H}$.

For a general $\mathcal{H}$, $\mathcal{H} \cong L^2[0,1] \otimes \mathcal{H}$. Use the same argument with $V_t \otimes I$ and $P_t \otimes I$ in place of V_t, P_t.

If $\mathcal{H}$ is infinite-dimensional, then N. Kuiper [Kui65] proved that $\mathcal{U}(\mathcal{H})$ is also contractible in the norm topology. The same is true for the unitary group of any properly infinite von Neumann algebra [BW76] and of the multiplier algebra of a stable σ-unital C*-algebra [CH87]. Note that the result fails if $dim(\mathcal{H}) = n < \infty$: the algebraic topology of $U(n, \mathbb{C})$ is fairly complicated.

I.4 Functional Calculus

We continue to follow the exposition of [Rie13] with some modernizations.

There is a very important procedure for applying functions to operators called *functional calculus*. Polynomials with complex coefficients can be applied to any operator in an obvious way. This procedure can be extended to holomorphic functions of arbitrary operators in a way described in II.1.5. For self-adjoint operators (and, more generally, for normal operators), this functional calculus can be defined for continuous functions and even bounded Borel measurable functions, as we now describe.

I.4.1 Functional Calculus for Continuous Functions

To describe the procedure for defining a continuous function of a self-adjoint operator, we first need a simple fact called the *Spectral Mapping Theorem* for polynomials:

I.4.1.1 PROPOSITION. Let $T \in \mathcal{L}(\mathcal{H})$, and let p be a polynomial with complex coefficients. Then $\sigma(p(T)) = \{p(\lambda) : \lambda \in \sigma(T)\}$.

The proof is an easy application of the Fundamental Theorem of Algebra: if $\lambda \in \mathbb{C}$, write $p(z) - \lambda = \alpha(z - \mu_1) \cdots (z - \mu_n)$; then

$$p(T) - \lambda I = \alpha(T - \mu_1 I) \cdots (T - \mu_n I)$$

and, since the $T - \mu_k I$ commute, $p(T) - \lambda I$ is invertible if and only if all the $T - \mu_k I$ are invertible.

Combining this result with I.2.6.10, we get:

I.4.1.2 COROLLARY. If T is self-adjoint and p is a polynomial with real coefficients, then $\|p(T)\| = \max\{|p(\lambda)| : \lambda \in \sigma(T)\}$.

(This is even true if p has complex coefficients (II.2.3.2)).

I.4.1.3 If T is self-adjoint and m, M are as in I.2.6.10, then polynomials with real coefficients are uniformly dense in the real-valued continuous functions on $[m, M]$ by the Stone-Weierstrass Theorem. If f is a real-valued continuous function on $\sigma(T)$ and (p_n) is a sequence of polynomials with real coefficients converging uniformly to f on $\sigma(T)$, then $(p_n(T))$ converges in norm to a self-adjoint operator we may call $f(T)$. The definition of $f(T)$ does not depend on the choice of the sequence (p_n). We have that $f \mapsto f(T)$ is an isometric algebra-isomorphism from the real Banach algebra $C_\mathbb{R}(\sigma(T))$, with supremum norm, onto the closed real subalgebra of $\mathcal{L}(\mathcal{H})$ generated by T. (This homomorphism extends to complex-valued continuous functions in a manner extending the application of polynomials with complex coefficients.) Since $f(T)$ is a limit of polynomials in T, $f(T) \in \{T\}''$; in particular, the $f(T)$ all commute with each other.

We have that $f(T) \geq 0$ if and only if $f \geq 0$ on $\sigma(T)$; more generally, we have $\sigma(f(T)) \subseteq [n, N]$, where

$$n = \min\{f(\lambda) : \lambda \in \sigma(T)\}, \quad N = \max\{f(\lambda) : \lambda \in \sigma(T)\}.$$

(Actually, with a bit more work (II.2.3.2) one can show the *Spectral Mapping Theorem* in this context: $\sigma(f(T)) = \{f(\lambda) : \lambda \in \sigma(T)\}$.)

I.4.2 Square Roots of Positive Operators

Applying this procedure to the function $f(t) = t^\alpha$, we get:

I.4.2.1 PROPOSITION. If $T \geq 0$, then there is a continuous function $\alpha \mapsto T^\alpha$ from $\mathbb{R}_+$ to $\mathcal{L}(\mathcal{H})_+$ satisfying $T^\alpha T^\beta = T^{\alpha+\beta}$ $\forall \alpha, \beta$ and $T^1 = T$. $T^\alpha \in \{T\}''$.

In particular, if $\alpha = 1/2$ we get a positive square root for T.

I.4.2.2 COROLLARY. Every positive operator has a positive square root.

In fact, the operator $T^{1/2}$ is the *unique* positive square root of T (II.3.1.2(vii)).

$T^{1/2}$ can be, and often is, explicitly defined using a sequence of polynomials converging uniformly to the function $f(t) = t^{1/2}$. For example, if $\|T\| \leq 1$ the sequence $p_0 = 0$, $p_{n+1}(t) = p_n(t) + \frac{1}{2}(t - p_n(t)^2)$ increases to $f(t)$ on $[0,1]$, and hence $p_n(T) \nearrow T^{1/2}$.

I.4.2.3 If T is self-adjoint, we define T_+, T_-, and $|T|$ to be $f(T)$, $g(T)$, and $h(T)$ respectively, where $f(t) = \max(t,0)$, $g(t) = -\min(t,0)$, and $h(t) = |t| = f(t) + g(t)$. We have $T_+, T_-, |T| \in \{T\}''$, $T_+, T_-, |T| \in \mathcal{L}(\mathcal{H})_+$, $T = T_+ - T_-$, $|T| = T_+ + T_-$, $T_+ T_- = 0$, $|T| = (T^2)^{1/2}$. T_+ and T_- are called the *positive* and *negative* parts of T, and $|T|$ is the *absolute value* of T.

If T is an arbitrary element of $\mathcal{L}(\mathcal{H})$, define $|T|$ to be $(T^*T)^{1/2}$.

I.4.3 Functional Calculus for Borel Functions

I.4.3.1 If T is self-adjoint, m, M as above, and f is a bounded nonnegative lower semicontinuous function on $[m, M]$, choose an increasing sequence f_n of nonnegative continuous functions on $[m, M]$ converging pointwise to f. Then $(f_n(T))$ is a bounded increasing sequence of positive operators, and hence converges strongly to a positive operator we may call $f(T)$. The definition of $f(T)$ is independent of the choice of the f_n, and $f(T) \in \{T\}''$. Similar considerations apply to decreasing sequences and upper semicontinuous functions.

I.4.3.2 If f is a bounded real-valued Borel measurable function on $[m, M]$, then f can be obtained by taking successive pointwise limits of increasing and decreasing sequences (perhaps transfinitely many times) beginning with continuous functions. We may then define $f(T)$ in the analogous way. This is again well defined; $f(T) \in \{T\}''$. See III.5.2.13 for a slick and rigorous treatment of Borel functional calculus. An alternate approach to defining $f(T)$ using the spectral theorem will be given in I.6.1.5.

I.5 Projections

Projections are the simplest non-scalar operators, and play a vital role throughout the theory of operators and operator algebras.

I.5.1 Definitions and Basic Properties

I.5.1.1 If $\mathcal{H}$ is a Hilbert space and $\mathcal{X}$ a closed subspace, then every $\xi \in \mathcal{H}$ can be uniquely written as $\eta + \zeta$, where $\eta \in \mathcal{X}$, $\zeta \in \mathcal{X}^\perp$. The function $\xi \mapsto \eta$ is a bounded self-adjoint idempotent operator $P_\mathcal{X}$, called the *(orthogonal) projection onto* $\mathcal{X}$. $P_\mathcal{X} \geq 0$, $P_\mathcal{X}^2 = P_\mathcal{X}$, $\|P_\mathcal{X}\| = 1$, and $\sigma(P_\mathcal{X}) = \{0,1\}$. Conversely, if $P \in \mathcal{L}(\mathcal{H})$ satisfies $P = P^* = P^2$, then P is a projection: $P = P_\mathcal{X}$, where $\mathcal{X} = \mathcal{R}(P)$. There is thus a one-one correspondence between projections in $\mathcal{L}(\mathcal{H})$ and closed subspaces of $\mathcal{H}$.

I.5.1.2 PROPOSITION. If $\mathcal{X}, \mathcal{Y}$ are closed subspaces of $\mathcal{H}$, then the following are equivalent:

(i) $P_\mathcal{X} \leq P_\mathcal{Y}$ (as elements of $\mathcal{L}(\mathcal{H})_+$).
(ii) $P_\mathcal{X} \leq \lambda P_\mathcal{Y}$ for some $\lambda > 0$.
(iii) $\mathcal{X} \subseteq \mathcal{Y}$.
(iv) $P_\mathcal{X} P_\mathcal{Y} = P_\mathcal{Y} P_\mathcal{X} = P_\mathcal{X}$.
(v) $P_\mathcal{Y} - P_\mathcal{X}$ is a projection (it is $P_{\mathcal{Y} \cap \mathcal{X}^\perp}$).

I.5.1.3 The set $\mathrm{Proj}(\mathcal{H})$ of projections in $\mathcal{L}(\mathcal{H})$ is a complete lattice (Boolean algebra). $P_\mathcal{X} \wedge P_\mathcal{Y} = P_{\mathcal{X} \cap \mathcal{Y}}$, $P_\mathcal{X} \vee P_\mathcal{Y} = P_{(\mathcal{X}+\mathcal{Y})^-}$, $P_\mathcal{X}^\perp = P_{\mathcal{X}^\perp} = I - P_\mathcal{X}$. $\bigwedge_i P_{\mathcal{X}_i} = P_{\cap \mathcal{X}_i}$ and $\bigvee_i P_{\mathcal{X}_i} = P_{(\sum \mathcal{X}_i)^-}$. If P and Q commute, then $P \wedge Q = PQ$ and $P \vee Q = P + Q - PQ$.

If (P_i) is an increasing net of projections, then $P_i \to \bigvee_i P_i$ strongly; if (P_i) is decreasing, then $P_i \to \bigwedge_i P_i$ strongly.

I.5.1.4 REMARK. Note that $\mathcal{L}(\mathcal{H})_+$ is not a lattice unless $\mathcal{H}$ is one-dimensional. For example, $P = \begin{bmatrix} 1 & 0 \\ 0 & 0 \end{bmatrix}$ and $Q = \frac{1}{2}\begin{bmatrix} 1 & 1 \\ 1 & 1 \end{bmatrix}$ have no least upper bound in $(\mathbb{M}_2)_+$. [The matrices $I = \begin{bmatrix} 1 & 0 \\ 0 & 1 \end{bmatrix}$ and $P + Q = \frac{1}{2}\begin{bmatrix} 3 & 1 \\ 1 & 1 \end{bmatrix}$ are upper bounds, but there is no T with $P, Q \leq T \leq I, P+Q$, i.e. $(\mathbb{M}_2)_+$ does not have the Riesz Interpolation Property.] Thus the supremum of P and Q in $\mathrm{Proj}(\mathcal{H})$ is not necessarily a least upper bound of P and Q in $\mathcal{L}(\mathcal{H})_+$.

However, if (P_i) is an increasing net of projections in $\mathcal{L}(\mathcal{H})$, then $P = \bigvee P_i$ is the least upper bound for $\{P_i\}$ in $\mathcal{L}(\mathcal{H})_+$, since $P_i \to P$ strongly, and for each $T \in \mathcal{L}(\mathcal{H})_+$, $\{S \in \mathcal{L}(\mathcal{H}) : 0 \leq S \leq T\}$ is strongly closed.

I.5.1.5 The lattice operations are not distributive in general: if P, Q, R are projections in $\mathcal{L}(\mathcal{H})$, then it is not true in general that $(P \vee Q) \wedge R = (P \wedge R) \vee (Q \wedge R)$ (this equality does hold if P, Q, R commute). But there is a weaker version called the *orthomodular law* which does hold in general:

I.5.1.6 PROPOSITION. Let P, Q, R be projections in $\mathcal{L}(\mathcal{H})$ with $P \perp Q$ and $P \leq R$. Then $(P + Q) \wedge R = P \vee (Q \wedge R)$.

I.5.1.7 As with unitaries (I.3.2.9), the strong, weak, σ-strong, σ-weak, strong-* topologies coincide on the set of projections on $\mathcal{H}$, using I.1.3.3, since if $P_i \to P$ weakly, for any $\xi \in \mathcal{H}$ we have

$$\|P_i\xi\|^2 = \langle P_i\xi, \xi\rangle \to \langle P\xi, \xi\rangle = \|P\xi\|^2.$$

The set of projections is strongly closed. The weak closure is the positive portion of the closed unit ball.

Projections in Tensor Products

Let $\mathcal{H}_1$ and $\mathcal{H}_2$ be Hilbert spaces, and P and Q projections in $\mathcal{L}(\mathcal{H}_1)$ and $\mathcal{L}(\mathcal{H}_2)$ respectively. Then $P \otimes Q$ is a projection in $\mathcal{L}(\mathcal{H}_1 \otimes \mathcal{H}_2)$: it is the projection onto the subspace $P\mathcal{H}_1 \otimes Q\mathcal{H}_2$.

I.5.1.8 PROPOSITION. Let $\{P_i : i \in I\}$ and $\{Q_j : j \in J\}$ be sets of projections in $\mathcal{L}(\mathcal{H}_1)$ and $\mathcal{L}(\mathcal{H}_2)$ respectively. If $\bigvee_i P_i = P$ in $\mathcal{L}(\mathcal{H}_1)$ and $\bigvee_j Q_j = Q$ in $\mathcal{L}(\mathcal{H}_2)$, then $\bigvee_{i,j} P_i \otimes Q_j = P \otimes Q$ in $\mathcal{L}(\mathcal{H}_1 \otimes \mathcal{H}_2)$.

PROOF: The span of $\{P_i\mathcal{H}_1 : i \in I\}$ is dense in $P\mathcal{H}_1$ and the span of $\{Q_j\mathcal{H}_2 : j \in J\}$ is dense in $Q\mathcal{H}_2$, so the span of

$$\{(P_i \otimes Q_j)(\mathcal{H}_1 \otimes \mathcal{H}_2) : i \in I, j \in J\}$$

is dense in $P\mathcal{H}_1 \otimes Q\mathcal{H}_2 = (P \otimes Q)(\mathcal{H}_1 \otimes \mathcal{H}_2)$.

I.5.2 Support Projections and Polar Decomposition

I.5.2.1 If $T \in \mathcal{L}(\mathcal{X}, \mathcal{Y})$, then the *right support projection* of T is the projection P_T onto $\mathcal{N}(T)^\perp$. P_T is the unique projection in $\mathcal{L}(\mathcal{X})$ with the property that $TP_T = T$ and $P_TS = 0$ whenever $S \in \mathcal{L}(\mathcal{X})$ and $TS = 0$, and is the smallest projection P such that $TP = T$. Similarly, the *left support projection* of T is the projection $Q_T \in \mathcal{L}(\mathcal{Y})$ onto the closure of the range of T, i.e. the right support projection of T^*. Q_T satisfies $Q_TT = T$ and $SQ_T = 0$ whenever $S \in \mathcal{L}(\mathcal{Y})$ and $ST = 0$.

If $T \in \mathcal{L}(\mathcal{H})$ is normal, then its left and right support projections coincide. If $T \geq 0$, then $T^\alpha \to P$ strongly as $\alpha \to 0$, so P_T is a strong limit of polynomials in T and therefore $P_T \in \{T\}''$. Also,

$$T(T + \epsilon I)^{-1} \to P_T$$

strongly as $\epsilon \to 0$. In general, $P_T = P_{T^*T}$ and $Q_T = Q_{TT^*} = P_{TT^*}$, so $(T^*T)^\alpha \to P_T$ and $(TT^*)^\alpha \to Q_T$ strongly as $\alpha \to 0$.

If P, Q are projections, then $P \vee Q$ is the support projection of $P + Q$; thus $P \vee Q$ and $P \wedge Q = I - [(I - P) \vee (I - Q)]$ are in $\{P, Q\}''$.

I.5.2.2 The operators $|T| = (T^*T)^{1/2}$ and T map $P_T\mathcal{X}$ one-one onto dense subspaces of $P_T\mathcal{X}$ and $Q_T\mathcal{Y}$ respectively. Also,

$$\||T|\xi\|^2 = \langle |T|\xi, |T|\xi \rangle = \langle T^*T\xi, \xi \rangle = \langle T\xi, T\xi \rangle = \|T\xi\|^2$$

for any ξ. Thus the map $|T|\xi \mapsto T\xi$ is well-defined, linear, and isometric, and extends to an isometry U from $P_T\mathcal{X}$ onto $Q_T\mathcal{Y}$. U may be regarded as an element of $\mathcal{L}(\mathcal{X}, \mathcal{Y})$ by setting $U = 0$ on $(P_T\mathcal{X})^\perp$. Then U satisfies $U^*U = P_T$, $UU^* = Q_T$, and $T = U|T|$, and is the unique element of $\mathcal{L}(\mathcal{X}, \mathcal{Y})$ with these properties. $T = U|T|$ is called the *(left) polar decomposition* of T. The operator T also has a right polar decomposition coming from the left polar decomposition of T^*; in fact,

$$TT^* = U|T|^2U^* = U(T^*T)U^*$$

and by iteration $(TT^*)^n = U(T^*T)^n U^*$ for all n, so

$$|T^*| = (TT^*)^{1/2} = U(T^*T)^{1/2}U^* = U|T|U^*$$

and hence $T = U|T| = |T^*|U$. The uniqueness of polar decomposition implies that if $T \in \mathcal{L}(\mathcal{H})$ with polar decomposition $T = U|T|$, and $V \in \mathcal{L}(\mathcal{H})$ is unitary, then the polar decomposition of VT is $(VU)|T|$.

$T(T^*T + \epsilon I_\mathcal{X})^{-1/2}$ converges strongly to U as $\epsilon \to 0$. So if $T \in \mathcal{L}(\mathcal{H})$, then $U \in \{T, T^*\}''$, and obviously $|T| \in \{T, T^*\}''$ also. If $T \in \mathcal{L}(\mathcal{X}, \mathcal{Y})$ and T^*T is invertible, then U is an isometry, and $U = T(T^*T)^{-1/2}$. So if $T \in \mathcal{L}(\mathcal{H})$ is invertible, then U (as well as $|T|$) is in the norm-closed algebra generated by T and T^*. Similarly, if $T \in \mathcal{L}(\mathcal{X}, \mathcal{Y})$ and TT^* is invertible, then U is a coisometry and $U = (TT^*)^{-1/2}T$. If T is invertible (or, more generally, one-one with dense range), then U is unitary.

I.5.2.3 In general, U is a partial isometry from $P_T\mathcal{X}$ to $Q_T\mathcal{X}$. A *partial isometry* is an operator $U \in \mathcal{L}(\mathcal{X}, \mathcal{Y})$ such that U^*U is a projection P. A partial isometry U is an isometry from $\mathcal{N}(U)^\perp$ onto $\mathcal{R}(U)$, and $P = U^*U = P_{\mathcal{N}(U)^\perp}$; UU^* is also a projection $Q = P_{\mathcal{R}(U)}$. The projections P and Q are called the *initial* and *final* projections, or *source* and *range* projections, of U.

There is a useful generalization of the construction of U:

I.5.2.4 PROPOSITION. Let $S, T \in \mathcal{L}(\mathcal{H})$ with $S^*S \leq T^*T$. Then there is a unique $W \in \mathcal{L}(\mathcal{H})$ with $W^*W \leq Q_T$ (hence $\|W\| \leq 1$), and $S = WT$. If $R \in \mathcal{L}(\mathcal{H})$ commutes with S, T, and T^*, then $RW = WR$.

PROOF: W is defined on $\mathcal{R}(T)$ by $W(T\xi) = S\xi$ (W is well defined since $\|S\xi\| \leq \|T\xi\|$ for all ξ). W extends to an operator on $Q_T\mathcal{H}$ by continuity; set $W = 0$ on $Q_T^\perp\mathcal{H}$. Then $W^*W \leq Q_T$ and $S = WT$. If $RT = TR$, $RT^* = T^*R$, and $RS = SR$, then for $\eta = T\xi \in \mathcal{R}(T)$ we have

$$WR\eta = WRT\xi = WTR\xi = SR\xi = RS\xi = RWT\xi = RW\eta$$

and thus $RW\eta = WR\eta$ for $\eta \in Q_T\mathcal{H}$. Since $R(TT^*) = (TT^*)R$, $RQ_T = Q_T R$, and thus R leaves $(I-Q_T)\mathcal{H}$ invariant; thus if $\eta \in Q_T^\perp\mathcal{H}$, then $WR\eta = RW\eta = 0$.

A simple matrix trick yields a multivariable version:

I.5.2.5 PROPOSITION. Let $S, T_1, \ldots, T_n \in \mathcal{L}(\mathcal{H})$ with $S^*S \le \sum_{k=1}^n T_k^* T_k$. Then there are unique $W_1, \ldots, W_n \in \mathcal{L}(\mathcal{H})$ with

$$\sum_{k=1}^n W_k^* W_k \le Q_{\sum T_k^* T_k}$$

(hence $\|W_k\| \le 1$), and $S = \sum_{k=1}^n W_k T_k$. If $R \in \mathcal{L}(\mathcal{H})$ commutes with S, T_k, and T_k^* ($1 \le k \le n$), then $RW_k = W_k R$ ($1 \le k \le n$).

PROOF: Identify $\mathcal{L}(\mathcal{H}^n)$ with $M_n(\mathcal{L}(\mathcal{H}))$, and set $\tilde{S} = diag(S, 0, \cdots, 0)$, and $\tilde{T}$ the matrix with first column $T_1, \ldots, T_n$ and other entries 0. Apply I.5.2.4 to obtain $\tilde{S} = \tilde{W}\tilde{T}$. Then $\tilde{W}$ has first row $W_1, \ldots, W_n$ and other entries 0.

I.5.2.6 If T is a bounded conjugate-linear operator on $\mathcal{H}$, and J is an involution (I.2.3.3), then $JT \in \mathcal{L}(\mathcal{H})$, and $T^*T = (JT)^*(JT)$, so $|T| = |JT|$ is well defined; if $JT = U|JT| = U|T|$ is the polar decomposition of JT, then $T = V|T|$ is a polar decomposition for T, where $V = JU$ is a "conjugate-linear partial isometry," i.e. V^*V and VV^* are projections which are naturally the right and left support projections for T. This construction is independent of the choice of J (if K is another involution, then JK is unitary). We also have $JT = |T^*J|U$ and $(JTT^*J)^n = J(TT^*)^n J$ for all n, so

$$|T^*J| = (JTT^*J)^{1/2} = J|T^*|J$$

and hence $T = J(JT) = |T^*|JU = |T^*|V$ as in the linear case. Thus bounded conjugate-linear operators have a well-defined and unique polar decomposition with analogous properties to those of linear operators.

I.6 The Spectral Theorem

The spectral theorem gives a complete description of self-adjoint (bounded) operators on a Hilbert space as generalized multiplication operators (I.2.4.3(i)). There are also versions of the theorem for normal operators and for unbounded self-adjoint operators. The version and exposition here comes essentially from [Rie13] with some modernizations.

I.6.1 Spectral Theorem for Bounded Self-Adjoint Operators

I.6.1.1 If T is a self-adjoint operator on a Hilbert space $\mathcal{H}$ and $\lambda \in \mathbb{R}$, let $E_{(\lambda,\infty)} = E_{(\lambda,\infty)}(T) = P_{(T-\lambda I)_+}$, the support projection of $(T-\lambda I)_+$ (I.4.2.3, I.5.2). The family $\{E_{(\lambda,\infty)} : \lambda \in \mathbb{R}\}$ is called the *spectral resolution* of T, and has the following properties:

$E_{(\lambda,\infty)} \geq E_{(\mu,\infty)}$ if $\lambda \leq \mu$

$(E_{(\lambda,\infty)})$ is strongly continuous from the right

If $\sigma(T) \subseteq [m, M]$, then $E_{(\lambda,\infty)} = I$ for $\lambda < m$ and $E_{(\lambda,\infty)} = 0$ for $\lambda \geq M$.

I.6.1.2 We then define $E_{[\lambda,\infty)} = \bigwedge_{\mu<\lambda} E_{(\mu,\infty)}$. We have that $E_{[\lambda,\infty)} \geq E_{(\lambda,\infty)}$, and $E_{[\lambda,\infty)} - E_{(\lambda,\infty)}$ is the projection onto the eigenspace $\{\xi \in \mathcal{H} : T\xi = \lambda\xi\}$, so for "most" λ we have $E_{[\lambda,\infty)} = E_{(\lambda,\infty)}$. Furthermore, for $\lambda \leq \mu$, set $E_{[\lambda,\mu]} = E_{[\lambda,\infty)} - E_{(\mu,\infty)}$, and similarly define $E_{(\lambda,\mu]}$, $E_{[\lambda,\mu)}$, and $E_{(\lambda,\mu)}$. More generally, if B is any Borel subset of $\sigma(T)$, E_B may be defined by successive infima and suprema. (See III.5.2.13 for another way to define the E_B.) E_B is called the *spectral projection* of T corresponding to B. The family $\{E_B\}$ satisfies:

(i) $E_A \leq E_B$ if $A \subseteq B$
(ii) $E_A \perp E_B$ if $A \cap B = \emptyset$
(iii) If $A \cap B = \emptyset$, then $E_{A \cup B} = E_A + E_B$
(iv) If $\bigcup_n B_n = B$, then $E_B = \bigvee_n E_{B_n}$
(v) $E_\emptyset = 0$, $E_{\sigma(T)} = I$
(vi) $\sigma(E_B T) \subseteq \bar{B} \cup \{0\}$.

A family of projections with properties (i)-(v) is called a *projection-valued measure* (or *resolution of the identity*) over $\sigma(T)$.

I.6.1.3 In addition, all E_B are strong limits of polynomials in T and are therefore in $\{T\}''$; in particular, all the spectral projections of T commute with each other and with T, and the range of E_B is invariant under T. (ii) and (vi) say that the range of E_B is the "part" of $\mathcal{H}$ on which T "looks like" a multiplication operator (with multiplicity) by a function taking values in B. In particular, $\lambda I \leq E_{[\lambda,\mu]} T \leq \mu I$ on the range of $E_{[\lambda,\mu]}$.

It follows easily that if $\sigma(T) \subseteq [m, M]$ and $\{\lambda_0 = m < \lambda_1 < \cdots < \lambda_n = M\}$ is a partition of $[m, M]$, then $\|T - \sum_{i=1}^n \lambda_i E_{[\lambda_{i-1},\lambda_i)}\| \leq \max\{|\lambda_i - \lambda_{i-1}|\}$. Thus we obtain:

I.6.1.4 THEOREM. [SPECTRAL THEOREM] If T is a (bounded) self-adjoint operator on a Hilbert space $\mathcal{H}$, and $\{E_B(T)\}$ are its spectral projections, then

$$T = \int_{-\infty}^{\infty} \lambda \, dE_\lambda = \int_{-\|T\|}^{\|T\|} \lambda \, dE_\lambda$$

under any reasonable interpretation of the integral (e.g. in the sense that the Riemann sum approximations to the integral converge in norm to T.)

One can also interpret functional calculus using the spectral resolution:

I.6.1.5 THEOREM. Let T be a self-adjoint operator on a Hilbert space $\mathcal{H}$, and f a bounded complex-valued Borel function on $\sigma(T)$. Then

$$f(T) = \int_{-\infty}^{\infty} f(\lambda) \, dE_\lambda = \int_{-\|T\|}^{\|T\|} f(\lambda) \, dE_\lambda$$

where the integral can be interpreted as a Riemann (or Riemann-Stieltjes) integral if f is (piecewise) continuous and a Lebesgue (Lebesgue-Stieltjes) integral otherwise.

I.6.2 Spectral Theorem for Normal Operators

I.6.2.1 There is also a version of the spectral theorem for normal operators. If N is a normal operator on $\mathcal{H}$, write $N = S+iT$ with S, T self-adjoint. Then, since S and T commute, the spectral projections for S and T all commute with each other and with S and T; the ranges of the spectral projections for S are invariant under T, and vice versa. Thus, if A and B are Borel subsets of $\sigma(S)$ and $\sigma(T)$ respectively, we may set $E_{A \times B}(N) = E_A(S) E_B(T)$. By successive orthogonal sums, infima, and suprema, we may then define a spectral projection $E_C(N)$ for any Borel subset C of $\sigma(N) \subseteq \mathbb{C}$. The family $\{E_C(N)\}$ is a projection-valued measure over $\sigma(N)$ (I.6.1.2), and we have:

I.6.2.2 THEOREM. [SPECTRAL THEOREM FOR NORMAL OPERATORS] Let N be a normal operator on a Hilbert space H, and $\{E_C(N)\}$ its spectral resolution. Then

$$N = \int_{\sigma(N)} \lambda \, dE_\lambda$$

where the integral can be interpreted as a Riemann integral over a rectangle in $\mathbb{C}$ (or as a Lebesgue integral over $\sigma(N)$).

I.6.2.3 Although there is no actual (scalar-valued) measure associated to a spectral resolution, there is a well-defined notion of a "set of measure zero": a Borel set B has N-measure zero if $E_B(N) = 0$. There is thus a notion of "N-almost everywhere." In fact, there is a well-defined measure class corresponding to the spectral "measure." No nonempty open set has zero measure.

I.6.2.4 If f is a bounded Borel function on $\sigma(N)$, then we may define the functional calculus element $f(N)$ by

$$f(N) = \int_{\sigma(N)} f(\lambda) \, dE_\lambda$$

where the integral can be interpreted as a Riemann integral if f is continuous and a Lebesgue integral in general. This functional calculus has the following properties:

(i) If p is a polynomial of two variables with complex coefficients, and $f(z) = p(z,\bar z)$, then $f(N) = p(N, N^*)$.
(ii) If $f_n \to f$ uniformly on $\sigma(N)$, then $f_n(N) \to f(N)$ in norm.
(iii) If (f_n) is a uniformly bounded sequence and $f_n \to f$ pointwise N-a.e. on $\sigma(N)$, then $f_n(N) \to f(N)$ strongly.
(iv) $\sigma(f(N)) \subseteq \{f(\lambda) : \lambda \in \sigma(N)\}^-$, with equality if f is continuous.
(v) $\|f(N)\|$ is the "essential supremum" of $|f|$ on $\sigma(N)$:

$$\|f(N)\| = \inf\{\alpha > 0 : E_{\{\lambda : |f(\lambda)| > \alpha\}}(N) = 0\}$$

and if f is continuous, then $\|f(N)\| = \sup\{|f(\lambda)| : \lambda \in \sigma(N)\}$.
(vi) $f(N)$ is self-adjoint if and only if $f(z) \in \mathbb{R}$ N-a.e.; $f(N)$ is a projection if and only if $f(z) = 0$ or 1 N-a.e. (i.e. if f is a characteristic function); $f(N)$ is unitary if and only if $|f(z)| = 1$ N-a.e.

A consequence of the Spectral Theorem is a precise realization of any normal operator as a multiplication operator:

I.6.2.5 THEOREM. Let T be a normal operator on a separable Hilbert space $\mathcal{H}$. Then, for $1 \leq n \leq \infty$, there are disjoint Borel subsets X_n of $\sigma(T)$, with $\cup X_n = \sigma(T)$, and finite regular Borel measures μ_n on X_n, such that T is unitarily equivalent to $\bigoplus_n M_z$ on $\bigoplus_n L^2(X_n, \mu_n, \mathcal{H}_n)$, where $\mathcal{H}_n$ is an n-dimensional Hilbert space and M_z is multiplication by $f(z) = z$. The X_n are uniquely determined up to sets of T-measure zero, and the μ_n are unique up to equivalence.

In connection with functional calculus, the following fact is significant. It was first proved by B. Fuglede [Fug50] in the case $M = N$, and the general result was obtained by C. Putnam [Put51]. The proof given here is from [Ros58].

I.6.2.6 THEOREM. Let $T, M, N \in \mathcal{L}(\mathcal{H})$, with M, N normal. If $TM = NT$, then $TM^* = N^*T$.

PROOF: If p is any polynomial with complex coefficients, then $Tp(M) = p(N)T$, so for any $\lambda \in \mathbb{C}$, $Te^{i\lambda M} = e^{i\lambda N}T$, i.e. $T = e^{-i\lambda N}Te^{i\lambda M}$. Thus, if $f(\lambda) = e^{-i\lambda N^*}Te^{i\lambda M^*}$, f is an entire function from $\mathbb{C}$ to $\mathcal{L}(\mathcal{H})$, and we have

$$f(\lambda) = e^{-i\lambda N^*}e^{-i\bar\lambda N}Te^{i\lambda M}e^{i\lambda M^*} = e^{-i(\lambda N^* + \bar\lambda N)}Te^{i(\bar\lambda M + \lambda M^*)}.$$

But $\lambda N^* + \bar\lambda N$ and $\bar\lambda M + \lambda M^*$ are self-adjoint for all λ, and thus $f(\lambda) = U(\lambda)TV(\lambda)$, where $U(\lambda), V(\lambda)$ are unitary. Thus $\|f(\lambda)\|$ is constant, and therefore f is constant by Liouville's Theorem. So $0 = f'(0) = -iN^*T + iTM^*$.

I.6.2.7 COROLLARY. [FUGLEDE] If $T, N \in \mathcal{L}(\mathcal{H})$, N is normal, and T commutes with N, then T commutes with N^* (alternatively, T^* commutes with N).

I.6.2.8 COROLLARY. If $T, N \in \mathcal{L}(\mathcal{H})$, N is normal, T commutes with N, and f is a bounded Borel function on $\sigma(N)$, then T and T^* commute with $f(N)$.

In particular, under the hypotheses, if $N = U|N|$ is the polar decomposition, then T (and also T^*) commutes with $Re(N)$, $Im(N)$, U, and $|N|$.

I.7 Unbounded Operators

It is beyond the scope of this volume to give a complete treatment of unbounded operators. However, at some points in the theory of von Neumann algebras, it is very useful to work with unbounded operators, so we briefly discuss the parts of the theory relevant to our applications.

I.7.1 Densely Defined Operators

I.7.1.1 DEFINITION. Let $\mathcal{H}$ be a Hilbert space. A *partially defined operator* on $\mathcal{H}$ with domain $\mathcal{D}$ consists of (1) a subspace $\mathcal{D}$ of $\mathcal{H}$ and (2) a linear transformation $T : \mathcal{D} \to \mathcal{H}$. A *densely defined operator* on $\mathcal{H}$ with domain $\mathcal{D}$ is a partially defined operator on $\mathcal{H}$ whose domain $\mathcal{D}$ is dense in $\mathcal{H}$.

One can also consider partially defined operators on real Hilbert spaces, or conjugate-linear partially defined operators on (complex) Hilbert spaces.

A partially defined operator (or densely defined operator) may be bounded or unbounded. We will be almost exclusively concerned with densely defined operators.

Note that the specification of the domain is a very important part of the definition of a partially defined operator (we usually write $\mathcal{D}(T)$ for the domain of T). We will sometimes omit explicit specification of the domain where the omission should cause no confusion (as long as the need to specify the domain is kept in mind). Thus a partially defined operator on $\mathcal{H}$ is technically not an operator on $\mathcal{H}$ in the sense of I.2.1.1 (sometimes called an *everywhere defined operator*) in general. A bounded densely defined operator on $\mathcal{H}$ extends uniquely to a bounded (everywhere defined) operator on $\mathcal{H}$, but we will distinguish between the operator and its extension.

I.7.1.2 If S, T are partially defined operators on $\mathcal{H}$, we say $S = T$ if $\mathcal{D}(S) = \mathcal{D}(T)$ and $S = T$ on $\mathcal{D}(S)$; we say $S \subseteq T$ if $\mathcal{D}(S) \subseteq \mathcal{D}(T)$ and $S = T$ on $\mathcal{D}(S)$ (T is called an *extension* of S). Define algebraic operations by letting $S + T$ be the partially defined operator whose domain is $\mathcal{D}(S + T) = \mathcal{D}(S) \cap \mathcal{D}(T)$ and which is $S + T$ on $\mathcal{D}(S+T)$; ST is the partially defined operator whose domain is

$$\mathcal{D}(ST) = \{\xi \in \mathcal{H} : \xi \in \mathcal{D}(T), T\xi \in \mathcal{D}(S)\}$$

and which is ST on $\mathcal{D}(ST)$. Thus a sum or product of densely defined operators is not necessarily densely defined. If T is a partially defined operator on $\mathcal{H}$ and $S \in \mathcal{L}(\mathcal{H})$, then $S+T$ and ST have the same domain as T (but TS may have a different domain, and need not be densely defined if T is). If $R, S \in \mathcal{L}(\mathcal{H})$, then $TR + TS \subseteq T(R+S)$, but equality does not hold in general (e.g. if $S = -R$). We do always have $(R+S)T = RT + ST$.

One must be careful in defining commutation relations among partially defined operators: $ST = TS$ implies $\mathcal{D}(ST) = \mathcal{D}(TS)$. This is a very strong requirement: T does not even commute with the 0 operator in this sense unless T is everywhere defined. A more useful notion if S is bounded is:

I.7.1.3 DEFINITION. Let T be a partially defined operator on $\mathcal{H}$, and $S \in \mathcal{L}(\mathcal{H})$. Then S and T are *permutable* if $ST \subseteq TS$. More generally, if $\mathcal{S} \subseteq \mathcal{L}(\mathcal{H})$, T is permutable with $\mathcal{S}$ if T is permutable with every $S \in \mathcal{S}$. In this case, we say T is *affiliated* with $\mathcal{S}'$, written $T \sim \mathcal{S}'$.

If P is a projection in $\mathcal{L}(\mathcal{H})$ which is permutable with T, then $PTP \subseteq (TP)P = TP$, but TP and PTP have the same domain, so $TP = PTP$. Also,

$$\mathcal{D}((I-P)T) = \mathcal{D}(T) = \mathcal{D}(T - TP) \subseteq \mathcal{D}(T(I-P))$$

and

$$(I-P)T\xi = T\xi - PT\xi = T\xi - TP\xi = T(I-P)\xi$$

for $\xi \in \mathcal{D}(T)$, so $(I-P)T \subseteq T(I-P)$ and therefore $T(I-P) = (I-P)T(I-P)$, and hence P reduces T, i.e. both P and $I-P$ map $\mathcal{D}(T)$ into itself, T maps $P\mathcal{D}(T)$ and $(I-P)\mathcal{D}(T)$ into themselves, and $T = PTP + (I-P)T(I-P)$ (in the strict sense). Conversely, if P reduces T in this sense, then P is permutable with T.

I.7.1.4 If T is a partially defined operator on $\mathcal{H}$ which is one-to-one on $\mathcal{D}(T)$, then there is an "inverse" T^{-1} with domain $\mathcal{D}(T^{-1}) = \mathcal{R}(T)$; $T^{-1}T$ and TT^{-1} are the identity maps on $\mathcal{D}(T)$ and $\mathcal{R}(T)$ respectively. If S and T are both one-one on their domains with "inverses" S^{-1} and T^{-1}, then ST is one-one on its domain and its "inverse" $(ST)^{-1}$ is precisely $T^{-1}S^{-1}$.

I.7.1.5 If T is a densely defined operator on $\mathcal{H}$, then a *core* for T is a subspace $\mathcal{D}$ of $\mathcal{D}(T)$ such that the graph $\Gamma(T|_\mathcal{D})$ is dense in the graph $\Gamma(T)$ (viewed as subspaces of $\mathcal{H} \oplus \mathcal{H}$). A core for T is obviously a dense subspace of $\mathcal{D}(T)$; note, however, that if T is unbounded and $\mathcal{D}$ is a dense subspace of $\mathcal{D}(T)$, then the graph of $T|_\mathcal{D}$ is not in general dense in the graph of T, even if T is self-adjoint (I.7.3.2), so the notion of core is more restrictive than just a dense subspace of $\mathcal{D}(T)$. If $\mathcal{T}$ is a set of densely defined operators, then a *core* for $\mathcal{T}$ is a subspace $\mathcal{D}$ of $\mathcal{D}(\mathcal{T}) = \bigcap_{T \in \mathcal{T}} \mathcal{D}(T)$ which is simultaneously a core for each $T \in \mathcal{T}$. (If $\mathcal{T}$ has a core, then $\mathcal{D}(\mathcal{T})$ is dense, but the converse is not true in general.) The notion of a core as a common domain for the operators in $\mathcal{T}$ is a useful one, even for a single operator, since there is often a dense proper subspace of $\mathcal{D}(T)$ on which T is simpler to describe than on all of $\mathcal{D}(T)$ (e.g. I.7.4.7).

I.7.2 Closed Operators and Adjoints

A partially defined operator is not continuous unless it is bounded. But there is a vestige of continuity in an important class of operators:

I.7.2.1 DEFINITION. A densely defined operator T on $\mathcal{H}$ is *closed* if its graph $\Gamma(T)$ is closed in $\mathcal{H} \times \mathcal{H}$. A densely defined operator T is *closable* (or *preclosed*) if T has a closed extension, i.e. if the closure of $\Gamma(T)$ is the graph of a (densely defined) operator (called the closure of T).

An everywhere defined closed operator is bounded by the Closed Graph Theorem. Conversely, a bounded densely defined operator is closable, and its closure is its unique everywhere defined extension. We will only consider closed (or sometimes closable) operators.

If T is closed and $S \in \mathcal{L}(\mathcal{H})$, then $S + T$ is closed. If T is closed and one-to-one with dense range, then T^{-1} is also closed. If T is closed and bounded below on $\mathcal{D}(T)$, then $\mathcal{R}(T)$ is closed.

Closable operators have a nice alternate characterization in terms of adjoints.

I.7.2.2 DEFINITION. Let T be a densely defined operator on $\mathcal{H}$. The *adjoint* T^* of T is the partially defined operator defined as follows. The domain $\mathcal{D}(T^*)$ is the set of $\eta \in \mathcal{H}$ for which there is a (necessarily unique) vector $\zeta \in \mathcal{H}$ such that $\langle \xi, \zeta \rangle = \langle T\xi, \eta \rangle$ for all $\xi \in \mathcal{D}(T)$ (i.e. $\eta \in \mathcal{D}(T^*)$ if and only if the linear functional $\xi \mapsto \langle T\xi, \eta \rangle$ is bounded on $\mathcal{D}(T)$ and thus extends to a bounded linear functional on $\mathcal{H}$.) Set $T^*\eta = \zeta$.

A densely defined operator T is *adjointable* if $\mathcal{D}(T^*)$ is dense in $\mathcal{H}$.

If T is densely defined and bounded, then T^* is everywhere defined and bounded, and is the usual adjoint of the closure of T. If T is a densely defined operator on $\mathcal{H}$, then $\mathcal{R}(T)^\perp \subseteq \mathcal{D}(T^*)$ and is the null space of T^*. If S, T are densely defined operators and $S \subseteq T$, then $T^* \subseteq S^*$ (so if T is adjointable, so is S). If T is adjointable, then T^* is also adjointable and $(T^*)^* \supseteq T$.

If T is adjointable, then it is easily seen that T^* is closed. Thus, if T is adjointable, then it is closable ($(T^*)^*$ is a closed extension). The converse is also true:

I.7.2.3 PROPOSITION. Let T be a densely defined operator on $\mathcal{H}$. Then T is closable if and only if T is adjointable. If T is closable, then $(T^*)^*$ is the closure of T.

PROOF: It suffices to show that if T is closed, then T is adjointable and $(T^*)^* = T$. Let V be the unitary operator on $\mathcal{H} \oplus \mathcal{H}$ defined by $V(\xi, \eta) = (\eta, -\xi)$. $V^2 = -I$, and $\Gamma(S^*) = [V\Gamma(S)]^\perp$ for any densely defined operator S on $\mathcal{H}$, and in particular $\Gamma(T^*) = [V\Gamma(T)]^\perp$. So, since V is unitary,

$$V\Gamma(T^*) = [V^2\Gamma(T)]^\perp = \Gamma(T)^\perp$$

$$[V\Gamma(T^*)]^\perp = \Gamma(T)^{\perp\perp} = \Gamma(T)$$

since T is closed. If $\eta \in \mathcal{D}(T^*)^\perp$, then $(0,\eta) \in [V\Gamma(T^*)]^\perp = \Gamma(T)$, so $\eta = 0$ and $\mathcal{D}(T^*)$ is dense. Furthermore, setting $S = T^*$ above, $\Gamma((T^*)^*) = [V\Gamma(T^*)]^\perp = \Gamma(T)$, $(T^*)^* = T$.

I.7.2.4 PROPOSITION. Let T be a closed operator on $\mathcal{H}$ which is one-to-one with dense range, and let T^{-1} be its "inverse," with domain $\mathcal{R}(T)$ (so T^{-1} is closed). Then T^* is one-one with dense range and $(T^*)^{-1} = (T^{-1})^*$.

PROOF: Since $\mathcal{N}(T^*) = \mathcal{R}(T)^\perp$ and $\mathcal{R}(T^*)^\perp = \mathcal{N}(T)$, T^* is one-one with dense range. Let U be the unitary in $\mathcal{L}(\mathcal{H} \times \mathcal{H})$ defined by $U(\xi,\eta) = (\eta,\xi)$, and let V be as in the proof of I.7.2.3. $U^2 = I$ and $UV = -VU$. Then $\Gamma(T^{-1}) = U\Gamma(T)$, and

$$\Gamma((T^{-1})^*) = [V\Gamma(T^{-1})]^\perp = [VU\Gamma(T)]^\perp = [UV\Gamma(T)]^\perp$$
$$= U\Gamma(T^*) = \Gamma((T^*)^{-1}).$$

I.7.2.5 DEFINITION. Let T be a densely defined operator on $\mathcal{H}$. The *spectrum* $\sigma(T)$ is the complement of the set of $\lambda \in \mathbb{C}$ such that $T - \lambda I$ is one-to-one on $\mathcal{D}(T - \lambda I) = \mathcal{D}(T)$ with dense range, and $(T - \lambda I)^{-1}$ is bounded (i.e. $T - \lambda I$ is bounded below on $\mathcal{D}(T)$ with dense range).

The spectrum of an unbounded densely defined operator need not be closed, bounded, or nonempty in general. If T is closed and $T - \lambda I$ is bounded below, then $T - \lambda I$ has closed range; hence if $\lambda \notin \sigma(T)$, then the range of $T - \lambda I$ is all of $\mathcal{H}$.

I.7.2.6 The results of this section hold also for densely defined operators on real Hilbert spaces, or conjugate-linear densely defined operators on (complex) Hilbert spaces, using the same arguments as in the bounded case (I.2.3.3).

I.7.3 Self-Adjoint Operators

I.7.3.1 DEFINITION. Let T be a densely defined operator on $\mathcal{H}$. Then T is *symmetric* if $T \subseteq T^*$, and T is *self-adjoint* if $T = T^*$.

A symmetric operator is closable (its closure is also symmetric) and a self-adjoint operator is closed. (Note, however, that the adjoint of a symmetric operator is not symmetric in general.)

I.7.3.2 EXAMPLES. Let $\mathcal{D}_2$ be the set of $f \in L^2[0,1]$ such that f is absolutely continuous and $f' \in L^2[0,1]$,

$$\mathcal{D}_1 = \{f \in \mathcal{D}_2 : f(0) = f(1)\}$$

$$\mathcal{D}_0 = \{f \in \mathcal{D}_2 : f(0) = f(1) = 0\}.$$

Then $\mathcal{D}_0$ is dense in $L^2[0,1]$, and $\mathcal{D}_k$ has codimension one in $\mathcal{D}_{k+1}$ ($k = 0, 1$). Let T_k be defined by $T_k f = f'$ with domain $\mathcal{D}_k$ ($k = 0, 1, 2$). Then each T_k is closed, T_1 is self-adjoint, and $T_0^* = T_2$, so T_0 is closed and symmetric, but not self-adjoint. $\mathcal{D}_0$ is dense in $\mathcal{D}(T_1)$, but $\Gamma(T_0)$ is not dense in $\Gamma(T_1)$. ($\mathcal{D}_2$ with its inner product inherited from its natural identification with $\Gamma(T_2)$ is the first Sobolev space.)

There exist closed symmetric operators with no self-adjoint extension (e.g. the inverse Cayley transform of the unilateral shift). On the other hand, a symmetric operator can have more than one self-adjoint extension, and they can be essentially different (e.g. different spectra).

I.7.3.3 If T is symmetric, then $\langle T\xi, \xi \rangle \in \mathbb{R}$ for all $\xi \in \mathcal{D}(T)$. Thus, if T is symmetric and $\lambda \in \mathbb{C} \setminus \mathbb{R}$, then $T - \lambda I$ is bounded below by $|\text{Im}(\lambda)|$ and hence is one-to-one and $(T - \lambda I)^{-1}$ is bounded. If T is closed and symmetric, then $T - \lambda I$ is closed and bounded below and hence has closed range. If T is self-adjoint, then $(T - \lambda I)^* = T - \bar{\lambda}I$ is also one-to-one, so $\mathcal{R}(T - \lambda I)$ is dense in $\mathcal{H}$ and therefore equal to $\mathcal{H}$ since it is closed; thus $(T - \lambda I)^{-1}$ is everywhere defined and bounded. So, if T is self-adjoint, $\sigma(T) \subseteq \mathbb{R}$. Conversely, if T is closed and symmetric and $\mathcal{R}(T - \lambda I)$ is dense for some $\lambda \in \mathbb{C} \setminus \mathbb{R}$, then T is self-adjoint.

The next proposition is an immediate corollary of I.7.2.4.

I.7.3.4 PROPOSITION. Let T be a self-adjoint densely defined operator on $\mathcal{H}$ which is one-one on $\mathcal{D}(T)$ with dense range. Then the "inverse" T^{-1}, with domain $\mathcal{R}(T)$, is self-adjoint.

I.7.3.5 THEOREM. Let T be a closed operator on $\mathcal{H}$. Then T^*T is densely defined and self-adjoint; $I + T^*T$ maps $\mathcal{D}(T^*T)$ one-to-one onto $\mathcal{H}$, and $(I + T^*T)^{-1}$ and $T(I + T^*T)^{-1}$ are everywhere defined, bounded, and of norm ≤ 1, and $(I + T^*T)^{-1} \geq 0$. Also, $\mathcal{D}(T^*T) = \mathcal{R}((I + T^*T)^{-1})$.

PROOF: Let $V \in \mathcal{L}(\mathcal{H} \times \mathcal{H})$ be as in the proof of I.7.2.3. Then $\Gamma(T) = [V\Gamma(T^*)]^\perp$, so if $\zeta \in \mathcal{H}$ there are unique $\xi \in \mathcal{D}(T)$ and $\eta \in \mathcal{D}(T^*)$ such that $(\zeta, 0) = (\xi, T\xi) + (T^*\eta, -\eta)$, i.e. $\zeta = \xi + T^*\eta$, $0 = T\xi - \eta$ (so $\eta = T\xi$), and

$$\|\xi\|^2 + \|\eta\|^2 \leq \|\xi\|^2 + \|T\xi\|^2 + \|T^*\eta\|^2 + \|\eta\|^2 = \|\zeta\|^2.$$

Set $R\zeta = \xi$ and $S\zeta = \eta$. Then R and S are everywhere defined and bounded, with $\|R\|, \|S\| \leq 1$, and $R + T^*S = I$ and $TR - S = 0$, so $S = TR$ and

$$(I + T^*T)R = I.$$

In particular, TR, $T^*S = T^*TR$, and $(I + T^*T)R$ are everywhere defined, i.e. $\mathcal{R}(R) \subseteq \mathcal{D}(T^*T)$. If $\xi \in \mathcal{H}$, then

$$\langle R\xi, \xi \rangle = \langle R\xi, (I + T^*T)R\xi \rangle = \langle R\xi, R\xi \rangle + \langle TR\xi, TR\xi \rangle \geq 0$$

and thus $R \geq 0$. R is clearly one-to-one and hence has dense range (I.7.1.1), so T^*T is densely defined. $I + T^*T$ is clearly one-to-one on its domain $\mathcal{D}(T^*T)$, and hence has an "inverse" defined on its range. From $(I + T^*T)R = I$, the range of $I + T^*T$ is all of $\mathcal{H}$, and therefore $(I + T^*T)^{-1}$ is defined everywhere and equals R. Since R is self-adjoint, its "inverse" $I + T^*T$ is self-adjoint by I.7.3.4, and hence T^*T is also self-adjoint; and $\mathcal{D}(T^*T) = \mathcal{D}(I+T^*T) = \mathcal{R}(R)$.

Note that $\mathcal{D}(T^*T)$ is a proper subset of $\mathcal{D}(T)$ in general ($\mathcal{D}(T^*T)$ is a core for T).

I.7.3.6 Many aspects of the theory of unbounded operators are somewhat counterintuitive. For example, if S and T are closed densely defined operators with $S \subseteq T$, then it is not true in general that $S^*S \subseteq T^*T$ (indeed, since both are self-adjoint, it would follow that $S^*S = T^*T$). In fact, if T_0, T_1, and T_2 are as in I.7.3.2, then $T_0 \subseteq T_1 \subseteq T_2$, but $T_0^*T_0$, $T_1^*T_1$, and $T_2^*T_2 = T_0T_0^*$ all have incomparable domains, although they all agree on the intersection of the domains, which is a dense subspace of $L^2[0,1]$.

I.7.3.7 A densely defined operator S is *positive* if S is self-adjoint and $\langle S\xi, \xi \rangle \geq 0$ for all $\xi \in \mathcal{D}(S)$ (the first condition does not follow from the second in the unbounded case). From I.7.3.5, if T is a closed operator, then T^*T is positive. If S is positive, then $\sigma(S) \subseteq [0, \infty)$.

I.7.4 The Spectral Theorem and Functional Calculus for Unbounded Self-Adjoint Operators

I.7.4.1 The simplest route to the Spectral Theorem for unbounded self-adjoint operators, and the approach taken by von Neumann in his original work [vN30a], is via the Cayley transform. For $t \in \mathbb{R}$, set $c(t) = \frac{t+i}{t-i}$. Then c maps $\mathbb{R}$ one-one onto $\mathbb{T} \setminus \{1\}$, and $c^{-1}(\lambda) = i\frac{\lambda+1}{\lambda-1}$.

If T is a self-adjoint operator on $\mathcal{H}$, let

$$c(T) = (T + iI)(T - iI)^{-1}.$$

Then $T + iI$ and $T - iI$ map $\mathcal{D}(T)$ one-one onto $\mathcal{H}$ (I.7.3.3), so $c(T)$ maps $\mathcal{H}$ one-one onto $\mathcal{H}$. Also, for $\xi \in \mathcal{D}(T)$, a simple computation shows that

$$\|(T + iI)\xi\| = \|(T - iI)\xi\|$$

so $c(T)$ is unitary. We also have that 1 is not an eigenvalue of $c(T)$. Conversely, if U is a unitary for which 1 is not an eigenvalue, so that $U - I$ is one-to-one, then $U - I$ has dense range (I.7.1.1) and it is easily verified that

$$c^{-1}(U) = i(U + I)(U - I)^{-1}$$

is a self-adjoint operator with domain $\mathcal{R}(U - I)$. If $1 \notin \sigma(U)$, then $c^{-1}(U)$ is everywhere defined and bounded. We have that $c^{-1}(c(T)) = T$ and $c(c^{-1}(U)) = U$; $c(T)$ is permutable with T, and commutes with any $S \in \mathcal{L}(\mathcal{H})$ which is permutable with T.

I.7.4.2 DEFINITION. The operator $c(T)$ is the *Cayley transform* of T, and $c^{-1}(U)$ is the *inverse Cayley transform* of U.

If E is a spectral projection of $c(T)$ corresponding to a Borel subset A of $\mathbb{T}$, then E is permutable with T; in fact, EU is a unitary on $E\mathcal{H}$, and it is easily verified that $c^{-1}(EU)$, which is a self-adjoint operator on $E\mathcal{H}$, agrees with $T|_{E\mathcal{H}}$. Furthermore, E commutes with any bounded operator which is permutable with T.

I.7.4.3 DEFINITION. E is called the *spectral projection* for T corresponding to $B = c^{-1}(A)$, denoted $E_B(T)$.

I.7.4.4 If A is bounded away from 1, so that $B = c^{-1}(A)$ is a bounded subset of $\mathbb{R}$, then $T|_{E\mathcal{H}}$ is bounded, and its spectrum is contained in $\bar{B}$; and if F is the spectral projection for $c(T)$ corresponding to $A' \subseteq A$, then F is the spectral projection in the usual sense for $T|_{E\mathcal{H}}$ corresponding to $c^{-1}(A')$. Thus, if we set $E_\lambda(T) = E_{c((-\infty,\lambda))}(U)$, we can by the ordinary Spectral Theorem (I.6.1.4) write $T|_{E\mathcal{H}} = \int_B \lambda \, dE(\lambda)$.

A byproduct of this argument is that

$$\sigma(T) \cap [-n, n] = \sigma(T|_{E_{[-n,n]}(T)\mathcal{H}})$$

is closed for each n, so $\sigma(T)$ is closed in $\mathbb{R}$.

So if (A_n) is an increasing sequence of Borel subsets of $\mathbb{T} \setminus \{1\}$, each bounded away from 1, with $\cup A_n = \mathbb{T} \setminus \{1\}$, so that $B_n = c^{-1}(A_n)$ is an increasing sequence of bounded Borel subsets of $\mathbb{R}$ with $\cup B_n = \mathbb{R}$, and $E_n = E_{A_n}(U) = E_{B_n}(T)$, then $\mathcal{H}_0 = \cup E_n \mathcal{H}$ is dense in $\mathcal{H}$, and is a core for T, i.e. T is the closure of $T|_{\mathcal{H}_0}$. So we obtain:

I.7.4.5 THEOREM. [SPECTRAL THEOREM FOR UNBOUNDED SELF-ADJOINT OPERATORS] Let T be a self-adjoint (densely defined) operator on a Hilbert space $\mathcal{H}$. Then there is a projection-valued measure $\{E_B\}$ on the Borel subsets of $\mathbb{R}$, consisting of projections permutable with T and commuting with all bounded operators permutable with T, for which

$$T = \int_{-\infty}^{\infty} \lambda \, dE(\lambda)$$

where the integral can be interpreted as an "improper Riemann integral" (in the precise sense described above).

Just as in the bounded case, there is a notion of "T-almost everywhere": a Borel subset B of $\mathbb{R}$ has T-measure 0 if $E_B(T) = 0$. The complement of $\sigma(T)$ has T-measure 0.

I.7.4.6 We can define functional calculus in analogy with the bounded case. Let $f: \mathbb{R} \to \mathbb{C}$ be Borel measurable. We want to define $f(T)$ for a self-adjoint T. $f(T)$ should be an operator which can be symbolically written

$$f(T) = \int_{-\infty}^{\infty} f(\lambda)\, dE(\lambda)$$

so we must properly interpret this expression; the only difficulty is with finding the right domain. Let (B_n) be an increasing sequence of bounded Borel subsets of $\mathbb{R}$ such that f is bounded on each B_n and $\cup B_n = \mathbb{R}$, and let $E_n = E_{B_n}(T)$. Then $f(T|_{E_n \mathcal{H}}) = \int_{B_n} f(\lambda)\, dE(\lambda)$ in the usual sense (I.6.2.4), and $f(T|_{E_{n+1}\mathcal{H}})$ agrees with $f(T|_{E_n\mathcal{H}})$ on $E_n\mathcal{H}$, so this family defines a densely defined operator $f(T)_0$ with domain $\mathcal{H}_0 = \cup E_n\mathcal{H}$. $(f(T)_0)^* \supseteq \bar{f}(T)_0$, so $f(T)_0$ is closable.

I.7.4.7 DEFINITION. $f(T)$ is the closure of $f(T)_0$.

The domain $\mathcal{H}_0$, and therefore the operator $f(T)_0$, depends on the choice of the E_n, but $f(T)$ is independent of the choice of the E_n. If T is bounded and f is bounded on $\sigma(T)$, then $f(T)$ agrees with the usual definition. Functional calculus has the following properties:

(i) If $f = g$ T-a.e. (in particular, if $f = g$ on $\sigma(T)$), then $f(T) = g(T)$.
(ii) $f(T)^* = \bar{f}(T)$; in particular, $f(T)$ is self-adjoint if (and only if) f is real-valued T-a.e.
(iii) $f(T)$ is bounded if (and only if) f is bounded T-a.e.; if f is bounded, $f(T)$ is normal and $\|f(T)\|$ is the T-essential supremum of f.
(iv) $f(T)$ is positive if (and only if) $f \geq 0$ T-a.e.
(v) $f(T)$ is unitary if (and only if) $|f| = 1$ T-a.e.
(vi) If (f_n) is a uniformly bounded sequence of functions converging pointwise T-a.e. to f, then $f_n(T) \to f(T)$ strongly.

Proof of (vi): Suppose $|f_n(t)| \leq K$ for all n and t. For each m and k, let

$$B_{m,k} = \{t : |f_n(t) - f(t)| \leq 1/m \text{ for all } n \geq k\}$$

and $E_{m,k} = E_{B_{m,k}}(T)$. For fixed m, $\mathbb{R} \setminus \cup_k B_{m,k}$ has T-measure 0; so $\cup_k E_{m,k}\mathcal{H}$ is dense in $\mathcal{H}$. Thus, if $\xi \in \mathcal{H}$, there is a k and $\eta \in E_{m,k}\mathcal{H}$ with $\|\xi - \eta\| \leq 1/m$, and then

$$\|(f_n(T) - f(T))\xi\| \leq \|f_n(T)(\xi - \eta)\|$$
$$+ \|(f_n(T) - f(T))\eta\| + \|f(T)(\eta - \xi)\| \leq \frac{2K+1}{m}$$

for all $n \geq k$.

Note that the exact domain of $f(T)$ is rather subtle; the domain in general depends on f, and need not contain or be contained in $\mathcal{D}(T)$. However, if $\{f_k\}$ is a countable set of functions, then there is a core for $\{f_k(T)\}$ which is also a core for T, namely $\cup E_{B_n}(T)\mathcal{H}$, where (B_n) is an increasing sequence of bounded Borel sets on which each of the f_k are individually bounded, and such that $\mathbb{R} \setminus \cup B_n$ has T-measure 0.

This functional calculus procedure also works, and is interesting, for unbounded Borel functions of bounded self-adjoint operators, and can be extended to unbounded functions of bounded normal operators.

I.7.4.8 An important special case of functional calculus is to define T_+, T_-, and $|T|$ for a self-adjoint operator T. These are positive operators. In addition, if T is positive, then T^α can be defined for any $\alpha > 0$, and in particular for $\alpha = 1/2$. Thus every positive operator has a square root. We have that $|T| = (T^2)^{1/2}$ for any self-adjoint T.

I.7.4.9 If T is a closed operator, then there is a partial isometry V from $[\mathcal{R}(T^*)]^-$ to $[\mathcal{R}(T)]^-$ such that $T = V|T| = |T^*|V$. (In particular, $\mathcal{D}(|T|) = \mathcal{D}(T)$.) The proof consists of taking a dense subspace $\mathcal{H}_0$ of $\mathcal{H}$ as above, using spectral projections for T^*T, and then noting that $\||T|\xi\| = \|T\xi\|$ for $\xi \in \mathcal{H}_0$ just as in the bounded case. Thus there is polar decomposition for closed operators. As in I.5.2.6, polar decomposition also works for closed conjugate-linear operators.

Note that if S and T are closed operators with $S \subseteq T$, then

$$\mathcal{D}(|S|) = \mathcal{D}(S) \subseteq \mathcal{D}(T) = \mathcal{D}(|T|)$$

but usually $|S| \not\subseteq |T|$ (I.7.3.6). For example, if T_0, T_1 are as in I.7.3.2, then $|T_0|$ has dense range since T_0 is one-to-one, but $|T_1|$ does not have dense range.

Finally, there is a more nontrivial application of functional calculus which is very important. Let H be a self-adjoint operator on $\mathcal{H}$, and for $t \in \mathbb{R}$ set $U_t = e^{itH}$. Then U_t is unitary, $U_{s+t} = U_s U_t$, and $t \mapsto U_t$ is strongly continuous ((U_t) is a *strongly continuous one-parameter group of unitaries*). The map $t \mapsto U_t$ is norm-continuous if and only if H is bounded. The converse is also true [Sto32]:

I.7.4.10 THEOREM. [STONE] Let (U_t) be a strongly continuous one-parameter group of unitaries on $\mathcal{H}$. Then there is a unique self-adjoint operator H on $\mathcal{H}$, called the *generator* of the group, such that $U_t = e^{itH}$ for all t. The domain of H consists of all vectors ξ for which $\lim_{t\to 0} t^{-1}(U_t - I)\xi$ exists, and $H\xi$ is this limit.

In fact, if (α_t) is a strongly continuous one-parameter automorphism group of $\mathcal{L}(\mathcal{H})$, then there is such a group of unitaries such that $\alpha_t = ad\, U_t$, i.e. there is a unique (up to adding a scalar) self-adjoint operator H such that $\alpha_t(T) = e^{itH} T e^{-itH}$ for all $t \in \mathbb{R}$ and $T \in \mathcal{L}(\mathcal{H})$.

I.8 Compact Operators

Some of the simplest, yet most important, classes of operators are the compact operators (often called *completely continuous* operators, particularly in earlier references), and the subclasses of finite-rank, trace-class, and Hilbert-Schmidt operators.

I.8.1 Definitions and Basic Properties

I.8.1.1 DEFINITION. Let $\mathcal{X}$ and $\mathcal{Y}$ be Banach spaces. An operator $T : \mathcal{X} \to \mathcal{Y}$ is *compact* if T sends bounded subsets of $\mathcal{X}$ to precompact subsets of $\mathcal{Y}$.

In other words, if B is a bounded subset of $\mathcal{X}$, then $\overline{T(B)}$ is a compact subset of $\mathcal{Y}$. Equivalently, whenever (ξ_n) is a bounded sequence of vectors in $\mathcal{X}$, then $(T\xi_n)$ has a convergent subsequence.

I.8.1.2 It is obvious that a compact operator is bounded, and the composition (in either order) of a compact operator with a bounded operator is compact. It is easily seen that a norm-limit of compact operators is compact. The set of compact operators from $\mathcal{X}$ to $\mathcal{Y}$, denoted $\mathcal{K}(\mathcal{X}, \mathcal{Y})$, is a closed subspace of $\mathcal{L}(\mathcal{X}, \mathcal{Y})$; $\mathcal{K}(\mathcal{H}) = \mathcal{K}(\mathcal{H}, \mathcal{H})$ is a closed two-sided ideal in $\mathcal{L}(\mathcal{H})$.

If $\mathcal{X}$ is a Hilbert space (or, more generally, a reflexive Banach space), $T \in \mathcal{K}(\mathcal{X}, \mathcal{Y})$, and B is the closed unit ball of $\mathcal{X}$, then it follows from I.1.3.2 and I.2.1.2 that $T(B)$ is already closed and therefore compact.

I.8.1.3 EXAMPLES.

(i) Any finite-rank bounded operator is compact. Conversely, if $T \in \mathcal{L}(\mathcal{X}, \mathcal{Y})$ is compact and bounded below, then $\mathcal{X}$ is finite-dimensional.

(ii) Let X and Y be measure spaces, and $k \in L^2(X \times Y)$. Define an operator $T : L^2(X) \to L^2(Y)$ by $(Tf)(y) = \int_X k(x,y)f(x)\,dx$. T is a compact operator, called the *integral kernel operator* with kernel k. (In fact, T is a Hilbert-Schmidt operator (I.8.5)).

A straightforward exercise shows:

I.8.1.4 PROPOSITION. Let $\mathcal{X}, \mathcal{Y}, \mathcal{Z}$ be Banach spaces, $T \in \mathcal{L}(\mathcal{X}, \mathcal{Y})$ compact, and (S_i) a uniformly bounded net of elements of $\mathcal{L}(\mathcal{Y}, \mathcal{Z})$ converging strongly to S. Then $S_i T \to ST$ in norm.

I.8.1.5 If $\mathcal{Y}$ is a Hilbert space and $\{\eta_i : i \in \Omega\}$ an orthonormal basis, for a finite subset $F \subseteq \Omega$ let P_F be the projection onto the span of $\{\eta_i : i \in F\}$. Then P_F is a finite-rank projection, and $P_F \to I$ strongly. So if $T \in \mathcal{L}(\mathcal{X}, \mathcal{Y})$ is compact, then $P_F T \to T$ in norm, i.e. T is a norm-limit of finite-rank operators. [This can fail if $\mathcal{Y}$ is not a Hilbert space [Enf73].] Thus $\mathcal{K}(\mathcal{X}, \mathcal{Y})$ is precisely the norm-closure of the set $\mathcal{C}(\mathcal{X}, \mathcal{Y})$ of bounded finite-rank operators.

As a corollary, the adjoint of a compact operator is compact. If T is compact, T^*T is too and so is $f(T^*T)$ for any continuous function f vanishing at 0. In particular, $|T| = (T^*T)^{1/2}$ is compact.

I.8.2 The Calkin Algebra

I.8.2.1 Since $\mathcal{K}(\mathcal{H})$ is a closed ideal in $\mathcal{L}(\mathcal{H})$ for a Hilbert space $\mathcal{H}$ (assumed infinite-dimensional), we may form the quotient Banach algebra $\mathcal{Q}(\mathcal{H}) = \mathcal{L}(\mathcal{H})/\mathcal{K}(\mathcal{H})$, called the *Calkin algebra of* $\mathcal{H}$. The term "Calkin algebra" without further qualification will mean the Calkin algebra of a separable, infinite-dimensional Hilbert space. We usually denote the quotient map from $\mathcal{L}(\mathcal{H})$ to $\mathcal{Q}(\mathcal{H})$ by π.

Since $\mathcal{K}(\mathcal{H})$ is self-adjoint, the involution $T \mapsto T^*$ on $\mathcal{L}(\mathcal{H})$ drops to a well-defined involution on $\mathcal{Q}(\mathcal{H})$. Actually, $\mathcal{Q}(\mathcal{H})$ is a C*-algebra (II.5.1.1).

Thus all the properties and constructions for general Banach algebras, or C*-algebras, are valid for $\mathcal{Q}(\mathcal{H})$; see Chapter 2. In this chapter we will need to use only the elementary fact that the group of invertible elements in a unital Banach algebra is open and contains the open ball of radius 1 around the identity (II.1.3).

I.8.2.2 If $T \in \mathcal{L}(\mathcal{H})$, we define the *essential spectrum* $\sigma_e(T)$ to be the spectrum of $\pi(T)$ in $\mathcal{Q}(\mathcal{H})$, i.e.

$$\sigma_e(T) = \{\lambda \in \mathbb{C} : \pi(T - \lambda I) \text{ is not invertible in } \mathcal{Q}(\mathcal{H})\}$$

Clearly $\sigma_e(T) \subseteq \sigma(T)$, and in fact $\sigma_e(T) \subseteq \cap_{K \in \mathcal{K}(\mathcal{H})} \sigma(T + K)$ (the inclusion is proper in general). By II.1.4.2, $\sigma_e(T)$ is a nonempty compact subset of $\mathbb{C}$.

I.8.3 Fredholm Theory

The theory of Fredholm operators and Fredholm index is not only important for applications, but (in retrospect) was a pioneering example of bringing algebraic topology into the theory of operator algebras (cf. II.8.4.32, II.8.4.36). In fact, it was one of the first examples of the process called "noncommutative topology," which underlies much of the modern theory of operator algebras.

I.8.3.1 DEFINITION. Let $\mathcal{X}$ and $\mathcal{Y}$ be Banach spaces and $T \in \mathcal{L}(\mathcal{X}, \mathcal{Y})$. Then T is a *Fredholm operator* if $\mathcal{N}(T)$ is finite-dimensional and $\mathcal{R}(T)$ has finite codimension. The *index* of T is $dim \, \mathcal{N}(T) - codim \, \mathcal{R}(T)$.

I.8.3.2 It follows easily from the Open Mapping Theorem that the range of a Fredholm operator is automatically closed. Thus, if $\mathcal{X}$ and $\mathcal{Y}$ are Hilbert spaces (the only situation we will consider), then $T \in \mathcal{L}(\mathcal{X}, \mathcal{Y})$ is Fredholm if and only if $\mathcal{R}(T)$ is closed and $\mathcal{N}(T)$ and $\mathcal{N}(T^*)$ are finite-dimensional, and then $index(T) = dim \, \mathcal{N}(T) - dim \, \mathcal{N}(T^*)$. If T is Fredholm, then T is bounded below on $\mathcal{N}(T)^\perp$; it follows that T^* is also Fredholm, and $index(T^*) = -index(T)$.

I.8.3.3 EXAMPLES.

(i) If $\mathcal{X}$ and $\mathcal{Y}$ are finite-dimensional, then every operator from $\mathcal{X}$ to $\mathcal{Y}$ is Fredholm with index $dim\ \mathcal{X} - dim\ \mathcal{Y}$.
(ii) If T is invertible, then T is Fredholm with index 0.
(iii) Let S be the unilateral shift (I.2.4.3). Then S is Fredholm with index -1. S^n is Fredholm with index $-n$, and $(S^*)^n$ is Fredholm with index n.

The parts of the following proposition are easy consequences of the definition of a compact operator.

I.8.3.4 PROPOSITION. Let $\mathcal{H}$ be a Hilbert space, $T \in \mathcal{K}(\mathcal{H})$, and $\lambda \in \mathbb{C}$, $\lambda \neq 0$. Then

(i) $\mathcal{N}(T - \lambda I)$ is finite-dimensional.
(ii) $T - \lambda I$ is bounded below on $\mathcal{N}(T - \lambda I)^\perp$, and hence $\mathcal{R}(T - \lambda I)$ is closed.
(iii) There is an n such that $\mathcal{N}((T - \lambda I)^n) = \mathcal{N}((T - \lambda I)^{n+1})$ [$T - \lambda I$ has *finite ascent*.] The smallest such n is called the *ascent* of $T - \lambda I$.

Applying the proposition to both $T - \lambda I$ and $T^* - \bar{\lambda}I = (T - \lambda I)^*$, we see that

I.8.3.5 COROLLARY. Let $\mathcal{H}$ be a Hilbert space, $T \in \mathcal{K}(\mathcal{H})$, and $\lambda \in \mathbb{C}$, $\lambda \neq 0$. Then

(i) $T - \lambda I$ is Fredholm.
(ii) There is an m such that $\mathcal{R}((T - \lambda I)^m) = \mathcal{R}((T - \lambda I)^{m+1})$ [$T - \lambda I$ has *finite descent*.]

The next theorem [Atk53] gives a useful alternate characterization of Fredholm operators.

I.8.3.6 THEOREM. [ATKINSON] Let $\mathcal{H}$ be an infinite-dimensional Hilbert space and $\pi : \mathcal{L}(\mathcal{H}) \to \mathcal{Q}(\mathcal{H}) = \mathcal{L}(\mathcal{H})/\mathcal{K}(\mathcal{H})$ the quotient map onto the Calkin algebra of $\mathcal{H}$. Then $T \in \mathcal{L}(\mathcal{H})$ is Fredholm if and only if $\pi(T)$ is invertible in $\mathcal{Q}(\mathcal{H})$.

PROOF: If T is Fredholm, then T is bounded below on $\mathcal{N}(T)^\perp$, so there is an $S \in \mathcal{L}(\mathcal{H})$ mapping $\mathcal{R}(T)$ to $\mathcal{N}(T)^\perp$ and zero on $\mathcal{R}(T)^\perp = \mathcal{N}(T^*)$, such that $ST = I - P_{\mathcal{N}(T)}$ and $TS = I - P_{\mathcal{N}(T^*)}$ (S is called a *quasi-inverse* for T; we have $TST = T$ and $STS = S$). Thus $\pi(S)$ is an inverse for $\pi(T)$.

Conversely, suppose $\pi(T)$ is invertible in $\mathcal{Q}(\mathcal{H})$, and let $S \in \mathcal{L}(\mathcal{H})$ such that $\pi(S) = \pi(T)^{-1}$. Then $ST = I + K$, $TS = I + L$ for $K, L \in \mathcal{K}(\mathcal{H})$. $\mathcal{N}(T) \subseteq \mathcal{N}(ST)$ which is finite-dimensional by I.8.3.5; $\mathcal{R}(T) \supseteq \mathcal{R}(TS)$ has finite codimension by I.8.3.2.

I.8.3.7 COROLLARY.

(i) If S and T are Fredholm and $K \in \mathcal{K}(\mathcal{H})$, then ST and $T+K$ are Fredholm.
(ii) The set $\mathcal{F}(\mathcal{H})$ of Fredholm operators on $\mathcal{H}$ is open.

The most important basic facts about the index function are:

I.8.3.8 THEOREM. Let $\mathcal{H}$ be a Hilbert space.

(i) If $S, T \in \mathcal{F}(\mathcal{H})$, then $index(ST) = index(S) + index(T)$.
(ii) The *index* is locally constant on $\mathcal{F}(\mathcal{H})$, and hence constant on connected components (path components) of $\mathcal{F}(\mathcal{H})$.
(iii) If $T \in \mathcal{F}(\mathcal{H})$ and $K \in \mathcal{K}(\mathcal{H})$, then $index(T + K) = index(T)$.

The proof of (i) is a straightforward but moderately involved calculation, and (ii) is a simple consequence of (i) and the fact that the invertible operators form an open set. For (iii), note that T and $T+K$ are connected by the path $\{T + tK : 0 \leq t \leq 1\}$ in $\mathcal{F}(\mathcal{H})$.

If T is a Fredholm operator of index 0 and K is a one-one map of $\mathcal{N}(T)$ onto $\mathcal{R}(T)^\perp$, then $S = T - K$ is invertible. Thus we obtain:

I.8.3.9 COROLLARY. A Fredholm operator T has index 0 if and only if $T = S + K$, where S is invertible and $K \in \mathcal{K}(\mathcal{H})$. (So, since $\mathcal{K}(\mathcal{H})$ and the set of invertible elements in $\mathcal{L}(\mathcal{H})$ are connected, *index* maps the connected components of $\mathcal{F}(\mathcal{H})$ one-one onto $\mathbb{Z}$.) In particular, if $T \in \mathcal{K}(\mathcal{H})$ and $\lambda \in \mathbb{C}$, $\lambda \neq 0$, then $T - \lambda I$ has index 0.

This last result is often stated in the following form, called the *Fredholm Alternative*:

I.8.3.10 COROLLARY. If $T \in \mathcal{K}(\mathcal{H})$ and $\lambda \in \mathbb{C}$, $\lambda \neq 0$, then $T - \lambda I$ is injective if and only if it is surjective. So if $\lambda \in \sigma(T)$, $\lambda \neq 0$, then λ is an eigenvalue of T.

So, under the hypotheses of the corollary, either the equation $(T-\lambda I)\xi = \eta$ has a unique solution for every η, or the homogeneous equation $(T - \lambda I)\xi = 0$ has a nontrivial (but finite-dimensional) space of solutions. This result, applied to integral kernel operators, is important in the theory of differential and integral equations.

I.8.3.11 If $T \in \mathcal{K}(\mathcal{H})$ and $\lambda \neq 0$, and $n > 0$ is the ascent of $T - \lambda I$ (I.8.3.4(iii)), let $\mathcal{H}_0 = \mathcal{N}((T - \lambda I)^n)$, $\mathcal{H}_1 = \mathcal{R}((T - \lambda I)^n)$. Then $\mathcal{H}_0$ and $\mathcal{H}_1$ are complementary closed subspaces of $\mathcal{H}$ invariant under $T - \lambda I$, $(T - \lambda I)|_{\mathcal{H}_0}$ is nilpotent, and $(T - \lambda I)|_{\mathcal{H}_1}$ is invertible. It follows that λ is an isolated point of $\sigma(T)$. Thus $\sigma(T)$ consists of 0 and a (possibly finite) sequence of eigenvalues converging to 0.

I.8.3.12 If $T \in \mathcal{L}(\mathcal{H})$, then the essential spectrum $\sigma_e(T)$ (I.8.2.2) is

$$\sigma_e(T) = \{\lambda \in \mathbb{C} : T - \lambda I \text{ is not Fredholm}\}.$$

It is not difficult to see that

$$\cap_{K \in \mathcal{K}(\mathcal{H})} \sigma(T + K) = \{\lambda \in \mathbb{C} : T - \lambda I \text{ is not Fredholm of index } 0\}.$$

The index is constant on connected components of $\mathbb{C} \setminus \sigma_e(T)$ and vanishes on $\mathbb{C} \setminus \sigma(T)$ and, in particular, on the unbounded component of $\mathbb{C} \setminus \sigma_e(T)$. Thus, if $\mathbb{C} \setminus \sigma_e(T)$ is connected, we have $\sigma_e(T) = \cap_{K \in \mathcal{K}(\mathcal{H})} \sigma(T + K)$.

I.8.3.13 Let $\mathcal{G}$ be the group of invertible elements of $\mathcal{Q}(\mathcal{H})$, and $\mathcal{G}_o$ the connected component (path component) of the identity in $\mathcal{G}$. I.8.3.8 implies that *index* is a homomorphism from $\mathcal{G}/\mathcal{G}_o$ to the additive group $\mathbb{Z}$. This map is a special case of the connecting map in the six-term exact sequence of K-theory (V.1.2.22) associated with the extension

$$0 \to \mathcal{K}(\mathcal{H}) \to \mathcal{L}(\mathcal{H}) \to \mathcal{Q}(\mathcal{H}) \to 0$$

and I.8.3.8 and I.8.3.9 imply that *index* is actually an isomorphism from $\mathcal{G}/\mathcal{G}_o \cong K_1(\mathcal{Q}(\mathcal{H}))$ onto $\mathbb{Z} \cong K_0(\mathcal{K}(\mathcal{H}))$.

I.8.3.14 Many of the properties of Fredholm operators carry over to semi-Fredholm operators. A bounded operator T on $\mathcal{H}$ is *semi-Fredholm* if T has closed range and either $\mathcal{N}(T)$ or $\mathcal{N}(T^*)$ is finite-dimensional. The index of a (non-Fredholm) semi-Fredholm operator is well defined as $\pm\infty$, and the index is additive where well defined (note, however, that the product of a semi-Fredholm operator of index $+\infty$ and a semi-Fredholm operator of index $-\infty$ is not semi-Fredholm in general.) A compact perturbation of a semi-Fredholm operator is semi-Fredholm of the same index. A version of Atkinson's theorem holds: an operator is semi-Fredholm if and only if its image in the Calkin algebra is left or right invertible.

I.8.3.15 Many of the results above, suitably restated, apply to Fredholm operators between different spaces.

I.8.4 Spectral Properties of Compact Operators

Compact operators behave almost like finite-rank operators with regard to spectral theory. The spectral theory of compact operators in its modern form was developed by F. Riesz, following important contributions of Fredholm and Hilbert.

I.8.4.1 By I.8.3.4, if $T \in \mathcal{K}(\mathcal{H})$, each nonzero point of $\sigma(T)$ is an eigenvalue of T, the corresponding eigenspace is finite-dimensional, and the eigenvalues form a finite set or a sequence converging to 0. (If $T \in \mathcal{K}(\mathcal{H})_+$, this also follows from the Spectral Theorem: for each $\lambda > 0$ the spectral projection $E_{(\lambda,\infty)}(T)$ has finite rank, since T is bounded below on its range.)

I.8.4.2 If $T \in \mathcal{K}(\mathcal{H})$, the nonzero eigenvalues of $|T|$, counted with multiplicity, can thus be arranged in a decreasing sequence

$$(\mu_0(T), \mu_1(T), \cdots)$$

called the *characteristic list* (or *list of characteristic numbers*) for T. (The characteristic list is finite if and only if T is finite-rank, in which case it is expanded into an infinite sequence by adding zeroes.) The characteristic lists for T and T^* are the same. By I.2.6.10, $\mu_0(T) = \|T\|$. The characteristic list is sometimes called the *eigenvalue list* of T, especially if $T \geq 0$.

The n'th characteristic number $\mu_n(T)$ can alternately be characterized as the distance from T to the (closed) set $\mathcal{C}_n(\mathcal{H})$ of operators in $\mathcal{L}(\mathcal{H})$ of rank $\leq n$ (cf. [GKn69]). From this we get several useful properties such as

$$|\mu_n(S) - \mu_n(T)| \leq \|S - T\|$$

$$\mu_{n+m}(S + T) \leq \mu_n(S) + \mu_m(T)$$

$$\mu_{n+m}(ST) \leq \mu_n(S)\mu_m T.$$

Denote the partial sum $\sum_{n=0}^{N-1} \mu_n(T)$ by $\sigma_N(T)$.

I.8.5 Trace-Class and Hilbert-Schmidt Operators

The trace of a finite-rank operator on a vector space is an important invariant in many contexts. The notion of trace can be extended to certain infinite-rank operators on Hilbert spaces, and this trace plays a crucial role in many applications, often acting as a "noncommutative integral."

I.8.5.1 Let $\mathcal{H}$ be a Hilbert space and $0 \leq T \in \mathcal{L}(\mathcal{H})$. We define the *trace* $Tr(T)$ to be $\sum_i \langle T\xi_i, \xi_i \rangle \in [0, \infty]$, where $\{\xi_i\}$ is an orthonormal basis for $\mathcal{H}$. If $S \in \mathcal{L}(\mathcal{H})$, then a simple calculation shows that $Tr(SS^*) = Tr(S^*S)$; so $Tr(T)$ is well defined independently of the choice of the orthonormal basis and satisfies $Tr(T) = Tr(U^*TU)$ for any unitary U. Also, obviously $Tr(T) \geq \|T\|$, $Tr(S+T) = Tr(S) + Tr(T)$ if $S, T \geq 0$, and $Tr(S) \leq Tr(T)$ if $0 \leq S \leq T$. We have that $Tr(T) < \infty$ if and only if T is compact and its characteristic list $(\mu_0(T), \mu_1(T), \cdots)$ satisfies $\sum_n \mu_n(T) < \infty$, in which case $Tr(T) = \sum_n \mu_n(T)$.

I.8.5.2 Tr is lower semicontinuous on $\mathcal{L}(\mathcal{H})_+$. In fact, if (T_j) is a net in $\mathcal{L}(\mathcal{H})_+$ converging weakly to T, then $\langle T_j\xi_i, \xi_i \rangle \to \langle T\xi_i, \xi_i \rangle$ for each i, so $Tr(T) \leq \liminf_j Tr(T_j)$. ($Tr$ is not continuous if $\mathcal{H}$ is infinite-dimensional, however: if P_n is a rank n projection and $T_n = \frac{1}{n}P_n$, then $T_n \to 0$ but $Tr(T_n) = 1$ for all n.)

I.8.5.3 DEFINITION. If $\mathcal{H}$ is a Hilbert space, then

$$\mathcal{L}^1(\mathcal{H})_+ = \{T \in \mathcal{L}(\mathcal{H})_+ : Tr(T) < \infty\}.$$

The set of *trace-class operators* on $\mathcal{H}$, denoted $\mathcal{L}^1(\mathcal{H})$, is the linear span of $\mathcal{L}^1(\mathcal{H})_+$.

$$\mathcal{L}^2(\mathcal{H}) = \{T \in \mathcal{L}(\mathcal{H}) : T^*T \in \mathcal{L}^1(\mathcal{H})\}$$

is the set of *Hilbert-Schmidt operators* on $\mathcal{H}$.

I.8.5.4 $\mathcal{L}^1(\mathcal{H})$ is a *-subspace of $\mathcal{K}(\mathcal{H})$, and Tr extends to a well-defined positive linear functional on $\mathcal{L}^1(\mathcal{H})$. A polarization argument shows that $\mathcal{L}^1(\mathcal{H})$ is a left ideal in $\mathcal{L}(\mathcal{H})$, and hence a two-sided ideal since it is *-closed. In particular, using polar decomposition we see that $T \in \mathcal{L}^1(\mathcal{H})$ if and only if $|T| \in \mathcal{L}^1(\mathcal{H})_+$. A similar calculation shows that $Tr(ST) = Tr(TS)$ for all $S \in \mathcal{L}(\mathcal{H}), T \in \mathcal{L}^1(\mathcal{H})$.

I.8.5.5 $\mathcal{L}^2(\mathcal{H})$ is closed under addition since

$$(S+T)^*(S+T) \leq (S+T)^*(S+T) + (S-T)^*(S-T) = 2(S^*S + T^*T)$$

for any S,T, and is thus a *-subspace of $\mathcal{K}(\mathcal{H})$. Since $T^*S^*ST \leq \|S\|^2 T^*T$ for any S,T, it follows that $\mathcal{L}^2(\mathcal{H})$ is also a two-sided ideal in $\mathcal{L}(\mathcal{H})$, and from the polarization identity

$$S^*T = \sum_{n=0}^{3} i^n (S - iT)^*(S - iT)$$

for any S,T we see that the product of any two elements of $\mathcal{L}^2(\mathcal{H})$ is in $\mathcal{L}^1(\mathcal{H})$. Conversely, every element of $\mathcal{L}^1(\mathcal{H})$ is such a product since

$$T \in \mathcal{L}^1(\mathcal{H}) \implies |T| \in \mathcal{L}^1(\mathcal{H})_+ \implies |T|^{1/2} \in \mathcal{L}^2(\mathcal{H})$$

so $T = U|T| = (U|T|^{1/2})|T|^{1/2}$ is a product of two elements of $\mathcal{L}^2(\mathcal{H})$. Since $S^2 \leq \|S\|S$ for $S \geq 0$, it follows that $|T|$ and hence T are also in $\mathcal{L}^2(\mathcal{H})$, i.e. $\mathcal{L}^1(\mathcal{H}) \subseteq \mathcal{L}^2(\mathcal{H})$. We also have $Tr(ST) = Tr(TS)$ for all $S, T \in \mathcal{L}^2(\mathcal{H})$.

As a result, $\langle S, T \rangle = Tr(T^*S)$ is an inner product on $\mathcal{L}^2(\mathcal{H})$. The norm $\|T\|_2 = [Tr(T^*T)]^{1/2}$ satisfies $\|T\|_2 \geq \|T\|$, so $\mathcal{L}^2(\mathcal{H})$ is complete (i.e. a Hilbert space) under this norm. If $S \in \mathcal{L}(\mathcal{H}), T \in \mathcal{L}^2(\mathcal{H})$, then $\|ST\|_2 \leq \|S\|\|T\|_2$.

I.8.5.6 It follows from the CBS inequality on $\mathcal{L}^2(\mathcal{H})$ that if $S, T \in \mathcal{L}^1(\mathcal{H})$,

$$Tr(|S+T|) \leq Tr(|S|) + Tr(|T|)$$

(even though it is not true that $|S+T| \leq |S| + |T|$ in general!) Thus $\|T\|_1 = Tr(|T|)$ is a norm on $\mathcal{L}^1(\mathcal{H})$ satisfying $\|ST\|_1 \leq \|S\|\|T\|_1$ for all $S \in \mathcal{L}(\mathcal{H})$, $T \in \mathcal{L}^1(\mathcal{H})$. Since $\|T\|_1 \geq \|T\|$, $\mathcal{L}^1(\mathcal{H})$ is complete (i.e. a Banach space) under $\|\cdot\|_1$.

I.8.5.7 It is clear that if $T \in \mathcal{K}(\mathcal{H})$ with characteristic list $(\mu_0(T), \mu_1(T), \cdots)$, then
$$T \in \mathcal{L}^1(\mathcal{H}) \iff \sum \mu_n(T) < \infty$$
and $\|T\|_1 = \sum \mu_n(T)$. Similarly,
$$T \in \mathcal{L}^2(\mathcal{H}) \iff \sum \mu_n(T)^2 < \infty$$
and $\|T\|_2 = (\sum \mu_n(T)^2)^{1/2}$. It is also easily verified that if $T \in \mathcal{L}^2(\mathcal{H})$ and $\{\xi_i\}$ is an orthonormal basis for $\mathcal{H}$, then
$$\|T\|_2 = \Big(\sum_{i,j} |\langle T\xi_i, \xi_j \rangle|^2\Big)^{1/2},$$
i.e. if T is represented by an infinite matrix, $\|T\|_2$ is the l^2-norm of the set of entries of the matrix.

The ideal $\mathcal{C}(\mathcal{H})$ of finite-rank operators is dense in both $\mathcal{L}^1(\mathcal{H})$ and $\mathcal{L}^2(\mathcal{H})$ in their respective topologies.

I.8.6 Duals and Preduals, σ-Topologies

The next theorem may be regarded as a noncommutative analog of the well-known fact that the Banach space dual $(c_o)^*$ of the Banach space c_o of sequences of complex numbers converging to zero, with supremum norm, is isometrically isomorphic to the space l^1 of summable sequences with its natural norm, and that the dual $(l^1)^*$ is the Banach space l^∞ of bounded sequences with supremum norm, using the natural pairings.

I.8.6.1 THEOREM.

(i) If $\mathcal{H}$ is a Hilbert space and ϕ is a bounded linear functional on the Banach space $(\mathcal{K}(\mathcal{H}), \|\cdot\|)$, then there is a unique $S \in \mathcal{L}^1(\mathcal{H})$ such that $\phi(T) = Tr(ST)$; and $\|S\|_1 = \|\phi\|$. So the Banach space dual of $\mathcal{K}(\mathcal{H})$ is isometrically isomorphic to $\mathcal{L}^1(\mathcal{H})$.

(ii) If $\mathcal{H}$ is a Hilbert space and ϕ is a bounded linear functional on the Banach space $(\mathcal{L}^1(\mathcal{H}), \|\cdot\|_1)$, then there is a unique $S \in \mathcal{L}(\mathcal{H})$ such that $\phi(T) = Tr(ST)$; and $\|S\| = \|\phi\|$. So the Banach space dual of $\mathcal{L}^1(\mathcal{H})$ is isometrically isomorphic to $\mathcal{L}(\mathcal{H})$.

Thus the second dual of $\mathcal{K}(\mathcal{H})$ is $\mathcal{L}(\mathcal{H})$.

I.8.6.2 DEFINITION. The *σ-weak operator topology* on $\mathcal{L}(\mathcal{H})$ is the weak-* topology from the identification of $\mathcal{L}(\mathcal{H})$ with $\mathcal{L}^1(\mathcal{H})^*$, i.e. the topology generated by the family of seminorms $\{\omega_T : T \in \mathcal{L}^1(\mathcal{H})\}$, where $\omega_T(S) = |Tr(ST)|$.

The word "operator" is often omitted in the name.

It is easily checked that the weak (weak operator) topology on $\mathcal{L}(\mathcal{H})$ is the topology generated by the family of seminorms $\{\omega_T : T \in \mathcal{C}(\mathcal{H})\}$, so the σ-weak topology is slightly stronger than the weak topology. The topologies coincide on bounded sets.

In fact, $\mathcal{L}^1(\mathcal{H})$ is the *unique* Banach space $\mathcal{X}$ such that $\mathcal{X}^*$ is isometrically isomorphic to $\mathcal{L}(\mathcal{H})$ (III.2.4.1). So the σ-weak topology is intrinsic to the Banach space structure of $\mathcal{L}(\mathcal{H})$.

I.8.6.3 PROPOSITION. The σ-*strong operator topology* on $\mathcal{L}(\mathcal{H})$ is the topology of pointwise convergence as left multiplication operators on $\mathcal{K}(\mathcal{H})$, i.e. $S_i \to S$ σ-strongly if $S_i T \to ST$ for all $T \in \mathcal{K}(\mathcal{H})$. The σ-strong topology is generated by the family of seminorms $\{\rho_T : T \in \mathcal{K}(\mathcal{H})\}$, where $\rho_T(S) = \|ST\|$.

The σ-*strong-* operator topology* on $\mathcal{L}(\mathcal{H})$ is similarly the topology of pointwise convergence as left or right multiplication operators on $\mathcal{K}(\mathcal{H})$, generated by the seminorms $\{\lambda_T, \rho_T : T \in \mathcal{K}(\mathcal{H})\}$, where ρ_T is as above and $\lambda_T(S) = \|TS\|$. Thus the σ-strong and σ-strong-* topologies are intrinsic to the Banach algebra structure of $\mathcal{L}(\mathcal{H})$.

The strong operator topology [resp. the strong-* topology] is generated by the seminorms $\{\rho_T : T \in \mathcal{C}(\mathcal{H})\}$ [resp. $\{\lambda_T, \rho_T : T \in \mathcal{C}(\mathcal{H})\}$], so the σ-topologies are slightly stronger; they agree on bounded sets (I.8.1.4). The σ-strong-* topology is the strict topology when $\mathcal{L}(\mathcal{H})$ is regarded as the multiplier algebra of $\mathcal{K}(\mathcal{H})$ (II.7.3).

I.8.6.4 PROPOSITION. Let $\mathcal{H}$ and $\mathcal{H}'$ be Hilbert spaces. Then the map

$$\mathcal{L}(\mathcal{H}) \to \mathcal{L}(\mathcal{H}) \otimes 1 \subseteq \mathcal{L}(\mathcal{H} \otimes \mathcal{H}')$$

is a homeomorphism for either the σ-weak or σ-strong topologies.

The proof is a simple and straightforward calculation. (Roughly speaking, it amounts to the fact that if $\mathcal{H}$ is infinite-dimensional and $\mathcal{H}'$ is separable, then $\mathcal{K}(\mathcal{H} \otimes \mathcal{H}') \cong \mathcal{K}(\mathcal{H})$.)

I.8.7 Ideals of $\mathcal{L}(\mathcal{H})$

I.8.7.1 The smallest nonzero ideal (unless otherwise qualified, "ideal" will mean "two-sided ideal") of $\mathcal{L}(\mathcal{H})$ is $\mathcal{C}(\mathcal{H})$. This is seen most easily by noting that every rank-one operator is of the form $\Theta_{\xi,\eta}$ for vectors $\xi, \eta \in \mathcal{H}$, where $\Theta_{\xi,\eta}(\zeta) = \langle \zeta, \eta \rangle \xi$. $T\Theta_{\xi,\eta} = \Theta_{T\xi,\eta}$ and $\Theta_{\xi,\eta} T = \Theta_{\xi,T^*\eta}$, so every nonzero ideal contains all rank-one operators, which span the finite-rank operators. (If $\mathcal{H}$ is finite-dimensional, of course, $\mathcal{C}(\mathcal{H}) = \mathcal{L}(\mathcal{H})$ is a simple algebra isomorphic to $M_n(\mathbb{C})$.)

$\mathcal{K}(\mathcal{H})$ is thus the smallest nonzero closed ideal in $\mathcal{L}(\mathcal{H})$.

I.8.7.2 If $\mathcal{H}$ is separable and infinite-dimensional, then $\mathcal{K}(\mathcal{H})$ is the only nontrivial closed ideal in $\mathcal{L}(\mathcal{H})$. If $\mathcal{H}$ is nonseparable, there are larger nontrivial closed ideals, for example the set of operators with separable range. In fact, if $\dim(\mathcal{H}) = \aleph_\alpha$, $\alpha \geq 0$, then for each ordinal $\beta \leq \alpha$ there is a closed ideal $\mathcal{K}_\beta$ consisting of the closure of the set of all operators for which the closure of the range has dimension $< \aleph_\beta$ (if $\beta > 0$ and is not a supremum of a countable number of strictly smaller ordinals, e.g. if β is not a limit ordinal, then it is unnecessary to take closure here.) Then $\mathcal{K}_0 = \mathcal{K}(\mathcal{H})$, and $\mathcal{K}_1$ is the ideal of operators with separable range. It is easily seen that every closed ideal of $\mathcal{L}(\mathcal{H})$ is of this form (observe that every closed ideal of $\mathcal{L}(\mathcal{H})$ is generated by projections), so the nontrivial closed ideals of $\mathcal{L}(\mathcal{H})$ form a chain $\{\mathcal{K}_\beta : 0 \leq \beta \leq \alpha\}$.

Nonclosed Ideals

I.8.7.3 There are many nonclosed ideals of $\mathcal{L}(\mathcal{H})$ intermediate between $\mathcal{C}(\mathcal{H})$ and $\mathcal{K}(\mathcal{H})$ (if $\mathcal{H}$ is infinite-dimensional). For example, there are $\mathcal{L}^1(\mathcal{H})$ and $\mathcal{L}^2(\mathcal{H})$. More generally, if $p \geq 1$, let

$$\mathcal{L}^p(\mathcal{H}) = \{T : |T|^p \in \mathcal{L}^1(\mathcal{H})\}.$$

Then $\mathcal{L}^p(\mathcal{H})$ is exactly the set of elements of $\mathcal{K}(\mathcal{H})$ with characteristic list $(\mu_1(T), \mu_2(T), \cdots)$ satisfying $\sum \mu_n(T)^p < \infty$; $\mathcal{L}^p(\mathcal{H})$ is an ideal, and has a norm

$$\|T\|_p = (\||T|^p\|_1)^{1/p} = \left(\sum \mu_n(T)^p\right)^{1/p}$$

making it a Banach space and satisfying $\|ST\|_p \leq \|S\|\|T\|_p$ for $S \in \mathcal{L}(\mathcal{H})$, $T \in \mathcal{L}^p(\mathcal{H})$. If $p, q \geq 1$, then $\mathcal{L}^p(\mathcal{H}) \subseteq \mathcal{L}^q(\mathcal{H})$ if and only if $p \leq q$. Also, if $\frac{1}{p} + \frac{1}{q} = 1$, the product of an element $S \in \mathcal{L}^p(\mathcal{H})$ and $\mathcal{L}^q(\mathcal{H})$ is in $\mathcal{L}^1(\mathcal{H})$ and the pairing $(S, T) \mapsto Tr(ST)$ identifies $(\mathcal{L}^p(\mathcal{H}), \|\cdot\|_p)^*$ isometrically with $(\mathcal{L}^q(\mathcal{H}), \|\cdot\|_q)$. These are called the *Schatten ideals*.

I.8.7.4 There are lots of other ideals. In fact, they can be characterized by *order ideals* in the space c_0^+ of decreasing sequences of nonnegative real numbers converging to zero, hereditary subsets closed under (coordinatewise) addition (informally, a set of all sequences going to zero at at least a certain rate); the ideal corresponding to an order ideal consists of all compact operators whose characteristic lists are in the order ideal.

I.8.7.5 Particularly important are the *symmetrically normed ideals*, an ideal $\mathcal{J}$ with a norm $\|\cdot\|_\mathcal{J}$ satisfying $\|T\|_\mathcal{J} \geq \|T\|$ (so $\mathcal{J}$ is complete with respect to $\|\cdot\|_\mathcal{J}$, $\|ST\|_\mathcal{J} = \|TS\|_\mathcal{J} \leq \|S\|\|T\|_\mathcal{J}$ for $S \in \mathcal{L}(\mathcal{H})$, $T \in \mathcal{J}$. The $\mathcal{L}^p(\mathcal{H})$ are examples of symmetrically normed ideals. Symmetrically normed ideals are in one-one correspondence with *norming functions* on the space c_{00}^+ of decreasing sequences of nonnegative real numbers which are eventually 0: a norming function is a function $\Phi : c_{00}^+ \to \mathbb{R}$ satisfying $\Phi(\alpha\xi) = \alpha\Phi(\xi)$ for

$\alpha \geq 0$, $\xi \in c_{00}^+$, $\Phi(\xi + \eta) \leq \Phi(\xi) + \Phi(\eta)$ for $\xi, \eta \in c_{00}^+$, $\Phi(\xi) > 0$ for $\xi \neq 0$, and $\Phi(\eta_1) = 1$, where $\eta_n = (1, \cdots, 1, 0, 0, \cdots)$ (n ones). Φ defines a norm on $\mathcal{C}(\mathcal{H})$ by $\|T\|_\Phi = \Phi((\mu_n(T)))$, and a "norm" on $\mathcal{K}(\mathcal{H})$ by $\|T\|_\Phi = \sup \|TP\|_\Phi$, where P runs over all finite-rank projections. Set

$$\mathcal{J}_\Phi = \{T \in \mathcal{K}(\mathcal{H}) : \|T\|_\Phi < \infty\}.$$

Then $\mathcal{J}_\Phi$ is a symmetrically normed ideal with norm $\|\cdot\|_\Phi$, and every symmetrically normed ideal is of this form. (For example, for $\mathcal{L}^p(\mathcal{H})$, the corresponding Φ is $\Phi((\xi_1, \xi_2, \cdots)) = (\sum \xi_n^p)^{1/p}$.) See [GKn69] for a full treatment.

I.8.7.6 Some other examples of symmetrically normed ideals, described in detail in [GKn69] or [Con94], are the ideals $\mathcal{L}^{(p,q)}(\mathcal{H})$ for $1 < p < \infty$, $1 \leq q \leq \infty$, defined by interpolation theory. We describe only the ideals with $q = 1$ or ∞, in which case p can also be 1 or ∞. If $p < \infty$, the ideal $\mathcal{L}^{(p,1)}(\mathcal{H})$, often denoted $\mathcal{L}^{p+}(\mathcal{H})$, is the *strong $\mathcal{L}^p$ space*

$$\mathcal{L}^{(p,1)}(\mathcal{H}) = \{T \in \mathcal{K}(\mathcal{H}) : \sum_{n=1}^{\infty} n^{1/p-1}\mu_{n-1}(T) < \infty\}$$

with norm $\|T\|_{p,1} = \sum n^{1/p-1}\mu_{n-1}(T)$. Note that $\mathcal{L}^{(1,1)}(\mathcal{H}) = \mathcal{L}^1(\mathcal{H})$. (More generally, $\mathcal{L}^{(p,p)}(\mathcal{H}) = \mathcal{L}^p(\mathcal{H})$ for any p.)

$$\mathcal{L}^{(\infty,1)}(\mathcal{H}) = \{T \in \mathcal{K}(\mathcal{H}) : \sum \frac{1}{n}\mu_{n-1}(T) < \infty\}$$

with norm $\|T\|_{\infty,1} = \sum \frac{1}{n}\mu_{n-1}(T)$. The ideal $\mathcal{L}^{(\infty,1)}(\mathcal{H})$ is called the *Macaev ideal*. The *weak $\mathcal{L}^p$ spaces*, denoted $\mathcal{L}^{p-}(\mathcal{H})$ or $\mathcal{L}^{(p,\infty)}(\mathcal{H})$, are the ideals which for $1 < p < \infty$ are given by

$$\mathcal{L}^{(p,\infty)}(\mathcal{H}) = \{T \in \mathcal{K}(\mathcal{H}) : \mu_n(T) = O(n^{-1/p})\}$$

with norm $\|T\|_{p,\infty} = \sup_{N \geq 1} N^{1/p-1}\sigma_N(T)$. For $p = 1$ the definition is slightly different:

$$\mathcal{L}^{(\infty,1)}(\mathcal{H}) = \{T \in \mathcal{K}(\mathcal{H}) : \sigma_N(T) = O(\log N)\}$$

with norm $\|T\|_{\infty,1} = \sup_{N \geq 2} \frac{1}{\log N}\sigma_n(T)$. For completeness, we set $\mathcal{L}^{(\infty,\infty)} = \mathcal{L}(\mathcal{H})$. Then we have that $\mathcal{L}^{(p,q)}(\mathcal{H}) \subseteq \mathcal{L}^{(p',q')}(\mathcal{H})$ if and only if $p < p'$ or $p = p'$ and $q \leq q'$.

Let $\mathcal{L}_0^{(p,q)}(\mathcal{H})$ be the closure of $\mathcal{C}(\mathcal{H})$ in $\mathcal{L}^{(p,q)}(\mathcal{H})$. Then $\mathcal{L}_0^{(p,q)}(\mathcal{H}) = \mathcal{L}^{(p,q)}(\mathcal{H})$ if $q < \infty$;

$$\mathcal{L}_0^{(1,\infty)}(\mathcal{H}) = \{T \in \mathcal{K}(\mathcal{H}) : \sigma_N(T) = o(\log N)\}$$

and $\mathcal{L}_0^{(\infty,\infty)}(\mathcal{H}) = \mathcal{K}(\mathcal{H})$. We have $\mathcal{L}^{(p,q)}(\mathcal{H})^* = \mathcal{L}^{(p',q')}(\mathcal{H})$ if $1 < p, q < \infty$, $\frac{1}{p} + \frac{1}{p'} = 1$, $\frac{1}{q} + \frac{1}{q'} = 1$; if $\frac{1}{p} + \frac{1}{p'} = 1$, then $\mathcal{L}_0^{(p,\infty)}(\mathcal{H})^* = \mathcal{L}^{(p',1)}(\mathcal{H})$ and $\mathcal{L}^{(p,1)}(\mathcal{H})^* = \mathcal{L}^{(p',\infty)}(\mathcal{H})$. $\mathcal{L}_0^{(1,\infty)}(\mathcal{H})^* = \mathcal{L}^{(\infty,1)}(\mathcal{H})$ and $\mathcal{L}^{(\infty,1)}(\mathcal{H})^* = \mathcal{L}^{(1,\infty)}(\mathcal{H})$. In each case, the pairing is the ordinary one: $(S,T) \mapsto Tr(ST)$.

I.8.7.7 There is an important class of functionals on $\mathcal{L}^{(1,\infty)}(\mathcal{H})$, called *Dixmier traces*, defined for positive T by $Tr_\omega(T) = \lim_\omega \frac{1}{\log N} \sigma_N(T)$, where $\lim_\omega$ is a suitable Banach limit. The technical details can be found in [Con94]. For any ω, $Tr_\omega(ST) = Tr_\omega(TS)$ for all $S \in \mathcal{L}(\mathcal{H})$, $T \in \mathcal{L}^{(1,\infty)}(\mathcal{H})$. The trace Tr_ω depends on the choice of ω, but there is a large class of operators called *measurable operators* for which $Tr_\omega(T)$ is independent of ω. The elements of $\mathcal{L}_0^{(1,\infty)}(\mathcal{H})$ are measurable, and Tr_ω vanishes on this ideal. Thus Tr_ω is a trace which is very different from the ordinary trace. There are other such traces, but the Dixmier trace is particularly important in noncommutative geometry, as described in [Con94].

I.9 Algebras of Operators

The main subject of this volume is the study of operator algebras. A *(concrete) operator algebra* is a *-subalgebra of $\mathcal{L}(\mathcal{H})$ which is topologically closed in a suitable sense (there is also a subject of non-self-adjoint operator algebras, but it is outside the scope of this volume.) A *concrete C^*-algebra* is a *-subalgebra of $\mathcal{L}(\mathcal{H})$ which is closed in the norm topology. An important example is $\mathcal{K}(\mathcal{H})$. This example shows that a concrete C*-algebra need not be unital. A *von Neumann algebra* is a (necessarily unital) *-subalgebra M of $\mathcal{L}(\mathcal{H})$ such that $M = M''$. A von Neumann algebra is weakly (hence strongly, norm-, $\cdots$) closed (and in particular is a concrete C*-algebra, albeit of a very special kind.)

I.9.1 Commutant and Bicommutant

One of the first, yet still one of the most crucial basic theorems of operator algebra theory is von Neumann's *Bicommutant Theorem* [vN30b]. We say a *-algebra A of operators on a Hilbert space $\mathcal{H}$ *acts nondegenerately* if $T\xi = 0$ for all $T \in A$ implies $\xi = 0$. Since A is a *-algebra, this is equivalent to the condition that the subspace

$$A\mathcal{H} = \mathrm{span}\{T\xi : T \in A, \xi \in \mathcal{H}\}$$

is dense in $\mathcal{H}$. If $I \in A$, then A obviously acts nondegenerately.

I.9.1.1 THEOREM. [BICOMMUTANT] Let A be a *-subalgebra of $\mathcal{L}(\mathcal{H})$ acting nondegenerately. Then A is σ-strongly dense in A''.

OUTLINE OF PROOF: Using I.2.5.4, the proof reduces to showing that if $\xi \in \mathcal{H}$ and $T \in A''$, there is a sequence (T_n) in A with $T_n\xi \to T\xi$. If $\mathcal{X} = \{S\xi : S \in A\}^-$, then $P_\mathcal{X} \in A'$ (since both $\mathcal{X}$ and $\mathcal{X}^\perp$ are invariant under A), so T commutes with $P_\mathcal{X}$, i.e. T leaves $\mathcal{X}$ invariant. It remains to show that $\xi \in \mathcal{X}$, so that $T\xi \in \mathcal{X}$. This is trivial if $I \in A$. In the general nondegenerate case, for each $S \in A$,

$$S[(I - P_{\mathcal{X}})\xi] = (I - P_{\mathcal{X}})[S\xi] = 0$$

since $S\xi \in \mathcal{X}$, so $(I - P_{\mathcal{X}})\xi = 0$.

The Bicommutant Theorem relates a topological closure property with the simple, natural, and purely algebraic property of being the bicommutant (or just commutant) of a *-closed set of operators.

I.9.1.2 Thus a unital (or just nondegenerate) *-subalgebra of $\mathcal{L}(\mathcal{H})$ is a von Neumann algebra if and only if it is σ-strongly closed.

There is an important technical strengthening of the bicommutant theorem [Kap51b]:

I.9.1.3 THEOREM. [KAPLANSKY DENSITY] Let A be a *-subalgebra of $\mathcal{L}(\mathcal{H})$ acting nondegenerately. Then the unit ball of A [resp. A_{sa}] is σ-strongly dense in the unit ball of A'' [resp. A''_{sa}].

The proof, while not difficult, uses material from succeeding chapters, so will not be given here; see, for example, [Ped79, 2.3.3] or [KR97a, 5.3.5].

I.9.1.4 There is a duality between a von Neumann algebra M and its commutant M', most of which will be described in Chapter III. The center $\mathcal{Z}(M)$ is $M \cap M' = (M \cup M')'$, and thus $\mathcal{Z}(M) = \mathcal{Z}(M')$ is also a von Neumann algebra. If M is a von Neumann algebra on $\mathcal{H}$ and $P \in M'$ is a projection, then $P\mathcal{H}$ is invariant under M and the restriction $PM = PMP$ of M to $P\mathcal{H}$ is a von Neumann algebra on $P\mathcal{H}$ whose commutant is $PM'P$ and whose center is $P\mathcal{Z}(M)$. Dually, if Q is a projection in M, then QMQ is a von Neumann algebra of operators on $Q\mathcal{H}$ with commutant $QM' = QM'Q$ and center $Q\mathcal{Z}(M)$.

I.9.1.5 A von Neumann algebra M is called a *factor* if $\mathcal{Z}(M) = \mathbb{C}I$. For example, $\mathcal{L}(\mathcal{H})$ is a factor. M is a factor if and only if M' is a factor. If M is a factor and P is a projection in M or M', then PMP is a factor.

I.9.2 Other Properties

I.9.2.1 Many of the results of this chapter are really facts about operator algebras in disguise. Here are some of the particularly important ones. If $T \in \mathcal{L}(\mathcal{H})$, write $C^*(T)$ for the C*-algebra generated by T, i.e. the norm-closure of the subalgebra generated by T and T^*; and $VN(T) = \{T, T^*\}''$ the von Neumann algebra generated by T. Of course, $C^*(T) \subseteq VN(T)$ (they are equal if and only if $VN(T)$ is finite-dimensional).

(i) If A is a concrete C*-algebra of operators, $T \in A$ is normal, and $f \in C_o(\sigma(T))$, then $f(T) \in A$, and $f \mapsto f(T)$ is an isometric *-isomorphism from $C_o(\sigma(T))$ onto $C^*(T) \subseteq A$. In particular, if $T = T^*$, then T^α, T_+, T_-, $|T|$ are all in $C^*(T)$.

(ii) If M is a von Neumann algebra on $\mathcal{H}$ and $\{P_i\}$ is a collection of projections in M, then $\bigwedge P_i$ and $\bigvee P_i$ are in M, i.e. the projections in M form a complete lattice.
(iii) If M is a von Neumann algebra on $\mathcal{H}$ and $T \in M$ with polar decomposition $T = U|T|$, then $U \in VN(T) \subseteq M$ (and of course by (i), $|T| \in VN(T) \subseteq M$). More generally, if $S, T \in M$ and $S^*S \leq T^*T$, then there is a $W \in M$ with $\|W\| \leq 1$ and $S = WT$.
(iv) If M is a von Neumann algebra and $T \in M$ is normal, then all the spectral projections of T are in $VN(T) \subseteq M$, and more generally $f(T) \in VN(T) \subseteq M$ for any bounded Borel function f on $\sigma(T)$ (and $f \mapsto f(T)$ maps the bounded Borel functions on $\sigma(T)$ onto $VN(T)$).

II
C*-Algebras

II.1 Definitions and Elementary Facts

In this section, we give the basic definitions of Banach algebras and C*-algebras, some of the important related terminology, and elementary facts about functional calculus and spectrum of elements. We consider only complex Banach algebras and C*-algebras; there is a similar theory of real C*-algebras (see, for example, [Goo82], [Sch93], or [Con01]).

The standard references for C*-algebra theory are the classic [Dix69b], [Sak71], [Arv76], [Ped79], [Mur90], [KR97a]–[KR97b], [Dav96], and [Tak02]–[Tak03b]. These references can be consulted for more details on the theory. There are also numerous more specialized books on various aspects of the theory, which are referenced in appropriate sections below.

II.1.1 Basic Definitions

II.1.1.1 DEFINITION. A *Banach algebra* is a (complex) algebra which is a Banach space under a norm which is submultiplicative ($\|xy\| \leq \|x\|\|y\|$ for all $x, y \in A$).

An *involution* on a Banach algebra A is a conjugate-linear isometric antiautomorphism of order two, usually denoted $x \mapsto x^*$. In other words, $(x+y)^* = x^* + y^*$, $(xy)^* = y^*x^*$, $(\lambda x)^* = \bar{\lambda} x^*$, $(x^*)^* = x$, $\|x^*\| = \|x\|$ for all $x, y \in A$, $\lambda \in \mathbb{C}$. A *Banach *-algebra* is a Banach algebra with an involution.

An *(abstract) C*-algebra* is a Banach *-algebra A satisfying the *C*-axiom*:

$$\|x^*x\| = \|x\|^2 \text{ for all } x \in A$$

The deceptively simple and innocuous C*-axiom turns out to be extremely powerful, forcing rigid structure on a C*-algebra. For example, it follows that the norm is completely determined by the algebraic structure and is thus unique (II.1.6.5), that *-homomorphisms of C*-algebras are automatically contractive (II.1.6.6), and that every C*-algebra can be isometrically represented as a concrete C*-algebra of operators (II.6.4.10). One obvious, but useful, consequence is that in a C*-algebra, $x^*x = 0 \Rightarrow x = 0$. [In fact, in a C*-algebra, $\sum x_j^* x_j = 0$ implies that each $x_j = 0$ (II.3.1.2(i), II.3.1.3).]

II.1.1.2 In many older references, abstract C*-algebras were called *B*-algebras*, with the name "C*-algebra" reserved for concrete C*-algebras. The term "C*-algebra," first introduced in [Seg47] (for concrete C*-algebras, but viewed in a somewhat abstract manner), did not become universal until well after the publication of [Dix69b] in 1964 (there were occasional references to "B*-algebras" in the literature at least as late as 1980). According to [DB86, p. 6], the "C" in "C*-algebra" originally meant "closed", and not, as commonly believed, "continuous", although the interpretation as standing for "continuous" is nicely in line with the modern point of view of C*-algebra theory as "noncommutative topology."

The issue of terminology is clouded by the fact that several different (but ultimately equivalent) axiom schemes have been used for C*-algebras over the years. For example, it is easily seen that the C*-axiom implies that the involution is isometric, so it is unnecessary to include isometry of the involution as an axiom. The C*-axiom has sometimes been replaced by the apparently weaker axiom that $\|x^*x\| = \|x^*\|\|x\|$ for all x. It turns out that this weaker axiom also implies isometry of the involution (a much harder result), so the weakened axiom is equivalent to the C*-axiom. See also II.3.1.4. See [DB86] for details about the C*-algebra axioms.

II.1.1.3 EXAMPLES.

(i) Any concrete C*-algebra of operators (I.9) is a C*-algebra with the usual operator norm and involution (I.2.3.1). In particular, $\mathcal{L}(\mathcal{H})$ and $\mathcal{K}(\mathcal{H})$ are C*-algebras for any Hilbert space $\mathcal{H}$. If $\mathcal{H}$ is n-dimensional, we obtain that the $n \times n$ matrices $\mathbb{M}_n = \mathcal{L}(\mathbb{C}^n)$ form a C*-algebra with the usual involution (conjugate transpose) and operator norm. We denote by $\mathbb{K}$ the C*-algebra of compact operators on a separable, infinite-dimensional Hilbert space. More generally, if $\mathcal{X}$ is any (complex) Banach space, then the algebra $\mathcal{L}(\mathcal{X})$ of bounded operators on $\mathcal{X}$ is a Banach algebra with the operator norm. $\mathcal{L}(\mathcal{X})$ does not have a natural involution in general (in fact, see [KM46]).

(ii) Let X be a locally compact Hausdorff space, and $C_o(X)$ the complex-valued continuous functions on X vanishing at infinity. Give $C_o(X)$ its usual pointwise operations and supremum norm. Define an involution by $f^*(x) = \overline{f(x)}$. Then $C_o(X)$ is a commutative C*-algebra. In fact, every commutative C*-algebra is of this form (II.2.2.4). $C_o(X)$ has a unit (identity) if and only if X is compact; in this case, we usually write $C(X)$. More generally, if B is a C*-algebra, then the set $C_o(X, B)$ of (norm-)continuous functions from X to B vanishing at infinity, with pointwise operations and supremum norm, is a C*-algebra. In particular,

$C_o(X, \mathbb{M}_n) \cong M_n(C_o(X))$ is a C*-algebra. (In fact, a matrix algebra over any C*-algebra is a C*-algebra (II.6.6); this is a special case of the tensor product for C*-algebras (II.9)).

(iii) Let G be a locally compact topological group with (left) Haar measure μ. Then $L^1(G, \mu)$ becomes a Banach *-algebra under convolution. It is not a C*-algebra unless G is trivial. This example will be treated in more detail in II.10.

Many more examples and constructions of interesting and important C*-algebras are given in II.8–II.10. Parts of these sections are elementary and can be consulted to give a general picture of the scope of the subject.

II.1.1.4 If A is a Banach algebra and I is a closed ideal ("ideal" will always mean "two-sided ideal" unless otherwise specified) in A, then the quotient norm makes A/I into a Banach algebra. If A is a Banach *-algebra and I is a *-ideal (i.e. closed under *), then A/I is a Banach *-algebra. It turns out that if A is a C*-algebra and I is a closed ideal in A, then I is automatically a *-ideal and A/I is a C*-algebra in the quotient norm (II.5.1.1).

II.1.2 Unitization

II.1.2.1 A Banach algebra, even a C*-algebra, need not be unital (e.g. $\mathbb{K}$, $C_o(X)$ for X noncompact). However, every nonunital Banach algebra A can be embedded in a unital Banach algebra $\tilde{A}$. Let $A^\dagger$ be $A \oplus \mathbb{C}$ with coordinatewise addition,
$$(a, \lambda)(b, \mu) = (ab + \lambda b + \mu a, \lambda \mu)$$
and $\|(a, \lambda)\| = \|a\| + |\lambda|$. [$A$ embeds via $a \mapsto (a, 0)$, and the unit is $(0, 1)$; we often write $a + \lambda 1$ for (a, λ).] If A is a Banach *-algebra, define $(a, \lambda)^* = (a^*, \bar{\lambda})$; then $A^\dagger$ becomes a (unital) Banach *-algebra. With this norm $A^\dagger$ is not a C*-algebra; but if A is a C*-algebra the operator norm on $A^\dagger$ as left multiplication operators on A, i.e.
$$\|(a, \lambda)\| = \sup\{\|ab + \lambda b\| : \|b\| = 1\}$$
is an equivalent C*-norm on $A^\dagger$. The algebra $A^\dagger$ contains A as a (closed) ideal, and $A^\dagger/A \cong \mathbb{C}$. If A is unital, $A^\dagger \cong A \oplus \mathbb{C}$ as C*-algebras (under the map $(a, \lambda) \mapsto (a - \lambda 1, \lambda)$). Set $\tilde{A} = A$ if A is unital, and $\tilde{A} = A^\dagger$ if A is nonunital.

II.1.2.2 As an example, if X is a locally compact noncompact Hausdorff space, it is easy to see that $\widetilde{C_o(X)} \cong C(X^\dagger)$, where $X^\dagger$ is the one-point compactification of X.

II.1.2.3 A bounded homomorphism between the Banach algebras A and B extends uniquely to a bounded unital homomorphism from $A^\dagger$ to $\tilde{B}$. A *-homomorphism between Banach *-algebras extends to a unital *-homomorphism between the unitizations.

II.1.2.4 The unitization process allows reduction of many aspects of the theory of Banach algebras or C*-algebras to the unital case. However, there are important reasons not to restrict attention to only the unital case. For example, we want to regard closed ideals in a C*-algebra as C*-algebras themselves. Many C*-algebras which arise in applications, such as C*-algebras of nondiscrete locally compact groups, are nonunital. Also, in many parts of the advanced theory of C*-algebras one needs to work with stable C*-algebras (II.6.6.12), or suspensions (II.5.5.10), which are always nonunital. Every C*-algebra has an approximate unit (II.4).

II.1.3 Power series, Inverses, and Holomorphic Functions

II.1.3.1 In a Banach space, every absolutely convergent infinite series converges. In particular, if A is a unital Banach algebra and $x \in A$ with $\|x\| < 1$, then $\sum_{n=0}^{\infty} x^n$ converges to an inverse for $1 - x$. Thus we obtain the following fundamental facts:

II.1.3.2 PROPOSITION. Let A be a unital Banach algebra. Then

(i) The invertible elements in A form an open set; if x is invertible, the open ball around x of radius $\|x^{-1}\|^{-1}$ is contained in the invertible elements of A. In particular, if $\|1 - y\| < 1$, then y is invertible.
(ii) Every maximal ideal in A is (norm-)closed.

II.1.3.3 There is a theory of holomorphic (analytic) functions from open sets in $\mathbb{C}$ taking values in a Banach space, which is nearly identical to the usual complex-valued theory. In particular, most of the standard theorems of complex analysis, such as the Cauchy Integral Formula, Liouville's Theorem, and the existence and radius of convergence of Taylor and Laurent expansions, have exact analogs in this setting. See [DS88a, III.14] for details.

II.1.4 Spectrum

II.1.4.1 DEFINITION. Let A be a Banach algebra, $x \in A$. The *spectrum* of x in A is

$$\sigma_A(x) = \{\lambda \in \mathbb{C} : x - \lambda 1 \text{ is not invertible in } \tilde{A}\}$$

If A is nonunital, then $0 \in \sigma_A(x)$ for every $x \in A$.

II.1.4.2 PROPOSITION. Let A be a Banach algebra, $x, y \in A$. Then

(i) $\sigma_A(x)$ is a nonempty compact subset of the plane.
(ii) $\max\{|\lambda| : \lambda \in \sigma_A(x)\} = \lim_{n \to \infty} \|x^n\|^{1/n} = \inf \|x^n\|^{1/n}$. This number is called the *spectral radius* of x, denoted $r(x)$.

(iii) If f is a polynomial with complex coefficients, then
$$\sigma_{\tilde{A}}(f(x)) = \{f(\lambda) : \lambda \in \sigma_A(x)\} .$$

(iv) $\sigma_A(xy) \cup \{0\} = \sigma_A(yx) \cup \{0\}$.

(v) If B is a Banach subalgebra of A and $x \in B$, then $\sigma_A(x) \cup \{0\} \subseteq \sigma_B(x) \cup \{0\}$ and $\partial(\sigma_B(x)) \cup \{0\} \subseteq \partial(\sigma_A(x)) \cup \{0\}$, where ∂ denotes the topological boundary in $\mathbb{C}$. In particular, the spectral radius of x in B is the same as the spectral radius in A (so the notation $r(x)$ is unambiguous).

The proofs of (iii) and (iv) are simple algebraic computations [cf. I.4.1.1, II.1.5.2 for (iii); for (iv), if $r = (1-xy)^{-1}$, then $xyr = rxy = r-1$, so $1+yrx = (1-yx)^{-1}$], and the other parts follow from straightforward applications of the theory of II.1.3, using the fact that the function $\lambda \mapsto (x-\lambda 1)^{-1}$ is holomorphic on the complement of the spectrum of x (called the *resolvent set* of x).

A simple consequence of (i) is:

II.1.4.3 COROLLARY. [GELFAND-MAZUR] The only (complex) Banach division algebra is $\mathbb{C}$.

II.1.5 Holomorphic Functional Calculus

If f is a polynomial with complex coefficients, without constant term (i.e. $f(0) = 0$), and x is an element of an algebra A, then there is an obvious way to apply f to x to obtain an element $f(x) \in A$. If A is a Banach algebra, there is a very important way of extending this procedure to holomorphic functions, called *functional calculus*.

II.1.5.1 If X is a compact subset of $\mathbb{C}$, denote by $H(X)$ the algebra of functions holomorphic in a neighborhood of X and vanishing at 0 if $0 \in X$, with functions identified if they agree on a neighborhood of X. Functional calculus gives a homomorphism from $H(\sigma_A(x))$ to the Banach subalgebra of A generated by x extending the map for polynomials. The image of f is denoted $f(x)$. The element $f(x)$ can be defined using the Cauchy Integral Formula, but in some cases (e.g. if f is entire) it is also given by a power series. If A is unital, functional calculus is also defined for holomorphic functions not vanishing at zero. Functional calculus has the following properties, which (along with the elementary definition for polynomials) determine it uniquely:

II.1.5.2 PROPOSITION. Let A be a Banach algebra and $x \in A$. Then

(i) For any $f \in H(\sigma_A(x))$, $\sigma_A(f(x)) = \{f(\lambda) : \lambda \in \sigma_A(x)\}$.
(ii) If $f \in H(\sigma_A(x))$ and $g \in H(f(\sigma_A(x))) = H(\sigma_A(f(x)))$, so $g \circ f \in H(\sigma_A(x))$, then $(g \circ f)(x) = g(f(x))$.

(iii) If $f_n, f \in H(\sigma_A(x))$ and $f_n \to f$ uniformly on a neighborhood of $\sigma_A(x)$, then $f_n(x) \to f(x)$.

(iv) If B is a Banach algebra and $\phi : A \to B$ a continuous (bounded) homomorphism, then $\phi(f(x)) = f(\phi(x))$ for any $f \in H(\sigma_A(x))$.

PROOF: (i): Suppose A is unital. Let f be analytic on $\mathcal{U}$ containing $\sigma_A(x)$. If $\lambda \in \sigma_A(x)$, then $f(z) - f(\lambda) = (z - \lambda)g(z)$ with g analytic on $\mathcal{U}$; then $f(x) - f(\lambda)1 = (x - \lambda 1)g(x)$, and since $x - \lambda 1$ and $g(x)$ commute and $x - \lambda 1$ is not invertible, $f(\lambda) \in \sigma_A(f(x))$. Conversely, if $\mu \notin \{f(\lambda) : \lambda \in \sigma_A(x)\}$, then $h(z) = (f(z) - \mu)^{-1}$ is analytic on $\{z \in \mathcal{U} : f(z) \neq \mu\}$, which contains $\sigma_A(x)$, and $h(x) = (f(x) - \mu 1)^{-1}$.

(ii)-(iv) are straightforward.

II.1.5.3 An especially important case of functional calculus uses the exponential function $f(z) = e^z$. If x is any element of a unital Banach algebra A, then $f(x)$ is defined and denoted e^x. The element e^x is given by the power series $\sum_{n=0}^{\infty} \frac{x^n}{n!}$. For any x, e^x is invertible, with inverse e^{-x}; $e^{x+y} = e^x e^y$ if x and y commute. If A is a Banach *-algebra, then $(e^x)^* = e^{(x^*)}$. Conversely, if $\sigma_A(x)$ is contained in a simply connected open set not containing 0, let $f(z)$ be a branch of the logarithm holomorphic in a neighborhood of $\sigma_A(x)$; then $y = f(x)$ satisfies $e^y = x$. In particular, if $\|1 - x\| < 1$, then there is a $y \in A$ with $\|y\| < \pi/2$ with $e^y = x$ (use the principal branch of log).

II.1.5.4 If A is a unital Banach algebra, write A^{-1} (also often written $GL_1(A)$) for the (open) set of invertible elements in A, and A_o^{-1} (or $GL_1(A)_o$) for the connected component of the identity in A^{-1}. Then

$$exp(A) = \{e^y : y \in A\} \subseteq A^{-1}$$

is path-connected, and by the above the subgroup of A^{-1} generated algebraically by $exp(A)$ is a connected open subgroup, hence equal to A_o^{-1}. In particular, A_o^{-1} is an open subgroup of A^{-1}, and every element of A_o^{-1} is a finite product of exponentials.

If A and B are unital Banach algebras and ϕ is a bounded homomorphism from A onto B, and $y \in B^{-1}$, then there is not necessarily an $x \in A^{-1}$ with $\phi(x) = y$:

II.1.5.5 EXAMPLE. Let $\mathbb{D}$ be the closed unit disk in $\mathbb{C}$, and $\mathbb{T}$ its boundary. There is a *-homomorphism $\phi : C(\mathbb{D}) \to C(\mathbb{T})$ given by restriction. But if $g \in C(\mathbb{T})$ with $g(z) = z$, then $g \in C(\mathbb{T})^{-1}$, but there is no $f \in C(\mathbb{D})^{-1}$ with $\phi(f) = g$.

The image in the Calkin algebra of a Fredholm operator of nonzero index is another example.

However:

II.1.5.6 PROPOSITION. *If $\phi : A \to B$ is a surjective bounded homomorphism of unital Banach algebras, then $\phi(A_o^{-1}) = B_o^{-1}$.*

This can be proved either by an application of the Open Mapping Theorem (I.2.1.4) or by noting that if $x \in A$, then $\phi(e^x) = e^{\phi(x)}$.

We now restrict our attention to C*-algebras. Certain types of elements have standard names arising from operator theory, reflecting the types of operators they become when the C*-algebra is represented as a concrete C*-algebra of operators:

II.1.5.7 DEFINITION. Let A be a C*-algebra and $x \in A$. Then x is

self-adjoint if $x = x^*$.
normal if $x^*x = xx^*$.
a *projection* if $x = x^* = x^2$.
a *partial isometry* if x^*x is a projection.

If A is unital, then x is

an *isometry* if $x^*x = 1$.
a *coisometry* if $xx^* = 1$.
unitary if $x^*x = xx^* = 1$.

II.1.5.8 The self-adjoint elements of A form a closed real vector subspace A_{sa} of A. (Note, however, that A_{sa} is not closed under multiplication unless A is commutative.) If A is unital, then the set of unitaries of A forms a group $\mathcal{U}(A)$. Every self-adjoint or unitary element is normal. If x is self-adjoint, then e^{ix} is unitary. 0 and 1 are projections; isometries (and, in particular, unitaries) are partial isometries. (Coisometries are too; in fact, it follows from II.2.3.5 that if x is a partial isometry, then so is x^*.)

II.1.5.9 If x is any element, then $a = (x + x^*)/2$ and $b = (x - x^*)/2i$ are self-adjoint, and $x = a + ib$; thus $A_{sa} + iA_{sa} = A$. The elements a and b are called the *real* and *imaginary* parts of x. It is obvious from their definitions that $\|a\|, \|b\| \leq \|x\|$.

II.1.6 Norm and Spectrum

II.1.6.1 It follows from the C*-axiom that every nonzero projection, and hence every nonzero partial isometry (in particular, every unitary) has norm 1. Thus, if x is unitary, then $\sigma_A(x) \subseteq \{\lambda : |\lambda| \leq 1\}$. Since $x^{-1} = x^*$ is also unitary and hence $\sigma_A(x^{-1}) = \{\lambda^{-1} : \lambda \in \sigma_A(x)\}$ is also contained in the unit disk, $\sigma_A(x)$ is actually contained in the unit circle.

II.1.6.2 If x is self-adjoint, then e^{ix} is unitary, so

$$\sigma_A(e^{ix}) = \{e^{i\lambda} : \lambda \in \sigma_A(x)\}$$

is contained in the unit circle, i.e. $\sigma_A(x) \subseteq \mathbb{R}$.

II.1.6.3 If x is self-adjoint, then from the C*-axiom $\|x^2\| = \|x\|^2$, and by iteration $\|x^{2^n}\| = \|x\|^{2^n}$ for all n. Thus $r(x) = \lim_{n\to\infty} \|x^{2^n}\|^{2^{-n}} = \|x\|$. More generally, if y is normal, then

$$r(y) = \lim_{n\to\infty} \|y^{2^n}\|^{2^{-n}} = \lim_{n\to\infty} \|(y^*)^{2^n} y^{2^n}\|^{2^{-n-1}}$$

$$= \lim_{n\to\infty} \|(y^*y)^{2^n}\|^{2^{-n-1}} = [r(y^*y)]^{1/2} = \|y^*y\|^{1/2} = \|y\| .$$

II.1.6.4 COROLLARY. A C*-algebra is a semisimple algebra.

PROOF: If A is a C*-algebra with radical R, and $x \in R$, then $x^*x \in R$; so x^*x is nilpotent, $r(x^*x) = 0$, $x^*x = 0$, $x = 0$.

II.1.6.5 COROLLARY. If A is a C*-algebra and $x \in A$, then $\|x\| = r(x^*x)^{1/2}$. So the norm on a C*-algebra is completely determined by its algebraic structure and is thus unique.

II.1.6.6 COROLLARY. If A is a Banach *-algebra, B a C*-algebra, and $\phi : A \to B$ a *-homomorphism, then $\|\phi\| \leq 1$.

PROOF: If $x \in A$, then $\sigma_B(\phi(x)) \cup \{0\} \subseteq \sigma_A(x) \cup \{0\}$. Thus

$$\|\phi(x)\|^2 = \|\phi(x^*x)\| = r(\phi(x^*x)) \leq r(x^*x) \leq \|x^*x\| \leq \|x\|^2 .$$

A *-algebra thus has at most one norm making it a (complete) C*-algebra; if it has one, it has no other (even incomplete) norm satisfying the C*-axiom (II.2.2.9). [A C*-algebra can have other C*-seminorms; and a *-algebra (e.g. a polynomial ring) can have many different (incomplete) norms satisfying the C*-axiom.]

II.1.6.7 COROLLARY. If B is a unital C*-subalgebra of a unital C*-algebra A and $x \in B$, then $\sigma_B(x) = \sigma_A(x)$. If B is a general C*-subalgebra of a general C*-algebra A, and $x \in B$, then $\sigma_B(x) \cup \{0\} = \sigma_A(x) \cup \{0\}$.

PROOF: This is true if $x = x^*$ by II.1.4.2(v). For general x, if $x - \lambda 1$ is invertible in A, then so are $(x - \lambda 1)^*(x - \lambda 1)$ and $(x - \lambda 1)(x - \lambda 1)^*$. Thus they are invertible in B, so $x - \lambda 1$ is left and right invertible in B.

There are unitary analogs of II.1.3.2(i) and II.1.5.6. If A is a unital C*-algebra, write $\mathcal{U}(A)$ for the unitary group of A and $\mathcal{U}(A)_o$ the connected component of the identity in $\mathcal{U}(A)$.

II.1.6.8 PROPOSITION. Let A be a unital C*-algebra. Then

(i) $\mathcal{U}(A)_o$ is a path-connected open subgroup of $\mathcal{U}(A)$, and every unitary in $\mathcal{U}(A)_o$ is a finite product of exponentials of the form e^{ix} for $x = x^*$.
(ii) If $\phi : A \to B$ is a surjective *-homomorphism, then $\phi(\mathcal{U}(A)_o) = \mathcal{U}(B)_o$.

Part (ii) follows from (i) and the fact that $\phi(A_{sa}) = B_{sa}$.

In fact, if u is a unitary in A and $\|u-1\| < 2$, then there is an $x \in A_{sa}$ with $u = e^{ix}$ and $\|x\| < \pi$. In this case, u is connected to 1 by a path $(u_t = e^{itx})$ of length $< \pi$.

If $\phi : A \to B$ is a surjective *-homomorphism of unital C*-algebras, then it is not true in general that $\phi(\mathcal{U}(A)) = \mathcal{U}(B)$ (II.1.5.5).

II.2 Commutative C*-Algebras and Continuous Functional Calculus

II.2.1 Spectrum of a Commutative Banach Algebra

II.2.1.1 If A is a unital commutative Banach algebra, let $\hat{A}$ be the set of (unital) homomorphisms from A to $\mathbb{C}$, and $\mathrm{Prim}(A)$ the set of maximal ideals of A. Every maximal ideal is closed (II.1.3) and hence every homomorphism is continuous. In fact:

II.2.1.2 PROPOSITION. If $\phi \in \hat{A}$ and $x \in A$, then $\phi(x) \in \sigma_A(x)$. Conversely, if $\lambda \in \sigma_A(x)$, then there is a $\phi \in \hat{A}$ with $\phi(x) = \lambda$. Thus, for any $\phi \in \hat{A}$, $x \in A$, $|\phi(x)| \leq r(x) \leq \|x\|$, and hence $\|\phi\| = 1$.

PROOF: $\phi(x - \phi(x)1) = 0$, so $x - \phi(x)1$ is not invertible. For the converse, $x - \lambda 1$ generates a proper ideal of A, which is contained in a maximal ideal I. Let $\phi : A \to A/I$ be the quotient map. $A/I \cong \mathbb{C}$ by II.1.4.3.

There is a useful consequence for general Banach algebras:

II.2.1.3 COROLLARY. Let B be a unital Banach algebra, $x, y \in B$. If $xy = yx$, then

$$\sigma_B(xy) \subseteq \sigma_B(x)\sigma_B(y) = \{\lambda\mu : \lambda \in \sigma_B(x), \mu \in \sigma_B(y)\}$$

and $\sigma_B(x+y) \subseteq \sigma_B(x) + \sigma_B(y)$.

PROOF: First suppose B is commutative. If $\gamma \in \sigma_B(xy)$, then there is a $\phi \in \hat{B}$ with $\phi(xy) = \gamma$. We have $\phi(x) \in \sigma_B(x)$, $\phi(y) \in \sigma_B(y)$, and $\gamma = \phi(x)\phi(y)$.

For a general B, let A be the closed subalgebra generated by $1, x, y$, and $(x - \lambda 1)^{-1}$, $(y - \mu 1)^{-1}$ for all $\lambda \notin \sigma_B(x)$, $\mu \notin \sigma_B(y)$. Then A is commutative, and $\sigma_A(x) = \sigma_B(x)$, $\sigma_A(y) = \sigma_B(y)$. We have

$$\sigma_B(xy) \subseteq \sigma_A(xy) \subseteq \sigma_A(x)\sigma_A(y) = \sigma_B(x)\sigma_B(y).$$

The proof for $\sigma_B(x+y)$ is almost identical.

II.2.1.4 We return to the case where A is a commutative unital Banach algebra. Then $\hat{A}$ may be identified with a closed subset of the unit ball of the dual space A^*, and can thus be given the weak-* topology (topology of pointwise convergence). So $\hat{A}$ is a compact Hausdorff space, called the *spectrum* of A. $\mathrm{Prim}(A)$ may be given the Jacobsen or hull-kernel topology (the closure of a set $\mathcal{J}$ is the set of all maximal ideals containing $\cap_{I \in \mathcal{J}} I$). $\mathrm{Prim}(A)$ is compact, but not necessarily Hausdorff (e.g. if A is the "disk algebra" of functions continuous on the closed unit disk in $\mathbb{C}$ and analytic on the open disk).

There is a continuous injective map $\Omega : \phi \mapsto Ker(\phi)$ from $\hat{A}$ to $\mathrm{Prim}(A)$. This map is surjective: if $I \in \mathrm{Prim}(A)$, then A/I is a Banach algebra in the quotient norm and also a field, hence is $\mathbb{C}$ by II.1.4.3.

II.2.2 Gelfand Transform

II.2.2.1 $C(\hat{A})$ is thus a commutative C*-algebra. There is a natural homomorphism $\Gamma : x \mapsto \hat{x}$ from A to $C(\hat{A})$ given by $\hat{x}(\phi) = \phi(x)$, called the *Gelfand transform*. The image obviously contains the constant functions and separates the points of $\hat{A}$.

II.2.2.2 If A is a C*-algebra, then every $\phi \in \hat{A}$ is a *-homomorphism [it suffices to show that $\phi(x) \in \mathbb{R}$ if $x = x^*$; but

$$\{\phi(x)\} = \sigma_{\mathbb{C}}(\phi(x)) \subseteq \sigma_A(x) \subseteq \mathbb{R}$$

by II.1.6.2.] Thus the Gelfand transform is a *-homomorphism.

II.2.2.3 If A is nonunital, let $\mathrm{Prim}(A)$ be as before, and $\hat{A}$ the set of nonzero homomorphisms to $\mathbb{C}$. Then $\mathrm{Prim}(A)$ and $\hat{A}$ may be homeomorphically identified with the open sets $\mathrm{Prim}(\tilde{A})\backslash\{A\}$ and $\hat{\tilde{A}}\backslash\{\tilde{A}/A\}$ respectively, and thus they are locally compact spaces. The Gelfand transform maps A to $C_\mathrm{o}(\hat{A})$.

II.2.2.4 THEOREM. [GELFAND-NAIMARK] If A is a commutative C*-algebra, then the Gelfand transform is an isometric *-isomorphism from A onto $C_\mathrm{o}(\hat{A})$. Furthermore, $\Omega : \hat{A} \to \mathrm{Prim}(A)$ is a homeomorphism.

(There are actually two "Gelfand-Naimark theorems," the other being II.6.4.10.)

PROOF: We may assume A is unital. For the first part, it remains only to show that Γ is isometric (then the range will be closed, hence all of $C_\mathrm{o}(\hat{A})$ by the Stone-Weierstrass Theorem). By the C*-axiom and II.1.6.6, it suffices to show that if $x = x^*$, then $\|\hat{x}\| \geq \|x\|$. By II.2.1.2 there is $\lambda \in \sigma_A(x)$ with $|\lambda| = \|x\|$, so there is a ϕ with $|\hat{x}(\phi)| = \|x\|$. It is a routine exercise to prove that $\mathrm{Prim}(C(X))$ is homeomorphic to X, and thus Ω is a homeomorphism for $C(X)$.

II.2 Commutative C*-Algebras and Continuous Functional Calculus

II.2.2.5 If X and Y are compact Hausdorff spaces, and $\phi : X \to Y$ is continuous, then ϕ defines a *-homomorphism $\hat{\phi} : C(Y) \to C(X)$ by $\hat{\phi}(f) = f \circ \phi$. The map $\hat{\phi}$ is injective [*resp.* surjective] if and only if ϕ is surjective [*resp.* injective]. Conversely, if $\psi : C(Y) \to C(X)$ is a *-homomorphism and $x \in X = \widehat{C(X)}$, then $\phi_x \circ \psi \in \widehat{C(Y)} = Y$; set $\check{\psi}(x) = \phi_x \circ \psi$. [Here ϕ_x is the homomorphism from $C(X)$ to $\mathbb{C}$ corresponding to x, i.e. $\phi_x(f) = f(x)$.] It is easy to check that $\check{\psi}$ is continuous. These maps are inverses of each other. In other words:

II.2.2.6 THEOREM. The correspondence $X \leftrightarrow C(X)$ is a contravariant category equivalence between the category of compact Hausdorff spaces and continuous maps and the category of commutative unital C*-algebras and unital *-homomorphisms.

II.2.2.7 There is a version of this theorem for locally compact spaces and general commutative C*-algebras. However, *-homomorphisms from $C_o(Y)$ to $C_o(X)$ do not correspond to continuous maps from X to Y, but rather to *proper* continuous maps from open subsets of X to Y. It is much cleaner to phrase things in terms of pointed compact spaces and basepoint-preserving continuous maps, identifying the locally compact space X with $(X^\dagger, \infty)$ and to the pointed compact space $(Y, *)$ associating the C*-algebra $C_o(Y \backslash \{*\})$.

II.2.2.8 THEOREM. The correspondence $(X, *) \leftrightarrow C_o(X \setminus \{*\})$ is a contravariant category equivalence between the category of pointed compact Hausdorff spaces and basepoint-preserving continuous maps and the category of commutative C*-algebras and *-homomorphisms.

An important immediate consequence is:

II.2.2.9 COROLLARY. Let A and B be C*-algebras, $\phi : A \to B$ an injective *-homomorphism. Then ϕ is isometric, i.e. $\|\phi(x)\| = \|x\|$ for all $x \in A$.

PROOF: If $x \in A$, then $\|\phi(x)\|^2 = \|\phi(x^*x)\|$, so replacing A and B by $C^*(x^*x)$ and $C^*(\phi(x^*x))$ respectively, it suffices to assume A and B are commutative, where the result is obvious from the theorem.

II.2.3 Continuous Functional Calculus

II.2.2.4 allows us to extend the holomorphic functional calculus of II.1.5 to continuous functions of normal elements. An immediate corollary is:

II.2.3.1 COROLLARY. Let A be a C*-algebra, and x a normal element of A. Then $C^*(x)$ is isometrically isomorphic to $C_o(\sigma_A(x))$ under an isomorphism which sends x to the function $f(t) = t$.

In fact, polynomials in x and x^* without constant term are uniformly dense in $C^*(x)$, and by the Stone-Weierstrass theorem polynomials in λ and

$\bar{\lambda}$ without constant term are dense in $C_o(\sigma_A(x))$. If p is a polynomial in two variables with complex coefficients and no constant term, then

$$\|p(x, x^*)\| = \max\{|p(\lambda, \bar{\lambda})| : \lambda \in \sigma_A(x)\}.$$

Thus, if f is a complex-valued function which is continuous on $\sigma_A(x)$, with $f(0) = 0$ if $0 \in \sigma_A(x)$, then there is a corresponding element $f(x) \in C^*(x)$. This extended functional calculus has the same properties as in II.1.5.2, with a stronger continuity property:

II.2.3.2 PROPOSITION. Let A be a C*-algebra and $x \in A$ a normal element. Then

(i) For any $f \in C_o(\sigma_A(x))$, $\sigma_A(f(x)) = \{f(\lambda) : \lambda \in \sigma_A(x)\}$.
(ii) If $f \in C_o(\sigma_A(x))$ and $g \in C_o(f(\sigma_A(x))) = C_o(\sigma_A(f(x)))$, so $g \circ f \in C_o(\sigma_A(x))$, then $(g \circ f)(x) = g(f(x))$.
(iii) If $f_n, f \in C_o(\sigma_A(x))$ and $f_n \to f$ uniformly on $\sigma_A(x)$, then $f_n(x) \to f(x)$.
(iv) If B is a C*-algebra and $\phi : A \to B$ a *-homomorphism, then $\phi(f(x)) = f(\phi(x))$ for any $f \in C_o(\sigma_A(x))$.

II.2.3.3 PROPOSITION. Let Y be a compact subset of $\mathbb{C}$, and (f_n) a sequence of elements of $C_o(Y)$ converging uniformly on Y to f. Let A be a C*-algebra, (x_n) a sequence of normal elements of A with $x_n \to x \in A$ and $\sigma_A(x_n) \subseteq Y$ (so $\sigma_A(x) \subseteq Y$). Then $f_n(x_n) \to f(x)$.

For the proof, let $\epsilon > 0$, and approximate f uniformly on Y within $\epsilon/4$ by a polynomial p; then

$$\|f_n(x_n) - f(x)\| \le$$
$$\|f_n(x_n) - f(x_n)\| + \|f(x_n) - p(x_n)\| + \|p(x_n) - p(x)\| + \|p(x) - f(x)\|.$$

If n is large enough that $|f_n - f|$ is uniformly less than $\epsilon/4$ on Y, the first term is less than $\epsilon/4$; the second and fourth terms are $< \epsilon/4$ by choice of p, and the third term goes to 0 as $n \to \infty$ by continuity of addition, multiplication, and involution.

This result is often used when (f_n) is a constant sequence.

From functional calculus we obtain:

II.2.3.4 COROLLARY. Let A be a C*-algebra, and x a normal element of A. Then

x is self-adjoint if and only if $\sigma_A(x) \subseteq \mathbb{R}$.
x is unitary if and only if $\sigma_A(x) \subseteq \{\lambda : |\lambda| = 1\}$.
x is a projection if and only if $\sigma_A(x) \subseteq \{0, 1\}$.

II.2.3.5 So if u is a partial isometry (i.e. u^*u is a projection), then by II.1.4.2(iv) and II.2.3.4 uu^* is also a projection. [There is a simpler direct proof: it is elementary to see that u is a partial isometry if and only if $u = uu^*u$. One direction is obvious; for the other, if u^*u is a projection and $y = u - uu^*u$, then $y^*y = 0$, so $y = 0$.] u^*u and uu^* are called the *initial* and *final* projections, or the *source* and *range* projections, of u, and u is called a partial isometry from u^*u to uu^*. (The terminology is inspired by I.5.2.3.)

II.2.3.6 PROPOSITION. Let $a = a^*$ in a C*-algebra A. Then there is a unique self-adjoint $b \in A$ with $b^3 = a$.

PROOF: We can take $b = f(a)$, where $f(t) = t^{1/3}$. For uniqueness, suppose $c = c^*$ and $c^3 = a$. Then c commutes with a, and hence with b which is a limit of polynomials in a. So $C^*(b, c)$ is commutative, and contains a. But a self-adjoint element of a commutative C*-algebra obviously (by the Gelfand transform) has a unique self-adjoint cube root.

The same is true for n'th roots, for any odd n. See II.3.1.2(vii) for an analogous statement for even roots.

II.3 Positivity, Order, and Comparison Theory

II.3.1 Positive Elements

II.3.1.1 DEFINITION. Let A be a C*-algebra. An element $x \in A$ is *positive* if $x = x^*$ and $\sigma(x) \subseteq [0, \infty)$. The set of positive elements in A is denoted A_+. If $x \in A_+$, we write $x \geq 0$.

Note that by II.1.6.7, the property of being positive is independent of the containing C*-algebra, i.e. if B is a C*-subalgebra of A, then $B_+ = B \cap A_+$.

A positive element of $C_o(X)$ is just a function taking only nonnegative real values. The following facts are obvious from II.2.2.4:

II.3.1.2 PROPOSITION. Let A be a C*-algebra and $x, y \in A$. Then

(i) If $x \geq 0$ and $-x \geq 0$, then $x = 0$.
(ii) If x is normal, then $x^*x \geq 0$. In particular, if $x = x^*$, then $x^2 \geq 0$.
(iii) If $x \geq 0$, then $\|x\| = \max\{\lambda : \lambda \in \sigma(x)\}$.
(iv) If $x = x^*$ and $\|x\| \leq 2$, then $x \geq 0$ if and only if $\|1 - x\| \leq 1$ (in $\tilde{A}$).
(v) If $x, y \geq 0$ and $xy = yx$, then $x + y$ and xy are positive.
(vi) If $x = x^*$, then there is a unique decomposition $x = x_+ - x_-$, where $x_+, x_- \geq 0$ and $x_+ x_- = 0$. We have $x_+, x_- \in C^*(x)$. [$x_+ = f(x)$ and $x_- = g(x)$, where $f(t) = \max(t, 0)$ and $g(t) = -\min(t, 0)$.] Thus every element of A is a linear combination of four positive elements.

(vii) Every positive element of a C*-algebra has a unique positive square root. More generally, if $x \geq 0$ and α is a positive real number, there is a positive element $x^\alpha \in C^*(x)_+$; these elements satisfy $x^\alpha x^\beta = x^{\alpha+\beta}$, $x^1 = x$, and $\alpha \mapsto x^\alpha$ is continuous. If x is invertible x^α is also defined for $\alpha \leq 0$.

(viii) $(x, \lambda) \geq 0$ in $A^\dagger$ if and only if $x = x^*$ and $\lambda \geq \|x_-\|$.

For (vii), set $x^\alpha = g_\alpha(x)$, where $g_\alpha(t) = t^\alpha$. Then $g_{\alpha^{-1}}(x^\alpha) = x$. If $b \in A_+$ with $g_{\alpha^{-1}}(b) = x$, then x commutes with b, hence x^α commutes with b; thus $b = x^\alpha$ in $C^*(x^\alpha, b) \subseteq A$.

The two crucial facts about positive elements are:

II.3.1.3 PROPOSITION. Let A be a C*-algebra. Then

(i) A_+ is a closed cone in A; in particular, if $x, y \geq 0$ then $x + y \geq 0$. [A *cone* C in a real or complex vector space is a subset closed under addition and under scalar multiplication by $\mathbb{R}_+$, often (but not always) with the property that $C \cap (-C) = \{0\}$.]

(ii) If $x \in A$, then $x^*x \geq 0$.

To prove (i), since $\mathbb{R}_+ A_+ \subseteq A_+$ it suffices to show that $A_+ \cap B$ is a closed convex set, where $B = B_1(A)$ is the closed unit ball in A. But $A_+ \cap B$ is the intersection of the closed convex sets A_{sa}, B, and $\{x : \|1 - x\| \leq 1\}$ (II.3.1.2(iv)).

The proof of property (ii) uses the fact that if $x = a + ib \in A$ with $a, b \in A_{sa}$, then $x^*x \geq 0$ if and only if $xx^* \geq 0$ by II.1.4.2(iv), and

$$x^*x + xx^* = 2a^2 + 2b^2 \geq 0$$

by (i) and II.3.1.2(ii). If $x^*x = c = c_+ - c_-$ as in II.3.1.2(vi) and $y = xc_-$, then $-y^*y = c_-^3 \geq 0$, so

$$yy^* = (y^*y + yy^*) + (-y^*y) \geq 0$$

and thus $y^*y \geq 0$. So $y^*y = 0$ by II.3.1.2(i), $c_- = 0$, $x^*x \geq 0$.

II.3.1.4 These properties were mysterious in the early days of the subject, and property (ii) was originally assumed as an axiom for C*-algebras (in the equivalent form that $1 + x^*x$ is invertible for all x). Property (i) was proved independently in [Fuk52] and [KV53], and (ii) was proved by Kaplansky (cf. [Sch52]). [Kap68] and [DB86] contain an account of the history of the development of the C*-axioms.

II.3.1.5 COROLLARY. If A is a C*-algebra and $a \in A_+$, then $x^*ax = (a^{1/2}x)^*(a^{1/2}x) \geq 0$ for any $x \in A$.

II.3.1.6 In light of II.3.1.5, the following useful *polarization identity*, valid for all x,y, gives a simplified proof of the last statement of II.3.1.2(vi) (in the nonunital case, II.3.2.1 is needed to show that $A = A^2$ for a C*-algebra A):

$$4y^*x = (x+y)^*(x+y) - (x-y)^*(x-y) + i(x+iy)^*(x+iy) - i(x-iy)^*(x-iy)$$

More generally, for any x, y, a we have

$$4y^*ax = \sum_{k=0}^{3} i^k (x+i^k y)^* a(x+i^k y) .$$

II.3.1.7 If A is a C*-algebra and $x \in A$, define $|x| = (x^*x)^{1/2} \in A_+$. If $x = x^*$, then $|x| = x_+ + x_- = f(x)$, where $f(t) = |t|$.

II.3.1.8 Because of II.3.1.3(i), it makes sense in a C*-algebra A to write $x \le y$ if $y - x \ge 0$. This defines a translation-invariant partial order on A (it is usually only used on A_{sa}). Note that $x \le y$ is well defined independent of the containing C*-algebra. If $x = x^*$ and $\sigma_A(x) \subseteq [\alpha, \beta]$, then $\alpha 1 \le x \le \beta 1$ (in $\tilde{A}$). If $a \le b$, then $x^*ax \le x^*bx$ for any x. If $x \in A$, $a \in A_+$, then $x^*ax \le x^*(\|a\|1)x = \|a\|x^*x$.

II.3.1.9 PROPOSITION. Let A be a C*-algebra, $x, y \in A$. Then

(i) $-(x^*x + y^*y) \le x^*y + y^*x \le x^*x + y^*y$
(ii) $(x+y)^*(x+y) \le 2(x^*x + y^*y)$. More generally, for any $x_1, \ldots, x_n$,

$$(x_1 + \cdots + x_n)^*(x_1 + \cdots + x_n) \le n(x_1^*x_1 + \cdots + x_n^*x_n)$$

(iii) If $0 \le x \le y$ and x is invertible, then y is invertible and $0 \le y^{-1} \le x^{-1}$.

PROOF: For (i), $(x+y)^*(x+y), (x-y)^*(x-y) \ge 0$. The first statement of (ii) is just $(x+y)^*(x+y) + (x-y)^*(x-y) = 2(x^*x + y^*y)$; for the general statement, if ω is a primitive n'th root of unity, then

$$n \sum_{j=1}^{n} x_j^* x_j = \sum_{j=1}^{n} \left[\left(\sum_{k=1}^{n} \omega^{j(k-1)} x_k \right)^* \left(\sum_{k=1}^{n} \omega^{j(k-1)} x_k \right) \right] .$$

(iii): If x is invertible, then $\epsilon 1 \le x$ for some $\epsilon > 0$, so $\epsilon 1 \le y$ and y is invertible by II.2.2.4, and $x^{-1}, y^{-1} \ge 0$. The inequality $y^{-1} \le x^{-1}$ is obvious from II.2.2.4 if x and y commute, and in particular if $y = 1$. For the general case, if $x \le y$, then

$$y^{-1/2}xy^{-1/2} \le y^{-1/2}yy^{-1/2} = 1$$
$$1 \le (y^{-1/2}xy^{-1/2})^{-1} = y^{1/2}x^{-1}y^{1/2}$$
$$y^{-1} = y^{-1/2}1y^{-1/2} \le y^{-1/2}(y^{1/2}x^{-1}y^{1/2})y^{-1/2} = x^{-1}.$$

II.3.1.10 PROPOSITION. Let A be a C*-algebra, $0 \leq x \leq y \in A$, and $0 < \alpha \leq 1$. Then $x^\alpha \leq y^\alpha$. (We say the function $t \mapsto t^\alpha$ is *operator-monotone*.)

PROOF: Let $S = \{\alpha \in (0, \infty) : t \mapsto t^\alpha$ is operator-monotone$\}$. Then $1 \in S$, and S is (topologically) closed in $(0, \infty)$ and also closed under multiplication. Two other observations:

(i) $\alpha \in S$ if and only if $0 \leq x \leq y$ and y is invertible implies $x^\alpha \leq y^\alpha$: For general x, y, $0 \leq x \leq y$, we then have $x^\alpha \leq (y + \epsilon 1)^\alpha$ (in $\tilde{A}$) for all $\epsilon > 0$, and thus $x^\alpha \leq \lim_{\epsilon \to 0}(y + \epsilon 1)^\alpha = y^\alpha$ (II.2.3.3).

(ii) If $x, y \geq 0$ and y is invertible, then

$$x^\alpha \leq y^\alpha \iff y^{-\alpha/2}x^\alpha y^{-\alpha/2} \leq 1$$

$$\iff \|y^{-\alpha/2}x^{\alpha/2}\|^2 = \|y^{-\alpha/2}x^\alpha y^{-\alpha/2}\| \leq 1.$$

Next we show $1/2 \in S$. If $0 \leq x \leq y$, and y is invertible, then $y^{-1/2}xy^{-1/2} \leq 1$, so $\|y^{-1/2}x^{1/2}\| \leq 1$. By II.1.4.2(iv),

$$r(y^{-1/4}x^{1/2}y^{-1/4}) = r(y^{-1/2}x^{1/2}) \leq \|y^{-1/2}x^{1/2}\| \leq 1$$

so $y^{-1/4}x^{1/2}y^{-1/4} \leq 1$, $x^{1/2} \leq y^{1/2}$.

Similarly, if $\alpha, \beta \in S$, we show $\gamma = (\alpha + \beta)/2 \in S$. If $0 \leq x \leq y$, and y is invertible, we have

$$r(y^{-\gamma/2}x^\gamma y^{-\gamma/2}) = r(y^{-\gamma}x^\gamma) = r(y^{-\alpha/2}x^\gamma y^{-\beta/2}) \leq \|(y^{-\alpha/2}x^{\alpha/2})(x^{\beta/2}y^{-\beta/2})\|$$

$$\leq \|y^{-\alpha/2}x^{\alpha/2}\|\|x^{\beta/2}y^{-\beta/2}\| \leq 1.$$

So S contains all of $(0, 1]$.

For an alternate proof, see [Ped79, 1.3.8].

II.3.1.11 The result can fail if $\alpha > 1$ unless x and y commute. There are counterexamples in $\mathbb{M}_2$, for example $x = \begin{bmatrix} 2 & 2 \\ 2 & 2 \end{bmatrix}$, $y = \begin{bmatrix} 3 & 0 \\ 0 & 6 \end{bmatrix}$; $x \leq y$, but $x^\alpha \not\leq y^\alpha$ for any $\alpha > 1$. Thus S is precisely $(0, 1]$.

II.3.1.12 While discussing inequalities, the following simple analog of Bessel's inequality is useful: if $a_1, \ldots, a_n$ are elements of a C*-algebra A, and $a_i^* a_j = 0$ for $i \neq j$, then for any $x_1, \ldots, x_n \in A$ we have

$$\left\|\sum_{i=1}^n a_i x_i\right\|^2 = \left\|\left(\sum_{i=1}^n a_i x_i\right)^* \left(\sum_{i=1}^n a_i x_i\right)\right\| = \left\|\sum_{i=1}^n x_i^* a_i^* a_i x_i\right\|$$

$$\leq \sum_{i=1}^n \|x_i^* a_i^* a_i x_i\| = \sum_{i=1}^n \|a_i x_i\|^2.$$

Orthogonality

II.3.1.13 Two projections (or, more generally, positive elements) p, q of a C*-algebra A are *orthogonal*, written $p \perp q$, if $pq = 0$. Orthogonal positive elements commute, and if $p \perp q$, then $\|p+q\| = \max(\|p\|, \|q\|)$. Orthogonality is usually used only for positive elements, but can be defined in general: if $x, y \in A$, then $x \perp y$ if $xy = yx = x^*y = yx^* = 0$ (hence $x^*y^* = y^*x^* = xy^* = y^*x = 0$ also).

Following [Ber72], we say $p, q \in A_+$ are *very orthogonal*, written $p \perp\!\!\!\perp q$, if $pxq = 0$ for all $x \in A$. This is equivalent to saying that the ideals generated by p and q are orthogonal. Unlike orthogonality, the notion that p and q are very orthogonal depends on the containing C*-algebra. If $p \perp\!\!\!\perp q$ in A, then $p \perp q$, even if A is nonunital ($pq = \lim pq^{1/n}q$; in fact, $p \perp\!\!\!\perp q$ in A if and only if $p \perp\!\!\!\perp q$ in $\tilde{A}$).

II.3.2 Polar Decomposition

Polar decomposition (I.5.2.2) cannot be done in general in a C*-algebra, but there is a weak version which is extremely useful. In fact, we have a weakened version of I.5.2.4.

II.3.2.1 PROPOSITION. *If A is a C*-algebra, $x \in A$, $a \in A_+$, $x^*x \leq a$, and $0 < \alpha < 1/2$, then there is a $u \in \overline{Aa} \subseteq A$ with $u^*u \leq a^{1-2\alpha}$ (hence $\|u\| \leq \|a^{1/2-\alpha}\|$), $uu^* \leq (xx^*)^{1-2\alpha}$, and $x = ua^\alpha$. In particular, $x = u(x^*x)^\alpha$ for some u with $u^*u = (x^*x)^{1-2\alpha}$ (and $\|u\| = \|(x^*x)^{1/2-\alpha}\|$). So every element of A can be written as a product of n elements of A, for any n.*

The proof consists of showing that the sequence $u_n = x[(1/n)+a]^{-1/2}a^{1/2-\alpha}$ is norm-convergent to an element u with $\|u\| \leq \|a^{1/2-\alpha}\|$ (see [Ped79, 1.4.4] for details). Then, if $y_n = 1 - [(1/n)+a]^{-1/2}a^{1/2}$, we have

$$\|x - u_n a^\alpha\|^2 = \|xy_n\|^2 = \|y_n x^* x y_n\| \leq \|y_n a y_n\| = \|a^{1/2}y_n\|^2 \to 0$$

since $f_n(t) = t^{1/2}[1 - (t/[(1/n)+t])^{1/2}] \to 0$ uniformly on $[0, \|a\|]$ by Dini's Theorem.

A closely related result, which can be used to give an alternate proof of II.3.2.1 (and in fact a slight generalization), will be discussed in III.5.2.16.

II.3.2.2 By applying the proposition to x^* and taking adjoints, we obtain the "right-handed version": if $xx^* \leq a$, then for $0 < \alpha < 1/2$ there is a $v \in \overline{aA} \subseteq A$ with $\|v\| \leq \|a^{1/2-\alpha}\|$ and $x = a^\alpha v$.

There is a closely related result about existence of "cube roots," with a simpler proof, which gives an alternate proof of the last statement of II.3.2.1:

II.3.2.3 PROPOSITION. Let A be a C*-algebra and $x \in A$. Then there is a unique $y \in A$ with $x = yy^*y$.

A proof of existence can be obtained from II.3.2.1: if $x = y(x^*x)^{1/3}$ with $y \in \overline{Ax^*x}$, then it can be shown that $y^*y = (x^*x)^{1/3}$. Uniqueness can also be shown directly. (See III.5.2.18 for a version of this argument using polar decomposition in the second dual.) But there is also a slick elementary proof of the proposition using functional calculus in 2×2 matrices (II.6.6.4).

II.3.2.4 If $0 \leq z \leq a$ in a C*-algebra A, we cannot in general write $z = a^{1/2}ca^{1/2}$ or $z = x^*ax$ for $x, c \in A$, even if A is commutative [consider $f(t) = t$ and $g(t) = t\sin^2(t^{-1})$ in $C([0,1])$], although this can always be done if A is a von Neumann algebra. But there is an approximate version:

II.3.2.5 COROLLARY. Let $0 \leq z \leq a$ in a C*-algebra A. Then

(i) For any $0 < \alpha < 1/2$ there is a $c \in A_+$ with $z = a^\alpha c a^\alpha$ and $\|c\| \leq \|a^{1-2\alpha}\|$.
(ii) There is a bounded sequence (x_n) in A with $x_n^* a x_n \to z$, and a bounded sequence (c_n) in A_+ with $a^{1/2} c_n a^{1/2} \to z$.

PROOF: Take $x = z^{1/2}$ in II.3.2.1, and find for each α a u_α with $z^{1/2} = u_\alpha a^\alpha = a^\alpha u_\alpha^*$ and $\|u_\alpha\| \leq \|a^{1/2-\alpha}\| \leq \max(1, \|a^{1/2}\|)$. Then $z = a^\alpha u_\alpha u_\alpha^* a^\alpha = u_\alpha a^{2\alpha} u_\alpha^*$. Set $k_n = 1/2 - 1/n$, $x_n = u_{k_n}^*$ and $c_n = u_{k_n} u_{k_n}^*$.

Riesz Decomposition

II.3.2.6 As a related matter, we do not have the usual Riesz Decomposition property in noncommutative C*-algebras: if $a, b, c \in A_+$ and $a \leq b + c$, there are not necessarily $a_1, a_2 \in A_+$ with $a_1 \leq b$, $a_2 \leq c$ and $a = a_1 + a_2$ (consider $b = \begin{bmatrix} 1 & 0 \\ 0 & 0 \end{bmatrix}$, $c = \begin{bmatrix} 0 & 0 \\ 0 & 1 \end{bmatrix}$, $a = \frac{1}{2}\begin{bmatrix} 1 & 1 \\ 1 & 1 \end{bmatrix}$ in $\mathbb{M}_2$). However, we have the following version of the Riesz Decomposition property in any C*-algebra:

II.3.2.7 PROPOSITION. Let A be a C*-algebra, $x, y_1, \ldots, y_n \in A$ with $x^*x \leq \sum_{j=1}^n y_j y_j^*$. Then there are $u_1, \ldots, u_n \in A$ with $u_j^* u_j \leq y_j^* y_j$ for each j, and $xx^* = \sum_{j=1}^n u_j u_j^*$.

The proof, similar to the proof of II.3.2.1, can be found in [Ped79, 1.4.10]. If A is a von Neumann algebra, there is a simplified proof using the argument of I.5.2.5. Note that the distinctions between x^*x and xx^*, and the $y_j^*y_j$ and $y_jy_j^*$, are merely for convenience, giving broader applicability; x and the y_j may be replaced in the statement by $(x^*x)^{1/2}$ and $(y_jy_j^*)^{1/2}$ respectively. [However, the distinction between $u_j^*u_j$ and $u_ju_j^*$ is essential, as the previous example shows.]

Well-Supported Elements

True polar decomposition can be done for invertible elements, and somewhat more generally:

II.3.2.8 DEFINITION. An element x in a C*-algebra A is *well-supported* if $\sigma_A(x^*x)\setminus\{0\}$ is closed (i.e. $\sigma_A(x^*x) \subseteq \{0\} \cup [\epsilon, \infty)$ for some $\epsilon > 0$).

The property of being well-supported is independent of the containing C*-algebra. Invertible elements and partial isometries are well-supported. The image of a well-supported element under a homomorphism is well-supported.

II.3.2.9 It follows from II.1.4.2(iv) that if x is well-supported, then so is x^*. From II.1.4.2(iii), if $x = x^*$ (or, more generally, if x is normal), then x is well-supported if and only if x^n is well-supported for some n.

If $f(0) = 0$ and $f(t) = 1$ for $t > 0$, then f is continuous on $\sigma_A(x^*x)$ if x is well-supported, and $p = f(x^*x)$ is a projection, called the *right support projection* of x. This projection has the properties that $xp = x$ and $py = 0$ whenever $xy = 0$. Similarly, $q = f(xx^*)$, the *left support projection* of x, satisfies $qx = x$ and $yq = 0$ whenever $yx = 0$. If x is normal, then its left and right support projections coincide. If $g(0) = 0$ and $g(t) = t^{-1/2}$ for $t > 0$, then $g(x^*x)$ is defined and $u = xg(x^*x)$ is a partial isometry with $u^*u = p$, $uu^* = q$, and $x = u(x^*x)^{1/2}$ gives a true polar decomposition for x. If x is invertible, then $p = q = 1$ and u is unitary; in this case, $g(x^*x) = (x^*x)^{-1/2}$, and if $x = va$ with v unitary and $a \geq 0$, then $v = u$ and $a = (x^*x)^{1/2}$. Also, if $x_n \to x$ with x_n, x invertible, and $x_n = u_n a_n$, $x = ua$, then $u_n \to u$ and $a_n \to a$.

II.3.2.10 Well-supported elements in a C*-algebra are precisely the elements which are *von Neumann regular* in the ring-theoretic sense of having a quasi-inverse (y is a quasi-inverse for x if $xyx = x$) [Goo91]. Well-supported elements are also precisely the elements which have closed range when the C*-algebra is represented as an algebra of operators on a Hilbert space (or a Hilbert module; cf. II.7.2.8). Thus well-supported elements are sometimes called *elements with closed range*. This term is also justified by the following observation:

II.3.2.11 PROPOSITION. Let A be a C*-algebra, $x \in A$. The following are equivalent:

(i) Ax is closed.
(ii) xA is closed.
(iii) x^*Ax is closed.
(iv) x is well-supported.

Convex Combinations of Unitaries

We now have enough machinery to obtain a sharpening, due to R. Kadison and G. Pedersen [KP85], of a result of B. Russo and H. Dye ([RD66]; cf. [Har72], [Gar84]). We begin with a simple observation:

II.3.2.12 PROPOSITION. Every element of a unital C*-algebra A is a linear combination of four unitaries. In fact, if $x = x^* \in A$ and $\|x\| \leq 2$, then x is a sum of two unitaries in $\mathcal{U}(A)_o$.

PROOF: Set $a = x/2$. Then $1 - a^2 \geq 0$ and $a \pm i(1-a^2)^{1/2}$ are unitaries. If $0 \leq t \leq 1$, then $(ta \pm i(1-t^2a^2)^{1/2})$ gives a path of unitaries from $\pm i1$ to $a \pm i(1-a^2)^{1/2}$, so $a \pm i(1-a^2)^{1/2} \in \mathcal{U}(A)_o$.

II.3.2.13 LEMMA. Let A be a unital C*-algebra, $y \in A$, $\|y\| < 1$, and $u \in \mathcal{U}(A)$. Then there are unitaries $w_1, w_2 \in A$ with $u + y = w_1 + w_2$. If $u \in \mathcal{U}(A)_o$, then w_1, w_2 may be chosen in $\mathcal{U}(A)_o$.

PROOF: Write $u+y = u(1+u^*y)$. Then $1+u^*y$ is invertible, $1+u^*y = v|1+u^*y|$ with $v \in \mathcal{U}(A)_o$, and $|1+u^*y|$ is a sum of two elements of $\mathcal{U}(A)_o$ by II.3.2.12.

II.3.2.14 THEOREM. If A is a unital C*-algebra, $x \in A$, $\|x\| < 1 - 2/n$, then there are $u_1, \ldots, u_n \in \mathcal{U}(A)_o$ with $x = (u_1 + \cdots + u_n)/n$.

PROOF: Let $y = (n-1)^{-1}(nx-1)$; then $\|y\| < 1$. By repeated use of II.3.2.13, we have

$$nx = 1+(n-1)y = u_1+w_1+(n-2)y = u_1+u_2+w_2+(n-3)y = \cdots = u_1+\cdots+u_n$$

for some $u_1, \ldots, u_n \in \mathcal{U}(A)_o$.

In fact, an element of norm $1 - 2/n$ in a unital C*-algebra A can also be written as a mean of n unitaries in A [Haa90]. This result is sharp: if v is a nonunitary isometry and $\lambda > 1 - 2/n$, then λv cannot be written as a convex combination of n unitaries in any containing C*-algebra [KP85]. The minimal number of unitaries necessary for an element $x \in A$ is closely related to the distance from x to the invertible elements of A ([Rør88], [Ped89b]; cf. II.3.2.19).

II.3.2.15 COROLLARY. [RUSSO-DYE] Let A be a unital C*-algebra. Then $co(\mathcal{U}(A)_o)$ (the convex hull of $\mathcal{U}(A)_o$) contains the open unit ball of A. Thus $\overline{co}(\mathcal{U}(A)) = \overline{co}(\mathcal{U}(A)_o)$ is the closed unit ball of A.

II.3.2.16 COROLLARY. Let A be a unital C*-algebra, B a normed vector space, and $\phi : A \to B$ a bounded linear map. Then $\|\phi\| = \sup \|\phi|_C\|$, where C runs over all unital commutative C*-subalgebras of A.

For the proof, note that $\|\phi\| = \sup\{\|\phi(u)\| : u \in \mathcal{U}(A)\}$, and $C^*(u)$ is commutative.

Not every element of the closed unit ball of a unital C*-algebra is a convex combination of unitaries. In fact, we have:

II.3.2.17 THEOREM. Let A be a C*-algebra. If A is nonunital, then there are no extreme points in the closed unit ball of A. If A is unital, then the extreme points of the closed unit ball of A are precisely the elements x such that
$$(1 - xx^*)A(1 - x^*x) = 0$$
(i.e. $(1 - xx^*) \perp\!\!\!\perp (1 - x^*x)$ (II.3.1.13)). Such an x is automatically a partial isometry. So every isometry or coisometry (in particular, every unitary) is an extreme point of the closed unit ball of A.

It is easy to see that an extreme point x in the unit ball of A is a partial isometry: if not (i.e. if x^*x is not a projection), write $x = u(x^*x)^{1/4}$ for some u, $\|u\| = 1$, and by functional calculus find a small positive a in $C^*(x^*x)$ such that $\|(x^*x)^{1/4}(1+a)\| = 1$ (where $1 \in \tilde{A}$ is used symbolically); then x is the average of $u(x^*x)^{1/4}(1 \pm a)$, a contradiction. So if x is extreme, $1 - xx^*$ and $1 - x^*x$ are projections in $\tilde{A}$.

Hence, if x is extreme and $a \in (1 - xx^*)A(1 - x^*x)$ with $\|a\| = 1$, then $ax^*x = 0$, so $ax^*xa^* = 0$, $ax^* = 0$, $a^*a \perp x^*x$, and similarly $a^*x = 0$, $aa^* \perp xx^*$. So
$$\|x \pm a\|^2 = \|(x^* \pm a^*)(x \pm a)\| = \|x^*x + a^*a\| = \max(\|x^*x\|, \|a^*a\|) = 1$$
and $x = [(x+a) + (x-a)]/2$, a contradiction; thus $(1 - xx^*)A(1 - x^*x) = 0$.

So if x is extreme, and if $y \in A$, then
$$(1 - x^*x)(1 - xx^*)y^*y(1 - xx^*)(1 - x^*x) = 0$$
and thus $y(1 - xx^*)(1 - x^*x) = 0$. Therefore $(1 - xx^*)(1 - x^*x) = 0$, $1 = x^*x + xx^* - (xx^*)(x^*x)$, and A is unital.

Showing that an x satisfying $(1 - xx^*)A(1 - x^*x) = 0$ is an extreme point is more difficult (cf. [Sak71, 1.6.4], [Ped79, 1.4.7]), but there is an elementary proof that a unitary u in a unital C*-algebra A is an extreme point in the unit ball. Since left multiplication by u is an isometry from A onto A, it suffices to show that 1 is an extreme point. The result holds in a commutative C*-algebra by the functional representation. In a general A, if $1 = (x + y)/2$ with $\|x\| = \|y\| = 1$, write $x = a + ib$, $y = c + id$; then $1 = (a + c)/2$ and $\|a\|, \|c\| \leq 1$, and $C^*(a, c)$ is commutative, so $a = c = 1$; thus x and y are normal, so $C^*(x, y)$ is commutative and $x = y = 1$.

There is a simple argument that an isometry $U \in \mathcal{L}(\mathcal{H})$ is an extreme point of the unit ball of $\mathcal{L}(\mathcal{H})$, which can be combined with the Gelfand-Naimark Theorem (II.6.4.10) to show that isometries (and coisometries) in general C*-algebras are extreme points. If $U = tS + (1-t)T$ with $\|S\| = \|T\| = 1$ and $0 < t < 1$, let $\xi \in \mathcal{H}$, $\|\xi\| = 1$. If $\eta = U\xi$, then $\|\eta\| = 1$, and η is an extreme point of the unit ball of $\mathcal{H}$ by the parallelogram law. But $\eta = t(S\xi) + (1-t)(T\xi)$, so $S\xi = T\xi = U\xi$. Since ξ is arbitrary, $S = T = U$.

II.3.2.18 Thus a nonunitary isometry is not a convex combination of unitaries. In fact, if A is unital, the convex hull of $\mathcal{U}(A)$ is the whole closed unit ball of A if and only if A has stable rank 1 (V.3.1.5) [Rør88]. See [PR88] and [HR93] for further related results. One simple related result is useful:

II.3.2.19 PROPOSITION. Let A be a unital C*-algebra, and u a nonunitary isometry in A. If z is an invertible element of A, then $\|u - z\| \geq 1$.

PROOF: Suppose z is invertible in A and $\|u - z\| < 1$. Set $x = zu^*$, $p = uu^*$, and $r = 1 - p \neq 0$. Then x is right invertible with right inverse $y = uz^{-1}$, and $xp = x$, so

$$\|p - pxp\| = \|p - px\| = \|p(p-x)\| = \|p(u-z)u^*\| \leq \|u-z\| < 1$$

and pxp is invertible in pAp. Since $py = y$, we have $(pxp)(pyr) = pxyr = pr = 0$, and since pxp is invertible in pAp, $pyr = 0$. But $(rxp)(pyr) = rxyr = r \neq 0$, a contradiction.

Since 0 is in the closure of the invertible elements of A, the distance from u to the invertible elements of A is exactly 1.

II.3.2.20 An interesting consequence of these theorems is that the (closed) unit ball of a unital C*-algebra is the closed convex hull of its extreme points, even though in general it is not compact in any natural topology.

Quasi-Invertible Elements

II.3.2.21 In [BP95], an element of a unital C*-algebra A of the form yuz, where y and z are invertible and u is an extreme point of the unit ball of A, is called a *quasi-invertible* element of A. The set of quasi-invertible elements of A is denoted A_q. Every left or right invertible element or extreme point of the unit ball is quasi-invertible. A quasi-invertible element x of A is well-supported, and the right and left support projections p and q of x satisfy $(1-p) \perp\!\!\!\perp (1-q)$; and conversely, every element of this form is quasi-invertible [if $x = u|x|$ is the polar decomposition, then u is an extreme point of the unit ball and $x = u(|x| + 1 - p)$]. Unlike invertibility (II.1.6.7), quasi-invertibility depends on the containing C*-algebra.

II.3.3 Comparison Theory for Projections

Just as in $\mathcal{L}(\mathcal{H})$, the order structure on projections in a C*-algebra has nice aspects not shared by general positive elements, summarized in the next proposition:

II.3.3.1 PROPOSITION. If p and q are projections in a C*-algebra A, then the following are equivalent:

(i) $p \leq q$.
(ii) $p \leq \lambda q$ for some $\lambda > 0$.
(iii) $pq = qp = p$.
(iv) $q - p$ is a projection.

If these conditions hold, we say p is a *subprojection* of q.

Note that unlike in $\mathcal{L}(\mathcal{H})$ (or in a von Neumann algebra), the projections in a C*-algebra do not form a lattice in general:

II.3.3.2 EXAMPLES.

(i) Let c be the C*-algebra of convergent sequences of complex numbers. Let $p^{(m)} = (p_n^{(m)})$ be the projection in c with $p_n^{(m)} = 1$ for $n \leq m$ odd, and $p_n^{(m)} = 0$ otherwise. Then the sequence $(p^{(m)})$ has no supremum in $\mathrm{Proj}(c)$.
(ii) Let A be the C*-algebra of convergent sequences in $\mathbb{M}_2$. Let $p = (p_n)$ and $q = (q_n)$ be the projections in A with $p_n = q_{2n-1} = diag(1,0)$ and
$$q_{2n} = \begin{bmatrix} 1 - n^{-1} & \sqrt{n^{-1} - n^{-2}} \\ \sqrt{n^{-1} - n^{-2}} & n^{-1} \end{bmatrix}.$$
Then $\{p, q\}$ has no supremum or infimum in $\mathrm{Proj}(A)$ (note that in $\mathbb{M}_2$, $p_{2n} \wedge q_{2n} = 0$ and $p_{2n} \vee q_{2n} = I$).
(iii) Let A be as in (ii), and let B be the (nonunital) C*-subalgebra of A of sequences converging to a multiple of $diag(1,0)$. Then $p, q \in B$, but $\{p, q\}$ has no upper bound in $\mathrm{Proj}(B)$.

It is very useful to have a notion of equivalence of projections:

II.3.3.3 DEFINITION. If p and q are projections in A, then p and q are *(Murray-von Neumann) equivalent (in A)*, written $p \sim q$, if there is a partial isometry $u \in A$ with $u^*u = p$, $uu^* = q$, and $p \precsim q$ if p is equivalent to a subprojection of q, i.e. if there is a partial isometry $u \in A$ with $u^*u = p$ and $uu^* \leq q$ (we say p is *subordinate* to q).

Although the notation does not reflect it, equivalence of projections is, of course, relative to a specified containing C*-algebra.

Unitarily equivalent projections are equivalent; the converse is false in general (in $\mathcal{L}(\mathcal{H})$, for example), but is "stably" true (V.1.1.2). Note that $p \precsim q$ and $q \precsim p$ do not together imply $p \sim q$ in general, e.g. in $\mathcal{O}_n$, $n > 2$ (II.8.3.3(ii), V.1.3.4), i.e. there is no Schröder-Bernstein theorem for equivalence of projections (although there is for von Neumann algebras (III.1.1.9)).

The next proposition is fundamental, and may be regarded as one of the starting points for the K-theory of C*-algebras. It has an analog for general Banach algebras [Bla98, 4.3.4].

II.3.3.4 PROPOSITION. Let A be a C*-algebra, p, q projections in A. If $\|p - q\| < 1$, then $p \sim q$. In fact, there is a unitary $v(p,q) \in \tilde{A}$ with $v(p,q)pv(p,q)^* = q$. The assignment $(p,q) \mapsto v(p,q)$ has the following properties:

(i) $v(p,p) = 1$ for all p.
(ii) $(p,q) \mapsto v(p,q)$ is jointly continuous in p and q; for any $\epsilon > 0$ there is a $\delta > 0$ such that $\|v(p,q) - 1\| < \epsilon$ for any p,q with $\|p - q\| < \delta$.
(iii) $(p,q) \mapsto v(p,q)$ is functorial in the sense that if $\phi : A \to B$ is a *-homomorphism, then $\tilde{\phi}(v(p,q)) = v(\phi(p), \phi(q))$, where $\tilde{\phi} : \tilde{A} \to \tilde{B}$ is the induced map.

PROOF: If $x = qp + (1-q)(1-p)$, then $1 - x^*x = 1 - xx^* = (p-q)^2$, so $\|1 - x^*x\| = \|1 - xx^*\| = \|p - q\|^2 < 1$ and x is invertible in $\tilde{A}$. Let $v(p,q)$ be the unitary in the polar decomposition of x, i.e. $v(p,q) = x(x^*x)^{-1/2}$, $x = v(p,q)(x^*x)^{1/2}$. We have $xp = qx = qp$, $px^* = x^*q = pq$, so $x^*xp = x^*qx = px^*x$, and so $(x^*x)^{1/2}$ commutes with p. Since $xpx^{-1} = qxx^{-1} = q$, we obtain $v(p,q)pv(p,q)^* = q$. Let $\gamma = \|p-q\|$. We have $x^*x = 1 - (p-q)^2 \leq 1$, so $\|x\| \leq 1$, and

$$\|x - 1\| = \|qp + (1-q)(1-p) - p - (1-p)\| \leq \|qp - p\| + \|(1-q)(1-p) - (1-p)\|$$

$$\leq \|q - p\| + \|(1-q) - (1-p)\| = 2\gamma.$$

Since $(1 - \gamma^2)1 \leq x^*x \leq 1$, we have

$$\|(x^*x)^{-1/2} - 1\| \leq \alpha := (1 - \gamma^2)^{-1/2} - 1$$

$$\|v(p,q) - 1\| = \|x(x^*x)^{-1/2} - x + x - 1\| \leq \|x[(x^*x)^{-1/2} - 1]\| + \|x - 1\| \leq \alpha + 2\gamma.$$

A similar argument shows the following useful technical result:

II.3.3.5 PROPOSITION. For every $\epsilon > 0$ there is a $\delta > 0$ such that, whenever A is a C*-algebra and p, q are projections in A with $\|pq - q\| < \delta$, then there is a projection $p' \in A$ with $q \leq p'$ and $\|p - p'\| < \epsilon$.

PROOF: Suppose $0 < \delta < 1$ and p, q are projections in a C*-algebra A with $\|pq - q\| < \delta$. Then $\|qpq - q\| < \delta$, so qpq is invertible in qAq with inverse a, and $q \leq a \leq (1-\delta)^{-1}q$. If $b = a^{1/2}$, then $\|q - b\| < \alpha := (1-\delta)^{-1/2} - 1$, and pq has polar decomposition $u(qpq)^{1/2}$, where $u = pqb$ is a partial isometry with $u^*u = q$. We have

$$\|u - q\| = \|u - pq + pq - q\| \leq \|pqb - pq\| + \|pq - q\|$$

$$= \|pq(b - q)\| + \|pq - q\| < \beta := \alpha + \delta$$

and $\|u^* - q\| < \beta$, so, since $u = uq$,

$$\|uu^* - q\| = \|uu^* - uq + uq - q\| \leq \|u(u^* - q)\| + \|(u - q)q\| < 2\beta$$

and $r = uu^* \leq p$. If $\epsilon > 0$ is given and β is small enough, then the unitary $v(q,r) \in \tilde{A}$ (II.3.3.4(ii)) satisfies $\|v(q,r) - 1\| < \epsilon/2$. Then, if $p' = v(q,r)pv(q,r)^*$, we have

$$\|p' - p\| \leq \|p' - v(q,r)p\| + \|v(q,r)p - p\|$$
$$= \|v(q,r)p[v(q,r)^* - 1]\| + \|[v(q,r) - 1]p\| < \epsilon$$

and $q = v(q,r)rv(q,r)^* \leq v(q,r)pv(q,r)^* = p'$.

II.3.4 Hereditary C*-Subalgebras and General Comparison Theory

II.3.4.1 DEFINITION. Let A be a C*-algebra. A C*-subalgebra B of A is *hereditary* if $0 \leq a \leq b$, $a \in A$, $b \in B$ implies $a \in B$.

II.3.4.2 PROPOSITION. Let A be a C*-algebra, B a hereditary C*-subalgebra, and $x \in A$. Then

(i) $BAB \subseteq B$ (so $BAB = B$ since $B = B^3$ (II.3.2.1)).
(ii) $\overline{x^*Ax} = \overline{(x^*x)A(x^*x)}$ is a hereditary C*-subalgebra of A containing x^*x, and is the smallest such algebra.
(iii) If A is separable, then every hereditary C*-subalgebra of A is of the form $\overline{hAh}$ for some $h \in A_+$.

PROOF: (i): First note that if $z \in B$, then $z^*Az \subseteq B$: it suffices to show $z^*A_+z \subseteq B_+$, which follows from II.3.1.8. Now let $x, y \in B$, and set $b = xx^* + y^*y$. Then there are $u, v \in B$ with $x = b^{1/4}v$ and $y = ub^{1/4}$ (II.3.2.1), so, for $a \in A$,

$$xay = b^{1/4}vaub^{1/4} \in b^{1/4}Ab^{1/4} \subseteq B.$$

(ii): $(x^*x)A(x^*x) \subseteq x^*Ax$, and $x^*Ax \subseteq (x^*x)^{1/4}A(x^*x)^{1/4}$ since $x = u(x^*x)^{1/4}$ for some u (II.3.2.1). $\overline{x^*Ax}$ is a C*-subalgebra of A.

$$(x^*x)^3 = x^*(xx^*)^2x \in x^*Ax,$$

so

$$(x^*x)^\alpha \in C^*((x^*x)^3) \subseteq \overline{(x^*x)A(x^*x)} \subseteq \overline{x^*Ax}$$

for any $\alpha > 0$. Thus we have

$$(x^*x)^{1/4}A(x^*x)^{1/4} \subseteq \overline{(x^*x)A(x^*x)}$$

by (i), so $\overline{x^*Ax} = \overline{(x^*x)A(x^*x)}$. It follows that $b \in \overline{x^*Ax}$ if and only if $\lim_{\alpha \to 0}(x^*x)^\alpha b(x^*x)^\alpha = b$. Thus, if $0 \leq a \leq b$ and $b \in \overline{x^*Ax}$, write $a = b^{1/4}cb^{1/4}$ (II.3.2.5) to see that $a \in \overline{x^*Ax}$.

(iii): Let (x_n) be a dense sequence in the unit ball of B, and let h be the element $\sum_{n=1}^{\infty} 2^{-n}x_n^*x_n$; then $B = \overline{hAh}$.

See II.5.3.2 and II.5.3.9 for additional comments about hereditary C*-subalgebras.

As with projections, it is useful to have some other notions of comparison between general positive elements in addition to the ordinary order $\leq$. Here are two other useful ones:

II.3.4.3 DEFINITION. Let A be a C*-algebra, $a, b \in A_+$. Then
(i) $b \ll a$ if $ab = b$.
(ii) $b \precsim a$ if there is a sequence (x_n) in A with $x_n^* a x_n \to b$ (equivalently, for any $\epsilon > 0$ there is an $x \in A$ with $\|x^* a x - b\| < \epsilon$).
(iii) $a \approx b$ if $a \precsim b$ and $b \precsim a$.

Of course, $\precsim$ and $\approx$ depend on the choice of A ($\ll$ is independent of the choice of the containing C*-algebra). These relations are transitive; $\precsim$ is reflexive, and $\approx$ is an equivalence relation. [It is not entirely trivial that $\precsim$ is transitive, since the sequence (x_n) cannot be chosen to be bounded in general; but if $b \precsim a$ and $c \precsim b$, and y satisfies $\|y^* by - c\| < \epsilon/2$, choose x with $\|x^* ax - b\| < \epsilon/2\|y\|^2$; then $\|y^* x^* axy - c\| < \epsilon$.] It is easy to see that if p and q are projections, this definition of $p \precsim q$ is equivalent to the previous one.

II.3.4.4 If $\|b\| \leq 1$, then $b \ll a \implies b \leq a$. We have $a \ll a$ if and only if a is a projection. If p and q are projections, then $p \ll q$ if and only if $p \leq q$. If $b \ll a$, then a and b obviously commute.

II.3.4.5 By II.3.2.5, $b \leq a \Rightarrow b \precsim a$. But $\precsim$ is a much weaker relation: $x^* ax \precsim a$ for any x, and if $b \in \overline{aAa}$, then $b \precsim a$, so $\overline{aAa} = \overline{bAb} \Rightarrow a \approx b$. In particular, $a \approx \alpha a \approx a^\alpha$ for any $\alpha > 0$. Also, by II.3.2.1 $x^* x \approx xx^*$ for any x. The relations $\precsim$ and $\approx$ measure the "width" (size of the support) of an element rather than its "height" or "location". The relation $b \precsim a$ roughly means that the hereditary C*-subalgebra generated by b is "smaller" than the one generated by a.

II.3.4.6 PROPOSITION. If $a, b, c \in A_+$, $b \ll a$, $\|a\| = 1$, and $\|a - c\| = \eta < 1$, then $(1 - \eta) b \leq b^{1/2} c b^{1/2} \leq (1 + \eta) b$, so $b^{1/2} c b^{1/2} \approx b$ and $b \precsim c$.

For the proof, note that $b^{1/2} c b^{1/2} = b^{1/2} (c - a) b^{1/2} + b^{1/2} a b^{1/2} \geq -\eta b + b$.

II.3.4.7 PROPOSITION.
(i) If $a, b, b_n \in A_+$, $b_n \precsim a$ for all n, and $b_n \to b$, then $b \precsim a$.
(ii) If $a_1, a_2, b_1, b_2 \in A_+$, $b_1 \precsim a_1$, $b_2 \precsim a_2$, and $a_1 \perp a_2$ (i.e. $a_1 a_2 = 0$), then $b_1 + b_2 \precsim a_1 + a_2$.

The proof of (i) is straightforward. For (ii), $b_i \precsim a_i^3$ for $i = 1, 2$, so choose x_n, y_n with $x_n^* a_1^3 x_n \to b_1$, $y_n^* a_2^3 y_n \to b_2$; then

$$(a_1 x_n + a_2 y_n)^* (a_1 + a_2)(a_1 x_n + a_2 y_n) \to b_1 + b_2.$$

II.3.4.8 There are several other very similar relations used in [Cun78], [Rør92], and other references (the same symbols have been used for distinct relations in various references, so the symbols we use do not necessarily coincide with those of the references):

Let A be a C*-algebra, $a, b \in A_+$. Then

$b \preceq a$ if there is a $z \in \tilde{A}$ with $z^*az = b$.
$b \preccurlyeq a$ if there are $x, y \in \tilde{A}$ with $b = xay$.
$b \precsim a$ if $b \leq z^*az$ for some $z \in \tilde{A}$.
$b \sim a$ if there is an $x \in A$ with $x^*x = a$, $xx^* = b$.
$b \simeq a$ if $b \preceq a$ and $a \preceq b$.

It is obvious that $\preceq$, $\preccurlyeq$, and $\precsim$ are transitive. The relations $\preceq$, $\preccurlyeq$, and $\precsim$ are practically the same, although there are nuances of differences between them. We have

$$b \preceq a \Longrightarrow b \preccurlyeq a \Longrightarrow b \precsim a \Longrightarrow b \precsim a$$

[if $0 \leq b = xay$ for $a \geq 0$, then $b \leq (x^* + y)^*a(x^* + y)$, so $b \preccurlyeq a \Longrightarrow b \precsim a$].
$b \precsim a \not\Longrightarrow b \preccurlyeq a$ (II.5.2.1(iii)); it appears to be unknown whether $b \preccurlyeq a$ implies $b \preceq a$, but is likely false in general. $\precsim$ is a "soft" version of each of these relations.

Using I.5.2.4, it can be shown that if A is a von Neumann algebra, then $\preceq$, $\preccurlyeq$, and $\precsim$ agree on A_+. Combining similar arguments with II.3.2.1, one can show:

II.3.4.9 PROPOSITION. Let A be a C*-algebra, $a, b \in A_+$. If $b \precsim a$, then $b^\beta \preceq a^\alpha$ for all α, β, $0 < \alpha < \beta$. In particular, $b \preceq a^\alpha$ for all α, $0 < \alpha < 1$.

The relation $\sim$ is transitive and thus an equivalence relation, although this is not obvious [Ped98]. It is obvious that $\simeq$ is an equivalence relation. We have $a \sim b \Longrightarrow a \approx b$ and $a \simeq b \Longrightarrow a \approx b$; $\simeq$ is a much "looser" equivalence relation than $\sim$ (e.g. $a \simeq \alpha a$ for any $\alpha > 0$), although $a \sim b$ does not quite imply $a \simeq b$ in general (it does in a von Neumann algebra). The relation $\approx$ is looser yet: we have $a \approx a^2$ for any a, but we rarely have $a \simeq a^2$ (only if a is well-supported, i.e. if $a \simeq p$ for a projection p).

II.3.4.10 There is a version of $\precsim$ for arbitrary elements: we say $x \precsim y$ if there are sequences (z_n), (w_n) with $z_n y w_n \to x$, and $x \approx y$ if $x \precsim y$ and $y \precsim x$. The relation $\preccurlyeq$ also obviously extends to general elements. It is not hard to see that these agree with the previous definitions if $x, y \in A_+$; we have $x \approx x^* \approx x^*x$ for any x. II.3.4.7 remains true in this context, provided that $a_1 \perp a_2$ in the C*-algebra sense, i.e. $a_1 a_2 = a_2 a_1 = a_1^* a_2 = a_1 a_2^* = 0$.

There are some useful functions for comparison theory, as well as other purposes (cf. [Cun78]):

II.3.4.11 DEFINITION. For $\epsilon > 0$, let f_ϵ be the continuous function on $\mathbb{R}$ which is 0 on $(-\infty, \epsilon]$, 1 on $[2\epsilon, \infty)$, and linear on $[\epsilon, 2\epsilon]$.

II.3.4.12 If $x \in A_+$, then the elements $f_\epsilon(x)$ have the following properties:

$0 \leq f_\epsilon(x)$ for all ϵ; $\|f_\epsilon(x)\| = 1$ if $\|x\| \geq 2\epsilon$.
$f_\epsilon(x) \ll f_\delta(x)$ for $0 < \delta \leq \epsilon/2$.
$\epsilon \mapsto f_\epsilon(x)$ is continuous.
$f_\epsilon(x) = xg_\epsilon(x) = g_\epsilon(x)^{1/2}xg_\epsilon(x)^{1/2} = x^{1/2}g_\epsilon(x)x^{1/2}$, where $g_\epsilon(t) = t^{-1}f_\epsilon(t)$; in particular, $f_\epsilon(x) \leq \|g_\epsilon(x)\|x$ and $f_\epsilon(x) \precsim x$.
If $h_\epsilon(t) = tf_\epsilon(t)$ for $t \geq 0$, then $h_\epsilon(x) = xf_\epsilon(x)$ and $f_\epsilon(x) = g_{\epsilon/2}(x)h_\epsilon(x)$, so $h_\epsilon(x) \simeq f_\epsilon(x)$ and hence $h_\epsilon(x) \approx f_\epsilon(x)$; as $\epsilon \to 0$,

$$h_\epsilon(x) = xf_\epsilon(x) = x^{1/2}f_\epsilon(x)x^{1/2} = f_\epsilon(x)^{1/2}xf_\epsilon(x)^{1/2} \to x.$$

If $x_n \to x$, then $f_\epsilon(x_n) \to f_\epsilon(x)$ for each ϵ (II.2.3.3).

II.3.4.13 It is also useful to consider the functions ℓ_ϵ, where $\ell_\epsilon(t) = \max(0, t - \epsilon)$; if $x \in A_+$, then $\ell_\epsilon(x) = (x - \epsilon 1)_+ \in A_+$ (computed in $\tilde{A}$). Then $\epsilon \mapsto \ell_\epsilon(x)$ is continuous, $\|\ell_\epsilon(x)\| = \max(0, \|x\| - \epsilon)$, and $\ell_\epsilon(x) \to x$ as $\epsilon \to 0$. We have $\|\ell_\epsilon(x) - h_\epsilon(x)\| \leq \epsilon$, and there are nonnegative continuous functions ϕ and ψ vanishing at 0 (depending on ϵ) such that $\ell_\epsilon = \phi h_\epsilon$ and $h_\epsilon = \psi \ell_\epsilon$, so $\ell_\epsilon(x) = \phi(x)h_\epsilon(x)$, $h_\epsilon(x) = \psi(x)\ell_\epsilon(x)$, and in particular $\ell_\epsilon(x) \simeq h_\epsilon(x) \simeq f_\epsilon(x)$.

The main technical virtue of the ℓ_ϵ is that $\ell_\delta \circ \ell_\epsilon = \ell_{\delta+\epsilon}$ for $\delta, \epsilon > 0$. There is no such simple relationship for the f_ϵ or h_ϵ, although $f_\delta(f_\epsilon(x)) \simeq f_\gamma(x)$ for $\gamma = \epsilon + \epsilon\delta$ (if $\delta \leq 1$).

II.3.4.14 If x is well-supported, then for any sufficiently small ϵ, $f_\epsilon(x^*x)$ and $f_\epsilon(xx^*)$ are the left and right support projections of x respectively. The elements $f_\epsilon(a)$ behave like "generalized support projections" of an element $a \in A_+$ and can often be used as a substitute for an (in general nonexistent, at least in A) actual support projection.

We have the following corollary of II.3.4.6:

II.3.4.15 PROPOSITION. If $x_n \to x$ in A_+, then, for any $\epsilon > 0$, $f_\epsilon(x_n) \precsim f_{\epsilon/2}(x)$ and $f_\epsilon(x) \precsim f_{\epsilon/2}(x_n)$ for all sufficiently large n.

This result can be improved to show that for any $0 < \delta < \epsilon$, $f_\epsilon(x_n) \precsim f_\delta(x)$ and $f_\epsilon(x) \precsim f_\delta(x_n)$ for all sufficiently large n [KR00, 2.5]. The proof uses the following version of II.3.4.6:

II.3.4.16 PROPOSITION. [Rør92, 2.2] If $x, y \in A_+$, $\|x - y\| < \epsilon$, then $f_\epsilon(x) \precsim y$.

PROOF: If $\delta = \|x - y\|$, then $x - \delta 1 \leq y$ (in $\tilde{A}$). So

$$(\epsilon - \delta)f_\epsilon(x) \leq f_\epsilon(x)^{1/2}(x - \delta 1)f_\epsilon(x)^{1/2} \leq f_\epsilon(x)^{1/2}yf_\epsilon(x)^{1/2}$$

so $f_\epsilon(x) \leq z^*yz$ with $z = (\epsilon - \delta)^{-1/2}f_\epsilon(x)^{1/2}$.

One can then show:

II.3.4.17 PROPOSITION. [Rør92, 2.4] Let A be a C*-algebra, $a, b \in A_+$. The following are equivalent:

(i) $b \precsim a$.
(ii) There are $x_n, y_n \in \tilde{A}$ with $x_n a y_n \to b$.
(iii) There are $x_n, y_n \in A$ with $x_n a y_n \to b$.
(iv) $f_\epsilon(b) \precsim a$ for all $\epsilon > 0$.
(v) For every $\epsilon > 0$, there is a $\delta > 0$ such that $f_\epsilon(b) \precsim f_\delta(a)$.

II.4 Approximate Units

While a C*-algebra need not be unital, it always contains an approximate unit (sometimes called an approximate identity) which can often be used as a substitute. Some, but not all, Banach algebras have approximate units in an analogous sense.

II.4.1 General Approximate Units

II.4.1.1 DEFINITION. Let A be a C*-algebra. An *approximate unit* for A is a net (h_λ) of positive elements of A of norm ≤ 1, indexed by a directed set Λ, such that $h_\lambda x \to x$ for all $x \in A$. If $h_\lambda \leq h_\mu$ for $\lambda \leq \mu$, the approximate unit (h_λ) is *increasing*.

An approximate unit (h_λ) is *idempotent* if each h_λ is a projection; it is *almost idempotent* if $h_\lambda \ll h_\mu$ for $\lambda < \mu$.

An approximate unit (h_λ) is *sequential* if $\Lambda = \mathbb{N}$; it is *continuous* if $\Lambda = (0, \infty)$ (or a cofinal subinterval) and $\lambda \mapsto h_\lambda$ is continuous.

In many references, approximate units are assumed to be increasing, but we will not do so since there are situations where this assumption is not natural. At the other extreme, the assumption that $\|h_\lambda\| \leq 1$ for all λ, or even that $\|h_\lambda\|$ is bounded, is not universal.

An approximate unit in a unital C*-algebra is just a net of (eventually) invertible positive elements in the unit ball converging to 1.

II.4.1.2 PROPOSITION. If (h_λ) is an approximate unit for a C*-algebra A, and $\alpha > 0$ is fixed, then for all $x \in A$:

(i) $xh_\lambda \to x$ and $h_\lambda x h_\lambda \to x$; if $x \geq 0$, then $x^{1/2} h_\lambda x^{1/2} \to x$.
(ii) $h_\lambda^\alpha x \to x$, i.e. (h_λ^α) is an approximate unit for A. [However, if $\alpha > 1$, (h_λ^α) need not be increasing even if (h_λ) is (II.3.1.11).]

For the first part of (i), $h_\lambda x^* \to x^*$; for the second part, note that if $x \geq 0$, then $h_\lambda x^{1/2} \to x^{1/2}$. For (ii), we have

$$\|h_\lambda x - x h_\lambda\| \leq \|h_\lambda x - x\| + \|x - x h_\lambda\| \to 0$$

so
$$\|h_\lambda^2 x - x\| \leq \|h_\lambda^2 x - h_\lambda x h_\lambda\| + \|h_\lambda x h_\lambda - x\| \to 0$$
and similarly $\|h_\lambda^{2^n} x - x\| \to 0$ for any n. If $\alpha > 0$, choose n so that $\alpha \leq 2^{n-1}$; for $\epsilon > 0$ we have $\|x^* h_\lambda^{2^n} x - x^* x\| < \epsilon$ for large λ, and since
$$x^* h_\lambda^{2^n} x \leq x^* h_\lambda^{2\alpha} x \leq x^* h_\lambda^\alpha x \leq x^* x$$
we have $\|x^* h_\lambda^\alpha x - x^* h_\lambda^{2\alpha} x\| < \epsilon$ and $\|x^* x - x^* h_\lambda^\alpha x\| < \epsilon$, so, for large λ,
$$\|x - h_\lambda^\alpha x\|^2 = \|(x^* - x^* h_\lambda^\alpha)(x - h_\lambda^\alpha x)\| = \|x^* x - 2x^* h_\lambda^\alpha x + x^* h_\lambda^{2\alpha} x\|$$
$$\leq \|x^* x - x^* h_\lambda^\alpha x\| + \|x^* h_\lambda^\alpha x - x^* h_\lambda^{2\alpha} x\| < 2\epsilon.$$

If A is a C*-algebra and I is a dense (two-sided) ideal in A, let Λ_I be the set of all positive elements of I of norm strictly less than 1. Give Λ_I its ordering as a subset of A_+.

II.4.1.3 PROPOSITION.

(i) Λ_I is a directed set.
(ii) Λ_I is an increasing approximate unit for A.

PROOF: (i): By II.3.1.9(iii), $h \mapsto (1-h)^{-1} - 1$ is an order-isomorphism of Λ_I onto I_+, which is a directed set. (The computation is done in $\tilde{A}$, but it is easy to check that the map and its inverse $h \mapsto 1 - (1+h)^{-1}$ map I into I by considering the quotient map $\tilde{A} \to \tilde{A}/I$.)

(ii): It suffices to show that for any $x \in A$, the decreasing net
$$\{x^*(1-h)x : h \in \Lambda_I\}$$
converges to 0, for then
$$\|(1-h)x\|^2 = \|x^*(1-h)^2 x\| \leq \|x^*(1-h)x\| \to 0.$$
Since Λ_I is dense in Λ_A [if $x \in A_+$, choose $y \in I$ close to $x^{1/2}$; then $y^* y \in I_+$ is close to x], we need only find for each x and ϵ an $h \in \Lambda_A$ with $\|x^*(1-h)x\| < \epsilon$. By II.3.1.2(vi), we may assume $x \geq 0$ and $\|x\| < 1$. But then $\|x(1-x^{1/n})x\| < \epsilon$ for sufficiently large n.

II.4.1.4 COROLLARY. If A is a C*-algebra and I is a dense (two-sided) ideal in A, then there is an increasing approximate unit for A contained in I. If A is separable, then the increasing approximate unit may be chosen to be sequential or continuous.

To construct an increasing sequential approximate unit for a separable A, let $\{x_n\}$ be a dense set in A, and inductively find h_n in Λ_I with $h_n \geq h_{n-1}$ and $\|h_n x_k - x_k\| < 1/n$ for $1 \leq k \leq n$. For a continuous approximate unit, set $h_t = (n+1-t)h_n + (t-n)h_{n+1}$ for $n \leq t \leq n+1$.

The following simple fact is worth noting:

II.4.1.5 PROPOSITION. Let A be a C*-algebra, (h_λ) an approximate unit for A, and p a projection in A. Then $ph_\lambda p$ is well-supported (II.3.2.8) for all sufficiently large λ.

The proof consists of noting that $ph_\lambda p \to p$, so $ph_\lambda p$ is invertible in pAp for sufficiently large λ.

II.4.2 Strictly Positive Elements and σ-Unital C*-Algebras

In the case where there is a countable approximate unit, we can do better.

II.4.2.1 PROPOSITION. If A is a C*-algebra and $h \in A_+$, the following are equivalent:

(i) hA is dense in A
(ii) hAh is dense in A
(iii) $(f_\epsilon(h))$ is an (increasing) approximate unit for A as $\epsilon \to 0$.

If $\|h\| \leq 1$, then these are also equivalent to

(iv) $(h^{1/n})$ is an (increasing) approximate unit for A.

[The condition in (iv) that $\|h\| \leq 1$ is not critical – it only insures that $(h^{1/n})$ is in the unit ball and increasing.]
An h satisfying these conditions is *strictly positive* in A.

II.4.2.2 If A has a unit, then h is strictly positive if and only if it is invertible. A positive element $h \in A$ is strictly positive in $C^*(h)$, and in $\overline{hAh}$. If h is strictly positive in A and $h \leq a \in A_+$, then a and h^α are strictly positive for any $\alpha > 0$.

II.4.2.3 A function in $C_o(X)$ is strictly positive in this sense if and only if it takes strictly positive values everywhere. There is a similar characterization of strictly positive elements in general C*-algebras: it follows from the Hahn-Banach Theorem and the results of II.6.3 that $h \in A_+$ is strictly positive if and only if $\phi(a) > 0$ for every state ϕ on A.

II.4.2.4 PROPOSITION. A C*-algebra contains a strictly positive element if and only if it has a countable approximate unit. (Such a C*-algebra is called *σ-unital*.)

PROOF: One direction is obvious from II.4.2.1. Conversely, suppose A has a countable approximate unit, which may be chosen sequential, say (h_n). Set $h = \sum_{n=1}^{\infty} 2^{-n} h_n$. Then for each n and m, $h^{1/m} \geq 2^{-n/m} h_n^{1/m} \geq 2^{-n/m} h_n$, so $(h^{1/m})$ is an approximate unit for A.

Of course, every unital C*-algebra is σ-unital. Every separable C*-algebra is σ-unital, but there are nonseparable nonunital σ-unital C*-algebras, as well as non-σ-unital C*-algebras: for example, $C_o(X)$ is σ-unital if and only if X is σ-compact.

II.4.2.5 COROLLARY. If A is a σ-unital C*-algebra, then A has a sequential increasing almost idempotent approximate unit, and has a continuous increasing approximate unit consisting of mutually commuting elements.

The next fact is useful in the theory of stable and real rank (V.3.1).

II.4.2.6 PROPOSITION. Let h be a strictly positive element in a C*-algebra A, and p a projection in A. Then php is invertible in pAp and hence well-supported.

Indeed, $ph^{1/n}p$ is invertible in pAp (and hence well-supported) for some $n = 2^k$ by II.4.1.5. Thus $ph^{2^{-k+1}}p \geq (ph^{2^{-k}}p)^2$ is also invertible in pAp, and proceeding inductively php is invertible in pAp.

II.4.3 Quasicentral Approximate Units

II.4.3.1 If I is an ideal in a C*-algebra A, it is frequently useful to have an approximate unit (h_λ) for I which asymptotically commutes with all elements of A, i.e. $\|[x, h_\lambda]\| \to 0$ for all $x \in A$, where $[x, y] = xy - yx$ is the usual commutator. Such an approximate unit is called *quasicentral* for A. Quasicentral approximate units were first introduced independently in [AP77] and [Arv77].

II.4.3.2 PROPOSITION. If A is a C*-algebra, I a closed ideal of A, and (h_λ) an [increasing] approximate unit for I, then there is an [increasing] approximate unit for I, contained in the convex hull of (h_λ), which is quasicentral for A. If (h_λ) is almost idempotent, the quasicentral approximate unit may be chosen almost idempotent. If A is separable, the quasicentral approximate unit may be chosen sequential or continuous.

The proof is based on the following two facts: (1) for any $x \in A$, $[x, h_\lambda] \to 0$ in the weak topology of A; and (2) a norm-closed convex subset of a Banach space is weakly closed.

II.4.3.3 Note that even if (h_λ) is idempotent, the quasicentral approximate unit cannot in general be chosen idempotent. And even if I is separable, if A is not separable the quasicentral approximate unit cannot in general be chosen countable. For example, if $\mathcal{H}$ is a separable infinite-dimensional Hilbert space, it can be shown that although $\mathcal{K}(\mathcal{H})$ has a sequential idempotent approximate unit, it does not have either a countable approximate unit or an idempotent approximate unit which is quasicentral for $\mathcal{L}(\mathcal{H})$. This is closely related to the notion of quasidiagonality, which will be discussed in V.4.2.

II.5 Ideals, Quotients, and Homomorphisms

Recall that "ideal" means "two-sided ideal" unless otherwise specified. We assume that all ideals (left, right, or two-sided) are closed under scalar

II.5 Ideals, Quotients, and Homomorphisms

multiplication; this is automatic in a unital C*-algebra or for a closed ideal (using an approximate unit for the algebra), but not in general (cf. II.5.2.1(v)).

II.5.1 Closed Ideals

II.5.1.1 PROPOSITION. Let I be a closed ideal in a C*-algebra A, and π the quotient map from A to the Banach algebra A/I. Then

(i) I is self-adjoint (so I is a C*-algebra and there is an induced involution on A/I).
(ii) I is a hereditary C*-subalgebra of A, and $x^*x \in I \Leftrightarrow xx^* \in I \Leftrightarrow x \in I$.
(iii) If (h_λ) is an approximate unit for I, then for all $x \in A$,

$$\|\pi(x)\| = \lim_\lambda \|x(1-h_\lambda)\| = \inf \|x(1-h_\lambda)\|.$$

(iv) A/I is a C*-algebra in the quotient norm.

PROOF: (i): If $x \in I$, then $(x^*x)^{1/n} \in I$ for all n.

$$\|x^* - (x^*x)^{1/n}x^*\|^2 = \|x - x(x^*x)^{1/n}\|^2$$
$$= \|(1-(x^*x)^{1/n})x^*x(1-(x^*x)^{1/n})\| \to 0.$$

(ii) follows immediately from (i) and II.3.2.1.
(iii): $\|\pi(x)\| = \inf\{\|x+y\| : y \in I\} \leq \inf \|x - xh_\lambda\|$. But if $y \in I$, then

$$\|x - xh_\lambda\| \leq \|(x+y)(1-h_\lambda)\| + \|y - yh_\lambda\| \leq \|x+y\| + \|y - yh_\lambda\|,$$

so $\limsup \|x - xh_\lambda\| \leq \|\pi(x)\|$.
(iv): For any $x \in A$,

$$\|\pi(x)\|^2 = \inf \|x(1-h_\lambda)\|^2 = \inf \|(1-h_\lambda)x^*x(1-h_\lambda)\|$$
$$\leq \inf \|x^*x(1-h_\lambda)\| = \|\pi(x^*x)\| = \|\pi(x)^*\pi(x)\|.$$

II.5.1.2 COROLLARY. Let A and B be C*-algebras, $\phi : A \to B$ a *-homomorphism. Then $\phi(A)$ is closed in B, and therefore a C*-subalgebra.

PROOF: Let $I = Ker(\phi)$. I is closed since ϕ is continuous (II.1.6.6), and ϕ induces an injective *-homomorphism from the C*-algebra A/I to B, which is an isometry by II.2.2.9.

II.5.1.3 COROLLARY. Let A be a C*-algebra, I a closed ideal of A, and B a C*-subalgebra of A. Then, for any $x \in B$,

$$\inf\{\|x+y\| : y \in I\} = \inf\{\|x+z\| : z \in B \cap I\}.$$

$B + I$ is closed in A (and is therefore a C*-subalgebra), and $(B+I)/I$ is isometrically isomorphic to $B/(B \cap I)$. In particular, if I and J are closed ideals in A, then $I + J$ is a closed ideal in A.

PROOF: The inclusion of B into A drops to a map from $B/(B \cap I)$ into A/I, which is injective, hence an isometry. The range is $(B+I)/I$, and $B+I$ is the inverse image in A of this C*-subalgebra of A/I.

II.5.1.4 Here are some other facts about closed ideals. Using II.3.2.7 and II.5.1.1(ii), one can show:

(i) If I and J are closed ideals of A, then $(I+J)_+ = I_+ + J_+$.

Using the fact that in a C*-algebra, every element is the product of two elements (II.3.2.1), we obtain:

(ii) If I and J are closed ideals in A, then $I \cap J = IJ$.
(iii) If I is a closed ideal in A and J is a closed ideal of I, then J is an ideal of A.

From (ii) and II.5.1.3, we can easily show:

(iv) If I, J, K are closed ideals of A, then $(I+K) \cap (J+K) = (I \cap J) + K$.

Indeed, if A is any ring, we have

$$(I+K)(J+K) \subseteq IJ + K \subseteq (I \cap J) + K \subseteq (I+K) \cap (J+K).$$

Let A be a C*-algebra, J a closed ideal, $\pi : A \to A/J$ the quotient map. If $y \in A/J$, then by definition of the quotient norm, for any $\epsilon > 0$ there is an $x \in A$ with $\pi(x) = y$ and $\|x\| < \|y\| + \epsilon$. But we can do better:

II.5.1.5 PROPOSITION. Let A be a C*-algebra, J a closed ideal in A, and $\pi : A \to A/J$ the quotient map. If $y \in A/J$, then there is an $x \in A$ with $\pi(x) = y$ and $\|x\| = \|y\|$. If $y = y^*$ [resp. $y \geq 0$], then we can choose x with $x = x^*$ [resp. $x \geq 0$].

PROOF: We may assume $\|y\| = 1$. If $y = y^*$, let z be any preimage of y and set $w = (z + z^*)/2$; if $y \geq 0$, let z be a preimage of $y^{1/2}$ and set $w = z^*z$. In the general case, let w be any preimage of y. In any case, $\pi(w) = y$.

If $f(t) = \min(t, 2)$, then $x = w(1+w^*w)^{-1}f(1+w^*w)$ (computed in $\tilde{A}$, but $x \in A$) is a preimage of y of norm 1, which is self-adjoint [positive] if y is self-adjoint [positive]. [Note that $x^*x = g(w^*w)$, where $g(t) = t(1+t)^{-2}f(1+t)^2$.]

In fact, we have the following [Ped79, 1.5.10]:

II.5.1.6 PROPOSITION. Let A be a C*-algebra, J a closed ideal in A, and $\pi : A \to A/J$ the quotient map. If $a \in A_+$ and $y \in A/J$ with $y^*y \leq \pi(a)$, then there is an $x \in A$ with $x^*x \leq a$ and $\pi(x) = y$.

II.5.2 Nonclosed Ideals

A C*-algebra can have many nonclosed ideals, and their structure can be enormously complicated. (See, for example, [GJ76] for a description of the ideal structure of $C(X)$.) Several of the results of II.5.1 can fail for nonclosed ideals: they need not be self-adjoint, a self-adjoint ideal need not be hereditary or even positively generated, and an ideal of an ideal of A need not be an ideal of A. (In a von Neumann algebra, it is an easy consequence of polar decomposition that every ideal is self-adjoint, positively generated, hereditary, and strongly invariant (II.5.2.3).)

II.5.2.1 EXAMPLES.

(i) Let $A = C([-1,1])$, $f \in A$ with $f(t) = t$ for $t \geq 0$, $f(t) = it$ for $t < 0$, I the principal ideal generated by f. Then I is not self-adjoint since $\bar{f} \notin I$.

(ii) Let $A = C([-1,1])$, $g \in A$ with $g(t) = t$ for all t, I the principal ideal generated by g. Then I is self-adjoint, but not positively generated: g is not a linear combination of positive elements of I.

(iii) Let $A = C([0,1])$, $f \in A$ with $f(t) = t$ for all t, I the principal ideal generated by f. Then I is self-adjoint and positively generated, but not hereditary: if $g(t) = t\sin^2(1/t)$, then $0 \leq g \leq f$ but $g \notin I$.

(iv) Let $A = C_b((0,1])$, $I = C_o((0,1])$, $J = \{f \in I : \lim_{t \to 0} f(t)/t \text{ exists }\}$. Then I is a closed ideal of A, and J is an ideal of I [note that if $f \in J$ and $g \in I$, then $\lim_{t \to 0} [f(t)g(t)]/t = [\lim_{t \to 0} f(t)/t][\lim_{t \to 0} g(t)] = 0.]$ But J is not an ideal of A: if $f(t) = t \in J$ and $h(t) = \sin(1/t) \in A$, then $fh \notin J$.

(v) Let A, I, J be as in (iv), and $K = \{f \in J : \lim_{t \to 0} f(t)/t \in \mathbb{Z}\}$. Then K is a ring-theoretic ideal of the C*-algebra I which is not closed under scalar multiplication.

(vi) [Ped69] Let A be the C*-algebra of sequences from $\mathbb{M}_2$ converging to 0, and let

$$I = \left\{ \left(\begin{bmatrix} a_n & b_n \\ c_n & d_n \end{bmatrix} \right) : |a_n| = O(1/n), \ |b_n|, |c_n|, |d_n| = o(1/n) \right\}.$$

Then I is a positively generated hereditary ideal of A (or of $\tilde{A}$), but if

$$x = \left(\begin{bmatrix} 0 & 0 \\ n^{-1/2} & 0 \end{bmatrix} \right),$$

then $x^*x \in I$, $xx^* \notin I$ (cf. II.5.1.1(ii)).

This example can be modified to take any combination of $|a_n|$, $|b_n|$, $|c_n|$, $|d_n|$ to be $O(1/n)$ and the remaining ones $o(1/n)$, giving examples of ideals in A which are not self-adjoint (using a_n and b_n or just b_n) or are self-adjoint but not positively generated (using b_n and c_n).

While we will be mostly concerned with closed ideals in C*-algebras, there are occasions where nonclosed ideals play an important role (e.g. in II.6.8.2). The ideals normally encountered are self-adjoint, positively generated, and hereditary. Such ideals can be described by their positive cones, and conversely the positive cones of such ideals can be nicely characterized:

II.5.2.2 PROPOSITION. Let A be a C*-algebra, and M a unitarily invariant hereditary cone in A_+ (unitarily invariant means $u^*xu \in M$ for any $x \in M$ and u unitary in $\tilde{A}$). Then $Span(M)$ is a self-adjoint, positively generated, hereditary ideal J of A with $M = J \cap A_+$.

PROOF: Suppose $b \in M$ and u is a unitary in $\tilde{A}$. Write $x = b^{1/2}$ and $y = b^{1/2}u$; then $4u^*b = \sum_{k=0}^{3} i^k(x+i^ky)^*(x+i^ky)$ (II.3.1.6), and by II.3.1.9

$$(x+i^ky)^*(x+i^ky) \leq 2(x^*x + y^*y) = 2(b + u^*bu) \in M$$

so $u^*b \in J$. By II.3.2.12, $ab \in J$ for all $a \in A$, so $AJ \subseteq J$. Since $J = J^*$, also $JA \subseteq J$. If $y = \sum_{k=1}^{n} \lambda_k x_k \geq 0$ for $x_1, \ldots, x_n \in M$, then

$$0 \leq y = y^* = \sum \bar{\lambda}_k x_k = (y+y^*)/2 = \sum (Re\,\lambda_k) x_k \leq \sum |\lambda_k| x_k \in M$$

so $y \in M$.

II.5.2.3 A hereditary cone M in A_+ is *strongly invariant* if $x^*x \in M \implies xx^* \in M$. A positively generated ideal I is strongly invariant if I_+ is strongly invariant. A strongly invariant hereditary cone is unitarily invariant, but the converse is not true in general (II.5.2.1(vi)).

There is one especially important and well-behaved ideal in any C*-algebra:

II.5.2.4 THEOREM. Let A be a C*-algebra. Then there is a dense ideal $Ped(A)$ in A which is contained in every dense ideal of A. $Ped(A)$ is called the *Pedersen ideal* of A, and has the following properties:

(i) $Ped(A)$ contains $A_+^c = A_+^f$, and is the ideal generated by this set, where

$$A_+^c = \{b \in A_+ : \exists a \in A_+ \text{ with } b \ll a\}$$

$$A_+^f = \{f(a) : a \in A_+, f \in C_o([0,\infty))_+ \text{ vanishes in a neighborhood of } 0\}.$$

In particular, $Ped(A)$ contains all projections (more generally, all well-supported elements) in A.

(ii) $Ped(A)$ is self-adjoint, hereditary, and spanned by its positive elements; $x^*x \in Ped(A) \iff xx^* \in Ped(A) \iff x \in Ped(A)$.

(iii) If $x_1, \ldots, x_n \in Ped(A)$, then the hereditary C*-subalgebra of A generated by $x_1, \ldots, x_n$ is contained in $Ped(A)$. In particular, $C^*(x_1, \ldots, x_n) \subseteq Ped(A)$.

See [Ped79, 5.6] for a discussion and proof. To see that $A_+^f = A_+^c$, if $b \ll a$, and $\|b\| = 1$, set $c = f_{1/2}(a)(b+1)/2$; then $c \geq 0$ (note that a and b commute) and $b = f_{1/2}(c)$, so $A_+^c \subseteq A_+^f$, and the other containment is obvious.

II.5.2.5 If A is unital, then $Ped(A) = A$, and this can even happen (but is unusual) if A is nonunital (e.g. V.2.3.6). It is difficult, if not impossible, to effectively describe the Pedersen ideal of a general C*-algebra. However, there are two illuminating examples where it can be easily described: $Ped(\mathcal{K}(\mathcal{H}))$ is the ideal of finite-rank operators on $\mathcal{H}$; and if X is a locally compact Hausdorff space, then $Ped(C_o(X)) = C_c(X)$, the ideal of functions of compact support.

II.5.2.6 If A is a C*-algebra and $a \in A_+$, then there is a sequence $a_n \in A_+^c \subseteq Ped(A)_+$ with $a = \sum a_n$ and $\|a_n\| \leq 2^{-n}$ for $n > 1$: let (h_n) be a sequence of nonnegative continuous functions on $(0, \infty)$, vanishing in a neighborhood of 0, with $\sum h_n(t) = t$ for all t and $\|h_n\| \leq 2^{-n}$ for $n > 1$ (e.g. $h_1(t) = tf_{1/4}(t)$, $h_n(t) = t(f_{2^{-n-1}}(t) - f_{2^{-n}}(t))$ for $n > 1$), and set $a_n = h_n(a)$.

II.5.2.7 If $\pi : A \to B$ is a quotient map (surjective *-homomorphism), then $\pi(Ped(A)) = Ped(B)$. This is obvious from II.5.2.4(i) since $\pi(A_+^f) = B_+^f$.

II.5.2.8 If B is a C*-subalgebra of A, then $Ped(A) \cap B$ is an ideal of B containing B_+^f, and hence it contains $Ped(B)$; but it may be strictly larger than $Ped(B)$ in general (consider the case where A is unital and B is an ideal in A).

The Pedersen ideal is a useful technical device in many arguments. Here is an example.

II.5.2.9 PROPOSITION. Let A be a C*-algebra, $\{J_i\}$ a set of ideals of A. Then $\cap J_i$ is dense in $\cap \bar{J}_i$.

PROOF: Let $x \in \cap \bar{J}_i$. Then, for any $\epsilon > 0$, $xf_\epsilon(|x|) \in Ped(\bar{J}_i) \subseteq J_i$ for each i, and $xf_\epsilon(|x|) \to x$.

Principal Ideals

If A is a C*-algebra and $a \in A$, then the "principal" ideal generated (algebraically) by a is

$$I_a = \{\sum_{j=1}^n y_j a z_j \; : \; y_j, z_j \in A\}.$$

Write J_a for the closure of I_a. If $a = a^*$, then I_a is self-adjoint, and by polarization (II.3.1.6) every element of I_a is a linear combination of elements of the form x^*ax, $x \in A$. In particular, every self-adjoint element of I_a has the form $\sum_{j=1}^{n} x_j^* a x_j - \sum_{k=1}^{m} y_k^* a y_k$.

If $a \in A_+$, then I_a is spanned by positive elements of the form x^*ax, $x \in A$. If $b \in (I_a)_+$, then it is plausible that b could be written as $\sum_{k=1}^{n} x_k^* a x_k$ (finite sum; cf. II.5.2.13). However, it is unknown, and probably false, that this can be done in general. But it is possible within a slightly smaller ideal. Let I_a^o be the linear span of $\{y^*az : y, z \in \overline{aA}\}$. Then $I_a^o \subseteq I_a$, and is dense in I_a since $a^\alpha \in I_a^o$ for all $\alpha > 1$ (thus $J_a = \overline{I_a^o}$ and $Ped(J_a) \subseteq I_a^o$); $I_a^o = I_a$ if $a \in I_a^o$ (and conversely if A is unital). This happens frequently, but not always: for example, if $a \in Ped(J_a)$ (e.g. if a is a projection or if J_a is algebraically simple), then $I_a = I_a^o = Ped(J_a)$.

II.5.2.10 PROPOSITION. Let A be a C*-algebra, and $a \in A_+$. If $b \geq 0$ is in the ideal I_a^o, then there are $x_1, \ldots, x_n$ in $\overline{aA} \subseteq J_a$ with $b = \sum_{k=1}^{n} x_k^* a x_k$.

PROOF: Write $b = \sum_{j=1}^{m} y_j^* a z_j$. Then

$$2b = b + b^* = \sum_{j=1}^{m}(y_j^* a z_j + z_j^* a y_j) \leq \sum_{j=1}^{m}(y_j^* a y_j + z_j^* a z_j)$$

$$= \sum_{k=1}^{n} w_k^* a w_k = \sum_{k=1}^{n}(w_k^* a^{1/2})(w_k^* a^{1/2})^*$$

with $w_k \in \overline{aA}$ (the inequality uses II.3.1.9(i)); and, applying II.3.2.7, we can write $b = \sum_{k=1}^{n} u_k u_k^*$, where $u_k^* u_k \leq a^{1/2} w_k w_k^* a^{1/2}$. Then by III.5.2.17, $u_k = x_k^* a^{1/2}$ for some $x_k \in \overline{aA}$, and $b = \sum_{k=1}^{n} x_k^* a x_k$.

II.5.2.11 COROLLARY. Let A be a C*-algebra, and $a \in A_+$. If $b \in Ped(J_a)_+$, then there are $x_1, \ldots, x_n$ in $\overline{aA} \subseteq J_a$ with $b = \sum_{k=1}^{n} x_k^* a x_k$. In particular, if p is a projection in J_a, then there are $x_1, \ldots, x_n$ in $\overline{aA}$ with $p = \sum_{k=1}^{n} x_k^* a x_k$.

The case of a projection has a simpler direct proof not requiring II.3.2.7.

There is a nice reformulation in terms of matrix algebras (a matrix algebra over a C*-algebra is again a C*-algebra by II.6.6):

II.5.2.12 COROLLARY. Let A be a C*-algebra, $a \in A_+$. If $b \in (J_a)_+$, then for any $\epsilon > 0$ there is an n such that $diag(f_\epsilon(b), 0, \ldots, 0) \precsim (a, a, \ldots, a)$ in $M_n(A)$.

Combining II.5.2.10 with II.5.2.6, we obtain:

II.5.2.13 COROLLARY. Let A be a C*-algebra, $a \in A_+$, and J_a the closed ideal of A generated by a. If $b \in (J_a)_+$, then there is a sequence (x_k) in $\overline{aA} \subseteq J_a$ with $b = \sum_{k=1}^{\infty} x_k^* a x_k$.

II.5.3 Left Ideals and Hereditary Subalgebras

II.5.3.1 If A is a C*-algebra and B a hereditary C*-subalgebra, then B_+ is a closed hereditary cone in A_+. If L is a closed left ideal of A, then $L \cap L^*$ is a C*-subalgebra B of A with $B_+ = L_+$. If C is a hereditary cone in A_+, set $L(C) = \{x \in A : x^*x \in C\}$.

II.5.3.2 PROPOSITION. Let A be a C*-algebra. Then

(i) If L is a closed left ideal in A, then $L \cap L^*$ is a hereditary C*-subalgebra of A, and hence L_+ is a hereditary cone in A_+.
(ii) If C is a closed hereditary cone in A_+, then $L(C)$ is a closed left ideal in A and $Span(C)$ is a hereditary C*-subalgebra of A.
(iii) If B is a hereditary C*-subalgebra of A, then $\overline{AB} = \overline{\{ab : a \in A, b \in B\}} = \{x \in A : x^*x \in B\}$ is a closed left ideal in A.
(iv) The maps $C \mapsto L(C)$ and $L \mapsto L_+$ are mutually inverse bijections between the set $\mathcal{L}$ of closed left ideals of A and the set $\mathcal{C}$ of closed hereditary cones in A_+; $L \mapsto L \cap L^*$ and $B \mapsto \overline{AB}$ are mutually inverse bijections between $\mathcal{L}$ and the set $\mathcal{H}$ of hereditary C*-subalgebras of A; and $B \mapsto B_+$ and $C \mapsto Span(C)$ are mutually inverse bijections between $\mathcal{H}$ and $\mathcal{C}$.

PROOF: (i): If $0 \le a \le b$ and $b \in B = L \cap L^*$, then $b^{1/4} \in B_+ = L_+$ and $a^{1/2} = ub^{1/4} \in L$ (II.3.2.1), so $a \in L$.

(ii): $L(C)$ is closed by continuity of multiplication, closed under left multiplication by A by II.3.1.8, and closed under addition by II.3.1.9(ii). If $a \in C$, then $a^2 \le \|a\|a$, so $a^2 \in C$, $a \in L(C)_+$. On the other hand, if $a \in C$, then $a^{1/2}f_\epsilon(a) \in C$ for all $\epsilon > 0$ since C is hereditary, so $a^{1/2} \in C$; hence if $b \in L(C)_+$, then $b^2 \in C$, so $b \in C$. Thus $L(C)_+ = C$, and $Span(C)$ is the hereditary C*-subalgebra $L(C) \cap L(C)^*$.

(iii): If $x \in \overline{AB}$, then $x^*x \in \overline{BAB} = B$, so $(x^*x)^{1/4} \in B$ and $x = u(x^*x)^{1/4} \in \overline{AB}$. The same argument shows that if $x \in A$, then $x^*x \in B$ if and only if $x \in \overline{AB}$. If $x = a_1b_1 + a_2b_2$ with $b_i \in B$, then

$$x^*x \le 2(\|a_1\|^2 b_1^* b_1 + \|a_2\|^2 b_2^* b_2) \in B$$

(II.3.1.8, II.3.1.9(ii)), so $x^*x \in B$, and $\overline{AB}$ is closed under addition.

(iv): It was shown in (ii) that if $C \in \mathcal{C}$, then $L(C)_+ = C$. Conversely, let $L \in \mathcal{L}$, and let $C = L_+$. If $x \in L$, then $x^*x \in C$, so $L \subseteq L(C)$; on the other hand, if $x^*x \in C$, then $(x^*x)^{1/4} \in C$ so $x = u(x^*x)^{1/4} \in L$, so $L = L(C)$ and the first statement is proved. For the third statement, if $B \in \mathcal{H}$, then $B = Span(B_+)$; conversely, if $C \in \mathcal{C}$, then $Span(C) = L(C) \cap L(C)^*$, so $Span(C)_+ = L(C)_+ = C$ by (ii). The second statement follows from the other two by noting that $(\overline{AB})_+ = B_+$: $B = B^2 \subseteq \overline{AB}$, and conversely if $x = ab \in (\overline{AB})_+$, then $x^2 = b^*a^*ab \le \|a\|^2 b^*b \in B_+$, so $x^2 \in B_+$, $x \in B_+$.

90   II C*-Algebras

II.5.3.3 One consequence worth noting is that if (h_λ) is an approximate unit for B, then (h_λ) is a right approximate unit for AB, i.e. every closed left ideal in a C*-algebra has a right approximate unit. (A closed left ideal need not have a left approximate unit.)

II.5.3.4 If B is a hereditary C*-subalgebra of A, let $K_B = \overline{Span(ABA)}$ be the closed ideal of A generated by B. Note that ABA is not closed under addition in general, and the generated ideal $Span(ABA)$ of A is not generally (topologically) closed, for example $A = \mathbb{K}$, $B = \mathbb{C}e_{11}$.

II.5.3.5 PROPOSITION. Let A be a C*-algebra, B a hereditary C*-subalgebra, K_B as above. Then $I \mapsto I \cap B = BIB$ and $J \mapsto \overline{Span(AJA)}$ are mutually inverse bijections between the closed ideals of A contained in K_B and the closed ideals of B.

The simple proof is similar to previous arguments.

Factorizations

The first statement in II.5.3.2(iii) is true even if B is not hereditary. This follows from a general factorization result for Banach modules over C*-algebras, a version of Cohen's Factorization Theorem for Banach algebras [Coh59] and Banach modules [Hew64].

II.5.3.6 DEFINITION. Let B be a C*-algebra. A *right Banach B-module* is a Banach space X which is a right B-module, for which there is a K satisfying $\|xb\| \leq K\|x\|\|b\|$ for all $x \in X$, $b \in B$. Left Banach B-modules are defined analogously.

If B is unital, we do not assume that $x1_B = x$ for all $x \in X$. A module with this property is called a *unital module*.

Caution: by convention, the term "Banach B-module" without "left" or "right" means a *B-bimodule* (II.5.5.15).

The K is not very important: a right Banach B-module has a natural structure as a unital right Banach $B^\dagger$-module, and has an equivalent norm with $K = 1$ for this extended action.

II.5.3.7 THEOREM. [Ped98, 4.1] Let B be a C*-algebra, X a right Banach B-module. For any y in the closed span Y of XB, and $\epsilon > 0$, there are $x \in Y$ and $b \in B_+$ with $y = xb$ and $\|x - y\| < \epsilon$. So $Y = XB$, and XB is a closed submodule of X. An analogous statement holds for left Banach B-modules.

PROOF: Y becomes a unital right Banach $\tilde{B}$-module in the evident way; let K be the corresponding constant, i.e. $\|yb\| \leq K\|y\|\|b\|$ for all $y \in Y$, $b \in \tilde{B}$. Let (h_λ) be an approximate unit for B. Then $yh_\lambda \to y$ for any $y \in Y$. Fix y and ϵ. Set $x_0 = y$ and $b_0 = 1 \in \tilde{B}$. Inductively let $h_n = h_{\lambda_n}$ for some λ_n with

$$\|x_{n-1} - x_{n-1}h_{\lambda_n}\| < 2^{-n}\epsilon/K$$

and let
$$b_n = b_{n-1} - 2^{-n}(1-h_n) \in \tilde{B}.$$
Then $b_n = 2^{-n}1 + \sum_{k=1}^{n} 2^{-k} h_k \geq 2^{-n}1$, so b_n is invertible in $\tilde{B}$ and $\|b_n^{-1}\| \leq 2^n$. Set $x_n = yb_n^{-1} \in Y$. We have
$$x_n - x_{n-1} = yb_{n-1}^{-1}(b_{n-1} - b_n)b_n^{-1} = 2^{-n} x_{n-1}(1-h_n) b_n^{-1}$$
so $\|x_n - x_{n-1}\| < 2^{-n}\epsilon$. Thus (x_n) and (b_n) are Cauchy sequences; let x and b be the limits. We have $b = \sum_{k=1}^{\infty} 2^{-k} h_k \in B_+$ and $\|x - y\| < \epsilon$, and $y = xb$ since $y = x_n b_n$ for all n.

II.5.3.8 COROLLARY. Let A be a C*-algebra, B a C*-subalgebra of A, and R a closed right ideal of A. If L is the closed linear span of RB, then $L = RB = LB$. In fact, every $y \in L$ can be written as xb, where $x \in L$, $b \in B_+$. In particular, the closed left ideal of A generated by B is AB.

PROOF: L is a right Banach B-module. Since L is the closure of the span of RB, if (h_λ) is an approximate unit for B we have $yh_\lambda \to y$ for every $y \in L$, and thus LB is dense in L. For the last statement, set $R = A$, and note that the span of AB is a left ideal in A and $B = B^2 \subseteq AB$.

This result is most frequently applied when R is a two-sided ideal of A. An important special case is when R is a C*-algebra and A is the multiplier algebra $M(R)$ (II.7.3).

II.5.3.9 COROLLARY. Let B be a C*-subalgebra of a C*-algebra A. Then the hereditary C*-subalgebra of A generated by B is equal to
$$BAB = \{x^* ax : x \in B, \ a \in A\} = \{bab : b \in B_+, \ a \in A\}.$$
In particular, if $x \in A$, then $\overline{x^* Ax} = \bigcup_{b \in C^*(x^*x)_+} bAb$.

PROOF: Let D be the hereditary C*-subalgebra of A generated by B. Then $AD = AB$ by II.5.3.2 and II.5.3.8. We have $BAB \subseteq DAD = D$ (II.3.4.2(i)). If $x \in D$, write $x = ycdz$ with $y, c, d, z \in D$. Then $(yc)^* = sb_1$ and $dz = tb_2$ for $s, t \in A$, $b_1, b_2 \in B_+$, so $x = b_1(s^*t)b_2$. If $b = (b_1^2 + b_2^2)^{1/4} \in B_+$, then $b_1 = bu$ and $b_2 = vb$ for some $u, v \in B \subseteq A$ (II.3.2.1), so $x = b(us^*tv)b$.

Full Elements and Hereditary Subalgebras

II.5.3.10 If A is a C*-algebra and $a \in A$, then a is *full* in A if the closed ideal of A generated by a is all of A. Similarly, if B is a subalgebra of A, then B is *full* in A if the closed ideal of A generated by B is all of A. A strictly positive element in A is full, so a σ-unital C*-algebra contains a full element. The converse (that a full positive element is strictly positive) is true if A is commutative (in particular, a full element in a unital commutative C*-algebra is invertible), but not in general (II.6.4.15).

II.5.3.11 PROPOSITION. Let A be a unital C*-algebra, B a full hereditary C*-subalgebra. Then B contains a full positive element.

PROOF: Since 1 is in the closed ideal of A generated by B, there are positive elements $c_1, \ldots, c_n \in B$ and elements $y_1, \ldots, y_n, z_1, \ldots, z_n \in A$ with $\sum_{k=1}^n y_k c_k z_k$ close to 1 and hence invertible. Then

$$n^{-1}\left(\sum_{k=1}^n y_k c_k z_k\right)^*\left(\sum_{k=1}^n y_k c_k z_k\right) \leq \sum_{k=1}^n z_k^* c_k y_k^* y_k c_k z_k \leq \sum_{k=1}^n \|y_k\|^2 z_k^* c_k^2 z_k = r$$

(II.3.1.9(ii)), so r is invertible. Let $a = \sum_{k=1}^n c_k^{1/2}$; then by II.3.2.1 we can write $c_k^2 = u_k a$, $1 = \sum_{k=1}^n \|y_k\|^2 r^{-1} z_k^* u_k a z_k$, so a is full.

Nonclosed Left Ideals and Hereditary *-Subalgebras

II.5.3.12 A nonclosed *-subalgebra (even an ideal) of a C*-algebra can be hereditary in the sense of II.3.4.1, yet fail to be generated by its positive elements (e.g. the last part of II.5.2.1(vi)). We say a *-subalgebra B of a C*-algebra A is *positively generated* if B is spanned by $B_+ = B \cap A_+$. If $x \in A$, then $x^* A x$ is a positively generated *-subalgebra of A, although not hereditary in general.

If B is a positively generated hereditary *-subalgebra of a C*-algebra A, and $b \in B_+$, then the C*-subalgebra $f_\epsilon(b) A f_\epsilon(b)$ of A is contained in B for any $\epsilon > 0$, and thus the union of the hereditary C*-subalgebras of B has dense intersection with B_+. Thus B has an approximate unit (h_λ) such that, for each λ, the hereditary C*-subalgebra of A generated by λ is contained in B.

II.5.3.13 Some, but not all, of the results of II.5.3.2 hold for nonclosed left ideals. Suppose $\mathfrak{N}$ is a left ideal in a C*-algebra A. By approximating the elements of a right approximate unit for the (norm-)closure of $\mathfrak{N}$ by elements of $\mathfrak{N}$, there is a right approximate unit in $\mathfrak{N}$. (In fact, with more care, it can be shown that any left ideal in a C*-algebra has an increasing right approximate unit; see e.g. [SZ79, 3.20].)

There are two *-subalgebras of A naturally associated to $\mathfrak{N}$, $\mathfrak{N} \cap \mathfrak{N}^*$ and

$$\mathcal{M}(\mathfrak{N}) = Span\{y^*x : x, y \in \mathfrak{N}\}$$

Of course, $\mathcal{M}(\mathfrak{N}) \subseteq \mathfrak{N} \cap \mathfrak{N}^*$; they coincide if $\mathfrak{N}$ is closed by II.5.3.2, but not in general. Using a right approximate unit for $\mathfrak{N}$, it is easily seen that the (norm-)closure of $\mathcal{M}(\mathfrak{N})$ (and hence also of $\mathfrak{N} \cap \mathfrak{N}^*$) is $L \cap L^*$, where L is the closure of $\mathfrak{N}$. By polarization, $\mathcal{M}(\mathfrak{N}) = Span\{x^*x : x \in \mathfrak{N}\}$, and in particular $\mathcal{M}(\mathfrak{N})$ is positively generated.

$\mathfrak{N} \cap \mathfrak{N}^*$ and $\mathcal{M}(\mathfrak{N})$ are not hereditary *-subalgebras of A in general, even if $\mathfrak{N}$ is a two-sided ideal (II.5.2.1(iii)). A condition on $\mathfrak{N}$ which insures that $\mathcal{M}(\mathfrak{N})$ is hereditary is:

II.5.3.14 DEFINITION. A left ideal $\mathfrak{N}$ in a C*-algebra A is *weighted* if $x_1, \ldots, x_n \in \mathfrak{N}$, $y \in A$, $y^*y \leq \sum_{k=1}^n x_k^* x_k$ implies $y \in \mathfrak{N}$.

Every left ideal in a von Neumann algebra is weighted (I.5.2.5). It follows from II.5.3.2 and II.5.3.15 that every norm-closed left ideal in a C*-algebra is weighted. The examples of II.5.2.1(i)–(v) are not weighted. The term "weighted" will be explained in II.6.7.9.

If $\mathfrak{M}$ is a positively generated hereditary *-subalgebra of a C*-algebra A, as before set $L(\mathfrak{M}) = \{x \in A : x^*x \in \mathfrak{M}\}$. It is obvious that $L(\mathfrak{M})$ is a weighted left ideal of A. Similarly, if $\mathfrak{C}$ is a hereditary cone in A, define $L(\mathfrak{C})$ in the same way. Essentially the same arguments as in the proof of II.5.3.2 show:

II.5.3.15 THEOREM. Let A be a C*-algebra. The maps $\mathfrak{N} \mapsto \mathcal{M}(\mathfrak{N})$, $\mathfrak{M} \mapsto L(\mathfrak{M})$, $\mathfrak{M} \mapsto \mathfrak{M}_+$, $\mathfrak{C} \mapsto Span(\mathfrak{C})$, $\mathfrak{N} \mapsto \mathcal{M}(\mathfrak{N})_+$, $\mathfrak{C} \mapsto L(\mathfrak{C})$ give mutually inverse bijections between the set of weighted left ideals of A, the set of positively generated hereditary *-subalgebras of A, and the set of hereditary cones in A.

If $\mathfrak{N}$ is a weighted left ideal in a C*-algebra A, even in a von Neumann algebra, then $\mathfrak{N} \cap \mathfrak{N}^*$ is not hereditary in general. However, it is "almost hereditary." Specifically, it is 2-hereditary:

II.5.3.16 DEFINITION. Let A be a C*-algebra, and $\mathfrak{M}$ a positively generated *-subalgebra. $\mathfrak{M}$ is α-*hereditary* $(\alpha > 0)$ if $x \in \mathfrak{M}_+$, $y \in A_+$, $y^\alpha \leq x^\alpha$ imply $y \in \mathfrak{M}$. $\mathfrak{M}$ is *completely hereditary* if it is α-hereditary for all $\alpha > 0$.

If $\mathfrak{M}$ is α-hereditary, it is β-hereditary for $\beta > \alpha$. $\mathfrak{M}$ is 1-hereditary if and only if it is positively generated and hereditary in the usual sense.

II.5.3.17 EXAMPLE. Let $0 < \alpha \leq 2$. Let A be a C*-algebra and $x \in A_+$. Set
$$\mathfrak{M} = \{y \in A : (y^*y)^{\alpha/2}, (yy^*)^{\alpha/2} \leq kx^\alpha \text{ for some } k > 0\}.$$
Then $\mathfrak{M}$ is a *-subalgebra of A (cf. the proof of II.5.3.2(iii)), and is α-hereditary. For suitable choice of A and x, $\mathfrak{M}$ is not β-hereditary for any $\beta < \alpha$.

II.5.4 Prime and Simple C*-Algebras

II.5.4.1 DEFINITION. A C*-algebra A is *simple* if it has no nontrivial closed ideals. A is *algebraically simple* if it contains no nontrivial ideals.

A simple unital C*-algebra is algebraically simple by II.1.3.2. A nonunital simple C*-algebra need not be algebraically simple (e.g. $\mathbb{K}$). The only examples of simple C*-algebras seen so far are $\mathbb{M}_n$ and $\mathbb{K}$ (and the Calkin algebra on a separable infinite-dimensional Hilbert space), but there are many others, and the study of simple C*-algebras is one of the main thrusts of the classification program [Rør02a].

If A is simple, then any hereditary C*-subalgebra of A is simple by II.5.3.5.

II.5.4.2 PROPOSITION. Let A be a simple C*-algebra, B a hereditary C*-subalgebra of $Ped(A)$ (II.5.2.4). Then B is algebraically simple.

PROOF: Let x, b be nonzero elements of B, with $b \geq 0$. It suffices to show that b is in the ideal of B generated by x. We have that $y = (xx^*)x(x^*x) \neq 0$, and $b^{1/3}$ is in the ideal of A generated by y (which is equal to $Ped(A)$). Thus there are elements $c_i, d_i \in A$ ($1 \leq i \leq n$) with $b^{1/3} = \sum_{i=1}^n c_i y d_i$; thus

$$b = \sum_{i=1}^n [b^{1/3} c_i x x^*] x [x^* x d_i b^{1/3}]$$

and the elements in brackets are in B (II.3.4.2(i)).

We have the following immediate consequence of II.5.2.10 and II.5.2.13:

II.5.4.3 PROPOSITION. Let A be a simple C*-algebra, $a, b \in A_+$ with $a \neq 0$. Then

(i) There is a sequence (x_k) in A with $b = \sum_{k=1}^\infty x_k^* a x_k$.
(ii) If $b \in Ped(A)$ (in particular, if A is unital or, more generally, algebraically simple), then there are $x_1, \ldots, x_n \in A$ with $b = \sum_{k=1}^n x_k^* a x_k$.

II.5.4.4 DEFINITION. A C*-algebra A is *prime* if, whenever J and K are ideals of A with $J \cap K = \{0\}$, either J or K is $\{0\}$ (i.e. $\{0\}$ is a prime ideal in A).

Any simple C*-algebra is prime. By II.5.2.9, A is prime if, whenever J and K are *closed* ideals of A with $J \cap K = \{0\}$, either J or K is $\{0\}$. A closed ideal I of A is a prime ideal if and only if A/I is a prime C*-algebra.

II.5.4.5 PROPOSITION. A C*-algebra A is prime if and only if, whenever $x, y \in A$ are nonzero elements, there is a $z \in A$ with $xzy \neq 0$.

PROOF: If J and K are nonzero ideals of A, let $x \in J$ and $y \in K$ be nonzero. Then, for any $z \in A$, $xzy \in J \cap K$. Conversely, if $x \neq 0$ and J is the ideal generated by x, let

$$K = \{y \in A : xzy = 0 \text{ for all } z \in A\}.$$

Then K is an ideal in A with $J \cap K = \{0\}$ [if $y \in J \cap K$, write $y^* y = \sum_{i=1}^n a_i x b_i$; then $(y^* y)^2 = \sum_{i=1}^n a_i x b_i y^* y = 0$.]

II.5.4.6 COROLLARY. If A is a concrete C*-algebra of operators acting irreducibly on a Hilbert space $\mathcal{H}$, then A is prime.

PROOF: If $S, T \in A$, $T \neq 0$, and $\|S\| = 1$, let $\xi, \eta \in \mathcal{H}$ with $T\xi \neq 0$ and $S\eta \neq 0$. Then by irreducibility there is an $R \in A$ with $\|RT\xi - \eta\| < \|S\eta\|$; so $\|SRT\xi - S\eta\| < \|S\eta\|$, $SRT\xi \neq 0$, $SRT \neq 0$.

II.5 Ideals, Quotients, and Homomorphisms

II.5.4.7 If J is a closed ideal in a C*-algebra B, set
$$J^\perp = \{x \in B : Jx = \{0\}\}.$$

$J^\perp$ is the *annihilator* of J. (This notation is consistent with the notation of Hilbert modules: $J^\perp$ is the orthogonal complement of J in the Hilbert module B.) $J^\perp$ is a closed ideal of B. J is *essential* in B if $J^\perp = \{0\}$; equivalently, J is essential if $J \cap I \neq \{0\}$ for every nonzero closed ideal I of B. If X is an open subset of a compact Hausdorff space Y, so that $C_o(X)$ is an ideal in $C(Y)$, then $C_o(X)^\perp$ is $C_o(Z)$, where Z is the interior of $Y \setminus X$, so $C_o(X)$ is essential in $C(Y)$ if and only if X is dense in Y.

A is prime if and only if every nonzero ideal of A is an essential ideal.

II.5.5 Homomorphisms and Automorphisms

II.5.5.1 If A and B are C*-algebras, denote by $\mathrm{Hom}(A, B)$ the set of *-homomorphisms from A to B. If A and B are unital, let $\mathrm{Hom}_1(A, B)$ be the set of unital *-homomorphisms from A to B. The set of *-automorphisms of a C*-algebra A is denoted $\mathrm{Aut}(A)$.

Unless otherwise specified, "homomorphism" and "automorphism" for C*-algebras will always mean "*-homomorphism" and "*-automorphism" respectively. (One occasionally, but rarely, considers homomorphisms of C*-algebras which are not *-homomorphisms; cf. [Gar65], [Chr81], [Pau02, Chapter 19]).

II.5.5.2 There is a natural composition from $\mathrm{Hom}(A, B) \times \mathrm{Hom}(B, C)$ to $\mathrm{Hom}(A, C)$, and also for Hom_1 if A, B, C are unital. $\mathrm{Hom}(A, A)$ is a monoid (semigroup with identity) and $\mathrm{Aut}(A)$ is a group; $\mathrm{Hom}_1(A, A)$ is also a monoid if A is unital. The identity element of each is the identity map id_A.

Topologies on Homomorphisms

II.5.5.3 There are two natural topologies on $\mathrm{Hom}(A, B)$:

The *norm topology*, induced from the operator norm on homomorphisms regarded as bounded operators from A to B.

The *point-norm topology*, induced by the pseudometrics
$$d_a(\phi, \psi) = \|\phi(a) - \psi(a)\|$$
for $a \in A$. This is the topology of pointwise (norm-)convergence.

II.5.5.4 PROPOSITION.

(i) Composition is jointly continuous for either the norm or point-norm topologies.
(ii) $\mathrm{Aut}(A)$ is a topological group in either the norm or point-norm topologies.

II.5.5.5 If B is a von Neumann algebra, one can also consider other topologies on $\mathrm{Hom}(A,B)$ such as the point-weak, point-σ-strong, etc. In fact, it follows easily from I.3.2.9 and II.3.2.12 that these topologies all coincide. But II.5.5.4 fails in general for these topologies, although there is a closely related topology under which $\mathrm{Aut}(B)$ is a topological group (III.3.2.1).

Homotopy Theory for Homomorphisms

The point-norm topology is in general much more useful than the norm topology, which is too strong for most purposes. A good example is homotopy of homomorphisms:

II.5.5.6 DEFINITION. Let ϕ_0, ϕ_1 be in $\mathrm{Hom}(A,B)$. A *homotopy* from ϕ_0 to ϕ_1 is a *-homomorphism ϕ from A to $C([0,1],B)$ (the continuous functions from $[0,1]$ to B with pointwise operations and sup norm) such that $\phi_0 = \pi_0 \circ \phi$, $\phi_1 = \pi_1 \circ \phi$, where $\pi_t : C([0,1],B) \to B$ is evaluation at t. Write $\phi_0 \simeq \phi_1$ if ϕ_0 and ϕ_1 are homotopic.

It is straightforward to check that this is an exact generalization of the notion of homotopy of continuous functions in topology: if $A = C(X)$ and $B = C(Y)$, then a homotopy of unital homomorphisms from A to B exactly corresponds to an ordinary homotopy between the corresponding continuous functions from Y to X.

II.5.5.7 PROPOSITION. If ϕ is a homotopy from ϕ_0 to ϕ_1, and ϕ_t is defined to be $\pi_t \circ \phi$ for $0 < t < 1$, then (ϕ_t) is a point-norm continuous path in $\mathrm{Hom}(A,B)$ from ϕ_0 to ϕ_1. Conversely, any point-norm continuous path (ϕ_t) defines a homotopy by $[\phi(a)](t) = \phi_t(a)$.

II.5.5.8 DEFINITION. Let A and B be C*-algebras. A *homotopically dominates* B if there are *-homomorphisms $\phi : A \to B$ and $\psi : B \to A$ such that $\phi \circ \psi \simeq id_B$. A and B are *homotopy equivalent* if there are $\phi : A \to B$ and $\psi : B \to A$ such that $\phi \circ \psi \simeq id_B$ and $\psi \circ \phi \simeq id_A$.

A C*-algebra A is *contractible* if it is homotopy equivalent to (or homotopy dominated by) the C*-algebra $\{0\}$, i.e. if there is a point-norm continuous path (ϕ_t) ($0 < t \leq 1$) of *-homomorphisms from A to A with $\phi_1 = id_A$ and $\lim_{t \to 0} \|\phi_t(x)\| = 0$ for all $x \in A$.

A C*-algebra is *subcontractible* if it is isomorphic to a C*-subalgebra of a contractible C*-algebra.

These notions are not exact analogs of the corresponding notions in topology, although $C_o(X)$ is contractible in the C*-sense if and only if the one-point compactification $X^\dagger$ is contractible to $\{\infty\}$ relative to $\{\infty\}$.

II.5 Ideals, Quotients, and Homomorphisms 97

II.5.5.9 Since any nonzero projection has norm one, a [sub]contractible C*-algebra cannot contain any nonzero projections. In particular, a [sub]contractible C*-algebra is always nonunital. If A is contractible, then $\tilde{A}$ is homotopy equivalent to $\mathbb{C}$, and conversely (if A is nonunital).

The most important examples of contractible and subcontractible C*-algebras are cones and suspensions:

II.5.5.10 DEFINITION. Let A be a C*-algebra (not necessarily unital). The *cone* over A, written $C(A)$ or often CA, is the C*-algebra $C_o((0,1], A)$ of continuous functions $f : [0,1] \to A$ with $f(0) = 0$. $[C(A) \cong C_o((0,1]) \otimes A$ (II.9.4.4).]
The *unital cone* $C_1(A)$ is $\widetilde{CA}$. If A is unital, $C_1(A)$ may be thought of as the set of $f : [0,1] \to A$ such that $f(0) \in \mathbb{C}1_A$.
The *suspension* of A, written $S(A)$ or often SA, is $C_o((0,1), A) \cong C_o(\mathbb{R}, A)$.
One can inductively define higher cones $C^n A = C(C^{n-1}A)$ and suspensions $S^n A = S(S^{n-1}A)$ for any n.

II.5.5.11 CA is contractible for any A: let $[\phi_{1-t}(f)](s)$ be 0 if $s \leq t$ and $f(s-t)$ if $t < s \leq 1$. SA is a C*-subalgebra of CA and hence subcontractible, but SA is not contractible in general (it is easy to see that $S\mathbb{C} \cong C_o(\mathbb{R})$ is not contractible, for example).

Automorphisms

II.5.5.12 DEFINITION.
(i) An automorphism α of a C*-algebra A is *inner* if there is a unitary $u \in \tilde{A}$ with $\alpha = Ad\ u$, i.e. $\alpha(x) = uxu^*$ for all $x \in A$. The set of all inner automorphisms of A is denoted $In(A)$; $In(A)$ is a normal subgroup of $Aut(A)$, and the map $\mathcal{U}(\tilde{A}) \to In(A)$ sending u to $Ad\ u$ is a homomorphism whose kernel is the center of $\mathcal{U}(\tilde{A})$.
(ii) $\phi, \psi \in Hom(A, B)$ are *unitarily equivalent* if there is a unitary $u \in \tilde{B}$ with $\psi = (Ad\ u) \circ \phi$.
(iii) $\phi, \psi \in Hom(A, B)$ are *approximately unitarily equivalent* if there is a net (u_i) of unitaries in $\tilde{B}$ with $(Ad\ u_i) \circ \phi \to \psi$ in the point-norm topology.
(iv) $\phi, \psi \in Hom(A, B)$ are *asymptotically unitarily equivalent* if there is a (norm-continuous) path (u_t) ($t \in [t_o, \infty)$) of unitaries in $\tilde{B}$ with $(Ad\ u_t) \circ \phi \to \psi$ in the point-norm topology.
(v) An automorphism of A is *approximately inner* [*resp. asymptotically inner*] if it is approximately unitarily equivalent [*resp.* asymptotically unitarily equivalent] to id_A.

II.5.5.13 REMARKS.

(i) Using multiplier algebras (II.7.3), one can similarly define multiplier unitary equivalence, multiplier approximate unitary equivalence, and multiplier asymptotic unitary equivalence as in II.5.5.12(ii)-(iv), replacing $\tilde{B}$ by $M(B)$, and multiplier inner automorphisms, multiplier approximately inner automorphisms, and multiplier asymptotically inner automorphisms as in II.5.5.12(i),(v) with $\tilde{A}$ replaced by $M(A)$. Denote the group of multiplier inner automorphisms of A by $\mathrm{Inn}(A)$; $\mathrm{Inn}(A)$ is a normal subgroup of $\mathrm{Aut}(A)$ containing $\mathrm{In}(A)$. If B is unital, there is of course no difference between inner and multiplier inner, etc. If B is nonunital, there is in general some difference between inner and multiplier inner, and between unitary equivalence and multiplier unitary equivalence, although the difference is not great and can often be overlooked. In at least many cases, approximate unitary equivalence and approximate multiplier unitary equivalence coincide even if B is nonunital. For example, if B is stable (II.6.6.12), then it is not hard to show that $\mathcal{U}(\tilde{B})$ is strictly dense in $\mathcal{U}(M(B))$, so the two relations coincide for homomorphisms from any A into B. If A is AF (II.8.2.2(iv)), then the two relations coincide for maps from A into any C*-algebra B. Asymptotic and asymptotic multiplier unitary equivalence similarly coincide in at least the most important cases.

(ii) If $t \mapsto u_t$ is norm-continuous [resp. strictly continuous (II.7.3)], then $t \mapsto \mathrm{Ad}\, u_t$ is norm-continuous [resp. point-norm-continuous].

(iii) If $\phi_1, \phi_2 : A \to B$, $\psi_1, \psi_2 : B \to C$, and the ϕ_i and ψ_i are unitarily equivalent [resp. approximately unitarily equivalent, asymptotically unitarily equivalent], then so are $\psi_1 \circ \phi_1$ and $\psi_2 \circ \phi_2$. A similar statement is true for multiplier unitary equivalence, etc., if ψ_1, ψ_2 are *strict* (II.7.3.13).

(iv) If A is separable and $\phi, \psi : A \to B$ are approximately unitarily equivalent, then there is a sequence (u_n) of unitaries in $\tilde{B}$ with $(\mathrm{Ad}\, u_n) \circ \phi \to \psi$ in the point-norm topology.

II.5.5.14 EXAMPLE. Every automorphism of $\mathcal{L}(\mathcal{H})$ is inner. For an automorphism must preserve rank-one projections, and then by an easy argument must be spatially implemented. Since an automorphism must preserve $\mathcal{K}(\mathcal{H})$, $\mathrm{Aut}(\mathcal{K}(\mathcal{H})) = \mathrm{Aut}(\mathcal{L}(\mathcal{H}))$ is isomorphic to the quotient of $\mathcal{U}(\mathcal{H})$ by its center $\mathbb{T}I$. The point-norm topology on $\mathrm{Aut}(\mathcal{K}(\mathcal{H}))$ is the quotient of the weak/strong/strong-* topology on $\mathcal{U}(\mathcal{H})$; the norm topologies on $\mathrm{Aut}(\mathcal{K}(\mathcal{H}))$ and $\mathrm{Aut}(\mathcal{L}(\mathcal{H}))$ coincide and are the quotient of the norm topology on $\mathcal{U}(\mathcal{H})$. The point-norm topology on $\mathrm{Aut}(\mathcal{L}(\mathcal{H}))$ is distinct from both of these (if $\mathcal{H}$ is infinite-dimensional), and is more difficult to describe (cf. [Haa75b]). Since $\mathcal{L}(\mathcal{H})$ is the multiplier algebra of $\mathcal{K}(\mathcal{H})$, every automorphism of $\mathcal{K}(\mathcal{H})$ is multiplier inner, i.e. $\mathrm{Inn}(\mathcal{K}(\mathcal{H})) = \mathrm{Aut}(\mathcal{K}(\mathcal{H}))$. However, $\mathrm{In}(\mathcal{K}(\mathcal{H}))$ is a smaller group (dense in $\mathrm{Aut}(\mathcal{K}(\mathcal{H}))$ in the point-norm topology).

Derivations

The subject of derivations on C*-algebras is a vast and important one which is treated in detail in [Bra86] (see also [Sak91] and [BR87]). We give only a quick overview here. The subject arises naturally not only in studying differentiability on manifolds and actions of Lie groups, but also in the mathematical formulation of quantum mechanics. Derivations are also important in the structure theory of operator algebras.

II.5.5.15 DEFINITION. Let A be a C*-algebra. A *Banach A-module* is an A-bimodule $\mathcal{X}$ which is a Banach space, such that there is a K with $\|ax\|, \|xa\| \leq K\|a\|\|x\|$ for all $a \in A$, $x \in \mathcal{X}$.

The term "Banach A-module" is conventionally used instead of the more strictly correct "Banach A-bimodule", leading to unfortunate confusion with the definition of a left or right Banach module (II.5.3.6). Standard examples of Banach A-modules are A itself, or any C*-algebra containing A, and A^* (or B^* for any B containing A). More generally, if $\mathcal{X}$ is a Banach A-module, then $\mathcal{X}^*$ has a natural structure as a Banach A-module by $[a\phi](x) = \phi(xa)$, $[\phi a](x) = \phi(ax)$ (note reversal of order); such a Banach A-module is called a *dual Banach A-module*. This dual module has the property that $\phi \mapsto a\phi$ and $\phi \mapsto \phi a$ are weak-* continuous maps from $\mathcal{X}^*$ to $\mathcal{X}^*$ for all $a \in A$.

If A is a von Neumann algebra, one also considers *normal Banach A-modules* (IV.2.5).

II.5.5.16 DEFINITION. Let A be a C*-algebra and $\mathcal{X}$ a Banach A-module. A *derivation* from A to $\mathcal{X}$ is a linear map δ from a dense *-subalgebra $\mathcal{A}$ of A to $\mathcal{X}$ satisfying $\delta(ab) = \delta(a)b + a\delta(b)$ for all $a, b \in \mathcal{A}$. $\mathcal{A}$ is called the *domain* of δ.

A derivation δ from A to $\mathcal{X}$ is *closed* if it is closed (I.7.2.1) as a partially defined operator from A to $\mathcal{X}$.

A derivation δ from A to $\mathcal{X}$ is *inner* if there is an $x \in \mathcal{X}$ with $\delta(a) = [a, x] = ax - xa$ for all $a \in \mathcal{A}$. An inner derivation is bounded.

Any linear combination of derivations (with the same domain) is a derivation. The closure of a closable derivation (i.e. a derivation which is a closable operator) is a derivation. There exist derivations which are not closable, although an everywhere-defined derivation is automatically closed (bounded) ([Sak60], [KR97a, 4.6.66]).

If $\mathcal{X}$ has an involution compatible with the action of A, a derivation δ from A to $\mathcal{X}$ has an adjoint δ^* with the same domain, defined by $\delta^*(a) = [\delta(a^*)]^*$. The derivation δ is *self-adjoint*, or a **-derivation*, if $\delta = \delta^*$. A general δ can be written as a linear combination of the *-derivations $\delta_r = (\delta + \delta^*)/2$ and $\delta_i = (\delta - \delta^*)/2i$, so the study of general derivations can be reduced to considering *-derivations; in some references, "derivation" means "*-derivation." Other authors prefer to work with antihermitian derivations, where $\delta^* = -\delta$. With the above notation, if δ is an arbitrary derivation, the derivations $\delta_a = i\delta_i$ and $\delta_b = -i\delta_r$ are antihermitian, and $\delta = \delta_a + i\delta_b$.

II.5.5.17 There are three flavors of derivations usually considered on C*-algebras, the second and third being generalizations of the first:

(i) Bounded *-derivations from a C*-algebra A to itself (or occasionally to a larger containing C*-algebra). Such a derivation is called a *bounded derivation on A*. The domain of a bounded derivation on A can (and will) be taken to be all of A.
There are classes of C*-algebras on which every bounded derivation is inner (as a derivation from A to $M(A)$ in the nonunital case): von Neumann algebras [Sak66] and more generally AW*-algebras [Ole74], simple C*-algebras [Sak68], and commutative C*-algebras and unital continuous-trace C*-algebras [AEPT76] (direct sums of these exhaust all separable C*-algebras with this property [Ell77]).

(ii) Unbounded *-derivations from a C*-algebra A to itself. Among other applications, these are important in the theory of one-parameter automorphism groups of A, in a manner analogous to Stone's theorem (I.7.4.10): if α is a (point-norm-)continuous action of $\mathbb{R}$ on A, let $\mathcal{A}$ be the set of all $a \in A$ for which the limit

$$\delta(a) = \lim_{t \to 0} \frac{\alpha_t(a) - a}{t}$$

exists; then it turns out that $\mathcal{A}$ is a dense *-subalgebra of A and δ is a closed *-derivation from A to A with domain $\mathcal{A}$, called the *generator* of α. Conversely, it is easy to check that if δ is a bounded *-derivation of A, then $\alpha_t = e^{it\delta}$ (computed in $\mathcal{L}(A)$) is an automorphism of A for each t, and $t \to \alpha_t$ is a continuous action of $\mathbb{R}$ on A, with generator δ. The same can be done for certain closed unbounded derivations, which can be intrinsically characterized. There is thus a one-one correspondence between such derivations and one-parameter groups of automorphisms.

(iii) Bounded derivations from A to a general (usually dual) Banach A-module $\mathcal{X}$ are important in the cohomology theory of operator algebras (see [KR71a]). A C*-algebra A is *amenable* if every bounded derivation from A to a dual Banach A-module is inner. This definition will be explained and justified in IV.3.3; it turns out that amenability is equivalent to nuclearity for C*-algebras. There is a von Neumann algebra version, discussed in IV.2.5.

II.6 States and Representations

We have gotten fairly far into the structure of C*-algebras without showing any direct connection with operator theory on Hilbert space except that algebras of operators provide some examples of C*-algebras. But the connection is a very close one and is important to understand, since it provides much of the motivation and applications of the theory. In addition, some structure facts can be easily proved using representation theory which cannot be done reasonably in any other way (e.g. II.6.6).

II.6.1 Representations

II.6.1.1 DEFINITION. A *representation* of a C*-algebra A is a *-homomorphism from A to $\mathcal{L}(\mathcal{H})$ for some Hilbert space $\mathcal{H}$.

Two representations π and ρ of A on $\mathcal{X}$ and $\mathcal{Y}$ respectively are *(unitarily) equivalent* if there is a unitary operator $U \in \mathcal{L}(\mathcal{X}, \mathcal{Y})$ with $U\pi(x)U^* = \rho(x)$ for all $x \in A$.

A *subrepresentation* of a representation π on $\mathcal{H}$ is the restriction of π to a closed invariant subspace of $\mathcal{H}$.

A representation is *irreducible* if it has no nontrivial closed invariant subspaces.

If π_i ($i \in \Omega$) is a representation of A on $\mathcal{H}_i$, then the *sum* $\oplus_i \pi_i$ of the π_i is the diagonal sum acting on $\oplus_i \mathcal{H}_i$. If each π_i is equivalent to a fixed representation ρ, then $\oplus_i \pi_i$ is a *multiple* or *amplification* of ρ by $card(\Omega)$.

II.6.1.2 A representation is always norm-decreasing (II.1.6.6) and hence continuous. Thus the kernel of a representation is a closed ideal. A representation with kernel 0 is called *faithful*. By II.2.2.9, a faithful representation is isometric.

II.6.1.3 If J is a closed ideal of A, then any representation of A/J gives a representation of A by composition with the quotient map, whose kernel contains J. Conversely, if π is a representation of A whose kernel contains J, then π drops to a representation of A/J.

II.6.1.4 If π is a representation of A on $\mathcal{H}$ and $\mathcal{X}$ is a closed subspace of $\mathcal{H}$, then the following are easily seen to be equivalent because $\pi(A)$ is a self-adjoint subset of $\mathcal{L}(\mathcal{H})$:

(i) $\mathcal{X}$ is invariant under $\pi(A)$.
(ii) $\mathcal{X}^\perp$ is invariant under $\pi(A)$.
(iii) $P_\mathcal{X} \in \pi(A)'$.

II.6.1.5 If $\mathcal{N}$ is the largest subspace of $\mathcal{H}$ on which $\pi(x) = 0$ for all $x \in A$, then $\mathcal{N}$ is closed, and $\mathcal{X} = \mathcal{N}^\perp$ is an invariant subspace, called the *essential subspace* of π. $\mathcal{X}$ is the closed span of $\{\pi(x)\xi : x \in A, \xi \in \mathcal{H}\}$. (Actually, by II.5.3.7, $\mathcal{X} = \{\pi(x)\xi : x \in A, \xi \in \mathcal{X}\} = \{\pi(x)\xi : x \in A, \xi \in \mathcal{H}\}$.) If A is unital, then $P_\mathcal{X} = \pi(1_A)$; in general, if (h_λ) is an approximate unit for A, then $\pi(h_\lambda) \to P_\mathcal{X}$ strongly. The representation π is *nondegenerate* (or *essential*) if $\mathcal{X} = \mathcal{H}$. An irreducible representation is always nondegenerate.

II.6.1.6 If J is a closed ideal in A, let $\mathcal{E}$ be the essential subspace of $\pi|_J$. Then $\mathcal{E}$ is invariant under $\pi(A)$, so $P_\mathcal{E} \in \pi(A)'$. So π decomposes into a sum of a representation essential on J and a representation which is zero on J (a representation of A/J). Conversely, if ρ is a nondegenerate representation of J on an $\mathcal{H}$, and (h_λ) is an approximate unit for J, then for any $a \in A$ the net $(\rho(ah_\lambda))$ converges strongly in $\mathcal{L}(\mathcal{H})$ to an operator we call $\rho_A(a)$ (cf. II.7.3.9), defining a representation of A extending ρ, and we have $\rho_A(A)'' = \rho(J)''$. (This ρ_A is the unique extension of ρ to a representation of A on $\mathcal{H}$.)

II.6.1.7 Note that for a nondegenerate representation π, the strong (or weak) closure of $\pi(A)$ is $\pi(A)''$ by the Bicommutant Theorem (I.9.1.1); for a general π, the strong closure of $\pi(A)$ is $P_\mathcal{X} \pi(A)''$, where $\mathcal{X}$ is the essential subspace of π. (Note that $P_\mathcal{X}$ is a central projection in $\pi(A)''$.)

Irreducible and Factor Representations

Since $\pi(A)'$ and $\pi(A)''$ are von Neumann algebras and are thus generated by their projections, we get the following fact from an application of the Bicommutant Theorem:

II.6.1.8 PROPOSITION. Let π be a representation of a C*-algebra A on a Hilbert space $\mathcal{H}$. Then the following are equivalent:

(i) π is irreducible.
(ii) $\pi(A)' = \mathbb{C}I$.
(iii) $\pi(A)'' = \mathcal{L}(\mathcal{H})$.
(iv) $\pi(A)$ is strongly dense in $\mathcal{L}(\mathcal{H})$.

It follows from II.6.1.6 that if π is an irreducible representation of A, and J is a closed ideal of A, then $\pi|_J$ is either zero or irreducible. More generally, we have:

II.6.1.9 PROPOSITION. Let B be a hereditary C*-subalgebra of a C*-algebra A, and let π be an irreducible representation of A on a Hilbert space $\mathcal{H}$. Let P be the projection onto the essential subspace of $\pi|_B$. Then B acts irreducibly on $P\mathcal{H}$.

PROOF: Let $\xi, \eta \in P\mathcal{H}$ with $\xi \neq 0$. Then for any $\epsilon > 0$ there is an $x \in A$ with $\pi(x)\xi = \pi(x)P\xi$ within $\epsilon/2$ of η, so $P\pi(x)P\xi$ is within $\epsilon/2$ of $P\eta = \eta$. Let (h_λ) be an approximate unit for B. Then $\pi(h_\lambda) \to P$ strongly, so $\|\pi(h_\lambda x h_\lambda)\xi - \eta\| < \epsilon$ for sufficiently large λ.

II.6.1.10 We say a representation π of A is a *factor representation* if $\pi(A)''$ is a factor (I.9.1.5). An irreducible representation, or any multiple of an irreducible representation, is a factor representation.

II.6.1.11 PROPOSITION. The kernel of a factor representation of a C*-algebra is a (closed) prime ideal.

PROOF: The kernel of any representation is closed. Let π be a factor representation of a C*-algebra A on $\mathcal{H}$. Replacing A by $A/\ker(\pi)$ we may assume π is faithful. Suppose there are nonzero ideals J and K of A with $J \cap K = \{0\}$. If (h_λ) and (k_μ) are increasing approximate units for J and K respectively, then $(\pi(h_\lambda))$ and $(\pi(k_\mu))$ converge strongly to the projections P and Q onto the closures of $\pi(J)\mathcal{H}$ and $\pi(K)\mathcal{H}$. Then P and Q are nonzero projections in $\pi(A)''$, and $P \perp Q$. Since $\pi(J)\mathcal{H}$ and $\pi(K)\mathcal{H}$ are invariant under $\pi(A)$, we also have $P, Q \in \pi(A)'$, a contradiction.

See III.1.1.8 for another proof.

There is a potential ambiguity in the definition of an irreducible representation. We say a representation is *algebraically irreducible* if it has no nontrivial invariant subspaces, closed or not. But R. Kadison [Kad57] showed the following:

II.6.1.12 THEOREM. [KADISON TRANSITIVITY] Let π be an irreducible representation of a C*-algebra A on a Hilbert space $\mathcal{H}$ and $\xi_1, \cdots, \xi_n, \eta_1, \cdots, \eta_n$ vectors in $\mathcal{H}$ with $\{\xi_1, \cdots, \xi_n\}$ linearly independent. Then there is an $x \in A$ such that $\pi(x)\xi_k = \eta_k$ for $1 \leq k \leq n$. In particular, π is algebraically irreducible.

This theorem is an immediate corollary of the next result:

II.6.1.13 THEOREM. Let π be an irreducible representation of a C*-algebra A on a Hilbert space $\mathcal{H}$, $\mathcal{X}$ a finite-dimensional subspace of $\mathcal{H}$, and $T \in \mathcal{L}(\mathcal{H})$. Then, for any $\epsilon > 0$, there is an $a \in A$ with $\|a\| < \|T\| + \epsilon$ and $\pi(a)|_\mathcal{X} = T|_\mathcal{X}$. If $T = T^*$ [resp. $T \geq 0$], we may choose $a = a^*$ [resp. $a \geq 0$].

For the proof, it follows from The Kaplansky Density Theorem (I.9.1.3) and II.5.1.5 (or its preceding comment) that there is a $a_1 \in A$ with $\|a_1\| \leq \|T\|$ and $\|(\pi(a_1) - T)|_\mathcal{X}\| < \epsilon/4$, with $a_1 = a_1^*$ [resp. $a_1 \geq 0$] if $T = T^*$ [resp. $T \geq 0$]. Inductively choose a_k ($k \geq 2$) of the same form with $\|a_k\| < 2^{-k}\epsilon$ and

$$\|(\sum_{k=1}^{n} \pi(a_k) - T)|_\mathcal{X}\| < 2^{-n-1}\epsilon.$$

Then $a = \sum_{k=1}^{\infty} a_k$ satisfies $\|a\| < \|T\| + \epsilon$ and $\pi(a)|_\mathcal{X} = T|_\mathcal{X}$.

One can actually obtain $\|a\| \leq \|T\|$ by a modified functional calculus argument [KR97a, 5.7.41].

II.6.2 Positive Linear Functionals and States

The connection between C*-algebras and Hilbert spaces is made via the notion of a state:

II.6.2.1 DEFINITION. Let A be a C*-algebra. A linear functional ϕ on A is *positive*, written $\phi \geq 0$, if $\phi(x) \geq 0$ whenever $x \geq 0$. A *state* on A is a positive linear functional of norm 1. Denote by $\mathcal{S}(A)$ the set of all states on A, called the *state space* of A.

II.6.2.2 A positive linear functional is automatically bounded [it suffices to show it is bounded on A_+ by II.3.1.2(vi); if it is not, for each n choose $x_n \geq 0$ with $\|x_n\| = 1$ and $\phi(x_n) \geq 4^n$, and then $x = \sum_{n=1}^{\infty} 2^{-n} x_n$ satisfies $\phi(x) \geq 2^{-n}\phi(x_n) \geq 2^n$ for all n, a contradiction.]

II.6.2.3 EXAMPLES.

(i) If A is a concrete C*-algebra of operators acting nondegenerately on $\mathcal{H}$ and $\xi \in \mathcal{H}$, and $\phi_\xi(x) = \langle x\xi, \xi \rangle$ for $x \in A$, then ϕ_ξ is a positive linear functional on A of norm $\|\xi\|^2$, so ϕ_ξ is a state if $\|\xi\| = 1$. Such a state is called a *vector state* of A. This example is the origin of the term "state": in the mathematical formulation of quantum mechanics, the states of a physical system are given by probability distributions (unit vectors in an L^2-space), and observables are self-adjoint operators; the value of the observable T on the state ξ is $\langle T\xi, \xi \rangle$ (cf. [BR87]).

(ii) By the Riesz Representation Theorem, there is a one-one correspondence between bounded linear functionals on $C_o(X)$ and finite regular complex Baire measures (complex Radon measures) on X. A bounded linear functional is positive if and only if the corresponding complex measure takes only nonnegative real values (i.e. is an ordinary measure). Thus the states on $C_o(X)$ are precisely given by the regular Baire probability measures on X. Any homomorphism from $C_o(X)$ to $\mathbb{C}$ is a state by II.2.1.2 and II.2.2.2.

A positive linear functional takes real values on self-adjoint elements, and thus $\phi(x^*) = \overline{\phi(x)}$ for any x if $\phi \geq 0$. So if ϕ is a positive linear functional on A, then ϕ defines a pre-inner product on A by $\langle x, y \rangle_\phi = \phi(y^*x)$. Thus we get the CBS inequality:

II.6.2.4 PROPOSITION. Let A be a C*-algebra, ϕ a positive linear functional on A, $x, y \in A$. Then $|\phi(y^*x)|^2 \leq \phi(x^*x)\phi(y^*y)$. More symmetrically, $|\phi(xy)|^2 \leq \phi(xx^*)\phi(y^*y)$.

II.6.2.5 PROPOSITION. Let A be a C*-algebra and ϕ a bounded linear functional on A. Then

(i) If $\phi \geq 0$, then $\|\phi\| = \sup\{\phi(x) : x \geq 0, \|x\| = 1\} = \lim \phi(h_\lambda)$ for any approximate unit (h_λ) for A (in particular, $\|\phi\| = \phi(1_A)$ if A is unital), and ϕ extends to a positive linear functional on $\tilde{A}$ by setting $\phi(1) = \|\phi\|$ (or $\phi(1) = t$ for any $t \geq \|\phi\|$). In particular, any state on A extends uniquely to a state on $\tilde{A}$.

(ii) If A is unital and $\phi(1_A) = \|\phi\|$, then $\phi \geq 0$.

For the first part of (i), if (h_λ) is an approximate unit for A, we have $\|\phi\| \geq \sup\{\phi(x) : x \geq 0, \|x\| = 1\} \geq \sup \phi(h_\lambda)$; but if $y \in A$, $\|y\| = 1$, $|\phi(y)|^2 > \|\phi\|^2 - \epsilon$, then, using II.6.2.4,

$$\|\phi\|^2 - \epsilon < |\phi(y)|^2 = \lim |\phi(h_\lambda^{1/2} y)|^2 \leq \liminf \phi(h_\lambda) \phi(y^* y) \leq \|\phi\| \liminf \phi(h_\lambda).$$

The second statement of (i) follows from the first and II.3.1.2(viii).

To prove (ii), if $x \geq 0$, restrict ϕ to $C^*(x, 1) \cong C(X)$; then ϕ corresponds to a complex measure μ on X, and $\mu(X) = |\mu|(X)$, so $\mu \geq 0$. A more elementary alternate argument can be based on II.1.6.3: if $\phi(1_A) = \|\phi\| = 1$, and $x \geq 0$, suppose $\phi(x) \notin [0, \infty)$. Then there is a $\lambda \in \mathbb{C}$ and $\rho \geq 0$ with $\sigma(x) \subseteq D = \{z \in \mathbb{C} : |z - \lambda| \leq \rho\}$, but $\phi(x) \notin D$. If $y = x - \lambda 1_A$, then y is normal and $\|y\| = r(y) \leq \rho$, so $|\phi(x) - \lambda| = |\phi(y)| \leq \|y\| \leq \rho$, a contradiction.

For a generalization of II.6.2.5 and another alternate argument for (ii), see II.6.9.4.

As a corollary, we obtain Kadison's inequality for states (cf. II.6.9.14):

II.6.2.6 COROLLARY. Let A be a C*-algebra, $x \in A$, ϕ a state on A. Then $|\phi(x)|^2 \leq \phi(x^* x)$.

PROOF: Extend ϕ to a state on $\tilde{A}$ and apply II.6.2.4 with $y = 1$.

II.6.2.7 Any sum or nonnegative scalar multiple of positive linear functionals on a C*-algebra A is positive. A weak-* limit of a net of positive linear functionals on A is positive. II.6.2.5(i) shows that if ϕ and ψ are positive linear functionals on A, then $\|\phi + \psi\| = \|\phi\| + \|\psi\|$. In particular, a convex combination of states is a state, so if A is unital, $\mathcal{S}(A)$ is a compact convex subset of the dual of A. If A is nonunital, then the states of $\tilde{A}$ are in natural one-one correspondence with the set of positive linear functionals on A of norm ≤ 1, and $\mathcal{S}(A)$ is locally compact but noncompact.

II.6.2.8 DEFINITION. A state of a C*-algebra A is *pure* if it is an extreme point of $\mathcal{S}(A)$. Denote the set of pure states of A by $\mathcal{P}(A)$.

II.6.2.9 If A is unital, then $\mathcal{S}(A)$ is the closed convex hull of the pure states; in general, the set of positive linear functionals of norm ≤ 1 is the closed convex hull of the pure states and the zero functional. The pure states of $C_o(X)$ are precisely the homomorphisms to $\mathbb{C}$, i.e. the states corresponding to point masses (II.6.2.3(ii)).

States can be defined more generally on operator systems:

II.6.2.10 DEFINITION. An *operator space* is a closed subspace of a C*-algebra.

An *operator system* is a closed self-adjoint subspace of a unital C*-algebra A which contains the unit of A.

(Sometimes operator spaces or operator systems are not required to be closed.) If X is an operator system, then a *positive linear functional* on X is a linear functional ϕ with $\phi(1) = \|\phi\|$. A positive linear functional ϕ on X is a *state* if $\phi(1) = 1$. Pure states are the extreme points of $\mathcal{S}(X)$, the state space of X.

The key feature of an operator system X in a C*-algebra A is that it is spanned by its positive cone $X_+ = X \cap A_+$. In fact, if $x \in X$, its real and imaginary parts are in X; and if $x = x^* \in X$, then $x = \|x\|1 - (\|x\|1 - x)$ and $\|x\|1, \|x\|1 - x \in X_+$. Positive linear functionals on X are precisely the linear functionals taking nonnegative values on X_+.

II.6.3 Extension and Existence of States

II.6.3.1 If A is a unital C*-algebra, X is an operator system in A, and ϕ is a state on X, then by the Hahn-Banach Theorem ϕ extends to a linear functional ψ on A of norm one. Since $\psi(1) = 1$, ψ is a state on A. If B is a (not necessarily unital) C*-subalgebra of a (not necessarily unital) C*-algebra A, and ϕ is a state on B, first extend ϕ to $\tilde{B}$, regard $\tilde{B}$ as a unital C*-subalgebra of $\tilde{A}$ and extend to $\tilde{A}$, and then restrict to A to get an extension of ϕ to a state on A.

II.6.3.2 If ϕ is a pure state on an operator system X in a unital A, or on a C*-subalgebra of a (not necessarily unital) C*-algebra A, then the set of extensions of ϕ to A is a weak-* compact convex subset of $\mathcal{S}(A)$, and any extreme point of this set (extreme points exist by the Krein-Milman Theorem) is a pure state of A. Thus we get:

II.6.3.3 PROPOSITION. Let A be a C*-algebra, $x \in A_{sa}$. Then there is a pure state ϕ on A with $|\phi(x)| = \|x\|$.

PROOF: Note that there is a pure state (homomorphism) ϕ on $C^*(x) \cong C_o(\sigma(x))$ with $|\phi(x)| = r(x) = \|x\|$, and extend ϕ to a pure state on A.

There is also a noncommutative version of the Jordan Decomposition Theorem for signed measures:

II.6.3.4 THEOREM. Let A be a C*-algebra, and ϕ a bounded self-adjoint linear functional on A. Then there are unique positive linear functionals ϕ_+ and ϕ_- on A with $\phi = \phi_+ - \phi_-$ and $\|\phi\| = \|\phi_+\| + \|\phi_-\|$.

Since every linear functional ϕ can be canonically written as $\phi_{re} + i\phi_{im}$ with ϕ_{re}, ϕ_{im} self-adjoint [define ϕ^* by $\phi^*(x) = \overline{\phi(x^*)}$, and let $\phi_{re} = (\phi + \phi^*)/2$,

$\phi_{im} = (\phi - \phi^*)/2i$], every bounded linear functional can be written canonically as a linear combination of four states. A simple consequence of this and the Hahn-Banach Theorem is:

II.6.3.5 COROLLARY. Let A be a C*-algebra, $x \in A$.

(i) If $\phi(x) = 0$ for all $\phi \in \mathcal{S}(A)$, then $x = 0$.
(ii) $x = x^*$ if and only if $\phi(x) \in \mathbb{R}$ for all $\phi \in \mathcal{S}(A)$.
(iii) $x \geq 0$ if and only if $\phi(x) \geq 0$ for all $\phi \in \mathcal{S}(A)$.

In fact, $\mathcal{S}(A)$ can be replaced by $\mathcal{P}(A)$ in this result.

II.6.4 The GNS Construction

The GNS construction, discovered independently by Gelfand and Naimark [GN43] and I. Segal [Seg47] (the essential idea appeared earlier in [Gel41]), while quite simple, is an "ingenious construction" (E. Hille [Hil43]) and one of the most fundamental ideas of the theory of operator algebras. It provides a method for manufacturing representations of C*-algebras.

II.6.4.1 Let A be a C*-algebra, ϕ a positive linear functional on A. Put a pre-inner product on A by $\langle x, y \rangle_\phi = \phi(y^*x)$. If

$$N_\phi = \{x \in A : \phi(x^*x) = 0\},$$

then N_ϕ is a closed left ideal of A, and $\langle \cdot, \cdot \rangle_\phi$ drops to an inner product on A/N_ϕ. If $a \in A$, let $\pi_\phi(a)$ be the left multiplication operator by a on A/N_ϕ, i.e. $\pi_\phi(a)(x + N_\phi) = ax + N_\phi$. Since

$$x^*a^*ax \leq \|a\|^2 x^*x$$

(II.3.1.8), $\pi_\phi(a)$ is a bounded operator and $\|\pi_\phi(a)\| \leq \|a\|$; so $\pi_\phi(a)$ extends to a bounded operator, also denoted $\pi_\phi(a)$, on the completion (Hilbert space) $\mathcal{H}_\phi = L^2(A, \phi)$ of A/N_ϕ. The representation π_ϕ is called the *GNS representation* of A associated with ϕ.

II.6.4.2 There is also a (unique) distinguished vector $\xi_\phi \in \mathcal{H}_\phi$ such that $\phi(a) = \langle \pi_\phi(a)\xi_\phi, \xi_\phi \rangle_\phi$ (and $\|\xi_\phi\|^2 = \|\phi\|$). If A is unital, then ξ_ϕ is just the image of 1_A in $\mathcal{H}_\phi$. In general, let $\tilde{\phi}$ be the unique extension of ϕ to $\tilde{A}$ with $\|\tilde{\phi}\| = \|\phi\|$; then from $\tilde{\phi}(1) = \lim \phi(h_\lambda) = \lim \phi(h_\lambda^2)$, where (h_λ) is an approximate unit for A, it follows that $\tilde{\phi}((1-h_\lambda)^2) \to 0$, so A/N_ϕ is dense in $\tilde{A}/N_{\tilde{\phi}}$ and thus $\mathcal{H}_\phi$ can be identified with $\mathcal{H}_{\tilde{\phi}}$, and ξ_ϕ may be taken to be $\xi_{\tilde{\phi}}$. This ξ_ϕ is a *cyclic vector* for π_ϕ, i.e. $\pi_\phi(A)\xi_\phi$ is dense in $\mathcal{H}_\phi$.

II.6.4.3 This construction is the inverse of II.6.2.3(i): if π is a representation of A on a Hilbert space $\mathcal{H}$ and ξ is a cyclic vector for π, and $\phi(a) = \langle \pi(a)\xi, \xi \rangle$, then $(\mathcal{H}_\phi, \pi_\phi, \xi_\phi)$ is unitarily equivalent to $(\mathcal{H}, \pi, \xi)$. Thus there is a precise one-one correspondence between positive linear functionals on A and (cyclic) representations of A with a specified cyclic vector.

II.6.4.4 EXAMPLE. Let $A = \mathbb{M}_n$, with Tr the (ordinary) trace on A. Then $\mathcal{H}_{\text{Tr}}$ can be identified with $\mathbb{M}_n$ with the Hilbert-Schmidt norm, with $\xi_{\text{Tr}} = I$. The action π_{Tr} is the ordinary one by left multiplication. $\mathbb{M}_n$ acting on itself in this way is called the *standard form* representation of $\mathbb{M}_n$. Its commutant is also isomorphic (or, more naturally, anti-isomorphic) to $\mathbb{M}_n$, acting by right multiplication. Each von Neumann algebra on a Hilbert space has a standard form generalizing this example (III.2.6.5).

II.6.4.5 If ϕ is a positive linear functional on A, then ϕ has a canonical extension to a positive linear functional on $\mathcal{L}(\mathcal{H}_\phi)$ of the same norm, also denoted ϕ, by $\phi(T) = \langle T\xi_\phi, \xi_\phi \rangle_\phi$.

The next simple result is a version of the Radon-Nikodym Theorem (the classical Radon-Nikodym theorem is essentially the special case where $A = L^\infty(X, \mu)$ for a finite measure space (X, μ), and $\phi(f) = \int f \, d\mu$). Other more difficult (and useful) versions of the Radon-Nikodym Theorem for von Neumann algebras appear in III.2.3.3 and III.4.7.5.

II.6.4.6 PROPOSITION. Let ϕ and ψ be positive linear functionals on A, with $\psi \leq \phi$ (i.e. $\psi(x) \leq \phi(x)$ for all $x \in A_+$). Then there is a unique operator $T \in \pi_\phi(A)' \subseteq \mathcal{L}(\mathcal{H}_\phi)$, with $0 \leq T \leq I$, such that $\psi(x) = \phi(T\pi_\phi(x)) = \langle T\pi_\phi(x)\xi_\phi, \xi_\phi \rangle_\phi$ for all $x \in A$.

PROOF: For $x, y \in A$, define $(\pi_\phi(x)\xi_\phi, \pi_\phi(y)\xi_\phi) = \psi(y^*x)$. Then $(\cdot, \cdot)$ extends to a bounded sesquilinear form (pre-inner product) on $\mathcal{H}_\phi$, and thus there is a $T \in \mathcal{L}(\mathcal{H}_\phi)$, $0 \leq T \leq I$, such that $(\eta, \zeta) = \langle T\eta, \zeta \rangle_\phi$ for all $\eta, \zeta \in \mathcal{H}_\phi$ (I.2.2.2). If $x, y, z \in A$, then

$$\langle T\pi_\phi(x)[\pi_\phi(z)\xi_\phi], \pi_\phi(y)\xi_\phi \rangle_\phi = \psi(y^*(xz)) = \psi((x^*y)^*z)$$

$$= \langle T\pi_\phi(z)\xi_\phi, \pi_\phi(x^*)[\pi_\phi(y)\xi_\phi] \rangle_\phi = \langle \pi_\phi(x)T[\pi_\phi(z)\xi_\phi], \pi_\phi(y)\xi_\phi \rangle_\phi$$

so, fixing x and letting y, z range over A, we conclude that $T\pi_\phi(x) = \pi_\phi(x)T$.

II.6.4.7 In fact, if $\psi \leq \phi$ and $\omega = \phi - \psi$, so that $\phi = \psi + \omega$, then π_ϕ is unitarily equivalent to the subrepresentation of $\pi_\psi \oplus \pi_\omega$ with cyclic vector $\xi_\psi \oplus \xi_\omega$, and T is projection onto the first coordinate, compressed to this subspace.

Because of the abundance of states on a C*-algebra described in (II.6.3.3), every C*-algebra has many representations. We can say even more:

II.6.4.8 PROPOSITION. If ϕ is a state of A, then π_ϕ is irreducible if and only if ϕ is a pure state.

PROOF: It follows immediately from II.6.4.6 and II.6.1.8 that if π_ϕ is irreducible, then any positive linear functional $\psi \leq \phi$ is a multiple of ϕ, and hence ϕ is pure. Conversely, if ϕ is pure, suppose there is a projection P in $\pi_\phi(A)'$, $P \neq 0, I$. Then $P\xi_\phi \neq 0$, for otherwise

$$0 = \pi_\phi(x) P \xi_\phi = P[\pi_\phi(x)\xi_\phi]$$

for all $x \in A$, so $P = 0$. Similarly, $(I - P)\xi_\phi \neq 0$. For $x \in A$, write

$$\phi_1(x) = \langle \pi_\phi(x) P\xi_\phi, P\xi_\phi \rangle_\phi = \phi(P\pi_\phi(x))$$

$$\phi_2(x) = \langle \pi_\phi(x)(I-P)\xi_\phi, (I-P)\xi_\phi \rangle_\phi = \phi((I-P)\pi_\phi(x)).$$

Then ϕ_1, ϕ_2 are positive linear functionals on A, $\phi = \phi_1 + \phi_2$, and $\phi_1 = \lambda \phi$, $\phi_2 = (1-\lambda)\phi$ for some λ, $0 < \lambda < 1$, because ϕ is pure. For any $\epsilon > 0$, there is an $x \in A$ with $\|\pi_\phi(x)\xi_\phi - P\xi_\phi\|_\phi < \epsilon$ and $\|\pi_\phi(x)\xi_\phi\|_\phi^2 = \phi(x^*x) = \|P\xi_\phi\|_\phi^2$. Then $\|(I-P)\pi_\phi(x)\xi_\phi\|_\phi = \|(I-P)(\pi_\phi(x)\xi_\phi - P\xi_\phi)\|_\phi < \epsilon$, and

$$(1-\lambda)\|P\xi_\phi\|_\phi^2 = (1-\lambda)\phi(x^*x) = \phi_2(x^*x) = \|(I-P)\pi_\phi(x)\xi_\phi\|_\phi^2 < \epsilon^2$$

contradicting $\lambda \neq 1$ and therefore the existence of P. Thus $\pi_\phi(A)' = \mathbb{C}I$.

II.6.4.9 COROLLARY. If A is a C*-algebra and $x \in A$, then there is an irreducible representation π of A with $\|\pi(x)\| = \|x\|$.

PROOF: Let $\pi = \pi_\phi$, where ϕ is a pure state of A with $\phi((xx^*)^2) = \|(xx^*)^2\| = \|x\|^4$ (II.6.3.3). Then

$$\|x\|^2 = \langle xx^*, xx^* \rangle_\phi^{1/2} = \|xx^*\|_\phi = \|\pi(x)x^*\|_\phi \leq \|\pi(x)\| \|x^*\|_\phi \leq \|\pi(x)\| \|x\| \ .$$

By considering sums of irreducible representations, we obtain

II.6.4.10 COROLLARY. If A is a C*-algebra, then A has a faithful representation, i.e. A is isometrically isomorphic to a concrete C*-algebra of operators on a Hilbert space $\mathcal{H}$. If A is separable, then $\mathcal{H}$ may be chosen to be separable.

This result is also called the "Gelfand-Naimark theorem" (cf. II.2.2.4).

II.6.4.11 PROPOSITION. Let A be a C*-algebra, B a C*-subalgebra of A, and ρ a representation of B on a Hilbert space $\mathcal{H}_0$. Then ρ can be extended to a representation π of A on a (possibly) larger Hilbert space $\mathcal{H}$, i.e. there is a Hilbert space $\mathcal{H}$ containing $\mathcal{H}_0$ and a representation π of A on $\mathcal{H}$ such that $\mathcal{H}_0$ is invariant under $\pi|_B$ and $\pi|_B$ agrees with ρ on $\mathcal{H}_0$. If ρ is irreducible, then π may be chosen irreducible.

Since every representation is a direct sum of cyclic representations, it suffices to show the result for ρ cyclic, i.e. ρ is the GNS representation corresponding to a state ψ of B. Extend ψ to a state ϕ of A (with ϕ pure if ψ is pure; see II.6.3.2), and let π be its GNS representation.

Using these results, facts about elements of C*-algebras can be deduced from corresponding facts about operators. The next result is an example.

II.6.4.12 THEOREM. [FUGLEDE] (cf. I.6.2.7) Let A be a C*-algebra, $x, y \in A$ with x normal. If y commutes with x (i.e. if $[x, y] = 0$), then y commutes with x^*. So y and y^* commute with $f(x)$ for any $f \in C_o(\sigma(x))$.

Various other results such as the next two can also be proved using irreducible representations.

II.6.4.13 PROPOSITION. Let A be a C*-algebra, $x \in A$. Then x is in the center $\mathcal{Z}(A)$ if and only if $\pi(x)$ is a scalar multiple of the identity for every irreducible representation π of A.

PROOF: If $x \notin \mathcal{Z}(A)$ and $[x, y] \neq 0$, then there is an irreducible representation π of A for which $\pi([x, y]) = [\pi(x), \pi(y)] \neq 0$.

II.6.4.14 PROPOSITION. Let A be a noncommutative C*-algebra. Then there is a nonzero $x \in A$ with $x^2 = 0$ (equivalently, $x^*x \perp xx^*$).

PROOF: If A is noncommutative, there is an irreducible representation π of A with $\pi(A)$ noncommutative and hence more than one-dimensional. Thus there is an $a = a^* \in A$ such that $\sigma(\pi(a))$ contains two distinct numbers λ and μ. Let f and g be continuous functions from $\mathbb{R}$ to $[0, 1]$, with disjoint suppports, and with $f(\lambda) = g(\mu) = 1$. Then $f(a)$ and $g(a)$ are orthogonal, and $\pi(f(a)) = f(\pi(a))$ and $\pi(g(a)) = g(\pi(a))$ are nonzero. Since $\pi(A)$ is prime (II.6.1.11), there is a $z \in \pi(A)$ with $\pi(f(a))z\pi(g(a)) \neq 0$ (II.5.4.5). If $y \in A$ with $\pi(y) = z$, then $x = f(a)yg(a) \neq 0$ and $x^2 = 0$. For the last statement, if $x^2 = 0$, then $(x^*x)(xx^*) = 0$; conversely, if $(x^*x)(xx^*) = 0$, then $[(x^*)^2x^2]^2 = 0$, so $(x^2)^*x^2 = 0$, $x^2 = 0$.

II.6.4.15 COROLLARY. Every noncommutative unital C*-algebra contains a full positive element which is not invertible.

PROOF: Let A be unital and noncommutative, and $x \in A$ with $\|x\| = 1$ but $x^2 = 0$. Fix $\epsilon > 0$ small enough that $z = f_\epsilon(xx^*)x = xf_\epsilon(x^*x) \neq 0$, $y =$

$xf_{\epsilon/2}(x^*x)$, $a = 1 - yy^*$. Let J be the closed ideal of A generated by a. Then $y^*y \in J$ since $y^*y \leq a$, and hence $yy^* \in J$. Also, $a \in J$, so $1 = a + yy^* \in J$, $J = A$. But a is not invertible since $a \perp zz^*$.

Extending States from Ideals

Using the GNS construction, we can obtain an extension of II.6.2.5(i) and a sharpening of II.6.3.1 in the case that the subalgebra is an ideal. This result is also closely related to II.6.1.6.

II.6.4.16 PROPOSITION. Let J be a closed ideal in a C*-algebra A, and ψ a positive linear functional on J. Then there is a unique extension of ψ to a positive linear functional ϕ on A with $\|\phi\| = \|\psi\|$. If (h_λ) is an increasing approximate unit for J, then $\phi(x) = \lim \psi(h_\lambda x) = \lim \psi(x h_\lambda)$ for all $x \in A$; if $x \in A_+$, then

$$\phi(x) = \lim \psi(x^{1/2} h_\lambda x^{1/2}) = \sup \psi(x^{1/2} h_\lambda x^{1/2}).$$

Any other positive linear functional ϕ' on A extending ψ satisfies $\phi' \geq \phi$.

PROOF: By II.6.3.1 there is a ϕ extending ψ with $\|\phi\| = \|\psi\|$, but we will not use this result. Instead we define ϕ directly as follows.

Let π_ψ be the GNS representation of J on $\mathcal{H}_\psi$ with cyclic vector ξ_ψ. Extend π_ψ to a representation π of A on $\mathcal{H}_\psi$ as in II.6.1.6. Then ξ_ψ defines a positive linear functional ϕ on A by $\phi(x) = \langle \pi(x)\xi_\psi, \xi_\psi \rangle$. We have $\|\phi\| = \|\xi_\psi\|^2 = \|\psi\|$, and $(\mathcal{H}_\phi, \pi_\phi, \xi_\phi)$ can be naturally identified with $(\mathcal{H}_\psi, \pi, \xi_\psi)$. Since $\pi(h_\lambda) \to I$ strongly, we have, for any $x \in A$,

$$\psi(h_\lambda x) = \langle \pi(h_\lambda)\pi(x)\xi_\psi, \xi_\psi \rangle \to \langle \pi(x)\xi_\psi, \xi_\psi \rangle = \phi(x)$$

and similarly $\psi(x h_\lambda) \to \phi(x)$ and, if $x \geq 0$, $\psi(x^{1/2} h_\lambda x^{1/2}) \to \phi(x)$.

If ϕ' is any extension of ψ to a positive linear functional on A, and $x \in A_+$, then $\phi'(x) \geq \phi'(x^{1/2} h_\lambda x^{1/2}) = \psi(x^{1/2} h_\lambda x^{1/2})$ for all λ, so $\phi'(x) \geq \phi(x)$. Thus $\omega = \phi' - \phi$ is a positive linear functional, and we have $\|\phi'\| = \|\phi\| + \|\omega\| = \|\psi\| + \|\omega\|$, so if $\|\phi'\| = \|\psi\|$, then $\omega = 0$, $\phi' = \phi$.

II.6.5 Primitive Ideal Space and Spectrum

It is common in mathematics (e.g. in algebraic geometry) to associate to a ring A a topological space of ideals of A, generally either the set of prime ideals (denoted Spec(A)) or the set of maximal ideals (denoted Maxspec(A)), with the hull-kernel topology (II.2.1.4). For C*-algebras, the most useful variant is to consider the primitive ideal space:

II.6.5.1 DEFINITION. Let A be a C*-algebra. A *primitive ideal* of A is an ideal which is the kernel of an irreducible representation of A. Denote by Prim(A) the set of primitive ideals of A. The topology on Prim(A) is the hull-kernel topology, i.e. $\{J_i\}^- = \{J : J \supseteq \cap J_i\}$.

II.6.5.2 A primitive ideal is closed by II.6.1.2 and prime by II.6.1.11. (In fact, the kernel of any factor representation is a closed prime ideal (II.6.1.11)). If A is separable, then every closed prime ideal of A is primitive (II.6.5.15). Recently, N. Weaver [Wea03] constructed a nonseparable prime C*-algebra with no faithful irreducible representation, so not every closed prime ideal in a nonseparable C*-algebra is primitive. It is not known whether the kernel of a factor representation of a nonseparable C*-algebra must be a primitive ideal (Weaver's example has no faithful factor representations, in fact no faithful cyclic representations).

II.6.5.3 By II.6.4.9, the intersection of the primitive ideals of a C*-algebra A is $\{0\}$. By considering irreducible representations of A/J, it follows that every closed ideal J of A is an intersection of primitive ideals. In particular, every maximal ideal is primitive. There is thus a natural one-one correspondence between the closed sets in $\mathrm{Prim}(A)$ and the closed ideals of A. If A is commutative, then every closed prime ideal is maximal and hence primitive, and $\mathrm{Prim}(A)$ agrees with the set defined in II.2.1.4 and may be identified with $\hat{A}$ as in II.2.2.4.

II.6.5.4 If J is a closed ideal in A, then we can identify $\mathrm{Prim}(A/J)$ with
$$\mathrm{Prim}_J(A) = \{K \in \mathrm{Prim}(A) : J \subseteq K\},$$
the corresponding closed set in $\mathrm{Prim}(A)$. If $K \in \mathrm{Prim}(A)$, $J \not\subseteq K$, and π is an irreducible representation of A with kernel K, then $\pi|J$ is an irreducible representation of J (II.6.1.6) with kernel $K \cap J$. So there is a map ρ_J from
$$\mathrm{Prim}^J(A) = \{K \in \mathrm{Prim}(A) : J \not\subseteq K\}$$
to $\mathrm{Prim}(J)$, defined by $\rho_J(K) = K \cap J$, which is surjective by II.6.1.6. Also, ρ_J is injective: if $K_1, K_2 \in \mathrm{Prim}(A)$ and $K_1 \cap J = K_2 \cap J \neq J$, then by II.5.1.4(iv) we have
$$(K_1 + K_2) \cap (J + K_2) = (K_1 \cap J) + K_2 = (K_2 \cap J) + K_2 = K_2$$
and, since K_2 is prime and $J + K_2 \neq K_2$, we have $K_1 + K_2 = K_2$, $K_1 \subseteq K_2$. Symmetrically, $K_2 \subseteq K_1$, so $K_1 = K_2$. Thus $\mathrm{Prim}(J)$ may be identified with $\mathrm{Prim}^J(A) = \mathrm{Prim}(A) \setminus \mathrm{Prim}_J(A)$, an open set in $\mathrm{Prim}(A)$.

II.6.5.5 If J is a closed ideal in a C*-algebra A and $x \in A$, write $\|x\|_J$ for the norm of x mod J. If $\{J_i\}$ is any collection of closed ideals in A and $J = \cap J_i$, then $\|\cdot\|_J = \sup_i \|\cdot\|_{J_i}$ (by II.1.6.5 it suffices to note that this formula defines a C*-norm on A/J); and if (J_i) is an increasing net of closed ideals with $J = [\cup J_i]^-$, then $\|\cdot\|_J = \inf_i \|\cdot\|_{J_i}$ (this follows immediately from the definition of the quotient norm). A simple consequence is:

II.6.5.6 PROPOSITION. Let A be a C*-algebra.

(i) If $x \in A$, define $\check{x} : \mathrm{Prim}(A) \to \mathbb{R}_+$ by $\check{x}(J) = \|x\|_J$. Then $\check{x}$ is lower semicontinuous.
(ii) If $\{x_i\}$ is a dense set in the unit ball of A, and $U_i = \{J \in \mathrm{Prim}(A) : \check{x}_i(J) > 1/2\}$, then $\{U_i\}$ forms a base for the topology of $\mathrm{Prim}(A)$.
(iii) If $x \in A$ and $\lambda > 0$, then $\{J \in \mathrm{Prim}(A) : \check{x}(J) \geq \lambda\}$ is compact (but not necessarily closed) in $\mathrm{Prim}(A)$.

PROOF: Part (i) is obvious. For (ii), let $J \in \mathrm{Prim}(A)$. A neighborhood of J is a set of the form $V = \{I \in \mathrm{Prim}(A) : K \not\subseteq I\}$ for a closed ideal K not contained in J. Choose $y \in K \backslash J$, and $0 < \epsilon \leq \frac{1}{2}\|y\|_J^2$, and let $x = f_\epsilon(y^*y)$. Then $x \in K$, and $\|x\| = \|x\|_J = 1$. Choose x_i with $\|x_i - x\| < 1/2$; then $J \in U_i$ and $U_i \subseteq V$.
To show (iii), let $\{K_i : i \in \Omega\}$ be a set of closed ideals in A, K the closed ideal generated by $\cup K_i$ (so $K = [\sum K_i]^-$), and
$$U_i = \{J \in \mathrm{Prim}(A) : K_i \not\subseteq J\}.$$
Then $\{U_i\}$ is an open cover of $F = \{J \in \mathrm{Prim}(A) : \check{x}(J) \geq \lambda\}$ if and only if $\|x\|_K < \lambda$. In this case, there are $i_1, \ldots, i_n$ such that if $K_0 = K_{i_1} + \cdots + K_{i_n}$, then $\|x\|_{K_0} < \lambda$, so $\{U_{i_1}, \ldots, U_{i_n}\}$ is a cover of F.

II.6.5.7 COROLLARY. $\mathrm{Prim}(A)$ is a locally compact space T_0-space which is compact if A is unital. If A is separable, then $\mathrm{Prim}(A)$ is second countable.

II.6.5.8 The function $\check{x}$ of II.6.5.6 is not continuous in general; in fact, if $\mathrm{Prim}(A)$ is not Hausdorff, then the $\check{x}$ cannot all be continuous since $\{\check{x} : x \in A\}$ separates the points of $\mathrm{Prim}(A)$. But if $\mathrm{Prim}(A)$ is Hausdorff, II.6.5.6(iii) shows that $\check{x}$ is continuous for all $x \in A$. If x is in the center of A, then $\check{x}$ is continuous (II.6.5.10).

II.6.5.9 EXAMPLES.

(i) Let $A = \tilde{\mathbb{K}}$. Then $\mathrm{Prim}(A) = \{0, \mathbb{K}\}$; $\{\mathbb{K}\}$ is a closed point, and $\{0\}$ is a dense open point. Thus $\mathrm{Prim}(A)$ is not T_1.
(ii) Let A be the C*-algebra of all sequences of elements of $\mathbb{M}_2$ which converge to a diagonal matrix. Then $\mathrm{Prim}(A)$ consists of a sequence of points simultaneously converging to two closed points at infinity. Thus $\mathrm{Prim}(A)$ is T_1 but not Hausdorff.
(iii) Let A be the C*-algebra of all sequences in $\tilde{\mathbb{K}}$ converging to a scalar multiple of 1. Then $\mathrm{Prim}(A)$ is the one-point compactification of a disjoint union of two-point spaces as in (i). This A gives a counterexample to problems 4.7.8, 4.7.9, and 10.10.11 of [Dix69b].

If A is commutative, then the Gelfand-Naimark Theorem (II.2.2.4) identifies elements of A with continuous functions on $\hat{A} = \mathrm{Prim}(A)$. There are various generalizations; the next result of Dauns and Hoffman [DH68] is one of the most important.

II.6.5.10 THEOREM. [DAUNS-HOFFMAN] Let A be a unital C*-algebra, and $\mathcal{Z}(A)$ its center. Then, for each $x \in \mathcal{Z}(A)$, the function $\check{x}$ of II.6.5.6 is continuous, and $x \mapsto \check{x}$ ($x \in \mathcal{Z}(A)_+$) extends to an isomorphism from $\mathcal{Z}(A)$ onto $C(\mathrm{Prim}(A))$.

The theorem also holds for nonunital A: if $M(A)$ is the multiplier algebra of A (II.7.3), then $\mathcal{Z}(M(A))$ is isomorphic to $C_b(\mathrm{Prim}(A))$, the bounded continuous complex-valued functions of $\mathrm{Prim}(A)$. See IV.1.6.7 for another "Dauns-Hoffman Theorem."

For the proof, let $Z = \mathcal{Z}(A)$. If π is an irreducible representation of A, then π maps Z into the scalars, so defines a homomorphism from Z to $\mathbb{C}$, i.e. an element of $\hat{Z}$. So if $J \in \mathrm{Prim}(A)$, then $\alpha(J) = J \cap Z \in \mathrm{Prim}(Z) \cong \hat{Z}$. The map $\alpha: \mathrm{Prim}(A) \to \mathrm{Prim}(Z)$ is obviously continuous from the way the hull-kernel topology is defined, and α is surjective since every irreducible representation of Z extends to an irreducible representation of A (II.6.4.11). Thus there is an induced homomorphism $\alpha_*: Z \cong C(\hat{Z}) \to C(\mathrm{Prim}(A))$, which is injective since α is surjective.

If $x \in Z \subseteq A$, and $\hat{x}$ is the corresponding function in $C(\hat{Z})$ (cf. II.2.2.4), then $\check{x} = |\alpha_*(\hat{x})| = |\hat{x} \circ \alpha|$, so $\check{x}$ is continuous. If $x \in Z_+$, then $\check{x} = \alpha_*(\hat{x})$.

To show that α_* is surjective, the key lemma in the proof given in [EO74] is the following, which is also useful for other purposes:

II.6.5.11 LEMMA. Let A be a C*-algebra, $a \in A_+$, f a continuous function from $\mathrm{Prim}(A)$ to $[0,1]$, and $\epsilon > 0$. Then there is a $b \in A_+$ such that

$$\|b - f(J)a\|_J < \epsilon$$

for all $J \in \mathrm{Prim}(A)$. If b' is another such element, then $\|b - b'\| < 2\epsilon$.

PROOF: Choose n such that $\frac{2\|a\|}{n} < \epsilon$, and for $1 \leq k \leq n$ let J_k be the ideal of A corresponding to the open set

$$\{J \in \mathrm{Prim}(A) : (k-1)/n < f(J) < (k+1)/n\}.$$

Then $J_1 + \cdots + J_n = A$, so by II.5.1.4 there are $a_k \in (J_k)_+$ with $a = a_1 + \cdots + a_n$. Set $b = \sum_k \frac{k}{n} a_k$. Then it is easily checked that b has the right properties. If b' is another element with the same properties, then $\|b - b'\|_J < 2\epsilon$ for all J, hence $\|b - b'\| < 2\epsilon$ by II.6.4.9.

In the situation of II.6.5.11, let b_n be a b corresponding to $\epsilon = 1/n$. Then (b_n) is a Cauchy sequence and thus converges to $b \in A$. So we obtain an exact version of II.6.5.11 which is equivalent to surjectivity of α_* (take $a = 1$; then $b \in Z$ by II.6.4.13):

II.6.5.12 COROLLARY. Let A be a C*-algebra, $a \in A_+$, and f a continuous function from $\mathrm{Prim}(A)$ to $[0,1]$. Then there is a unique $b \in A_+$ such that $b = f(J)a \mod J$ for all $J \in \mathrm{Prim}(A)$.

II.6.5.13 For a general C*-algebra A we define $\hat{A}$ to be the set of unitary equivalence classes of irreducible representations of A. There is a natural map from $\hat{A}$ onto $\mathrm{Prim}(A)$ sending π to $\ker(\pi)$. This map is not injective in general, i.e. there can be inequivalent representations with the same kernel. We will return to this in IV.1.5.7. There is also an obvious map from $\mathcal{P}(A)$ onto $\hat{A}$, sending a pure state to its GNS representation; the composite map from $\mathcal{P}(A)$ to $\mathrm{Prim}(A)$ is continuous and open when $\mathcal{P}(A)$ is given the relative weak-* topology. $\hat{A}$ can be given a unique topology making the maps $\mathcal{P}(A) \to \hat{A}$ and $\hat{A} \to \mathrm{Prim}(A)$ continuous and open, but $\hat{A}$ is not T_0 unless the map to $\mathrm{Prim}(A)$ is injective (and hence a homeomorphism).

By II.6.1.6, if J is a closed ideal in A, we may identify $\hat{J}$ and $\mathrm{Prim}(J)$ with open subsets of $\hat{A}$ and $\mathrm{Prim}(A)$ respectively; the complements are naturally identified with $\widehat{A/J}$ and $\mathrm{Prim}(A/J)$. We may do the same with $\mathcal{P}(J)$ since if $\phi \in \mathcal{P}(A)$, then either $\phi|_J = 0$ or $\phi|J$ is a pure state of J.

A theorem of G. Choquet (cf. [Dix69b, B14]) says that the set of extreme points of a compact convex set in a locally convex Hausdorff topological vector space is a Baire space (i.e. the intersection of a countable family of dense open sets is dense). Since $\mathcal{P}(A)$ is such a set for A unital, and $\mathcal{P}(A)$ is a dense open subset of $\mathcal{P}(\tilde{A})$ for nonunital A, we get:

II.6.5.14 THEOREM. Let A be a C*-algebra. Then $\mathcal{P}(A)$, $\hat{A}$, and $\mathrm{Prim}(A)$ are Baire spaces.

II.6.5.15 COROLLARY. Let A be a separable C*-algebra. An ideal of A is primitive if and only if it is closed and prime.

PROOF: A primitive ideal in any C*-algebra is closed and prime. If A is separable and J is a closed prime ideal in A, by replacing A by A/J we may assume $J = \{0\}$. $\{0\}$ is an intersection of countably many primitive ideals $\{J_n\}$; each J_n is essential in A since A is prime, so the corresponding open set $\mathrm{Prim}(J_n)$ is dense in $\mathrm{Prim}(A)$. If $\{0\}$ is not primitive, then $\cap_n \mathrm{Prim}(J_n) = \emptyset$, contradicting II.6.5.14.

II.6.5.16 Another way of describing the topology on $\hat{A}$ is to fix a Hilbert space $\mathcal{H}$ and let $\mathrm{Irr}_{\mathcal{H}}(A)$ be the set of irreducible representations of A on $\mathcal{H}$. $\mathrm{Irr}_{\mathcal{H}}(A)$ has a natural topology, the topology of elementwise convergence, and the map from $\mathrm{Irr}_{\mathcal{H}}(A)$ to $\hat{A}$ is continuous and open, i.e. $\hat{A}$ has the relative quotient topology (more care must be taken in describing the topology if A has irreducible representations of different dimensions; cf. IV.1.4.7. The difficulty can be avoided by stabilization (II.6.6.11) if A is separable, since there are natural homeomorphisms $[A \otimes \mathbb{K}]\hat{\,} \cong \hat{A}$ and $\mathrm{Prim}(A \otimes \mathbb{K}) \cong \mathrm{Prim}(A)$ (II.7.6.13)). This procedure also yields a Borel structure (the *Mackey Borel structure*) on $\hat{A}$, the quotient of the natural Borel structure on $\mathrm{Irr}_{\mathcal{H}}(A)$ induced by its topology, which is finer than the Borel structure on $\hat{A}$ induced by its topology, and which is more useful than the topology if $\hat{A}$ is not T_0 (e.g.

the Mackey Borel structure separates points of $\hat{A}$). See IV.1.5.12 for more discussion of the Mackey Borel structure and its relation to the topology.

II.6.6 Matrix Algebras and Stable Algebras

As an example of a basic fact which can be proved easily by representation theory but for which there seems to be no decent space-free proof (there is a quick proof using Hilbert modules (II.7.2.6)), we observe that a matrix algebra over a C*-algebra is also a C*-algebra.

II.6.6.1 If $\mathcal{H}$ is a Hilbert space, then $\mathcal{L}(\mathcal{H}^n)$ is naturally isomorphic to the matrix algebra $M_n(\mathcal{L}(\mathcal{H}))$. (More generally, $\mathcal{L}(\mathcal{H}^n, \mathcal{H}^m)$ can be identified with the $m \times n$ matrices over $\mathcal{L}(\mathcal{H})$.) So if A is a concrete C*-algebra of operators on $\mathcal{H}$, then the matrix algebra $M_n(A)$ acts naturally as a concrete C*-algebra of operators on $\mathcal{H}^n$. The adjoint of a matrix (a_{ij}) is the matrix (b_{ij}), where $b_{ij} = a_{ji}^*$.

II.6.6.2 Thus, if A is a C*-algebra, we need only to take a faithful representation π of A on a Hilbert space $\mathcal{H}$; then π defines entrywise a faithful representation $\pi^{(n)}$ of $M_n(A)$ (with involution defined as above) on $\mathcal{H}^n$, and thus $M_n(A)$ is a C*-algebra with the induced operator norm.

II.6.6.3 Of course, by general theory, this operator norm is the unique C*-norm on $M_n(A)$ with this involution. In the future, we will often use matrix algebras over C*-algebras, and will always implicitly use this involution and norm. Note that there is no explicit formula for the norm of a matrix in terms of the entries in general. (There is, however, a simple estimate: if $a = (a_{ij})$ is a matrix, then

$$\max_{i,j} \|a_{ij}\| \leq \|a\| \leq \sum_{i,j} \|a_{ij}\|$$

since it is easily seen that if b is a matrix with exactly one nonzero entry x, then $\|b\| = \|x\|$.)

II.6.6.4 As an example of the use of matrices, we give a proof of II.3.2.3 (cf. [RW98, 2.31]). Let x be an element of a C*-algebra A. Note that if b is a self-adjoint (or normal) element in a unital C*-algebra B, and u is a unitary in B, then $f(u^*bu) = u^*f(b)u$ for any $f \in C(\sigma(b))$; thus, if $u^*bu = -b$, then $u^*b^{1/3}u = (-b)^{1/3} = -b^{1/3}$. Applying this to $b = \begin{bmatrix} 0 & x^* \\ x & 0 \end{bmatrix}$ and $u = \begin{bmatrix} 1 & 0 \\ 0 & -1 \end{bmatrix}$ in $M_2(\tilde{A})$, we obtain that $b^{1/3} = \begin{bmatrix} 0 & y^* \\ y & 0 \end{bmatrix}$ for some $y \in A$, and $x = yy^*y$. For uniqueness, if $x = zz^*z$, let $c = \begin{bmatrix} 0 & z^* \\ z & 0 \end{bmatrix}$; then c is self-adjoint and $c^3 = b$, so $c = b^{1/3}$ by II.2.3.6.

Order and Norm

There is a tight relationship between the order structure and norm in a C*-algebra: II.3.1.2(iv) is an example. The next result is another example.

II.6.6.5 PROPOSITION. Let A be a unital C*-algebra, and $x \in A$. Then $\begin{bmatrix} 1 & x \\ x^* & 1 \end{bmatrix} \geq 0$ in $M_2(A)$ if and only if $\|x\| \leq 1$.

The proof is a simple inner product calculation if $A \subseteq \mathcal{L}(\mathcal{H})$ (cf. [Pau02, 3.1]); for the general case, just choose a faithful representation of A. There is also a simple and elegant space-free proof [ER00, 1.3.2]. A similar proof shows more generally:

II.6.6.6 PROPOSITION. Let A be a unital C*-algebra, $x, y \in A$. Then $\begin{bmatrix} 1 & x \\ x^* & y \end{bmatrix} \geq 0$ in $M_2(A)$ if and only if $x^*x \leq y$.

There is also a tight relationship with the algebraic structure:

II.6.6.7 THEOREM. [Wal03] Let A be a unital C*-algebra, $u, v \in \mathcal{U}(A)$. If $x \in A$, then $\begin{bmatrix} 1 & u & x \\ u^* & 1 & v \\ x^* & v^* & 1 \end{bmatrix} \geq 0$ in $M_3(A)$ if and only if $x = uv$.

Tensor Product Notation

II.6.6.8 $M_n(A)$ is isomorphic to $A \otimes \mathbb{M}_n = A \otimes_{\mathbb{C}} \mathbb{M}_n$, and it is convenient to use tensor product notation in matrix algebras. In $\mathbb{M}_n$, let e_{ij} be the matrix with 1 in the (i, j)'th entry and zeroes elsewhere, and write $a \otimes e_{ij}$ for the element of $M_n(A)$ with a in the (i, j)'th entry and zeroes elsewhere. The e_{ij} are called the *standard matrix units* in $\mathbb{M}_n$. (There is a similar set of standard matrix units in $\mathbb{K}$.)

Stable Algebras

II.6.6.9 As in II.6.6.1, $\mathcal{L}(\mathcal{H}^\infty)$ can be identified with an algebra of infinite matrices over $\mathcal{L}(\mathcal{H})$ (although it is impossible to give an explicit description of which matrices give bounded operators). If A is a concrete C*-algebra of operators on $\mathcal{H}$, let $M_\infty(A)$ denote the infinite matrices over A with only finitely many nonzero entries. Then $M_\infty(A)$ acts naturally as a *-algebra of bounded operators on $\mathcal{H}^\infty$, and its closure in $\mathcal{L}(\mathcal{H}^\infty)$ is denoted $A \otimes \mathbb{K}$.

II.6.6.10 If A is an (abstract) C*-algebra, we can form $M_\infty(A)$ in the same manner, and by choosing a faithful representation π of A, $M_\infty(A)$ may be identified with $M_\infty(\pi(A))$ and thus given a norm. Using the uniqueness of norm on $M_n(A)$ for each n, it is easily seen that the norm on $M_\infty(A)$ does not depend on the choice of π.

II.6.6.11 DEFINITION. The completion of $M_\infty(A)$ is called the *stable algebra* of A, denoted $A \otimes \mathbb{K}$.

This is also a special case of a tensor product of C*-algebras (II.9.4.2).

II.6.6.12 DEFINITION. A C*-algebra A is *stable* if $A \cong A \otimes \mathbb{K}$. Two C*-algebras A and B are *stably isomorphic* if $A \otimes \mathbb{K} \cong B \otimes \mathbb{K}$.

II.6.6.13 PROPOSITION. $\mathcal{K}(\mathcal{H}) \otimes \mathrm{M}_n = \mathcal{K}(\mathcal{H}^n)$ and $\mathcal{K}(\mathcal{H}) \otimes \mathbb{K} = \mathcal{K}(\mathcal{H}^\infty)$. In particular,
$$\mathbb{K} \cong M_n(\mathbb{K}) \cong \mathrm{M}_n \otimes \mathbb{K} \cong \mathbb{K} \otimes \mathrm{M}_n$$
for all n, and $\mathbb{K}$ is isomorphic to $\mathbb{K} \otimes \mathbb{K}$. In fact, the map from $\mathbb{K}$ to $\mathbb{K} \otimes \mathrm{M}_n$ or $\mathbb{K} \otimes \mathbb{K}$ given by $x \to x \otimes e_{11}$ is homotopic to an isomorphism.

The first statement is almost obvious. To prove the last statement, by a judicious choice of bases it suffices to find a strongly continuous path of isometries on a separable Hilbert space linking the identity to an isometry with infinite codimension. The operators V_t on $L^2[0,1]$ defined in I.3.2.10, for, say, $1/2 \leq t \leq 1$, do the trick.

II.6.6.14 Similarly, it is easily verified that $(A \otimes \mathbb{K}) \otimes \mathbb{K} \cong A \otimes \mathbb{K}$ for any A, so the stable algebra $A \otimes \mathbb{K}$ of a C*-algebra A is a stable C*-algebra, and A is stably isomorphic to $A \otimes \mathbb{K}$. Two stable C*-algebras are stably isomorphic if and only if they are isomorphic. In the same way, A and $M_n(A)$ are stably isomorphic for any A and n. Stably isomorphic C*-algebras are "the same up to 'size';" a stable C*-algebra has uniformly infinite "size." (This must be taken with a grain of salt in the non-σ-unital case; but can be made precise in a nice way for σ-unital C*-algebras (II.7.6.11, II.7.6.13)).

II.6.7 Weights

Weights are one of the two most important generalizations of positive linear functionals.

II.6.7.1 DEFINITION. A *weight* on a C*-algebra A is a function $\phi : A_+ \to [0, \infty]$ such that $\phi(0) = 0$, $\phi(\lambda a) = \lambda \phi(a)$ for $a \in A_+$ and $\lambda > 0$, and $\phi(a+b) = \phi(a) + \phi(b)$ for all $a, b \in A_+$. The weight ϕ is *densely defined* if $\{x \in A_+ : \phi(x) < \infty\}$ is dense in A_+, and ϕ is *faithful* if $\phi(x) = 0$ implies $x = 0$.

II.6.7.2 EXAMPLES.

(i) Any positive linear functional defines a continuous finite weight. Every weight which is finite everywhere comes from a positive linear functional (and hence is continuous).
(ii) Tr is a lower semicontinuous weight on $\mathcal{L}(\mathcal{H})$.

(iii) Any sum of weights is a weight. Any sum of lower semicontinuous weights is a lower semicontinuous weight. In particular, any sum of positive linear functionals is a lower semicontinuous weight. (The converse is also true (III.2.2.18)).

(iv) The function ϕ defined by $\phi(0) = 0$, $\phi(x) = \infty$ for $x \neq 0$ is a (degenerate) lower semicontinuous weight.

(v) Let X be a locally compact Hausdorff space and μ a Borel measure on X, and set $\phi(f) = \int_X f \, d\mu$ for $f \in C_o(X)_+$. Then ϕ is a weight on $C_o(X)$, which is lower semicontinuous by Fatou's lemma.

Every lower semicontinuous weight on a commutative C*-algebra is of this form. To see this, if ϕ is a lower semicontinuous weight on $C_o(X)$ and U is an open set in X, set $\mu(U) = \sup \phi(f)$, where $f : X \to [0, 1]$ is supported in U. If V is the union of all U such that $\mu(U) < \infty$, then the Riesz Representation Theorem gives an extension of μ to a locally finite Borel measure on V. We can set $\mu(Y) = \infty$ for any nonempty $Y \subseteq X \setminus V$. Then $\phi(f) = \int_X f \, d\mu$ for any $f \in C_o(X)_+$.

The correspondence between weights and measures is not one-to-one, however: two measures give the same weight if and only if they agree on the σ-compact open sets.

(vi) Let I be a hereditary *-subalgebra (e.g. an ideal) in A, and ψ a positive linear functional on I (not necessarily bounded). Define $\phi(x)$ to be $\psi(x)$ if $x \in I_+$ and $\phi(x) = \infty$ if $x \in A_+ \setminus I_+$. Then ϕ is a weight on A which is not lower semicontinuous in general (e.g. if I is dense and ψ is bounded).

(vii) As an example of (vi), let $A = c_o$ (I.8.6), and $I = \{(a_n) : (na_n) \text{ is bounded}\}$; let ω be a free ultrafilter on $\mathbb{N}$, and for $a = (a_n) \in I$ set $\psi(a) = \lim_\omega na_n$. Since ϕ vanishes on c_{oo}, the dense ideal of sequences which are eventually 0, the only positive linear functional θ on A with $\theta(a) \leq \phi(a)$ for all $a \in A_+$ is $\theta = 0$.

II.6.7.3 We can do constructions with weights generalizing the description of the trace-class and Hilbert-Schmidt operators (I.8.5.3) and the GNS construction (II.6.4). If ϕ is a weight on A and $0 \leq a \leq b$, then $\phi(a) \leq \phi(b) = \phi(a) + \phi(b - a)$. Thus

$$\mathfrak{N}_\phi = \{x \in A : \phi(x^*x) < \infty\}$$

(sometimes written D_ϕ^2) is a left ideal of A containing

$$N_\phi = \{x \in A : \phi(x^*x) = 0\}$$

($\mathfrak{N}_\phi$ and N_ϕ are closed under addition by the argument in the proof of II.5.3.2(iii)). N_ϕ is closed if ϕ is lower semicontinuous, but $\mathfrak{N}_\phi$ is not closed in general unless ϕ is continuous. ϕ extends to a positive linear functional, also denoted ϕ, on the span $\mathfrak{M}_\phi$ (sometimes written D_ϕ^1) of $\{a \in A_+ : \phi(a) < \infty\}$. Since by polarization $y^*x \in \mathfrak{M}_\phi$ for any $x, y \in \mathfrak{N}_\phi$, ϕ defines a pre-inner product on $\mathfrak{N}_\phi$ as in the GNS construction (i.e. $\langle x, y \rangle_\phi = \phi(y^*x)$) which drops to an inner product on $\mathfrak{N}_\phi/N_\phi$. A acts by bounded operators on this space by left multiplication, thus giving a GNS representation π_ϕ on the completion $\mathcal{H}_\phi$.

II.6.7.4 There is no natural cyclic vector ξ_ϕ in this case, however. In fact, the GNS representation from a weight is not even nondegenerate in general unless the weight is lower semicontinuous (e.g. the example of II.6.7.2(vii)).

II.6.7.5 Example II.6.7.2(iv) shows that we must make further restrictions to get interesting weights. It is tempting to say that a (lower semicontinuous) weight ϕ on A is *semifinite* if, for every $x \in A_+$,

$$\phi(x) = \sup\{\phi(y) : y \in (\mathfrak{M}_\phi)_+, y \leq x\}.$$

It turns out that this definition is too restrictive unless ϕ is a trace. A better definition comes from the following:

II.6.7.6 PROPOSITION. Let ϕ be a lower semicontinuous weight on a C*-algebra A. The following are equivalent:

(i) For every approximate unit (h_λ) for $\mathfrak{M}_\phi$ and $x \in A_+$,

$$\phi(x) \leq \liminf_\lambda \phi(h_\lambda x h_\lambda).$$

(ii) For every approximate unit (h_λ) for $\mathfrak{M}_\phi$ and $x \in A_+$ with $\phi(x) = \infty$, $\lim_\lambda \phi(h_\lambda x h_\lambda) = \infty$.
(iii) For every approximate unit (h_λ) for $\mathfrak{M}_\phi$ and $x \in A_+$ with $\phi(x) = \infty$, $\sup_\lambda \phi(h_\lambda x h_\lambda) = \infty$.

The weight ϕ is *semifinite* if it satisfies these conditions.

For the proof, (i) $\Leftrightarrow$ (ii) is clear, since the condition of (i) is automatically satisfied for $x \in \mathfrak{M}_\phi$ by lower semicontinuity. (ii) $\Longrightarrow$ (iii) is trivial; the converse is not entirely trivial since $(h_\lambda x h_\lambda)$ and hence $\phi(h_\lambda x h_\lambda)$ is not increasing in general, but (iii) $\Longrightarrow$ (ii) follows easily by passing to a subnet of the given approximate unit, which is again an approximate unit.

II.6.7.7 If ϕ is semifinite, then $\mathfrak{M}_\phi^\perp = \{0\}$; the converse is unclear. If ϕ is lower semicontinuous and $\mathfrak{N}_\phi$ is dense in A, then ϕ is semifinite. The converse is false in general, but see III.2.2.20. If ϕ satisfies the condition of II.6.7.5, and $x \in A_+$ with $\phi(x) = \infty$, for any $K > 0$ there is a $y \in (\mathfrak{M}_\phi)_+$ with $y \leq x$ and $\phi(y) > K$. If (h_λ) is an approximate unit for $\mathfrak{M}_\phi$, then $\phi(h_\lambda y h_\lambda) > K$ for some λ by lower semicontinuity. Since $h_\lambda y h_\lambda \leq h_\lambda x h_\lambda$, $\phi(h_\lambda x h_\lambda) > K$, so ϕ is semifinite by II.6.7.6(iii).

II.6.7.8 If ϕ is a faithful semifinite weight (e.g. a faithful state), then π_ϕ is a faithful representation. Note, however, that π_ϕ can be faithful even if ϕ is not faithful: for example, if A is simple, then π_ϕ is always faithful (unless ϕ is degenerate, i.e. takes only the values 0 and ∞). If ϕ is semifinite, then π_ϕ is faithful if and only if N_ϕ contains no nonzero ideal.

II.6.7.9 If ϕ is a weight on A, then the left ideal $\mathfrak{N}_\phi$ is a weighted left ideal (II.5.3.14). Conversely, if $\mathfrak{N}$ is a weighted left ideal of a C*-algebra A, and $\mathfrak{M} = \mathcal{M}(\mathfrak{N})$ (II.5.3.13), define a weight ϕ on A by fixing a positive linear functional f on A, and setting $\phi(x) = f(x)$ if $x \in \mathfrak{M}_+$, $\phi(x) = \infty$ if $x \in A_+ \setminus \mathfrak{M}_+$. Then $\mathfrak{N} = \mathfrak{N}_\phi$. Thus the weighted left ideals of a C*-algebra A are precisely the "ideals of definition" of weights on A, justifying the term "weighted."

II.6.8 Traces and Dimension Functions

The most important special kinds of weights (or states) on a C*-algebra are the traces, which play a crucial role in several places in the structure theory.

II.6.8.1 DEFINITION. A *trace* on a C*-algebra A is a weight τ on A satisfying $\tau(x^*x) = \tau(xx^*)$ for all $x \in A$. A *tracial state* (or *normalized trace*) is a state which is a trace.

There is some nonuniformity of terminology concerning traces: in some references, "trace" means "tracial state."

II.6.8.2 It is immediate that if τ is a trace on A, then $\tau(u^*xu) = \tau(x)$ for any $x \in A_+$ and u unitary in $\tilde{A}$, and that $\mathfrak{M}_\tau$, $\mathfrak{N}_\tau$, and N_τ are ideals of A; by polarization $\tau(xy) = \tau(yx)$ for any $x, y \in \mathfrak{N}_\tau$. If τ is lower semicontinuous, then N_τ is a closed ideal of A. In particular, a nonzero lower semicontinuous trace on a simple C*-algebra is automatically faithful.

II.6.8.3 EXAMPLES.

(i) Any weight on a commutative C*-algebra is a trace, and every state is a tracial state.
(ii) Tr is a trace on $\mathcal{L}(\mathcal{H})$ (or on $\mathcal{K}(\mathcal{H})$). $\tau = \frac{1}{n} \cdot \text{Tr}$ is a tracial state on $\mathbb{M}_n$.
(iii) If τ is a trace on A, define τ_n on $M_n(A)$ by

$$\tau_n((x_{ij})) = \sum_{i=1}^{n} \tau(x_{ii}).$$

Then τ_n is a trace on $M_n(A)$. If τ is a tracial state, then so is $\frac{1}{n}\tau_n$. Conversely, if σ is a trace on $M_n(A)$, for $x \in A_+$ set $\tau(x) = \sigma(diag(x, 0, \cdots, 0))$; then τ is a trace on A and $\sigma = \tau_n$.
(iv) The degenerate weight of II.6.7.2(iv) is a trace.
(v) The Dixmier trace (I.8.7.7) on $\mathbb{K}$ (or on $\mathcal{L}(\mathcal{H})$) is a trace which is not lower semicontinuous.

II.6.8.4 If τ is a trace on A and u is a unitary in $\tilde{A}$, then $\tau(u^*xu) = \tau(x)$ for every $x \in A_+$ (τ is *unitarily invariant*). An interesting question is whether the converse holds: is a unitarily invariant weight necessarily a trace? The following example shows that the answer is no in general; but a lower semicontinuous unitarily invariant weight is a trace (II.6.8.7), as is any unitarily invariant linear functional.

II.6.8.5 EXAMPLE. Let A and I be as in II.5.2.1(vi). If $a = \left(\begin{bmatrix} a_n & b_n \\ c_n & d_n \end{bmatrix} \right)$ is in A_+, define $\tau(a) = a_1 + d_1$ if $a \in I$ and $\tau(x) = \infty$ if $a \notin I$. Then it is easily seen that τ is unitarily invariant since I_+ is unitarily invariant. But if $x = \left(\begin{bmatrix} 0 & 0 \\ n^{-1/2} & 0 \end{bmatrix} \right)$, then $\tau(x^*x) = 1$ and $\tau(xx^*) = \infty$.

II.6.8.6 PROPOSITION. Let A be a C*-algebra, I an ideal of A (not necessarily closed), and ϕ a linear functional on I such that $\phi(u^*xu) = \phi(x)$ for all $x \in I$ and all unitaries $u \in \tilde{A}$. Then $\phi(xy) = \phi(yx)$ for all $x \in I$, $y \in A$.

PROOF: If u is a unitary in $\tilde{A}$, and $x \in I$, then

$$\phi(ux) = \phi(u^*(ux)u) = \phi(xu).$$

If $y \in A$, write y as a linear combination of unitaries (II.3.2.12) to obtain $\phi(xy) = \phi(yx)$.

II.6.8.7 PROPOSITION. A lower semicontinuous unitarily invariant weight is a trace.

PROOF: Let ϕ be a lower semicontinuous unitarily invariant weight on A. Then $\mathfrak{M}_\phi$ is a two-sided ideal of A (II.5.2.2), and ϕ defines a linear functional on $\mathfrak{M}_\phi$, also denoted ϕ. Let (h_λ) be a right approximate unit for $\mathfrak{N}_\phi$ (II.5.3.3). Suppose $x \in \mathfrak{N}_\phi$, i.e. $\phi(x^*x) < \infty$. Then $xh_\lambda^2 x^* \to xx^*$, so $\phi(xx^*) \leq \liminf \phi(xh_\lambda^2 x^*)$. But since $x, h_\lambda, (x^*x)^{1/2} \in \mathfrak{N}_\phi$, we have $xh_\lambda, (x^*x)^{1/2}h_\lambda \in \mathfrak{M}_\phi$, and therefore by II.6.8.6, for all λ,

$$\phi(xh_\lambda^2 x^*) = \phi(h_\lambda x^* x h_\lambda) = \phi((x^*x)^{1/2} h_\lambda^2 (x^*x)^{1/2}) \leq \phi(x^*x)$$

and so $x^* \in \mathfrak{N}_\phi$ and $\phi(xx^*) \leq \phi(x^*x)$. By the symmetric argument $\phi(xx^*) = \phi(x^*x)$. (Actually, for all λ, $\phi(xh_\lambda^2 x^*) \leq \phi(xx^*)$, so $\phi(xx^*) = \lim_\lambda \phi(xh_\lambda^2 x^*)$, and similarly $\phi(x^*x) = \lim_\lambda \phi((x^*x)^{1/2} h_\lambda^2 (x^*x)^{1/2})$.)

II.6.8.8 We usually restrict attention to semifinite traces. For C*-algebras, we most frequently consider densely defined lower semicontinuous traces (in the von Neumann algebra case, σ-weakly densely defined), which are automatically semifinite.

II.6.8.9 PROPOSITION. Let τ be a lower semicontinuous trace on a C*-algebra A. The following are equivalent:

(i) τ is semifinite.
(ii) For every $x \in A_+$, $\tau(x) = \sup\{\tau(y) : y \in (\mathfrak{M}_\tau)_+, y \leq x\}$.
(iii) $\mathfrak{M}_\tau$ is an essential ideal (II.5.4.7) in A.
(iv) $\mathfrak{N}_\tau$ is an essential ideal in A.

This follows easily from II.6.7.6 and the observation that if τ is semifinite, then
$$\tau(h_\lambda x h_\lambda) = \tau(x^{1/2} h_\lambda^2 x^{1/2}) \to \tau(x)$$
for every $x \in A_+$ and every approximate unit (h_λ) for $\mathfrak{M}_\phi$, since $x^{1/2} h_\lambda^2 x^{1/2} \leq x$.

II.6.8.10 The set $\mathcal{T}(A)$ of tracial states of A is a closed convex subset of $\mathcal{S}(A)$, and hence is a compact convex set if A is unital.

II.6.8.11 THEOREM. Let A be a unital C*-algebra. Then $\mathcal{T}(A)$ is a Choquet simplex.

See [Sak71, 3.1.18] for a proof.
$\mathcal{S}(A)$ is a Choquet simplex if and only if A is commutative (and unital). (See [AS03] for a complete description of the structure of $\mathcal{S}(A)$ as a compact convex set.)

There is a close connection between traces and dimension functions:

II.6.8.12 DEFINITION. A *dimension function* on a C*-algebra A is a function $d : A_+ \to [0, \infty]$ such that $d(x) \leq d(y)$ if $x \precsim y$ (II.3.4.3) and $d(x+y) = d(x) + d(y)$ if $x \perp y$, and such that d extends to a function on $M_n(A)_+$ with the same properties. A dimension function d is *normalized* if $\sup\{d(x) : x \in A_+\} = 1$.

Dimension functions measure the "size of the support" of the elements of the algebra. Dimension functions are often defined just on the projections of the algebra, particularly in the von Neumann algebra case; such a dimension function can be extended to all elements by setting $d(x) = d(p_x)$, where p_x is the support projection of x. The technical condition that a dimension function extend to matrix algebras is automatic in most cases; in general, it suffices that it extend to 2×2 matrices [BH82].

II.6.8.13 A lower semicontinuous trace τ on A defines a dimension function d_τ. The definition can be made in several ways. Most elegant is to proceed as follows: if $B \cong C_o(X)$ is a commutative C*-subalgebra of A, τ defines a lower semicontinuous weight on $C_o(X)$ and hence a Borel measure μ on X. If $f \in C_o(X)_+$, let
$$d_\tau(f) = \mu(\{x \in X : f(x) \neq 0\}).$$

It is easily checked that this gives a well-defined dimension function on A, which is lower semicontinuous. Explicit formulas for d_τ, which can be used as alternate definitions, are

$$d_\tau(a) = \lim_{n\to\infty} \tau(a^{1/n}) = \lim_{\epsilon\to 0} \tau(f_\epsilon(a))$$

(II.3.4.11). If τ is a tracial state, then d_τ is normalized.

II.6.8.14 The procedure can be (almost) reversed. If d is a dimension function on A, and $B \cong C_o(X)$ is a commutative C*-subalgebra of A, then d defines a finitely additive measure μ on the algebra of subsets of X generated by the σ-compact open sets, where if U is open and σ-compact, $\mu(U) = d(f)$, where f is any function in $C_o(X)_+$ nonzero precisely on U. If d is lower semicontinuous, then μ extends to a countably additive Borel measure on X and hence a lower semicontinuous weight τ on B. The definition of τ on all of A_+ is unambiguous, and τ is a lower semicontinuous quasitrace:

II.6.8.15 DEFINITION. A *quasitrace* on a C*-algebra A is a function $\tau : A_+ \to [0,\infty]$ such that $\tau(x^*x) = \tau(xx^*)$ for all $x \in A$, $\tau(\lambda a) = \lambda\tau(a)$ for all $\lambda \geq 0$ and $a \in A_+$, and $\tau(a+b) = \tau(a) + \tau(b)$ if a and b are commuting elements of A_+, and such that τ extends to a map $M_n(A)_+ \to [0,\infty]$ with the same properties. τ is *normalized* if $\sup\{\tau(a) : 0 \leq a \leq 1\} = 1$. Denote by $QT(A)$ the set of normalized quasitraces on A. $QT(A)$ is a convex set, which is compact in the topology of elementwise convergence if A is unital.

II.6.8.16 One of the oldest, most famous, and most important open general structure questions for C*-algebras is whether every quasitrace is a trace (i.e. additive on A_+). The general question can be reduced to the case of AW*-factors (III.1.8.3) [BH82]. U. Haagerup showed that the answer is yes for well-behaved C*-algebras:

II.6.8.17 THEOREM. Every quasitrace on an exact C*-algebra (II.9.6.6, IV.3) is a trace.

The first proof of this theorem was in [Haa92], which remains unpublished; see [HT99] for a proof using random matrices. Actually, these references only treat the case of normalized quasitraces on unital C*-algebras; E. Kirchberg (cf. [Kir97]) observed that the general case follows from this.

It is at least true that $QT(A)$ is a Choquet simplex if A is unital, which is metrizable if A is separable [BH82]. $T(A)$ is a face in $QT(A)$.

II.6.9 Completely Positive Maps

The other crucial generalization of positive linear functionals is the notion of completely positive map.

If A and B are C*-algebras and $\phi : A \to B$ is a function, let

$$\phi^{(n)} : M_n(A) \to M_n(B)$$

be the induced map obtained by applying ϕ entrywise.

II.6.9.1 DEFINITION. Let A, B be C*-algebras, and $\phi : A \to B$ a linear function. ϕ is *positive* (written $\phi \geq 0$) if $a \in A_+$ implies $\phi(a) \in B_+$. ϕ is *n-positive* if $\phi^{(n)} : M_n(A) \to M_n(B)$ is positive, and ϕ is *completely positive* if it is n-positive for all n.

Positive, n-positive, and completely positive maps can more generally be defined on operator systems.

A map $\phi : A \to B$ is an *order embedding* if ϕ is isometric and $\phi(x) \geq 0$ if and only if $x \geq 0$ (i.e. if ϕ and $\phi^{-1} : \phi(A) \to A$ are both positive contractions). The map ϕ is a *complete order embedding* if $\phi^{(n)}$ is an order embedding for all n.

In some contexts, completely positive maps (or completely positive contractions) are the appropriate morphisms between C*-algebras.

II.6.9.2 A positive map is automatically bounded (the proof is essentially the same as for positive linear functionals). In fact, a unital positive map is a contraction (II.6.9.4).

II.6.9.3 EXAMPLES.

(i) Any *-homomorphism ϕ is positive, since $a \geq 0 \Rightarrow a = x^*x$ for some $x \in A$, so $\phi(a) = \phi(x)^*\phi(x) \geq 0$. Since $\phi^{(n)}$ is also a *-homomorphism, it follows that a *-homomorphism is completely positive.

(ii) If $\phi : A \to B$ is n-positive and $v \in B$, then $\psi : A \to B$ defined by $\psi(x) = v^*\phi(x)v$ is also n-positive. ψ is called the *compression* of ϕ by v. In particular, any compression of a *-homomorphism is completely positive.

(iii) An $(n+1)$-positive map is n-positive, but the converse is not true in general. The simplest example is the transpose map τ on $\mathbb{M}_2$. It is easily seen that τ is positive. But τ is not 2-positive. For example,

$$a = \begin{bmatrix} 1 & 0 & 0 & 1 \\ 0 & 0 & 0 & 0 \\ 0 & 0 & 0 & 0 \\ 1 & 0 & 0 & 1 \end{bmatrix} \geq 0, \text{ but } \tau^{(2)}(a) = \begin{bmatrix} 1 & 0 & 0 & 0 \\ 0 & 0 & 1 & 0 \\ 0 & 1 & 0 & 0 \\ 0 & 0 & 0 & 1 \end{bmatrix} \not\geq 0.$$

[$M_2(\mathbb{M}_2) \cong \mathbb{M}_4$, and $\tau^{(2)}$ takes the transpose in each 2×2-subblock.]

(iv) A positive map ϕ from $\mathbb{C}^n$ to a C*-algebra B is just a map of the form

$$\phi(\lambda_1, \ldots, \lambda_n) = \lambda_1 b_1 + \cdots + \lambda_n b_n$$

where $b_1, \ldots, b_n \in B_+$. If $\|\phi(1)\| \leq 1$, i.e. $\sum b_k \leq 1$, then $\|\phi\| \leq 1$ by I.2.6.13. In fact, any positive map from $\mathbb{C}^n$ to any B is completely positive (II.6.9.10).

(v) A (completely) positive unital map from $\mathbb{C}^n$ to $C(X)$ is of the form

$$(\lambda_1, \ldots, \lambda_n) \mapsto \sum \lambda_k f_k$$

where $f_1, \ldots, f_n$ are nonnegative functions summing to 1, i.e. a partition of unity in X. If $g_1, \ldots, g_m$ are finitely many elements of $C(X)$ and $\epsilon > 0$, choose a finite open cover $\mathcal{U}$ of X such that each g_k varies by less than $\epsilon/2$ on each $U \in \mathcal{U}$, and let $\{f_1, \ldots, f_n\}$ be a partition of unity subordinate to $\mathcal{U}$. Then the corresponding (completely) positive unital map from $\mathbb{C}^n$ into $C(X)$ has elements $(\lambda_{k1}, \ldots, \lambda_{kn}) \in \mathbb{C}^n$ with $\|(\lambda_{k1}, \ldots, \lambda_{kn})\| = \|g_k\|$ and $\|g_k - \phi(\lambda_{k1}, \ldots, \lambda_{kn})\| < \epsilon$ for all k.

II.6.9.4 PROPOSITION. [RD66, Cor. 1]) Let A and B be unital C*-algebras, and $\phi : A \to B$ a unital linear map. Then ϕ is positive if and only if ϕ is a contraction.

PROOF: Suppose ϕ is a contraction. Let π be a faithful unital representation of B on $\mathcal{H}$. For any unit vector ξ in $\mathcal{H}$, define $\sigma : A \to \mathbb{C}$ by $\sigma(x) = \langle \pi(\phi(x))\xi, \xi \rangle$. Then σ is a linear functional on A, $\|\sigma\| \leq 1$, and $\sigma(1) = 1$, so σ is a state on A (II.6.2.5(ii)); so if $x \in A_+$, $0 \leq \sigma(x) = \langle \pi(\phi(x))\xi, \xi \rangle$. Since ξ is arbitrary, $\pi(\phi(x)) \geq 0$.

We give an alternate argument using an idea which first appeared in [Kad51]. We first show that ϕ is self-adjoint. If not, choose $a \in A$, $a = a^*$, $\|a\| = 1$, such that $\phi(a) = b + ic$, $b = b^*$, $c = c^*$, $c \neq 0$. Replacing a by $-a$ if necessary, we may assume there is $0 < \lambda \in \sigma(c)$. Then for any $n > 0$ we have

$$\lambda + n \leq \|c + n1_B\| \leq \|b + i(c + n1_B)\| = \|\phi(a + in1_A)\| \leq \|a + in1_A\| = (1+n^2)^{1/2}$$

which is a contradiction for sufficiently large n. Positivity now follows from II.3.1.2(iv).

Now suppose ϕ is positive. By II.6.9.2, ϕ is bounded; and from II.3.2.16 we have that the norm of ϕ is the supremum of its restrictions to unital commutative C*-subalgebras of A. We may thus assume A is commutative, i.e. $A = C(X)$. By the construction in II.6.9.3(v), $\|\phi\|$ can be approximated by $\|\phi \circ \psi\|$, where ψ is a unital positive map from $\mathbb{C}^n$ to $C(X)$; thus we may assume $A = \mathbb{C}^n$. But a unital positive map from $\mathbb{C}^n$ to B is a contraction by I.2.6.13.

II.6.9.5 PROPOSITION. Let $\phi : A \to B$ be a positive contraction. Then the extended unital map $\tilde{\phi} : A^\dagger \to \tilde{B}$ is positive.

PROOF: Suppose $a + \lambda 1 \geq 0$ in $A^\dagger$. Then $a = a^*$ and $\lambda \geq 0$. If $\lambda = 0$, then $\tilde{\phi}(a + \lambda 1) = \phi(a) \geq 0$. If $\lambda > 0$, then $-\lambda^{-1}a \leq 1$, $\|\lambda^{-1}a\| \leq 1$, $\|\phi(\lambda^{-1}a)\| \leq 1$, $\phi(a) \geq -\lambda 1$, $\tilde{\phi}(a + \lambda 1) = \phi(a) + \lambda 1 \geq 0$.

II.6.9.6 It is not obvious that a positive linear functional is a completely positive map to $\mathbb{C}$, but by the GNS construction a state ϕ on A is the compression of a *-homomorphism ($\pi_\phi : A \to \mathcal{L}(\mathcal{H}_\phi)$) by a projection onto a one-dimensional subspace (spanned by ξ_ϕ). Thus a positive linear functional is completely positive. (Another argument and a generalization will be given below.)

W. Stinespring [Sti55] was the first to study completely positive maps in detail, and he proved the following dilation theorem, which is a generalization of the GNS construction, and which shows that every completely positive map is the compression of a *-homomorphism:

II.6.9.7 THEOREM. [STINESPRING] Let $\phi : A \to \mathcal{L}(\mathcal{H})$ be a completely positive map. Then there is a Hilbert space $\mathcal{H}_\phi$, a representation π_ϕ of A on $\mathcal{H}_\phi$, and $V_\phi \in \mathcal{L}(\mathcal{H}, \mathcal{H}_\phi)$ with $\|V_\phi\|^2 = \|\phi\|$, such that $\phi(a) = V_\phi^* \pi_\phi(a) V_\phi$ for all $a \in A$, and such that $(\mathcal{H}_\phi, \pi_\phi, V_\phi)$ are canonical and minimal in the sense that if $\mathcal{H}'$ is another Hilbert space with a representation ρ of A, and $W \in \mathcal{L}(\mathcal{H}, \mathcal{H}')$ with $\phi(a) = W^* \rho(a) W$ for all $a \in A$, then there is an isometry $U : \mathcal{H}_\phi \to \mathcal{H}'$ onto a subspace invariant under ρ and intertwining π_ϕ and ρ, and such that $W = UV_\phi$.

If A and ϕ are unital, then $\mathcal{H}_\phi$ can be chosen to contain $\mathcal{H}$, and $\phi(a) = P_\mathcal{H} \pi(a) P_\mathcal{H}$.

The proof is just a few lines (plus a routine check of many details): define a pre-inner product on the algebraic tensor product $A \odot \mathcal{H}$ (over $\mathbb{C}$) by

$$\langle x \otimes \xi, y \otimes \eta \rangle_\phi = \langle \phi(y^* x)\xi, \eta \rangle_\mathcal{H}$$

and then divide out by vectors of length 0 and complete to get $\mathcal{H}_\phi$, and let $\pi_\phi(a)$ be left multiplication

$$\pi_\phi(a)(x \otimes \xi) = ax \otimes \xi.$$

If (h_λ) is an approximate unit for A, then $(h_\lambda \otimes \xi)$ is a Cauchy net in $\mathcal{H}_\phi$ for any ξ; call the limit $V\xi$. If $(\mathcal{H}', \rho, W)$ are as in the statement of the theorem, define $U(x \otimes \xi) = \rho(x)(W\xi)$.

The great contribution of Stinespring (in connection with this result) was to observe that complete positivity of ϕ is precisely what is needed to make $\langle \cdot, \cdot \rangle_\phi$ positive semidefinite. This can be conveniently summarized in the following proposition, one direction of which is the crucial technical point in the proof of the theorem, and the converse a corollary of the theorem.

II.6.9.8 PROPOSITION. Let A be a C*-algebra, and $\phi : A \to \mathcal{L}(\mathcal{H})$ a linear map. Then ϕ is completely positive if and only if, for every $a_1, \ldots, a_n \in A$ and $\xi_1, \ldots, \xi_n \in \mathcal{H}$,

$$\sum_{i,j=1}^n \langle \phi(a_i^* a_j)\xi_j, \xi_i \rangle \geq 0.$$

PROOF: For the direction used in the proof of Stinespring's Theorem, let $a \in M_n(A)$ be the matrix with first row $(a_1, \ldots, a_n)$ and other entries 0.

Then $a^*a \geq 0$, and $(a^*a)_{ij} = a_i^* a_j$. If ϕ is completely positive, then the matrix $T \in M_n(\mathcal{L}(\mathcal{H})) \cong \mathcal{L}(\mathcal{H}^n)$ with $T_{ij} = \phi(a_i^* a_j)$ is positive. Conversely, if ϕ satisfies the condition in the statement, the construction in the proof of Stinespring's Theorem works to give a dilation of ϕ to a *-homomorphism, so ϕ is completely positive.

By breaking up $\mathcal{H}$ into a direct sum of cyclic subspaces, we can rephrase this result as follows:

II.6.9.9 PROPOSITION. Let A be a C*-algebra, and $\phi : A \to \mathcal{L}(\mathcal{H})$ a self-adjoint linear map. Then ϕ is completely positive if and only if, for every $a_1, \ldots, a_n$ and $x_1, \ldots, x_n$ in A and $\xi \in \mathcal{H}$,

$$\sum_{i,j=1}^{n} \langle \phi(x_i^*)\phi(a_i^* a_j)\phi(x_j)\xi, \xi \rangle \geq 0.$$

In fact, it suffices to let ξ run over a set $\{\xi_i | i \in I\}$ such that the closed span of $\{\phi(x)\xi_i | x \in A, i \in I\}$ is $\mathcal{H}$.

II.6.9.10 An examination of the proof of Stinespring's Theorem shows that if $\mathcal{H}$ is n-dimensional, then it suffices to assume that ϕ is n-positive, and it then follows from the conclusion that ϕ is actually completely positive. Since $\phi : A \to B$ is m-positive if and only if $\rho \circ \phi$ is m-positive for all $\rho \in \hat{B}$, we conclude that if every irreducible representation of B has dimension $\leq n$, then any n-positive map from a C*-algebra A into B is completely positive. This is also true if all irreducible representations of A have dimension $\leq n$ [Sti55]. (Conversely, if B is a C*-algebra and any n-positive map from any C*-algebra A into B is completely positive, then every irreducible representation of B has dimension $\leq n$ [Tom82].)

II.6.9.11 In fact, there is a one-one correspondence between completely positive maps from a C*-algebra A into $\mathbb{M}_n$ and positive linear functionals on $M_n(A)$. If σ is a positive linear functional on $M_n(A)$, and $a \in A$, define $\phi_\sigma(a)_{ij} = n\sigma(a \otimes e_{ij})$; then ϕ_σ is a completely positive map from A to $\mathbb{M}_n$. Conversely, if $\phi : A \to \mathbb{M}_n$ is completely positive, define σ_ϕ by

$$\sigma_\phi((a_{ij})) = \frac{1}{n} \sum_{i,j} \phi(a_{ij})_{ij}.$$

Then σ_ϕ is a positive linear functional on $M_n(A)$. The two assignments are easily checked to be mutually inverse. If A is unital and ϕ is unital, then σ_ϕ is a state (this is the reason for the convention of including the n and $1/n$ in the formulas); however, if σ is a state, then ϕ_σ is not necessarily unital.

More generally, the same procedure gives a one-one correspondence between cp-maps from A to $M_n(B)$ and cp-maps from $M_n(A)$ to B, for any C*-algebra B.

Combining this correspondence with II.6.3, we get the finite-dimensional case of the following theorem [Arv77]:

II.6.9.12 THEOREM. [ARVESON EXTENSION] Let A be a C*-algebra, X an operator system in A, $\mathcal{H}$ a Hilbert space, and $\phi : X \to \mathcal{L}(\mathcal{H})$ a completely positive map. Then ϕ extends to a completely positive map from A into $\mathcal{L}(\mathcal{H})$, of the same norm.

If X is a C*-subalgebra B of A, the general case of Arveson's Extension Theorem follows easily from Stinespring's Theorem: let $(\mathcal{H}_\phi, \pi_\phi, V_\phi)$ be the Stinespring dilation of $\phi : B \to \mathcal{L}(\mathcal{H})$, extend the representation π_ϕ of B to a representation ρ of A on a Hilbert space $\mathcal{H}'$ containing $\mathcal{H}_\phi$, and define $\psi(a) = V_\phi^* P_{\mathcal{H}_\phi} \rho(a) V_\phi$; ψ is an extension of ϕ to A.

II.6.9.13 A C*-algebra C for which the statement of II.6.9.12 is true with $\mathcal{L}(\mathcal{H})$ replaced by C is called an *injective C*-algebra* (it is an injective object in the category of C*-algebras and completely positive maps). Thus, any finite direct sum of copies of $\mathcal{L}(\mathcal{H})$'s (in particular, any finite-dimensional C*-algebra) is injective. The description and classification of injective C*-algebras which act on separable Hilbert spaces is one of the great achievements of the theory of operator algebras; see IV.2.1.

Another useful consequence of Stinespring's Theorem is *Kadison's inequality*:

II.6.9.14 PROPOSITION. Let $\phi : A \to B$ be a completely positive contraction. Then $\phi(x^*x) \geq \phi(x)^*\phi(x)$ for all $x \in A$.

Indeed, by embedding B in $\mathcal{L}(\mathcal{H})$ for some $\mathcal{H}$ and letting $(\mathcal{H}_\phi, \pi_\phi, V_\phi)$ be the Stinespring dilation, we have

$$\phi(x^*x) = V_\phi^* \pi_\phi(x^*x) V_\phi = V_\phi^* \pi_\phi(x)^* \pi_\phi(x) V_\phi$$

$$\geq V_\phi^* \pi_\phi(x)^* V_\phi V_\phi^* \pi_\phi(x) V_\phi = \phi(x)^* \phi(x)$$

since $V_\phi V_\phi^* \leq I$.

The Kadison inequality actually holds for 2-positive contractions [Cho74]: if $\phi : A \to B$ is a 2-positive contraction, extend ϕ to a unital 2-positive map from $A^\dagger$ to $\tilde{B}$. If $x \in A$, then $\begin{bmatrix} 1 & x \\ x^* & x^*x \end{bmatrix} \geq 0$ by II.6.6.6, so $\begin{bmatrix} 1 & \phi(x) \\ \phi(x)^* & \phi(x^*x) \end{bmatrix} \geq 0$. Apply II.6.6.6 again.

II.6.9.15 Conversely, a positive map $\phi : A \to B$ satisfying the Kadison inequality must be a contraction: if $\|\phi\| > 1$ choose $x \in A$ with $\|x\| = 1$ and $1 < \|\phi(x)\| \leq \|\phi\| < \|\phi(x)\|^2$; then

$$\|\phi(x)\|^2 = \|\phi(x)^*\phi(x)\| \leq \|\phi(x^*x)\| \leq \|\phi\|$$

since $\|x^*x\| = 1$, a contradiction.

Complete Order Isomorphisms

II.6.9.16 Recall that a map $\phi : A \to B$ is an *order embedding* if ϕ is an isometry and $\phi(x) \geq 0$ if and only if $x \geq 0$, and ϕ is a *complete order embedding* if $\phi^{(n)}$ is an order embedding for all n. A surjective [complete] order embedding is a *[complete] order isomorphism*. If A and B are unital, a complete order isomorphism must be unital; and a unital bijection ϕ is a [complete] order isomorphism if and only if both ϕ and ϕ^{-1} are [completely] positive (II.6.9.4).

If $\phi : A \to B$ is a complete order isomorphism (it is enough that $\phi^{(2)}$ be an order isomorphism), then ϕ and ϕ^{-1} satisfy Kadison's inequality, i.e. $\phi(x^*x) = \phi(x)^*\phi(x)$ for all x. Then, using polarization (II.3.1.6), we obtain:

II.6.9.17 THEOREM. Let A and B be C*-algebras, and $\phi : A \to B$ a complete order isomorphism. Then ϕ is an algebraic *-isomorphism.

PROOF: Let $x, y \in A$. Then

$$4\phi(y)^*\phi(x) = \sum_{k=0}^{3} i^k (\phi(x)+i^k\phi(y))^*(\phi(x)+i^k\phi(y)) = \sum_{k=0}^{3} i^k \phi(x+i^k y)^* \phi(x+i^k y)$$

$$= \sum_{k=0}^{3} i^k \phi((x+i^k y)^*(x+i^k y)) = \phi\left(\sum_{k=0}^{3} i^k (x+i^k y)^*(x+i^k y)\right) = 4\phi(y^*x) \ .$$

Thus, the algebraic structure of a C*-algebra is completely determined by its complete order structure (or just its 2-order structure). An explicit derivation of the algebraic structure from the complete order structure can be obtained from II.6.6.7, giving an alternate proof of the theorem. [There is a technicality in obtaining the theorem from II.6.6.7, since it is not clear how to characterize unitaries order-theoretically; but a complete order isomorphism from A to B gives a complete order isomorphism from $C_1(A)$ to $C_1(B)$, and the unitaries in these C*-algebras are precisely the extreme points of the unit balls. Note that it is sufficient to prove multiplicativity on unitaries by II.3.2.15.]

Note that the order structure on a C*-algebra is not sufficient to determine its algebraic structure: for example, if A is any C*-algebra, then A and A^{op} are order-isomorphic by the identity map, but need not be *-isomorphic (IV.1.7.16).

In fact, we have:

II.6.9.18 PROPOSITION. Let $\phi : A \to B$ be a completely positive contraction, and $x \in A$. If $\phi(x^*x) = \phi(x)^*\phi(x)$, then $\phi(yx) = \phi(y)\phi(x)$ for all $y \in A$.

The proof consists of applying Kadison's inequality to $\begin{bmatrix} x & y^* \\ 0 & 0 \end{bmatrix}$ (and thus works if ϕ is just 4-positive). See e.g. [Pau02, 3.18] for details.

Matrix Ordered Spaces

II.6.9.19 If X is an operator space in A (II.6.2.10), then the set of $n \times n$ matrices with elements in X can be naturally identified with an operator space in $M_n(A)$, which is an operator system if X is an operator system. Thus, if X is an operator system, $M_n(X)$ has an induced order structure (which in general depends on how X is embedded in A), and it makes sense to consider completely positive maps from X to a C*-algebra B (or to an operator system Y).

II.6.9.20 The situation can be abstracted [CE77a]. A *matrix ordered space* is a Banach space X with involution such that $M_n(X)$ is equipped with a norm $\|\cdot\|_n$ for each n, and a partial order defined by a set $M_n(X)_+ \subseteq M_n(X)_{sa}$ (using the induced involution), such that $c^*[M_m(X)_+]c \subseteq M_n(X)_+$ for any $c \in M_{m,n}(\mathbb{C})$.

Every operator system is a matrix ordered space as above. It makes sense to talk about completely positive maps between matrix ordered spaces in the obvious way.

If X and Y are matrix ordered spaces, then $\mathcal{L}(X,Y)$ can be partially ordered by taking the completely positive maps as the positive cone. If $x \in M_n(X)$, define $\Theta(x) : \mathbb{M}_n \to X$ by $\Theta(x)(a) = \sum_{i,j} a_{ij} x_{ij}$. Then $\Theta : M_n(X) \to \mathcal{L}(\mathbb{M}_n, X)$ is an order isomorphism.

II.6.9.21 Besides C*-algebras themselves, the most important matrix ordered spaces are duals of C*-algebras; indeed, one of the main motivations for developing the theory of matrix ordered spaces was to provide a setting to treat C*-algebras and their duals on an equal footing as ordered spaces. If X is a matrix ordered space (e.g. C*-algebra) with dual space X^*, then $M_n(X^*)$ can be identified with the *-vector space dual of $M_n(X)$ and can thus be ordered by the dual cone; X^* then becomes a matrix ordered space.

If X and Y are matrix ordered spaces, denote by $CP(X,Y) \subseteq \mathcal{L}(X,Y)$ the set of completely positive maps from X to Y.

Completely Bounded Maps

An operator space also inherits a matricial structure from its containing C*-algebra. The resulting matricial spaces can be axiomatized [Rua88] (cf. [ER00]).

II.6.9.22 If $\phi : X \to Y$ is a linear map between operator spaces X and Y, then ϕ is *completely bounded* if $\|\phi\|_{cb} = \sup_n \|\phi^{(n)}\|$ is finite. The map ϕ is a *complete contraction* if $\|\phi\|_{cb} \leq 1$.

A completely positive map ϕ (between C*-algebras or operator systems) is completely bounded, and $\|\phi\|_{cb} = \|\phi\|$. Conversely, a unital complete contraction between operator systems is completely positive (II.6.9.4).

132   II C*-Algebras

The elementary theory of completely bounded maps can be largely reduced to the completely positive case using the following dilation technique of V. Paulsen [Pau02]. The proof is similar to II.6.6.5.

II.6.9.23 PROPOSITION. Let A and B be unital C*-algebras, X an operator space in A, and $\phi : X \to B$ a complete contraction. Set

$$Y = \left\{ \begin{bmatrix} \lambda 1 & x^* \\ x & \mu 1 \end{bmatrix} : \lambda, \mu \in \mathbb{C}, x \in X \right\}.$$

Y is an operator system in $M_2(A)$, and $\psi : Y \to M_2(B)$ is completely positive, where

$$\psi\left(\begin{bmatrix} \lambda 1 & x^* \\ x & \mu 1 \end{bmatrix} \right) = \begin{bmatrix} \lambda 1 & \phi(x)^* \\ \phi(x) & \mu 1 \end{bmatrix}.$$

II.6.9.24 One can obtain completely bounded analogs of Stinespring's Theorem [Pau84] and Arveson's Extension Theorem [Wit84]. See [Pau02] for details. See also [ER00] for a complete discussion of the theory of operator spaces and completely bounded maps.

II.6.10 Conditional Expectations

Conditional expectations from a C*-algebra onto a C*-subalgebra play an important role in several parts of the theory, and are especially important in the theory of subfactors.

II.6.10.1 DEFINITION. Let B be a C*-subalgebra of a C*-algebra A. A *conditional expectation* from A to B is a completely positive contraction $\theta : A \to B$ such that $\theta(b) = b$ and such that $\theta(bx) = b\theta(x)$ and $\theta(xb) = \theta(x)b$ for all $x \in A$, $b \in B$ (θ is B-*linear*).

A conditional expectation is an idempotent map from A to A (a projection onto B) of norm one. The converse is also true [Tom57]:

II.6.10.2 THEOREM. Let A be a C*-algebra, B a C*-subalgebra, and $\theta : A \to B$ a projection of norm 1. Then θ is a conditional expectation.

PROOF: (cf. [Str81, 9.1]) We first prove the module property in the case where A is unital and B is generated by projections; the general case can be reduced to this by considering the second dual $\theta^{**} : A^{**} \to B^{**}$ (III.5.2.10). Let p be a projection in B, and $q = 1 - p$. Then pb and qb are in B for any $b \in B$, and in particular $\theta(p\theta(x)) = p\theta(x)$ and $\theta(q\theta(x)) = q\theta(x)$ for all $x \in A$. Using II.3.1.12, we obtain, for any $\lambda \in \mathbb{R}$ and $x \in A$,

$$(1+\lambda)^2 \|q\theta(px)\|^2 = \|q\theta(px + \lambda q\theta(px))\|^2 \leq \|px + \lambda q\theta(px)\|^2$$

$$\leq \|px\|^2 + \|\lambda q\theta(px)\|^2 = \|px\|^2 + \lambda^2 \|q\theta(px)\|^2$$

Thus $(1+2\lambda)\|q\theta(px)\|^2 \leq \|px\|^2$ for all λ, so $q\theta(px) = 0$, $\theta(px) = p\theta(px)$. Interchanging p and q in the argument, we obtain $p\theta(qx) = 0$, $p\theta(x) = p\theta(px)$. Thus $p\theta(x) = \theta(px)$ for any $x \in A$ and projection $p \in B$. Since B is generated by its projections, $\theta(bx) = b\theta(x)$ for all $x \in A$, $b \in B$.

Taking $x = 1$, we obtain that $\theta(x)$ is a projection in B which is a unit for B. Thus θ is a unital contraction from A to B and hence positive by II.6.9.4; in particular, it is self-adjoint, from which it follows that $\theta(xb) = \theta(x)b$ for all $x \in X$, $b \in B$.

Regard $B \subseteq \mathcal{L}(\mathcal{H})$ for some $\mathcal{H}$ by a faithful representation. If $a_1, \ldots, a_n$ and $x_1, \ldots, x_n$ are in A and $\xi \in \mathcal{H}$, we have, using

$$\sum_{i,j=1}^n \theta(x_i)^* a_i^* a_j \theta(x_j) = (\sum_{i=1}^n a_i \theta(x_i))^* (\sum_{i=1}^n a_i \theta(x_i)) \geq 0$$

and positivity of θ,

$$\sum_{i,j=1}^n \langle \theta(x_i)^* \theta(a_i^* a_j) \theta(x_j)\xi, \xi \rangle = \sum_{i,j=1}^n \langle \theta(\theta(x_i)^* a_i^* a_j \theta(x_j))\xi, \xi \rangle$$

$$= \langle \theta(\sum_{i,j=1}^n \theta(x_i)^* a_i^* a_j \theta(x_j))\xi, \xi \rangle \geq 0.$$

Thus θ is completely positive by II.6.9.9.

II.6.10.3 COROLLARY. Let A be a C*-algebra, B a C*-subalgebra, and $\theta : A \to B$ an idempotent positive B-linear map. Then θ is a conditional expectation.

PROOF: Expanding $\theta((x-\theta(x))^*(x-\theta(x))) \geq 0$ for $x \in A$ shows that θ satisfies Kadison's inequality and is thus a contraction (II.6.9.15). Apply II.6.10.2, or simply use the last part of its proof to conclude that θ is completely positive.

II.6.10.4 EXAMPLES.

(i) Let $(X, \mathcal{A}, \mu)$ be a probability measure space, and let $\mathcal{B}$ be a sub-σ-algebra of $\mathcal{A}$. Let $A = L^\infty(X, \mathcal{A}, \mu)$ and $B = L^\infty(X, \mathcal{B}, \mu)$. Then an ordinary conditional expectation from A to B in the sense of probability theory is a conditional expectation in the sense of II.6.10.1.

(ii) If A is unital and $B = \mathbb{C}1$, a conditional expectation of A onto B is just a state on A.

(iii) If B is a C*-subalgebra of A isomorphic to $\mathcal{L}(\mathcal{H})$ (or a finite direct sum of copies of such algebras), then the identity map from B to B extends to a conditional expectation from A onto B by II.6.9.12. In particular, if B is a finite-dimensional C*-subalgebra of a C*-algebra A, there is always a conditional expectation from A onto B. (See II.6.10.13 for a more general result.)

(iv) Let B be a finite-dimensional C*-subalgebra of a C*-algebra A, and $\mathcal{U}(B)$ its unitary group. For $x \in A$, define

$$\theta(x) = \int_{\mathcal{U}(B)} u^* x u \, d\mu(u)$$

where μ is normalized Haar measure on $\mathcal{U}(B)$. Then θ is a conditional expectation from A onto $B' \cap A$ (or $1_B(B' \cap A)$ if B is not a unital subalgebra of A).

(v) As a generalization of (iv), let G be a compact topological group acting as automorphisms $\{\alpha_g\}$ of a C*-algebra A, with $g \mapsto \alpha_g$ continuous for the point-norm topology, and let A^G (often written A^α) be the fixed-point algebra

$$A^G = \{x \in A : \alpha_g(x) = x \text{ for all } g \in G\}.$$

Define

$$\theta(x) = \int_G \alpha_g(x) \, d\mu(g)$$

where μ is normalized Haar measure on G. Then θ is a conditional expectation from A onto A^G.

See IV.2.2.14 for a closely related result.

(vi) Let X and Y be compact Hausdorff spaces, and $\pi : X \to Y$ a (necessarily finite) covering map. Regard $B = C(Y)$ as a C*-subalgebra of $A = C(X)$ by identifying g with $g \circ \pi$. For $f \in A$, define $\theta(f) \in B$ by

$$[\theta(f)](y) = \frac{1}{|\pi^{-1}(\{y\})|} \sum_{x \in \pi^{-1}(\{y\})} f(x).$$

Then θ is a conditional expectation. (If X is path-connected, then this example is a special case of (v) using the group of deck transformations.)

(vii) Other important examples come from tensor products; see II.9.7.1.

II.6.10.5 If A is a C*-algebra and B a C*-subalgebra, there is not generally a conditional expectation from A onto B; in fact, the existence of a conditional expectation is fairly exceptional. For example, if $\theta : \tilde{A} \to A$ is a conditional expectation, then for $x \in A$ we have

$$x = \theta(x) = \theta(1x) = \theta(1)x$$

so $\theta(1)$ is a unit for A and A is unital. If $A = C(X)$ is commutative and unital and $B = C(Y)$ a unital C*-subalgebra, the embedding $C(Y) \subseteq C(X)$ corresponding to a map f from X onto Y, a necessary (but not sufficient) condition for the existence of a faithful conditional expectation from A onto

B is that the cardinality of the fibers $f^{-1}(\{y\})$ $(y \in Y)$ must be lower semicontinuous, which implies that f is an open mapping. Nonfaithful conditional expectations exist in slightly greater generality; in the commutative case, a nonfaithful conditional expectation is the composition of a faithful conditional expectation and a quotient map. (Note that the first part of [Str81, 10.16] is in error.)

The simplest example of this type where there fails to be a conditional expectation is obtained by letting $X = \mathbb{Z} \cup \{\pm\infty\}$, so $A = C(X)$ is the C*-algebra of sequences $(x_n)_{n \in \mathbb{Z}}$ such that $\lim_{n \to +\infty} x_n$ and $\lim_{n \to -\infty} x_n$ exist, and B is the C*-subalgebra of sequences (x_n) such that $\lim_{n \to +\infty} x_n = \lim_{x \to -\infty} x_n$.

The C*-subalgebras of $\mathcal{L}(\mathcal{H})$ admitting conditional expectations are described in IV.2.1.4.

Conditional Expectations with Invariant State

It is clear from II.6.10.4(ii) that conditional expectations onto subalgebras are not unique in general. There are situations, however, where conditional expectations (sometimes with additional natural properties) are unique. The most important instance is when there is a faithful state commuting with the expectation. In this case, there is a nice geometric interpretation of the expectation.

II.6.10.6 Let A be a C*-algebra, B a C*-subalgebra, ϕ a faithful state on A, $\psi = \phi|_B$. Then the GNS Hilbert space $\mathcal{H}_\psi$ can be identified with a closed subspace of $\mathcal{H}_\phi$. Let ι be the natural embedding of A into $\mathcal{H}_\phi$.

II.6.10.7 PROPOSITION. Suppose $\theta : A \to B$ is a conditional expectation such that $\phi = \phi \circ \theta$. Then, if A is identified with $\iota(A)$, θ is the restriction to A of the orthogonal projection from $\mathcal{H}_\phi$ onto $\mathcal{H}_\psi$.

PROOF: Let Θ be the operator on $\iota(A)$ defined by θ, i.e. $\Theta(\iota(x)) = \iota(\theta(x))$. Since Θ is an idempotent map from $\iota(A)$ onto $\iota(B)$, it suffices to show that Θ is bounded and $\|\Theta\| \leq 1$. But, if $x \in A$,

$$\|\Theta(\iota(x))\|^2 = \|\iota(\theta(x))\|^2 = \phi(\theta(x)^*\theta(x)) \leq \phi(\theta(x^*x)) = \phi(x^*x) = \|\iota(x)\|^2$$

since $\theta(x)^*\theta(x) \leq \theta(x^*x)$ by Kadison's inequality (II.6.9.14).

In the above situation, ϕ is obviously completely determined by ψ and θ as $\phi = \psi \circ \theta$. It is not quite so obvious, but follows immediately from II.6.10.7, that θ is completely determined by ϕ:

II.6.10.8 COROLLARY. Let A be a C*-algebra, B a C*-subalgebra, and ϕ a faithful state on A. Then there is at most one conditional expectation $\theta : A \to B$ with $\phi = \phi \circ \theta$.

II.6.10.9 In the situation of the corollary, the θ does not always exist, even if there is a conditional expectation from A onto B. For example, if $A = \mathbb{M}_2$ and B the C*-subalgebra of diagonal matrices, there is a unique conditional expectation from A to B, the map which replaces the off-diagonal elements with 0's. [If θ is a conditional expectation, then

$$\theta(e_{12})^*\theta(e_{12}) \leq \theta(e_{12}^*e_{12}) = e_{22}$$

and also

$$\theta(e_{12})^*\theta(e_{12}) = \theta(e_{12})\theta(e_{12})^* \leq \theta(e_{12}e_{12}^*) = e_{11}$$

so $\theta(e_{12}) = 0$.] If ϕ is a faithful state on $\mathbb{M}_2$ which does not vanish on e_{12} (cf. II.6.3.5(i)), then there is no θ with $\phi = \phi \circ \theta$.

See III.4.7.7 for a criterion for existence in the von Neumann algebra setting.

These results and arguments generalize to weights:

II.6.10.10 PROPOSITION. Let A be a C*-algebra, B a C*-subalgebra, ϕ a faithful weight on A with $\mathfrak{N}_\phi$ dense in A. Then

(i) There is at most one conditional expectation $\theta : A \to B$ with $\phi = \phi \circ \theta$.
(ii) If θ is such a conditional expectation, then θ agrees on $\iota(\mathfrak{N}_\phi)$ with the orthogonal projection from $\mathcal{H}_\phi$ onto the closure of $\iota(\mathfrak{N}_\phi \cap B)$.

Idempotent Maps Onto Subspaces

It is sometimes useful to consider idempotent completely positive contractions whose range is not a C*-subalgebra. Such maps retain some properties of conditional expectations:

II.6.10.11 THEOREM. [CE77a, 3.1] Let A be a C*-algebra, $\theta : A \to A$ an idempotent ($\theta \circ \theta = \theta$) completely positive contraction. Then

(i) $\theta(xy) = \theta(x\theta(y)) = \theta(\theta(x)y)$ for all $x, y \in A$.
(ii) $\theta(A)$ is completely order isomorphic to a C*-algebra B; in fact, the multiplication $x \cdot y = \theta(xy)$ makes $\theta(A)$ into a C*-algebra with its involution and norm inherited from A.

There is a complementary result [CE76a, 4.1]:

II.6.10.12 THEOREM. Let A and B be C*-algebras, ω a complete order embedding of B into A, and C the C*-subalgebra of A generated by $\omega(B)$. Then

(i) There is a *-homomorphism $\phi : C \to B$ such that $\phi \circ \omega$ is the identity map on B.
(ii) If θ is any idempotent completely positive contraction from A onto $\omega(B)$, then $(\omega^{-1} \circ \theta)|_C$ is a *-homomorphism.

II.6.10.13 Finally, note that by II.6.9.12, if A and B are C*-algebras with B finite-dimensional (or, more generally, injective), and ω is a complete order embedding of B into A, then $\omega^{-1} : \omega(B) \to B$ extends to a completely positive contraction $\phi : A \to B$, and $\theta = \omega \circ \phi$ is an idempotent completely positive contraction from A onto $\omega(B)$. By II.6.10.12(ii), if C is the C*-subalgebra of A generated by $\omega(B)$, then $\phi|_C$ is a *-homomorphism.

II.7 Hilbert Modules, Multiplier Algebras, and Morita Equivalence

One of the most useful developments in modern operator algebra theory is the notion of Hilbert module, which simultaneously generalizes Hilbert spaces and C*-algebras and makes possible clean and unified treatments of several important aspects of the subject. Hilbert modules were introduced by I. Kaplansky [Kap53] in the commutative case and W. Paschke [Pas73] and M. Rieffel [Rie74] in general (see also [Saw68]), and the theory was further greatly developed by Kasparov [Kas80a].

General references: [Lan95], [RW98], [MT05].

II.7.1 Hilbert Modules

We will consider (left or right) B-modules over a C*-algebra B, which have a B-valued "inner product". All modules will be assumed to be vector spaces under a compatible scalar multiplication. Unlike the case of Hilbert spaces ($\mathbb{C}$-modules), or more generally modules over commutative algebras, it will be necessary to carefully distinguish between the "left theory" and the "right theory," although the two theories are formally identical with only a straightforward notational translation. We will concentrate our attention on right modules with just the comment that all results have exact analogs in the left module case. There are several reasons why right modules are the natural ones to work with: for example, operators can be written on the left; and almost all the literature (except about Morita equivalence) uses right modules exclusively.

II.7.1.1 DEFINITION. Let B be a C*-algebra and $\mathcal{E}$ a right B-module. A *B-valued pre-inner product* on $\mathcal{E}$ is a function $\langle \cdot, \cdot \rangle : \mathcal{E} \times \mathcal{E} \to B$ with the following properties for $\xi, \eta, \zeta \in \mathcal{E}$, $b \in B$, $\lambda \in \mathbb{C}$:

(i) $\langle \xi, \eta + \zeta \rangle = \langle \xi, \eta \rangle + \langle \xi, \zeta \rangle$ and $\langle \xi, \lambda \eta \rangle = \lambda \langle \xi, \eta \rangle$
(ii) $\langle \xi, \eta b \rangle = \langle \xi, \eta \rangle b$
(iii) $\langle \eta, \xi \rangle = \langle \xi, \eta \rangle^*$
(iv) $\langle \xi, \xi \rangle \geq 0$ (as an element of B).

If $\langle \xi, \xi \rangle = 0$ implies $\xi = 0$, then $\langle \cdot, \cdot \rangle$ is called a *B-valued inner product*.

A right B-module with a B-valued pre-inner product is called a *(right) pre-Hilbert B-module*.

II.7.1.2 If $\mathcal{E}$ is a left A-module, there is an analogous definition of an A-valued inner product with (i) replaced by linearity in the first variable and (ii) by the condition $\langle a\xi, \eta \rangle = a\langle \xi, \eta \rangle$. Conditions (iii) and (iv) remain the same. A left A-module with an A-valued inner product in this sense is called a *left pre-Hilbert A-module*. The term "pre-Hilbert module" without the word "left" will always mean a right pre-Hilbert module.

II.7.1.3 It follows from II.7.1.1(ii) and (iii) that $\langle \xi b, \eta \rangle = b^* \langle \xi, \eta \rangle$, i.e. $\langle \cdot, \cdot \rangle$ is "conjugate-linear" in the first variable. (It is truly conjugate-linear with respect to scalar multiplication.) The inner product in a left pre-Hilbert module is "conjugate-linear" in the second variable.

Just as in the scalar case, there is a CBS inequality and an induced seminorm $\|\xi\|^2 = \|\langle \xi, \xi \rangle\|$ on a pre-Hilbert B-module:

II.7.1.4 PROPOSITION. Let $\mathcal{E}$ be a pre-Hilbert B-module, $\xi, \eta \in \mathcal{E}$, $b \in B$. Then

(i) $\langle \xi, \eta \rangle^* \langle \xi, \eta \rangle \leq \|\langle \xi, \xi \rangle\| \langle \eta, \eta \rangle$ as elements of B [CBS INEQUALITY]
(ii) $\|\xi + \eta\| \leq \|\xi\| + \|\eta\|$
(iii) $\|\xi b\| \leq \|\xi\| \|b\|$.

These and similar results can be easily proved using the fact that if $\mathcal{E}$ is a pre-Hilbert B-module and ϕ is a state on B, then $(\xi, \eta)_\phi = \phi(\langle \eta, \xi \rangle)$ is an ordinary pre-inner product on $\mathcal{E}$. To prove the CBS inequality, it suffices to show that
$$\phi(\langle \xi, \eta \rangle^* \langle \xi, \eta \rangle) \leq \|\langle \xi, \xi \rangle\| \phi(\langle \eta, \eta \rangle)$$
for every state ϕ on B (II.6.3.5(iii)). Given ϕ, use II.3.1.8 and the ordinary CBS inequality $|\phi(\langle \zeta, \eta \rangle)|^2 \leq \phi(\langle \zeta, \zeta \rangle) \phi(\langle \eta, \eta \rangle)$ with $\zeta = \xi \langle \xi, \eta \rangle$. The inequality can also be proved directly in a similar manner to the usual CBS inequality (cf. [Lan95, 1.1]).

So the "seminorm" is a true seminorm, and is a norm if the pre-inner product is an inner product. $N_\mathcal{E} = \{\xi : \|\xi\| = 0\}$ is a submodule of $\mathcal{E}$, and the pre-inner product and seminorm drop to an inner product and norm on the quotient module $\mathcal{E}/N_\mathcal{E}$. $\mathcal{E}$ is called a *Hilbert B-module* if the seminorm is a norm and $\mathcal{E}$ is complete.

Unlike the case of Hilbert spaces, we do not have equality in (iii) in general (II.7.1.13).

A routine argument shows:

II.7.1.5 PROPOSITION. The completion of a pre-Hilbert B-module (with elements of norm zero divided out) is a Hilbert B-module.

II.7.1.6 In some applications, one must consider pre-Hilbert modules over an (incomplete) pre-C*-algebra B. The completion of such a module is a Hilbert module over the completion of B.

II.7.1.7 EXAMPLES.

(i) A left pre-Hilbert $\mathbb{C}$-module is just a vector space with a pre-inner product in the ordinary sense. A left Hilbert $\mathbb{C}$-module is a Hilbert space. A pre-Hilbert $\mathbb{C}$-module is a vector space with a "pre-inner product" which is conjugate-linear in the first variable (and scalar multiplication written on the right). There is an obvious identification with ordinary Hilbert spaces: if $\mathcal{H}$ is a Hilbert space, define the *conjugate Hilbert space* $\overline{\mathcal{H}}$ to be $\mathcal{H}$ with its usual addition and inner product, but with right scalar multiplication defined by $\xi\alpha = \bar{\alpha}\xi$ for $\xi \in \mathcal{H}$, $\alpha \in \mathbb{C}$.

(ii) B itself, regarded as a right B-module in the usual way, has a B-valued inner product defined by $\langle x, y \rangle = x^*y$, and is a Hilbert B-module. More generally, any right ideal of B becomes a pre-Hilbert B-module in this manner, and is complete if and only if it is closed. Similarly, if A is a C*-algebra regarded as a left A module, then A (or any closed left ideal of A) is a left Hilbert A-module with the inner product $\langle x, y \rangle = xy^*$.

(iii) If V is a complex vector bundle over a compact Hausdorff space X with a Hermitian structure (i.e. a locally trivial bundle of finite-dimensional Hilbert spaces over X), then the set $\Gamma(V)$ of continuous sections has a natural structure as a $C(X)$-module, and the pointwise inner product makes it into a Hilbert $C(X)$-module. (The local triviality and finite-dimensionality are not essential here.)

(iv) Any (algebraic) direct sum $\bigoplus \mathcal{E}_i$ of [pre-]Hilbert B-modules becomes a [pre-]Hilbert B-module with inner product

$$\langle \oplus \xi_i, \oplus \eta_i \rangle = \sum_i \langle \xi_i, \eta_i \rangle.$$

A finite direct sum of Hilbert B-modules is complete, but an infinite direct sum will not be.

(v) If $\mathcal{E}$ is a Hilbert B-module, the completion of a direct sum of a sequence of copies of $\mathcal{E}$ is denoted $\mathcal{E}^\infty$ (we can also form $\mathcal{E}^n$). B^∞ is usually denoted $\mathcal{H}_B$.

$$\mathcal{H}_B = \{(x_1, x_2, \cdots) : x_n \in B, \sum x_n^* x_n \text{ converges in } B\}$$

with inner product

$$\langle (x_1, x_2, \cdots), (y_1, y_2, \cdots) \rangle = \sum x_n^* y_n.$$

(An important technical point: $\mathcal{H}_B$ is more than just the sequences (x_n) for which $\sum \|x_n\|^2 < \infty$ in general.)

II.7.1.8 In many respects, Hilbert modules behave much like Hilbert spaces, and constructions and arguments can be carried over almost verbatim. But there is one dramatic exception: orthogonality. Orthogonality in a Hilbert module can be defined in the same way as in a Hilbert space, but is relatively rare in general: for example, it is easy to find examples of a proper closed B-submodule $\mathcal{F}$ of a Hilbert B-module $\mathcal{E}$ such that $\mathcal{F}^\perp = \{0\}$, and thus $\mathcal{F}^{\perp\perp}$ is strictly larger than $\mathcal{F}$. (In fact, it is relatively rare that a closed submodule has a complementary closed submodule or even a closed complementary vector subspace.) Thus most arguments using orthogonality must be discarded in the Hilbert module case.

Since B may be nonunital, it is not obvious that a Hilbert B-module $\mathcal{E}$ satisfies $\mathcal{E} = \mathcal{E}B = \{\xi b : \xi \in \mathcal{E}, b \in B\}$, but is true:

II.7.1.9 PROPOSITION. If $\mathcal{E}$ is a Hilbert B-module, then $\mathcal{E}B = \mathcal{E}$. If $\xi \in \mathcal{E}$, then $\|\xi\| = \sup\{\|\langle \xi, \eta b \rangle\| : \eta \in \mathcal{E}, b \in B, \|\eta\| = \|b\| = 1\}$.

PROOF: By II.5.3.7, it suffices to show that if (h_λ) is an approximate unit for B, then for any ξ, $\xi h_\lambda \to \xi$, so $\mathcal{E}B$ is dense in $\mathcal{E}$. But

$$\langle \xi - \xi h_\lambda, \xi - \xi h_\lambda \rangle = \langle \xi, \xi \rangle - h_\lambda \langle \xi, \xi \rangle - \langle \xi, \xi \rangle h_\lambda + h_\lambda \langle \xi, \xi \rangle h_\lambda \to 0.$$

One can alternatively show the following, which is strongly reminiscent of II.3.2.1 (and, from the right point of view, essentially the same result):

II.7.1.10 PROPOSITION. Let $\mathcal{E}$ be a Hilbert B-module, $\xi \in \mathcal{E}$, $\|\xi\| = 1$. Then for any α, $0 < \alpha < 1/2$, there is a $\eta \in \mathcal{E}$ with $\|\eta\| = 1$ and $\xi = \eta[\langle \xi, \xi \rangle]^\alpha$.

The proof is very similar to the proof of II.3.2.1: η is the limit of the norm-convergent sequence

$$\xi[(1/n) + b]^{-1/2} b^{1/2-\alpha}$$

where $b = \langle \xi, \xi \rangle$.

See [RW98, 2.31] for a related result with a slick proof.

A significant fact is G. Kasparov's Stabilization or Absorption Theorem [Kas80a] (first proved in the commutative case by J. Dixmier and A. Douady [DD63]; cf. also [Bro77]), which says that $\mathcal{H}_B$ is the "universal" Hilbert B-module:

II.7 Hilbert Modules, Multiplier Algebras, and Morita Equivalence

II.7.1.11 THEOREM. [STABILIZATION OR ABSORPTION] Let B be a C*-algebra, and $\mathcal{E}$ a countably generated Hilbert B-module. Then $\mathcal{E} \oplus \mathcal{H}_B \cong \mathcal{H}_B$.

The best proof, due to J. Mingo and W. Phillips [MP84], uses polar decomposition; see [Bla98, 13.6.2] or [Lan95, 6.2].

Note, however, that $\mathcal{H}_B$ is not itself countably generated unless B is σ-unital (II.4.2.4). There are some technical complications in the theory of Hilbert B-modules when B is not σ-unital, so in applications of the theory we often restrict to the σ-unital case (which of course includes the case of separable B).

II.7.1.12 There is another technical complication in Hilbert module theory which turns out to be a blessing in disguise. If $\mathcal{E}$ is a Hilbert B-module, then $\{\langle \xi, \eta \rangle : \xi, \eta \in \mathcal{E}\}$ is not generally closed under addition, and its closed linear span $J_\mathcal{E}$ is an ideal of B which is not all of B in general. If $J_\mathcal{E} = B$, then $\mathcal{E}$ is said to be a *full* Hilbert B-module. $\mathcal{E}$ can be regarded as a full Hilbert $J_\mathcal{E}$-module, and $\mathcal{E} J_\mathcal{E} = \mathcal{E}$ by II.7.1.9. The Hilbert $J_\mathcal{E}$-module structure is the natural one on $\mathcal{E}$, and the B-module structure on $\mathcal{E}$ can be recovered from the $J_\mathcal{E}$-module structure. In fact, if $J_\mathcal{E}$ is an ideal in a C*-algebra D, then the Hilbert $J_\mathcal{E}$-module structure extends uniquely to a Hilbert D-module structure [by a calculation as in the proof of II.7.1.9, if (h_λ) is an approximate unit for $J_\mathcal{E}$, then for any $\xi \in \mathcal{E}$ and $d \in D$, $(\xi h_\lambda d)$ is a Cauchy net in $\mathcal{E}$ and its limit can be called ξd.] This observation leads to the theory of multiplier algebras via Hilbert modules (II.7.3).

II.7.1.13 Note that if $\mathcal{E}$ is a Hilbert B-module, $\xi \in \mathcal{E}$, $b \in B$, then we do not have equality in II.7.1.4(iii) in general (e.g. in example II.7.1.7(ii)). It is not even true that if $b \in B$, then $\xi b = 0$ for all $\xi \in \mathcal{E}$ implies $b = 0$; it only implies that b is orthogonal to $J_\mathcal{E}$ (cf. II.7.3). If $\mathcal{E}$ is a full Hilbert B-module and $\xi b = 0$ for all $\xi \in \mathcal{E}$, then $b = 0$.

II.7.2 Operators

The appropriate analogs of bounded operators for Hilbert modules are the adjointable operators.

II.7.2.1 DEFINITION. Let B be a C*-algebra and $\mathcal{E}$ and $\mathcal{F}$ Hilbert B-modules. A function $T : \mathcal{E} \to \mathcal{F}$ is an *adjointable operator* if there is a function $T^* : \mathcal{F} \to \mathcal{E}$ such that

$$\langle T\xi, \eta \rangle_\mathcal{F} = \langle \xi, T^*\eta \rangle_\mathcal{E}$$

for all $\xi \in \mathcal{E}$, $\eta \in \mathcal{F}$. The operator T^* is called the *adjoint* of T. Denote by $\mathcal{L}(\mathcal{E}, \mathcal{F})$ the set of adjointable operators from $\mathcal{E}$ to $\mathcal{F}$. Write $\mathcal{L}(\mathcal{E})$ for $\mathcal{L}(\mathcal{E}, \mathcal{E})$.

Some experts justifiably object to the term "adjointable" as a language abomination, but the term is unfortunately now well established.

An adjointable operator is automatically $\mathbb{C}$-linear and even B-linear (a simple calculation using the linearity of the inner products) and bounded (by the Closed Graph Theorem). Not every bounded module morphism is adjointable, however; for example, the inclusion of a noncomplemented submodule is not adjointable. $\mathcal{L}(\mathcal{E}, \mathcal{F})$ is a Banach space under the usual operator norm; and it is easily checked that $\mathcal{L}(\mathcal{E})$ is a C*-algebra with this norm and involution (the proof is essentially identical to the Hilbert space case).

Since the definition of $\mathcal{L}(\mathcal{E}, \mathcal{F})$ is given without explicit reference to the B-module structure, the same set of adjointable operators is obtained no matter which C*-algebra $\mathcal{E}$ and $\mathcal{F}$ are regarded as Hilbert modules over (cf. II.7.1.12) so long as the inner products on $\mathcal{E}$ and $\mathcal{F}$ are unchanged.

By definition of the operator norm, if $T \in \mathcal{L}(\mathcal{E}, \mathcal{F})$ and $\xi \in \mathcal{E}$, then $\|T\xi\| \leq \|T\|\|\xi\|$. But there is a more precise inequality which follows from the fact that $\mathcal{L}(\mathcal{E})$ is a C*-algebra:

II.7.2.2 PROPOSITION. Let $\mathcal{E}$ be a Hilbert B-module, $T \in \mathcal{L}(\mathcal{E})$, and $\xi \in \mathcal{E}$. Then $\langle T\xi, T\xi \rangle \leq \|T\|^2 \langle \xi, \xi \rangle$ as elements of B.

Since $S = \|T\|^2 I - T^*T \geq 0$ in $\mathcal{L}(\mathcal{E})$, we have

$$\|T\|^2 \langle \xi, \xi \rangle - \langle T\xi, T\xi \rangle = \langle S\xi, \xi \rangle = \langle S^{1/2}\xi, S^{1/2}\xi \rangle \geq 0.$$

This is also an easy corollary of II.7.1.10. We may assume $\|\xi\| = 1$; for $\alpha < 1/2$ let $\eta_\alpha \in \mathcal{E}$ with $\|\eta_\alpha\| = 1$ and $\xi = \eta_\alpha b^\alpha$, where $b = \langle \xi, \xi \rangle$. Then

$$\langle T\xi, T\xi \rangle = b^\alpha \langle T\eta_\alpha, T\eta_\alpha \rangle b^\alpha \leq \|T\|^2 b^{2\alpha}$$

and the inequality follows by letting $\alpha \to 1/2$. (This argument also works if T is just a bounded module-homomorphism.)

II.7.2.3 COROLLARY. Let $\mathcal{E}$ be a Hilbert B-module, $T \in \mathcal{L}(\mathcal{E})$, and ϕ a state on B. Define the pre-inner product $(\cdot, \cdot)_\phi$ on $\mathcal{E}$ as in II.7.1.4, and let $\mathcal{H}_\phi$ be the completion (Hilbert space). Then T extends to a bounded operator T_ϕ on $\mathcal{H}_\phi$, with $\|T_\phi\| \leq \|T\|$.

II.7.2.4 In analogy with the Hilbert space case, there are special "rank-one" operators in $\mathcal{L}(\mathcal{E}, \mathcal{F})$. If $\eta \in \mathcal{E}, \xi \in \mathcal{F}$, define $\Theta_{\xi,\eta}$ by $\Theta_{\xi,\eta}(\zeta) = \xi \langle \eta, \zeta \rangle$. Then $\Theta^*_{\xi,\eta} = \Theta_{\eta,\xi}$, so $\Theta_{\xi,\eta} \in \mathcal{L}(\mathcal{E}, \mathcal{F})$. If $T \in \mathcal{L}(\mathcal{F}, \mathcal{G})$, then $T\Theta_{\xi,\eta} = \Theta_{T\xi,\eta}$; and $\Theta_{\xi,\eta} T = \Theta_{\xi, T^*\eta}$ for $T \in \mathcal{L}(\mathcal{G}, \mathcal{E})$. Let $\mathcal{K}(\mathcal{E}, \mathcal{F})$ be the closed linear span of $\{\Theta_{\xi,\eta} : \eta \in \mathcal{E}, \xi \in \mathcal{F}\}$, and $\mathcal{K}(\mathcal{E}) = \mathcal{K}(\mathcal{E}, \mathcal{E})$. The set $\mathcal{K}(\mathcal{E})$ is a closed ideal in the C*-algebra $\mathcal{L}(\mathcal{E})$. The set $\mathcal{K}(\mathcal{E}, \mathcal{F})$ is often regarded as the set of "compact" operators from $\mathcal{E}$ to $\mathcal{F}$, although these operators will generally not be compact in the sense of I.8.1.1. In fact, we can sometimes have $I \in \mathcal{K}(\mathcal{E})$ (II.7.2.6(i)) (I of course denotes the identity map on $\mathcal{E}$; $I \in \mathcal{L}(\mathcal{E})$ and $I^* = I$); since $\mathcal{K}(\mathcal{E})$ is an ideal in $\mathcal{L}(\mathcal{E})$, this happens if and only if $\mathcal{K}(\mathcal{E}) = \mathcal{L}(\mathcal{E})$.

Note that the $\mathcal{K}(\mathcal{E})$-valued inner product $\langle \xi, \eta \rangle = \Theta_{\xi,\eta}$ makes $\mathcal{E}$ into a full left Hilbert $\mathcal{K}(\mathcal{E})$-module (cf. II.7.6.5(ii)).

A simple but useful technical fact (cf. [Bla98, 13.6.3], [Lan95, 6.7]) is:

II.7.2.5 PROPOSITION. Let $\mathcal{E}$ be a Hilbert B-module. Then $\mathcal{E}$ is countably generated if and only if $\mathcal{K}(\mathcal{E})$ is σ-unital.

II.7.2.6 EXAMPLES.

(i) If B is regarded as a Hilbert B-module, and $x, y \in B$, then
$$\Theta_{x,y} z = x \langle y, z \rangle = x y^* z$$
for any z, so $\Theta_{x,y}$ is left multiplication by xy^*. It is easily checked that $\|\Theta_{x,y}\| = \|xy^*\|$. Since $B = B^2$, we conclude that $\mathcal{K}(B) \cong B$, where B acts on itself by left multiplication. If B is unital, then $\Theta_{1,1} = I$, so $\mathcal{K}(B) = \mathcal{L}(B) \cong B$. (If B is nonunital, then of course $\mathcal{K}(B) \neq \mathcal{L}(B)$ since $I \in \mathcal{L}(B) \setminus \mathcal{K}(B)$.)

(ii) If $\mathcal{E}$ and $\mathcal{F}$ are Hilbert B-modules, then $\mathcal{L}(\mathcal{E}^n, \mathcal{F}^m)$ and $\mathcal{K}(\mathcal{E}^n, \mathcal{F}^m)$ can be naturally identified with the set of $m \times n$ matrices over $\mathcal{L}(\mathcal{E}, \mathcal{F})$ and $\mathcal{K}(\mathcal{E}, \mathcal{F})$ respectively, with the usual involution. So $\mathcal{L}(\mathcal{E}^n) \cong M_n(\mathcal{L}(\mathcal{E}))$ and $\mathcal{K}(\mathcal{E}^n) \cong M_n(\mathcal{K}(\mathcal{E}))$.

(iii) If $\mathcal{E}$ is a Hilbert B-module, then $\mathcal{L}(\mathcal{E}^\infty)$ can be identified with an algebra of infinite matrices over $\mathcal{L}(\mathcal{E})$, although it is difficult if not impossible to characterize which matrices give elements of $\mathcal{L}(\mathcal{E}^\infty)$. $\mathcal{K}(\mathcal{E}^\infty)$ is the closure of $M_\infty(\mathcal{K}(\mathcal{E}))$ (II.6.6.9), so is isomorphic to $\mathcal{K}(\mathcal{E}) \otimes \mathbb{K}$ as defined in II.6.6.11. In particular, $\mathcal{K}(\mathcal{H}_B) \cong B \otimes \mathbb{K}$.

(iv) If $\mathcal{E}, \mathcal{F}$ are Hilbert B-modules, then $\mathcal{K}(\mathcal{E}, \mathcal{F})$ is a left $\mathcal{K}(\mathcal{F})$-module by left multiplication, and has a $\mathcal{K}(\mathcal{F})$-valued inner product defined by $\langle S, T \rangle = ST^*$, making it a left Hilbert $\mathcal{K}(\mathcal{F})$-module.

II.7.2.7 Unlike in $\mathcal{L}(\mathcal{H})$, operators in $\mathcal{L}(\mathcal{E})$ do not have support projections or polar decomposition in general. In fact, since $\mathcal{L}(\mathcal{E})$ can be any unital C*-algebra, these should not be expected. From a Hilbert module point of view, the problem is that the closure of the range of an adjointable operator need not have an orthogonal complement. Even an isometric B-module map need not be adjointable (if the range is uncomplemented). It is true, although a nontrivial result [Lan95, 3.5], that an isometric B-module map T from a Hilbert B-module $\mathcal{E}$ *onto* a Hilbert B-module $\mathcal{F}$ is adjointable and unitary ($T^*T = I_\mathcal{E}$, $TT^* = I_\mathcal{F}$). There is also a result due to A. Miščenko [Miš79] (cf. [Lan95, 3.2]):

II.7.2.8 THEOREM. Let $\mathcal{E}$ and $\mathcal{F}$ be Hilbert B-modules, and $T \in \mathcal{L}(\mathcal{E}, \mathcal{F})$ with closed range. Then T^* also has closed range; $\mathcal{N}(T)^\perp = \mathcal{R}(T^*)$, $\mathcal{R}(T^*)^\perp = \mathcal{N}(T)$ (and hence $\mathcal{N}(T^*)^\perp = \mathcal{R}(T)$, $\mathcal{R}(T)^\perp = \mathcal{N}(T^*)$), and T has left and right support projections and polar decomposition.

Compare with II.3.2.10: an element of $\mathcal{L}(\mathcal{E})$ which is well-supported in the sense of II.3.2.8 has closed range. Conversely, it follows from the Open Mapping Theorem that if T has closed range, then T is bounded below on $\mathcal{N}(T)^\perp$ and hence is well-supported.

II.7.2.9 $\mathcal{L}(\mathcal{E})$ for a Hilbert B-module $\mathcal{E}$ (and, more generally, $\mathcal{L}(\mathcal{E},\mathcal{F})$) has various other topologies analogous to the topologies on $\mathcal{L}(\mathcal{H})$: weak, σ-weak, strong, σ-strong, strong-*, and σ-strong-*. The last two are the most important. The strong-* topology on $\mathcal{L}(\mathcal{E})$ is the topology of pointwise *-convergence on $\mathcal{E}$, and is generated by the seminorms $T \mapsto \|T\xi\|$ and $T \mapsto \|T^*\xi\|$ for $\xi \in \mathcal{E}$; the σ-strong-* is the topology of pointwise convergence as left and right multipliers of $\mathcal{K}(\mathcal{E})$, and is generated by the seminorms $T \mapsto \|TS\|$ and $T \mapsto \|ST\|$ for $S \in \mathcal{K}(\mathcal{E})$. The term *strict topology* is often used for these topologies; unfortunately, this term is used in the literature for both topologies, sometimes even by the same author. The two topologies agree on bounded sets, so the ambiguity is not too serious; but for consistency we will use the term "strict topology" only to mean the σ-strong-* topology. (The term comes from the theory of multiplier algebras (II.7.3), where it definitely means the σ-strong-* topology.)

Just as in $\mathcal{L}(\mathcal{H})$, multiplication is separately strictly continuous, and jointly strictly continuous on bounded sets.

It is easily shown that if (H_λ) is an approximate unit for $\mathcal{K}(\mathcal{E})$, then $H_\lambda \to I$ strictly, and it follows that $\mathcal{K}(\mathcal{E},\mathcal{F})$ is strictly dense in $\mathcal{L}(\mathcal{E},\mathcal{F})$ for any $\mathcal{E},\mathcal{F}$. There is also a version of the Kaplansky Density Theorem [Lan95, 1.4]: if A is a *-subalgebra of $\mathcal{L}(\mathcal{E})$ with strict closure D, then the unit ball of A is strictly dense in the unit ball of D.

II.7.3 Multiplier Algebras

If A is a C*-algebra (especially a nonunital one), we now examine the ways in which A can be embedded as an ideal in a unital C*-algebra D. If $A = C_o(X)$ is commutative, this corresponds to the ways in which the locally compact Hausdorff space X can be embedded as an open set in a compact Hausdorff space Y. There is a minimal way to do this: let $Y = X^\dagger$ be the one-point compactification. The C*-analog is to take $D = A^\dagger$ (II.1.2.1). In the commutative case, there is a "maximal" compactification, the Stone-Čech compactification βX. There is a C*-analog here, too, called the multiplier algebra of A.

II.7.3.1 THEOREM. *Let A be a C*-algebra. Then there is a unital C*-algebra $M(A)$ containing A as an essential ideal (II.5.4.7), which is universal in the sense that whenever A sits as an ideal in a C*-algebra D, the identity map on A extends uniquely to a *-homomorphism from D into $M(A)$ with kernel $A^\perp$ (II.5.4.7). (Thus $M(A)$ is unique up to isomorphism over A.) $M(A)$ is called the multiplier algebra of A.*

II.7.3.2 There are several ways of constructing $M(A)$ and proving the theorem. The original construction and proof, due to R. Busby [Bus68], used double centralizers. A *double centralizer* of a C*-algebra A is a pair (L,R) of linear maps from A to A satisfying $xL(y) = R(x)y$ for all $x,y \in A$ [think of L

II.7 Hilbert Modules, Multiplier Algebras, and Morita Equivalence 145

and R as left and right multiplication by an element of $M(A)$.] More generally, a *left centralizer* of A is a linear map $L : A \to A$ satisfying $L(xy) = L(x)y$ for all $x, y \in A$. Right centralizers are defined analogously.

II.7.3.3 PROPOSITION. [Ped79, 3.12.2] Every left (or right) centralizer of a C*-algebra is bounded. PROOF: Suppose not. Let L be an unbounded left multiplier of A, and $x_n \in A$ with $\|x_n\| < 1/n$ but $\|L(x_n)\| > n$. Let $a = \sum x_n x_n^*$. Then $x_n = a^{1/3} u_n$ for some $u_n \in A$ with $\|u_n\| \leq \|a^{1/6}\|$ (II.3.2.2), so

$$\|L(x_n)\| = \|L(a^{1/3})u_n\| \leq \|L(a^{1/3})\|\|a^{1/6}\|$$

for all n, a contradiction. The proof for right centralizers is similar.

II.7.3.4 If (L, R) is a double centralizer of A, then L is a left centralizer [if (h_λ) is an approximate unit for A, and $x, y \in A$, then $h_\lambda L(xy) = R(h_\lambda)xy = h_\lambda L(x)y$ for each λ], and similarly R is a right centralizer. Thus L and R are necessarily bounded; and it is automatic that $\|L\| = \|R\|$. Let

$$(L_1, R_1)(L_2, R_2) = (L_2 L_1, R_1 R_2)$$

and $(L, R)^* = (R^*, L^*)$, where $T^*(x) = T(x^*)^*$. Then the set of double centralizers with coordinatewise addition and this multiplication, and operator norm, is a C*-algebra $M(A)$. A embeds via $a \mapsto (L_a, R_a)$, where $L_a(x) = ax$ and $R_a(x) = xa$. If A is an ideal of D, then in the same way an element $d \in D$ defines $(L_d, R_d) \in M(A)$.

II.7.3.5 A second traditional approach is to take a faithful nondegenerate representation π of A on $\mathcal{H}$, and take $M(A)$ to be the idealizer

$$\{T \in \mathcal{L}(\mathcal{H}) : T\pi(A) \subseteq \pi(A), \pi(A)T \subseteq \pi(A)\}.$$

This approach, whose proof follows easily from II.6.1.6, will be subsumed in II.7.3.9. But there is an interesting consequence:

II.7.3.6 PROPOSITION. [Ped79, 3.12.3] Let A be a concrete C*-algebra of operators acting nondegenerately on a Hilbert space $\mathcal{H}$. If L is a left centralizer of A, then there is a unique $x \in A''$ with $L(y) = xy$ for all $y \in A$.

PROOF: Let (h_λ) be an increasing approximate unit for A. Then the net $(L(h_\lambda))$ is bounded (II.7.3.3), and hence has a weak limit point $x \in A''$. If $y \in A$, then $L(h_\lambda)y = L(h_\lambda y) \to L(y)$ in norm, hence weakly; but xy is a weak limit point of $(L(h_\lambda)y)$, so $L(y) = xy$. Uniqueness is obvious, since if $L(y) = zy$ for all $y \in A$, then $(x - z)h_\lambda = 0$ for all λ. But $h_\lambda \to 1$ strongly.

II.7.3.7 COROLLARY. $\mathcal{L}(\mathcal{H})$ is σ-strongly complete.

PROOF: Let (x_i) be a σ-strong Cauchy net in $\mathcal{L}(\mathcal{H})$. Then for each $y \in \mathcal{K}(\mathcal{H})$, $(x_i y)$ is a norm Cauchy net. Define $L(y)$ to be the limit. Then L is a left multiplier of $\mathcal{K}(\mathcal{H})$, so there is an $x \in \mathcal{L}(\mathcal{H})$ with $L(y) = xy$ for all $y \in \mathcal{K}(\mathcal{H})$, and $x_i \to x$ σ-strongly.

Multiplier Algebras Via Hilbert Modules

II.7.3.8 The most elegant approach to multiplier algebras is to use Hilbert modules. If $\mathcal{E}$ is a Hilbert B-module, then a *-homomorphism π from a C*-algebra A into $\mathcal{L}(\mathcal{E})$ is *nondegenerate* if $\pi(A)\mathcal{E}$ is dense in $\mathcal{E}$. This is easily shown to be equivalent to the statement that $\pi(h_\lambda) \to I$ strictly for some (hence every) approximate unit for A. (Actually, if π is nondegenerate, then $\pi(A)\mathcal{E} = \mathcal{E}$ (II.5.3.7).)

II.7.3.9 THEOREM. Let $\mathcal{E}$ be a Hilbert B-module, D a C*-algebra with closed ideal A, and $\pi : A \to \mathcal{L}(\mathcal{E})$ a nondegenerate *-homomorphism. Then π extends uniquely to a *-homomorphism from D into $\mathcal{L}(\mathcal{E})$. If π is faithful on A, then the kernel of the extended map is $A^\perp$. In particular, the idealizer of $\pi(A)$ in $\mathcal{L}(\mathcal{E})$ is isomorphic to the multiplier algebra of A.

The proof (cf. II.6.1.6) consists of observing that, if $d \in D$, $a_1, \cdots, a_n \in A$, $\xi_1, \cdots, \xi_n \in \mathcal{E}$, and (h_λ) is an approximate unit for A, then

$$\|\sum_i \pi(da_i)\xi_i\| = \lim_\lambda \|\sum_i \pi(dh_\lambda a_i)\xi_i\|$$

$$= \lim_\lambda \|\pi(dh_\lambda)\sum_i \pi(a_i)\xi_i\| \leq \|d\|\|\sum_i \pi(a_i)\xi_i\|$$

so $\sum_i \pi(a_i)\xi_i \mapsto \sum_i \pi(da_i)\xi_i$ is well defined and bounded, and has an adjoint $\sum_i \pi(a_i)\xi_i \mapsto \sum_i \pi(d^*a_i)\xi_i$.

Applying the theorem to $A = \mathcal{K}(\mathcal{E})$, we get:

II.7.3.10 COROLLARY. If $\mathcal{E}$ is a Hilbert B-module, then $\mathcal{L}(\mathcal{E})$ is the multiplier algebra of $\mathcal{K}(\mathcal{E})$.

II.7.3.11 Theorem II.7.3.1 may be viewed as the special case of this where $\mathcal{E} = A$ viewed as a Hilbert A-module and A is identified with $\mathcal{K}(A)$. The *strict topology* on $M(A)$ is the strict topology in the previous sense when $M(A)$ is identified with $\mathcal{L}(A)$ (actually, in this case the strong-* and σ-strong-* topologies coincide, so there is no ambiguity.) The strict topology is generated by the seminorms $x \mapsto \|ax\|$ and $x \mapsto \|xa\|$ for $a \in A$, i.e. $x_i \to x$ strictly in $M(A)$ if and only if $ax_i \to ax$ and $x_ia \to xa$ in norm for all $a \in A$.

In fact, $M(A)$ is the strict completion of A [Bus68], and in particular is strictly complete. Thus, for example, $\mathcal{L}(\mathcal{H})$ is complete in the σ-strong-* topology. The proof is almost identical to the one in II.7.3.7.

II.7.3.12 EXAMPLES.

(i) If A is unital, then $M(A) = A$, and the strict topology is the norm topology. $M(A)$ is always unital, so $M(A) \neq A$ if A is nonunital. (In fact, $M(A)$ is much larger than A in general: it is always nonseparable if A is nonunital.)

(ii) $M(\mathbb{K}) \cong \mathcal{L}(\mathcal{H})$. The strict topology is the σ-strong-* operator topology.

(iii) If X is a locally compact Hausdorff space, then $M(C_o(X)) \cong C(\beta X)$, which is isomorphic to the C*-algebra $C_b(X)$ of bounded continuous complex-valued functions on X (in fact, a good construction of βX is as the primitive ideal space of this C*-algebra.) On bounded sets, the strict topology is the topology of uniform convergence on compact subsets of X.

(iv) [APT73, 3.4] Generalizing (ii), if B is any C*-algebra and X is a locally compact Hausdorff space, then $M(C_o(X, B))$ is isomorphic to the C*-algebra of strictly continuous functions from βX to $M(B)$.

II.7.3.13 The main statement of Theorem II.7.3.9 remains true in greater generality: the *-homomorphism π is called *strict* if, for an approximate unit (h_λ) for A, the net $(\pi(h_\lambda))$ converges strictly to an operator $P \in \mathcal{L}(\mathcal{E})$ (which is necessarily a projection). The extended homomorphism from D goes into $P\mathcal{L}(\mathcal{E})P$. But the conclusion of Theorem II.7.3.9 does not hold for a general *-homomorphism π from A to $\mathcal{L}(\mathcal{E})$. In general, a *-homomorphism from A to B does not extend to a *-homomorphism from $M(A)$ to $M(B)$ (i.e. $A \mapsto M(A)$ is not functorial). The problems are that a *-homomorphism need not be strictly continuous, and the submodule $(\pi(A)\mathcal{E})^-$ need not be complemented. For example, the inclusion of a nonunital hereditary C*-subalgebra A of a separable unital C*-algebra B does not extend to a *-homomorphism from $M(A)$ into $M(B) = B$.

II.7.3.14 If p is a projection in $M(A)$, the hereditary C*-subalgebra pAp of A is called a *corner*. Corners are the "nicely embedded" hereditary C*-subalgebras. The corner pAp has a *complementary corner* $(1-p)A(1-p)$. The inclusion of pAp into $M(A)$ is strict, so by the extended version of II.7.3.9 we get that $M(pAp) \cong pM(A)p$.

II.7.4 Tensor Products of Hilbert Modules

There is a way of tensoring Hilbert modules which generalizes the tensor product of Hilbert spaces. In fact, there are two appropriate notions which are both useful. Here we discuss the "internal tensor product"; there is also an "external tensor product" (cf. [Bla98, 13.5] or [Lan95, §4]).

II.7.4.1 If $\mathcal{E}$ is a Hilbert A-module and $\mathcal{F}$ a Hilbert B-module, and $\phi : A \to \mathcal{L}(\mathcal{F})$ is a *-homomorphism, then $\mathcal{F}$ can be regarded as an $A - B$ bimodule via ϕ, and we can form the algebraic tensor product $\mathcal{E} \odot_A \mathcal{F}$ of $\mathcal{E}$ and $\mathcal{F}$ over A as the quotient of the tensor product vector space $\mathcal{E} \odot_\mathbb{C} \mathcal{F}$ by the subspace spanned by

$$\{\xi a \otimes \eta - \xi \otimes \phi(a)\eta : \xi \in \mathcal{E}, \eta \in \mathcal{F}, a \in A\}.$$

Then $\mathcal{E} \odot_A \mathcal{F}$ is a right B-module in the obvious way: $(\xi \otimes \eta)b = \xi \otimes \eta b$. Define a B-valued pre-inner product on $\mathcal{E} \odot_A \mathcal{F}$ by

$$\langle \xi_1 \otimes \eta_1, \xi_2 \otimes \eta_2 \rangle = \langle \eta_1, \phi(\langle \xi_1, \xi_2 \rangle_A) \eta_2 \rangle_B.$$

(One must check that this pre-inner product is well defined and positive; in fact, it is actually an inner product. Note the similarity with the Stinespring construction.) The completion of $\mathcal{E} \odot_A \mathcal{F}$ with respect to this inner product is a Hilbert B-module called the *(internal) tensor product* of $\mathcal{E}$ and $\mathcal{F}$, denoted $\mathcal{E} \otimes_\phi \mathcal{F}$ (or just $\mathcal{E} \otimes_A \mathcal{F}$ when there is no ambiguity about the map ϕ).

The tensor product is associative and distributive over direct sums in the obvious way.

II.7.4.2 EXAMPLES.

(i) If $\mathcal{F}$ is any Hilbert B-module, then $\mathbb{C}^n \otimes_\mathbb{C} \mathcal{F} \cong \mathcal{F}^n$. If $\mathcal{H}$ is a separable infinite-dimensional Hilbert space, regarded as a Hilbert $\mathbb{C}$-module, then $\mathcal{H} \otimes_\mathbb{C} \mathcal{F} \cong \mathcal{F}^\infty$. (The homomorphism from $\mathbb{C}$ to $\mathcal{L}(\mathcal{F})$ is the obvious unital one.) In particular, $\mathbb{C}^n \otimes_\mathbb{C} B \cong B^n$ and $\mathcal{H} \otimes_\mathbb{C} B \cong \mathcal{H}_B$.

(ii) If $\mathcal{F}$ is a Hilbert B-module and $\phi : A \to \mathcal{L}(\mathcal{F})$ is a nondegenerate *-homomorphism, then $A \otimes_\phi \mathcal{F} \cong \mathcal{F}$. Similarly, $A^n \otimes_\phi \mathcal{F} \cong \mathcal{F}^n$, $\mathcal{H}_A \otimes_\phi \mathcal{F} \cong \mathcal{F}^\infty$.

II.7.4.3 If $S \in \mathcal{L}(\mathcal{E})$, then there is a natural operator $S \otimes I \in \mathcal{L}(\mathcal{E} \otimes_\phi \mathcal{F})$, defined by

$$(S \otimes I)(\xi \otimes \eta) = S\xi \otimes \eta.$$

Thus there is a *-homomorphism from $\mathcal{L}(\mathcal{E})$ into $\mathcal{L}(\mathcal{E} \otimes_\phi \mathcal{F})$. (Note that this homomorphism does not send $\mathcal{K}(\mathcal{E})$ into $\mathcal{K}(\mathcal{E} \otimes_\phi \mathcal{F})$ in general.) However, unlike in the Hilbert space case, if $T \in \mathcal{L}(\mathcal{F})$, there is no reasonable definition of $I \otimes T \in \mathcal{L}(\mathcal{E} \otimes_\phi \mathcal{F})$ in general unless T commutes with $\phi(A)$.

Correspondences

II.7.4.4 If A and B are C*-algebras and $\mathcal{E}$ is a Hilbert B-module, a homomorphism $\phi : A \to \mathcal{L}(\mathcal{E})$ can be regarded as a "generalized homomorphism" from A to B. Such a tuple $(A, B, \mathcal{E}, \phi)$ is called a *correspondence* (or *C*-correspondence*) from A to B. An important example is a (not necessarily strict) homomorphism from A to $M(B)$. In fact, in general $\mathcal{K}(\mathcal{E})$ is a C*-algebra which is a "rescaled" version of an ideal of B (a precise statement uses Morita equivalence; see II.7.6.5(ii) and II.7.6.13), and ϕ is a homomorphism from A to $M(\mathcal{K}(\mathcal{E}))$.

The correspondence $(A, B, \mathcal{E}, \phi)$ is often abbreviated as the bimodule ${}_A\mathcal{E}_B$; the Hilbert B-module structure is understood. Note that there is no left Hilbert A-module structure in general. An $A - B$-imprimitivity bimodule (II.7.6.1) is a special kind of correspondence.

Correspondences can be composed: if $(A, B, \mathcal{E}, \phi)$ and $(B, D, \mathcal{F}, \psi)$ are correspondences from A to B and B to D respectively, then

$$(A, D, \mathcal{E} \otimes_\psi \mathcal{F}, \phi \otimes I)$$

is a correspondence from A to D which is the natural composition. This composition generalizes ordinary composition of homomorphisms once certain natural identifications are made. Imprimitivity bimodules are isomorphisms under this composition.

Correspondences from A to B can also be added:

$$(A, B, \mathcal{E}, \phi) + (A, B, \mathcal{F}, \psi) = (A, B, \mathcal{E} \oplus \mathcal{F}, \phi \oplus \psi).$$

In order to obtain a well-behaved theory (e.g. to have associativity of composition), one must identify isomorphic correspondences, where an isomorphism is an isometric bimodule isomorphism. For ordinary morphisms, this amounts to identifying unitarily equivalent homomorphisms. There is thus a category whose objects are C*-algebras and whose morphisms are isomorphism classes of correspondences.

Correspondences from A to B are closely related to the construction of the group $KK(A, B)$ (V.1.4.4) which is roughly a group of equivalence classes of formal differences of correspondences from A to B; cf. [Bla98, 17.6.2, 18.4.2(c)].

Interesting C*-algebras, generalizing both Cuntz-Krieger algebras and crossed products, can be associated with correspondences; see [Pim97] and [Kat04b].

II.7.5 The Generalized Stinespring Theorem

II.7.5.1 There is a version of Stinespring's Theorem for Hilbert modules, which is perhaps the most general version of the GNS construction. The statement and proof are nearly the same as for the original Stinespring theorem, with the technical details handled through the general theory developed in the last few sections. The statement must be slightly modified to avoid the difficulty of uncomplemented submodules. The appropriate notion is that of a strict completely positive map: if A and B are C*-algebras, $\mathcal{E}$ is a Hilbert B-module, and $\phi : A \to \mathcal{L}(\mathcal{E})$ is completely positive, we say that ϕ is *strict* if $(\phi(h_\lambda))$ converges strictly in $\mathcal{L}(\mathcal{E})$ for some (hence every) approximate unit (h_λ) for A (cf. II.7.3.13).

II.7.5.2 Theorem. [Generalized Stinespring] [Kas80a] Let A and B be C*-algebras, $\mathcal{E}$ a Hilbert B-module, and $\phi : A \to \mathcal{L}(\mathcal{E})$ be a strict completely positive map. Then there is a Hilbert B-module $\mathcal{E}_\phi$, a *-homomorphism π_ϕ from A to $\mathcal{L}(\mathcal{E}_\phi)$, and $V_\phi \in \mathcal{L}(\mathcal{E}, \mathcal{E}_\phi)$ such that $\phi(a) = V_\phi^* \pi_\phi(a) V_\phi$ for all $a \in A$, and $\pi_\phi(A) V_\phi \mathcal{E}$ is dense in $\mathcal{E}_\phi$, and such that $(\mathcal{E}_\phi, \pi_\phi, V_\phi)$ are canonical and minimal in the sense that if $\mathcal{E}'$ is another Hilbert B-module with a representation ρ of A, and $W \in \mathcal{L}(\mathcal{E}, \mathcal{E}')$ with $\phi(a) = W^* \rho(a) W$ and $\rho(A) W \mathcal{E}$ is dense in $\mathcal{E}'$, then there is a unitary $U \in \mathcal{L}(\mathcal{E}_\phi, \mathcal{E}')$ intertwining π_ϕ and ρ, and such that $W = UV_\phi$.

If ϕ is nondegenerate (in particular, if A and ϕ are unital), then $\mathcal{E}$ can be identified with a complemented submodule of $\mathcal{E}_\phi$, and $\phi(a) = P_\mathcal{E} \pi(a) P_\mathcal{E}$.

Combining this result with the Stabilization Theorem, we obtain:

II.7.5.3 Corollary. Let A and B be C*-algebras, with A separable and B σ-unital, and let $\phi : A \to \mathcal{L}(\mathcal{H}_B)$ be a completely positive contraction. Then there is a faithful *-homomorphism $\pi = \begin{bmatrix} \pi_{11} & \pi_{12} \\ \pi_{21} & \pi_{22} \end{bmatrix} : A \to M_2(\mathcal{L}(\mathcal{H}_B))$ with $\pi_{11} = \phi$. If ϕ is nondegenerate, π may be chosen nondegenerate.

II.7.6 Morita Equivalence

It sometimes happens that if A and B are C*-algebras, there are $A - B$-bimodules which are simultaneously left Hilbert A-modules and right Hilbert B-modules, and these lead to a very important equivalence relation on C*-algebras called Morita equivalence. Such modules also appear in the theory of induced representations of locally compact groups; in fact, this was the main original motivation for the theory (cf. [Rie74], [RW98]).

II.7.6.1 Definition. Let A and B be C*-algebras. An $A - B$-*imprimitivity bimodule* (or *equivalence bimodule*) is an $A - B$-bimodule $\mathcal{E}$ which is simultaneously a full left Hilbert A-module under an A-valued inner product ${}_A\langle \cdot, \cdot \rangle$ and a full right Hilbert B-module under a B-valued inner product $\langle \cdot, \cdot \rangle_B$, satisfying

$$ {}_A\langle \xi, \eta \rangle \zeta = \xi \langle \eta, \zeta \rangle_B $$

for all $\xi, \eta, \zeta \in \mathcal{E}$.

II.7.6.2 Proposition. Let A and B be C*-algebras, and $\mathcal{E}$ an $A - B$-imprimitivity bimodule. Then, for all $\xi, \eta \in \mathcal{E}$, $a \in A$, $b \in B$, $\langle a\xi, \eta \rangle_B = \langle \xi, a^*\eta \rangle_B$ and ${}_A\langle \xi b^*, \eta \rangle = {}_A\langle \xi, \eta b \rangle$. Thus L_a, defined by $L_a \xi = a\xi$, is an adjointable operator on the right Hilbert B-module $\mathcal{E}_B$ with $L_a^* = L_{a^*}$, and similarly for right multiplication R_b by b on the left Hilbert A-module ${}_A\mathcal{E}$; hence L_a and R_b are bounded, and $\langle a\xi, a\xi \rangle_B \leq \|a\|^2 \langle \xi, \xi \rangle_B$ and ${}_A\langle \xi b, \xi b \rangle \leq \|b\|^2 {}_A\langle \xi, \xi \rangle$.

II.7 Hilbert Modules, Multiplier Algebras, and Morita Equivalence

PROOF: If $\zeta \in \mathcal{E}$, then

$$\zeta \langle a\xi, \eta \rangle_B = {}_A\langle \zeta, a\xi \rangle \eta = {}_A\langle \zeta, \xi \rangle a^* \eta = \zeta \langle \xi, a^* \eta \rangle_B$$

and thus $\langle a\xi, \eta \rangle_B = \langle \xi, a^* \eta \rangle_B$ by II.7.1.13 since $\mathcal{E}$ is a full Hilbert B-module. The argument for the other equality is similar. Thus $L_a \in \mathcal{L}(\mathcal{E}_B)$, $L_a^* = L_{a^*}$, and $a \to L_a$ is a *-homomorphism from the C*-algebra A to the C*-algebra $\mathcal{L}(\mathcal{E}_B)$; so $\|L_a\| \le \|a\|$. The inequality $\langle a\xi, a\xi \rangle_B \le \|a\|^2 \langle \xi, \xi \rangle_B$ then follows from II.7.2.2, and the arguments for b and R_b are analogous.

II.7.6.3 The two inner products define norms $\|\cdot\|_A$ and $\|\cdot\|_B$ on $\mathcal{E}$. For any $\xi \in \mathcal{E}$,

$$\|\xi\|_B^6 = \|\langle \xi, \xi \rangle_B\|^3 = \|\langle \xi, \xi \rangle_B^3\| = \|\langle \xi \langle \xi, \xi \rangle_B, \xi \langle \xi, \xi \rangle_B \rangle_B\|$$
$$= \|\langle {}_A\langle \xi, \xi \rangle \xi, {}_A\langle \xi, \xi \rangle \xi \rangle_B\| \le \|{}_A\langle \xi, \xi \rangle\|^2 \|\langle \xi, \xi \rangle_B\| = \|\xi\|_A^4 \|\xi\|_B^2$$

and the symmetric argument shows $\|\xi\|_A^6 \le \|\xi\|_B^4 \|\xi\|_A^2$, so $\|\xi\|_A = \|\xi\|_B$, i.e. the two norms coincide.

II.7.6.4 One also sometimes needs to consider pre-imprimitivity bimodules over pre-C*-algebras (e.g. in the theory of induced representations of groups). For such incomplete modules, the conclusions of II.7.6.2 (especially the boundedness of L_a and R_b) are not automatic, and must be assumed as part of the definition (in many references, some or all of these assumptions are also explicitly included in the definition of an imprimitivity bimodule). With these assumptions, the completion is an imprimitivity bimodule over the completions of the algebras.

II.7.6.5 EXAMPLES.

(i) B is a $B-B$-imprimitivity bimodule under the inner products $\langle x, y \rangle_B = x^* y$ and ${}_B\langle x, y \rangle = xy^*$.

(ii) If $\mathcal{E}$ is a Hilbert B-module, then $\mathcal{E}$ becomes a full left Hilbert $\mathcal{K}(\mathcal{E})$-module with the natural left action and ${}_{\mathcal{K}(\mathcal{E})}\langle \xi, \eta \rangle = \Theta_{\eta,\xi}$. If $\mathcal{E}$ is a full Hilbert B-module, then $\mathcal{E}$ is a $\mathcal{K}(\mathcal{E}) - B$-imprimitivity bimodule. In particular, $\mathcal{H}_B$ is a $B \otimes \mathbb{K} - B$-imprimitivity bimodule and B^n is an $M_n(B) - B$-imprimitivity bimodule. Conversely, if $\mathcal{E}$ is an $A - B$-imprimitivity bimodule, then the map ${}_A\langle \xi, \eta \rangle \mapsto \Theta_{\eta,\xi}$ extends to a *-isomorphism from A onto $\mathcal{K}(\mathcal{E}_B)$.

(iii) Let $p \in M(B)$ be a projection such that the ideal $Span(BpB)$ is dense in B; let $\mathcal{E} = pB$ and $A = pBp$ (A is called a *full corner* in B.) Define an A-valued inner product as in (i). Then $\mathcal{E}$ is an $A - B$-imprimitivity bimodule. Bp has similar inner products making it a $B - A$-imprimitivity bimodule.

(iv) Let B be a C*-algebra, $\mathcal{E} = M_{m,n}(B)$ the set of $m \times n$ matrices over B. Then $\mathcal{E}$ has a natural structure as an $M_m(B) - M_n(B)$-imprimitivity bimodule by the same formulas as in (i) and (iii).

(v) If $\mathcal{E}$ is an A, B-imprimitivity bimodule, let $\mathcal{E}^*$ be the "dual module": $\mathcal{E}^*$ is $\{\xi^* : \xi \in \mathcal{E}\}$ with $\xi^* + \eta^* = (\xi + \eta)^*$ and $\alpha\xi^* = [\bar{\alpha}\xi]^*$, made into a $B - A$ bimodule by $b\xi^*a = [a^*\xi b^*]^*$. With the inner products ${}_B\langle \xi^*, \eta^* \rangle = \langle \eta, \xi \rangle_B (= \langle \xi, \eta \rangle_B^*)$ and $\langle \xi^*, \eta^* \rangle_A = {}_A\langle \eta, \xi \rangle$, the bimodule $\mathcal{E}^*$ is a $B - A$ imprimitivity bimodule. In example (i), $B^* \cong B$ via $x \mapsto x^*$, and similarly in (iii) $(pB)^* \cong Bp$, and in (iv) $M_{m,n}(B)^* \cong M_{n,m}(B)$. In (ii), $\mathcal{E}^* \cong \mathcal{K}(\mathcal{E}, B)$.

All these examples are basically the same. In fact, they are all special cases of a general construction (II.7.6.9).

II.7.6.6 If $\mathcal{E}$ is an $A-B$-imprimitivity bimodule, $\mathcal{F}$ a Hilbert C-module, and $\phi : B \to \mathcal{L}(\mathcal{F})$, then the Hilbert C-module $\mathcal{E} \otimes_\phi \mathcal{F}$ has a natural left action of A via adjointable operators. Similarly, if $\mathcal{E}$ is a left Hilbert A-module, $\mathcal{F}$ a $B - C$-imprimitivity bimodule, and $\psi : B \to \mathcal{L}(\mathcal{E})$, then C acts by adjointable operators on the left Hilbert module $\mathcal{E} \otimes_\psi \mathcal{F}$. If $\mathcal{E}$ is an $A - B$-imprimitivity bimodule and $\mathcal{F}$ a $B - C$-imprimitivity bimodule, then these actions are compatible and make $\mathcal{E} \otimes_\phi \mathcal{F} \cong \mathcal{E} \otimes_\psi \mathcal{F}$ into an $A-C$-imprimitivity bimodule, usually written $\mathcal{E} \otimes_B \mathcal{F}$.

II.7.6.7 PROPOSITION. Let $\mathcal{E}$ be an $A - B$-imprimitivity bimodule. Then $\xi^* \otimes \eta \mapsto \langle \xi, \eta \rangle_B$ gives an isomorphism of $\mathcal{E}^* \otimes_A \mathcal{E}$ with B as Hilbert B-modules (and, in fact, as $B - B$ imprimitivity bimodules).

II.7.6.8 DEFINITION. Let A and B be C*-algebras. A and B are *Morita equivalent* if there is an $A - B$-imprimitivity bimodule.

Morita equivalence is symmetric and reflexive by II.7.6.5(i) and (v), and transitive by the tensor product construction.

This relation was originally called "strong Morita equivalence" in [Rie74] because it is not exactly the C*-analog of Morita equivalence in algebra [Mor58]. But the word "strong" is now customarily omitted.

II.7.6.9 THEOREM. Let A and B be C*-algebras. Then A and B are Morita equivalent if and only if they are isomorphic to complementary full corners in a C*-algebra C, i.e. if there is a projection $p \in M(C)$ with $Span(CpC)$ and $Span(C(1-p)C)$ dense in C and $pCp \cong A$, $(1-p)C(1-p) \cong B$.

One direction is clear from II.7.6.5(iii): if there is such a C, then A and B are Morita equivalent to C via pC and $(1-p)C$ respectively (and $pC(1-p)$ is an $A - B$-imprimitivity bimodule). Conversely, if $\mathcal{E}$ is an $A - B$-imprimitivity bimodule, we form the *linking algebra* of A and B with respect to $\mathcal{E}$:

$$C = \begin{bmatrix} A & \mathcal{E} \\ \mathcal{E}^* & B \end{bmatrix} = \left\{ \begin{bmatrix} a & \xi \\ \eta^* & b \end{bmatrix} : a \in A, \xi, \eta \in \mathcal{E}, b \in B \right\}$$

with involution $\begin{bmatrix} a & \xi \\ \eta^* & b \end{bmatrix}^* = \begin{bmatrix} a^* & \eta \\ \xi^* & b^* \end{bmatrix}$ and multiplication

$$\begin{bmatrix} a_1 & \xi_1 \\ \eta_1^* & b_1 \end{bmatrix} \begin{bmatrix} a_2 & \xi_2 \\ \eta_2^* & b_2 \end{bmatrix} = \begin{bmatrix} a_1 a_2 + {}_A\langle \xi_1, \eta_2 \rangle & a_1\xi_2 + \xi_1 b_2 \\ \eta_1^* a_2 + b_1\eta_2^* & \langle \eta_1, \xi_2 \rangle_B + b_1 b_2 \end{bmatrix}.$$

C acts as a *-algebra of adjointable operators on the Hilbert B-module $\mathcal{E} \oplus B$ via $\begin{bmatrix} a & \xi \\ \eta^* & b \end{bmatrix} \begin{bmatrix} \zeta \\ z \end{bmatrix} = \begin{bmatrix} a\zeta + \xi z \\ \langle \eta, \zeta \rangle_B + bz \end{bmatrix}$, and the operator norm is equivalent to $\left\| \begin{bmatrix} a & \xi \\ \eta^* & b \end{bmatrix} \right\| = \|a\| + \|\xi\| + \|\eta\| + \|b\|$, so C is complete (a C*-algebra) with respect to the operator norm. The projection $p = \begin{bmatrix} I & 0 \\ 0 & 0 \end{bmatrix}$ defines a multiplier of C. Fullness is a routine computation.

In general, the linking algebra C depends up to isomorphism on the choice of $\mathcal{E}$ and not just on A and B. $\mathcal{E}$ can be recovered from C as $pC(1-p)$.

II.7.6.10 Since A is Morita equivalent to $A \otimes \mathbb{K}$, it is obvious that if A and B are stably isomorphic, then A and B are Morita equivalent. The converse is not true, however: if $\mathcal{H}$ is a nonseparable Hilbert space, then $\mathcal{K}(\mathcal{H})$ is Morita equivalent to $\mathbb{C}$, but not stably isomorphic (and much more complicated examples exist: for example, if A is a simple unital C*-algebra with a strictly increasing net of projections indexed by the first uncountable ordinal, such as the quotient of $\mathcal{L}(\mathcal{H})$ for nonseparable $\mathcal{H}$ by a maximal ideal, there is a hereditary C*-subalgebra B of A which is Morita equivalent to A but $A \otimes \mathcal{K}(\mathcal{H}') \not\cong B \otimes \mathcal{K}(\mathcal{H}')$ for any Hilbert space $\mathcal{H}'$.) However, the σ-unital case (including the separable case) is very nice, due to the next theorem [BGR77] (cf. [Bro77]):

II.7.6.11 THEOREM. [BROWN-GREEN-RIEFFEL] Let A and B be σ-unital C*-algebras. Then $A \otimes \mathbb{K} \cong B \otimes \mathbb{K}$ if (and only if) A and B are Morita equivalent.

Indeed, if $\mathcal{E}$ is an $A-B$-imprimitivity bimodule, it suffices to show that $\mathcal{H}_A \otimes_A \mathcal{E} \cong \mathcal{H}_B$ (as Hilbert B-modules); for then, since $\mathcal{H}_A \otimes_A \mathcal{E} \cong \mathcal{E}^\infty$, we have

$$A \otimes \mathbb{K} \cong \mathcal{K}(\mathcal{E}^\infty) \cong \mathcal{K}(\mathcal{H}_B) \cong B \otimes \mathbb{K}.$$

But since B is σ-unital, $\mathcal{E}^*$ is a countably generated Hilbert A-module (II.7.2.5), and thus $\mathcal{E}^* \oplus \mathcal{H}_A \cong \mathcal{H}_A$ by the Stabilization Theorem (II.7.1.11). Thus

$$\mathcal{H}_A \otimes_A \mathcal{E} \cong \mathcal{H} \otimes_\mathbb{C} \mathcal{H}_A \otimes_A \mathcal{E} \cong \mathcal{H} \otimes_\mathbb{C} (\mathcal{H}_A \oplus \mathcal{E}^*) \otimes_A \mathcal{E}$$

$$\cong (\mathcal{H} \otimes_\mathbb{C} \mathcal{H}_A \otimes_A \mathcal{E}) \oplus (\mathcal{H} \otimes_\mathbb{C} \mathcal{E}^* \otimes_A \mathcal{E}) \cong (\mathcal{H}_A \otimes_A \mathcal{E}) \oplus (\mathcal{H} \otimes_\mathbb{C} B) \cong \mathcal{H}_B$$

by the Stabilization Theorem, since $\mathcal{E}^* \otimes_A \mathcal{E} \cong B$ (II.7.6.7) and $\mathcal{H}_A \otimes_A \mathcal{E}$ is countably generated (because $\mathcal{H}_A$ and $\mathcal{E}$ are countably generated by II.7.2.5).

As a corollary of the proof, we obtain the following variation of the Stabilization Theorem. This is essentially the version of the Stabilization Theorem proved in [DD63] for B commutative; see [Bro77] and [MP84, 1.9] for the general case.

II.7.6.12 COROLLARY. If B is a σ-unital C*-algebra and $\mathcal{E}$ is a countably generated full Hilbert B-module, then $\mathcal{E}^\infty \cong \mathcal{H}_B$ (as Hilbert B-modules).

PROOF: $\mathcal{E}$ may be regarded as a $\mathcal{K}(\mathcal{E}) - B$-imprimitivity bimodule, and $\mathcal{K}(\mathcal{E})$ is σ-unital (II.7.2.5), so $\mathcal{E}^\infty \cong \mathcal{H}_{\mathcal{K}(\mathcal{E})} \otimes_{\mathcal{K}(\mathcal{E})} \mathcal{E} \cong \mathcal{H}_B$.

This version of the Stabilization Theorem implies the usual version if B is σ-unital: if $\mathcal{E}$ is a countably generated Hilbert B-module, then $(\mathcal{E} \oplus \mathcal{H}_B)^\infty \cong \mathcal{H}_B$, so

$$\mathcal{H}_B \cong (\mathcal{E} \oplus \mathcal{H}_B)^\infty \cong (\mathcal{E} \oplus \mathcal{H}_B) \oplus (\mathcal{E} \oplus \mathcal{H}_B)^\infty \cong \mathcal{E} \oplus \mathcal{H}_B \oplus \mathcal{H}_B \cong \mathcal{E} \oplus \mathcal{H}_B .$$

Morita Equivalence and Representations

II.7.6.13 One important feature of a Morita equivalence is that it gives a functorial correspondence between representations of the algebras. If $\mathcal{E}$ is an $A - B$-imprimitivity bimodule, $\mathcal{F}$ a Hilbert C-module, and $\phi : B \to \mathcal{L}(\mathcal{F})$ (i.e. a representation of B on $\mathcal{F}$), then $\mathcal{E}$ defines a representation of A on $\mathcal{E} \otimes_\phi \mathcal{F}$; and the process is reversible by tensoring on the left with $\mathcal{E}^*$ (using II.7.6.7). In particular, if $C = \mathbb{C}$ (i.e. ϕ is an ordinary Hilbert space representation), we obtain an induced representation of A on the Hilbert space $\mathcal{E} \otimes_\phi \mathcal{F}$. This correspondence sends irreducible representations to irreducible representations, so one gets natural homeomorphisms between $\hat{A}$ and $\hat{B}$, and also Prim(A) and Prim(B). (There is a more direct correspondence between closed ideals of A and those of B: if I is a closed ideal of A, regard I as a right Hilbert A-module and let $\mathcal{F}$ be the right Hilbert B-module $I \otimes_A \mathcal{E}$, and let $J = J_\mathcal{F}$ (the closed linear span of $\langle \mathcal{F}, \mathcal{F} \rangle_B$); then J is the ideal of B corresponding to I, and $\mathcal{F}$ is an $I - J$-imprimitivity bimodule.) This is a precise formulation of the statement (II.6.6.14) that Morita equivalent C*-algebras are "the same up to 'size'."

II.8 Examples and Constructions

So far only the most elementary examples of C*-algebras have appeared. In this section, we will discuss various constructions which will yield an enormous number of interesting C*-algebras. Two other important constructions, tensor products and crossed products, will be discussed in the next two sections.

The book [Dav96] has a much more detailed analysis of standard examples and constructions in C*-algebra theory.

II.8.1 Direct Sums, Products, and Ultraproducts

II.8.1.1 If $\{A_1, \cdots, A_n\}$ is a finite set of C*-algebras, there is a natural notion of direct sum or direct product: $A_1 \oplus \cdots \oplus A_n = A_1 \times \cdots \times A_n$ is the ordinary algebraic direct sum with norm $\|(a_1, \cdots, a_n)\| = \max \|a_i\|$.

If $\{A_i : i \in \Omega\}$ is an infinite set of C*-algebras, there are separate notions of direct sum and product which do not coincide with the algebraic ones.

II.8.1.2 DEFINITION.

$$\prod_{i \in \Omega} A_i = \{(a_i) : \|(a_i)\| = \sup_i \|a_i\| < \infty\}$$

$$\bigoplus_{i \in \Omega} A_i = \{(a_i) : \|a_i\| \to 0 \text{ as } i \to \infty\}$$

in the sense that for every $\epsilon > 0$ there are only finitely many i for which $\|a_i\| \geq \epsilon$. $\bigoplus A_i$ is the closure in $\prod A_i$ of the algebraic direct sum (Ω-tuples with only finitely many nonzero entries).

II.8.1.3 It is easily checked that $\prod A_i$ is a C*-algebra, and $\bigoplus A_i$ is a closed ideal in $\prod A_i$. The quotient C*-algebra is an analog of the Calkin algebra which is useful in various constructions. The multiplier algebra of $\bigoplus A_i$ can be naturally identified with $\prod M(A_i)$ (cf. II.7.3.12).

Direct sums and products are useful in many arguments. For example, approximate versions of exact results can sometimes be obtained. The next result is an approximate version of Fuglede's theorem (II.6.4.12), which is interesting even in the self-adjoint case where Fuglede's theorem is trivial.

II.8.1.4 PROPOSITION. Let X be a compact subset of $\mathbb{C}$ and $f : X \to \mathbb{C}$ a continuous function. Then for any $\epsilon > 0$ there is a $\delta > 0$ such that, whenever x, y are elements of a C*-algebra A with x normal, $\sigma(x) \subseteq X$, $\|y\| \leq 1$, and $\|[x,y]\| < \delta$, then $\|[f(x),y]\| < \epsilon$ (in $\tilde{A}$).

While a direct computational proof can be given (see e.g. [Arv77, p. 332] for the self-adjoint case, and [Dav96, Exercises II.8–II.9] for other special cases), it is more satisfying to give the following argument which entirely avoids computations. Suppose the conclusion is false. Then there is an X, f, and $\epsilon > 0$ such that there exist unital C*-algebras A_n and $x_n, y_n \in A_n$ satisfying the hypotheses, with $\|[x_n, y_n]\| < 1/n$ but $\|[f(x_n), y_n]\| \geq \epsilon$ for all n. Let $A = \prod A_n$, $J = \bigoplus A_n$, $x = (\cdots x_n \cdots)$ (note that $\{\|x_n\|\}$ is bounded since x_n is normal and $\sigma(x_n) \subseteq X$), $y = (\cdots y_n \cdots)$, and $\pi : A \to A/J$ the quotient map. Then $[\pi(x), \pi(y)] = 0$, so $[f(\pi(x)), \pi(y)] = 0$ (II.6.4.12). But $f(\pi(x)) = \pi(f(x)) = \pi(\cdots f(x_n) \cdots)$, and

$$\|[\pi(\cdots f(x_n) \cdots), \pi(\cdots y_n \cdots)]\| = \limsup \|[f(x_n), y_n]\| \geq \epsilon,$$

a contradiction.

II.8.1.5 COROLLARY. Let J be a closed ideal in a C*-algebra A, and (h_λ) an approximate unit for J which is quasicentral for A. Then, for any $\alpha > 0$, (h_λ^α) is an approximate unit for J which is quasicentral for A.

Ultraproducts

II.8.1.6 If ω is a free ultrafilter on Ω, then

$$J_\omega = \left\{(a_i) \in \prod A_i : \lim_\omega \|a_i\| = 0\right\}$$

is a closed ideal in $\prod A_i$. The quotient is called the *ultraproduct* of the A_i with respect to ω.

II.8.1.7 These constructions are frequently used with $\Omega = \mathbb{N}$ and all A_i the same A. The product of a sequence of copies of A is usually denoted $l^\infty(A)$ and the direct sum $c_o(A)$. The ultraproduct with respect to a free ultrafilter ω on $\mathbb{N}$ is denoted A_ω.

II.8.2 Inductive Limits

II.8.2.1 One of the simplest and most useful constructions of C*-algebras is the inductive limit (sometimes called direct limit). An *inductive system* of C*-algebras is defined just as in the purely algebraic setting: a collection $\{(A_i, \phi_{ij}) : i, j \in \Omega, i \leq j\}$, where Ω is a directed set, the A_i are C*-algebras, and ϕ_{ij} is a *-homomorphism from A_i to A_j with $\phi_{ik} = \phi_{jk} \circ \phi_{ij}$ for $i \leq j \leq k$. Each ϕ_{ij} is norm-decreasing, so there is a naturally induced C*-seminorm on the algebraic direct limit defined by

$$\|a\| = \lim_{j > i} \|\phi_{ij}(a)\| = \inf_{j > i} \|\phi_{ij}(a)\|$$

for $a \in A_i$; the completion of the algebraic direct limit (with elements of seminorm 0 divided out) is a C*-algebra called the *inductive limit* of the system, written $\varinjlim(A_i, \phi_{ij})$, or just $\varinjlim A_i$ if the ϕ_{ij} are understood. There is a natural *-homomorphism ϕ_i from A_i to the inductive limit.

If all the connecting maps are injective (and hence isometric), the algebraic direct limit may be thought of as the "union" of the A_i, and the inductive limit as the completion of this union. In general, the inductive limit of a system (A_i, ϕ_{ij}) may be naturally regarded as a C*-subalgebra of $(\prod A_i)/(\bigoplus A_i)$.

II.8.2.2 EXAMPLES.

(i) An inductive system (A_i, ϕ_{ij}) where each $A_i = C(X_i)$ is unital and commutative, and the connecting maps ϕ_{ij} are unital, exactly corresponds to an ordinary inverse system (X_i, f_{ij}) of compact Hausdorff spaces. The inductive limit $\varinjlim(A_i, \phi_{ij})$ is naturally isomorphic to $C(X)$, where X is the usual inverse limit $\varprojlim(X_i, f_{ij})$.

(ii) Let I be a set and $\{A_i : i \in I\}$ a collection of C*-algebras. Let Ω be the collection of finite subsets of I, directed by inclusion, and for each $F \in \Omega$ let $B_F = \bigoplus_{i \in F} A_i$. If $F \subseteq G \subseteq I$, define $\phi_{F,G} : B_F \to B_G$ by setting the additional coordinates zero. Then $\varinjlim(B_F, \phi_{F,G}) \cong \bigoplus_{i \in I} A_i$.

(iii) Let $\phi_{n,n+k} : \mathbb{M}_n \to \mathbb{M}_{n+k}$ be the (nonunital) *-homomorphism obtained by adding k rows and columns of zeroes. Then $\varinjlim(\mathbb{M}_n, \phi_{n,n+k})$ is isomorphic to $\mathbb{K}$.

(iv) Let $A_n = \mathbb{M}_{2^n}$, and let $\phi_{n,n+1} : A_n \to A_{n+1}$ be defined by $\phi_{n,n+1}(a) = diag(a,a)$ (this is called an *embedding of multiplicity* 2). $\phi_{n,n+1}$ is unital. The inductive limit is called the *UHF algebra of type* 2^∞, or the *CAR algebra* (this C*-algebra also arises from representations of the Canonical Anticommutation Relations of mathematical physics). Similarly, we could take $A_n = \mathbb{M}_{3^n}$ and $\phi_{n,n+1}$ an embedding of multiplicity 3 to obtain a UHF algebra of type 3^∞; more generally, the multiplicities could be varied at each step. The structure of these UHF algebras will be discussed in V.1.1.16(iv). More generally, we will consider the class of *AF algebras*, inductive limits of sequences of finite-dimensional C*-algebras.

II.8.2.3 If A_0 is a dense *-subalgebra of a C*-algebra A, there is no good relationship between A_0 and the (closed) ideals of A in general; A may have nonzero ideals whose intersection with A_0 is zero. But if A_0 is a union of C*-subalgebras of A (e.g. if $A = \varinjlim A_i$ and A_0 is the image of the algebraic direct limit), then the relationship is very nice:

II.8.2.4 PROPOSITION. Let A be a C*-algebra, $\{A_i\}$ a collection of C*-subalgebras such that $A_0 = \cup A_i$ is a dense *-subalgebra of A. If J is any closed ideal of A, then $J \cap A_0 = \cup_i (J \cap A_i)$ is dense in J.

PROOF: Let K be the closure in A of $J \cap A_0$; then K is a closed ideal of A contained in J, and $K \cap A_i = J \cap A_i$ for all i. Let $\pi : A \to A/K$ be the quotient map, and set $\bar{A} = A/K$, $\bar{A}_i = \pi(A_i)$, $\bar{J} = \pi(J)$. Let $\rho : \bar{A} \to \bar{A}/\bar{J}$ be the quotient map. Then $\rho|_{\bar{A}_i}$ is injective since $\bar{J} \cap \bar{A}_i = \{0\}$ for all i; hence ρ is isometric on each $\bar{A}_i$ and hence on all of $\bar{A}$; so ρ is injective, $\bar{J} = \{0\}$, $J = K$.

II.8.2.5 COROLLARY. Let $A = \varinjlim(A_i, \phi_{ij})$. If each A_i is simple, then A is simple.

In particular, the UHF algebras (II.8.2.2) are infinite-dimensional simple unital C*-algebras.

Using II.8.2.4, many simple inductive limits of sequences of nonsimple C*-algebras can also be constructed (cf. II.10.4.12(ii), IV.1.4.23). C*-algebras of this sort are the basic objects of study in large parts of the classification program [Rør02a].

II.8.2.6 There is a generalized inductive limit construction using inductive systems in which the connecting maps are not necessarily *-homomorphisms but are only asymptotically multiplicative, which is playing a role in the advanced theory (V.4.3).

II.8.3 Universal C*-Algebras and Free Products

Universal constructions have been playing an increasingly important role in the theory of C*-algebras in recent years. Many important C*-algebras can be simply and naturally expressed as universal C*-algebras on sets of generators and relations.

II.8.3.1 Suppose a set $\mathcal{G} = \{x_i : i \in \Omega\}$ of generators and a set $\mathcal{R}$ of relations are given. The relations can be of a very general nature, but are usually algebraic relations among the generators and their adjoints, or more generally of the form

$$\|p(x_{i_1}, \cdots, x_{i_n}, x_{i_1}^*, \cdots, x_{i_n}^*)\| \leq \eta$$

where p is a polynomial in $2n$ noncommuting variables with complex coefficients and $\eta \geq 0$. The only restriction on the relations is that they must be realizable among operators on a Hilbert space and they must (at least implicitly) place an upper bound on the norm of each generator when realized as an operator. A *representation* of $(\mathcal{G}|\mathcal{R})$ has the obvious meaning: a set $\{T_i : i \in \Omega\}$ of bounded operators on a Hilbert space $\mathcal{H}$ satisfying the relations. A representation of $(\mathcal{G}|\mathcal{R})$ defines a *-representation of the free *-algebra $\mathcal{A}$ on the set $\mathcal{G}$. For $x \in \mathcal{A}$, let

$$\|x\| = \sup\{\|\pi(x)\| : \pi \text{ a representation of } (\mathcal{G}|\mathcal{R})\}.$$

If this supremum is finite for all $x \in \mathcal{A}$ (it is enough to check this on the generators), it defines a C*-seminorm on $\mathcal{A}$, and the completion (with elements of seminorm 0 divided out) is called the *universal C*-algebra on* $(\mathcal{G}|\mathcal{R})$, denoted $C^*(\mathcal{G}|\mathcal{R})$. We may also consider the universal unital C*-algebra on a set of generators and relations by adding an additional generator 1 with relations $1 = 1^* = 1^2$ and $x1 = 1x = x$ for each other generator.

This construction is extremely general; indeed, every C*-algebra can be written (in an uninteresting way) as the universal C*-algebra on a suitable set of generators and relations. But the interesting examples are, of course, ones which come from simple (especially finite) presentations. Sometimes the universal property is described informally when there is a straightforward translation into a precise description.

II.8.3.2 EXAMPLES.

(i) There is no "universal C*-algebra generated by a single self-adjoint element," since there is no bound on the norm of the element. But there is a "universal C*-algebra generated by a single self-adjoint element of norm one," with $\mathcal{G} = \{x\}$ and $\mathcal{R} = \{x = x^*, \|x\| \leq 1\}$. This C*-algebra is isomorphic to $C_0([-1,1])$, the continuous functions on $[-1,1]$ vanishing at zero, via functional calculus. Similarly, the "universal unital

C*-algebra generated by a self-adjoint element of norm one" is isomorphic to $C([-1, 1])$. There is a "universal C*-algebra generated by a single positive element of norm one," isomorphic to $C_o((0, 1])$, and a "universal C*-algebra generated by a single normal element of norm one," isomorphic to $C_0(\mathbb{D})$, where $\mathbb{D}$ is the unit disk in $\mathbb{C}$.

(ii) There is a "universal C*-algebra generated by a single unitary," since a unitary automatically has norm one. We may take $\mathcal{G} = \{u, 1\}$,

$$\mathcal{R} = \{1 = 1^* = 1^2, u1 = 1u = u, u^*u = uu^* = 1\}.$$

$C^*(\mathcal{G}|\mathcal{R}) \cong C(\mathbb{T})$, the continuous functions on the circle. Similarly, the universal C*-algebra generated by n commuting unitaries is isomorphic to $C(\mathbb{T}^n)$, continuous functions on the n-torus.

(iii) If G is a group, there is a "universal C*-algebra generated by a group of unitaries isomorphic to G," with generators and relations from any presentation of G and additional relations making each generator a unitary. The corresponding C*-algebra is called the *group C*-algebra* of (discrete) G, denoted $C^*(G)$. The representations of $C^*(G)$ are in canonical one-one correspondence with the unitary representations of G. For example, $C^*(\mathbb{Z}) \cong C(\mathbb{T})$ and $C^*(\mathbb{Z}^n) \cong C(\mathbb{T}^n)$. Another important example is $\mathbb{F}_n$, the free group on n generators; $C^*(\mathbb{F}_n)$ is the universal C*-algebra generated by n unitaries. To see that the group G actually embeds in the unitary group of $C^*(G)$, consider the *left regular representation* λ of G on $l^2(G)$ induced by G acting on itself by left translation. The representation λ extends to a representation of $C^*(G)$, which is not faithful in general; the quotient $\lambda(C^*(G))$ is called the *reduced group C*-algebra* of G, denoted $C_r^*(G)$. The group C*-algebra construction generalizes to locally compact topological groups and will be discussed in detail in II.10.

(iv) Let $\mathcal{G} = \{e_{ij} : 1 \leq i, j \leq n\}$,

$$\mathcal{R} = \{e_{ij}^* = e_{ji}, e_{ij}e_{kl} = \delta_{jk}e_{il} : 1 \leq i, j, k, l \leq n\},$$

where δ_{jk} is the Kronecker symbol. The relations imply that each generator is a partial isometry, hence of norm 1, so the universal C*-algebra exists. $C^*(\mathcal{G}, \mathcal{R}) \cong \mathbb{M}_n$. A set of nonzero elements $\{f_{ij}\}$ in a C*-algebra A satisfying the relations is called a *set of matrix units* of type $\mathbb{M}_n$ in A, and generate a (possibly nonunital) C*-subalgebra isomorphic to $\mathbb{M}_n$. There is a similar description of $\mathbb{K}$ as the universal C*-algebra generated by an (infinite) set of matrix units.

More generally, if D is a finite-dimensional C*-algebra, then D is isomorphic to a direct sum $\oplus_{r=1}^m \mathbb{M}_{n_r}$ of matrix algebras (this can be proved by noting that every C*-algebra is semisimple (II.1.6.4) and applying Wedderburn's theorem, or by regarding D as a von Neumann algebra and using either the central decomposition (III.1.6.4) or just some elementary linear algebra.) The natural matrix units

$$\{e_{ij}^{(r)} : 1 \leq r \leq m, 1 \leq i, j \leq n_r\}$$

satisfy the relations $e_{ij}^{(r)*} = e_{ji}^{(r)}$, $e_{ij}^{(r)} e_{kl}^{(s)} = \delta_{rs}\delta_{jk}e_{il}$ for all i,j,k,l,r,s. Conversely, the universal C*-algebra generated by generators satisfying these relations is isomorphic to D. Such a set of generators is called a *set of matrix units of type D*.

(v) There is a "universal C*-algebra generated by a single isometry," called the *Toeplitz algebra* and denoted $\mathcal{T}$. If v is the generator, then $\mathcal{T}$ has an identity v^*v. If $p = v^*v - vv^*$, and $e_{ij} = v^{i-1}p(v^*)^{j-1}$, then it is easily verified that the span of the e_{ij} is an essential ideal in $\mathcal{T}$ isomorphic to $\mathbb{K}$. Modulo this ideal v is unitary, so $\mathcal{T}/\mathbb{K} \cong C(\mathbb{T})$. $\mathcal{T} \cong C^*(S)$, where S is the unilateral shift (I.2.4.3(ii)).

In addition to these basic examples, here are some more classes which play a very important role in the theory:

II.8.3.3 EXAMPLES.

(i) Let θ be a real number (usually in the unit interval), and let A_θ be the universal C*-algebra generated by two unitaries u,v, with the relation $vu = e^{2\pi i\theta}uv$. A_θ is called a *rotation algebra*, and in the particularly important case where θ is irrational, an *irrational rotation algebra*. An irrational rotation algebra is simple. Rotation algebras can be alternately described as crossed products (II.10.4.12). More generally, a *noncommutative torus* is a universal C*-algebra generated by unitaries $u_1, \cdots, u_n$, with $u_j u_i = e^{2\pi i\theta_{ij}} u_i u_j$ for $\theta_{ij} \in \mathbb{R}$. Noncommutative tori arise in applications (e.g. the quantum Hall effect in physics [Con94], and string theory [CDS98]) and are a natural setting for noncommutative geometry.

(ii) Set $\mathcal{G} = \{s_1, \cdots, s_n\}$,

$$\mathcal{R} = \left\{ s_i^* s_i = 1, \sum_{j=1}^n s_j s_j^* = 1 : 1 \leq i \leq n \right\}.$$

$C^*(\mathcal{G}|\mathcal{R})$ is the *Cuntz algebra* O_n, the universal (unital) C*-algebra generated by n isometries whose range projections are mutually orthogonal and add up to the identity. O_n is simple, and in fact is *purely infinite*: if $x \in O_n$, $x \neq 0$, then there are $a, b \in O_n$ with $axb = 1$. There is also an O_∞, the universal C*-algebra generated by a sequence of isometries with mutually orthogonal range projections; O_∞ is also purely infinite and hence simple. The simplicity of these algebras means that any set of isometries satisfying the relations generates a C*-algebra isomorphic to O_n.

(iii) As a generalization of (ii), let A be an $n \times n$ matrix of zeroes and ones, and let O_A be the universal C*-algebra generated by partial isometries $s_1, \cdots, s_n$ with relations

$$s_i^* s_i = \sum_{j=1}^{n} A_{ij} s_j s_j^*$$

and with $s_i s_i^* \perp s_j s_j^*$ for $i \neq j$ (the last relations are often automatic from the first ones). The range projections of the s_i add up to an identity, so O_A is unital. If A has all ones, $O_A \cong O_n$. O_A is simple (and purely infinite) if and only if A is irreducible (the corresponding directed graph is connected) and not a permutation matrix. The O_A arise in the study of topological Markov chains [CK80], and are called *Cuntz-Krieger algebras*. More generally, if A is a row-finite infinite matrix of zeroes and ones (every row has only finitely many ones), corresponding to a directed graph in which every vertex is the source of only finitely many edges, one can form a universal C*-algebra O_A as above. If A is infinite, O_A is nonunital. Such graph C*-algebras have been extensively studied; see, for example, [KPR98].

(iv) Let $SU(2) = \left\{ \begin{bmatrix} a & -\bar{c} \\ c & \bar{a} \end{bmatrix} : a, c \in \mathbb{C}, |a|^2 + |c|^2 = 1 \right\}$ be the special unitary group. Topologically, $SU(2) \cong S^3$. Define two coordinate functions α, γ by

$$\alpha\left(\begin{bmatrix} a & -\bar{c} \\ c & \bar{a} \end{bmatrix}\right) = a \, , \, \gamma\left(\begin{bmatrix} a & -\bar{c} \\ c & \bar{a} \end{bmatrix}\right) = c \, .$$

Then $\mathcal{G} = \{\alpha, \gamma\}$ generates $C(SU(2))$, and α and γ satisfy the relations

$$\mathcal{R} = \{\alpha^*\alpha + \gamma^*\gamma = 1, \alpha\alpha^* + \gamma\gamma^* = 1, \gamma^*\gamma = \gamma\gamma^*, \gamma\alpha = \alpha\gamma, \gamma^*\alpha = \alpha\gamma^*\}$$

(the relations are written deliberately in a less than optimal form for this example, in preparation for (v); the last relation follows from the previous ones in a C*-algebra by Fuglede's theorem, and the first two relations imply that $\alpha^*\alpha = \alpha\alpha^*$, so α and γ are commuting normal elements). It is not difficult to see that $C^*(\mathcal{G}|\mathcal{R})$ is naturally isomorphic to $C(SU(2)) \cong C(S^3)$.

(v) Let $q \in [-1, 1] \setminus \{0\}$. Similarly to (iv), let $\mathcal{G} = \{\alpha, \gamma\}$,

$$\mathcal{R} = \{\alpha^*\alpha + \gamma^*\gamma = 1, \alpha\alpha^* + q^2\gamma\gamma^* = 1, \gamma^*\gamma = \gamma\gamma^*, q\gamma\alpha = \alpha\gamma, q\gamma^*\alpha = \alpha\gamma^*\}$$

Then $A_q = C^*(\mathcal{G}|\mathcal{R})$ is a noncommutative C*-algebra if $q \neq 1$. It has a comultiplication $\Delta : A_q \to A_q \otimes A_q$ defined by

$$\Delta(\alpha) = \alpha \otimes \alpha + q\gamma^* \otimes \gamma \, , \quad \Delta(\gamma) = \gamma \otimes \alpha + \alpha^* \otimes \gamma$$

making it into a compact quantum group in the sense of Woronowicz [Wor87] (II.10.8.21). A_q is a "quantization" of the group $SU(2)$, often denoted $SU_q(2)$.

Free Products

II.8.3.4 If $\{A_i\}$ is a collection of C*-algebras, the *free product* $*A_i = *\{A_i\}$ (write $A*B$ for $*\{A,B\}$) is the universal C*-algebra generated by copies of the A_i with no additional relations. Free products tend to be "large" and pathological (e.g. they are rarely nuclear). For example, $\mathbb{C}*\mathbb{C}$ is the infinite-dimensional nonunital universal C*-algebra generated by two projections (this example is actually nuclear, and can be explicitly described as a subhomogeneous C*-algebra (IV.1.4.2)). If the A_i are unital, one can also form the *unital free product* $*_\mathbb{C} A_i$ generated by unital copies of the A_i. For example, $\mathbb{C}^2 *_\mathbb{C} \mathbb{C}^2$ is isomorphic to $\mathbb{C}*\mathbb{C}$ with unit adjoined; and if G and H are groups and $G*H$ is their free product, then $C^*(G*H) \cong C^*(G) *_\mathbb{C} C^*(H)$.

II.8.3.5 More generally, if D is a C*-algebra, regarded as a C*-subalgebra of each A_i via an embedding (or just *-homomorphism) ϕ_i, we may form the *amalgamated free product* $*_D A_i = *_{\phi_i} A_i$ as the universal C*-algebra generated by "copies" of the A_i with the copies of D identified, i.e. $\phi_i(d) = \phi_j(d)$ for all i,j and all $d \in D$. It is a somewhat nontrivial fact [Bla78, 3.1] that each A_i embeds in the amalgamated free product (if each ϕ_i is injective). The group C*-algebra of an amalgamated free product of groups can be constructed as an amalgamated free product of the group C*-algebras.

II.8.3.6 There is also a notion of reduced free product (or reduced amalgamated free product) of C*-algebras with respect to states. Reduced free products are very important in free probability theory, and increasingly in other aspects of operator algebra theory; see [Voi00].

Stable Relations and Semiprojectivity

The notions of projectivity and semiprojectivity are noncommutative analogs of the topological notions of absolute retract (AR) and absolute neighborhood retract (ANR) respectively. These notions were originally used in the development of shape theory for C*-algebras, first introduced by E. Effros and J. Kaminker [EK86]. Semiprojective C*-algebras have rigidity properties which make them conceptually and technically important in several other aspects of C*-algebra theory; this is reflected especially in the work of T. Loring and his coauthors on lifting problems (see, for example, [Lor97]). It is not too easy for a C*-algebra to be semiprojective, but there does seem to be a reasonable supply of such algebras.

II.8.3.7 DEFINITION. A separable C*-algebra A is *semiprojective* if, for any C*-algebra B, increasing sequence $\langle J_n \rangle$ of (closed two-sided) ideals of B, with $J = [\cup J_n]^-$, and *-homomorphism $\phi : A \to B/J$, there is an n and a *-homomorphism $\psi : A \to B/J_n$ such that $\phi = \pi \circ \psi$, where $\pi : B/J_n \to B/J$ is the natural quotient map.

A ϕ for which such a ψ exists is said to be *partially liftable*. If there is a $\psi : A \to B$ with $\phi = \pi \circ \psi$, then ϕ is *liftable*; if every homomorphism from A is liftable, A is said to be *projective*.

The definition of semiprojectivity first appeared in this form in [Bla85b] (a somewhat different, less restrictive, definition previously appeared in [EK86]).

This definition can be applied in various categories; we will use the category of general C*-algebras and *-homomorphisms unless otherwise specified. For comparison and motivation, we note:

II.8.3.8 PROPOSITION. Let $A = C(X)$ be a unital commutative C*-algebra, with X a compact metrizable space. Then A is projective [*resp.* semiprojective] in the category of unital commutative C*-algebras if and only if X is an AR [*resp.* ANR].

The proof is a simple exercise. Note, however, that if X is an AR, $C(X)$ is not even semiprojective in general in the category of general C*-algebras (or the general unital category) (II.8.3.16(x)). In fact, it is likely that if X is a compact metrizable space, then $C(X)$ is semiprojective in the general (or unital general) category if and only if X is an ANR and $dim(X) \leq 1$ (a proof of this could probably be assembled from known results).

II.8.3.9 For convenience, we have only defined semiprojectivity for separable C*-algebras (although the same definition makes sense also for nonseparable C*-algebras, it is probably not the appropriate one). Thus all semiprojective C*-algebras will implicitly be separable. In the definition, B is not required to be separable; however, it is easily seen that the definition of semiprojectivity does not change if in II.8.3.7 we make any or all of the following restrictions:

(i) B is separable.
(ii) ϕ is surjective.
(iii) ϕ is injective.

(II.5.1.3 is needed to prove (i) and (ii); for (iii), replace B by $A \oplus B$ and J_n by $0 \oplus J_n$.)

II.8.3.10 Suppose A is projective (in the general category). Let $B = CA$, $\pi : CA \to A$ evaluation at 0, and $\phi : A \to A$ the identity. Then ϕ lifts to $\psi : A \to CA$, showing that A is contractible. If A is projective in the unital category, a similar argument shows that A is homotopy equivalent to $\mathbb{C}$. (But not every contractible C*-algebra is projective or even semiprojective: CA is always contractible, but is rarely semiprojective; a necessary, but not sufficient, condition is that A be semiprojective.)

The prototype example of a projective C*-algebra is $C_o((0,1])$, the universal C*-algebra generated by a single positive element of norm 1. Projectivity follows immediately from II.5.1.5.

We now list, mostly without proof, several simple facts about semiprojective C*-algebras.

II.8.3.11 PROPOSITION. [Bla85b, 2.18] Let B, J_n, and J be as in II.8.3.7, and let $q_1, \ldots, q_k$ be mutually orthogonal projections in B/J. Then for sufficiently large n, there are mutually orthogonal projections $p_1, \ldots, p_k$ in B/J_n with $\pi(p_j) = q_j$ for all j. If B (and hence B/J) is unital and $q_1 + \cdots + q_k = 1$, then we may choose the p_j so that $p_1 + \cdots + p_k = 1$.

The proof is a simple induction, with a functional calculus argument to prove the case $k = 1$.

II.8.3.12 COROLLARY. [Bla85b, 2.16] If A is unital, then the definition of semiprojectivity for A does not change if in II.8.3.7 B and ϕ are required to be unital. In particular, $\mathbb{C}$ is semiprojective.

Thus a unital C*-algebra which is semiprojective in the unital category is semiprojective in the general category.

Note that $\mathbb{C}$ is projective in the unital category. However, $\mathbb{C}$ is not projective in the general category: a *-homomorphism from $\mathbb{C}$ to B/J is effectively just a choice of projection in B/J, and projections do not lift from quotients in general.

II.8.3.13 PROPOSITION. [Bla85b, 2.23] Let B, J_n, J be as in II.8.3.7. Let v be a partial isometry in B/J, and set $q_1 = v^*v$, $q_2 = vv^*$. Suppose there are projections $p_1, p_2 \in B/J_n$ for some n with $\pi(p_j) = q_j$. Then, after increasing n if necessary, there is a partial isometry $u \in B/J_n$ with $\pi(u) = v$ and $p_1 = u^*u$, $p_2 = uu^*$.

II.8.3.14 PROPOSITION. ([Bla85b, 2.19], [Lor97]) A finite direct sum of semiprojective C*-algebras is semiprojective.

II.8.3.15 PROPOSITION. ([Bla85b, 2.28-2.29], [Lor97]) If A is semiprojective, then $M_n(A)$ is semiprojective for all n. If A is semiprojective, then any unital C*-algebra Morita equivalent to A is also semiprojective.

The unital cases of II.8.3.14 and II.8.3.15 are simple consequences of II.8.3.11 and II.8.3.13, but the nonunital cases are more delicate.

II.8.3.16 EXAMPLES. Simple repeated applications of II.8.3.11-II.8.3.15 show that the following C*-algebras are semiprojective:

(i) $\mathbb{M}_n = M_n(\mathbb{C})$, and more generally any finite-dimensional C*-algebra. (Semiprojectivity of $\mathbb{M}_n$ was first proved, although not explicitly stated, by J. Glimm [Gli60].)
(ii) $C(\mathbb{T})$, where $\mathbb{T}$ is a circle (the universal C*-algebra generated by one unitary).
(iii) Generalizing (ii), $C^*(\mathbb{F}_n)$, the full C*-algebra of the free group on n generators for n *finite* (the universal C*-algebra generated by n unitaries).

(iv) The Toeplitz algebra $\mathcal{T}$ (the universal C*-algebra generated by an isometry).
(v) The Cuntz-Krieger algebras O_A for a finite square $0-1$ matrix A (II.8.3.3(iii)), and in particular the Cuntz algebras O_n ($n \neq \infty$).
(vi) The universal unital C*-algebra U_n^{nc} generated by elements $\{x_{ij} : 1 \leq i, j \leq n\}$, with the $n \times n$ matrix (x_{ij}) a unitary. Many variations of this example are possible, including some which are interesting compact quantum groups.

Some potential or actual non-examples are:

(vii) $C^*(\mathbb{F}_\infty)$, the universal C*-algebra generated by a sequence of unitaries. The problem is that, in the setting of II.8.3.7 with B and ϕ unital, the n might have to be increased each time an additional generator is partially lifted. In fact, $C^*(\mathbb{F}_\infty) = \varinjlim(C^*(\mathbb{F}_n))$ violates the conclusion of II.8.3.17, so is not semiprojective.
(viii) The Cuntz algebra O_∞, the universal C*-algebra generated by a sequence of isometries with mutually orthogonal range projections, has the same potential difficulty as $C^*(\mathbb{F}_\infty)$. However, it turns out that O_∞ *is* semiprojective [Bla04b]. In fact, it seems likely that every (separable) purely infinite simple nuclear C*-algebra with finitely generated K-theory is semiprojective (see [Bla04b] and [Spi01] for partial results), and it appears possible that these and the finite-dimensional matrix algebras exhaust the semiprojective simple C*-algebras.
(ix) $\mathbb{K}$ is not semiprojective. In fact, it follows easily from II.8.3.17 that no infinite-dimensional AF algebra is semiprojective.
(x) $C(\mathbb{T}^n)$ for $n \geq 2$ is the universal C*-algebra generated by n commuting unitaries. Commutation relations are difficult to lift in general, and it can be shown that $C(\mathbb{T}^n)$ ($n \geq 2$) fails to satisfy the conclusion of II.8.3.17 and is thus not semiprojective. Similarly, $C([0,1]^n)$ is not semiprojective for $n > 1$.

Other interesting examples of semiprojective C*-algebras are given in [Lor97].

One of the most important features of semiprojective C*-algebras is the following approximate factorization property. The proof uses a mapping telescope construction, and is a straightforward noncommutative adaptation of the proof of the corresponding property for ANR's.

II.8.3.17 PROPOSITION. [Bla85b, 3.1] Let A be a semiprojective C*-algebra, and $(B_n, \beta_{m,n})$ be an inductive system of C*-algebras. Let

$$B = \varinjlim(B_n, \beta_{m,n}).$$

If $\phi : A \to B$ is a homomorphism, then for all sufficiently large n there are homomorphisms $\phi_n : A \to B_n$ such that $\beta_n \circ \phi_n$ is homotopic to ϕ and converges pointwise to ϕ as $n \to \infty$, where β_n is the standard map from B_n to B.

II.8.3.18 This property can be phrased in terms of stable relations. Let $(\mathcal{G}|\mathcal{R})$ be a set of generators and relations, which to avoid complications we will take to be finite. If $\mathcal{G} = \{x_1, \ldots, x_k\}$, then a set $\{y_1, \ldots, y_k\}$ in a C*-algebra A *approximately satisfies the relations within* $\delta > 0$ if, whenever

$$\|p(x_{i_1}, \cdots, x_{i_n}, x_{i_1}^*, \cdots, x_{i_n}^*)\| \leq \eta$$

is a relation in $\mathcal{R}$, we have

$$\|p(y_{i_1}, \cdots, y_{i_n}, y_{i_1}^*, \cdots, y_{i_n}^*)\| < \eta + \delta.$$

II.8.3.19 DEFINITION. The finite set $(\mathcal{G}|\mathcal{R})$ is *weakly stable* (or has *weakly stable relations*) if for any $\epsilon > 0$ there is a $\delta > 0$ such that, whenever A is a C*-algebra and $\{y_1, \ldots, y_k\} \subseteq A$ approximately satisfy the relations within δ, there are $x_1, \ldots, x_k \in A$ exactly satisfying the relations with $\|x_j - y_j\| < \epsilon$ for $1 \leq j \leq k$.

II.8.3.20 It is easily seen that if $C^*(\mathcal{G}|\mathcal{R})$ is semiprojective, then $(\mathcal{G}|\mathcal{R})$ is weakly stable. The converse is not quite true.

The situation can be rephrased: for $\delta > 0$, let $\mathcal{R}_\delta$ be the set of relations in $\mathcal{R}$, "softened" by replacing each η by $\eta + \delta$. Then there is a natural homomorphism π_δ from $C^*(\mathcal{G}|\mathcal{R}_\delta)$ to $C^*(\mathcal{G}|\mathcal{R})$. The relations $\mathcal{R}$ are weakly stable if there is an approximate right inverse for π_δ for sufficiently small δ. It follows easily from the definition that if $C^*(\mathcal{G}|\mathcal{R})$ is semiprojective, then there is an exact right inverse for π_δ for small δ. If there is an exact right inverse for π_δ for some δ, the relations are said to be *stable*. It turns out that if $\mathcal{R}$ is finite and stable, then $C^*(\mathcal{G}|\mathcal{R})$ is semiprojective, so in the finitely presented case semiprojectivity is equivalent to stable relations.

See [Lor97] for a complete discussion of these and related matters.

II.8.3.21 An important special case concerns approximate matrix units. If D is a finite-dimensional C*-algebra and $\{æ_{ij}^{(r)}\}$ is a set of elements of a C*-algebra A, indexed like a set of matrix units of type D (II.8.3.2(iv)), then $\{æ_{ij}^{(r)}\}$ is a set of *approximate matrix units of type D within δ* if the $æ_{ij}^{(r)}$ approximately satisfy the matrix unit relations within δ. It then follows that

II.8.3.22 PROPOSITION. Let D be a finite-dimensional C*-algebra. Then for any $\epsilon > 0$ there is a $\delta > 0$ such that, whenever A is a C*-algebra and $\{æ_{ij}^{(r)}\}$ is a set of approximate matrix units of type D within δ in A, then there is a set $\{e_{ij}^{(r)}\}$ of exact matrix units of type D in A with $\|e_{ij}^{(r)} - æ_{ij}^{(r)}\| < \epsilon$ for all i, j, r.

This result is used, for example, in the theory of AF algebras. Recall (II.8.2.2(iv)) that an AF algebra is a C*-algebra which is isomorphic to an inductive limit of a sequence of finite-dimensional C*-algebras. Here is a "local" description:

II.8.3.23 DEFINITION. A C*-algebra A is an *AF algebra in the local sense* (*local AF algebra*) if, for every $x_1, \ldots, x_n \in A$ and $\epsilon > 0$, there is a finite-dimensional C*-subalgebra B and elements $y_1, \ldots, y_n \in B$ with $\|x_i - y_i\| < \epsilon$ for $1 \leq i \leq n$.

An AF algebra is obviously an AF algebra in the local sense, but the converse is not obvious since the finite-dimensional C*-subalgebras are not nested in general. Using II.8.3.22, one can easily show:

II.8.3.24 COROLLARY. *Every separable local AF algebra is an AF algebra. In particular, any countable inductive limit of AF algebras is an AF algebra.*

One could relax the definition of an AF algebra to allow inductive limits of finite-dimensional C*-algebras over arbitrary (directed) index sets. An AF algebra in this sense is an AF algebra in the local sense, and a separable AF algebra in this sense is an AF algebra in the sense of II.8.2.2. The equivalence of AF and local AF is unclear in the nonseparable case, and both definitions have been used for nonseparable AF algebras. Nonseparable AF algebras are somewhat mysterious, and separability is usually included in the definition of an AF algebra.

II.8.4 Extensions and Pullbacks

The theory of extensions of C*-algebras should properly be regarded as a part of K-theory; indeed, it is the starting point for "K-homology" and bivariant K-theory, as will be described in V.1.4. But it is appropriate to discuss the elementary aspects in this section, as a source of examples and also to serve as an introduction to the use of homological algebra ideas in C*-algebras.

Extension theory is important in many contexts, since it describes how more complicated C*-algebras can be constructed out of simpler "building blocks". Some of the most important applications of extension theory are:

(i) Structure of type I C*-algebras, group C*-algebras, and crossed products.
(ii) Classification of essentially normal operators.
(iii) Index theory for elliptic pseudodifferential operators.
(iv) Using associated homological invariants to distinguish between C*-algebras (often simple C*-algebras).

Extensions

II.8.4.1 DEFINITION. Let A and B be C*-algebras. An *extension of A by B* is a short exact sequence

$$0 \longrightarrow B \xrightarrow{j} E \xrightarrow{q} A \longrightarrow 0$$

of C*-algebras ("exact" means j is injective, q is surjective, and $ker(q) = im(j)$).

One could simply regard the C*-algebra E as the extension, but the exact sequence language gives a much better theory since it keeps track of how the ideal B and the quotient $A \cong E/B$ are related to E.

There is some nonuniformity of terminology concerning extensions: sometimes such a sequence is called an extension of B by A. We have adopted what seems to have become the dominant terminology (it is standard in homological algebra), especially since it matches up nicely with the notation of bivariant K-theory.

The goal of extension theory is, given A and B, to describe and classify all extensions of A by B up to a suitable notion of equivalence.

II.8.4.2 There is always at least one extension $E = A \oplus B$. If B is unital, this is the only extension (up to strong isomorphism; cf. II.8.4.12). Thus extension theory is only interesting if B is nonunital. The most interesting case is when B is stable.

It is most important to study *essential* extensions, ones where B is an essential ideal in E (II.5.4.7). Essential extensions only occur if B is nonunital.

II.8.4.3 EXAMPLES.

(i) Let $A = \mathbb{C}$, $B = C_o((0,1))$. There are four possible choices of E : $C_o((0,1)) \oplus \mathbb{C}$, $C_o((0,1])$, $C_o([0,1))$, and $C(S^1)$. Each has an obvious associated exact sequence. The last three are essential. It is not clear at this point whether we should regard the extensions corresponding to $C_o((0,1])$ and $C_o([0,1))$ as being the "same" or "different".

(ii) Let $\mathcal{T}$ be the Toeplitz algebra (II.8.3.2(v)). Then $\mathcal{T}$ is an essential extension of $C(\mathbb{T})$ by $\mathbb{K}$.

The Busby Invariant

The key to analyzing extensions is the so-called Busby invariant. R. Busby [Bus68] was the first to study extensions of C*-algebras. The Busby invariant is based on an earlier, purely algebraic construction of Hochschild.

II.8.4.4 Given an extension $0 \to B \to E \to A \to 0$, B sits as an ideal of E. Hence there is an associated *-homomorphism σ from E into $M(B)$ (II.7.3.1), which is injective if and only if B is essential in E. If we compose σ with the quotient map $\pi : M(B) \to Q(B) = M(B)/B$, we obtain a *-homomorphism τ from $E/B \cong A$ to $Q(B)$. $Q(B)$ is called the *outer multiplier algebra* of B.

II.8.4.5 DEFINITION. τ is the *Busby invariant* of the extension $0 \to B \to E \to A \to 0$.

The Busby invariant τ is injective if and only if B is essential in A.

II.8.4.6 EXAMPLES.

(i) In the situation of II.8.4.3(i), $M(B) \cong C(\beta\mathbb{R})$, and $Q(B) \cong C(\beta\mathbb{R} \setminus \mathbb{R})$. $\beta\mathbb{R} \setminus \mathbb{R}$ has two components. The Busby invariant corresponding to the four extensions is the map from $\mathbb{C}$ to $Q(B)$ sending 1 respectively to 0, the characteristic function of the component at $+\infty$, the characteristic function of the component at $-\infty$, and 1.

(ii) In II.8.4.3(ii), if $\mathcal{T} = C^*(S)$, the Busby invariant is the map from $C(\mathbb{T})$, regarded as the universal C*-algebra generated by a single unitary u, to the Calkin algebra $\mathcal{Q} = Q(\mathbb{K})$ sending u to the unitary $\pi(S)$.

II.8.4.7 An extension can be recovered from its Busby invariant (II.8.4.10(i)). We often identify an extension with its Busby invariant, so an extension is frequently regarded as a *-homomorphism into an outer multiplier algebra instead of as an exact sequence.

Pullbacks

We now discuss a useful general algebraic construction which works for C*-algebras.

II.8.4.8 Suppose A_1, A_2, B are C*-algebras, and ϕ_i is a *-homomorphism from A_i to B. We seek a C*-algebra P and *-homomorphisms ψ_i from P to A_i making the following diagram commutative:

$$\begin{array}{ccc} P & \xrightarrow{\psi_2} & A_2 \\ \psi_1 \downarrow & & \downarrow \phi_2 \\ A_1 & \xrightarrow{\phi_1} & B \end{array}$$

and which is universal in the sense that if C is any C*-algebra and $\omega_i : C \to A_i$ satisfies $\phi_1 \circ \omega_1 = \phi_2 \circ \omega_2$, then there is a unique *-homomorphism $\theta : C \to P$ such that $\omega_i = \psi_i \circ \theta$.

Any such P is obviously unique up to isomorphism commuting with the ψ_i.

Such a P exists. One way of constructing P is as

$$\{(a_1, a_2) \mid \phi_1(a_1) = \phi_2(a_2)\} \subseteq A_1 \oplus A_2.$$

II.8.4.9 DEFINITION. P is called the *pullback* of (A_1, A_2) along (ϕ_1, ϕ_2).

II.8.4.10 EXAMPLES.

(i) Let $0 \to B \to E \to A \to 0$ be a short exact sequence of *-algebras. Form the Busby invariant $\tau : A \to Q(B)$. Then E is naturally isomorphic to the pullback of $(A, M(B))$ along (τ, π).

(ii) Let $\phi : A \to B$ be a *-homomorphism. Let $\pi_0 : C_o([0,1), B) \to B$ be evaluation at 0. Then the pullback of $(A, C_o([0,1), B))$ along (ϕ, π_0) is called the *mapping cone* of ϕ, denoted C_ϕ.

Mapping cones are an analog of the corresponding construction in topology, and are important in deriving exact sequences in bivariant K-theory.

See [Ped99] and [ELP99] for more interesting examples, applications, and generalizations of pullbacks.

II.8.4.11 The pullback construction shows that every *-homomorphism from A to $M(B)$ is the Busby invariant of an extension of A by B. In particular, if B is stable, then $Q(B)$ contains a (unital) copy of $\mathcal{Q}$ and hence of $\mathcal{L}(\mathcal{H})$ for separable $\mathcal{H}$, and thus if A is separable there are essential extensions of A by B.

Equivalence of Extensions

II.8.4.12 To have a reasonable classification of the extensions of A by B, we need a suitable notion of equivalence. There are several obvious candidates. Throughout, we fix A and B, and consider two extensions

$$0 \longrightarrow B \xrightarrow{j_1} E_1 \xrightarrow{q_1} A \longrightarrow 0$$

and

$$0 \longrightarrow B \xrightarrow{j_2} E_2 \xrightarrow{q_2} A \longrightarrow 0$$

with associated Busby invariants τ_1 and τ_2.

(i) Strong isomorphism (called "strong equivalence" in [Bus68] and [Ros82]): there is a *-isomorphism γ making the following diagram commute:

$$\begin{array}{ccccccccc} 0 & \longrightarrow & B & \xrightarrow{j_1} & E_1 & \xrightarrow{q_1} & A & \longrightarrow & 0 \\ & & {\scriptstyle id}\downarrow & & {\scriptstyle \gamma}\downarrow & & {\scriptstyle id}\downarrow & & \\ 0 & \longrightarrow & B & \xrightarrow{j_2} & E_2 & \xrightarrow{q_2} & A & \longrightarrow & 0 \end{array}$$

(ii) Weak isomorphism (called "weak equivalence" in [Bus68] and [Ros82]): there are *-isomorphisms α, β, γ making the following diagram commute:

$$\begin{array}{ccccccccc} 0 & \longrightarrow & B & \xrightarrow{j_1} & E_1 & \xrightarrow{q_1} & A & \longrightarrow & 0 \\ & & {\scriptstyle \alpha}\downarrow & & {\scriptstyle \gamma}\downarrow & & {\scriptstyle \beta}\downarrow & & \\ 0 & \longrightarrow & B & \xrightarrow{j_2} & E_2 & \xrightarrow{q_2} & A & \longrightarrow & 0 \end{array}$$

(iii) Strong (unitary) equivalence: there is a unitary $u \in M(B)$ such that $\tau_2(a) = \pi(u)\tau_1(a)\pi(u)^*$ for all $a \in A$.

(iv) Weak (unitary) equivalence: there is a unitary $v \in Q(B)$ such that $\tau_2(a) = v\tau_1(a)v^*$ for all $a \in A$.
(v) Homotopy equivalence: the homomorphisms $\tau_i : A \to Q(B)$ are homotopic.

It follows from the uniqueness of pullbacks that two extensions are strongly isomorphic if and only if their Busby invariants coincide. Thus the Busby invariant exactly determines the strong isomorphism class of an extension.

It turns out that strong and weak isomorphism are not very tractable equivalence relations on extensions in general; the other relations are much more amenable to analysis by methods of noncommutative topology. The most useful equivalence relation is actually a stabilized version of (iii)-(v) (II.8.4.18, II.8.4.27).

Addition of Extensions

In this subsection, we assume B is stable. Then for any A there is an additive structure on the set of strong (or weak) equivalence classes of extensions of A by B. This additive structure is the C*-version of a standard notion from homological algebra. An additive structure can also sometimes be defined even if B is not stable, as long as $B \cong M_2(B)$.

II.8.4.13 Fix an isomorphism of $\mathbb{K}$ with $M_2(\mathbb{K})$ (II.6.6.13); this isomorphism induces an isomorphism $B \cong M_2(B)$ and hence isomorphisms $M(B) \cong M_2(M(B))$, $Q(B) \cong M_2(Q(B))$. These isomorphisms are called *standard isomorphisms* and are uniquely determined up to unitary equivalence (and up to homotopy).

II.8.4.14 DEFINITION. If τ_1, τ_2 are extensions of A by B, then the *sum* $\tau_1 \oplus \tau_2$ is the extension whose Busby invariant is

$$\tau_1 \oplus \tau_2 : A \to Q(B) \oplus Q(B) \subseteq M_2(Q(B)) \cong Q(B)$$

where the last isomorphism is a standard isomorphism.

We have cheated slightly here: the sum of two extensions is well defined only up to strong equivalence. Thus we should really define the sum on strong equivalence classes, giving a binary operation on the set of strong equivalence classes.

II.8.4.15 PROPOSITION. Addition is a well defined binary operation on the set $\mathbf{Ext}(A, B)$ of strong equivalence classes of extensions of A by B, and is associative and commutative. So $\mathbf{Ext}(A, B)$ is a commutative semigroup.

It is convenient to define $\mathbf{Ext}(A, B)$ for general B to be $\mathbf{Ext}(A, B \otimes \mathbb{K})$.

Split and Semisplit Extensions

II.8.4.16 DEFINITION. An extension

$$0 \longrightarrow B \xrightarrow{j} E \xrightarrow{q} A \longrightarrow 0$$

is *split* if there is a *-homomorphism $s : A \to E$ such that $q \circ s = id_A$. The map s is called a *splitting*, or *cross section*, for the extension.

It turns out to be natural to regard split extensions as "trivial" extensions.

II.8.4.17 If τ is the Busby invariant of an extension, then the extension is split if and only if there is a *-homomorphism $\sigma : A \to M(B)$ such that $\tau = \pi \circ \sigma$. In particular, if B is stable, the set of split extensions is a subsemigroup of **Ext**(A, B), which is nonempty since the extension $A \oplus B$ is split.

II.8.4.18 DEFINITION. $Ext(A, B)$ is the quotient semigroup of **Ext**(A, B) by the subsemigroup of classes of trivial (split) extensions.

In other words, if τ_1 and τ_2 are extensions of A by B, then $[\tau_1] = [\tau_2]$ in $Ext(A, B)$ if and only if there are trivial extensions τ_1' and τ_2' such that $\tau_1 \oplus \tau_1'$ and $\tau_2 \oplus \tau_2'$ are strongly equivalent, i.e. the equivalence relation in $Ext(A, B)$, called *stable equivalence*, is the one generated by strong equivalence and addition of trivial extensions.

$Ext(A, B)$ is an abelian monoid (semigroup with identity) whose identity is the class of trivial extensions.

II.8.4.19 For fixed B, $Ext(A, B)$ is obviously contravariantly functorial in A. It is less obvious, but true, that for fixed A, $Ext(A, B)$ is covariantly functorial in B (cf. [Bla98, 15.9]).

II.8.4.20 Not much is known in general about the semigroup $Ext(A, B)$. It can be very pathological. Even if A is separable, it is not known in general whether the semigroup has cancellation. There is a separable C*-algebra A such that $Ext(A, \mathbb{K})$ is not a group [And78].

We can, however, give a nice description of the invertible classes. In good cases it turns out that every class is invertible, i.e. $Ext(A, B)$ is an abelian group.

II.8.4.21 The key observation is as follows. Suppose B is stable and τ is an invertible extension, i.e. there is an extension τ^{-1} such that $\tau \oplus \tau^{-1}$ is trivial. (It is customary to write τ^{-1} instead of $-\tau$ for the inverse extension even though the operation is written additively.) Then $\tau \oplus \tau^{-1} : A \to M_2(Q(B))$ lifts to a *-homomorphism

$$\phi = \begin{bmatrix} \phi_{11} & \phi_{12} \\ \phi_{21} & \phi_{22} \end{bmatrix} : A \to M_2(M(B)).$$

ϕ_{11} (and also ϕ_{22}), being the compression of a *-homomorphism, must be a completely positive contraction from A to $M(B)$, and $\pi \circ \phi_{11} = \tau$. So if τ is invertible, then τ has a completely positive lifting to $M(B)$.

The converse is true by the Generalized Stinespring Theorem (II.7.5.2) if A is separable: if τ has a completely positive contractive lifting ϕ_{11} to $M(B)$, then ϕ_{11} can be dilated to a homomorphism

$$\phi = (\phi_{ij}) : A \to M_2(M(B)).$$

$\pi \circ \phi_{11} = \tau$ is a homomorphism, so (it is easily checked) $\pi \circ \phi_{22}$ is also a homomorphism from A to $Q(B)$. $\pi \circ \phi_{22}$ is an inverse for τ. Thus we have proved

II.8.4.22 THEOREM. If A is separable and B is stable, an extension $\tau : A \to Q(B)$ defines an invertible element of $Ext(A, B)$ if and only if τ (stably) lifts to a completely positive contraction from A to $M(B)$.

II.8.4.23 DEFINITION. An extension

$$0 \longrightarrow B \xrightarrow{j} E \xrightarrow{q} A \longrightarrow 0$$

is *semisplit* if there is a completely positive contraction $s : A \to E$ such that $q \circ s = id_A$. The map s is called a *cp-splitting*, or *cp-cross section*, for the extension.

II.8.4.24 COROLLARY. An extension τ of A by B defines an invertible element of $Ext(A, B)$ if and only if $\tau \oplus \tau'$ is semisplit for some trivial τ'.

II.8.4.25 If A is separable and nuclear (II.9.4), then every extension of A by B is semisplit (IV.3.2.5), and hence $Ext(A, B)$ is a group for all B.

II.8.4.26 Let A be separable and B stable. If we consider the group $Ext(A, B)^{-1}$ of invertible elements of $Ext(A, B)$, we may express the elements as pairs (ϕ, P), where ϕ is a *-homomorphism from A to $M(B)$ and P is a projection in $M(B)$ which commutes with $\phi(A) \mod B$. A pair is trivial if and only if P actually commutes with $\phi(A)$. Since ϕ and P are only determined up to "compact perturbation" (modulo B), we must regard two pairs which agree mod B as identical. Strong equivalence corresponds to unitary equivalence of pairs, and sum to direct sum of pairs. Thus the group $Ext(A, B)$ is isomorphic to the quotient of the semigroup of equivalence classes of such pairs, under the equivalence relation generated by unitary equivalence and "compact perturbation", with direct sum, modulo the subsemigroup of classes of exact (trivial) pairs.

II.8.4.27 Using bivariant K-theory, it can be shown that the relation of "stabilized homotopy" (the equivalence relation generated by homotopy (II.5.5.6) and addition of trivial extensions) coincides with the relation in $Ext(A, B)^{-1}$ for A, B separable.

II.8.4.28 An extension τ is *absorbing* if $\tau \oplus \tau'$ is strongly equivalent to τ for any trivial extension τ'. It turns out that if A and B are separable, and at least one is nuclear, then there is an absorbing extension in each class in $Ext(A, B)$, which is unique up to strong equivalence. The existence and uniqueness of absorbing extensions follows from the Weyl-von Neumann Theorems of Voiculescu and Kasparov:

II.8.4.29 THEOREM. [NONCOMMUTATIVE WEYL-VON NEUMANN THEOREM] [Voi76] Let A be a separable C*-algebra, $\mathcal{H}$ a separable Hilbert space, π and ρ faithful nondegenerate representations of A on $\mathcal{H}$ with $\pi(A) \cap \mathcal{K}(\mathcal{H}) = \rho(A) \cap \mathcal{K}(\mathcal{H}) = \{0\}$. Then π and ρ are unitarily equivalent mod $\mathcal{K}(\mathcal{H})$: there is a unitary $U \in \mathcal{L}(\mathcal{H})$ such that $U^*\pi(x)U - \rho(x) \in \mathcal{K}(\mathcal{H})$ for all $x \in A$. If $x_1, \ldots, x_n \in A$ and $\epsilon > 0$, such a unitary exists with the additional property that $\|U^*\pi(x_j)U - \rho(x_j)\| < \epsilon$ for $1 \leq j \leq n$.

This result generalizes the classical Weyl-von Neumann Theorem (II.8.4.38), hence the name. Kasparov [Kas80a] further generalized the result to Hilbert modules.

II.8.4.30 COROLLARY. If A is a separable C*-algebra, then all nonunital split essential extensions of A by $\mathbb{K}$ are equivalent and absorbing.

There is also a version for unital extensions.

Essentially Normal Operators

Brown, Douglas, and Fillmore ([BDF73]; [BDF77]) made the first careful study of Ext-groups in the case $B = \mathbb{C}$; they were almost exclusively interested in the case $A = C(X)$ for a compact metrizable space X.

II.8.4.31 DEFINITION. If X is a locally compact Hausdorff space, then

$$Ext(X) = Ext(C_o(X), \mathbb{C}) = Ext(C_o(X), \mathbb{K}).$$

Since $C_o(X)$ is nuclear (II.9.4.4), $Ext(X)$ is a group for all X.

II.8.4.32 BDF proved that $X \mapsto Ext(X)$ is a homotopy-invariant *covariant* functor on the category of compact metrizable spaces, and proved Bott Periodicity and the six-term cyclic exact sequence, along with the special case of Voiculescu's theorem (II.8.4.30) on absorbing extensions of $C(X)$ by $\mathbb{K}$. The main theoretical consequence of these general results is that $Ext(X) \cong K_1(X)$, the first K-homology group of X. (See Chapter 5). Other important consequences include the classification of essentially normal operators and the structure of some naturally occurring C*-algebras.

II.8.4.33 Let $\mathcal{H}$ be a separable Hilbert space. An operator $T \in \mathcal{L}(\mathcal{H})$ is *essentially normal* if $T^*T - TT^* \in \mathbb{K}$. In other words, if $t = \pi(T)$ is the image in the Calkin algebra, then T is essentially normal if t is normal. The basic question studied by BDF was: under what conditions can T be written as $N + K$, where N is normal and K compact? More generally, given two essentially normal operators T_1 and T_2, under what conditions is T_1 unitarily equivalent to a compact perturbation of T_2?

II.8.4.34 For the second question, one obvious necessary condition is that T_1 and T_2 have the same essential spectrum (I.8.2.2). So the question may be rephrased: given a compact subset X of $\mathbb{C}$, denote by $EN(X)$ the set of essentially normal operators with essential spectrum X. Given $T_1, T_2 \in EN(X)$, under what conditions is T_1 unitarily equivalent to a compact perturbation of T_2?

II.8.4.35 The problem is translated into an extension problem by noting that if $T \in EN(X)$, then $C^*(t, 1) \cong C(X)$; so if we set $A(T) = C^*(T, \mathbb{K}, 1)$, $A(T)$ corresponds naturally to an extension of $C(X)$ by $\mathbb{K}$. The question of whether T_1 is unitarily equivalent to a compact perturbation of T_2 is exactly the question of whether the corresponding extensions are strongly equivalent, i.e. whether they represent the same element of $Ext(X)$.

The main theorem in the classification of essentially normal operators is the following:

II.8.4.36 THEOREM. If $X \subseteq \mathbb{C}$, $Ext(X) \cong [\mathbb{C} \setminus X, \mathbb{Z}]$, the group of homotopy classes of continuous functions of compact support from $\mathbb{C} \setminus X$ to $\mathbb{Z}$. Thus $Ext(X) \cong \prod \mathbb{Z}$, with one factor for each bounded component of $\mathbb{C} \setminus X$. The isomorphism sends the class of $T \in EN(X)$ to $(\cdots index(T - \lambda_n 1) \cdots)$ (I.8.3.1), where λ_n is in the n-th bounded component of $\mathbb{C} \setminus X$. (This *index* is constant on connected components of $\mathbb{C} \setminus X$ (I.8.3.8) and vanishes on the unbounded component.)

This is actually a special case of the Universal Coefficient Theorem (V.1.5.8).

II.8.4.37 COROLLARY. An essentially normal operator T can be written $T = N + K$, with N normal and K compact, if and only if $index(T - \lambda 1) = 0$ for all λ not in $\sigma_e(T)$. If $T_1, T_2 \in EN(X)$, then T_1 is unitarily equivalent to a compact perturbation of T_2 if and only if $index(T_1 - \lambda 1) = index(T_2 - \lambda 1)$ for all $\lambda \notin X$.

II.8.4.38 COROLLARY. If X is a compact subset of $\mathbb{C}$ with connected complement, then any essentially normal operator with essential spectrum X can be written as (normal) + (compact), and any two essentially normal operators with essential spectrum X are unitarily equivalent up to compact perturbation. In particular, any $T \in EN(X)$ is a compact perturbation of a diagonalizable normal operator (one with an orthonormal basis of eigenvectors).

In the case $X \subseteq \mathbb{R}$ (the case of essentially self-adjoint operators), II.8.4.38 applies and is the classical Weyl-von Neumann Theorem. D. Berg [Ber71] later showed that any two normal operators with the same essential spectrum are unitarily equivalent up to compact perturbation; Voiculescu's theorem (II.8.4.29) may be regarded as a generalization of this fact.

II.8.4.39 The surjectivity of the map from $Ext(X)$ to $[\mathbb{C} \setminus X, \mathbb{Z}]$ provides many examples of essentially normal, nonnormal operators. For example, let X be the "Hawaiian earring" formed as the union of all the circles of radius $1/n$ centered at $(1/n, 0)$. Then $Ext(X)$ is the full direct product of a countable number of copies of $\mathbb{Z}$; thus $Ext(X)$ is uncountable. To obtain an explicit essentially normal operator corresponding to the sequence $(k_n) \in \prod_\mathbb{N} \mathbb{Z}$, take $\bigoplus \frac{1}{n}(I + S^{d_n})$, where $d_n = k_n - k_{n-1}$ ($k_{-1} = 0$), S is the unilateral shift, $S^d = S^{*|d|}$ for $d < 0$, and $S^0 = U$, the bilateral shift.

II.8.5 C*-Algebras with Prescribed Properties

It is frequently useful to be able to construct examples of C*-algebras with a prescribed set of properties. Sometimes there is a ready example which is not separable, but it is worthwhile to have a general method for obtaining a separable example.

In this section, we provide a general method for constructing separable C*-subalgebras of C*-algebras with prescribed properties, so that the subalgebra inherits all of the properties. Thus if a nonseparable C*-algebra with a prescribed set of properties of the proper type can be found, so can a separable one.

Some of the properties mentioned here, for which this method is useful, will only be carefully introduced in subsequent sections.

II.8.5.1 DEFINITION. A property P of C*-algebras is *separably inheritable (SI)* if

(i) whenever A is a C*-algebra with property P and B is a separable C*-subalgebra of A, then there is a separable C*-subalgebra E of A which contains B and which has property P.
(ii) whenever (A_n, ϕ_n) is an inductive system of separable C*-algebras with injective connecting maps, and each A_n has property P, so does $\varinjlim A_n$.

This definition can be used either within the category of general C*-algebras and homomorphisms, or the category of unital C*-algebras and unital homomorphisms.

II.8.5.2 EXAMPLES.

(i) Any property which is inherited by all subalgebras and inductive limits is an (SI) property. For example:
"A is finite" (III.1.3.1).
(ii) Any property of the following forms is an (SI) property:
"A contains an element of the form ..."
"A contains a separable C*-subalgebra isomorphic to ..."
"A contains a separable C*-subalgebra which has (or does not have) the property ..."
So, for example, the properties "A contains a separable nonnuclear C*-subalgebra" and "A is infinite" (III.1.3.1) are (SI) properties.
(iii) The property "A is unital" is not an (SI) property in the general category, since it is not preserved under inductive limits.

The idea of the proof of the following proposition can be used to prove that many other more interesting properties are (SI) properties.

II.8.5.3 PROPOSITION. Let $P_1, P_2, \ldots$ be a sequence of (SI) properties. Then $(P_1 \wedge P_2 \wedge \cdots)$ is an (SI) property.

PROOF: Let A be a C*-algebra with properties $P_1, P_2, \ldots$, and B a separable C*-subalgebra. Define a sequence of separable C*-subalgebras

$$B \subseteq \tilde{B}_{1,1} \subseteq \tilde{B}_{2,1} \subseteq \tilde{B}_{2,2} \subseteq \cdots \subseteq \tilde{B}_{n,1} \subseteq \tilde{B}_{n,2} \subseteq \cdots \subseteq \tilde{B}_{n,n} \subseteq \tilde{B}_{n+1,1} \subseteq \cdots$$

of A such that $\tilde{B}_{n,k}$ has property P_k, by successive application of the (SI) property of each P_k. Set $\tilde{B} = [\cup \tilde{B}_{n,k}]^-$. $\tilde{B}$ is a separable C*-subalgebra of A, and $\tilde{B}$ has property P_k for each k since $\tilde{B} = \varinjlim_{n \geq k} \tilde{B}_{n,k}$.

A similar iteration argument is used in the following proof. We say that a unital C*-algebra A has *stable rank 1* if the invertible elements of A are dense in A, and a general A has stable rank one if $\tilde{A}$ does. A complete treatment of the theory of stable rank for C*-algebras is found in V.3.1.

II.8.5.4 PROPOSITION. The property "A has stable rank 1" is an (SI) property.

PROOF: Let A be a C*-algebra with stable rank 1; assume for simplicity that A is unital (the nonunital case is an easy variation). Then the invertible elements are dense in A. Let B be a separable C*-subalgebra of A. Choose a countable dense set $x_1, x_2, \ldots$ in B. For each n, let (x_{nk}) be a sequence of invertible elements in A with $x_{nk} \to x_n$, and let $\tilde{B}_1$ be the C*-subalgebra of A generated by $\{x_{nk}\}$. Then $\tilde{B}_1$ is separable, contains B, and has the property that the closure of the invertible elements of $\tilde{B}_1$ contains B. Similarly, construct a separable $\tilde{B}_2$ containing $\tilde{B}_1$ such that the closure of the invertibles in $\tilde{B}_2$ contains $\tilde{B}_1$. Continue the construction inductively, and let $\tilde{B} = [\cup \tilde{B}_n]^-$. $\tilde{B}$ is separable and has stable rank 1.

It is obvious that stable rank 1 is preserved under inductive limits.

II.8.5.5 In an almost identical way, the following properties may be proved to be (SI) properties:

"A has stable rank $\leq n$" (V.3.1.2)
"A has real rank $\leq n$" (V.3.2.1)
"A has connected unitary group" (in the unital category)
"$M_n(A)$ has connected unitary group" (in the unital category)
"$K_1(A) = 0$" (V.1.2.2)
"A has unique tracial state" (in the unital category)
"A has [strict] cancellation" (V.2.4.13)
"$K_0(A)$ is [weakly] unperforated" (V.2.4.15)
"$K_0(A)$ has the Riesz interpolation property" (V.2.4.20)
"A is an AF algebra" (II.8.2.2(iv))

As another example, we give a proof for "A has a unique tracial state." (See [Phi04] for a different proof.) If A is a unital C*-algebra with unique tracial state τ_0 and B is a separable unital C*-subalgebra with trace space $\mathcal{T}(B)$, let $\tau_1 = \tau_0|_B \in \mathcal{T}(B)$. If D is a separable C*-subalgebra of A containing B, let $\mathcal{T}_D(B)$ be the set of tracial states on B which extend to traces on D. Then $\mathcal{T}_D(B)$ is closed in $\mathcal{T}(B)$, and $D \mapsto \mathcal{T}_D(B)$ reverses inclusion. If $\tau \in \cap_D \mathcal{T}_D(B)$, for each D let τ_D be a tracial state on D extending τ, and σ_D an extension of τ_D to a state on A. Then any weak-* limit point of (σ_D) is a tracial state on A extending τ. Thus $\cap_D \mathcal{T}_D(B) = \{\tau_1\}$. Since $\mathcal{T}(B)$ is second countable (compact and metrizable), there is a sequence (D_n) with $\cap_n \mathcal{T}_{D_n}(B) = \{\tau_1\}$. Let $\tilde{B}_1$ be the C*-subalgebra of A generated by $\cup D_n$; then $\tilde{B}_1$ is separable, and the only tracial state on B which extends to a trace on $\tilde{B}_1$ is τ_1. Iterate the process to get

$$B \subseteq \tilde{B}_1 \subseteq \cdots \subseteq \tilde{B}_n \subseteq \tilde{B}_{n+1} \subseteq \cdots$$

such that the only tracial state on $\tilde{B}_n$ which extends to a trace on $\tilde{B}_{n+1}$ is τ_0 (restricted to $\tilde{B}_n$). If $\tilde{B} = [\cup \tilde{B}_n]^-$, then $\tilde{B}$ is separable and has unique trace.

II.8.5.6 THEOREM. [Bla78, 2.2] The property "A is simple" is an (SI) property.

The proof of this theorem follows the same general scheme, but is slightly more complicated technically. The proof uses the following fact, which is of independent interest (and possibly somewhat surprising at first glance since there can be uncountably many closed ideals in a separable C*-algebra):

II.8.5.7 PROPOSITION. Let B be a separable C*-algebra. Then there is a countable subset S of B such that if J is any closed ideal (not necessarily proper) of B, then $S \cap J$ is dense in J.

PROOF: First note that since $\operatorname{Prim}(B)$ is second countable (II.6.5.7), there is a countable set $\{U_n\}$ of open sets, containing $\emptyset$ and $\operatorname{Prim}(B)$, closed under finite unions and intersections, forming a base for the topology. U_n corresponds to a closed ideal J_n, and the J_n have the property that if J is a (not necessarily proper) closed ideal of B, then $\cup\{J_n : J_n \subseteq J\}$ is dense in J. Thus we can take a countable dense set S_n in J_n and let $S = \cup S_n$.

We now give the proof of II.8.5.6.

PROOF: Let A be a simple C*-algebra, and B a separable C*-subalgebra. Let $S = \{0, x_1, x_2, \dots\}$ be a countable dense set in B as in II.8.5.7. Since A is simple, for each i, j, n there is a finite set $\{y_{ijk}^{(n)}, z_{ijk}^{(n)} : 1 \leq k \leq r\} \subseteq A$ (where r depends on i, j, n) such that

$$\left\| x_j - \sum_{k=1}^{r} y_{ijk}^{(n)} x_i z_{ijk}^{(n)} \right\| < \frac{1}{n}.$$

Let $Ƀ_1$ be the C*-subalgebra of A generated by B and all the $y_{ijk}^{(n)}, z_{ijk}^{(n)}$. Then $Ƀ_1$ is separable, and if J is any closed ideal of $Ƀ_1$, either $J \cap B$ is 0, or it contains x_i for some i, in which case it contains all the x_j, so contains B. Iterate the construction to get an increasing sequence $(Ƀ_n)$, where if J is any closed ideal of $Ƀ_{n+1}$, then $J \cap Ƀ_n$ is 0 or $Ƀ_n$. If $Ƀ = [\cup Ƀ_n]^-$, then $Ƀ$ is separable, and simple by II.8.2.4.

The class of simple C*-algebras is closed under inductive limits by II.8.2.5.

II.8.5.8 There are many more (SI) properties. For example, nuclearity is an (SI) property (IV.3.1.9).

II.8.5.9 EXAMPLE. An example of an application of this theory is to show that there exists an infinite-dimensional separable simple unital C*-algebra C which has unique trace, stable rank 1, real rank 0, connected unitary group, trivial K_1, totally ordered K_0, and which cannot be embedded in a nuclear C*-algebra (hence is not exact; cf. IV.3.4.18). Let B be the full C*-algebra of the free group on two generators (II.8.3.2(iii)); B cannot be embedded in a nuclear C*-algebra by II.9.6.6. B is residually finite-dimensional (V.2.1.11), and hence can be embedded in the full direct product of a sequence of matrix algebras, which can in turn be embedded in a II_1 factor A. Construct a separable $Ƀ$ in between which has all the above properties.

II.9 Tensor Products and Nuclearity

The theory of tensor products of C*-algebras is fraught with a surprising number of technical complications, but the theory ends up in a rather satisfactory form, and behaves very nicely for a large and natural class of C*-algebras called "nuclear C*-algebras," which also perhaps surprisingly arise in several other aspects of the subject. Tensor products of C*-algebras were first studied by T. Turumaru [Tur52]; in this and some other early references they were called "direct products."

II.9.1 Algebraic and Spatial Tensor Products

II.9.1.1 If A and B are C*-algebras, we can form their algebraic tensor product $A \odot B$ over $\mathbb{C}$. $A \odot B$ has a natural structure as a *-algebra with multiplication
$$(a_1 \otimes b_1)(a_2 \otimes b_2) = a_1 a_2 \otimes b_1 b_2$$
and involution $(a \otimes b)^* = a^* \otimes b^*$. We want to show the existence of a C*-norm on $A \odot B$ and determine the extent to which it is unique. If γ is a C*-norm on $A \odot B$, we will write $A \otimes_\gamma B$ for the completion.

II.9.1.2 $A \odot B$ has the usual universal property for bilinear maps. As an algebra, it has the universal property that whenever π_A and π_B are *-homomorphisms from A and B respectively, to a complex *-algebra C, such that $\pi_A(A)$ and $\pi_B(B)$ commute, then there is a unique *-homomorphism π from $A \odot B$ to C such that $\pi(a \otimes b) = \pi_A(a)\pi_B(b)$ for all $a \in A$, $b \in B$. Taking $C = \mathcal{L}(\mathcal{H})$, we get *-representations of $A \odot B$ and hence induced C*-seminorms.

The Minimal or Spatial Tensor Product

II.9.1.3 A standard way to generate such representations is via tensor products of Hilbert spaces: if π_A and π_B are representations of A and B on Hilbert spaces $\mathcal{H}_1$ and $\mathcal{H}_2$ respectively, we can form the representation $\pi = \pi_A \otimes \pi_B$ of $A \odot B$ on $\mathcal{H}_1 \otimes \mathcal{H}_2$ by $\pi(a \otimes b) = \pi_A(a) \otimes \pi_B(b)$ (I.2.5.2). If π_A and π_B are faithful, then it is not difficult to show (cf. I.2.5.5) that $\pi_A \otimes \pi_B$ is faithful on $A \odot B$, so $A \odot B$ has at least one C*-norm. Also, for any π_A and π_B we have $\|(\pi_A \otimes \pi_B)(\sum_{i=1}^n a_i \otimes b_i)\| \leq \sum_{i=1}^n \|a_i\|\|b_i\|$, so the norm
$$\left\|\sum_{i=1}^n a_i \otimes b_i\right\|_{\min} = \sup \left\|(\pi_A \otimes \pi_B)\left(\sum_{i=1}^n a_i \otimes b_i\right)\right\|$$
(over all representations π_A of A and π_B of B) is finite and hence a C*-norm, called the *spatial* norm on $A \odot B$; it is also called the *minimal C*-norm* because it indeed turns out to be the smallest C*-norm on $A \odot B$ (II.9.5.1). (A consequence of the minimality of $\|\cdot\|_{\min}$ is that if π_A and π_B are any faithful representations of A and B respectively, and $x \in A \odot B$, then $\|(\pi_A \otimes \pi_B)(x)\| = \|x\|_{\min}$, i.e. the spatial norm is independent of the faithful representations chosen.) The completion of $A \odot B$ with respect to this norm is written $A \otimes_{\min} B$ and called the *minimal* or *spatial* tensor product of A and B.

II.9.2 The Maximal Tensor Product

There is also a maximal C*-norm on $A \odot B$. Indeed, every *-representation of $A \odot B$ comes from a pair of commuting representations of A and B as above:

II.9.2.1 THEOREM. Let A and B be C*-algebras, and π a nondegenerate representation of $A \odot B$ on a Hilbert space $\mathcal{H}$. Then there are unique nondegenerate representations π_A of A and π_B of B on $\mathcal{H}$ such that $\pi(a \otimes b) = \pi_A(a)\pi_B(b) = \pi_B(b)\pi_A(a)$ for all a, b. If π is a factor representation, then π_A and π_B are also factor representations.

The first statement is obvious if A and B are unital ($\pi_A(a) = \pi(a \otimes 1_B)$, $\pi_B(b) = \pi(1_A \otimes b)$). In the general case, if (h_λ) and (k_μ) are approximate units for A and B respectively, then for any $a \in A$ and $b \in B$ the nets $\pi(a \otimes k_\mu)$ and $\pi(h_\lambda \otimes b)$ converge in the strong operator topology to operators we call $\pi_A(a)$ and $\pi_B(b)$ respectively, and these define representations with the right properties (the proof is essentially an application of II.9.3.2 to vector states). To prove the last statement, we have $\pi_A(A)'' \subseteq \pi_B(B)'$, so

$$\pi_A(A)' \cap \pi_A(A)'' \subseteq \pi_A(A)' \cap \pi_B(B)' \subseteq \pi(A \odot B)'.$$

Also, by construction $\pi_A(A) \subseteq \pi(A \odot B)''$, so $\pi_A(A)'' \subseteq \pi(A \odot B)''$,

$$\pi_A(A)' \cap \pi_A(A)'' \subseteq \pi(A \odot B)' \cap \pi(A \odot B)'' = \mathbb{C}I.$$

II.9.2.2 COROLLARY. Let A and B be C*-algebras. Then every C*-seminorm γ on $A \odot B$ extends uniquely to a C*-seminorm on $\tilde{A} \odot \tilde{B}$, and satisfies $\gamma(a \otimes b) \leq \|a\|\|b\|$ for all $a \in A$, $b \in B$.

Thus most questions about tensor products can be reduced to the unital case.

II.9.2.3 So $\|\pi(\sum_{i=1}^n a_i \otimes b_i)\| \leq \sum_{i=1}^n \|a_i\|\|b_i\|$ for any representation π of $A \odot B$, and hence

$$\left\|\sum_{i=1}^n a_i \otimes b_i\right\|_{\max} = \sup \left\|\pi\left(\sum_{i=1}^n a_i \otimes b_i\right)\right\|$$

where the supremum is taken over all representations, is a (finite) C*-norm on $A \odot B$ which is the largest possible C*-norm. The completion is denoted $A \otimes_{\max} B$, and called the *maximal tensor product* of A and B.

II.9.2.4 If $\mathcal{X}$ and $\mathcal{Y}$ are Banach spaces, the (semi)norm defined on $\mathcal{X} \odot \mathcal{Y}$ by

$$\|\xi\|_\wedge = \inf\left\{\sum \|x_i\|\|y_i\| : \xi = \sum x_i \otimes y_i\right\}$$

is obviously the largest possible subcross seminorm ($\|x \otimes y\| \leq \|x\|\|y\|$ for all x, y), and is in fact a cross norm [Gro55]. The completion, denoted $\mathcal{X} \hat{\otimes} \mathcal{Y}$, is called the *projective tensor product* (it has a universal property analogous to projectivity). If A and B are C*-algebras, we obviously have $\|\cdot\|_{\max} \leq \|\cdot\|_\wedge$; equality rarely holds (i.e. $\|\cdot\|_\wedge$ is rarely a C*-norm).

II.9.2.5 It is obvious that $A \otimes_{\max} B$ and $A \otimes_{\min} B$ are unital if (and only if) both A and B are unital. In this case, there are natural embeddings $A \to A \otimes 1_B \subseteq A \otimes_\gamma B$ and $B \to 1_A \otimes B \subseteq A \otimes_\gamma B$ for any C*-norm γ on $A \odot B$. If A and B are unital C*-algebras, then $A \otimes_{\max} B$ can be described as the universal unital C*-algebra generated by unital copies of A and B which commute; it is harder to describe $A \otimes_{\max} B$ as a universal C*-algebra in the nonunital case since A and B do not naturally embed.

II.9.2.6 Both $\otimes_{\max}$ and $\otimes_{\min}$ are associative and commutative in the obvious senses: $(a \otimes b) \otimes c \mapsto a \otimes (b \otimes c)$ gives isomorphisms

$$(A \otimes_{\max} B) \otimes_{\max} C \cong A \otimes_{\max} (B \otimes_{\max} C)$$

$$(A \otimes_{\min} B) \otimes_{\min} C \cong A \otimes_{\min} (B \otimes_{\min} C)$$

and $a \otimes b \mapsto b \otimes a$ gives $A \otimes_{\max} B \cong B \otimes_{\max} A$ and $A \otimes_{\min} B \cong B \otimes_{\min} A$.

II.9.3 States on Tensor Products

II.9.3.1 $\|\cdot\|_{\max}$ and $\|\cdot\|_{\min}$ can be alternately described using appropriate positive linear functionals on $A \odot B$, where a positive linear functional on $A \odot B$ is a linear functional f such that $f(x^*x) \geq 0$ for all $x \in A \odot B$. The set of positive linear functionals on $A \odot B$ is denoted $(A \odot B)^*_+$. The set $(A \odot B)^*_+$ is a cone in the algebraic dual $(A \odot B)^d$ which is closed in the topology of pointwise convergence (weak^d topology).

If A and B are unital, a *state* on $A \odot B$ is an element $\phi \in (A \odot B)^*_+$ with $\phi(1 \otimes 1) = 1$. The set of states, denoted $\mathcal{S}(A \odot B)$, is a weak^d-compact convex set.

II.9.3.2 PROPOSITION. *If A and B are C*-algebras and $f \in (A \odot B)^*_+$, then f extends to a positive linear functional $\tilde{f}$ on $\tilde{A} \odot \tilde{B}$.*

PROOF: Let (h_λ) be an increasing approximate unit for A. If $b \in B_+$, and $f_b(a) = f(a \otimes b)$, then f_b is a positive linear functional on A and hence bounded; thus $\lim(f(h_\lambda \otimes b)) = \sup(f(h_\lambda \otimes b))$ exists, and we may set $\tilde{f}(1 \otimes b)$ to be the limit. Writing a general $b \in B$ as a linear combination of positive elements, we then get a linear functional $\tilde{f}$ on $\tilde{A} \odot B$ with the property that

$$\tilde{f}(x) = \lim(f((h_\lambda \otimes 1)x)) = \lim(f((h_\lambda^{1/2} \otimes 1)x(h_\lambda^{1/2} \otimes 1)))$$

for all $x \in \tilde{A} \odot B$. Positivity of $\tilde{f}$ follows from the fact that if $x \in \tilde{A} \odot B$, then

$$\tilde{f}(x^*x) = \lim(f([x(h_\lambda^{1/2} \otimes 1)]^*[x(h_\lambda^{1/2} \otimes 1)])).$$

Then $\tilde{f}$ can be extended to $\tilde{A} \odot \tilde{B}$ in the same manner.

In particular, if $f \in (A \odot B)^*_+$, then

$$\sup\{f(a \otimes b) : a \in A_+, b \in B_+, \|a\| = \|b\| = 1\}$$

is finite. We say f is a *state* of $A \odot B$ if this supremum is 1; this definition agrees with the previous one if A and B are unital.

II.9.3.3 There is a GNS representation corresponding to a positive linear functional. An element of $(A \odot B)^*_+$ defines a pre-inner product on $A \odot B$, and $A \odot B$ acts by left multiplication. But there is a technicality in obtaining a representation of $A \odot B$ (and hence $A \otimes_{\max} B$) on the completion: it must be shown that $A \odot B$ acts by bounded operators. This follows immediately from the next proposition (the nonunital case follows from the unital one by II.9.3.2). We say $x \in A \odot B$ is *algebraically positive*, written $0 \leq_a x$, if $x = \sum_{k=1}^n z_k^* z_k$ for some $z_k \in A \odot B$. Write $x \leq_a y$ if $0 \leq_a y - x$. If $x \leq_a y$, then $z^* x z \leq_a z^* y z$ for any $z \in A \odot B$, and if $x_k \leq_a y_k$ ($1 \leq k \leq n$), then $\sum x_k \leq_a \sum y_k$.

II.9.3.4 PROPOSITION. Let A and B be unital C*-algebras and $x = x^* \in A \odot B$. Then there is a $\lambda > 0$ such that $x \leq_a \lambda(1 \otimes 1)$.

PROOF: First note that if $a \in A_+$, $b \in B_+$, then $0 \leq_a a \otimes b = (a^{1/2} \otimes b^{1/2})^2$, and
$$a \otimes b + (\|a\|1 - a) \otimes b + \|a\|1 \otimes (\|b\|1 - b) = \|a\|\|b\|(1 \otimes 1)$$
so $a \otimes b \leq_a \|a\|\|b\|(1 \otimes 1)$. Also note that $(A \odot B)_{sa} = A_{sa} \odot B_{sa}$: if $x = x^* = \sum_{k=1}^n a_k \otimes b_k$, then
$$x = \frac{1}{4}\left(\sum_{k=1}^n [(a_k + a_k^*) \otimes (b_k + b_k^*) - i(a_k - a_k^*) \otimes i(b_k - b_k^*)]\right).$$

Finally, if $x = \sum_{k=1}^n a_k \otimes b_k$ with $a_k = a_k^*$, $b_k = b_k^*$, then
$$x \leq_a \sum_{k=1}^n (a_{k+} + a_{k-}) \otimes (b_{k+} + b_{k-}) \leq_a \left(\sum_{k=1}^n \|a_k\|\|b_k\|\right)(1 \otimes 1).$$

Thus $\mathcal{S}(A \odot B)$ can be identified with $\mathcal{S}(A \otimes_{\max} B)$; the identification is an affine homeomorphism for the weak^d and weak-* topologies.

II.9.3.5 If ϕ and ψ are positive linear functionals on A and B respectively, then it can be checked using the complete positivity of ϕ and ψ (II.6.9.6) that the functional $\phi \otimes \psi$ defined by
$$(\phi \otimes \psi)\left(\sum_i a_i \otimes b_i\right) = \sum_i \phi(a_i)\psi(b_i)$$
is well defined and positive (this also follows easily by considering the representation $\pi_\phi \otimes \pi_\psi$ and the vector $\xi_\phi \otimes \xi_\psi$). Let $(A \odot B)^*_{++}$ be the subset of $(A \odot B)^*_+$ of pointwise limits of convex combinations of such product functionals (one can just use multiples of products of pure states). Then for $x \in A \odot B$ we have:
$$\|x\|^2_{\max} = \sup\left\{\frac{\phi(y^* x^* x y)}{\phi(y^* y)} : \phi \in (A \odot B)^*_+, y \in A \odot B, \phi(y^* y) \neq 0\right\}$$

$$\|x\|_{\min}^2 = \sup\left\{\frac{\phi(y^*x^*xy)}{\phi(y^*y)} : \phi \in (A \odot B)^*_{++}, y \in A \odot B, \phi(y^*y) \neq 0\right\}$$

$$= \sup\left\{\frac{(\phi \otimes \psi)(y^*x^*xy)}{(\phi \otimes \psi)(y^*y)} : \phi \in \mathcal{P}(A), \psi \in \mathcal{P}(B), y \in A \odot B, (\phi \otimes \psi)(y^*y) \neq 0\right\}.$$

II.9.3.6 So $\mathcal{S}(A \otimes_{\min} B)$ [resp. $(A \odot B)^*_{++}$] can be identified with the closure of $\mathcal{S}(A \odot B) \cap (A^* \odot B^*)$ in $\mathcal{S}(A \odot B)$ [resp. the closure of $(A \odot B)^*_+ \cap (A^* \odot B^*)$ in $(A \odot B)^*_+$.] More generally, if γ is any cross norm on $A \odot B$, then $\mathcal{S}(A \otimes_\gamma B)$ can be identified with the set of states on $A \odot B$ which are continuous for γ. This subset completely determines γ by the Hahn-Banach Theorem.

II.9.3.7 There is a standard identification $f \leftrightarrow T_f$ of $(A \odot B)^d$ with the set $\text{Hom}(A, B^d)$ of linear maps from A to B^d, where $[T_f(a)](b) = f(a \otimes b)$. Recall that B^* has a structure as a matrix ordered space (II.6.9.20).

II.9.3.8 PROPOSITION.

(i) Let $f \in (A \odot B)^d$. Then $f \in (A \odot B)^*_+$ if and only if T_f is a completely positive map from A to $B^* \subseteq B^d$.

(ii) The map $f \to T_f$ is an affine homeomorphism from $(A \odot B)^*_+$ with the weak^d-topology onto $CP(A, B^*)$ with the topology of pointwise weak-* convergence.

II.9.4 Nuclear C*-Algebras

II.9.4.1 A C*-algebra A is called *nuclear* if, for every C*-algebra B, there is a unique C*-norm on $A \odot B$. If A is nuclear, we often just write $A \otimes B$ for $A \otimes_{\max} B = A \otimes_{\min} B$.

The term "nuclear" is in analogy with the notion of a nuclear space in Grothendieck's theory of tensor products of topological vector spaces [Gro55], although an infinite-dimensional C*-algebra is never a nuclear space in the sense of Grothendieck.

Section IV.3 contains a more detailed study of nuclear C*-algebras.

II.9.4.2 EXAMPLES. Let A be any C*-algebra. Then $\mathbb{M}_n \odot A$ can be identified with $M_n(A)$ in the standard way: if $\{e_{ij}\}$ are the standard matrix units of $\mathbb{M}_n$ (II.6.6.8), then

$$\left(\sum_{i,j=1}^n e_{ij} \otimes a_{ij}\right) \mapsto (a_{ij})$$

is an isomorphism. Since $M_n(A)$ is a C*-algebra under the norm of II.6.6.2, this norm is the unique C*-norm on $\mathbb{M}_n \odot A$, and in particular agrees with $\|\cdot\|_{\max}$ and $\|\cdot\|_{\min}$, and

$$\mathbb{M}_n \otimes_{\max} A = \mathbb{M}_n \otimes_{\min} A = \mathbb{M}_n \odot A$$

(written $\mathbb{M}_n \otimes A$). So $\mathbb{M}_n$ is nuclear. Similarly, it is easily seen that $A \otimes \mathbb{K}$ in the sense of II.6.6.11 agrees with $A \otimes_{\max} \mathbb{K}$ and $A \otimes_{\min} \mathbb{K}$ in the tensor product sense (although it is not the same as $A \odot \mathbb{K}$ in general), so $\mathbb{K}$ is nuclear.

II.9.4.3 There is a partial converse to this example. Let $\{e_{ij} : 1 \leq i, j \leq n\}$ be a set of matrix units of type $\mathbb{M}_n$ (II.8.3.2(iv)) in a C*-algebra A. Then the e_{ij} generate a (possibly nonunital) C*-subalgebra isomorphic to $\mathbb{M}_n$. If A is unital and $\sum_i e_{ii} = 1$, then for $a \in A$ set $a_{ij} = e_{1i} a e_{j1} \in e_{11} A e_{11}$; then $a \mapsto (a_{ij})$ is an isomorphism from A onto $M_n(e_{11} A e_{11}) \cong \mathbb{M}_n \otimes e_{11} A e_{11}$.

Another important example is:

II.9.4.4 THEOREM. [Tak64] Let $A = C_o(X)$ be commutative. Then A is nuclear, and for any C*-algebra B, $A \otimes B$ can be identified with $C_o(X, B)$ (II.1.1.3(ii)) under the map $(f \otimes b)(x) = f(x)b$. In particular, if $B = C_o(Y)$ is also commutative, then $C_o(X) \otimes C_o(Y) \cong C_o(X \times Y)$ under the identification

$$(f \otimes g)(x, y) = f(x) g(y).$$

To show A is nuclear, we may assume for simplicity that A and B are unital (II.9.2.1). Let ϕ be any pure state of $A \otimes_{\max} B$, and π_A and π_B the restrictions to A and B (II.9.2.1) of the irreducible GNS representation π_ϕ. Since A is commutative, it follows from II.9.2.1 that $\pi_A(A) \subseteq \mathbb{C}I$, and so $\pi_\phi(a \otimes b) = \chi(a) \pi_B(b)$ for all $a \in A$, $b \in B$, where $\chi \in \mathcal{P}(A) = \hat{A}$. Therefore $\phi(a \otimes b) = \chi(a) \psi(b)$ for some $\psi \in \mathcal{P}(B)$ ($\psi(b) = \langle \pi_\phi(1 \otimes b) \xi_\phi, \xi_\phi \rangle$), i.e. $\phi = \chi \otimes \psi$. Thus $\mathcal{P}(A \otimes_{\max} B) \cong X \times \mathcal{P}(B)$, and every pure state of $A \otimes_{\max} B$ factors through $A \otimes_{\min} B$, so $\|\cdot\|_{\max} = \|\cdot\|_{\min}$ on $A \odot B$. We also obtain that $\mathrm{Prim}(A \otimes_{\max} B) \cong X \times \mathrm{Prim}(B)$, and it is routine to verify the isomorphism with $C_o(X, B)$. Now let γ be a C*-norm and π the quotient map from $A \otimes_{\max} B$ onto $A \otimes_\gamma B$. π corresponds to a closed set in $\mathrm{Prim}(A \otimes_{\max} B)$; if this subset is proper, since the topology is the product topology there are ideals I in A and J in B such that $\pi|_{I \odot J} = 0$. But if a and b are nonzero elements of I and J respectively, $\gamma(a \otimes b) \neq 0$ since γ is a norm, a contradiction.

II.9.4.5 The class of nuclear C*-algebras is closed under most of the standard operations: extensions, inductive limits, tensor products (all elementary observations) and quotients (a deep theorem, cf. IV.3.1.13). It is thus not so easy to give examples of nonnuclear C*-algebras. The most elementary examples are $C^*(\mathbb{F}_2)$ and $C^*_r(\mathbb{F}_2)$:

II.9.4.6 THEOREM. Let λ and ρ be the left and right regular representations of the free group $\mathbb{F}_2$, and $\pi : C^*(\mathbb{F}_2) \to C^*_r(\mathbb{F}_2)$ the quotient map (II.8.3.2(iii)). Then

(i) [Tak64] The representation of $C_r^*(\mathbb{F}_2) \odot C_r^*(\mathbb{F}_2)$ on $l^2(\mathbb{F}_2)$ induced by (λ, ρ) is not continuous for $\|\cdot\|_{\min}$, i.e. $\|\cdot\|_{\max} \neq \|\cdot\|_{\min}$ on $C_r^*(\mathbb{F}_2) \odot C_r^*(\mathbb{F}_2)$.

(ii) The representation of $C^*(\mathbb{F}_2) \odot C_r^*(\mathbb{F}_2)$ on $l^2(\mathbb{F}_2)$ induced by $(\lambda \circ \pi, \rho)$ is not continuous for $\|\cdot\|_{\min}$, i.e. $\|\cdot\|_{\max} \neq \|\cdot\|_{\min}$ on $C^*(\mathbb{F}_2) \odot C_r^*(\mathbb{F}_2)$.

(iii) [Was90] However, the representation of $C^*(\mathbb{F}_2) \odot C^*(\mathbb{F}_2)$ on $l^2(\mathbb{F}_2)$ induced by $(\lambda \circ \pi, \rho \circ \pi)$ *is* continuous for $\|\cdot\|_{\min}$.

See [Was94, Chapter 3] for a discussion and proof.

It is also true that $\mathcal{L}(\mathcal{H})$ is nonnuclear if $\mathcal{H}$ is infinite-dimensional. In fact, "most" C*-algebras are nonnuclear, as is clear from the characterizations of IV.3.1.12.

II.9.4.7 A is nuclear if and only if $(A \odot B)^*_{++} = (A \odot B)^*_+$ for every B. This has an alternate description which motivates the connection between nuclearity and completely positive approximations:

II.9.4.8 PROPOSITION. Let π and ρ be representations of C*-algebras A and B on $\mathcal{H}$ and $\mathcal{H}'$ respectively, giving a representation $\pi \otimes \rho$ of $A \odot B$ (which extends to $A \otimes_{\min} B$) on $\mathcal{H} \otimes \mathcal{H}'$. Let ξ be a unit vector in $\mathcal{H} \odot \mathcal{H}'$, and ϕ the corresponding vector state on $A \odot B$. Then the map $T_\phi : A \to B^*$ of II.9.3.7 is a completely positive finite-rank contraction. Every completely positive finite-rank contraction from A to B^* arises in this manner.

II.9.4.9 COROLLARY. Let A and B be C*-algebras. Identify $(A \odot B)^*_+$ with $CP(A, B^*)$ as in II.9.3.7. Then $(A \odot B)^*_{++}$ is the closure in $CP(A, B^*)$ (with the topology of pointwise weak-* convergence) of the finite-rank completely positive maps. In particular, A is nuclear if and only if the finite-rank maps are dense in $CP(A, B^*)$ for every B.

In fact, nuclearity can be characterized by certain identity maps being approximable by completely positive finite-rank contractions; see IV.3.1.5.

II.9.5 Minimality of the Spatial Norm

II.9.5.1 The proof that $\|\cdot\|_{\min}$ is the smallest C*-norm on $A \odot B$ for arbitrary A and B is similar to II.9.4.4, reducing the general problem to the commutative case. Indeed, if A, B are unital and γ is a C*-norm on $A \odot B$, and $\phi \in \mathcal{P}(A)$, $\psi \in \mathcal{P}(B)$, by II.9.3.1 it suffices to show that $\phi \otimes \psi$ extends to a state on $A \otimes_\gamma B$. Let $\mathcal{S}_\gamma$ be the subset of $\mathcal{P}(A) \times \mathcal{P}(B)$ of pairs which can be extended; $\mathcal{S}_\gamma$ is weak-* closed. If $\phi \in \mathcal{P}(A)$ and $b \in B_+$, let C be the commutative C*-subalgebra of B generated by b, and ω a pure state on C with $\omega(b) = \|b\|$. Since $\|\cdot\|_\gamma = \|\cdot\|_{\min}$ on $A \otimes_\gamma C$, $\phi \otimes \omega$ is a pure state on $A \otimes_\gamma C$, and extends to a pure state θ on $A \otimes_\gamma B$. Then $\psi = \theta|_{1 \otimes B}$ is a pure state of B, and $\theta = \phi \otimes \psi$ [if $y \in B$, $0 \leq y \leq 1$, then $\phi_1(x) = \theta(x \otimes y)$ and $\phi_2(x) = \theta(x \otimes (1-y))$ are positive linear functionals on A, and $\phi = \phi_1 + \phi_2$, so by purity $\phi_1 = \psi(y)\phi$.] Thus

$$\{\psi \in \mathcal{P}(B) : \phi \otimes \psi \in \mathcal{S}_\gamma\}$$

is norming for B, and hence equals $\mathcal{P}(B)$ by the Hahn-Banach Theorem.

II.9.5.2 Thus, for any C*-norm γ on $A \odot B$, we have $\|\cdot\|_{\min} \leq \gamma \leq \|\cdot\|_{\max}$ and therefore γ satisfies

$$\gamma(a \otimes b) = \|a\|\|b\|$$

for all $a \in A$, $b \in B$ (γ is a *cross norm*). In fact, $\|\cdot\|_{\min}$ is the smallest C*-seminorm γ on $A \odot B$ satisfying $\gamma(a \otimes b) \neq 0$ for all nonzero $a \in A$, $b \in B$. A consequence of this minimality is:

II.9.5.3 COROLLARY. If A and B are simple C*-algebras, then $A \otimes_{\min} B$ is simple.

II.9.6 Homomorphisms and Ideals

II.9.6.1 If $\phi : A_1 \to A_2$ and $\psi : B_1 \to B_2$ are *-homomorphisms, there is a natural induced *-homomorphism

$$\phi \otimes \psi : A_1 \odot B_1 \to A_2 \odot B_2$$

which induces *-homomorphisms

$$\phi \otimes_{\max} \psi : A_1 \otimes_{\max} B_1 \to A_2 \otimes_{\max} B_2$$

$$\phi \otimes_{\min} \psi : A_1 \otimes_{\min} B_1 \to A_2 \otimes_{\min} B_2$$

(some details must be checked to show existence of these homomorphisms). Both homomorphisms are usually written $\phi \otimes \psi$.

II.9.6.2 The maximal and minimal norms have some obvious permanence properties. If A is a C*-subalgebra of A_1, and B is any C*-algebra, then the natural inclusion of $A \odot B$ into $A_1 \otimes_{\min} B$ extends to an isometric embedding of $A \otimes_{\min} B$ into $A_1 \otimes_{\min} B$. The corresponding statement for $\otimes_{\max}$ is false, however: since a C*-subalgebra of a nuclear C*-algebra is not necessarily nuclear (IV.3.5.7), suppose A is a nonnuclear C*-subalgebra of a nuclear C*-algebra A_1, and B is a C*-algebra for which $A \otimes_{\max} B \neq A \otimes_{\min} B$; then the induced map from $A \otimes_{\max} B$ into $A_1 \otimes_{\max} B = A_1 \otimes_{\min} B$ is not isometric. But if J is a (closed) ideal of A, then from II.6.1.6 it follows that the natural map from $J \otimes_{\max} B$ to $A \otimes_{\max} B$ is isometric for any B. Hence, if $A \otimes_{\max} B = A \otimes_{\min} B$, then $J \otimes_{\max} B = J \otimes_{\min} B$. In particular, we have:

II.9.6.3 PROPOSITION. A closed ideal in a nuclear C*-algebra is a nuclear C*-algebra.

II.9.6.4 DEFINITION.

(i) A C*-algebra A is *quasinuclear* if, whenever $B \subseteq B_1$ are C*-algebras, the "inclusion" map $A \otimes_{\max} B \to A \otimes_{\max} B_1$ is isometric.
(ii) A C*-algebra A is *seminuclear* if, whenever $A \subseteq A_1$ and B are C*-algebras, the "inclusion" map $A \otimes_{\max} B \to A_1 \otimes_{\max} B$ is isometric.

Any nuclear C*-algebra is quasinuclear and seminuclear. It is clear from the argument of the previous paragraph that a C*-subalgebra of a nuclear C*-algebra is quasinuclear or seminuclear if and only if it is nuclear.

Actually, it is a (deep) fact that any quasinuclear C*-algebra is nuclear (IV.3.1.12). But there are seminuclear C*-algebras which are not nuclear: for example, any injective C*-algebra is seminuclear (use II.9.7.1).

II.9.6.5 Maximal tensor products commute with arbitrary inductive limits, and minimal tensor products commute with inductive limits where the connecting maps are *injective*. If $A = \varinjlim(A_i, \phi_{ij})$ is an inductive system of C*-algebras, and B is any C*-algebra, then $(A_i \otimes_{\max} B, \phi_{ij} \otimes id_B)$ and $(A_i \otimes_{\min} B, \phi_{ij} \otimes id_B)$ are inductive systems; the inductive limit of the first system is isomorphic to $A \otimes_{\max} B$, and if the ϕ_{ij} are injective, then the inductive limit of the second system is $A \otimes_{\min} B$ (some details need to be checked for $\otimes_{\max}$). If the ϕ_{ij} are injective, then the connecting maps in the minimal tensor product system are also injective, but the ones in the maximal tensor product system need not be injective. Minimal tensor products do *not* commute with inductive limits with noninjective connecting maps in general.

II.9.6.6 If J is a closed ideal in B, then the natural quotient map from $A \odot B$ to $A \odot (B/J)$ extends to quotient maps

$$\pi_{\max} : A \otimes_{\max} B \to A \otimes_{\max} (B/J)$$

$$\pi_{\min} : A \otimes_{\min} B \to A \otimes_{\min} (B/J).$$

The kernel of $\pi_{\max}$ is exactly $A \otimes_{\max} J$ (regarded as an ideal in $A \otimes_{\max} B$), but the kernel of $\pi_{\min}$ can be strictly larger than $A \otimes_{\min} J$ (II.9.4.6). In the language of exact sequences, if

$$0 \to J \to B \to B/J \to 0$$

is an exact sequence of C*-algebras, and A is a C*-algebra, then

$$0 \to A \otimes_{\max} J \to A \otimes_{\max} B \to A \otimes_{\max} (B/J) \to 0$$

is exact (i.e. maximal tensor product with A is an *exact functor*), but

$$0 \to A \otimes_{\min} J \to A \otimes_{\min} B \to A \otimes_{\min} (B/J) \to 0$$

is not exact in general (i.e. minimal tensor product with A is not an exact functor). If A has the property that minimal tensor product with A is an exact functor, then A is called an *exact C*-algebra*. Every nuclear C*-algebra is exact, but not every exact C*-algebra is nuclear. A C*-subalgebra of an exact C*-algebra is exact (IV.3.4.3); thus any C*-subalgebra of a nuclear C*-algebra is exact (but not necessarily nuclear). A deep theorem of E. Kirchberg (IV.3.4.18) says that every separable exact C*-algebra can be embedded as a C*-subalgebra of a nuclear C*-algebra.

There are C*-algebras which are not exact, for example $C^*(\mathbb{F}_2)$ and any C*-algebra containing it (e.g. $\mathcal{L}(\mathcal{H})$ for infinite-dimensional $\mathcal{H}$): if J is the kernel of the quotient map $\pi : C^*(\mathbb{F}_2) \to C^*_r(\mathbb{F}_2)$, then the sequence

$$0 \to C^*(\mathbb{F}_2) \otimes_{\min} J \to C^*(\mathbb{F}_2) \otimes_{\min} C^*(\mathbb{F}_2) \to C^*(\mathbb{F}_2) \otimes_{\min} C^*_r(\mathbb{F}_2) \to 0$$

is not exact by II.9.4.6.

See IV.3 for more details about exact C*-algebras.

II.9.6.7 It is not easy to describe the ideal structure of $A \otimes_{\min} B$, let alone $A \otimes_{\max} B$, in general. The map $(I, J) \mapsto I \otimes B + A \otimes J$ gives a continuous injective map from $\mathrm{Prim}(A) \times \mathrm{Prim}(B)$ onto a dense subspace of $\mathrm{Prim}(A \otimes_{\min} B)$, which is a bijective homeomorphism if A and B are separable and one is exact (IV.3.4.25).

II.9.6.8 As a related matter, if A and B are C*-algebras and C, D are C*-subalgebras of A and B respectively, then

$$C \otimes_{\min} D \subseteq (A \otimes_{\min} D) \cap (C \otimes_{\min} B)$$

(as subsets of $A \otimes_{\min} B$). It is a difficult and subtle question whether equality holds in general, which turns out to have a negative answer. But if one of the subalgebras is hereditary, it is true and elementary to prove:

II.9.6.9 PROPOSITION. Let A and B be C*-algebras, C a C*-subalgebra of A, J a hereditary C*-subalgebra of B (e.g. a closed ideal). Then

$$(A \otimes_{\min} J) \cap (C \otimes_{\min} B) = C \otimes_{\min} J.$$

PROOF: We have $C \otimes_{\min} J \subseteq (A \otimes_{\min} J) \cap (C \otimes_{\min} B) \subseteq (\tilde{A} \otimes_{\min} J) \cap (C \otimes_{\min} B)$ (as subsets of $\tilde{A} \otimes_{\min} B$). Let (h_λ) be an approximate unit for J. If

$$x \in (\tilde{A} \otimes_{\min} J) \cap (C \otimes_{\min} B)$$

then $(1 \otimes h_\lambda)x(1 \otimes h_\lambda) \to x$. But

$$(1 \otimes h_\lambda)(C \odot B)(1 \otimes h_\lambda) \subseteq C \odot J$$

so

$$(1 \otimes h_\lambda)(C \otimes_{\min} B)(1 \otimes h_\lambda) \subseteq C \otimes_{\min} J.$$

Thus $(1 \otimes h_\lambda)x(1 \otimes h_\lambda) \in C \otimes_{\min} J$, $x \in C \otimes_{\min} J$.

More generally, if $D \subseteq B$ and (ψ_λ) is a net of completely positive contractions (or even uniformly completely bounded maps) from B to D such that $\psi_\lambda(x) \to x$ for all $x \in D$, a virtually identical proof shows that

$$(A \otimes_{\min} D) \cap (C \otimes_{\min} B) = C \otimes_{\min} D.$$

This condition is always satisfied if D is nuclear (IV.3.1.8).

If C is simple and exact, then for any A, B, and $D \subseteq B$ it turns out that $(A \otimes_{\min} D) \cap (C \otimes_{\min} B) = C \otimes_{\min} D$. Conversely, if C is simple and non-exact, then there exist A, B, D such that equality fails. Details will be found in a forthcoming book on exact C*-algebras by E. Kirchberg and S. Wassermann. See [Hur79] or [Kye84, 3.1] for a counterexample to a closely related question, the C*-analog of III.4.5.9.

II.9.7 Tensor Products of Completely Positive Maps

II.9.7.1 Using the Stinespring dilation, it is easily shown from the homomorphism result that if $\phi : A_1 \to A_2$ and $\psi : B_1 \to B_2$ are completely positive contractions, then the map $\phi \otimes \psi$ from $A_1 \odot B_1$ to $A_2 \odot B_2$ given by

$$(\phi \otimes \psi)\left(\sum x_i \otimes y_i\right) = \sum \phi(x_i) \otimes \psi(y_i)$$

is well defined and extends to a completely positive contraction, also denoted $\phi \otimes \psi$, from $A_1 \otimes_{\min} B_1$ to $A_2 \otimes_{\min} B_2$. If ϕ and ψ are conditional expectations, so is $\phi \otimes \psi$.

As an important special case, if ϕ is a state on A, then for any B the *right slice map* R_ϕ from $A \otimes_{\min} B$ to B defined by

$$R_\phi\left(\sum x_i \otimes y_i\right) = \sum \phi(x_i) y_i$$

is a completely positive contraction; if A is unital, R_ϕ is a conditional expectation from $A \otimes_{\min} B$ to $1 \otimes B$. Similarly, if ψ is a state on B, there is a left slice map $L_\psi : A \otimes_{\min} B \to A$.

There is an analogous maximal tensor product of completely positive maps. We first state a tensor product version of Stinespring's Theorem (cf. [Was94]).

II.9.7.2 LEMMA. Let A and B be C*-algebras, and $\phi : A \to \mathcal{L}(\mathcal{H})$ and $\psi : B \to \mathcal{L}(\mathcal{H})$ be completely positive contractions with $\phi(a)\psi(b) = \psi(b)\phi(a)$ for all $a \in A$, $B \in B$. Then there is a Hilbert space $\mathcal{H}'$, commuting representations π of A and ρ of B on $\mathcal{H}'$, and $V \in \mathcal{L}(\mathcal{H}, \mathcal{H}')$, with $\|V\| \leq 1$, such that $\phi(a) = V^*\pi(a)V$ and $\psi(b) = V^*\rho(b)V$ for all $a \in A$, $b \in B$.

The proof is almost identical to the proof of II.6.9.7: put a pre-inner product on $A \odot B \odot \mathcal{H}$ by

$$\langle a_1 \otimes b_1 \otimes \xi, a_2 \otimes b_2 \otimes \eta \rangle = \langle \phi(a_2^* a_1) \psi(b_2^* b_1) \xi, \eta \rangle_\mathcal{H}$$

and let $\pi(a)(x \otimes y \otimes \xi) = ax \otimes y \otimes \xi$, $\rho(b)(x \otimes y \otimes \xi) = x \otimes by \otimes \xi$.

II.9.7.3 COROLLARY. If A_1, A_2, B_1, B_2 are C*-algebras and $\phi : A_1 \to A_2$ and $\psi : B_1 \to B_2$ are completely positive contractions, then the map $\phi \otimes \psi$ from $A_1 \odot B_1$ to $A_2 \odot B_2$ given by

$$(\phi \otimes \psi)\left(\sum x_i \otimes y_i\right) = \sum \phi(x_i) \otimes \psi(y_i)$$

extends to a completely positive contraction, also denoted $\phi \otimes \psi$, from $A_1 \otimes_{\max} B_1$ to $A_2 \otimes_{\max} B_2$.

PROOF: Represent $A_2 \otimes_{\max} B_2$ faithfully on $\mathcal{H}$, and regard ϕ and ψ as mapping into $\mathcal{L}(\mathcal{H})$. Apply II.9.7.2; π and ρ define a representation σ of $A_1 \otimes_{\max} B_1$ on $\mathcal{H}'$. Set $\phi \otimes \psi(x) = V^*\sigma(x)V$ for $x \in A_1 \otimes_{\max} B_1$.

II.9.8 Infinite Tensor Products

II.9.8.1 The definition of maximal or minimal tensor product can be extended in an obvious way to finite sets of C*-algebras. Infinite tensor products of *unital* C*-algebras can also be defined as follows. We describe the infinite minimal tensor product, which is the one most frequently used; the infinite maximal tensor product is formally identical. If $\{A_i : i \in \Omega\}$ is a set of C*-algebras, then for every finite $\mathcal{F} = \{i_1, \cdots, i_n\} \subseteq \Omega$ set

$$B_\mathcal{F} = A_{i_1} \otimes_{\min} \cdots \otimes_{\min} A_{i_n}.$$

If $\mathcal{F} \subseteq \mathcal{G}$, then there is a natural isomorphism $B_\mathcal{G} \cong B_\mathcal{F} \otimes_{\min} B_{\mathcal{G} \setminus \mathcal{F}}$. If all the A_i are unital, there is a natural inclusion of $B_\mathcal{F}$ into $B_\mathcal{G}$ by $x \mapsto x \otimes 1_{\mathcal{G} \setminus \mathcal{F}}$. Thus the collection of $B_\mathcal{F}$ form an inductive system, and we define the *infinite tensor product* $\bigotimes_{i \in \Omega}^{\min} A_i$ to be the inductive limit. The "min" is usually omitted, especially if the A_i are nuclear, in which case $\bigotimes_{i \in \Omega} A_i$ is also nuclear. If each A_i is simple, then $\bigotimes_{i \in \Omega} A_i$ is simple.

II.9.8.2 EXAMPLE. Let $\Omega = \mathbb{N}$, $A_i = \mathbb{M}_2$ for all i. Then $\bigotimes_{i \in \mathbb{N}} A_i = \bigotimes_\mathbb{N} \mathbb{M}_2$ is isomorphic to the CAR algebra (II.8.2.2). $\bigotimes_\mathbb{N} \mathbb{M}_3$ is isomorphic to the UHF algebra of type 3^∞.

II.9.8.3 One can more generally form an infinite tensor product $\bigotimes (A_i, p_i)$ with respect to projections $p_i \in A_i$, by embedding $B_\mathcal{F}$ into $B_\mathcal{G}$ by

$$x \mapsto x \otimes (\otimes_{i \in \mathcal{G} \setminus \mathcal{F}} p_i).$$

The resulting C*-algebra depends strongly on the choice of the p_i. See [Bla77a] for a description of the structure of such tensor products.

II.10 Group C*-Algebras and Crossed Products

Groups arise widely in mathematics, physics and many applications, because they are the mathematical structures which encode symmetries of the systems under consideration. We examine here how to incorporate symmetries into the context of operator algebras, via crossed products.

The crossed product construction, in its various forms and generalizations, has proved to be one of the most important ideas in operator algebras, both for internal structure theory and for applications.

The idea is: given a C*-algebra A, a (locally compact topological) group G, and a (continuous) action α of G as *-automorphisms of A, construct a larger C*-algebra $A \rtimes_\alpha G$, called the *crossed product*, containing A and a group of unitaries isomorphic to G, so that the unitaries implement the action. (This is strictly correct only if A is unital and G discrete; in general, A and G sit in the multiplier algebra $M(A \rtimes_\alpha G)$.) $A \rtimes_\alpha G$ encodes the "dynamical system" (A, G, α). In particular, the covariant representations of (A, G, α) on Hilbert spaces are in natural one-one correspondence with the representations of $A \rtimes_\alpha G$.

Two key cases, which have been historically important and which motivate the usefulness of such a construction, are:

(i) Let (X, G, α) be an ordinary topological dynamical system, i.e. X is a locally compact Hausdorff space, G a locally compact topological group, and α a continuous action on X (continuous means the corresponding map from $X \times G$ to X is continuous). The term "dynamical system" is most often used when G is $\mathbb{R}$ or $\mathbb{Z}$. Then there is a natural induced action of G on the C*-algebra $C_o(X)$, and the crossed product $C_o(X) \rtimes_\alpha G$ encodes the dynamical system (X, G, α). Important applications of operator algebras in both mathematics and physics come from this construction.

(ii) If G is a locally compact group, let G act trivially on $\mathbb{C}$. The crossed product $\mathbb{C} \rtimes_\alpha G$ is called the *group C*-algebra* of G, denoted $C^*(G)$, and encodes the unitary representation theory of G. Group C*-algebras have proved to be a powerful tool in representation theory, and group representations have been one of the principal applications of operator algebras since the early days of the subject.

Crossed products are not only important in applications, but are also the source of very interesting examples of C*-algebras. In addition, some of the fundamental structure of operator algebras, particularly von Neumann algebras, involves the use of crossed products.

Various generalizations of the crossed product construction have also turned out to be important, particularly the association of a C*-algebra to a locally compact topological groupoid, and recently the notion of a "quantum group."

The structure of group C*-algebras and crossed products is enormously complicated (it includes the entire theory of unitary representations of locally compact groups), and even some basic aspects of the theory are still not completely understood and are the subject of active current research.

This is a vast subject, and we will be able to only hit the high points of the basic theory. We will describe the particularly important case of crossed products by abelian groups in somewhat more detail. In fact, the three most important groups in the theory of crossed products are $\mathbb{Z}_2$, $\mathbb{Z}$, and $\mathbb{R}$ (in approximate increasing order of both difficulty and importance).

II.10.1 Locally Compact Groups

We briefly review the most important basic features of the theory of locally compact topological groups. A full treatment can be found in many books on topological groups and/or harmonic analysis, such as [HR63].

Haar Measure

II.10.1.1 If G is a locally compact (Hausdorff topological) group, then there is a nonzero Radon measure μ on the Baire sets of G which is left invariant, i.e. $\mu(tE) = \mu(E)$ for every $t \in G$ and Baire set E, where $tE = \{ts : s \in E\}$. The measure μ is unique up to scalar multiple and is called *(a) left Haar measure* on G. The measure μ has the properties that $\mu(U) > 0$ for every nonempty open set U, and $\mu(K) < \infty$ for every compact set K, and μ is (inner and outer) regular, so that $C_c(G) \subseteq L^1(G, \mu)$ and is dense in $L^1(G, \mu)$.

There is often a natural normalization for μ, e.g. when G is discrete (so μ is just counting measure) or compact (where we usually take $\mu(G) = 1$).

If G is a Lie group, left Haar measure is a Riemannian measure given by a differential form of top dimension; hence $C_c^\infty(G)$ is a dense subspace of $L^1(G)$.

The lack of a Haar measure for groups which are not locally compact is a major obstacle to developing a unified representation theory for them.

II.10.1.2 Left Haar measure is not right invariant in general. However, if $t \in G$ is fixed, then the measure μ_t defined by $\mu_t(E) = \mu(Et)$ is left invariant, hence is $\Delta_G(t)\mu$ for some $\Delta_G(t) \in \mathbb{R}_+$. The function $\Delta_G : G \to \mathbb{R}_+$ is a continuous homomorphism called the *modular function* of G. The measure ν defined by $\nu(E) = \int_E \Delta_G(t)\, d\mu(t)$ is right invariant, called *right Haar measure* on G. (Thus Δ_G is the Radon-Nikodym derivative $\left[\frac{d\nu}{d\mu}\right]$.) Right Haar measure is unique up to a scalar multiple, and has properties analogous to those of left Haar measure.

G is *unimodular* if $\Delta_G \equiv 1$, i.e. $\mu = \nu$. The class of unimodular groups includes all groups which are discrete, compact, or abelian.

Group Algebras

If G is a group, the *group algebra* of G, denoted $\mathbb{C}G$, the algebra of formal (finite) linear combinations of elements of G, with convolution as multiplication, is an important tool, particularly when G is finite. When G is locally compact, an analogous Banach *-algebra structure on $L^1(G)$ is most useful.

II.10.1.3 Let μ be a fixed left Haar measure on G. If $f, g \in L^1(G, \mu)$, define the *convolution* $f * g$ by

$$[f * g](t) = \int_G f(s) g(s^{-1} t) \, d\mu(s).$$

$L^1(G, \mu)$ becomes a Banach algebra under this multiplication. If f^* is defined by $f^*(t) = \Delta_G(t^{-1}) \bar{f}(t^{-1})$, then $f \mapsto f^*$ is an involution on $L^1(G, \mu)$ making it into a Banach *-algebra.

If a different left Haar measure $\mu' = \alpha \mu$ is chosen, $L^1(G, \mu') \cong L^1(G, \mu)$ by the map $f \mapsto \alpha f$; thus the group algebra $L^1(G)$ is (up to isomorphism) independent of the choice of left Haar measure.

$C_c(G)$ is a dense *-subalgebra of $L^1(G)$; if G is discrete, then $C_c(G) = \mathbb{C}G$. If G is finite, then $L^1(G) = \mathbb{C}G$. If G is a Lie group, then $C_c^\infty(G)$ is also a dense *-subalgebra.

II.10.1.4 $L^1(G)$ is unital if and only if G is discrete (the unit is the characteristic function of the identity element). But $L^1(G)$ always has an approximate unit. Let Λ be the net of compact neighborhoods of the identity in G, directed by reverse inclusion, and for each $\lambda \in \Lambda$ let h_λ be a nonnegative function in $C_c(G)$ supported in λ, with $\int_G h_\lambda = 1$; then (h_λ) is an approximate unit of norm one in $L^1(G)$.

II.10.1.5 More generally, let $M(G)$ be the set of finite complex Radon measures on G. Define convolution and adjoint on $M(G)$ by identifying $M(G)$ with $C_o(G)^*$ by the Riesz Representation Theorem and defining, for $m, n \in M(G)$, $f \in C_o(G)$,

$$\int_G f(s) \, d(m * n)(s) = \int_G \int_G f(ts) \, dn(s) dm(t)$$

$$\int_G f(s) \, d(m^*)(s) = [\int_G \bar{f}(s^{-1}) \, dm(s)]^-$$

Then $M(G)$ is a unital Banach *-algebra, and if $L^1(G)$ is identified via the Radon-Nikodym theorem with the subset of $M(G)$ consisting of measures absolutely continuous with respect to μ, then $L^1(G)$ is an ideal in $M(G)$ ($M(G)$ is the "multiplier algebra" of $L^1(G)$). $M(G)$ is called the *measure algebra* of G.

Amenability

The notion of amenability for groups, and its generalizations to operator algebras, is of fundamental importance. Here we give only the definition and list some elementary properties; amenability will be further discussed in IV.3.3

II.10 Group C*-Algebras and Crossed Products 195

and IV.3.5. A full development of the theory of amenable groups can be found in [Gre69] (see also [Ped79, 7.3]); see [Pie84] for a more modern treatment, and [Pat88] for amenability in a broader context.

II.10.1.6 Let $C_b(G)$ denote the C*-algebra of bounded continuous functions on G (whose maximal ideal space is the Stone–Čech compactification of G). Then G acts by left translation on $C_b(G)$, though this action is often not strongly continuous. By a *mean* on G is meant a state of $C_b(G)$. G is *amenable* if there is a left translation-invariant mean on G. (This use of means is the origin of the term "amenable".) There are actually several equivalent senses in which an amenable group has an invariant mean.

II.10.1.7 A more concrete condition equivalent to amenability is *Følner's condition*: for every compact set C of G and every $\epsilon > 0$ there exists a ("big") compact set D of G such that

$$\frac{\mu(D \triangle (tD))}{\mu(D)} < \epsilon$$

for every $t \in C$, where $\triangle$ denotes "symmetric difference."

II.10.1.8 Every compact group is amenable (take Haar measure), as is every solvable locally compact group (hence all abelian ones). Important examples of non-amenable groups are (discrete) free groups on two or more generators, and $SL(n, \mathbb{R})$ for $n \geq 2$.

There are many equivalent characterizations of amenable groups, most of which have to do with the existence of approximately invariant objects in certain actions of the group. The next theorem summarizes some of the most important ones for discrete groups; others will be given later using the group C*-algebra.

II.10.1.9 THEOREM. Let G be a (discrete) group. Then the following are equivalent:

(i) There is a (left, right, etc.) invariant mean on $C_b(G)$ $(= l^\infty(G))$.
(ii) There is a net μ_i of asymptotically invariant probability measures on G.
(iii) G satisfies the Følner condition.
(iv) G does not admit a paradoxical (Banach-Tarski) decomposition.
(v) The trivial representation of G is weakly contained in the (left) regular representation λ.

II.10.1.10 DEFINITION. A *paradoxical decomposition* of a (discrete) group G is a collection
$$\{A_1, \ldots, A_n, B_1, \ldots, B_m\}$$
of disjoint subsets of G and elements $t_1, \ldots, t_n, s_1, \ldots, s_m$ of G such that

$$\{t_1 A_1, \ldots, t_n A_n\} \quad \text{and} \quad \{s_1 B_1, \ldots, s_m B_m\}$$

are partitions of G.

The standard example of a group with a paradoxical decomposition is the free group $\mathbb{F}_2$ on two generators a and b: let A_1 be the set of (reduced) words beginning with a, A_2 the words beginning with a^{-1}, B_1 the words beginning with b, and B_2 the words beginning with b^{-1}; set $t_1 = s_1 = e$, $t_2 = a$, $s_2 = b$. Thus $\mathbb{F}_2$ is not amenable.

Paradoxical decompositions are closely related to the Banach-Tarski paradox, which was of great interest to many leading mathematicians of the twentieth century, including von Neumann ([Neu28], [Neu29]), who observed (i) $\implies$ (iv) of II.10.1.9 (if G has a paradoxical decomposition, consider the characteristic functions of the A_i and B_j to conclude that G cannot have a left invariant mean). A full discussion of the Banach-Tarski paradox and related matters can be found in [Wag93].

Pontrjagin Duality

II.10.1.11 Suppose G is a locally compact *abelian* group. Let $\hat{G}$ be the set of all continuous homomorphisms from G to the circle group $\mathbb{T}$ (these are called the *characters* of G). Then $\hat{G}$ is an abelian group under pointwise multiplication, and is a topological group under the compact-open topology (topology of uniform convergence on compact sets). $\hat{G}$ is called the *(Pontrjagin) dual group* of G.

II.10.1.12 EXAMPLES.

(i) $\hat{\mathbb{Z}} \cong \mathbb{T}$ via $n \mapsto z^n$ for fixed $z \in \mathbb{T}$, and $\hat{\mathbb{T}} \cong \mathbb{Z}$ via $z \mapsto z^n$ for fixed $n \in \mathbb{Z}$. Similarly, $(\mathbb{Z}^n)\hat{} \cong \mathbb{T}^n$ and $(\mathbb{T}^n)\hat{} \cong \mathbb{Z}^n$.
(ii) $\hat{\mathbb{R}} \cong \mathbb{R}$ via $s \mapsto \gamma_s$, where $\gamma_s(t) = e^{ist}$.
(iii) If G is finite, then $\hat{G} \cong G$ (but the isomorphism is not natural).

The next theorem summarizes the main results about $\hat{G}$.

II.10.1.13 THEOREM. Let G be a locally compact abelian group, and $\hat{G}$ its dual group. Then

(i) $\hat{G}$ is locally compact.
(ii) $\hat{G}$ is second countable if and only if G is second countable.
(iii) $\hat{G}$ is compact [*resp.* discrete] if and only if G is discrete [*resp.* compact].
(iv) $\hat{\hat{G}} \cong G$; in fact, the natural map $t \mapsto \hat{\hat{t}}$, where $\hat{\hat{t}}(\gamma) = \gamma(t)$, is a topological isomorphism from G onto $\hat{\hat{G}}$.

If $\gamma \in \hat{G}$ and $t \in G$, we often write $\gamma(t)$ as $\langle \gamma, t \rangle$ (or $\langle t, \gamma \rangle$) to emphasize the symmetry between G and $\hat{G}$.

II.10 Group C*-Algebras and Crossed Products

There are many finer aspects to Pontrjagin duality: for example, if G is discrete, then $\hat{G}$ is connected [resp. totally disconnected] if and only if G is torsion-free [resp. torsion].

Pontrjagin duality can be generalized to nonabelian groups using Hopf C*-algebras; see II.10.8.13.

Fourier-Plancherel Transform

II.10.1.14 We continue to let G be a locally compact abelian group. If $f \in L^1(G) = L^1(G, \mu)$, we define the *Fourier-Plancherel transform* $\hat{f}$ on $\hat{G}$ by

$$\hat{f}(\gamma) = \int_G \overline{\langle \gamma, t \rangle} f(t) \, d\mu(t).$$

It is easy to verify that $(f * g)\hat{} = \hat{f}\hat{g}$, and $(f^*)\hat{} = (\hat{f})^-$ (complex conjugate).

More generally, if $m \in M(G)$, define $\hat{m}$ by $\hat{m}(\gamma) = \int_G \overline{\langle \gamma, t \rangle} \, dm(t)$.

II.10.1.15 THEOREM.

(i) [RIEMANN-LEBESGUE LEMMA] If $f \in L^1(G)$, then $\hat{f} \in C_o(\hat{G})$.
(ii) If $f \in L^1(G) \cap L^2(G)$, then $\hat{f} \in L^2(\hat{G})$.
(iii) [PLANCHEREL THEOREM] If Haar measure σ on $\hat{G}$ is suitably normalized, then $f \mapsto \hat{f}$ ($f \in L^1(G) \cap L^2(G)$) extends to an isometry from $L^2(G, \mu)$ onto $L^2(\hat{G}, \sigma)$.

II.10.2 Group C*-Algebras

II.10.2.1 Suppose π is a *(strongly continuous unitary) representation* of the locally compact group G on a Hilbert space $\mathcal{H}$, i.e. a homomorphism from G to $\mathcal{U}(\mathcal{L}(\mathcal{H}))$ which is continuous for the strong operator topology (or any of the other operator topologies; cf. I.3.2.9). If $f \in L^1(G)$, the operator

$$\pi(f) = \int_G f(t) \pi(t) \, d\mu(t)$$

(see e.g. [DS88a] for information about Banach space valued integrals) is well defined and bounded, in fact $\|\pi(f)\| \leq \|f\|_1$; and it is easy to check that $f \to \pi(f)$ is a nondegenerate *-homomorphism from the Banach *-algebra $L^1(G)$ to $\mathcal{L}(\mathcal{H})$ (a *representation* of $L^1(G)$). This representation of $L^1(G)$ is called the *integrated form* of π.

More generally, if $m \in M(G)$, define $\pi(m) = \int_G \pi(t) \, dm(t)$; then $m \mapsto \pi(m)$ is a unital *-homomorphism from $M(G)$ to $\mathcal{L}(\mathcal{H})$.

II.10.2.2 THEOREM. Every representation of $L^1(G)$ on a Hilbert space arises from a (strongly continuous unitary) representation of G; thus there is a one-one correspondence between the representation theories of G and $L^1(G)$.

The proof uses the approximate unit (h_λ) of II.10.1.4. If $t \in G$, let $_th_\lambda \in L^1(G)$ be defined by $_th_\lambda(s) = h_\lambda(t^{-1}s)$; then it can be shown that if π is a representation of $L^1(G)$ on $\mathcal{H}$, then $(\pi(_th_\lambda))$ converges strongly to a unitary we can call $\pi(t)$, and that $t \to \pi(t)$ is a strongly continuous unitary representation of G giving π.

One could try to use $C_c(G)$ or $M(G)$ in place of $L^1(G)$ in this correspondence, but the result is not true in general since both $C_c(G)$ and $M(G)$ can have additional representations: a representation of $C_c(G)$ need not be bounded, and a representation of $M(G)$ need not be nondegenerate on $L^1(G)$.

II.10.2.3 DEFINITION. Let G be a locally compact group. The *(full) group C*-algebra* $C^*(G)$ is the universal enveloping C*-algebra of the Banach *-algebra $L^1(G)$. In other words, if $f \in L^1(G)$, define $\|f\| = \|f\|_{C^*(G)}$ by

$$\|f\| = \sup\{\|\pi(f)\| : \pi \text{ a representation of } L^1(G)\}$$

and let $C^*(G)$ be the completion of $(L^1(G), \|\cdot\|)$.

It is easy to see that this definition agrees (up to an obvious isomorphism) with the definition in II.8.3.2(iii) if G is discrete.

II.10.2.4 It follows immediately that there is a natural one-one correspondence between the (strongly continuous unitary) representations of G and the (nondegenerate) representations of $C^*(G)$. If π is a representation of G and therefore of $C^*(G)$, then $\pi(G)'' = \pi(C^*(G))''$; hence, in particular, π is an irreducible [*resp.* factor] representation of G if and only if it is an irreducible [*resp.* factor] representation of $C^*(G)$.

The Reduced Group C*-Algebra

II.10.2.5 To show that $\|\cdot\|$ is actually a norm on $L^1(G)$, consider the *left regular representation* λ of G on $L^2(G)$ defined by

$$[\lambda(t)\phi](s) = \phi(t^{-1}s).$$

It is not hard to see that if $0 \neq f \in L^1(G)$, then $\lambda(f) \neq 0$.

II.10.2.6 DEFINITION. The completion of $L^1(G)$ with respect to the norm $\|f\|_r = \|\lambda(f)\|$ is called the *reduced group C*-algebra* $C_r^*(G)$. $C_r^*(G)$ is isomorphic to the closure of $\lambda(L^1(G))$ in $\mathcal{L}(L^2(G))$. There is a natural quotient map from $C^*(G)$ onto $C_r^*(G)$.

The quotient map from $C^*(G)$ to $C_r^*(G)$ is an isomorphism if and only if G is amenable. $C^*(G)$ is never a simple C*-algebra unless G is trivial; in fact, $C^*(G)$ always has a one-dimensional quotient coming from the trivial representation of G. But $C_r^*(G)$ is sometimes simple, e.g. when G is a free group on more than one generator ([Pow75], [PS79], [Ake81], [Avi82], [dlH85], [BN88], [HR89], [Béd91]). See III.2.5.20.

II.10.2.7 PROPOSITION. If G is a locally compact abelian group, then the natural quotient map from $C^*(G)$ to $C_r^*(G)$ is an isomorphism, and the Fourier-Plancherel transform extends to an isomorphism from $C^*(G)$ onto $C_o(\hat{G})$ (with its usual pointwise operations).

In fact, the characters of G are precisely the irreducible representations of G, hence of $C^*(G)$; and it is easy to check that the weak-* topology on the set of characters from $C^*(G)^*$ agrees with the compact-open topology on $\hat{G}$.

II.10.2.8 $C^*(G)$ is unital if and only if G is discrete. In general, there is a natural group $\{u_t : t \in G\}$ of unitaries in $M(C^*(G))$ isomorphic to G: u_t is the strict limit of $({}_t h_\lambda)$. The map $t \to u_t$ is strictly continuous.

II.10.2.9 $C^*(G)$ is a separable C*-algebra if and only if G is second countable. Such a G is sometimes called "separable" in the literature, but this term is inconsistent with standard terminology in topology, since a locally compact group which is separable in the topological sense of having a countable dense set is not in general second countable (countable base for the topology).

II.10.2.10 The C*-algebra structure of $C^*(G)$ is not sufficient to recover G. For example, if G is any finite abelian group with n elements, then $C^*(G)$ is isomorphic to $\mathbb{C}^n$. But $C^*(G)$ has additional structure making it into a Hopf C*-algebra, which completely determines G (II.10.8.11).

II.10.3 Crossed products

In this subsection, we describe the construction and basic properties of covariant systems and crossed products. We restrict attention to locally compact groups for lack of a good theory which is broader (the existence of Haar measure is the fundamental tool needed).

Covariant Systems and Representations

II.10.3.1 DEFINITION. A *covariant system* is a triple (A, G, α) consisting of a C*-algebra A, a locally compact group G, and a homomorphism α of G into $\text{Aut}(A)$, which is continuous for the point-norm topology (*continuous action* of G on A). Another term commonly used for covariant systems is *C*-dynamical system*.

(A, G, α) is a *separable C*-dynamical system* if A is separable and G is second countable.

The term "C*-dynamical system" comes from the fact that if (A, G, α) is a covariant system with $A \cong C_o(X)$ commutative, then (A, G, α) comes from a topological dynamical system (X, G, α) as in II.10(i).

II.10.3.2 If (A, G, α) is a covariant system, then α extends to a homomorphism from G to $\mathrm{Aut}(M(A))$ such that $t \mapsto \alpha_t(x)$ is strictly continuous for all $x \in M(A)$. However, $(M(A), G, \alpha)$ is not a C*-dynamical system in general because the extended α is not generally point-norm continuous.

II.10.3.3 PROPOSITION. Let (A, G, α) be a C*-dynamical system, and $m \in M(G)$. Then the formula
$$\alpha_m(x) = \int_G \alpha_t(x) \, dm(t)$$
defines a norm-decreasing *-homomorphism from $M(G)$ to $\mathcal{L}(A)$ with involution $T \mapsto T^*$, where $T^*(x) = [T(x^*)]^*$. If $f \in L^1(G)$, then
$$\alpha_f(x) = \int_G f(t)\alpha_t(x) \, d\mu(t).$$

Just as representations of C*-algebras and unitary representations of groups are of central importance in many applications, so representations of covariant systems are of basic importance. Experience shows that the required property is that under the representation the action α should be given by conjugation by unitary operators corresponding to the group elements. Thus:

II.10.3.4 DEFINITION. Let (A, G, α) be a covariant system. A *covariant representation* of (A, G, α) is a pair of representations (π, ρ) of A and G respectively on the *same* Hilbert space such that the "covariance relation"
$$\rho(t)\pi(a)\rho(t)^* = \pi(\alpha_t(a))$$
holds for all $a \in A$ and $t \in G$. We require that π be non-degenerate, and that ρ be strongly continuous (as is normal for unitary representations).

The Covariance Algebra and (Full) Crossed Product

II.10.3.5 We now define the covariance algebra $L^1(G, A)$ of a C*-dynamical system, the analog (generalization) of $L^1(G)$. To avoid the question of which functions from G to A should be regarded as measurable (it turns out they are what one would expect), it is convenient to begin with $C_c(G, A)$ and then complete.

Define a multiplication (convolution) and involution on $C_c(G, A)$ by
$$[f * g](t) = \int_G f(s)\alpha_s(g(s^{-1}t)) \, d\mu(s)$$
$$f^*(t) = \Delta_G(t^{-1})\alpha_t(f(t^{-1})^*)$$

Then with $\|f\|_1 = \int_G \|f(t)\| \, d\mu(t)$, $C_c(G, A)$ becomes a *-normed algebra. Let $L^1(G, A)$ be the completion.

Elements of $L^1(G, A)$ can be naturally identified with integrable functions from G to A. In particular, if $f \in L^1(G)$, $x \in A$, then $t \mapsto f(t)x$ can be naturally identified with an element of $L^1(G, A)$, denoted $f \otimes x$. The linear span of such elements is dense.

II.10.3.6 If (π, ρ) is a covariant representation of (A, G, α) on $\mathcal{H}$, then there is an associated representation (nondegenerate *-homomorphism to $\mathcal{L}(\mathcal{H})$) $\pi \rtimes \rho$ of $L^1(G, A)$: if $f \in L^1(G, A)$, set

$$[\pi \rtimes \rho](f) = \int_G \pi(f(t))\rho(t) \, d\mu(t).$$

We have that $\|[\pi \rtimes \rho](f)\| \leq \|f\|_1$. The representation $\pi \rtimes \rho$ of $L^1(G, A)$ is called the *integrated form* of (π, ρ).

Just as in II.10.2.2, every representation of $L^1(G, A)$ arises from a covariant representation of (A, G, α), which can be recovered using an approximate unit (another argument is given below in II.10.3.11); thus there is a natural one-one correspondence between the representation theory of $L^1(G, A)$ and the covariant representations of (A, G, α).

We may also put a norm (cf. II.10.3.13) on $L^1(G, A)$ by

$$\|f\| = \sup\{\|[\pi \rtimes \rho](f)\| : (\pi, \rho) \text{ a covariant representation of } (A, G, \alpha)\}$$

II.10.3.7 DEFINITION. The completion of $L^1(G, A)$ with respect to this norm (the universal enveloping C*-algebra of $L^1(G, A)$) is the *(full) crossed product* of (A, G, α), denoted $A \rtimes_\alpha G$ or sometimes $C^*(A, G, \alpha)$.

The term "crossed product" without further qualification will always mean "full crossed product." Various other notations are also sometimes used for the crossed product, such as $C^*(G, A, \alpha)$, $G \ltimes_\alpha A$, $A \times_\alpha G$.

If $A \cong C_o(X)$ is commutative, $A \rtimes_\alpha G$ is often called a *transformation group C*-algebra* (II.10.4.1).

II.10.3.8 There is a natural one-one correspondence between the (nondegenerate) representations of $A \rtimes_\alpha G$ and the covariant representations of (A, G, α). If (π, ρ) is a covariant representation of (A, G, α), then

$$[(\pi \rtimes \rho)(A \rtimes_\alpha G)]'' = [\pi(A) \cup \rho(G)]''.$$

II.10.3.9 $A \rtimes_\alpha G$ is unital if and only if A is unital and G is discrete. In general, there is an embedding $\sigma : A \to M(A \rtimes_\alpha G)$ which maps into $A \rtimes_\alpha G$ if G is discrete: for any $x \in A$ set

$$(xf)(t) = xf(t) \quad \text{and} \quad (fx)(t) = f(t)\alpha_t(x)$$

for $f \in C_c(G, A)$. It is easily verified that in this way x determines an element $\sigma(x)$ of the multiplier algebra of $A \rtimes_\alpha G$, and we obtain a *-homomorphism of

A into $M(A \rtimes_\alpha G)$, which is seen to be injective, so isometric. If G is discrete, this specializes to an embedding of A into $A \rtimes_\alpha G$ itself, given by sending x to the function which has value x at the identity element of G and 0 elsewhere. (This motivates the formulas given above for non-discrete G.)

II.10.3.10 In much the same way, there is a group $\{u_t\}$ of unitaries in $M(A \rtimes_\alpha G)$ isomorphic to G: for any $t \in G$ set

$$(tf)(s) = \alpha_t(f(t^{-1}s)) \quad \text{and} \quad (ft)(s) = f(st^{-1})\Delta_G(t^{-1})$$

for $f \in C_c(G, A)$. It is easily verified that in this way t determines a unitary element u_t of $M(A \rtimes_\alpha G)$, and we obtain an injective homomorphism of G into the unitary group of $M(A \rtimes_\alpha G)$ which is continuous for the strict topology (II.7.3.11). If G is discrete and A is unital, this specializes to a homomorphism from G into $A \rtimes_\alpha G$ given by sending $t \in G$ to the function which has value 1_A at t and 0 elsewhere. (This motivates the formulas given above for the more general case.) It is easily checked that the homomorphisms of A and G into $M(A \rtimes_\alpha G)$ satisfy the covariance relation of II.10.3.8, suitably interpreted.

II.10.3.11 These embeddings give an alternate proof of the fact that representations of $A \rtimes_\alpha G$ arise from covariant representations of (A, G, α): suppose that ω is a non-degenerate representation of $A \rtimes_\alpha G$. Then by II.6.1.6 it extends to a representation of $M(A \rtimes_\alpha G)$, which can then be restricted to A and G viewed as included in $M(A \rtimes_\alpha G)$. This gives a covariant representation of (A, G, α), and one can check that the integrated form of this covariant representation is exactly the original representation ω.

II.10.3.12 There is also an induced homomorphism ρ from $C^*(G)$ to the multiplier algebra $M(A \rtimes_\alpha G)$ (into $A \rtimes_\alpha G$ if A is unital) extending the map

$$f \mapsto \int_G f(t) u_t \, d\mu(t)$$

for $f \in L^1(G)$; if $x \in A$, $y \in C^*(G)$, then

$$\sigma(x)\rho(y) \in A \rtimes_\alpha G$$

and linear combinations of such elements are dense in $A \rtimes_\alpha G$.

The map ρ is, however, not injective in general. For example, in II.10.4.3, the map ρ always factors through $C_r^*(G)$.

The Reduced Crossed Product

The definition of the norm on a crossed product algebra does not give much indication of how to compute it, since it is usually quite difficult to exhibit all covariant representations. But there is an analog of the regular representation and reduced group C*-algebra in the setting of crossed products. This reduced crossed product construction can be used to show that $\|\cdot\|$ is a norm on $L^1(G, A)$, and that the map $t \mapsto u_t$ from G to $\mathcal{U}(M(A \rtimes_\alpha G))$ is injective.

II.10.3.13 Suppose (A, G, α) is a C*-dynamical system. Let π be a (nondegenerate) representation, usually assumed faithful, of A on a Hilbert space $\mathcal{H}$. Define a representations π_α and λ of A and G, respectively, on $L^2(G, \mathcal{H}) \cong \mathcal{H} \otimes L^2(G)$ by

$$([\pi_\alpha(x)]\xi)(t) = \alpha_t(x)(\xi(t))$$
$$[\lambda(t)\xi](s) = \xi(t^{-1}s)$$

Then (π_α, λ) is a covariant representation of (A, G, α), and it is easily seen that the corresponding representation $\pi_\alpha \rtimes \lambda$ of $L^1(G, A)$ is faithful. Define, for $f \in L^1(G, A)$,

$$\|f\|_r = \|[\pi_\alpha \rtimes \lambda](f)\| \leq \|f\|.$$

It is easy to show that $\|\cdot\|_r$ is independent of the choice of π, provided only that it is faithful; thus $\|\cdot\|_r$ is a well-defined norm on $L^1(G, A)$.

II.10.3.14 DEFINITION. The completion of $L^1(G, A)$ with respect to $\|\cdot\|_r$ is called the *reduced crossed product* of (A, G, α), denoted $A \rtimes_\alpha^r G$ or $C_r^*(A, G, \alpha)$.

There is a natural quotient map from $A \rtimes_\alpha G$ onto $A \rtimes_\alpha^r G$. This map is injective if G is amenable (or, more generally, if α is an amenable action [AD02], which can occur even if G is not amenable (cf. II.10.4.3)).

II.10.3.15 EXAMPLES.

(i) Let ι be the trivial action of G on a C*-algebra A, i.e. $\iota_t(x) = x$ for all $t \in G$, $x \in A$. Then

$$A \rtimes_\iota G \cong A \otimes_{\max} C^*(G) \quad \text{and} \quad A \rtimes_\iota^r G \cong A \otimes_{\min} C_r^*(G).$$

Thus it is sometimes useful to think of $A \rtimes_\alpha G$ for non-trivial α as being some kind of "twisted" tensor product, and there has been substantial exploration of how properties of $\otimes_{\max}$ extend to general crossed product C*-algebras.

(ii) Crossed product algebras are closely related to semidirect products of groups. Indeed if α is an action of a locally compact group H by automorphisms of another locally compact group N, then H acts on $C^*(N)$ by transport of structure. Denote this action still by α. Then, if G is the semidirect product $N \rtimes_\alpha H$, there are natural isomorphisms

$$C^*(G) \cong C^*(N) \rtimes_\alpha H \quad \text{and} \quad C_r^*(G) \cong C_r^*(N) \rtimes_\alpha^r H.$$

The C*-algebra of a general group extension may be constructed as a twisted crossed product (II.10.7.6).

(iii) Examples with A commutative will be discussed in more detail in II.10.4.12, but one very simple example is fundamental. Let X be an n-point space and $G = \mathbb{Z}_n$ acting on X by cyclic permutations. Then it is easily seen that $C(X) \rtimes_\alpha G \cong \mathbb{M}_n$. This example generalizes to an arbitrary locally compact group acting on itself by left translation (II.10.4.3).

(iv) Fix $n > 1$. For each $k \in \mathbb{Z}$, let $A_k = \mathbb{M}_n$. For $k \geq 0$, let $p_k = 1_{\mathbb{M}_n} \in A_k$, and for $k < 0$ let $p_k = e_{11} \in A_k$. Let $A = \bigotimes_{\mathbb{Z}} (A_k, p_k)$ (II.9.8.3). Then $\mathbb{Z}$ acts on A by shifting the tensor product factors. It is not difficult to see that $A \rtimes_\sigma \mathbb{Z} \cong O_n \otimes \mathbb{K}$. More generally, if A is any $\{0,1\}$-matrix, then $O_A \otimes \mathbb{K}$ is the crossed product of an AF algebra by a shift-type automorphism [CE81].

Invariance Under Conjugacy

II.10.3.16 Different actions can produce isomorphic crossed products. If α and β are *conjugate* actions of G on A, in the sense that there is an automorphism γ of A with $\beta_t = \gamma^{-1} \circ \alpha_t \circ \gamma$ for all $t \in G$, then the map $\phi : C_c(G, A) \to C_c(G, A)$ defined by $(\phi f)(t) = \gamma(f(t))$ extends to isomorphisms $A \rtimes_\beta G \cong A \rtimes_\alpha G$ and $A \rtimes_\beta^r G \cong A \rtimes_\alpha^r G$.

Similarly, if θ is an automorphism of G, and $\beta_t = \alpha_{\theta(t)}$ for all $t \in G$, then the map $\psi : C_c(G, A) \to C_c(G, A)$ given by $(\psi f)(t) = \lambda f(\theta(t))$, where $\lambda \in \mathbb{R}_+$ is the factor by which left Haar is multiplied under θ, gives isomorphisms $A \rtimes_\beta G \cong A \rtimes_\alpha G$ and $A \rtimes_\beta^r G \cong A \rtimes_\alpha^r G$.

II.10.3.17 There is a more important invariance. Suppose α and β are actions of G on A which agree up to inner automorphisms, i.e. for each t there is a unitary w_t in $M(A)$ such that $\beta_t(x) = w_t \alpha_t(x) w_t^*$ for all $x \in A$. Then, for any $s, t \in G$,
$$w_{st} = z(s,t) w_s \alpha_s(w_t)$$
for some unitary $z(s,t) \in \mathcal{Z}(M(A))$.

To obtain an isomorphism between $A \rtimes_\beta G$ and $A \rtimes_\alpha G$, we need a little more:

(i) The map $t \to w_t$ is measurable.
(ii) For each $s, t \in G$, $z(s,t) = 1$, i.e. $w_{st} = w_s \alpha_s(w_t)$.

Then the map $\phi : L^1(G, A) \to L^1(G, A)$ with $(\phi f)(t) = f(t) w_t$ extends to isomorphisms $A \rtimes_\beta G \cong A \rtimes_\alpha G$ and $A \rtimes_\beta^r G \cong A \rtimes_\alpha^r G$.

II.10.3.18 DEFINITION. Actions α and β of a locally compact group G on a C*-algebra A are *outer conjugate* if there is a $\gamma \in \text{Aut}(A)$ such that $\gamma^{-1} \circ \beta_t \circ \gamma$ and α_t agree modulo inner automorphisms for each $t \in G$.

The actions α and β are *cocycle conjugate* if there is a $\gamma \in \text{Aut}(A)$ and $t \mapsto w_t \in \mathcal{U}(M(A))$, satisfying II.10.3.17 (i) and (ii), such that
$$\gamma^{-1} \circ \beta_t \circ \gamma(x) = w_t \alpha_t(x) w_t^*$$
for all $t \in G$, $x \in A$.

As above, if α and β are cocycle conjugate, then $A \rtimes_\beta G \cong A \rtimes_\alpha G$ and $A \rtimes_\beta^r G \cong A \rtimes_\alpha^r G$. If α and β are outer conjugate, then $A \rtimes_\beta G$ is isomorphic to a twisted crossed product of A by G under α (II.10.7.6).

II.10.3.19 There is a slightly stronger equivalence on actions which is extremely important. If α and β are actions of G on A, and (w_t) can be chosen in $\mathcal{U}(M(A))$ with $\beta_t = (Ad\ w_t) \circ \alpha_t$, satisfying the cocycle condition II.10.3.17(ii), and such that $t \mapsto w_t$ is strictly continuous, then α and β are *exterior equivalent*.

The following characterization of exterior equivalence is known as Connes' 2×2 matrix trick (actually half of Connes' trick – the other half is in the proof of III.4.6.3), first used to show that the modular automorphism group of a von Neumann algebra is unique up to exterior equivalence.

II.10.3.20 PROPOSITION. Let α and β be actions of G on a C*-algebra A. Then α and β are exterior equivalent if and only if there is an action γ of G on $M_2(A)$ such that $\gamma_t(diag(x,y)) = diag(\alpha_t(x), \beta_t(y))$ for all $x, y \in A$, $t \in G$.

PROOF: If α and β are exterior equivalent via (w_t), define γ by

$$\gamma_t\left(\begin{bmatrix} a & b \\ c & d \end{bmatrix}\right) = \begin{bmatrix} \alpha_t(a) & \alpha_t(b)w_t^* \\ w_t\alpha_t(c) & \beta_t(d) \end{bmatrix}.$$

Conversely, if γ exists, extend γ to $M(M_2(A)) \cong M_2(M(A))$ as in II.10.3.2, and for each t set $m_t = \gamma_t\left(\begin{bmatrix} 0 & 0 \\ 1 & 0 \end{bmatrix}\right)$. Since $m_t^* m_t = diag(1,0)$ and $m_t m_t^* = diag(0,1)$ for all t, m_t must be of the form $\begin{bmatrix} 0 & 0 \\ w_t & 0 \end{bmatrix}$ for some unitary w_t in $M(A)$, and it is easily checked that the w_t have the right properties (cf. [Ped79, 8.11.2]).

II.10.4 Transformation Group C*-Algebras

II.10.4.1 We now consider the special case of covariant systems (A, G, α) in which A is commutative, so that $A \cong C_o(X)$ for some locally compact space X. Then α gives an action as homeomorphisms of the pure state space of A, which is X. We denote this action again by α, so that

$$(\alpha_t f)(x) = f(\alpha_{t^{-1}}(x))$$

for $f \in A$, $t \in G$ and $x \in X$. Then α is seen to be a jointly continuous action, so that, by definition, (X, G, α) is a transformation group. Conversely, if (X, G, α) is a transformation group, then α defines an action of G on $A = C_o(X)$ such that (A, G, α) is a covariant system. In this case the crossed product algebra $A \rtimes_\alpha G$ is often called a *transformation group C*-algebra*. It can be denoted by $C^*(G, X, \alpha)$. One also has the corresponding reduced transformation group C*-algebra.

II.10.4.2 Probably the most fundamental example comes from the action of a locally compact group G on itself by left translation. An exact analog of the left regular representation of G gives an action λ of G on $C_o(G)$: if $t \in G$, $f \in C_o(G)$, set $[\lambda_t(f)](s) = f(t^{-1}s)$. Then

II.10.4.3 THEOREM. For any locally compact group G, $C_o(G) \rtimes_\lambda G$ is isomorphic to $\mathcal{K}(L^2(G))$.

This theorem is almost obvious if G is discrete, since there are obvious matrix units in this case. In the general case, if $\{f_i\}$ is an orthonormal basis for $L^2(G)$ contained in $C_c(G)$, then the elements $f_i \otimes f_j$ form a set of matrix units in the crossed product, whose linear combinations are dense.

The general result is a special case of Takai duality (II.10.5.2) and its noncommutative generalization, although the result long predates the duality theorem: the result for $\mathbb{R}$ (in the form of II.10.4.5) is known as the *Stone-von Neumann Theorem* ([Sto30], [vN31]), and the general result was proved in [Mac49] and [Loo52] (cf. [Rie72], [Ros04]).

Note that since $\mathcal{K}(L^2(G))$ is simple, the homomorphism from $C_o(G) \rtimes_\lambda G$ onto the reduced crossed product must be an isomorphism, regardless of whether G is amenable.

II.10.4.4 Suppose now that G is abelian. Then $C_o(G) \cong C^*(\hat{G})$ (II.10.2.7), and representations of $C_o(G)$ correspond to unitary representations of $\hat{G}$. Thus covariant representations of $(C_o(G), G, \lambda)$ correspond to pairs (ρ, U) of unitary representations of G and $\hat{G}$ respectively which satisfy a version of the covariance relation. It is easily seen that this version is

$$U_u \rho_t = \langle t, u \rangle \rho_t U_u$$

for $t \in G$ and $u \in \hat{G}$. This relation is exactly the Weyl form of the Heisenberg commutation relations, especially when $G = \mathbb{R}^n$.

It follows that representations of the Weyl form of the Heisenberg commutation relations correspond to representations of $C_o(G) \rtimes_\lambda G \cong \mathcal{K}(L^2(G))$. Since every representation of the latter is unitarily equivalent to a direct sum of copies of its canonical (irreducible) representation on $L^2(G)$ (IV.1.2.2), we obtain:

II.10.4.5 COROLLARY. [STONE-VON NEUMANN THEOREM] Every representation of the Weyl form of the Heisenberg commutation relations for the abelian group G is unitarily equivalent to a direct sum of copies of the canonical representation on $L^2(G)$, for which G acts by translation and $\hat{G}$ acts by pointwise multiplication by characters.

```
              Buchhandlung Dr. Kohl
                  Am Kornmarkt
               55543 Bad Kreuznach
                Tel.: 0671 / 8340470


Ihre Kundennummer : 6437
Dietmar Wolter
Guldenbachstr. 8
55444 Schweppenhausen


Operator Algebras
978-3-540-28486-4                   112,30 1



         Total: 1              112,30 EUR

         Bar:                  120,00 EUR
         ─────────────────────────────────
         Zurück:                  7,70 EUR

Betrag enthält 7,35 EUR MWSt:
1:  7,00% =     7,35  Netto:  104,95
USt-Idnr.: DE 148095562
05.04.07  17:02:23  82-3-2004


       Vielen Dank für Ihren Einkauf
```

II.10.4.6 One generalization of the above theorem concerns free proper actions. An action of G on X is said to be *free* if $\alpha_t(x) = x$ for some $x \in X$ implies $t = e_G$, the identity element of G.

Thus the action of G on the orbit of any point looks like the action of G on itself by left translation. The action α is said to be *proper* if for any compact subset C of G the set
$$\{t \in G : \alpha_t(C) \cap C \neq \emptyset\}$$
is compact (i.e. the map $G \times X \to X \times X$ sending (t, x) to $(x, \alpha_t(x))$ is a proper map). Let X/α denote the space of orbits of points of X with the quotient topology. When α is proper this quotient topology is Hausdorff and locally compact. (The converse is not true in general.) The algebra $C_o(X/\alpha)$ can be viewed as a subalgebra of $C_b(X)$ consisting of functions which are constant on orbits. Thus $C_o(X)$, and in particular $C_c(X)$, can be viewed as right modules over $C_o(X/\alpha)$. We turn $C_c(X)$ into a pre-Hilbert $C_o(X/\alpha)$-module (II.7.1) by defining on it a $C_o(X/\alpha)$-valued inner product given by
$$\langle \xi, \eta \rangle(\dot{x}) = \int_G \bar{\xi}(\alpha_{t^{-1}}(x))\eta(\alpha_{t^{-1}}(x))dt,$$
where $\dot{x}$ denotes the orbit through x.

II.10.4.7 Now $C_o(X)$ acts on $C_c(X)$ by pointwise multiplication, and the corresponding operators are adjointable (II.7.2.1) for the above inner product. The action of G on $C_c(X)$ by α is also adjointable, and is "unitary" once it is adjusted by the modular function. These actions satisfy the covariance relation of II.10.3.8, and their "integrated form" gives a representation of $C_c(G, C_o(X))$ with its convolution product and involution defined earlier. Once $C_c(X)$ is completed to form a Hilbert $C_o(X/\alpha)$-module, $\mathcal{E}$, this representation extends to a homomorphism of $C_o(X) \rtimes_\alpha G$ into $\mathcal{L}(\mathcal{E})$.

II.10.4.8 The "rank-one operators" (II.7.2.4) are seen to be given by elements of $C_o(X) \rtimes_\alpha G$, namely,
$$\Theta_{\xi,\eta}(t, x) = \Delta(t)^{-1}\xi(x)\bar{\eta}(\alpha_t^{-1}(x)).$$

These rank-one operators generate an ideal in $C_o(X) \rtimes_\alpha G$, which is Morita equivalent (II.7.6.8) to $C_o(X/\alpha)$. (For any proper action the full and reduced crossed products coincide [Phi89].)

When α is free as well as proper, one finds that this ideal is all of $C_o(X) \rtimes_\alpha G$ (and conversely). Thus we obtain:

II.10.4.9 THEOREM. Let α be a free and proper action of G on X. Then $C_o(X) \rtimes_\alpha G$ is Morita equivalent to $C_o(X/\alpha)$.

II.10.4.10 EXAMPLE. Let α be the action of the two-element group $\mathbb{Z}_2$ on the unit sphere S^2 in $\mathbb{R}^3$ which takes each point to its negative (i.e. its antipodal point), so that S^2/α is the corresponding projective space. Then $C(S^2) \rtimes_\alpha \mathbb{Z}_2$ is Morita equivalent to $C(S^2/\alpha)$. One can see that $C(S^2) \rtimes_\alpha \mathbb{Z}_2$ is a continuous field (IV.1.6.1) of 2×2 matrix algebras over S^2/α, and so is a homogeneous (IV.1.4.1) C*-algebra, but that it is not isomorphic to $M_2(C(S^2/\alpha))$. This illustrates why the looser notion of Morita equivalence is in many situations more useful than isomorphism in elucidating the structure of C*-algebras.

II.10.4.11 The above theorem points to another very important role for crossed products and their generalizations. There are many situations in which understanding the interaction between an action α on X and analytical or geometrical structure on X is best seen in terms of features of X/α, for example its algebraic topology. But suppose now that α, while free, is far from proper.

II.10.4.12 EXAMPLES.

(i) *Irrational rotation algebras* (cf. II.8.3.3(i)). Let α denote the action of the group $\mathbb{Z}$ on the circle, $\mathbb{T}$, by rotations through multiples of a fixed angle $2\pi\theta$ where θ is irrational. This action is free, but every orbit is dense, and the quotient topology on the orbit space $\mathbb{T}/\alpha$ has only two open sets. Thus the quotient topology is useless for most purposes. But the crossed product $C(\mathbb{T}) \rtimes_\alpha \mathbb{Z}$ is still well-defined, and is a simple unital C*-algebra called an *irrational rotation C*-algebra*. One can ask when two irrational rotation C*-algebras are isomorphic. One sees easily that each is isomorphic to one with θ in the interval $[0, 1/2]$, and then by using K-theoretic techniques [Rie81] one finds that different θ's in that interval give non-isomorphic algebras. Moreover, different θ's give algebras which are Morita equivalent exactly if the θ's are in the same orbit under the action of $SL(2, \mathbb{Z})$ by linear fractional transformations.

This definition is equivalent to the one in II.8.3.3(i): if u is the unitary $u(z) = z$ in $C(\mathbb{T})$, and v is the generator of $\mathbb{Z} \subseteq C(\mathbb{T}) \rtimes_\alpha \mathbb{Z}$, then $vu = e^{2\pi i\theta} uv$, and $C(\mathbb{T}) \rtimes_\alpha \mathbb{Z}$ is the universal C*-algebra generated by unitaries satisfying this relation.

(ii) *Bunce–Deddens algebras.* Let $\{m_k\}$ be a strictly increasing sequence of positive integers such that m_k divides m_{k+1} for each k. Let G_k be the subgroup of the circle group $\mathbb{T}$ consisting of the m_k-th roots of unity. Note that $G_k \subset G_{k+1}$ for each k. Let $G = \cup G_k$, with the discrete topology. Then G is a dense subgroup of $\mathbb{T}$ consisting of torsion elements. Let α be the action of G on $\mathbb{T}$ by translation. Again the action is free, but every orbit is dense, so that the topology on $\mathbb{T}/\alpha$ is useless. The crossed product $C(\mathbb{T}) \rtimes_\alpha G$ is a simple unital C*-algebra called a *Bunce–Deddens C*-algebra*. It is easily seen that $C(\mathbb{T}) \rtimes_\alpha G$ is the inductive limit (II.8.2) of the $C(\mathbb{T}) \rtimes_\alpha G_k$'s (cf. IV.1.4.23). The isomorphism classes of Bunce-Deddens algebras are classified just as the UHF algebras, using generalized integers (V.1.1.16(iv)).

By taking Fourier-Plancherel transform, $\hat{G}$ is a compact totally disconnected torsion-free group (isomorphic to the additive group of p-adic integers if $m_{k+1}/m_k = p$ for all k) containing a dense subgroup isomorphic to $\hat{\mathbb{T}} \cong \mathbb{Z}$, and the Bunce-Deddens algebra is isomorphic to $C(\hat{G}) \rtimes_\beta \mathbb{Z}$, where β is translation by $\mathbb{Z} \subseteq \hat{G}$.

(iii) The UHF algebras can similarly be described as transformation group C*-algebras. For each $k \in \mathbb{N}$ let G_k be a finite cyclic group. Then $G = \bigoplus G_k$ is a dense subgroup of $\Gamma = \prod G_k$, and thus G acts on Γ by translation. If G is given the discrete topology, then $C(\Gamma) \rtimes_\alpha G$ is a UHF algebra whose generalized integer is $\prod |G_k|$. If each G_k is $\mathbb{Z}_2$, the crossed product is the CAR algebra.

II.10.4.13 II.10.4.12(i) suggests that when α is a free action on X which is not proper, so that X/α can be highly singular as in the above examples, the crossed product $C_o(X) \rtimes_\alpha G$ may well carry much of the information which one would have found on X/α if it were not singular. This turns out to be true in practice. Thus crossed product algebras, and their generalizations, can serve as "desingularizations" of "bad" spaces.

II.10.4.14 When the action α is not free matters are, of course, more complicated. We give just one simple but important example, as a hint of what might happen more generally. Let H be a proper closed subgroup of a locally compact group G. Then the coset space G/H with the quotient topology is a locally compact space, which carries an evident action, α, of G by left translation. This action is not free, but the orbit space has just one point. We will construct a Morita equivalence of $C_o(G/H) \rtimes_\alpha G$ with another algebra, namely $C^*(H)$. As imprimitivity bimodule we take a completion of $C_c(G)$. For simplicity we assume that G and H are unimodular, but the general case can be handled by inserting modular functions in various places below. We let the convolution algebra $C_c(H)$ act on the right on $C_c(G)$ by convolution (by using the Haar measure on H to view the elements of $C_c(H)$ as finite measures on G). We define a $C_c(H)$-valued inner product on $C_c(G)$ by

$$\langle f, g \rangle = (f^* * g)|_H.$$

The completion, $\mathcal{E}$, of $C_c(G)$ is then a right Hilbert $C^*(H)$-module. We view functions in $C_o(G/H)$ as functions on G which are constant on cosets of H, so that $C_o(G/H)$ acts by pointwise multiplication on $C_c(G)$. This action together with the action of G by left translation on $C_c(G)$ satisfy the "covariance relation" (II.10.3.8), and their integrated form gives an injective homomorphism of $C_o(G/H) \rtimes_\alpha G$ into $\mathcal{L}(\mathcal{E})$. The rank-one operators for the $C^*(H)$-valued inner product are seen to be in the range of this homomorphism, and their linear span is seen to be dense in it, so that $C_o(G/H) \rtimes_\alpha G \cong \mathcal{K}(\mathcal{E})$. Thus

II.10.4.15 THEOREM. Let H be a closed subgroup of the locally compact group G. Then $C_o(G/H) \rtimes_\alpha G$ is Morita equivalent to $C^*(H)$.

II.10.4.16 In particular, every unitary representation of H gives a representation of $C_o(G/H) \rtimes_\alpha G$, and so a unitary representation of G. These representations of G are just those "induced" from H, and the above theorem is just a C*-algebraic formulation of Mackey's imprimitivity theorem which says that a representation of G is induced from H exactly if it admits a "system of imprimitivity", that is, a representation of $C_o(G/H)$ which satisfies the covariance relation with it.

Actions by Compact Groups

II.10.4.17 In the previous section we saw that if a transformation group involves a proper action, then the corresponding crossed product has some favorable properties. It is not clear how best to define "proper" actions on non-commutative C*-algebras (cf. [Rie90]), but since the action of any *compact* group on any space is proper, it is reasonable to expect that actions of compact groups on non-commutative spaces should also always be "proper", whatever the right definition.

In line with this idea, we proceed in the following way which parallels the situation for proper actions. Let (A, G, α) be a covariant system with G compact. Because Haar measure is now finite, we can normalize it to give G mass 1, and then we can average with it. Thus, as in II.6.10.4(v), we can define a conditional expectation, ϕ, of A onto its fixed point algebra, A^G, by

$$\phi(a) = \int_G \alpha_t(a) dt.$$

Since ϕ is clearly faithful, A^G is not trivial, though it may be one-dimensional. Recall (II.7.3.14) that a corner of a C*-algebra B is a hereditary subalgebra of form pBp for some projection $p \in M(B)$. It is Morita equivalent to the ideal $\overline{BpB}$.

II.10.4.18 THEOREM. [Ros79] Let (A, G, α) be a covariant system with G compact. Then A^G is naturally isomorphic with a corner of $A \rtimes_\alpha G$, and so is Morita equivalent to an ideal in $A \rtimes_\alpha G$.

PROOF: It suffices to let p be the function on G which has constant value 1, viewed as an element of $M(A \rtimes_\alpha G)$ as discussed in II.10.3.12. Simple calculations show that $p(A \rtimes_\alpha G)p$ consists of functions with values in A^G which are constant on G, and that these form a subalgebra of $A \rtimes_\alpha G$ isomorphic to A^G.

II.10.4.19 When A is commutative, the ideal generated by p can be seen to coincide with the ideal which entered into the discussion of proper actions in II.10.4.8. As indicated there, this ideal is all of $A \rtimes_\alpha G$ exactly when the action is free. It is not clear how best to define freeness of an action on a noncommutative C*-algebra. The above suggests that when G is compact one possibility is to require that this ideal be all of $A \rtimes_\alpha G$. This possibility, as well as others, have been extensively explored in [Phi87].

II.10.5 Takai Duality

II.10.5.1 We now consider what happens when G is commutative (but need not be compact). In this case the dual group $\hat{G}$ is useful in understanding the structure of $A \rtimes_\alpha G$. The reason for this is that there is a natural action, $\hat{\alpha}$, called the *dual action*, of $\hat{G}$ on $A \rtimes_\alpha G$, defined at the level of functions by

$$(\hat{\alpha}_\gamma(f))(t) = \langle t, \gamma \rangle f(t)$$

for $\gamma \in \hat{G}$. It is then natural to ask about the structure of $(A \rtimes_\alpha G) \rtimes_{\hat{\alpha}} \hat{G}$. The answer was given by H. Takai [Tak75], following a corresponding theorem of Takesaki about crossed products for von Neumann algebras (III.3.2.8).

II.10.5.2 THEOREM. [TAKAI DUALITY] With notation as above, we have

$$(A \rtimes_\alpha G) \rtimes_{\hat{\alpha}} \hat{G} \cong A \otimes \mathcal{K}(L^2(G)).$$

Under this isomorphism the second dual action $\hat{\hat{\alpha}}$ of G becomes the action $\alpha \otimes \lambda$ on $A \otimes \mathcal{K}(L^2(G))$, where λ is the action on $\mathcal{K}(L^2(G))$ consisting of conjugating by the representation of left translation on $L^2(G)$.

In the case in which $A = \mathbb{C}$ we are back to the Heisenberg commutation relations discussed in II.10.4.3.

II.10.5.3 Suppose that I is an ideal in A which is carried into itself by α. Then $I \rtimes_\alpha G$ is an ideal in $A \rtimes_\alpha G$, and for G abelian this latter ideal is carried into itself by $\hat{\alpha}$. One can use the Takai duality theorem to prove the converse, so that:

II.10.5.4 THEOREM. Let (A, G, α) be a covariant system with G abelian. Then passing from I to $I \rtimes_\alpha G$ gives a bijection between the α-invariant ideals of A and the $\hat{\alpha}$-invariant ideals of $A \rtimes_\alpha G$.

It takes more complicated techniques to get more information about the general ideals of $A \rtimes_\alpha G$, and there are no completely general results known; some of the known results are discussed in the next section.

II.10.5.5 There is a version of Takai duality for nonabelian groups and, more generally, for actions of certain Hopf C*-algebras (II.10.8.14).

II.10.6 Structure of Crossed Products

The structure of crossed product C*-algebras (and their generalizations) is in general very complicated. There is by now a vast literature dealing with various facets of this matter; see e.g. [GR79], [Kis80], [Ech93], [RW93], [Pac94]. To give the flavor we will simply state here one fundamental result, without even giving full definitions, much less an indication of the fairly complicated proof.

II.10.6.1 Let (A, G, α) be a covariant system. Then α determines an action of G on $\mathrm{Prim}(A)$. We say that two points in $\mathrm{Prim}(A)$ are in the same *quasi-orbit* if the closures of their G-orbits in $\mathrm{Prim}(A)$ coincide. (Recall that $\mathrm{Prim}(A)$ is locally compact, but perhaps not Hausdorff.) Let $J \in \mathrm{Prim}(A)$, and let G_J denote the stability subgroup of J in G. Then we can form $A \rtimes_\alpha G_J$. Using $A \rtimes_\alpha G_J$ there is a way of "inducing" J to a primitive ideal, $\mathrm{Ind}(J)$, of $A \rtimes_\alpha G$. This inducing construction uses representations of A whose kernel is J, and generalizes the construction of the regular representations given in II.10.3.13, in a way which can be viewed as a special case of the construction in II.10.4.14. Two elements in $\mathrm{Prim}(A)$ which are in the same quasi-orbit will induce to the same element of $\mathrm{Prim}(A \rtimes_\alpha G)$.

II.10.6.2 THEOREM. [Gre78] Assume that (A, G, α) is separable. If G is amenable, then every primitive ideal of $A \rtimes_\alpha G$ is an induced primitive ideal. Moreover if G acts freely on $\mathrm{Prim}(A)$, then the inducing process establishes a bijection between $\mathrm{Prim}(A \rtimes_\alpha G)$ and the quasi-orbits in $\mathrm{Prim}(A)$. In particular, if G acts freely and every orbit is dense, then $A \rtimes_\alpha G$ is simple.

II.10.6.3 In particular, we can conclude from this theorem that the irrational rotation algebras and the Bunce–Deddens algebras are simple. But since for both these examples A and G are commutative, the proof can be much simplified.

II.10.7 Generalizations of Crossed Product Algebras

Driven by the needs of various mathematical situations, many generalizations of crossed product C*-algebras have been developed and studied extensively. We briefly indicate some of them here.

Cocycles

II.10.7.4 Let G be a locally compact group, and let c be a 2-cocycle on G with values in the circle group $\mathbb{T}$, that is, c is a function on $G \times G$ satisfying

$$c(r, st)c(s, t) = c(r, s)c(rs, t).$$

These arise, for example, when dealing with projective representations of G — so important in quantum physics. Then the definition of convolution of functions on G can be generalized to

$$(f *_c g)(t) = \int f(s)g(s^{-1}t)c(s, s^{-1}t)ds.$$

Some care must be taken due to the fact that in important situations it is not possible to arrange that c be continuous—it is only measurable. But one obtains a C*-algebra, $C^*(G, c)$, as well as $C_r^*(G, c)$, often called a "twisted" group algebra.

II.10.7.5 EXAMPLE. *Noncommutative tori.* Let θ be a skew-adjoint operator on $\mathbb{R}^n$ for its usual inner product. Define a cocycle c_θ on $\mathbb{Z}^n$ by

$$c_\theta(u, v) = \exp(2\pi i \langle \theta u, v \rangle).$$

The resulting C*-algebras, $A_\theta = C^*(\mathbb{Z}^n, c_\theta)$, are called *noncommutative tori* (cf. II.8.3.3(i)). They arise in many situations. The irrational rotation algebras can be put in this form with $n = 2$ by taking Fourier transforms in the $\mathbb{T}$ variable.

Twisted Crossed Products

II.10.7.6 Given a covariant system (A, G, α), one can consider in addition a 2-cocycle with values in the unitary group of $M(A)$, satisfying suitable relations with α. Then one can again define a twisted convolution of A-valued functions to obtain C*-algebras $C^*(G, A, \alpha, c)$ and their reduced forms. Since general 2-cocycles are closely related to group extensions, these twisted group algebras can also often be defined (without mentioning the cocycle) in terms of extensions of G [Gre78], as well as in a framework of C*-algebras "graded" over G [Fel69].

Groupoids

II.10.7.7 Groups are effective in dealing with the global symmetries of mathematical objects. But in recent years there has been rapidly increasing use of groupoids for dealing with local symmetries, and for other purposes. A groupoid, G, is like a group except that it has many units (i.e. identity elements). For each element there is a left unit and a right unit. The (associative) product xy of $x, y \in G$ is only defined when the left unit of y coincides with the right unit of x. Each element is required to have an inverse. The elements which have a given unit u as simultaneously both left unit and right unit form a group, which can be considered to be the group of symmetries at u.

Groupoids which have a locally compact topology are of great importance, as are groupoids which are manifolds ("Lie groupoids"). In both cases there are important examples (such as holonomy groupoids) where the topology is not Hausdorff, though in a specific sense there are many Hausdorff subsets.

II.10.7.8 Unlike locally compact groups, groupoids do not automatically have a "Haar measure". One must include it as an axiom if needed (and check its existence for examples being studied). To do analysis on groupoids it is needed. A "Haar measure" is actually a field of measures, indexed by the units. It is usually far from unique. The measure μ^u associated to a unit u is supported on the set G^u of elements whose left unit is u. The field $\{\mu^u\}$ is required to be continuous in the sense that for any $f \in C_c(G)$ the function f^0 defined on the space G^0 of units by

$$f^0(u) = \int_{G^u} f(y) d\mu^u(y)$$

must be in $C_c(G^0)$. The field $\{\mu^u\}$ must also be left invariant in a natural sense.

II.10.7.9 When all this is properly formulated (see [Ren80] for details), one can again define convolution, by

$$(f * g)(x) = \int_{G^{r(x)}} f(y)g(y^{-1}x) d\mu^{r(x)}(y),$$

where $r(x)$ denotes the left unit ("range") of x. There is then a C*-norm defined in terms of representations of groupoids, and so one obtains the groupoid C*-algebra $C^*(G)$, and its reduced version, $C_r^*(G)$.

One can then consider generalizations of the various variations of crossed products considered above, so as to get, for example, groupoid C*-algebras twisted by cocycles, and crossed products for groupoids acting on C*-algebras.

Many familiar C*-algebras which appear in other guises can also be usefully realized as groupoid C*-algebras. This is true, for example, of the Cuntz algebras and, more generally, the graph C*-algebras (II.8.3.3(iii)).

II.10.8 Duality and Quantum Groups

Duality for Nonabelian Groups

There are two distinct, but related, types of duality for locally compact abelian groups, Pontrjagin duality and the more recent Takesaki-Takai duality. Both can be generalized to nonabelian groups, and farther.

II.10.8.10 The idea in generalizing Pontrjagin duality is to identify a "dual object" to a nonabelian group. This was first done by T. Tannaka [Tan39] and M. Krein [Kre49] for compact groups, and significant contributions to the theory were made by many mathematicians, notably I. Segal [Seg50], W. Stinespring [Sti59], G. Kac [Kac63], [Kac65], [VK74], P. Eymard [Eym64], N. Tatsuuma [Tat67], J. Ernest [Ern67], M. Takesaki [Tak72], M. Enock and J.-M. Schwartz [ES75], and E. Kirchberg [Kir77b]. We will not describe the evolution and various points of view used in duality theory (see e.g. [HR70] for a complete treatment of Tannaka-Krein duality); we will only outline one modern point of view using operator algebras.

II.10.8.11 If G is a finite group, then the finite-dimensional C*-algebra $C(G)$ has a structure as a commutative Hopf *-algebra, with comultiplication $\Delta : C(G) \to C(G) \odot C(G) \cong C(G \times G)$ defined by

$$[\Delta(f)](s,t) = f(st)$$

and counit $\epsilon : C(G) \to \mathbb{C}$ and antipode $S : C(G) \to C(G)$ given by

$$\epsilon(f) = f(e_G) , \quad [S(f)](t) = f(t^{-1}) .$$

The Hopf *-algebra $C(G)$ with these operations is usually denoted $K(G)$.

The group algebra $\mathbb{C}G$, which coincides with $C^*(G)$ and also $\mathfrak{L}(G)$ (III.3.3.1), has a natural structure as a cocommutative Hopf *-algebra via:

$$\Delta(u_t) = u_t \otimes u_t , \quad \epsilon(u_t) = 1 , \quad S(u_t) = u_{t^{-1}}$$

for $t \in G$ and u_t the corresponding unitary in $\mathfrak{L}(G)$, extended by linearity.

If G is abelian, then $K(G) \cong \mathbb{C}\hat{G}$ (so $\mathbb{C}G \cong K(\hat{G})$) as Hopf *-algebras.

II.10.8.12 A finite-dimensional Hopf *-algebra A which is not necessarily either commutative or cocommutative can thus be regarded as a generalization of a finite group. But there is another crucial ingredient present in the group case: a *Haar state*, a state $h : A \to \mathbb{C}$ satisfying $(h \otimes id)(\Delta(a)) = h(a)1$ for all $a \in A$. To insure existence of a Haar state (which is automatically unique, tracial, and faithful, thus giving a C*-norm on A via the GNS representation), it must be assumed that A satisfies $a^*a = 0 \Longrightarrow a = 0$. Such a Hopf *-algebra is now usually called a *finite quantum group*.

It is not hard to show using II.2.2.4 that a commutative finite quantum group is of the form $K(G)$ for a finite group G; it is similarly easy to show that a cocommutative quantum group is of the form $\mathbb{C}G$.

If A is a finite quantum group, then its dual space A^* also has a natural structure as a finite quantum group, denoted $\hat{A}$, and $\hat{\hat{A}} \cong A$. If $A \cong K(G)$ as above, then $\hat{A} \cong \mathbb{C}G$. There is thus a duality theory for finite quantum groups. See, for example, [KT99, §2].

II.10.8.13 If G is a general locally compact group, then $L^\infty(G)$ and $\mathfrak{L}(G)$ can be made into coinvolutive Hopf-von Neumann algebras in more or less the same way as for finite groups. (If G is infinite, these are technically not Hopf algebras, since the comultiplication maps M into $M \bar{\otimes} M$ and not into $M \odot M$.) If G is abelian, then $\mathfrak{L}(G) \cong L^\infty(\hat{G})$ (so $L^\infty(G) \cong \mathfrak{L}(\hat{G})$), so $L^\infty(G)$ and $\mathfrak{L}(G)$ can be regarded as duals of each other.

These Hopf-von Neumann algebras also have a *left Haar weight* induced by left Haar measure. A Hopf-von Neumann algebra with a left Haar weight is called a *Kac algebra*. The left Haar weight on a Kac algebra is unique up to a scalar multiple, and is faithful and semifinite; it is a trace if and only if it is also a right Haar weight (for example, the left Haar weight on $\mathfrak{L}(G)$ is a trace if and only if G is unimodular).

There is a duality theory for Kac algebras generalizing Pontrjagin duality and the duality of II.10.8.12 [ES75]; under this duality $\mathfrak{L}(G)$ is the dual of $L^\infty(G)$ (and vice versa) even if G is not abelian. See [ES92], [Tak73], or [Str81, §18] for a more complete treatment, which is closely related to modular theory for von Neumann algebras.

II.10.8.14 It is more delicate to obtain a duality at the group C*-algebra level. One of the many technical difficulties is that the natural comultiplication on $C_o(G)$ does not map into $C_o(G) \otimes C_o(G)$, but only into the multiplier algebra. One is led to study "Hopf C*-algebras" (or, perhaps more correctly, "C*-bialgebras"), where the comultiplication is a map from the algebra A to $M(A \otimes A)$, with a Haar weight (*Kac C*-algebras*). It is not obvious which cross norm to use on $A \odot A$; it turns out that the minimal cross norm is the best choice for a duality theory. Hopf C*-algebras were first considered in [Iór80] and Kac C*-algebras in [Val85].

A Kac C*-algebra has an enveloping Kac algebra obtained by taking the weak closure in the GNS representation from the Haar weight. Conversely, it was shown in [EV93] that every Kac algebra M contains a unique Kac C*-subalgebra A such that M is the enveloping Kac algebra of A (A is characterized by the properties that A is closed under the antipode, the comultiplication maps A into $M(A \otimes A)$, realized as the idealizer of $A \otimes A$ in $M \bar{\otimes} M$, and the restriction of the Haar weight to A is densely defined). There is thus a natural one-one correspondence between Kac C*-algebras and Kac algebras. In the cases above, the Kac C*-subalgebra of $L^\infty(G)$ is $C_o(G)$, and the Kac C*-subalgebra of $\mathfrak{L}(G)$ is $C_r^*(G)$.

From this correspondence, one can obtain a duality theory for Kac C*-algebras.

See [VVD01] for an approach to defining Hopf C*-algebras without the assumption of a Haar weight.

Crossed Products and Takesaki-Takai Duality

Takesaki-Takai duality can also be generalized to nonabelian groups and, even more generally, to Kac algebras and Kac C*-algebras. We will describe how this works for finite groups; the general group case is quite similar but with topological and measure-theoretic technical complications. This duality theorem was first proposed in the case G compact by J. Roberts [Rob76] and carried out in the general case by M. Landstad ([Lan79], [Lan77]), Y. Nakagami [Nak77], and S. Stratila, D. Voiculescu, and L. Zsido [SVZ77]. For the general Kac algebra case, see [Eno77], [ES80], [SVZ76], [Zsi78]. See [CM84], [BM85], and [VDZ00] for purely algebraic versions.

II.10.8.15 An *action* of a Hopf *-algebra (A, Δ) on a *-algebra N is a *-linear map $\delta : N \to N \odot A$ such that $(\delta \otimes id) \circ \delta = (id \otimes \Delta) \circ \delta$ as maps from N to $N \odot A \odot A$. We similarly define an action if A is a Hopf-von Neumann algebra, N is a von Neumann algebra, and $\delta : A \to N \bar{\otimes} A$ is bounded and σ-weakly continuous, or if N is a C*-algebra, A is a Hopf C*-algebra, and $\delta : N \to M(N \otimes A)$ is bounded.

The term "action" is conventionally used instead of "coaction", but an algebra N with an A-action in this sense is called an A-*comodule*. What may be called a coaction of a Hopf algebra (A, Δ) on a unital algebra N is a map $\psi : A \odot N \to N$ making N a left A-module, with the properties that $\psi(a \otimes xy) = \sum_j \psi(b_j \otimes x)\psi(c_j \otimes y)$ for $a \in A$, $x, y \in N$, and $\Delta(a) = \sum_j b_j \otimes c_j$, and $\psi(a \otimes 1) = \epsilon(a)1$, where $\epsilon : A \to \mathbb{C}$ is the counit. An algebra with an A-coaction in this sense is called an A-*module algebra*. See [Swe69] for details.

II.10.8.16 If G is a finite group, then the actions of $K(G)$ are in natural one-one correspondence with the actions of G: if N is a *-algebra and σ is a homomorphism from G into the group of *-automorphisms of N, then σ induces an action δ_σ of $K(G)$ on N by identifying $N \odot K(G)$ with $C(G, N)$ in the standard way and letting $[\delta_\sigma(x)](t) = \sigma_t(x)$ for $x \in N$. It is easily seen that every action of $K(G)$ arises in this manner. If G is just locally compact, there is a similar action of $C_o(G)$ if N is a C*-algebra with a G-action:

$$\delta(x) \in C_b(G, N) \cong N \otimes C_b(G) \subseteq M(N \otimes C_o(G))$$

and there is also an analogous action of $L^\infty(G)$ on a von Neumann algebra N with a G-action as in III.3.2.2.

Returning to the case of G finite, if N is a C*-algebra we may regard $K(G)$ as represented on $L^2(G)$ by multiplication operators, and identify $N \rtimes_\sigma G$ with the subalgebra of $N \odot \mathcal{L}(L^2(G))$ generated by $\delta(N) \cup (1 \otimes K(G))$.

II.10.8.17 There is an action γ of the Hopf *-algebra $\mathfrak{L}(G)$ on $\mathcal{L}(L^2(G))$ defined as follows. The comultiplication on $\mathfrak{L}(G)$ is defined by

$$\Delta(x) = W^*(x \otimes 1)W$$

where $\mathfrak{L}(G)$ is regarded as a subalgebra of $\mathcal{L}(L^2(G))$ and W is the unitary on $L^2(G) \otimes L^2(G) \cong L^2(G \times G)$ defined by $[Wf](s, t) = f(s, st)$. It can be checked that this formula for any $x \in \mathcal{L}(L^2(G))$ gives an element of $\mathcal{L}(L^2(G)) \odot \mathfrak{L}(G)$, and defines an action of $\mathfrak{L}(G)$. If G is abelian, it is not hard to see that this coincides with the action of $K(\hat{G})$ on $\mathcal{L}(L^2(G))$ coming from the action of $\hat{G}$ dual to the action of G on $\mathcal{L}(L^2(G))$ by the left regular representation.

From this we get an action $\beta = id \otimes \gamma$ of $\mathfrak{L}(G)$ on $N \odot \mathcal{L}(L^2(G))$ in the case where G acts on N. It is easy to see that β maps the subalgebra $N \rtimes_\sigma G$ into $(N \rtimes_\sigma G) \odot \mathfrak{L}(G)$ and hence induces an action of $\mathfrak{L}(G)$ on $N \rtimes_\sigma G$, called the *dual action* and denoted $\hat{\delta}$. The Takesaki-Takai duality theorem then becomes:

II.10.8.18 PROPOSITION. The subalgebra of $N \odot \mathcal{L}(L^2(G)) \odot \mathcal{L}(L^2(G))$ generated by $\hat{\delta}(N \rtimes_\sigma G)$ and $1 \otimes 1 \otimes K(G)$ (where $K(G)$ acts as multiplication operators on $L^2(G)$) is isomorphic to $N \odot \mathcal{L}(L^2(G))$.

II.10.8.19 In the general case of a Kac algebra A acting via δ on a von Neumann algebra N, there is a W*-crossed product $N\bar\rtimes_\delta A$, defined to be the von Neumann subalgebra of $N\bar\otimes A$ generated by $\delta(N)\cup(1\otimes A)$, and dual action $\hat\delta$ of $\hat A$ on $N\bar\rtimes_\delta A$; the second crossed product $(N\bar\rtimes_\delta A)\bar\rtimes_{\hat\delta}\hat A$ is isomorphic to $N\bar\otimes\mathcal{L}(\mathcal{H})$, where A acts on $\mathcal{H}$ in standard form (III.2.6). The second dual action can also be nicely described. There is a C*-algebra (Takai duality) version of this result.

Quantum Groups

II.10.8.20 Kac algebras (or Kac C*-algebras) are not the last word in generalizing locally compact groups via Hopf algebras, however. There are other Hopf-algebra-type objects which are "quantizations" of groups; these objects have come to be known as *quantum groups*. Many of these can be defined and studied without reference to operator algebras. See, for example, [Dri87], [Kas95], [KS97], and [Lus93].

Many of the interesting and important quantum groups, especially noncompact ones, are formed by deforming a Lie group (i.e. deforming the multiplication in the associated commutative Hopf algebra) in the direction of a Poisson bracket. These are called *deformation quantizations*.

While mathematicians seem to know a quantum group when they see one, and there are many examples and constructions which everyone agrees give quantum groups, coming up with a satisfactory definition of a quantum group has been a slow and difficult process which has not yet reached a conclusion. Operator algebraists, beginning with S. Woronowicz, have championed a C*-algebra approach for the locally compact case. A recent article ([KT99], [KT00]) surveys this approach in detail, so we will only hit a few selected high points.

II.10.8.21 If Kac algebras (or Kac C*-algebras) are generalizations of locally compact groups, then a Kac algebra (or Kac C*-algebra) in which the Haar weight is a state, such as $L^\infty(G)$ or $C(G)$ for G compact (a compact group) or $\mathfrak{L}(G)$ or $C_r^*(G)$ for G discrete (the dual of a discrete group), should correspond to a generalized compact group.

Woronowicz discovered that there are natural compact quantum groups which do not correspond to Kac algebras in any reasonable sense. An example is the quantum $SU(2)$ (II.8.3.3(v)). The problem is that the counit and antipode may be unbounded and not everywhere defined on the C*-algebra; the antipode is invertible but not involutive in general. Woronowicz gave a definition of a compact quantum group which is now generally accepted: a unital C*-algebra A with a *-homomorphism $\Delta : A \to A \otimes A$ (minimal tensor product), such that $(id\otimes\Delta)\circ\Delta = (\Delta\otimes id)\circ\Delta$, and such that $\Delta(A)(A\otimes 1)$ and $\Delta(A)(1\otimes A)$ are dense in A. Every finite quantum group is a compact quantum group. Every compact quantum group has a unique Haar state, which is however not faithful or tracial in general. The commutative [resp. cocommutative] compact quantum groups with faithful Haar state are precisely the C*-algebras $C(G)$ for G compact [resp. $C_r^*(G)$ for G discrete] with the usual structure.

The (co)representation theory of compact quantum groups can be studied using the multiplicative unitaries of S. Baaj and G. Skandalis [BS93]. Baaj and Skandalis associated quantum group-like objects to their multiplicative unitaries and obtained a duality theorem, although the theory must be extended [Wor96] to cover (apparently) general quantum groups.

Woronowicz proved a generalization of the Tannaka-Krein duality theorem (that a compact quantum group is effectively determined by its tensor category of finite-dimensional unitary corepresentations), using tensor C*-categories, which provide close connections between quantum groups and quantum field theory, knots, and subfactors (cf. [Jon90], [Wen88]).

II.10.8.22 A discrete quantum group should be the "dual" of a compact quantum group. P. Podles and Woronowicz [PW90] proposed a definition of a discrete quantum group, and a different approach was proposed in [ER94] and [VD96]. A. van Daele [VD98] then provided a general framework for unifying the theories of compact and discrete quantum groups, a purely algebraic version of multiplier Hopf algebras with invariant state, which essentially includes the theory of compact quantum groups, and in which there is a duality theory. Van Daele's category includes far more than just compact and discrete quantum groups.

The connection is that within every compact quantum group A there is a unique dense *-subalgebra $\mathcal{A}$ which is maximal with respect to the property that the comultiplication Δ maps $\mathcal{A}$ into $\mathcal{A}\odot\mathcal{A}$. This $\mathcal{A}$ is one of the objects considered by van Daele.

II.10.8.23 E. Kirchberg proposed the first reasonable definition of a locally compact quantum group. T. Masuda and Y. Nakagami [MN94] built on this work and developed a duality theory in the von Neumann algebra setting; however, the axiom scheme was very complex.

Only recently has an apparently satisfactory definition been given for a general locally compact quantum group, by J. Kustermans and S. Vaes [KV00] (see [MNW03] for another related attempt). In this definition, existence of left and right Haar weights must be included as an axiom.

II.10.8.24 Recently, the notion of a "quantum groupoid" has been explored. There are close connections with the theory of subfactors. See [Eno03] for a survey.

III
Von Neumann Algebras

Recall (I.9) that a von Neumann algebra is a *-subalgebra M of $\mathcal{L}(\mathcal{H})$ for a Hilbert space $\mathcal{H}$, satisfying $M = M''$. A von Neumann algebra is unital, weakly closed, and contains an abundance of projections.

Every von Neumann algebra is, of course, a C*-algebra, and all the results and techniques of Chapter II apply to von Neumann algebras; but it is not usually useful to think of von Neumann algebras merely as C*-algebras. In some respects, their C*-algebra structure is extremely well-behaved (e.g. comparison theory and ideal structure), while in other ways they are pathological as C*-algebras (for example, an infinite-dimensional von Neumann algebra is always nonseparable and almost always nonnuclear).

Even the predominant philosophy behind the theory of von Neumann algebras is different. While it is common to view C*-algebras as "noncommutative topological spaces," it is usual to regard von Neumann algebras as "noncommutative measure spaces." For example, if (X, μ) is a locally finite measure space, then $M = L^\infty(X, \mu)$ acts naturally as a von Neumann algebra of operators on $L^2(X, \mu)$ (in fact, $M = M'$, i.e. M is a *maximal abelian subalgebra (masa)* in $\mathcal{L}(L^2(X, \mu))$). By II.2.2.4, $M \cong C(Y)$ for some compact Hausdorff space Y, but this is rarely the best way to view M (Y is a huge non-first-countable extremally disconnected or Stonean space in general). Indeed, every commutative von Neumann algebra looks like an L^∞ algebra acting by multiplication on L^2 of a locally finite measure space, with multiplicity (III.1.5.18).

There is a whole range of techniques which are special to von Neumann algebras, many of which are motivated by ideas from measure theory. Most fall into two general groups: (1) algebraic arguments using the fact that the projections in a von Neumann algebra generate the algebra in a very strong sense and form a complete lattice; and (2) analytic/topological arguments using normal linear functionals and the σ-weak topology (which arise from the algebra's structure as a dual space). In this chapter, we will discuss the basic structure of von Neumann algebras and the techniques for working with them.

Von Neumann algebras, originally called "rings of operators," were the first operator algebras to be systematically studied, in the seminal papers of Murray and von Neumann [MVN36]-[MvN43]. The theory was already regarded as well developed when a series of spectacular advances, beginning with the Tomita-Takesaki Modular Theory in the late 1960's and culminating in Connes' classification of injective factors, ushered in the "modern era" of the subject.

A word on notation: even though von Neumann algebras are concrete algebras of operators, it is customary (but, unfortunately, far from universal) to use lower-case letters to refer to elements of the algebra, and we will do so. We will also use 1 to denote the identity element.

Besides the standard references listed at the beginning of Chapter 2, some of which (especially [KR97a]-[KR97b] and [Tak02]-[Tak03b]) concentrate on von Neumann algebras, other standard references for von Neumann algebras are the classic [Dix69a] and [SZ79]. As in the C*-algebra case, there are also numerous specialized volumes, some of which are cited below.

III.1 Projections and Type Classification

III.1.1 Projections and Equivalence

Recall (I.9.2.1) that the projections in a von Neumann algebra form a complete lattice. The way this is most often used in the theory is:

III.1.1.1 PROPOSITION. Let M be a von Neumann algebra on a Hilbert space $\mathcal{H}$, and let $\{p_i\}$ be a set of mutually orthogonal projections in M. Then the net of finite sums of the p_i converges strongly to $\bigvee_i p_i$, which is usually denoted $\sum_i p_i$.

III.1.1.2 Recall also (II.3.3.3) that projections p, q in a von Neumann algebra (or C*-algebra) M are *(Murray-von Neumann) equivalent*, written $p \sim q$, if there is a partial isometry $u \in M$ with $u^*u = p$, $uu^* = q$, and p is *subordinate* to q ($p \precsim q$) if $p \sim q' \leq q$. (These relations depend, of course, on the choice of the algebra M containing p and q.) Equivalence is additive in a von Neumann algebra: if $\{p_i, q_i\}$ is a set of projections in M with $p_i \perp p_j$ and $q_i \perp q_j$ for $i \neq j$ and $p_i \sim q_i$ for all i, then

$$\sum_i p_i \sim \sum_i q_i$$

(the partial isometries can be added in the strong operator topology); similarly, if $p_i \precsim q_i$ for all i, then $\sum_i p_i \precsim \sum_i q_i$. If $x \in M$ and $x = u|x|$ is its polar decomposition, then $u, |x| \in M$ and the source and range projections (right and left support projections) $p_x = u^*u$ and $q_x = uu^*$ are equivalent in M. If x is normal, then $p_x = q_x$ is the support projection of x, denoted $s(x)$.

III.1.1.3 PROPOSITION. Let p, q be projections in a von Neumann algebra M. Then
$$(p \vee q) - p \sim q - (p \wedge q).$$

In fact, $(p \vee q) - p$ and $q - (p \wedge q)$ are the source and range projections of $q(1-p)$ respectively.

III.1.1.4 PROPOSITION. Let p, q be (nonzero) projections in a von Neumann algebra M. If $qMp \neq 0$, then there are nonzero projections $p' \leq p$ and $q' \leq q$ with $p' \sim q'$.

If $qxp \neq 0$, then the source and range projections of qxp do the trick.

III.1.1.5 Every projection p in a von Neumann algebra M has a *central carrier* (or *central support projection*) z_p, the smallest projection in the center $\mathcal{Z}(M)$ containing p as a subprojection (z_p exists since $\mathcal{Z}(M)$ is itself a von Neumann algebra). By III.1.1.4,
$$1 - z_p = \bigvee \{q \in M : qMp = 0\}.$$

Projections p and q have nonzero equivalent subprojections if and only if $z_p z_q \neq 0$. In particular, if $q \precsim p$, then $z_q \leq z_p$.

III.1.1.6 PROPOSITION. Let M be a von Neumann algebra, $p, \{q_i : i \in I\}$ projections in M, with the q_i mutually orthogonal and all orthogonal to p, and $q_i \precsim p$ for all i. If $z = p + \sum_i q_i$ is in $\mathcal{Z}(M)$, then $z = z_p$.
PROOF: We have $p \leq z_p$ and $q_i \leq z_{q_i} \leq z_p$ for all i, so $z \leq z_p$. On the other hand, $p \leq z$, and z_p is the smallest central projection dominating p, so $z_p \leq z$.

III.1.1.7 PROPOSITION. If M is a von Neumann algebra and $x, y \in M$ with $yMx = 0$, then there is a projection z in the center $\mathcal{Z}(M)$ such that $x \in zM$ (i.e. $xz = x$) and $y \in (1-z)M$ (i.e. $yz = 0$).

We can take $z = z_{q_x}$; note that $p_y M q_x = 0$, so $p_y \leq 1 - z_{q_x}$.

From this, we get a quick proof of II.6.1.11:

III.1.1.8 COROLLARY. Let A be a concrete C*-algebra of operators on a Hilbert space $\mathcal{H}$. If $M = A''$ is a factor, then A is prime.
PROOF: We will use II.5.4.5. Let $x, y \in A$. If $xAy = 0$, then $xMy = 0$ by separate weak continuity of multiplication; so III.1.1.7 implies that either x or y is 0.

III.1.1.9 PROPOSITION. [SCHRÖDER-BERNSTEIN] Let p, q be projections in a von Neumann algebra M. If $p \precsim q$ and $q \precsim p$, then $p \sim q$.

The proof is very similar to the usual Schröder-Bernstein Theorem of set theory.

III.1.1.10 PROPOSITION. [GENERALIZED COMPARABILITY] Let p, q be projections in a von Neumann algebra M. Then there is a central projection $z \in \mathcal{Z}(M)$ such that $pz \precsim qz$ and $q(1-z) \precsim p(1-z)$.

The proof is a typical maximality or exhaustion argument: let $\{p_i, q_i\}$ be a maximal family of projections such that $p_i \leq p$, $q_i \leq q$, $p_i \sim q_i$ for all i and $p_i \perp p_j$, $q_i \perp q_j$ for $i \neq j$. Set

$$p' = \sum p_i, \; q' = \sum q_i, \; r = p - p', \; s = q - q'.$$

By maximality and III.1.1.4 we have $sMr = 0$, so there is a central projection z with $rz = 0$ and $sz = s$. Then

$$pz = p'z \sim q'z \leq qz$$

and similarly

$$q(1-z) = q'(1-z) \sim p'(1-z) \leq p(1-z).$$

III.1.1.11 COROLLARY. [COMPARABILITY] Let p, q be projections in a factor M. Then either $p \precsim q$ or $q \precsim p$.

Compare the next result with III.1.1.6:

III.1.1.12 PROPOSITION. Let p be a projection in a von Neumann algebra M, and $\{q_i : i \in I\}$ a maximal set of mutually orthogonal projections in M with $q_i \precsim p$ for all i. Then $\sum_i q_i = z_p$.

PROOF: Set $q = \sum_i q_i$. Since $q_i \leq z_{q_i} \leq z_p$ for all i, we have $q \leq z_p$. Set $r = z_p - q$. There is a central projection $z \in z_p M$ (i.e. $z \leq z_p$) with $zr \precsim zp$ and $(z_p - z)p \precsim (z_p - z)r$. Then $(z_p - z)p$ is equivalent to a subprojection of $(z_p - z)r$, hence of r, so $(z_p - z)p = 0$ by maximality. But

$$(z_p - z)p = z_p p - zp = p - zp,$$

i.e. $p = zp$, $p \leq z$, so $z = z_p$. But then $r = zr \precsim zp = p$, so $r = 0$ by maximality.

Ideals in von Neumann Algebras

III.1.1.13 The ideal structure of a von Neumann algebra is relatively simple compared to that of a general C*-algebra. In particular, the weakly closed ideals are very easy to describe. If M is a von Neumann algebra on $\mathcal{H}$ and I is an ideal which is closed in any of the standard operator topologies (weak, σ-strong-*, etc.), then I is a von Neumann algebra on its essential subspace $\mathcal{X}$, which is an invariant subspace for M; thus $p = P_{\mathcal{X}}$ is a projection in $\mathcal{Z}(M)$, $p \in I$, and $I = Mp$ (and thus is weakly closed). Thus the weakly closed ideals in M are precisely the principal ideals generated by central projections. The ideal Mp has a complementary ideal $M(1-p) \cong M/Mp$, and

$$M \cong Mp \oplus M(1-p).$$

If L is a weakly closed left ideal in M, and (h_λ) is an increasing right approximate unit in L, then (h_λ) converges strongly to a projection p (which is the supremum of the support projections of the h_λ). Thus $p \in L$, and $L = Mp$. Thus the weakly closed left ideals of M are precisely the principal left ideals generated by projections in M. Similarly, every weakly closed right ideal of M is of the form pM for a (unique) projection $p \in M$.

Norm-closed ideals in a von Neumann algebra are not necessarily weakly closed (e.g. $\mathcal{K}(\mathcal{H})$ in $\mathcal{L}(\mathcal{H})$), and their structure is more complicated, especially if $\mathcal{Z}(M)$ is large. The structure of the norm-closed ideals in a factor can be precisely described, however (III.1.7.11).

III.1.1.14 If L is a norm-closed left ideal in a von Neumann algebra M, with weak closure Mp, and (h_λ) is an increasing right approximate unit in L, then as above $h_\lambda \to p$ strongly. If $\mathfrak{N}$ is any left ideal in M with norm-closure L, then by approximating h_λ by an element of $\mathfrak{N}$ of norm 1, we obtain a net k_λ in the unit ball of $\mathfrak{N}$ converging strongly to p. If $x \in Mp$, then (xk_λ) is a bounded net in $\mathfrak{N}$ converging strongly to x. Thus, using the joint strong continuity of multiplication on bounded sets, we obtain a simple analog of the Kaplansky Density Theorem:

III.1.1.15 PROPOSITION. Let M be a von Neumann algebra, $\mathfrak{N}$ a left ideal in M with weak closure Mp. Then the unit ball of $\mathcal{M}(\mathfrak{N})$ (II.5.3.13) is σ-strongly dense in the unit ball of pMp.

III.1.1.16 More generally, suppose $\mathfrak{N}_1$ and $\mathfrak{N}_2$ are left ideals in a von Neumann algebra M, with weak closures Mp and Mq respectively. By the same argument, the σ-strong closure of

$$\mathfrak{N}_2^* \mathfrak{N}_1 = Span\{y^*x : x \in \mathfrak{N}_1, y \in \mathfrak{N}_2\} \subseteq \mathfrak{N}_1 \cap \mathfrak{N}_2^*$$

is qMp (the *Span* is even unnecessary). In particular, if $\mathfrak{N}_1$ and $\mathfrak{N}_2$ are σ-weakly dense in M, so is $\mathfrak{N}_2^*\mathfrak{N}_1$.

III.1.2 Cyclic and Countably Decomposable Projections

III.1.2.1 We say a projection p in a von Neumann algebra is *countably decomposable* if any collection of mutually orthogonal nonzero subprojections of p is countable. Any subprojection of a countably decomposable projection is countably decomposable. A von Neumann algebra M is countably decomposable if 1_M is countably decomposable. Any von Neumann algebra on a separable Hilbert space is countably decomposable, but there are countably decomposable von Neumann algebras not isomorphic to a von Neumann algebra on a separable Hilbert space (III.3.1.5). Similarly, if $\aleph$ is any cardinal, we can define $\aleph$-decomposability in the obvious manner. More generally, we say a von Neumann algebra M is *locally countably decomposable* if there is a set $\{z_i\}$ of mutually orthogonal central projections in M with $\sum z_i = 1_M$ and Mz_i countably decomposable. Every commutative von Neumann algebra is isomorphic to L^∞ of a locally finite measure space (III.1.5.18), hence is locally countably decomposable (see III.1.2.6 for a generalization).

III.1.2.2 The interplay and duality between a von Neumann algebra M on $\mathcal{H}$ and its commutant M' is the heart of the spatial theory of von Neumann algebras. The fundamental tool in analyzing this interplay is the notion of a cyclic projection, a generalization of the idea of a cyclic vector or representation.

If $\mathcal{S}$ is a subset of $\mathcal{L}(\mathcal{H})$ and $\xi \in \mathcal{H}$, write $[\mathcal{S}\xi]$ for the closed linear span of $\{S\xi : S \in \mathcal{S}\}$. Recall that if A is a *-algebra of operators on $\mathcal{H}$ and $\xi \in \mathcal{H}$, then ξ is *cyclic for* A (or *cyclic under* A) if $[A\xi] = \mathcal{H}$. We also say that ξ is *separating for* A if $x \in A$, $x\xi = 0$ implies $x = 0$, i.e. $x \mapsto x\xi$ is injective.

III.1.2.3 DEFINITION. Let M be a von Neumann algebra on $\mathcal{H}$, and q a projection in M'. Then q is *cyclic under* M (or *cyclic in* M') if there is a $\xi \in \mathcal{H}$ such that $[M\xi] = q\mathcal{H}$ (i.e. if the representation of M on $q\mathcal{H}$ is cyclic). ξ is called a *cyclic vector* (or *generating vector*) for q (under M).
If p is a projection in M, a vector $\xi \in \mathcal{H}$ is *separating* for (p, M) if $p\xi = \xi$ and $x \in M$, $x\xi = 0$ implies $xp = 0$.

A vector ξ is cyclic [resp. separating] for M in the usual sense if and only if it is cyclic for $1_{M'}$ under M [resp. separating for $(1, M)$].

We most commonly consider a von Neumann algebra M and projection $p \in M$; p is cyclic in M if it is cyclic under M'. Likewise, we sometimes consider vectors which are separating for M'. The application of these definitions at will to both M and M' can be slightly confusing, but is the cornerstone of the spatial analysis of von Neumann algebras.

III.1.2.4 PROPOSITION. Let M be a von Neumann algebra on $\mathcal{H}$, p a projection in M, and $\xi \in \mathcal{H}$ with $p\xi = \xi$. Then ξ is separating for (p, M) if and only if ξ is cyclic for p (under M'). In particular, ξ is cyclic for M if and only if it is separating for M'.
PROOF: Suppose ξ is cyclic for p under M', and $x \in M$. If $x\xi=0$, then $xy\xi = yx\xi = 0$ for all $y \in M'$, so

$$x[M'\xi] = xp\mathcal{H} = \{0\}$$

and $xp = 0$. Conversely, if ξ is separating for (p, M), let q be the projection from $\mathcal{H}$ onto $[M'\xi]$. Then $q \in (M')' = M$, and since $p(y\xi) = yp\xi = y\xi$ for all $y \in M'$, $q \leq p$. $(p-q)\xi = 0$, so $p - q = (p-q)p = 0$, $q = p$.

III.1.2.5 The following facts are nearly immediate:
(i) If p is cyclic in M with cyclic vector ξ, and $q \in M$ with $q \leq p$, then q is cyclic in M with cyclic vector $q\xi$.
(ii) If p is any nonzero projection in M, then there is a nonzero projection $q \in M$, $q \leq p$, which is cyclic in M [let $0 \neq \xi \in \mathcal{H}$ with $p\xi = \xi$, and let q be the projection onto $[M'\xi]$.] Thus by a simple maximality argument, p is the sum of a set of mutually orthogonal cyclic projections.
(iii) A cyclic projection is countably decomposable [if p is cyclic in M with cyclic vector ξ, and $\{q_i : i \in I\}$ is a set of mutually orthogonal nonzero subprojections of p in M, then $q_i\xi \neq 0$ for all i (III.1.2.4), and
$$\sum_i \|q_i\xi\|^2 \leq \|\xi\|^2,$$
so I is countable].
(iv) If (z_n) is a sequence of mutually orthogonal *central* projections in M, and each z_n is cyclic in M with cyclic vector ξ_n of norm one, then $\sum_n z_n$ is cyclic in M (or under M) with cyclic vector $\sum_n n^{-1}\xi_n$. Thus if $\mathcal{Z}(M)$ is countably decomposable, then there is a largest projection in $\mathcal{Z}(M)$ which is cyclic in M (or under M).

Note, however, that even a sum of two noncentral cyclic projections need not be cyclic: for example, if $M = \mathcal{L}(\mathcal{H})$, then a projection in M is cyclic if and only if it has rank ≤ 1.

Putting (ii) for $p = 1$ together with (iii), we get:

III.1.2.6 PROPOSITION. If M is a von Neumann algebra, then there is a set $\{p_i\}$ of mutually orthogonal, countably decomposable projections in M with $\sum p_i = 1$.

III.1.3 Finite, Infinite, and Abelian Projections

III.1.3.1 DEFINITION. A projection p in a C*-algebra A is

abelian if pAp is commutative.
finite if $p \sim p' \leq p$ implies $p' = p$.
infinite if it is not finite.
properly infinite if $p \sim p_1$, $p \sim p_2$ with $p_1, p_2 \leq p$ and $p_1 \perp p_2$.

A unital C*-algebra A is finite [*resp.* infinite, properly infinite] if 1_A is finite [*resp.* infinite, properly infinite].

By this definition, 0 is properly infinite (it is also finite). Although counterintuitive, it is convenient for the statement of decomposition results to regard 0 as both finite and properly infinite.

Any projection equivalent to a finite [*resp.* abelian, infinite, properly infinite] projection is finite [*resp.* abelian, infinite, properly infinite].

An abelian projection is obviously finite. If p is finite [resp. abelian] and $q \precsim p$, then q is finite [resp. abelian]. The finite projections in $\mathcal{L}(\mathcal{H})$ are the finite-rank projections, and the abelian projections are the rank one projections.

III.1.3.2 PROPOSITION. If p is an infinite projection in a C*-algebra A, then there is a sequence of mutually orthogonal equivalent nonzero subprojections of p. The converse is true if A is a von Neumann algebra.

PROOF: If p is infinite and $u \in A$ with

$$u^*u = p, \ uu^* = p_1 \lneq p,$$

let $q_1 = p - p_1$ and $q_n = uq_{n-1}u^*$ for $n > 1$. Then (q_n) is a sequence of mutually orthogonal equivalent nonzero subprojections of p. Conversely, if A is a von Neumann algebra and (q_n) is such a sequence, set $q = \sum_{n=1}^{\infty} q_n$; then

$$p = q + (p - q) \sim \sum_{n=2}^{\infty} q_n + (p - q) \lneq p,$$

so p is infinite.

The converse can fail in a C*-algebra, for example $A = \tilde{\mathbb{K}}$, $p = 1$. It is, however, true in a simple C*-algebra (V.2.3.1).

Properly Infinite Projections

We begin with a variation of III.1.3.2. See III.1.3.5 and V.2.3.1 for closely related results.

III.1.3.3 PROPOSITION. Let A be a properly infinite (unital) C*-algebra. Then there is a sequence of isometries in A with mutually orthogonal ranges, i.e. A contains a unital copy of O_∞ (II.8.3.3(ii)).

If u and v are isometries with mutually orthogonal ranges (i.e. $u^*v = 0$), then u, vu, v^2u, ... form such a sequence.

III.1.3.4 PROPOSITION. If p is an infinite projection in a von Neumann algebra M, then there is a central projection $z \in \mathcal{Z}(M)$ with pz nonzero and properly infinite.

PROOF: Let $\{q_n\}$ be as in III.1.3.2. Then this set can be expanded to a maximal set $\{q_i : i \in \Omega\}$ of mutually orthogonal equivalent subprojections of p, and if $q = \sum_{i \in \Omega} q_i$, then by breaking Ω into three disjoint subsets $\{i_0\}, \Omega', \Omega''$ with $Card(\Omega') = Card(\Omega'') = Card(\Omega)$ we get projections $q' = \sum_{i \in \Omega'} q_i$ and $q'' = \sum_{i \in \Omega''} q_i$ with $q' \perp q''$, $q = q_{i_0} + q' + q''$,

$$q \sim q' \sim q'' \sim q' + q_{i_0}.$$

If $p_0 = p - q$, then by III.1.1.10 and maximality there is a central projection z with $p_0 z \neq 0$ and $p_0 z \precsim q_{i_0} z$, so

$$pz = qz + p_0 z \precsim q'z + q_{i_0} z \precsim qz,$$

$pz \precsim q''z$ and pz is properly infinite.

III.1.3.5 It is then an easy consequence that a projection p in a von Neumann algebra M is properly infinite if and only if for each central projection z, pz is either 0 or infinite. It follows from the Schröder-Bernstein Theorem that if p is properly infinite, then $p = p_1 + p_2$, where $p_1 \perp p_2$ and $p_1 \sim p_2 \sim p$, i.e. a properly infinite projection can be "halved." By continuing to subdivide, p can be "divided by n" for any n. Also, we can generate a sequence (q_n) of mutually orthogonal subprojections of p each equivalent to p; if $q = \sum q_n$, then $q \sim p$ by III.1.1.9. If $u^*u = p$ and $uu^* = q$, set $p_n = u^* q_n u$; then (p_n) is a sequence of mutually orthogonal projections, each equivalent to p, with $\sum p_n = p$. Thus p can also be "divided by $\aleph_0$."

III.1.3.6 PROPOSITION. Let M be a properly infinite, locally countably decomposable von Neumann algebra, and p a projection in M with central support 1. Then there is a sequence (p_n) of mutually orthogonal projections in M, each equivalent to p, with $\sum p_n = 1$.

PROOF: Let $\{z_i\}$ be a maximal family of mutually orthogonal nonzero central projections in M such that for each i, there is a sequence (p_{in}) of mutually orthogonal subprojections of z_i, each equivalent to pz_i, with $\sum_n p_{in} = z_i$. It suffices to show that $\sum z_i = 1_M$, for then $p_n = \sum_i p_{in}$ does the trick. If $z = 1 - \sum z_i \neq 0$, then $q = pz \neq 0$ since p has central support 1. By replacing Mz by a direct summand, we may assume that $N = Mz$ is countably decomposable; q has central support 1 in N. Since 1_N is properly infinite, then by III.1.3.5 there is a sequence of mutually orthogonal projections in N each equivalent to q. Choose a maximal family $\{q_i\}$ containing such a sequence; then $\{q_i\}$ is countably infinite. Let $r = \sum q_i$. By generalized comparability, there is a central projection $z' \in N$ with $(1_N - r)z' \precsim qz'$ and $q(1_N - z') \precsim (1_N - r)(1_N - z')$. By maximality of $\{q_i\}$, $z' \neq 0$. Fix an index i_0. Then

$$z' = (1_N - r)z' + rz' \sim (1_N - r)z' + \sum_{i \neq i_0} q_i z' \precsim q_{i_0} + \sum_{i \neq i_0} q_i z' = rz' \leq z'$$

so by III.1.1.9 $rz' \sim z'$, contradicting maximality of $\{z_i\}$.

Combining the previous two results, we get:

III.1.3.7 COROLLARY. Let M be a locally countably decomposable von Neumann algebra, and p a properly infinite projection in M with central support 1. Then $p \sim 1$.

Finite Projections

We now turn our attention to finite projections.

III.1.3.8 PROPOSITION. Let M be a von Neumann algebra and p a projection in M. Then the following are equivalent:

(i) p is finite.
(ii) Whenever q, r are subprojections of p with $q \sim r$, then $p - q \sim p - r$.
(iii) Whenever q, r are subprojections of p with $q \precsim r$, then $p - r \precsim p - q$.

PROOF: (i) $\implies$ (ii): If $p - q \not\sim p - r$, then by generalized comparability there is a nonzero central projection z such that $pz - qz$ is equivalent to a proper subprojection s of $pz - rz$ (or vice versa). Then

$$p = (pz - qz) + qz + p(1 - z)$$

is equivalent to the proper subprojection $s + rz + p(1 - z)$, so p is infinite.
(ii) $\implies$ (iii): If $q \sim s \leq r$, then $p - q \sim p - s \geq p - r$.
(iii) $\implies$ (i): If p is equivalent to a proper subprojection p_1, set $q = p$, $r = p_1$. Then $p - r \not\precsim p - q$.

Note that (ii) $\implies$ (iii) $\implies$ (i) is valid in a general C*-algebra. But (i) $\not\implies$ (ii) in a general C*-algebra, even in a matrix algebra over a commutative C*-algebra. This is the cancellation question in nonstable K-theory, which will be discussed in V.2.4.13.

III.1.3.9 PROPOSITION. Let p, q be projections in a von Neumann algebra M. If p and q are finite, then $p \vee q$ is finite.

PROOF: Suppose p and q are finite and $p \vee q$ is infinite. Using III.1.1.3 and replacing q by $(p \vee q) - p$, we reduce to the case where $p \perp q$, and then using III.1.3.4 we may assume $p \vee q = p + q$ is properly infinite. By halving there are r, s with $r \perp s$, $p + q = r + s$, $r \sim s \sim p + q$. By generalized comparability we may assume $r \wedge p \precsim s \wedge q$. Then

$$r = (r - r \wedge p) + (r \wedge p) \precsim (r \vee p - p) + (s \wedge q) \leq q,$$

so r is finite, a contradiction.

Note that in a general C*-algebra A, if p, q are finite projections in A with $p \perp q$, then $p + q$ is not necessarily finite (V.2.1.6).

A supremum of abelian projections (in a von Neumann algebra) is not necessarily abelian, and an infinite supremum of finite projections is not finite in general. An arbitrary supremum of finite [resp. abelian] *central* projections is finite [resp. abelian].

III.1.3.10 If p is an abelian projection, then it follows easily from I.9.1.4 that p is a minimal projection among all projections with central carrier z_p, and every subprojection of p is of the form pz for some central projection z. If q is any projection with $z_p \leq z_q$, then $p \precsim q$ by generalized comparability.

III.1.4 Type Classification

III.1.4.1 A consequence of III.1.3.4 and a simple maximality argument is that if M is a von Neumann algebra and p is a projection in M, then there is a central projection $z \in M$ such that pz is finite and $p(1-z)$ is properly infinite. In particular, taking $p = 1$, there is a unique central projection $z_f \in M$ such that z_f is finite and $1 - z_f$ is properly infinite. Mz_f is called the *finite part* of M and $M(1 - z_f)$ the *properly infinite part*.

$$M \cong Mz_f \oplus M(1 - z_f).$$

(Note, however, that Mz_f does not contain all finite projections of M in general; for example, if $M = \mathcal{L}(\mathcal{H})$, then $z_f = 0$.)

III.1.4.2 DEFINITION. Let p be a projection in a von Neumann algebra M. Then p is *semifinite* if every nonzero subprojection of p contains a nonzero finite subprojection, and p is *purely infinite* (or *type III*) if p contains no nonzero finite subprojections. M is semifinite if 1_M is semifinite, and purely infinite (or Type III) if 1_M is purely infinite.

(For convenience, 0 is both semifinite and purely infinite.)

III.1.4.3 If p is a finite projection in a von Neumann algebra M, then by generalized comparability any projection with the same central support is semifinite, and in particular z_p itself is semifinite. Then another maximality argument shows that every von Neumann algebra M contains a central projection z_{sf} such that z_{sf} is semifinite and $1 - z_{sf}$ purely infinite. Mz_{sf} is the *semifinite part* of M and Mz_{III} the *Type III part*, where $z_{III} = 1 - z_{sf}$. $z_{sf} = \bigvee \{p : p \text{ is finite}\}$. The finite part of M is contained in the semifinite part; thus M is a direct sum of a finite part, a properly infinite semifinite part, and a purely infinite part. (It is not yet obvious that the purely infinite part can be nonzero; but examples will be given in III.3.)

Semifinite von Neumann Algebras

III.1.4.4 We will now examine the semifinite part more carefully. We say a projection p in a von Neumann algebra is *continuous* if it contains no nonzero abelian subprojection, and *discrete* if it contains no nonzero continuous subprojection (i.e. every nonzero subprojection contains a nonzero abelian subprojection). Any purely infinite projection is continuous, and it turns out that some von Neumann algebras also have nonzero finite continuous projections. (Again, for convenience, 0 is both discrete and continuous.)

III.1.4.5 If $\{p_i\}$ is a family of abelian projections which is maximal with respect to the condition that the central carriers z_{p_i} are pairwise orthogonal, then $z_I = \sum z_{p_i}$ is a discrete central projection and $1 - z_I$ is continuous.

$$z_I = \bigvee \{p : p \text{ is discrete}\} = \bigvee \{p : p \text{ is abelian}\}.$$

Mz_I is called the *discrete* or *Type I* part of M.

III.1.4.6 We then let $z_{II} = 1 - z_I - z_{III}$; Mz_{II} is called the *Type II* part of M. z_{II} is the largest continuous semifinite projection. We can further decompose the Type II part of M into a finite and properly infinite part: set $z_{II_1} = z_{II}z_f$ and $z_{II_\infty} = z_{II}(1 - z_f)$; Mz_{II_1} and Mz_{II_∞} are called the *Type II_1* and *Type II_∞* parts of M. z_{II_1} is the largest finite continuous central projection, and z_{II_∞} the largest properly infinite semifinite continuous projection.

III.1.4.7 Thus M can be canonically written as

$$M = Mz_I \oplus Mz_{II_1} \oplus Mz_{II_\infty} \oplus Mz_{III}.$$

Some of these central summands may be zero. If $z_I = 1$, then M is called a *(pure) Type I von Neumann algebra*; Type II_1, Type II_∞, and Type III von Neumann algebras are defined similarly. Thus every von Neumann algebra is a direct sum of von Neumann algebras of pure type. If M is a factor, then M is either Type I, Type II_1, Type II_∞, or Type III. ($\mathcal{L}(\mathcal{H})$ is an example of a type I factor; examples of factors of the latter three types will be given in III.3).

We may further break up the Type I part of M into finite and infinite parts as with the Type II part; the central projection corresponding to the properly infinite Type I part is denoted z_{I_∞}. There is a more refined decomposition and structure description possible for Type I von Neumann algebras (III.1.5). There is also a further decomposition of Type III factors into Type III_λ, $0 \leq \lambda \leq 1$ (III.4.6.13).

III.1.5 Tensor Products and Type I von Neumann Algebras

Matrix Units

III.1.5.1 We first discuss a generalization of II.9.4.3. Suppose $\{p_i : i \in \Omega\}$ is a set of mutually orthogonal equivalent projections in $\mathcal{L}(\mathcal{H})$, with $\sum p_i = 1$. Fix an i_0, and for each j let u_j be a partial isometry from p_j to p_{i_0}. Set $e_{ij} = u_i^* u_j$. Then $\{e_{ij}\}$ is a *set of matrix units* in $\mathcal{L}(\mathcal{H})$ of type $Card(\Omega)$, and the von Neumann algebra N generated by $\{e_{ij}\}$ is a subalgebra of $\mathcal{L}(\mathcal{H})$ isomorphic to $\mathcal{L}(\mathcal{H}_1)$, where $\mathcal{H}_1$ has dimension $Card(\Omega)$. Furthermore, if $\{\eta_k : k \in K\}$ is an orthonormal basis for $\mathcal{H}_2 = p_{i_0}\mathcal{H}$, and for each i, k we set $\zeta_{i,k} = e_{ii_0}\eta_k$, then $\{\zeta_{i,k}\}$ is an orthonormal basis for $\mathcal{H}$, and if $\{\xi_i\}$ is an orthonormal basis for $\mathcal{H}_1$, then $\xi_i \otimes \eta_k \mapsto \zeta_{i,k}$ is a unitary from $\mathcal{H}_1 \otimes \mathcal{H}_2$ onto $\mathcal{H}$ carrying $\mathcal{L}(\mathcal{H}_1) \otimes I$ onto N. The relative commutant $N' \cap M$ is naturally isomorphic to $p_i M p_i$ for any i.

III.1.5.2 A type I factor always has a set of matrix units. If N is a type I factor and p is a nonzero abelian projection in N, let $\{p_i\}$ be a maximal set of mutually orthogonal projections in N equivalent to p. (Actually $\{p_i\}$ is just a maximal set of mutually orthogonal nonzero abelian projections in N since every nonzero abelian projection in N is equivalent to p by III.1.3.10; we also have that pNp is one-dimensional by III.1.3.10.) By maximality, comparability, and III.1.3.10, $\sum p_i = 1$, and, if $\{e_{ij}\}$ is a corresponding set of matrix units, it is not difficult to see that $\{e_{ij}\}$ generates N as a von Neumann algebra. So we have:

III.1.5.3 PROPOSITION. Let N be a type I factor on a Hilbert space $\mathcal{H}$. Then there is an isomorphism $\mathcal{H} \cong \mathcal{H}_1 \otimes \mathcal{H}_2$ such that $N \cong \mathcal{L}(\mathcal{H}_1) \otimes I$. In particular, every type I factor is isomorphic to $\mathcal{L}(\mathcal{H}_1)$ for some Hilbert space $\mathcal{H}_1$.

Tensor Products of von Neumann Algebras

III.1.5.4 There is a notion of spatial tensor product of von Neumann algebras. If M_i is a von Neumann algebra on $\mathcal{H}_i$ ($i = 1, 2$), then $M_1 \odot M_2$ acts naturally on $\mathcal{H}_1 \otimes \mathcal{H}_2$, and the weak closure is a von Neumann algebra denoted $M_1 \bar{\otimes} M_2$ (this is larger than the norm closure $M_1 \otimes_{\min} M_2$ if M_1, M_2 are infinite-dimensional).

It follows easily from I.2.5.4 and the essential uniqueness of the representation of a von Neumann algebra (III.2.2.8) that $M_1 \bar{\otimes} M_2$ depends up to isomorphism only on M_1 and M_2 and not on the way they are represented as von Neumann algebras, i.e. if $M_i \cong N_i$ ($i = 1, 2$), then $M_1 \bar{\otimes} M_2 \cong N_1 \bar{\otimes} N_2$.

III.1.5.5 PROPOSITION. If N is a type I factor on a Hilbert space $\mathcal{H}$, and $\mathcal{H} = \mathcal{H}_1 \otimes \mathcal{H}_2$ as in III.1.5.3, then

$$\mathcal{L}(\mathcal{H}) \cong \mathcal{L}(\mathcal{H}_1) \bar{\otimes} \mathcal{L}(\mathcal{H}_2).$$

If N is a subalgebra of a von Neumann algebra M, then

$$M \cong \mathcal{L}(\mathcal{H}_1) \bar{\otimes} (N' \cap M) \cong \mathcal{L}(\mathcal{H}_1) \bar{\otimes} pMp$$

where p is a one-dimensional projection in N.

The first statement follows from I.2.5.4, since

$$[\mathcal{L}(\mathcal{H}_1) \bar{\otimes} \mathcal{L}(\mathcal{H}_2)]' = (\mathcal{L}(\mathcal{H}_1) \otimes I) \cap (I \otimes \mathcal{L}(\mathcal{H}_2)) = \mathbb{C}(I \otimes I)$$

and the other statement is a straightforward consequence.

III.1.5.6 EXAMPLES.

(i) Let $L^\infty(X_i, \mu_i)$ act on $L^2(X_i, \mu_i)$ by multiplication ($i = 1, 2$). Then

$$L^\infty(X_1, \mu_1) \bar{\otimes} L^\infty(X_2, \mu_2) \cong L^\infty(X_1 \times X_2, \mu_1 \times \mu_2)$$

acting on

$$L^2(X_1, \mu_1) \otimes L^2(X_2, \mu_2) \cong L^2(X_1 \times X_2, \mu_1 \times \mu_2).$$

(ii) If M is properly infinite, then there is a sequence (p_n) of mutually orthogonal projections in M, each equivalent to 1, with $\sum p_n = 1$ (III.1.3.5). Thus M is unitarily equivalent to $\mathcal{L}(\mathcal{H})\bar{\otimes}M$, where $\mathcal{H}$ is a separable infinite-dimensional Hilbert space.

(iii) Let M be a properly infinite locally countably decomposable von Neumann algebra on $\mathcal{H}$, and p a projection in M with central support 1. Then by III.1.3.6, there is a sequence (p_n) of mutually orthogonal projections in M which are all equivalent to p, with $\sum p_n = 1$. Let $\{e_{ij}\}$ be the corresponding set of matrix units (III.1.5.1). Then $\{e_{ij}\}$ generates a type I factor $N \subseteq M$, with $N' \cap M$ naturally isomorphic to pMp. If $\mathcal{H} = \mathcal{H}_1 \otimes \mathcal{H}_2$ is the decomposition corresponding to N, then $\mathcal{H}_1$ is infinite-dimensional and separable, and $\mathcal{H}_2$ can be identified with $p\mathcal{H}$, and the corresponding decomposition of M is $\mathcal{L}(\mathcal{H}_1)\bar{\otimes}pMp$.

The following rephrasing of I.5.1.8 is useful:

III.1.5.7 PROPOSITION. Let M_1, M_2 be von Neumann algebras on $\mathcal{H}_1$ and $\mathcal{H}_2$ respectively. Let $\{p_i : i \in I\}$ be a set of mutually orthogonal projections in M_1, and $\{q_j : j \in J\}$ a set of mutually orthogonal projections in M_2. Then
$$\{p_i \otimes q_j : i \in I, j \in J\}$$
is a set of mutually orthogonal projections in $M_1\bar{\otimes}M_2$, and
$$\sum_{i,j}(p_i \otimes q_j) = \left(\sum_i p_i\right) \otimes \left(\sum_j q_j\right).$$

Combining this result with III.1.1.6, we obtain:

III.1.5.8 COROLLARY. Let M_1, M_2 be von Neumann algebras on $\mathcal{H}_1$ and $\mathcal{H}_2$ respectively. Let p be projection in M_1, and q a projection in M_2. Then the central support $z_{p\otimes q}$ of $p \otimes q$ in $M_1\bar{\otimes}M_2$ is $z_p \otimes z_q$.

III.1.5.9 If $\mathcal{H} = \mathcal{H}_1 \otimes \mathcal{H}_2$, then
$$(\mathcal{L}(\mathcal{H}_1) \otimes I)' = I \otimes \mathcal{L}(\mathcal{H}_2)$$
(I.2.5.4), and thus
$$(M\bar{\otimes}\mathcal{L}(\mathcal{H}_2))' = M'\bar{\otimes}I$$
and therefore
$$(M \otimes I)' = M'\bar{\otimes}\mathcal{L}(\mathcal{H}_2)$$
if M is a von Neumann algebra on $\mathcal{H}_1$. If M_i is a von Neumann algebra on $\mathcal{H}_i$ ($i = 1, 2$), then it turns out to be true that $(M_1\bar{\otimes}M_2)' = M'_1\bar{\otimes}M'_2$. This is not difficult to prove if the M_i are semifinite, but is a surprisingly deep theorem in general (III.4.5.8). The original proof was one of the first major applications of the Tomita-Takesaki theory (and the problem was one of the principal motivations for developing the theory). There is now a more elementary proof [KR97b, 11.2].

There is a simple proof of an important partial result:

III.1.5.10 PROPOSITION. If M_i is a factor on $\mathcal{H}_i$ ($i = 1, 2$), then $M_1 \bar{\otimes} M_2$ is a factor.

PROOF: $[(M_1 \bar{\otimes} M_2) \cap (M_1 \bar{\otimes} M_2)']'$ contains both $M_1 \otimes I$ and $M_1' \otimes I$, hence also
$$(M_1 \cup M_1')'' \otimes I = \mathcal{L}(\mathcal{H}_1) \otimes I.$$
Similarly, it contains $I \otimes \mathcal{L}(\mathcal{H}_2)$, and hence
$$[(M_1 \bar{\otimes} M_2) \cap (M_1 \bar{\otimes} M_2)']' = \mathcal{L}(\mathcal{H}_1 \otimes \mathcal{H}_2).$$

III.1.5.11 There are von Neumann algebra analogs of the maximal tensor product (the *binormal tensor product*) and of nuclearity (semidiscreteness/injectivity), which are discussed in IV.2.

Structure of Type I von Neumann Algebras

III.1.5.12 Now we can describe the structure of Type I von Neumann algebras. If n is a cardinal, either finite or infinite, we say a Type I von Neumann algebra is of *Type* I_n if 1 is a sum of n equivalent abelian projections. By the above, if M is Type I_n, then $M \cong \mathcal{L}(\mathcal{H}_n) \bar{\otimes} Z_n$, where $\mathcal{H}_n$ is an n-dimensional Hilbert space and Z_n is a commutative von Neumann algebra. The Type I_1 von Neumann algebras are precisely the commutative ones. A Type I_n factor is isomorphic to $\mathbb{M}_n$.

If M is a Type I von Neumann algebra, we will show that for each n, there is a central projection z_{I_n} in M such that $M z_{I_n}$ is type I_n and $\sum_n z_{I_n} = 1$. The key property of a Type I von Neumann algebra is that it contains an abelian projection with central carrier 1.

III.1.5.13 Although a unified approach is possible, it is convenient to treat the finite and infinite parts separately. First suppose M is finite and Type I. Let p be an abelian projection of central carrier 1. Let z_{I_1} be the largest central projection in M such that $z_{I_1} \leq p$; then $1 - z_{I_1}$ is the central carrier of $1 - p$. If p_2 is an abelian projection orthogonal to p and with central carrier $1 - z_{I_1}$, let z_{I_2} be the largest central subprojection of $1 - z_{I_1}$ with $z_{I_2} \leq p + p_2$; $1 - z_{I_1} - z_{I_2}$ is the central carrier of $1 - p - p_2$. Let p_3 be an abelian projection orthogonal to $p + p_2$ with central carrier $1 - z_{I_1} - z_{I_2}$, and define z_{I_3} as before. Continue inductively to get z_{I_n} for all finite n. We must have $\sum_n z_{I_n} = 1$, since otherwise $1 - \sum z_{I_n}$ contains a sequence of mutually equivalent orthogonal nonzero projections and is thus infinite by III.1.3.2 (or, alternately, by III.1.5.1 $M(1 - \sum z_{I_n})$ contains a copy of $\mathcal{L}(\mathcal{H})$ for an infinite-dimensional $\mathcal{H}$).

Since non-Type-I von Neumann algebras contain matrix subalgebras of arbitrarily large order, the last argument gives a useful characterization of finite Type I von Neumann algebras of bounded degree:

III.1.5.14 PROPOSITION. Let M be a von Neumann algebra, and $n \in \mathbb{N}$. The following are equivalent:

(i) M is a direct sum of Type I_m von Neumann algebras for $m \leq n$.
(ii) M does not contain $n+1$ mutually equivalent, mutually orthogonal non-zero projections.
(iii) M does not contain a (not necesarily unital) C*-subalgebra isomorphic to $\mathbb{M}_{n+1}$.

III.1.5.15 The properly infinite part is easy in the countably decomposable case, but slightly trickier in general. Suppose M is properly infinite and Type I. If M is locally countably decomposable and p is an abelian projection with central support 1, then the result follows from III.1.3.6 and III.1.5.5. In the general case, let $\{z_i\}$ be a maximal family of mutually orthogonal central projections such that every set of mutually orthogonal and equivalent projections in Mz_i is countable, and let $z_{I_{\aleph_0}} = \sum z_i$. It is routine to show that $Mz_{I_{\aleph_0}}$ contains a sequence (p_n) of equivalent orthogonal abelian projections with $\sum p_n = z_{I_{\aleph_0}}$, and that if $z_{I_{\aleph_0}} \neq 1$, then $M(1 - z_{I_{\aleph_0}})$ contains an uncountable set of mutually orthogonal equivalent projections with central carrier $1 - z_{I_{\aleph_0}}$. One can proceed transfinitely in a similar way to define $z_{I_\aleph}$ for uncountable cardinals $\aleph$.

It is customary to use the term "Type I_∞" to denote any properly infinite Type I von Neumann algebra, even if not homogeneous. (This distinction occurs only in the case of von Neumann algebras which are not locally countably decomposable.) Thus a Type I_∞ factor is one isomorphic to $\mathcal{L}(\mathcal{H})$ for any infinite-dimensional $\mathcal{H}$.

Commutative von Neumann Algebras

A commutative von Neumann algebra is Type I. We can give a measure-theoretic description of commutative von Neumann algebras.

III.1.5.16 PROPOSITION. Let Z be a commutative von Neumann algebra on $\mathcal{H}$ with a cyclic and separating vector. Then there is a finite measure space (X, μ) such that Z is unitarily equivalent to $L^\infty(X, \mu)$ acting by multiplication on $L^2(X, \mu)$, and Z is a masa (i.e. $Z = Z'$).

PROOF: $Z \cong C(X)$ for a compact Hausdorff space X. Suppose Z has a cyclic and separating vector ξ of norm one. The state ϕ on Z given by ξ is faithful, and corresponds to a probability measure μ on X by the Riesz Representation Theorem, and the GNS representation of Z from ϕ is equivalent to the identity representation. Since Z is weakly closed, it follows easily that the image of Z under the GNS representation from ϕ is all of $L^\infty(X, \mu)$. To see that $L^\infty(X, \mu)$ is a masa on $L^2(X, \mu)$, let $T \in L^\infty(X, \mu)'$ and $\chi \in L^2(X, \mu)$ the constant function 1, and let $f = T\chi \in L^2$. Then f is not obviously bounded, but M_f is a densely defined operator on L^2 whose domain contains the dense subspace L^∞. If $g \in L^\infty \subseteq L^2$, then

$$Tg = T(g\chi) = TM_g\chi = M_gT\chi = M_gf = gf = M_fg$$

so M_f is bounded and thus f is (essentially) bounded, and $T = M_f$ on all of L^2.

III.1.5.17 PROPOSITION. Let Z be a commutative von Neumann algebra on $\mathcal{H}$. Then Z' is Type I.

PROOF: Break up $\mathcal{H}$ into a direct sum of cyclic subspaces under Z; on each subspace a direct summand of Z acts faithfully and cyclically. If q is the projection onto one of these cyclic subspaces, then $q \in Z'$, and $qZ'q = Zq$ since Zq is a masa on $q\mathcal{H}$. Thus q is an abelian projection in Z'.

Combining this with the previous structure results (decomposing Z' into a direct sum of von Neumann algebras of Type I_n), we obtain:

III.1.5.18 COROLLARY. Let Z be a commutative von Neumann algebra on $\mathcal{H}$. Then $\mathcal{H}$ can be uniquely decomposed into $\bigoplus_n \mathcal{H}_n$ ($1 \leq n \leq \infty$) such that each $\mathcal{H}_n$ is invariant under Z and there is a locally finite measure space (X_n, μ_n) such that $Z_{\mathcal{H}_n}$ is spatially isomorphic to the n-fold amplification of $L^\infty(X_n, \mu_n)$ acting on $L^2(X_n, \mu_n)$. In particular, $Z \cong L^\infty(X, \mu)$ for some locally finite measure space (X, μ).

A useful consequence is:

III.1.5.19 COROLLARY. Let Z be a countably decomposable commutative von Neumann algebra on $\mathcal{H}$. Then Z is a masa in $\mathcal{L}(\mathcal{H})$ if and only if Z has a cyclic and separating vector.

III.1.5.20 An important consequence of these structure theorems, along with III.1.5.3, is that if M is a Type I von Neumann algebra on a Hilbert space $\mathcal{H}$, then M' is also Type I. This is also true for Type II and Type III von Neumann algebras (III.2.6.12).

III.1.6 Direct Integral Decompositions

There is a way of decomposing a general von Neumann algebra as a "measurable direct sum" of factors, generalizing the structure of Type I von Neumann algebras (by III.1.5.12 and III.1.5.17, a Type I_n von Neumann algebra looks like $L^\infty(X, \mu, \mathcal{L}(\mathcal{H}_n))$ acting on $L^2(X, \mu, \mathcal{H}_n)$, where (X, μ) is a locally finite measure space). The theory only works well on separable Hilbert spaces, so for simplicity we will restrict to this case. The technical details of the theory are considerable, and require some of the results from succeeding sections. We only outline the construction, referring the reader to [KR97b], [Dix69a], or [Sak71] for details.

III.1.6.1 If (X,μ) is a standard measure space, let $\mathcal{H}_x$ be a separable Hilbert space for each (or for almost all) $x \in X$, with inner product $\langle \cdot, \cdot \rangle_x$. A *measurable field* (or *integrable field*; *square-integrable field* would be a better name) is a vector subspace $\Gamma \subseteq \prod_x \mathcal{H}_x$, closed under multiplication by $L^\infty(X,\mu)$, such that $x \mapsto \langle \xi(x), \eta(x) \rangle_x$ is measurable for all $\xi, \eta \in \Gamma$, and such that

$$\int_X \langle \xi(x), \xi(x) \rangle_x \, d\mu(x) < \infty$$

for all $\xi \in \Gamma$ (so

$$\langle \xi, \eta \rangle = \int_X \langle \xi(x), \eta(x) \rangle_x \, d\mu(x)$$

is a pre-inner product on Γ), which is generated as an $L^\infty(X,\mu)$-module by a countable subset $\{\xi_n\}$ of Γ such that $\overline{\mathrm{span}}\{\xi_n(x)\} = \mathcal{H}_x$ for almost all x. The completion of Γ (with sections agreeing a.e. identified) is a separable Hilbert space $\mathcal{H}$ which can be identified with a space of equivalence classes of measurable sections of the field $(\mathcal{H}_x)$; we write

$$\mathcal{H} = \int_X^\oplus \mathcal{H}_x \, d\mu(x)$$

and call $\mathcal{H}$ the *direct integral* of the Hilbert spaces $\mathcal{H}_x$.

III.1.6.2 If $T_x \in \mathcal{L}(\mathcal{H}_x)$, then (T_x) is a *measurable field* of bounded operators if $(T_x \xi(x))$ is a measurable section for each measurable section ξ. If $\|T_x\|$ is uniformly bounded, then (T_x) defines an operator $T \in \mathcal{L}(\mathcal{H})$ in the obvious way, and $\|T\|$ is the essential supremum of $\|T_x\|$. Such a T is called *decomposable* and is written

$$T = \int_X^\oplus T_x \, d\mu(x).$$

$L^\infty(X,\mu)$ acts via decomposable operators by the formula

$$f \mapsto \int_X^\oplus f(x) I_x \, d\mu(x).$$

The image is called the *algebra of diagonalizable operators* of the field.

III.1.6.3 THEOREM. Let M be a von Neumann algebra on a separable Hilbert space $\mathcal{H}$, and let Z be a commutative von Neumann subalgebra of M'. Then there is a standard measure space (X,μ), a measurable field of Hilbert spaces $(\mathcal{H}_x)$ over (X,μ), and a unitary

$$U : \mathcal{H} \to \int_X^\oplus \mathcal{H}_x \, d\mu(x)$$

carrying Z onto the set of diagonalizable operators, and such that

$$UTU^* = \int_X^\oplus T_x \, d\mu(x)$$

is measurable for each $T \in Z' \supseteq M$. Furthermore, if, for each x, M_x is the von Neumann subalgebra of $\mathcal{L}(\mathcal{H}_x)$ generated by $\{T_x : T \in M\}$, then a measurable operator

$$S = \int_X^\oplus S_x \, d\mu(x)$$

satisfies $U^*SU \in (M \cup Z)''$ if and only if $S_x \in M_x$ a.e. (Note that $(M \cup Z)'' = M$ if $Z \subseteq \mathcal{Z}(M)$.) We write

$$M = \int_X^\oplus M_x \, d\mu(x).$$

If Z contains $\mathcal{Z}(M)$, then M_x is a factor for almost all x. If Z is a masa in M', then $M_x = \mathcal{L}(\mathcal{H}_x)$ a.e.

III.1.6.4 The most natural choice of Z is $\mathcal{Z}(M)$, for then not only is M_x a factor a.e., but the operators in M' are also decomposable,

$$M' = \int_X^\oplus (M')_x \, d\mu(x)$$

and $(M')_x = (M_x)'$ a.e. This is called the *central decomposition* of M and is the sense in which every von Neumann algebra on a separable Hilbert space is a direct integral of factors. Each central projection of M corresponds to a measurable subset of X in the central decomposition. In particular, the finite, semifinite, properly infinite, Type I_n, Type II_1, Type II_∞, and Type III parts respect the central decomposition.

III.1.6.5 Direct integrals respect tensor products in the following sense: if M_i is a von Neumann algebra on separable $\mathcal{H}_i$ ($i = 1, 2$), with commutative $Z_i \subseteq M_i'$, and

$$M_i = \int_{X_i}^\oplus (M_i)_x \, d\mu_i(x)$$

is the corresponding decomposition, then the decomposition of $M_1 \bar\otimes M_2$ with respect to $Z_1 \bar\otimes Z_2$ is

$$\int_{X_1 \times X_2}^\oplus N_{(x,y)} \, d(\mu_1 \times \mu_2)(x, y)$$

where $\mathcal{H}_{(x,y)} \cong (\mathcal{H}_1)_x \otimes (\mathcal{H}_2)_y$ and $N_{(x,y)} \cong (M_1)_x \bar\otimes (M_2)_y$ a.e. If $Z_i = \mathcal{Z}(M_i)$, then

$$Z_1 \bar\otimes Z_2 = \mathcal{Z}(M_1 \bar\otimes M_2)$$

(III.4.5.10), so the central decomposition respects tensor products.

III.1.7 Dimension Functions and Comparison Theory

Comparison of projections in a von Neumann algebra is completely determined by the dimension functions (II.6.8.12) on the algebra. In this subsection, we will concentrate on the case of a factor, where the dimension function is unique; the general case is similar but technically more complicated, and can be described by a "center-valued dimension function." The general case is subsumed in the theory of traces (III.2.5).

Let M be a factor. The idea is that there is a dimension function d (II.6.8.12) from $\mathrm{Proj}(M)$ to $[0, \infty]$ (plus other infinite cardinals in the non-countably-decomposable case), unique up to normalization, such that $p \precsim q$ if and only if $d([p]) \leq d([q])$. The range of d has only a few possibilities, and determines the type of M. Furthermore, d extends to all of M_+ by setting $d(x) = d([p_x])$, and the extension is lower semicontinuous.

It is simplest to describe the dimension function in the different cases separately. First of all, we will have $d([p]) < \infty$ if and only if p is finite. If M is not countably decomposable, to define $d(p)$ for infinite p we need:

III.1.7.1 PROPOSITION. Let p be a projection in a factor M. If p is not countably decomposable, the cardinality of all maximal sets of mutually orthogonal countably decomposable nonzero subprojections of p is the same.

See [KR97b, 6.3.9] for a proof.

If p is infinite (and hence properly infinite since M is a factor), set $d([p]) = \aleph$, where $\aleph$ is the cardinal of an infinite set of mutually equivalent and mutually orthogonal nonzero countably decomposable subprojections of p. It is straightforward to check (using e.g. III.1.3.5) that this definition has the right properties for properly infinite projections. This completes the definition of d if M is Type III. Of course, if M is countably decomposable, then the only infinite cardinal appearing is $\aleph_0$, usually denoted ∞, and all properly infinite nonzero projections in M are equivalent (III.1.3.7).

III.1.7.2 If M is type I, then all nonzero abelian projections in M are equivalent, and it is customary to normalize d so that $d([p]) = 1$ if $p \neq 0$ is abelian. The function d is then uniquely determined by orthogonal additivity: if $M \cong \mathcal{L}(\mathcal{H})$, then $d([p])$ is the rank of p, or, alternatively, $d([p]) = n$ if and only if pMp is Type I_n.

III.1.7.3 The Type II case requires more work. First consider the case where M is Type II_1. It is customary to normalize in this case so that $d([1]) = 1$. Since M contains no abelian projections, every nonzero projection in M contains two equivalent orthogonal nonzero subprojections. Thus, if $\{p_i, q_i\}$ is a maximal family of mutually orthogonal projections in M with $p_i \sim q_i$ for each i, and $p = \sum p_i$, $q = \sum q_i$, then $p \sim q$ and $p + q = 1$. Set $p_{1/2} = p$. Similarly, there is a $p_{1/4} \leq p_{1/2}$ with $p_{1/4} \sim (p_{1/2} - p_{1/4})$. If $u^*u = p$, $uu^* = q$, set $q_{1/4} = u p_{1/4} u^*$; then

$$q_{1/4} \sim p_{1/4} \sim (q - q_{1/4}).$$

Set $p_{3/4} = p_{1/2} + q_{1/4}$. Proceeding in this way, we get a chain (p_α) for dyadic rational $\alpha \in [0,1]$, with $p_0 = 0$, $p_1 = 1$, and $(p_\beta - p_\alpha) \sim p_{\beta-\alpha}$ for $\alpha < \beta$. For any $\alpha \in [0,1]$, let p_α be the supremum of the p_β as β runs over all dyadic rationals $\leq \alpha$. Of course, $p_\alpha \leq p_\beta$ if $\alpha \leq \beta$; and by finiteness $p_\alpha \precsim p_\beta$ only if $\alpha \leq \beta$. It is clear from the construction that any dimension function d on M with $d(1) = 1$ must have $d(p_\alpha) = \alpha$ for all α.

III.1.7.4 LEMMA. If p is any nonzero projection of M, then $p_\alpha \precsim p$ for all sufficiently small positive α.

PROOF: If not, by comparability $p \precsim p_\alpha$ for all $\alpha > 0$. Then for each n there is a subprojection of $p_{2^{-n+1}} - p_{2^{-n}}$ equivalent to p, which by III.1.3.2 contradicts finiteness of 1_M.

III.1.7.5 COROLLARY. For any $\alpha < 1$, $p_\alpha = \bigwedge \{p_\beta : \beta > \alpha\}$.

PROOF: If $p'_\alpha = \bigwedge \{p_\beta : \beta > \alpha\}$, then

$$p'_\alpha - p_\alpha \leq p_\beta - p_\gamma \sim p_{\beta - \gamma}$$

for any dyadic rational β, γ with $\gamma < \alpha < \beta$.

III.1.7.6 Now let p be any projection in M, and set $\alpha = \sup\{\beta : p_\beta \precsim p\}$. By comparability, $\alpha = \inf\{\beta : p \precsim p_\beta\}$. Then $p \sim p_\alpha$; for either $p \precsim p_\alpha$ or vice versa, and if $q \leq p_\alpha$, $q \sim p$, then by III.1.3.8 $p_\alpha - q \precsim p_\alpha - p_\beta \sim p_{\alpha-\beta}$ for all $\beta < \alpha$, so $p_\alpha - q = 0$ by III.1.7.4, and the other case is similar. Thus we can (and must) set $d(p) = \alpha$. The function d is the unique real-valued function on Proj(M) satisfying

$$d([p + q]) = d([p]) + d([q]) \text{ if } p \perp q$$

and $d([1]) = 1$; and furthermore, d satisfies

$$d([\vee p_i]) = \sup d([p_i])$$

for any increasing net (p_i) of projections in M.

III.1.7.7 We can then extend d to M_+ by setting $d(x) = d([p_x])$. It is clear that $d(x + y) = d(x) + d(y)$ if $x \perp y$, and it is easily verified that $d(x) \leq d(y)$ if $x \precsim y$ in the sense of II.3.4.3. It is also clear that $d(x) = \sup_{\epsilon > 0} d(f_\epsilon(x))$ (II.3.4.11) for all $x \in M_+$; it then follows from II.3.4.15 that d is norm-lower semicontinuous on M_+, and in fact d is σ-weakly lower semicontinuous.

III.1.7.8 Now suppose M is Type II$_\infty$. If p is any finite nonzero projection in M, then pMp has a dimension function d_p defined as above, with $d_p(p) = 1$. Fix a finite projection p_0, and for any finite q set $d(q) = d_{p_0}(r)$ if $q \sim r \leq p_0$ and $d(q) = d_q(s)^{-1}$ if $p_0 \sim s \leq q$. Then d is a dimension function on all of the finite projections of M, normalized so that $d(p_0) = 1$, and is the unique such dimension function. The number $d(q)$ can be characterized as

$$\sup\{m/n : m \cdot [q] \leq n \cdot [p_0]\} = \inf\{m/n : n \cdot [p_0] \leq m \cdot [q]\}$$

where $n \cdot [p] = [p_1 + \cdots + p_n]$, with $p_i \sim p$ and $p_i \perp p_j$ for $i \neq j$. The dimension function d can be extended to all positive elements of M as before.

We summarize the construction in a theorem. To avoid verbosity about infinite cardinals, we state the theorem only for the countably decomposable case.

III.1.7.9 THEOREM. Let M be a countably decomposable factor. Then there is a dimension function d, which is unique up to normalization, such that $p \precsim q$ if and only if $d([p]) \leq d([q])$. The range of d is

$\{0, 1, \ldots, n\}$ if M is Type I$_n$
$\{0, 1, 2, \ldots, \infty\}$ if M is Type I$_\infty$
$[0, 1]$ if M is Type II$_1$
$[0, \infty]$ if M is Type II$_\infty$
$\{0, \infty\}$ if M is Type III.

Furthermore, the associated dimension function on M_+ defined by $d(x) = d([p_x])$ is σ-weakly lower semicontinuous.

III.1.7.10 It turns out that if M is semifinite, then there is a trace τ on M such that $d([p]) = \tau(p)$ for any projection p. The existence of a *quasitrace* (II.6.8.15) τ corresponding to d is straightforward (II.6.8.14), and was first proved by Murray and von Neumann [MVN36]; but the linearity of τ was elusive and not proved until [MvN37]. We will discuss the trace in III.2.5.

Norm-Closed Ideals in Factors

An easy consequence of the existence of a dimension function on factors is a description of the norm-closed ideals of a factor; the general case is almost identical to the case of a Type I factor described in I.8.7.2.

First note that if I is an ideal in a von Neumann algebra M, and $x \in I_+$, then $E_{[\epsilon, \infty)}(x) \in I$ for all $\epsilon > 0$. Thus, if I is norm-closed, it is the norm-closed linear span of its projections. Also, if p and q are projections in M with $q \precsim p$, then $q = upu^*$ for some $u \in M$, so if $p \in I$, then $q \in I$. Since the dimension function determines comparability of projections, if M is a factor and $d(1_M) = \aleph_\alpha$, every norm-closed ideal of M is of the form $\mathcal{K}_\beta = \mathcal{K}_\beta(M)$ for some $\beta \leq \alpha$, where $\mathcal{K}_\beta$ is the closed linear span of the projections $p \in M$ with $d(p) < \aleph_\beta$. So we obtain:

III.1.7.11 PROPOSITION. Let M be a factor. Then
(i) If M is finite (Type I_n or Type II_1), then M is a simple C*-algebra.
(ii) If M is Type III and countably decomposable, then M is a simple C*-algebra.
(iii) If M is Type I_∞ or Type II_∞ and countably decomposable, then M has exactly one nontrivial norm-closed ideal $\mathcal{K}_0$, the closure of the set of elements of finite trace.
(iv) If M is Type I_∞ or Type II_∞ and $d(1_M) = \aleph_\alpha$, then the nontrivial norm-closed ideals of M are $\{\mathcal{K}_\beta : 0 \leq \beta \leq \alpha\}$.
(v) If M is Type III and $d(1_M) = \aleph_\alpha$, $\alpha \geq 1$, then the nontrivial norm-closed ideals of M are $\{\mathcal{K}_\beta : 1 \leq \beta \leq \alpha\}$.
(vi) If M is properly infinite, then $M \cong N \bar{\otimes} \mathcal{L}(\mathcal{H})$ for some infinite-dimensional $\mathcal{H}$ and some N which is a simple C*-algebra; $\dim(\mathcal{H})$ is uniquely determined by M, and the ideals of M are in one-one correspondence with those of $\mathcal{L}(\mathcal{H})$, with the exception of $\mathcal{K}_0$ if M is Type III.

In (vi), unless $\beta = 0$ and M is Type III, the ideal $\mathcal{K}_\beta(M)$ is the ideal of $N \bar{\otimes} \mathcal{L}(\mathcal{H})$ generated by $N \otimes \mathcal{K}_\beta(\mathcal{L}(\mathcal{H}))$ (norm-closure of $N \odot \mathcal{K}_\beta(\mathcal{L}(\mathcal{H}))$), but is in general larger than this C*-tensor product. For example, if M is a countably decomposable Type II_∞ factor, $M \cong N \bar{\otimes} \mathcal{L}(\mathcal{H})$ with N Type II_1 with tracial state τ, and $\{e_{ij}\}$ is a set of matrix units in $\mathcal{L}(\mathcal{H})$, let p_n be a projection in N with $\tau(p_n) = 2^{-n}$ for each n, and $q_n = p_n \otimes e_{nn} \in M$. If $q = \sum_n q_n$, then q is a finite projection in M, hence in $\mathcal{K}_0(M)$, but $q \notin N \otimes \mathcal{K}(\mathcal{H})$.

Note also that the N in (vi) is not completely determined by M in the II_∞ case since there are II_1 factors with nontrivial fundamental group (III.3.3.14).

III.1.8 Algebraic Versions

Most of the arguments in this section are valid in much greater generality. An attempt was begun in the early 1950's, led by Kaplansky [Kap51a], to give an algebraic characterization of which C*-algebras are isomorphic to von Neumann algebras (such a C*-algebra is called a *W*-algebra*, where the W stands for "weak"), postulating properties making the arguments of this section valid.

III.1.8.1 DEFINITION. An *AW*-algebra* is a C*-algebra M such that every maximal commutative C*-subalgebra of M is generated by projections, and that the projections of M form a complete lattice.

There are several equivalent ways of phrasing this definition. One of the most useful involves annihilators. If S is a subset of a C*-algebra (or ring) M, then the *right annihilator* of S in M is $\{y \in M : xy = 0 \,\forall x \in S\}$.

III.1.8.2 THEOREM. A C*-algebra M is an AW*-algebra if and only if the right annihilator of any subset of M is generated by a projection, i.e. is of the form pM for a projection p. (One could similarly work with left annihilators.)

III.1.8.3 Every W*-algebra is an AW*-algebra (the right annihilator of any subset is a weakly closed right ideal). An example of a commutative AW*-algebra which is not a W*-algebra was given by Dixmier [Dix51]: let M be the *-algebra of bounded Borel functions on $\mathbb{R}$, and define

$$\|f\| = \inf\{\alpha > 0 : \{t : |f(t)| > \alpha\} \text{ is of first category}\}$$

(dividing out by the functions of seminorm 0). There are even "wild" (non-W*) AW*-factors of Type III ([Tak78], [Sai79]); one of the outstanding open questions of operator algebra theory is whether there exist wild AW*-factors of type II_1 (see II.6.8.16 for interesting reformulations of this problem).

III.1.8.4 In fact, W*-algebras can be abstractly characterized as AW*-algebras with sufficiently many normal linear functionals (III.2.4.5) (the wild AW*-algebras in the previous paragraph have no nonzero normal linear functionals). W*-algebras also have an abstract characterization as C*-algebras which are dual Banach spaces (III.2.4.2).

III.1.8.5 It is almost, but not quite, true that an AW*-algebra which acts on a separable Hilbert space is a W*-algebra. Dixmier's commutative example (III.1.8.3) can be faithfully represented on a separable Hilbert space [Wri76]. But the commutative case is essentially the only exception:

III.1.8.6 THEOREM. [Wri80] Let M be an AW*-algebra on a separable Hilbert space. If $\mathcal{Z}(M)$ is a W*-algebra, then M is a W*-algebra. In particular, any AW*-factor which can be (faithfully) represented on a separable Hilbert space is a W*-algebra.

In fact, an AW*-factor of type II_1 with a faithful state must be a W*-algebra [Wri76]. This is, however, false for Type III AW*-factors [Sai79].

III.1.8.7 All the arguments of III.1.1, III.1.3, III.1.4, and III.1.7 (except III.1.7.10), and some of those in III.1.5, are valid for AW*-algebras. Many are valid in the purely algebraic setting of Baer *-rings; see [Ber72].

III.2 Normal Linear Functionals and Spatial Theory

Let M be a von Neumann algebra on $\mathcal{H}$. Since $\mathcal{L}(\mathcal{H})$ is the Banach space dual of $\mathcal{L}^1(\mathcal{H})$ (I.8.6.1) and M is weak-* (σ-weakly) closed, M is the Banach space dual of $\mathcal{L}^1(\mathcal{H})/M^\perp$, called the *predual* M_* of M (the terminology is justified since it turns out that M_* is the unique Banach space X such that $X^* \cong M$ (III.2.4.1)). When M_* is identified with a closed subspace of the dual space $M^* = (M_*)^{**}$, the linear functionals in M_* are called the *normal linear functionals* on M. In this section, we will describe these linear functionals and the representations of M.

III.2.1 Normal and Completely Additive States

There are several potentially different natural continuity conditions for linear functionals on a von Neumann algebra. In this subsection we show that these notions are all actually the same.

III.2.1.1 DEFINITION. Let M be a von Neumann algebra on $\mathcal{H}$, and ϕ a state on M. Then ϕ is *completely additive* if, whenever $\{p_i\}$ is a collection of mutually orthogonal projections of M,

$$\phi\left(\sum_i p_i\right) = \sum_i \phi(p_i).$$

III.2.1.2 DEFINITION. Let ϕ be a (bounded) linear functional on a von Neumann algebra M. Then ϕ is *normal* if, whenever (x_i) is a bounded increasing net in M_+ with $x = \sup x_i$, we have $\phi(x) = \lim \phi(x_i)$.

III.2.1.3 The importance of complete additivity and normality are that they are intrinsic to the algebraic structure and independent of the way M is represented as an algebra of operators (it turns out that the σ-strong and σ-weak topologies on M are also independent of the representation (III.2.2.12), but the strong and weak operator topologies are not). A σ-strongly continuous linear functional is obviously normal, and a normal state clearly completely additive. A linear functional ϕ is normal if and only if, whenever (x_i) is a decreasing net in M_+ converging to 0 strongly, $\lim \phi(x_i) = 0$. It follows easily that if ϕ is normal and $y \in M$, then $_y\phi$ and ϕ_y, defined by $_y\phi(x) = \phi(yx)$ and $\phi_y(x) = \phi(xy)$, are normal; and that every normal linear functional is a linear combination of normal states.

The fundamental technical result is:

III.2.1.4 THEOREM. Let M be a von Neumann algebra on $\mathcal{H}$ and ϕ a (bounded) linear functional on M. Then the following are equivalent:

(i) There are sequences $(\xi_n), (\eta_n)$ of vectors in $\mathcal{H}$ with $\sum_n \|\xi_n\|^2 < \infty$ and $\sum_n \|\eta_n\|^2 < \infty$, and $\phi(x) = \sum_n \langle x\xi_n, \eta_n\rangle$ for all $x \in M$.
(ii) ϕ is σ-weakly continuous on M.
(iii) ϕ is σ-strongly continuous on M.
(iv) ϕ is weakly continuous on the unit ball of M.
(v) ϕ is strongly continuous on the unit ball of M.
(vi) ϕ is normal.

If ϕ is a state, these are also equivalent to:

(vii) There is an orthogonal sequence (ξ_n) of vectors in $\mathcal{H}$ with $\sum_n \|\xi_n\|^2 = 1$ and $\phi(x) = \sum_n \langle x\xi_n, \xi_n\rangle$ for all $x \in M$.
(viii) ϕ is completely additive.

Normal linear functionals are not weakly or strongly continuous on M in general; in fact, the weakly continuous linear functionals and the strongly continuous linear functionals are precisely the linear functionals as in (i) with a finite sum. To prove that every strongly continuous linear functional ϕ has such a representation (such a linear functional is obviously weakly continuous), the inverse image of the unit disk in $\mathbb{C}$ under ϕ contains a strong neighborhood of 0, so there are vectors $\xi_1, \cdots, \xi_n \in \mathcal{H}$ such that $\|x\xi_i\| \leq 1$ for all i implies $|\phi(x)| \leq 1$. Set

$$\xi = \xi_1 \oplus \cdots \oplus \xi_n \in \mathcal{H}^{(n)}$$

and $\tilde{\mathcal{H}}$ the closed span of $\{(x \otimes I)\xi\}$ in $\mathcal{H}^{(n)}$, and define a bounded linear functional ψ on $\tilde{\mathcal{H}}$ by

$$\psi((x \otimes I)\xi) = \phi(x).$$

Apply the Riesz-Frèchet Theorem (I.1.3.1) to obtain $\eta = (\eta_1, \cdots, \eta_n) \in \tilde{\mathcal{H}}$ with $\psi(x) = \langle (x \otimes I)\xi, \eta \rangle$. The proof of (iii) $\Longrightarrow$ (i) is nearly identical.

The implications (i) $\Longrightarrow$ (ii) $\Longrightarrow$ (iii) $\Longrightarrow$ (vi), and (vi) $\Longrightarrow$ (viii) and (vii) $\Longrightarrow$ (i) if ϕ is a state, are trivial; (ii) $\Longrightarrow$ (iv) and (iii) $\Longrightarrow$ (v) follow from the fact that the weak and σ-weak, and the strong and σ-strong, topologies agree on bounded sets, and (iv) $\Longrightarrow$ (ii) and (v) $\Longrightarrow$ (iii) follow from an application of the Krein-Smulian Theorem.

The next lemma is the key observation in the proof of (ii) $\Longrightarrow$ (vii).

III.2.1.5 LEMMA. Let M be a von Neumann algebra on $\mathcal{H}$, and ϕ be a σ-weakly continuous state on M. Then ϕ extends to a σ-weakly continuous state on $\mathcal{L}(\mathcal{H})$.

PROOF: Choose (ξ_n), (η_n) in $\mathcal{H}$ as in (i). Let $\mathcal{H}'$ be a separable infinite-dimensional Hilbert space with orthonormal basis $\{\zeta_n\}$, and set $\xi = \sum \xi_n \otimes \zeta_n$, $\eta = \sum \eta_n \otimes \zeta_n$ in $\mathcal{H} \otimes \mathcal{H}'$; then, for $x \in M$, $\phi(x) = \langle (x \otimes 1)\xi, \eta \rangle$. If $x \geq 0$, we have

$$0 \leq \phi(x) = \langle (x \otimes 1)\xi, \eta \rangle = \langle (x \otimes 1)\eta, \xi \rangle \leq \langle (x \otimes 1)(\xi + \eta), \xi + \eta \rangle,$$

so by II.6.4.6 there is a $T \in (M \otimes 1)'$ such that

$$\phi(x) = \langle T(x \otimes 1)(\xi + \eta), \xi + \eta \rangle = \langle (x \otimes 1)T^{1/2}(\xi + \eta), T^{1/2}(\xi + \eta) \rangle$$

for all $x \in M$. Define a linear functional ψ on $\mathcal{L}(\mathcal{H} \otimes \mathcal{H}')$ (and, in particular, on $\mathcal{L}(\mathcal{H}) \otimes 1$) by

$$\psi(S) = \langle ST^{1/2}(\xi + \eta), T^{1/2}(\xi + \eta) \rangle.$$

Then ψ is positive and σ-weakly continuous (in fact, weakly continuous). Since $\psi(1) = \phi(1) = 1$, ψ is a state. Since the restriction of the weak or σ-weak topology on $\mathcal{L}(\mathcal{H} \otimes \mathcal{H}')$ to $\mathcal{L}(\mathcal{H}) \otimes 1 \cong \mathcal{L}(\mathcal{H})$ is the σ-weak topology on $\mathcal{L}(\mathcal{H})$, the result follows.

To show that (ii) $\implies$ (vii) for a state ϕ, extend ϕ to a σ-weakly continuous state ψ on $\mathcal{L}(\mathcal{H})$. By I.8.6.2 there is thus an $S \in \mathcal{L}^1(\mathcal{H})$, $S \geq 0$, such that $\psi(T) = \text{Tr}(ST)$ for all $T \in \mathcal{L}(\mathcal{H})$. Let $(\mu_1, \mu_2, \cdots)$ be the characteristic list (eigenvalue list) of S, with corresponding unit eigenvectors ζ_n, and set $\xi_n = \mu_n^{1/2}\zeta_n$. Then, for $x \in M$,

$$\phi(x) = \text{Tr}(xS) = \sum \langle xS\zeta_n, \zeta_n \rangle = \sum \langle x\mu_n \zeta_n, \zeta_n \rangle = \sum \langle x\xi_n, \xi_n \rangle.$$

Thus to finish the proof of the theorem it suffices to show (viii) $\implies$ (v) since every normal linear functional is a linear combination of normal states.

The key technical lemma in the proof of (viii) $\implies$ (v) is:

III.2.1.6 LEMMA. Let ϕ be a completely additive state on a von Neumann algebra M on a Hilbert space $\mathcal{H}$. Then there is a nonzero projection $p \in M$ and a unit vector $\xi \in \mathcal{H}$ such that $\phi(x) \leq \langle x\xi, \xi \rangle$ for all $x \in (pMp)_+$ (and hence ϕ_p is weakly continuous).

PROOF: Let ξ be a unit vector in $\mathcal{H}$, and set $\psi(x) = \langle x\xi, \xi \rangle$. Let (q_i) be a maximal family of mutually orthogonal projections in M such that $\psi(q_i) < \phi(q_i)$ for all i. If $q = \sum_i q_i$, then $\psi(q) < \phi(q)$ since ϕ and ψ are completely additive; hence $q \neq 1$ since $\phi(1) = \psi(1) = 1$. Set $p = 1 - q$. Then, for every projection $r \leq p$, $\phi(r) \leq \psi(r)$, so $\phi(x) \leq \psi(x)$ for all $x \in (pMp)_+$ since every such x is a norm-limit of nonnegative linear combinations of subprojections of p.

III.2.1.7 Now to prove (viii) $\implies$ (v), let ϕ be completely additive, and let $\{p_i : i \in \Omega\}$ be a maximal mutually orthogonal family of projections in M such that there is a vector ξ_i such that $\phi(x) \leq \langle x\xi_i, \xi_i \rangle$ for all $x \in (p_i M p_i)_+$. If $q = 1 - \sum_i p_i \neq 0$, apply III.2.1.6 to qMq on $q\mathcal{H}$ to obtain a contradiction. Thus $\sum_i p_i = 1$, and so $\sum_i \phi(p_i) = 1$. For each i and each $x \in M$, we have

$$\phi(p_i x^* x p_i) \leq \langle p_i x^* x p_i \xi_i, \xi_i \rangle = \|x p_i \xi_i\|^2.$$

Now let (x_j) be a net in the unit ball of M converging strongly to 0. Given $\epsilon > 0$, choose a finite subset F of Ω such that $\phi(1-p) < \epsilon^2/4$, where $p = \sum_{i \in F} p_i$. If j is sufficiently large, $\sum_{i \in F} \|x_j p_i \xi_i\| < \epsilon/2$. Then, applying the CBS inequality, we have:

$$|\phi(x_j)| \leq |\phi(x_j(1-p))| + \sum_{i \in F} |\phi(x_j p_i)|$$

$$\leq \phi(x_j^* x_j)^{1/2} \phi(1-p)^{1/2} + \sum_{i \in F} \phi(p_i x_j^* x_j p_i)^{1/2} \leq \phi(1-p)^{1/2} + \sum_{i \in F} \|x_j p_i \xi_i\| < \epsilon.$$

III.2.1.8 It follows that M_* consists precisely of the (bounded) linear functionals on M satisfying the conditions of III.2.1.4. Note that M_* depends only on the C*-algebra structure of M and not on the way it is represented as an algebra of operators.

Because $\mathcal{K}(\mathcal{H})^* = \mathcal{L}^1(\mathcal{H}) = \mathcal{L}(\mathcal{H})_*$, we obtain:

III.2.1.9 PROPOSITION. Every bounded linear functional ψ on $\mathcal{K}(\mathcal{H})$ extends uniquely to a normal linear functional ϕ on $\mathcal{L}(\mathcal{H})$, and $\|\phi\| = \|\psi\|$. If ψ is a state, the normal extension is the unique state extension of ψ to $\mathcal{L}(\mathcal{H})$ (II.6.4.16).

It is also worth noting the following immediate consequence of III.2.1.5:

III.2.1.10 COROLLARY. Let $N \subseteq M$ be von Neumann algebras. Then every normal state on N extends to a normal state on M.

Singular States

III.2.1.11 DEFINITION. Let M be a von Neumann algebra, and ϕ a state on M. Then ϕ is *singular* (on M) if there is no nonzero normal positive linear functional ψ on M with $\psi \leq \phi$.

Note that if $N \subseteq M$ are von Neumann algebras, ϕ is a state on M, and $\phi|_N$ is singular on N, then ϕ is singular on M (since N is a unital subalgebra of M).

III.2.1.12 PROPOSITION. Let ϕ be a state on $\mathcal{L}(\mathcal{H})$. Then the following are equivalent:

(i) ϕ is singular on $\mathcal{L}(\mathcal{H})$.
(ii) $\phi|_{\mathcal{K}(\mathcal{H})} = 0$.
(iii) $\phi(p) = 0$ for every rank-one projection $p \in \mathcal{L}(\mathcal{H})$.

PROOF: (ii) $\implies$ (iii) is trivial, and (iii) $\implies$ (ii) is nearly trivial since linear combinations of rank-one projections are norm-dense in $\mathcal{K}(\mathcal{H})$. For (iii) $\implies$ (i), suppose ψ is a nonzero normal positive linear functional on $\mathcal{L}(\mathcal{H})$ with $\psi \leq \phi$, and let $\{p_i\}$ be a maximal family of orthogonal rank-one projections in $\mathcal{L}(\mathcal{H})$. Then $0 < \psi(1) = \sum_i \psi(p_i)$, so $0 < \psi(p_i) \leq \phi(p_i)$ for some i. For (i) $\implies$ (ii), suppose $\omega = \phi|_{\mathcal{K}(\mathcal{H})} \neq 0$, and let ψ be the unique normal extension of ω to $\mathcal{L}(\mathcal{H})$ (III.2.1.9). Then $\psi \leq \phi$ by II.6.4.16.

III.2.2 Normal Maps and Isomorphisms of von Neumann Algebras

Normal Maps

III.2.2.1 If M and N are von Neumann algebras, then a *-homomorphism (or, more generally, a positive map) $\phi : M \to N$ is *normal* if, whenever (x_i) is a bounded increasing net in M_+, then

$$\phi(\sup x_i) = \sup \phi(x_i).$$

It is obvious that a *-isomorphism from M onto N is a lattice-isomorphism from M_+ onto N_+ and hence normal.

III.2.2.2 PROPOSITION. Let M and N be von Neumann algebras, and $\phi : M \to N$ a completely positive map. The following are equivalent:

(i) ϕ is normal.
(ii) ϕ is continuous for the σ-weak topologies on M and N.
(iii) ϕ is continuous for the σ-strong topologies on M and N.
(iv) $\phi|_{B_1(M)}$ is continuous for the σ-strong topologies on $B_1(M)$ and N.

PROOF: (i) $\iff$ (ii): Since the σ-weak topology on N is the weakest topology making all elements of N_* continuous, ϕ is σ-weakly continuous if and only if $f \circ \phi \in M_*$ for all $f \in N_*$; it is also obvious that ϕ is normal if and only if $f \circ \phi$ is normal for every normal state f on N. The result then follows from III.2.1.4.
(ii) $\implies$ (iii): By scaling we may assume $\|\phi\| \leq 1$. Suppose $x_i \to 0$ σ-strongly. Then $x_i^* x_i \to 0$ σ-weakly. If ϕ is continuous for the σ-weak topologies, then $\phi(x_i^* x_i) \to 0$ σ-weakly. Since $\phi(x_i)^* \phi(x_i) \leq \phi(x_i^* x_i)$, $\phi(x_i)^* \phi(x_i) \to 0$ σ-weakly, so $\phi(x_i) \to 0$ σ-strongly.
(iii) $\implies$ (iv) is trivial.
(iv) $\implies$ (i): If (x_i) is a bounded increasing net in M with $x = \sup x_i$, then $x_i \to x$ σ-strongly, so $\phi(x_i) \to \phi(x)$ σ-strongly. Since $\phi(x_i) \to \sup \phi(x_i)$ strongly, $\phi(x) = \sup \phi(x_i)$.

III.2.2.3 PROPOSITION. Let M be a von Neumann algebra, ϕ a normal state on M, and π_ϕ the GNS representation. Then π_ϕ is a normal representation, i.e. a normal *-homomorphism from M to $\mathcal{L}(\mathcal{H}_\phi)$.

In fact, π_ϕ is continuous for the σ-weak and σ-strong topologies.

The Normal Stinespring Theorem

We have a normal version of Stinespring's theorem (II.6.9.7):

III.2.2.4 THEOREM. [NORMAL STINESPRING THEOREM] Let M be a von Neumann algebra, and $\phi : M \to \mathcal{L}(\mathcal{H})$ be a normal completely positive map. Then there is a Hilbert space $\mathcal{H}_\phi$, a normal representation π_ϕ of M on $\mathcal{H}_\phi$, and $V_\phi \in \mathcal{L}(\mathcal{H}, \mathcal{H}_\phi)$ with $\|V_\phi\|^2 = \|\phi\|$, such that

$$\phi(a) = V_\phi^* \pi_\phi(a) V_\phi$$

for all $a \in M$, and such that $(\mathcal{H}_\phi, \pi_\phi, V_\phi)$ are canonical and minimal in the sense that if $\mathcal{H}'$ is another Hilbert space with a normal representation ρ of M, and $W \in \mathcal{L}(\mathcal{H}, \mathcal{H}')$ with $\phi(a) = W^* \rho(a) W$ for all $a \in M$, then there is an isometry $U : \mathcal{H}_\phi \to \mathcal{H}'$ onto a subspace invariant under ρ and intertwining π_ϕ and ρ, and such that $W = UV_\phi$.

If A and ϕ are unital, then $\mathcal{H}_\phi$ can be chosen to contain $\mathcal{H}$, and $\phi(a) = P_\mathcal{H} \pi(a) P_\mathcal{H}$.

An examination of the proof of II.6.9.7 shows that if ϕ is σ-strongly continuous, then so is the representation π_ϕ defined there.

Tensor Products of Normal Completely Positive Maps

III.2.2.5 Using III.2.2.4, as in the C*-case we immediately deduce the existence of tensor products of normal completely positive contractions: if M_1, M_2, N_1, N_2 are von Neumann algebras and $\phi : M_1 \to M_2$ and $\psi : N_1 \to N_2$ are normal completely positive contractions, then the map $\phi \otimes \psi$ from $M_1 \odot N_1$ to $M_2 \odot N_2$ given by

$$(\phi \otimes \psi)\left(\sum x_i \otimes y_i\right) = \sum \phi(x_i) \otimes \psi(y_i)$$

is well defined and extends to a normal completely positive contraction, also denoted $\phi \otimes \psi$, from $M_1 \bar{\otimes} N_1$ to $M_2 \bar{\otimes} N_2$. If ϕ and ψ are conditional expectations, so is $\phi \otimes \psi$.

III.2.2.6 As an important special case, there is also a normal version of slice maps (II.9.7.1). Let M and N be von Neumann algebras on $\mathcal{H}_1$ and $\mathcal{H}_2$ respectively, and let ϕ be a normal state on M. Then the slice map R_ϕ defined by

$$R_\phi\left(\sum x_i \otimes y_i\right) = \sum \phi(x_i) y_i$$

extends to a normal conditional expectation from $M \bar{\otimes} N$ to N (and similarly for a left slice map L_ψ for a normal state ψ on N).

III.2.2.7 Another important special case is, of course, the tensor product of normal states: if ϕ and ψ are normal states on M and N, then there is a corresponding normal state $\phi \otimes \psi$ on $M \bar{\otimes} N$. In this case, there is an obvious alternate way to view $\phi \otimes \psi$, as the vector state corresponding to $\xi_\phi \otimes \xi_\psi$.

Other versions of Stinespring's theorem and tensor products of normal maps can be found in IV.2.3.3 and IV.2.3.4. Tensor products of normal semifinite weights are discussed in III.2.2.30.

Isomorphisms of von Neumann Algebras

The following corollary of III.2.1.4 shows that an algebraic isomorphism of von Neumann algebras is "almost spatial." An *amplification* of a representation π of a C*-algebra A on $\mathcal{H}$ is a representation $\pi \otimes 1$ of A on $\mathcal{H} \otimes \mathcal{H}'$ for some $\mathcal{H}'$, and a *reduction* of π is the restriction of π to in invariant subspace $\mathcal{X}$ such that $P_\mathcal{X}$ has central support 1 in $\pi(A)'$.

III.2.2.8 THEOREM. Let M and N be von Neumann algebras on Hilbert spaces $\mathcal{H}$ and $\mathcal{H}'$. If π is an algebraic *-isomorphism from M onto N, regarded as a representation of M (which is necessarily normal by III.2.2.1), then π is equivalent to a reduction of an amplification of the identity representation of M. Thus there is a Hilbert space $\mathcal{H}''$ such that $M \otimes 1$ on $\mathcal{H} \otimes \mathcal{H}''$ is unitarily equivalent to $N \otimes 1$ on $\mathcal{H}' \otimes \mathcal{H}''$. If $\mathcal{H}$ and $\mathcal{H}'$ are separable, $\mathcal{H}''$ may be chosen separable.

PROOF: Since π is a direct sum of cyclic normal representations (countably many if $\mathcal{H}'$ is separable), i.e. GNS representations from normal states, it suffices to show that if ϕ is a normal state of M, then π_ϕ is equivalent to a subrepresentation of a countable amplification of the identity representation of M. Let $\{\xi_n\}$ be as in III.2.1.4, and let $\mathcal{H}''$ be a separable Hilbert space with orthonormal basis $\{\eta_n\}$. Then π_ϕ is equivalent to the cyclic subrepresentation of the amplification $M \otimes 1$ on $\mathcal{H} \otimes \mathcal{H}''$ with cyclic vector $\sum \xi_n \otimes \eta_n$.

From III.1.3.7 we obtain:

III.2.2.9 COROLLARY. If M and N are isomorphic von Neumann algebras, and M' and N' are properly infinite and locally countably decomposable, then M and N are unitarily equivalent.

III.2.2.10 DEFINITION. If π and ρ are normal representations of a von Neumann algebra M, write $\pi \precsim \rho$ if π is equivalent to a subrepresentation of ρ. If $\pi(M)$ acts on $\mathcal{H}$ and z is a central projection in M, write π_z for the subrepresentation of π on $\pi(z)\mathcal{H}$.

III.2.2.11 COROLLARY. Let M be a von Neumann algebra, and π and ρ normal representations of M. Then there is a central projection $z \in \mathcal{Z}(M)$ such that $\pi_z \precsim \rho_z$ and $\rho_{1-z} \precsim \pi_{1-z}$.

PROOF: Let σ be a faithful normal representation of M on $\mathcal{H}$ of sufficiently high multiplicity that $\pi \precsim \sigma$ and $\rho \precsim \sigma$, i.e. there are projections $p, q \in \sigma(M)'$ with $\sigma|_{p\mathcal{H}}$ and $\sigma|_{q\mathcal{H}}$ equivalent to π and ρ respectively. Apply generalized comparability to p and q.

A more detailed discussion of when algebraic isomorphisms of von Neumann algebras are spatial is in III.2.6.

III.2.2.12 COROLLARY. Let M and N be von Neumann algebras, and $\phi : M \to N$ an (algebraic) *-isomorphism. Then ϕ is a homeomorphism from M to N for either the σ-weak topologies or the σ-strong topologies.

This is also a corollary of III.2.2.2.

III.2.2.13 A *-isomorphism $\phi : M \to N$ of von Neumann algebras is not weakly or strongly continuous in general, but it is if M' and N' are both properly infinite, in which case the weak and σ-weak topologies coincide on M and N, as do the strong and σ-strong. But we do have:

III.2.2.14 PROPOSITION. Let M and N be von Neumann algebras, and $\phi : M \to N$ an (algebraic) *-isomorphism. Then ϕ is a homeomorphism from $B_1(M)$ (closed unit ball) to $B_1(N)$ for either the weak, strong, or strong-* operator topologies.

PROOF: The result for the weak or strong topology follows immediately from III.2.2.12 and the agreement of the weak and σ-weak topologies, and the strong and σ-strong, on bounded sets, and the result for the strong-* topologies follows almost immediately. An alternate proof for the strong or strong-* topologies can be based on I.3.2.8.

III.2.2.15 It should be noted that every infinite-dimensional von Neumann algebra has non-normal representations, e.g. the GNS representation from a nonnormal state, or an irreducible representation of a non-Type-I factor. Thus a concrete C*-algebra of operators which is algebraically *-isomorphic to a von Neumann algebra is not necessarily weakly closed. (Such representations generally only occur on nonseparable Hilbert spaces.)

Characterization of Strong Topologies

III.2.2.16 We also obtain a useful characterization of the strong or σ-strong topology on bounded sets. If ϕ is a state on a C*-algebra A, we can define a seminorm $\|\cdot\|_{2,\phi}$ on A by

$$\|x\|_{2,\phi} = \phi(x^*x)^{1/2}$$

(this is the seminorm induced by the GNS inner product $\langle x, y \rangle_\phi = \phi(y^*x)$).

III.2.2.17 PROPOSITION. Let M be a von Neumann algebra, and $\{\phi_i : i \in I\}$ a separating family of normal states on M. Then the topology generated by $\{\|\cdot\|_{2,\phi_i} : i \in I\}$ agrees with the strong (or σ-strong) topology on bounded subsets of M. In particular, a bounded net (x_j) in M converges to 0 (σ-) strongly if and only if $\phi_i(x_j^*x_j) \to 0$ for all i.

This follows almost immediately from I.3.1.2, since $\pi = \oplus_i \pi_{\phi_i}$ is a faithful normal representation of M, and

$$\{y\xi_{\phi_i} : y \in \pi(M)', i \in I\}$$

is total. For each y and i, if (x_j) is a net in M and $\phi_i(x_j^*x_j)^{1/2} \to 0$, then

$$\|\pi(x_j)y\xi_{\phi_i}\| = \|y\pi(x_j)\xi_{\phi_i}\| \leq \|y\|\|\pi(x_j)\xi_{\phi_i}\| = \|y\|\phi(x_j^*x_j)^{1/2} \to 0 .$$

Conversely, if $x_j \to 0$ strongly, then $\|\pi(x_j)\xi_{\phi_i}\| = \phi_i(x_j^*x_j)^{1/2} \to 0$ for each i.

Normal Weights

A weight ϕ on a von Neumann algebra M is *normal* if, whenever (x_i) is an increasing net in M_+ converging strongly to x, $\phi(x) = \lim \phi(x_i)$. A normal weight is completely additive on projections, but the converse is not true for unbounded weights in general.

The next theorem gives alternate characterizations of normality.

III.2.2.18 THEOREM. Let ϕ be a weight on a von Neumann algebra M. Then the following are equivalent:

(i) ϕ is normal.
(ii) ϕ is σ-weakly lower semicontinuous.
(iii) ϕ is the supremum of a family of normal linear functionals on M.
(iv) ϕ is the supremum of an upward directed family of normal linear functionals on M.
(v) ϕ is the sum of a family of normal linear functionals on M.

The implications (v) $\iff$ (iv) $\implies$ (iii) $\implies$ (ii) $\implies$ (i) are obvious. (i) $\implies$ (iii) was proved in [Haa75a], and (iii) $\implies$ (iv) then follows from a result of F. Combes [Com68]:

III.2.2.19 THEOREM. Let ϕ be a normal weight on a von Neumann algebra M. Then
$$\{f \in M_* : (1+\epsilon)f \leq \phi \text{ for some } \epsilon > 0\}$$
is upward directed.

This result is closely related in spirit to the fact that if A is a C*-algebra, then $\{x \in A_+ : \|x\| < 1\}$ is upward directed (II.4.1.3(i)). See [Str81] for details of the proof of this theorem and III.2.2.18.

If ϕ is a weight on a von Neumann algebra M, then the σ-weak closure of $\mathfrak{N}_\phi$ is of the form Mp for a projection p, and the σ-weak closure of $\mathfrak{M}_\phi$ is then pMp (III.1.1.15).

III.2.2.20 PROPOSITION. A normal weight ϕ on a von Neumann algebra M is semifinite if and only if $\mathfrak{N}_\phi$ (and hence $\mathfrak{M}_\phi$) is σ-weakly dense in M.

PROOF: If $\mathfrak{M}_\phi$ is σ-weakly dense in M, and (h_λ) is an increasing approximate unit for $\mathfrak{M}_\phi$, then $h_\lambda \to 1$ σ-strongly, so $h_\lambda x h_\lambda \to x$ for any $x \in M$ by joint strong continuity of multiplication on bounded sets. Thus, if $x \in M_+$,
$$\phi(x) \leq \liminf_\lambda \phi(h_\lambda x h_\lambda)$$
by σ-strong lower semicontinuity of ϕ. Conversely, if ϕ is semifinite, and the σ-weak closure of $\mathfrak{M}_\phi$ is pMp, then
$$\mathfrak{M}_\phi^\perp = (1-p)M(1-p) = \{0\}$$
(II.6.7.7).

The conclusion of this result is usually taken as the definition of a normal semifinite weight.

III.2.2.21 So if ϕ is a normal semifinite weight, $\mathfrak{M}_\phi$ (and hence $\mathfrak{N}_\phi \cap \mathfrak{N}_\phi^*$) is σ-strongly dense in M. Actually, if ϕ and ψ are normal semifinite weights on M, then

$$\mathfrak{N}_\psi^* \mathfrak{N}_\phi = Span\{y^*x : x \in \mathfrak{N}_\phi, y \in \mathfrak{N}_\psi\} \subseteq \mathfrak{N}_\phi \cap \mathfrak{N}_\psi^*$$

is σ-strongly dense in M (III.1.1.16).

Semifiniteness is very important for normal weights, but can be easily arranged:

III.2.2.22 PROPOSITION. Let ϕ be a normal weight on a von Neumann algebra M. Then there is a normal semifinite weight ψ on M with $\mathfrak{N}_\psi \supseteq \mathfrak{N}_\phi$, and such that the inclusion of $\mathfrak{N}_\phi$ into $\mathfrak{N}_\psi$ induces an isometry from $\mathcal{H}_\phi$ onto $\mathcal{H}_\psi$ (and hence the GNS representations π_ϕ and π_ψ are unitarily equivalent). The formula for ψ is $\psi(x) = \phi(pxp)$ (where the σ-weak closure of $\mathfrak{N}_\phi$ is Mp), and ψ agrees with ϕ on $\mathfrak{M}_\phi$.

PROOF: If $\psi(x) = \phi(pxp)$, then $\mathfrak{N}_\psi = \mathfrak{N}_\phi \oplus M(1-p)$ and $N_\psi = N_\phi \oplus M(1-p)$, so ψ has the stated properties and is clearly normal and semifinite.

Support Projections of States and Weights

III.2.2.23 Let ϕ be a normal weight (e.g. a normal state) on a von Neumann algebra M, and let C be the set of $x \in M_+$ with $\phi(x) = 0$. By normality, if $x \in C$, then its support projection $s(x) \in C$. If p and q are projections in C, then $p \vee q = s(p+q) \in C$, so the set of projections in C is upward directed. If q is the supremum, then $q \in C$ by normality, and $p = 1 - q$ satisfies $\phi(x) = \phi(pxp)$ for all $x \in M_+$. This projection p is called the *support projection*, or *carrier*, of ϕ, denoted $s(\phi)$.

III.2.2.24 Let

$$N_\phi = \{x \in M : \phi(x^*x) = 0\}.$$

Then N_ϕ is a left ideal of M. If ϕ is a state, from the CBS inequality $x \in N_\phi$ if and only if $\phi(yx) = 0$ for all $y \in M$. Thus N_ϕ is σ-weakly closed, and hence $N_\phi = Mq$ for some projection $q \in M$. This is the same q as above, so $p = s(\phi) = 1 - q$. Since, for any $x \in M$, $\phi(qx) = \overline{\phi(x^*q)} = 0$, it follows that

$$\phi(x) = \phi(px) = \phi(xp) = \phi(pxp)$$

for all $x \in M$; and $\phi|_{pMp}$ is faithful. If π_ϕ is the GNS representation of M, then the kernel of π_ϕ is $M(1 - z_{s(\phi)})$; thus π_ϕ is faithful on $Mz_{s(\phi)}$.

III.2.2.25 More generally, if ϕ is a positive map from M to another von Neumann algebra, we can define N_ϕ in the same way; N_ϕ is a norm-closed left ideal of M which is σ-weakly closed if ϕ is normal; in this case ϕ has a support projection $s(\phi)$ as above. If ϕ is completely positive, then by the Kadison inequality we have $\phi(yx) = 0$ for all $x \in N_\phi$, $y \in M$, so if ϕ is normal and $p = s(\phi)$, then, as above,

$$\phi(x) = \phi(px) = \phi(xp) = \phi(pxp)$$

for all $x \in M$; and $\phi|_{pMp}$ is faithful.

Faithful Weights and States

III.2.2.26 If M is a von Neumann algebra, let $\{\phi_i : i \in \Omega\}$ be a family of normal linear functionals which is maximal with respect to the condition that the supports p_i are mutually orthogonal. On M_+, let $\psi = \sum_i \phi_i$. Then ψ is a faithful normal semifinite weight on M, and $\mathfrak{N}_\psi \cap \mathfrak{N}_\psi^*$ contains $\cup_{F \in \mathcal{F}(\Omega)} p_F M p_F$, where $\mathcal{F}(\Omega)$ is the directed set of finite subsets of Ω and $p_F = \sum_{i \in F} p_i$. (Such a faithful normal semifinite weight is called *strictly semifinite*; the condition is equivalent to the property that the restriction of ϕ to M^ϕ (III.4.6.1) is semifinite (III.4.7.9). Not every faithful normal semifinite weight is strictly semifinite.) Thus every von Neumann algebra has a faithful normal semifinite weight.

If M is countably decomposable and $\{\phi_n : n \in \mathbb{N}\}$ is a maximal family of normal states on M with mutually orthogonal supports, set $\phi = \sum_n 2^{-n} \phi_n$. Then ϕ is a faithful normal state on M. Thus we obtain:

III.2.2.27 PROPOSITION. Let M be a von Neumann algebra. Then the following are equivalent:

(i) M has a faithful state.
(ii) M has a faithful normal state.
(iii) M is countably decomposable.
(iv) The strong (or σ-strong) operator topology, restricted to the unit ball of M, is first countable.
(v) The strong (or σ-strong) operator topology, restricted to the unit ball of M, is metrizable.

The implication (iii) $\implies$ (ii) results from the previous discussion, and (ii) $\implies$ (i) $\implies$ (iii) and (v) $\implies$ (iv) are trivial. To prove (ii) $\implies$ (v), let ϕ be a faithful normal state on M. Then by III.2.2.17 the norm

$$\|x\|_{2,\phi} = \phi(x^*x)^{1/2}$$

gives the strong operator topology on bounded subsets of M. For (iv) $\implies$ (iii), if $\{p_i\}$ is an uncountable set of mutually orthogonal nonzero projections in M, the net of finite sums of the p_i converges strongly to $\sum p_i$, but no sequence of finite sums converges strongly to $\sum p_i$, so the strong operator topology on the unit ball of M is not first countable.

III.2.2.28 The strong-* (or σ-strong-*) topology on the unit ball of a countably decomposable M is also metrizable via the metric

$$d^*(x,y) = \|x - y\|_{2,\phi} + \|x^* - y^*\|_{2,\phi}$$

where ϕ is a faithful normal state, and conversely.

The σ-weak topology on the unit ball of a countably decomposable von Neumann algebra is not metrizable in general. In fact, it is metrizable if and only if the von Neumann algebra has separable predual, a straightforward consequence of the fact that a compact metrizable space is second countable.

Note that if M is infinite-dimensional, none of the weak, σ-weak, strong, σ-strong, strong-*, σ-strong-* topologies are metrizable (or first countable) on all of M: if $\{p_n\}$ is a sequence of mutually orthogonal nonzero projections in M, then 0 is in the σ-strong-* closure of $\{\sqrt{n}p_n\}$, but no subsequence converges weakly to 0 by Uniform Boundedness; cf. I.1.3.4.

States and Weights on Tensor Products

We can form tensor products of normal states. Let M and N be von Neumann algebras, with normal states ϕ and ψ respectively. We can form the tensor product state $\phi \otimes \psi$ on $M \bar{\otimes} N$ as in III.2.2.7.

III.2.2.29 PROPOSITION. The support projection $s(\phi \otimes \psi)$ is $s(\phi) \otimes s(\psi)$. In particular, if ϕ and ψ are faithful, so is $\phi \otimes \psi$.

PROOF: We have that $[(1-s(\phi))\otimes 1] \perp s(\phi\otimes\psi)$ and $[1\otimes(1-s(\psi))] \perp s(\phi\otimes\psi)$, so

$$s(\phi \otimes \psi) \leq [s(\phi) \otimes 1] \wedge [1 \otimes s(\psi)] = s(\phi) \otimes s(\psi).$$

So

$$z_{s(\phi\otimes\psi)} \leq z_{s(\phi)\otimes s(\psi)} = z_{s(\phi)} \otimes z_{s(\psi)}$$

(III.1.5.8) [this also follows easily from III.2.2.24 and the fact that $\pi_{\phi\otimes\psi} = \pi_\phi \otimes \pi_\psi$.] We may thus cut down to $M z_{s(\phi)}$ and $N z_{s(\psi)}$, i.e. we may assume that π_ϕ and π_ψ are faithful. So $\pi_{\phi\otimes\psi} = \pi_\phi \otimes \pi_\psi$ is also faithful, i.e. $z_{s(\phi\otimes\psi)} = z_{s(\phi)\otimes s(\psi)} = z_{s(\phi)} \otimes z_{s(\psi)}$.

It is easily seen that $s(\phi)\mathcal{H}_\phi = [\pi_\phi(M)'\xi_\phi]$, and similarly $s(\psi)\mathcal{H}_\psi = [\pi_\psi(M)'\xi_\psi]$,

$$s(\phi \otimes \psi)\mathcal{H}_{\phi\otimes\psi} = [\pi_{\phi\otimes\psi}(M\bar{\otimes}N)'\xi_{\phi\otimes\psi}] = [(\pi_\phi(M)\bar{\otimes}\pi_\psi(N))'(\xi_\phi \otimes \xi_\psi)]$$

$$\supseteq [(\pi_\phi(M)'\bar{\otimes}\pi_\psi(N)')(\xi_\phi \otimes \xi_\psi)] = (s(\phi) \otimes s(\psi))(\mathcal{H}_\phi \otimes \mathcal{H}_\psi)$$

so $s(\phi) \otimes s(\psi) \leq s(\phi \otimes \psi)$.

Note that this proof can be streamlined by applying the (rather deep) Commutation Theorem for Tensor Products (III.4.5.8).

III.2.2.30 If ϕ and ψ are normal semifinite weights on M and N respectively, then a tensor product weight can be constructed on $M\bar\otimes N$, which is normal and semifinite and satisfies $s(\phi \otimes \psi) = s(\phi) \otimes s(\psi)$. The details, involving left Hilbert algebras, are rather delicate, however (cf. III.4.4.5); see [Tak03a, VIII.4] for a complete treatment.

III.2.3 Polar Decomposition for Normal Linear Functionals and the Radon-Nikodym Theorem

The next result, due to S. Sakai [Sak58], may be regarded as giving a polar decomposition for normal bounded linear functionals on a von Neumann algebra.

III.2.3.1 PROPOSITION. Let M be a von Neumann algebra and ψ a normal linear functional on M. Then there is a positive linear functional ϕ on M, with $\|\phi\| = \|\psi\|$, and a partial isometry $u \in M$, such that $\psi(x) = \phi(xu)$ and $\phi(x) = \psi(xu^*)$ for all $x \in M$. The ϕ is uniquely determined, and denoted $|\psi|$; u may be chosen so that u^*u is the support of ϕ, and such u is also unique.

We prove only the existence. We may assume $\|\psi\| = 1$. Let

$$K = \{a \in M : \psi(a) = 1 = \|a\|\}.$$

Since ψ is σ-weakly continuous, K is a nonempty σ-weakly compact subset of M. Let v be an extreme point of K; then v is an extreme point of the unit ball of M, and therefore a partial isometry (II.3.2.17). Set $\phi = \psi_v$, i.e. $\phi(x) = \psi(xv)$ for all x. Then $\|\phi\| \leq 1 = \phi(1)$, so ϕ is a state. If q is the support of ϕ, then $q \leq vv^*$ since

$$\phi(vv^*) = \psi(vv^*v) = \psi(v) = 1.$$

Set $u = v^*q$; u is a partial isometry with $u^*u = q$ and

$$\psi(xu^*) = \psi(xqv) = \phi(xq) = \phi(x)$$

for all x. In particular, $\psi(u^*) = 1$. Set $p = uu^*$. To show that $\psi(x) = \phi(xu)$ for all x, it suffices to show that $\psi(x(1-p)) = 0$ for all x. If $\psi(x(1-p)) = \beta > 0$ for some x, $\|x\| = 1$, then, for any n,

$$n + \beta = \psi(nu^* + x(1-p)) \leq \|nu^* + x(1-p)\|$$

$$= \|[nu^* + x(1-p)][nu + (1-p)x^*]\|^{1/2} = \|n^2 q + x(1-p)x^*\|^{1/2} \leq \sqrt{n^2+1}$$

which is a contradiction for large n (cf. the proof of II.6.9.4).

III.2.3.2 PROPOSITION. Let ϕ be a normal positive linear functional on a von Neumann algebra M, and $b \in M$. Then $|\phi_b| \leq \|b\|\phi$.

PROOF: Let u be as in III.2.3.1, i.e. $|\phi_b|(x) = \phi_b(xu^*) = \phi(xu^*b)$ and $\phi(xb) = \phi_b(x) = |\phi_b|(xu)$ for all $x \in M$. Set $a = u^*b$; then $|\phi_b| = \phi_a$ is positive, and thus, for all $x \in M$,

$$\phi(xa) = \phi_a(x) = (\phi_a)^*(x) = \overline{\phi_a(x^*)} = \overline{\phi(x^*a)} = \overline{\phi((a^*x)^*)} = \phi(a^*x)$$

and hence $\phi(xa^2) = \phi(a^*xa)$ for all x, so $\phi_{a^2} \geq 0$. Similarly, $\phi(xa^{2^{n+1}}) = \phi((a^{2^n})^*x(a^{2^n}))$ for all n. Then, if $x \geq 0$, by the CBS inequality,

$$|\phi_b|(x) = \phi(xa) = \phi(x^{1/2}(x^{1/2}a)) \leq \phi(x)^{1/2}\phi(a^*xa)^{1/2} = \phi(x)^{1/2}\phi(xa^2)^{1/2}$$

$$\leq \phi(x)^{1/2}\phi(x)^{1/4}\phi(xa^4)^{1/4} \leq \cdots \leq \phi(x)^{1-2^{-n}}\phi(xa^{2^n})^{2^{-n}}$$

$$\leq \phi(x)^{1-2^{-n}}(\|\phi\|\|x\|\|a\|^{2^n})^{2^{-n}} \to \|a\|\phi(x) \leq \|b\|\phi(x) \text{ as } n \to \infty.$$

Generalizing the classical Radon-Nikodym Theorem of measure theory to the situation of weights on a von Neumann algebra has proved to be an important and difficult problem, intimately tied up with the deep structure of the algebras. In this section, we will only prove the following special case, due to Sakai [Sak65], which is a crucial technical result in the spatial theory of von Neumann algebras. Compare this result to II.6.4.6 (which is used in the proof).

III.2.3.3 THEOREM. Let M be a von Neumann algebra, and ϕ and ψ normal positive linear functionals on M, with $\psi \leq \phi$. Then there is a unique $t \in M$ with $0 \leq t \leq 1$, such that $\psi(x) = \phi(txt)$ for all $x \in M$, and with the support projection of t equal to the support of ψ.

PROOF: By cutting down by the support of ϕ, we may assume ϕ is faithful, and identify M with $\pi_\phi(M)$. By II.6.4.6, there is an $r \in M'$, $0 \leq r \leq 1$, such that

$$\psi(x) = \langle rx\xi_\phi, \xi_\phi\rangle = \langle xs\xi_\phi, s\xi_\phi\rangle$$

where $s = r^{1/2}$. Let ω be the linear functional on M' defined by $\omega(y) = \langle ys\xi_\phi, \xi_\phi\rangle$. Then $\omega = (\phi')_s$, where $\phi'(y) = \langle y\xi_\phi, \xi_\phi\rangle$. By III.2.3.2, $|\omega| \leq \phi'$, so there is a $t \in M$, $0 \leq t \leq 1$, such that $|\omega|(y) = \langle yt\xi_\phi, \xi_\phi\rangle$ for $y \in M'$. Write $\omega = |\omega|_u$, $|\omega| = \omega_{u^*}$ for $u \in M'$ as in III.2.3.1. Then, for all $y \in M'$,

$$\langle t\xi_\phi, y\xi_\phi\rangle = \langle y^*t\xi_\phi, \xi_\phi\rangle = |\omega|(y^*) = \omega(y^*u^*) = \langle y^*u^*s\xi_\phi, \xi_\phi\rangle = \langle u^*s\xi_\phi, y\xi_\phi\rangle$$

so since ξ_ϕ is cyclic for M' we have $t\xi_\phi = u^*s\xi_\phi$. Also,

$$\langle uu^*s\xi_\phi, y\xi_\phi\rangle = \langle y^*uu^*s\xi_\phi, \xi_\phi\rangle = |\omega|(y^*u) = \omega(y^*) = \langle y^*s\xi_\phi, \xi_\phi\rangle = \langle s\xi_\phi, y\xi_\phi\rangle$$

for $y \in M'$, and hence $ut\xi_\phi = uu^*s\xi_\phi = s\xi_\phi$. Then, for $x \in M$,

$$\psi(x) = \langle xs\xi_\phi, s\xi_\phi \rangle = \langle xs\xi_\phi, ut\xi_\phi \rangle$$
$$= \langle xu^*s\xi_\phi, t\xi_\phi \rangle = \langle xt\xi_\phi, t\xi_\phi \rangle = \langle txt\xi_\phi, \xi_\phi \rangle = \phi(txt).$$

The result is often stated by letting $h = t^2$, i.e. there is a unique $h \in M$ with $0 \leq h \leq 1$, such that $\psi(x) = \phi(h^{1/2}xh^{1/2})$ for all $x \in M$, and with the support projection of h equal to the support of ψ. This h is an appropriate "Radon-Nikodym derivative" $\left[\frac{d\psi}{d\phi}\right]$.

The analogous statement for weights is false in general. A version of the Radon-Nikodym theorem for weights will be discussed in III.4.7.5.

III.2.4 Uniqueness of the Predual and Characterizations of W*-Algebras

Suppose M is a von Neumann algebra, and $\mathcal{X}$ is a (complex) Banach space such that $\mathcal{X}^*$ is isometrically isomorphic to M. Regard $\mathcal{X}$ as a subspace of $\mathcal{X}^{**} \cong M^*$ in the usual way.

III.2.4.1 THEOREM. Under this identification, $\mathcal{X} = M_*$; so there is a unique predual for every von Neumann algebra.

The proof of this theorem uses the weak-* topology on M as the dual of $\mathcal{X}$ (sometimes called the $\sigma(M, \mathcal{X})$-topology). The first step is to prove that M_{sa} is weak-* closed. By the Krein-Smulian Theorem, it suffices to prove that $M_{sa} \cap B_1(M)$ is weak-* closed, where $B_1(M)$ is the (norm-)closed unit ball of M (which is weak-* compact by Alaoglu's Theorem). If (x_j) is a net in $M_{sa} \cap B_1(M)$ converging weak-* to $a + ib$ with $a, b \in M_{sa}$, then, if $b \neq 0$, for suitably large positive or negative n we have

$$\|a + i(b + n1)\| \geq \|b\| + |n| > \sqrt{1 + n^2} \geq \|x_j + in1\|$$

so $(x_j + in1)$ cannot converge to $a + i(b + n1)$ since $B_{\sqrt{1+n^2}}(M)$ is weak-* closed, a contradiction.

$$M_+ \cap B_1(M) = (M_{sa} \cap B_1(M)) \cap (1 - M_{sa} \cap B_1(M))$$

is also weak-* closed, so M_+ is weak-* closed. It follows that $\mathcal{X}$ is the linear span of the states in $\mathcal{X}$, and the states in $\mathcal{X}$ are norming on M_{sa}. By the weak-* compactness of $B_1(M)$, every bounded increasing net in M_+ has a convergent subnet, and it follows that every state in $\mathcal{X}$ is normal; so $\mathcal{X} \subseteq M_*$. An application of the Hahn-Banach Theorem then shows that $\mathcal{X} = M_*$.

Characterization of W*-Algebras

There is also an elegant characterization of W*-algebras, due to Sakai [Sak56]:

III.2.4.2 THEOREM. Let A be a C*-algebra. If there is a (complex) Banach space $\mathcal{X}$ such that $\mathcal{X}^*$ is isometrically isomorphic to A, then A is a W*-algebra.

By a slight refinement of the argument in the proof of III.2.4.1, such a C*-algebra has a representation as a concrete C*-algebra M of operators with the property that every masa in M is monotone complete. The proof can then be finished using the next theorem [Ped72], which is of independent interest:

III.2.4.3 THEOREM. Let M be a unital C*-subalgebra of $\mathcal{L}(\mathcal{H})$ with the property that every masa in M is monotone closed. Then M is a von Neumann algebra.

In fact, it suffices to assume that M contains all spectral projections of each of its self-adjoint elements and that the sum of every family of mutually orthogonal projections in M is in M. The second condition is automatic from the first if the Hilbert space is separable.

III.2.4.4 If ϕ is a normal state on an AW*-algebra A and π_ϕ is the GNS representation, then $\pi_\phi(A)$ satisfies the hypotheses of III.2.4.3 and is thus a von Neumann algebra. So an AW*-algebra with a separating family of normal states is a W*-algebra, i.e. we get:

III.2.4.5 THEOREM. Let A be a C*-algebra. Then A is a W*-algebra if and only if every maximal commutative C*-subalgebra of A is monotone complete and A has a separating family of normal states.

III.2.5 Traces on von Neumann Algebras

Perhaps the most crucial structure fact about a finite von Neumann algebra M is the existence of a center-valued trace on M, from which it follows that M has a separating family of normal tracial states.

III.2.5.1 DEFINITION. Let A be a von Neumann algebra (or a unital C*-algebra), with center $Z = \mathcal{Z}(A)$. A *center-valued state* on A is a conditional expectation from A onto Z. A center-valued state θ on A is a *center-valued trace* if $\theta(xy) = \theta(yx)$ for all $x, y \in A$.

III.2.5.2 PROPOSITION. Let M be a von Neumann algebra with center Z. Then M has a normal center-valued state.

PROOF: The von Neumann algebra Z' is type I (III.1.5.20), and contains M (and also M'). Let p be an an abelian projection in Z' with central support 1. Then the map $\psi : Z \to pZ = pZ'p$ defined by $\psi(x) = px$ is an algebraic isomorphism and a σ-weak homeomorphism. Define $\phi : Z' \to Z$ by $\phi(x) = \psi^{-1}(pxp)$; then ϕ is a normal center-valued state.

Compare this result to IV.2.2.10 (cf. also IV.2.1.8).

A von Neumann algebra with a center-valued trace must be finite, and a center-valued trace on a von Neumann algebra must be faithful (if $x \in M_+$ and $\theta(x) = 0$, then $x = 0$). This is most easily seen using the notion of a monic projection:

III.2.5.3 DEFINITION. Let M be a von Neumann algebra, z a central projection, and $n \in \mathbb{N}$. A projection $p \in M$ is (n,z)-*monic* if there are mutually orthogonal projections $p_1, \ldots, p_n \in M$, each equivalent to p, with $z = p_1 + \cdots + p_n$. The projection p is *monic* if it is (n,z)-monic for some n and z.

A monic projection is also called *fundamental* in some references if the n is a power of 2.

III.2.5.4 EXAMPLES.
(i) If M is Type I_n, then any abelian projection $p \in M$ is (n, z_p)-monic.
(ii) If p is properly infinite and z_p is countably decomposable, then p is $(1, z_p)$-monic (and (n, z_p)-monic for any n) by III.1.3.6.
(iii) If M is Type II_1, then by an argument similar to the one in III.1.7.3, for any n and z there is an (n, z)-monic projection, and any two are equivalent.

In fact, for a given n and z, any two (n, z)-monic projections are equivalent.

III.2.5.5 PROPOSITION. If M is a finite von Neumann algebra, then every projection in M is a sum of a mutually orthogonal family of monic projections.

To prove this, it suffices to show that every nonzero projection $p \in M$ contains a nonzero monic subprojection; the result then follows from a maximality argument. One may furthermore assume that M is either Type I_n or Type II_1. The Type I_n case follows easily from III.2.5.4(i), and the argument in the Type II_1 case is similar to the one in III.1.7.4, using III.2.5.4(iii).

III.2.5.6 PROPOSITION. Let M be a von Neumann algebra with center-valued trace θ. Then M is finite, and θ is faithful.

For the proof, first note that if $p \precsim q$, then $\theta(p) \leq \theta(q)$. Thus, if p is (n, z)-monic, we must have $\theta(p) = \frac{1}{n}z$. If z is a properly infinite central projection in M, then z is $(2, z)$-monic by III.1.3.5, so $z = \theta(z) = \frac{1}{2}z$, a contradiction unless $z = 0$. Thus M is finite. If x is a nonzero element of M_+, then there is a nonzero projection p with $p \leq mx$ for some m, and a nonzero monic projection $q \leq p$; then $0 < \theta(q) \leq \theta(p) \leq m\theta(x)$.

The Main Theorem

III.2.5.7 THEOREM. Let M be a finite von Neumann algebra. Then M has a unique center-valued trace θ. Furthermore, θ has the following properties:

(i) θ is normal, and hence σ-weakly continuous (III.2.2.2).
(ii) θ is faithful, i.e. if $x \in M_+$ and $\theta(x) = 0$, then $x = 0$.
(iii) For any $x \in M$, $\theta(x)$ is the unique central element in the norm-closed convex hull of $\{uxu^* : u \in \mathcal{U}(M)\}$.
(iv) Every tracial state on M is of the form $f \circ \theta$, where f is a state on $\mathcal{Z}(M)$, i.e. every state on $\mathcal{Z}(M)$ extends uniquely to a tracial state on M.

III.2.5.8 COROLLARY. Every semifinite von Neumann algebra has a faithful normal semifinite trace. Every finite von Neumann algebra has a separating family of normal tracial states. Every countably decomposable finite von Neumann algebra (in particular, every II_1 factor) has a faithful normal tracial state.

PROOF: To prove the last statement, just note that if $\mathcal{Z}(M)$ is countably decomposable, then it has a faithful normal state. For the first two statements, every semifinite von Neumann algebra is a direct sum of algebras of the form $M \bar{\otimes} \mathcal{L}(\mathcal{H})$, where M is a countably decomposable finite von Neumann algebra (cf. III.1.5.6(iii)); such an algebra has a faithful normal semifinite tensor product trace.

III.2.5.9 We now discuss the proof of III.2.5.7. Once a normal center-valued trace θ is shown to exist, it must be faithful by III.2.5.6. Uniqueness of the center-valued trace also follows easily from a similar argument: if θ' is any center-valued trace on M, and p any projection in M, write $p = \sum_i p_i$, where p_i is (n_i, z_i)-monic; then, since $\theta(p_i) = \theta'(p_i) = \frac{1}{n_i} z_i$ for all i, and since θ is normal, we have

$$\theta(p) = \sum_i \theta(p_i) = \sum_i \theta'(p_i) \leq \theta'(p).$$

Since also
$$1 - \theta(p) = \theta(1 - p) \leq \theta'(1 - p) = 1 - \theta'(p)$$
we conclude that $\theta(p) = \theta'(p)$. By spectral theory we then obtain that $\theta(x) = \theta'(x)$ for all $x \in M_{sa}$.

The proof of (iv) is nearly identical, at least for normal tracial states (a bit more argument is needed for non-normal traces): if τ is a normal tracial state, then $\tau(p) = \tau \circ \theta(p)$ for every monic projection p, hence for every projection p by III.2.5.5, hence $\tau(x) = \tau \circ \theta(x)$ for all $x = x^*$ by spectral theory.

There are several proofs of the existence of θ. Most are based on the following approximate version. This result was proved by Murray and von Neumann [MvN37] in the factor case, and Dixmier [Dix49] observed that the argument works essentially verbatim in general.

III.2.5.10 LEMMA. Let M be a finite von Neumann algebra, and $\epsilon > 0$. Then there is a normal center-valued state ψ with the property that $\psi(xx^*) \leq (1+\epsilon)\psi(x^*x)$ for all $x \in M$.

The proof of the lemma proceeds in three steps. The first step, which requires most of the work, is to show that if ϕ is any normal center-valued state (such exist by III.2.5.2), then there is a projection $p \in M$ with $\phi(p) \neq 0$ and
$$\phi(xx^*) \leq (1+\epsilon)\phi(x^*x)$$
for all $x \in pMp$. To do this, first let $q = 1 - \sum q_i$, where $\{q_i\}$ is a maximal family of mutually orthogonal projections with $\phi(q_i) = 0$; then ϕ is faithful on qMq, and $\phi(q) = 1$. Then let $\{e_i, f_i\}$ be a maximal family of subprojections of q such that $e_i \perp e_j$ and $f_i \perp f_j$ for $i \neq j$ and $e_i \sim f_i$ and $\phi(e_i) > \phi(f_i)$ for all i (where ">" means "$\geq$ and not equal to" in C). If $e = q - \sum e_i$ and $f = q - \sum f_i$, then (unless ϕ is already a trace) $\phi(f) > \phi(e) \geq 0$, so $f \neq 0$, and $e \neq 0$ since $e \sim f$ (III.1.1.2), so $\phi(e) \neq 0$. If μ is the smallest number such that $\phi(e') \leq \mu\phi(f')$ whenever $e' \leq e$, $f' \leq f$, $e' \sim f'$, then $0 \neq \phi(e) \leq \mu\phi(f)$, so $\mu > 0$; and there is $e' \leq e$, $f' \leq f$, $e' \sim f'$, $(1+\epsilon)\phi(e') \not\leq \mu\phi(f')$, and by cutting down by a suitable central projection we can make $(1+\epsilon)\phi(e') > \mu\phi(f')$. Let $\{e_i'', f_i''\}$ be a maximal family such that $e_i'' \leq e'$, $f_i'' \leq f'$, $e_i'' \sim f_i''$, $(1+\epsilon)\phi(e_i'') \leq \mu\phi(f_i'')$. Then
$$0 \neq p = e' - \sum e_i'' \sim f'' = f' - \sum f_i''$$
and if $p_1, p_2 \leq p$, $p_1 \sim p_2$, there is $g \leq f''$, $g \sim p_1$, and we have
$$\phi(p_2) \leq \mu\phi(g) \leq (1+\epsilon)\phi(p_1).$$
By spectral theory this implies $\phi(xx^*) \leq (1+\epsilon)\phi(x^*x)$ whenever $x \in pMp$.

The second step is to show that there is a nonzero central projection z and a normal center-valued state ψ_z on zM with $\psi_z(xx^*) \leq (1+\epsilon)\psi(x^*x)$ for all $x \in zM$. Let p be as in the first part. We may assume p is monic by III.2.5.5; let $p_i, \ldots, p_n$ be a set of mutually orthogonal projections with $\sum p_i = z$ and p_i equivalent to p via u_i. A simple computation shows that ψ_z defined by $\psi_z(x) = \sum_{i=1}^{n} \phi(u_i x u_i^*)$ is the desired center-valued state.

The third step is a straightforward exhaustion argument.

III.2.5.11 The first existence proof, in [MvN37], valid on its face only for factors (although it can be modified for the general case), first constructs the quasitrace τ as in II.6.8.14, and then it must be shown that τ is linear. If ψ is as in III.2.5.10, then it follows from a simple argument that $|\psi(p) - \tau(p)| < \epsilon$ for every monic projection p, and thus (using normality of ψ and τ)
$$|\psi(x) - \tau(x)| < \epsilon\|x\|$$
for all $x \in M_+$. Therefore τ is ϵ-linear for all ϵ.

III.2.5.12 The second proof, by Dixmier [Dix49], was the first to handle the non-factor case, and also established property (iii) of III.2.5.7. This proof will be discussed in more detail in III.2.5.15.

III.2.5.13 The simplest existence proof using III.2.5.10 is due to Kadison ([Kad55], [Kad61]; see [KR97b, 8.2]). If (a_n) is a sequence of positive numbers strictly decreasing to 1, and ψ_n is a center-valued state as in III.2.5.10 with $\epsilon = a_n - 1$, a simple computation shows that $a_m^2 \psi_m - \psi_n$ is a positive map for $m < n$, hence completely positive by II.6.9.10, and so it follows from II.6.9.4 that (ψ_n) is norm-convergent to a center-valued state θ which must be a normal center-valued trace.

III.2.5.14 The most elegant existence proof, and the only standard proof not using III.2.5.10, is the argument of F. Yeadon [Yea71]. It suffices to show that every normal state on $Z = \mathcal{Z}(M)$ extends to a normal tracial state on M; then, by the above proof, the extension is unique, and the map ζ from Z_* to M_* sending a normal state to its tracial extension is a linear isometry; the dual map $\theta = \zeta^*$ from $(M_*)^* = M$ to $(Z_*)^* = Z$ is a surjective idempotent map of norm one, and hence a center-valued trace.

For each $u \in \mathcal{U}(M)$, let $T_u : M_* \to M_*$ be defined by

$$[T_u(f)](x) = f(u^*xu).$$

Then $\{T_u : u \in \mathcal{U}(M)\}$ is a group of isometries. Fix an $f \in M_*$, and let K_f be the norm-closed convex hull of $\{T_u(f) : u \in \mathcal{U}(M)\}$. Then each $g \in K_f$ satisfies $g|_Z = f|_Z$, and each T_u maps K_f onto itself. The crucial use of finiteness in the proof is in showing that K_f is weakly compact (the weak topology is the topology of pointwise convergence on M). Thus the Ryll-Nardzewski fixed-point theorem [RN67] applies (the "noncontracting" hypothesis is trivial) to yield a $\tau \in K_f$ with $T_u(\tau) = \tau$ for all $u \in \mathcal{U}(M)$, so τ is a tracial state with $\tau|_Z = f|_Z$. Since the restriction map from M_* to Z_* is surjective (III.2.1.10), this completes the argument.

The Dixmier Property

III.2.5.15 We now outline the existence proof of Dixmier [Dix49], which also establishes property (iii). It is worthwhile to name the relevant properties since they are used not only in the study of von Neumann algebras, but also for many C*-algebras such as certain reduced group C*-algebras.

III.2.5.16 DEFINITION. Let A be a unital C*-algebra with center $\mathcal{Z}(A)$. A has the *Dixmier property* if, for every $x \in A$, the norm-closed convex hull $\mathcal{K}_x$ of $\{uxu^* : u \in \mathcal{U}(A)\}$ has nonempty intersection with $\mathcal{Z}(A)$. A has the *strong Dixmier property* if $\mathcal{K}_x \cap \mathcal{Z}(A)$ consists of exactly one point for all $x \in A$.

III.2.5.17 PROPOSITION. A C*-algebra with the strong Dixmier property has a center-valued trace.

Indeed, $x \mapsto \phi(x) \in \mathcal{K}_x \cap \mathcal{Z}(A)$ is a linear contraction which is the identity on $\mathcal{Z}(A)$, and hence a center-valued state. If $u \in \mathcal{U}(A)$, then $\phi(uxu^*) = \phi(x)$ for all x, and it follows that $\phi(x^*x) = \phi(xx^*)$ for all x (and hence by polarization $\phi(xy) = \phi(yx)$ for all x, y) by an argument nearly identical to the proof of II.6.8.7.

III.2.5.18 THEOREM. Every von Neumann algebra has the Dixmier property.

The main step in the proof is to show that given $x \in M_{sa}$, there is a $u \in \mathcal{U}(M)$ and $y \in \mathcal{Z}(M)$ with $\|(x + uxu^*)/2 - y\| \leq \frac{3}{4}\|x\|$. In fact, if $\|x\| = 1$ and p is the support projection of x_+, choose a central projection z such that $pz \precsim (1-p)z$ and $(1-p)(1-z) \precsim p(1-z)$ and partial isometries v, w with $v^*v = pz$, $vv^* \leq (1-p)z$, $w^*w = (1-p)(1-z)$, $ww^* \leq p(1-z)$. Then the element

$$u = v + v^* + w + w^* + ((1-p)z - vv^*) + (p(1-z) - ww^*)$$

is unitary, and it and $y = (1 - 2z)/4$ work. The proof is then finished by iterating the construction.

III.2.5.19 THEOREM. Every finite von Neumann algebra has the strong Dixmier property.

Because of III.2.5.18, it is only necessary to show that if M is finite and $x \in M_{sa}$, then $\mathcal{K}_x$ cannot contain more than one point. But if $y, z \in \mathcal{K}_x$ and a center-valued state ψ is chosen as in III.2.5.10 with $\epsilon < \|y - z\|$, a contradiction is easily obtained.

III.2.5.20 It is shown in [HZ84] that every simple unital C*-algebra has the Dixmier property and every simple unital C*-algebra with unique tracial state has the strong Dixmier property. It has also been shown that $C_r^*(G)$ has the strong Dixmier property for certain G, and hence is simple with unique trace; cf. [Pow75], [PS79], [Ake81], [dlH85], [BN88], [HR89], [Béd91]. See [Avi82] for a related condition on reduced free products of C*-algebras.

Characterization of Finite Projections

We can use the existence of a center-valued trace to describe the σ-strong topology on bounded subsets of a finite von Neumann algebra M and show that the involution is σ-strongly continuous. The first result is an immediate corollary of III.2.2.17 and III.2.5.8:

III.2.5.21 PROPOSITION. Let M be a finite von Neumann algebra, and $\{\tau_i : i \in I\}$ be a separating family of normal tracial states on M (e.g. the set of all normal tracial states on M). Then the topology generated by the seminorms $\{\|\cdot\|_{2,\tau_i} : i \in I\}$ (III.2.2.17) agrees with the σ-strong topology on bounded subsets of M.

III.2.5.22 COROLLARY. If M is a finite von Neumann algebra, then the involution is σ-strongly continuous on bounded subsets of M, and hence the σ-strong and σ-strong-* topologies coincide on bounded subsets of M.

PROOF: If (x_j) is a bounded net in M converging σ-strongly to 0, then, for each tracial state τ on M,

$$\|x_j^*\|_{2,\tau} = \tau(x_j x_j^*)^{1/2} = \tau(x_j^* x_j)^{1/2} \to 0$$

so $x_j^* \to 0$ σ-strongly.

III.2.5.23 PROPOSITION. Let M be a von Neumann algebra, and p a projection in M. The following are equivalent:

(i) p is finite.
(ii) The involution is σ-strongly continuous on bounded subsets of pMp.
(iii) The function $x \mapsto px^*$ is σ-strongly continuous on bounded subsets of M.

PROOF: (i) $\Longrightarrow$ (ii) is III.2.5.22. For (iii) $\Longrightarrow$ (i), suppose p is not finite. By cutting down by a central projection, we can assume that p is properly infinite. Then there is a sequence (q_n) of mutually orthogonal subprojections of p, each equivalent to p (III.1.3.5); let v_n be a partial isometry with $v_n^* v_n = q_n$ and $v_n v_n^* = p$. Then $v_n \to 0$ σ-strongly, but $v_n^* = p v_n^* \not\to 0$ σ-strongly.
For (ii) $\Longrightarrow$ (iii), suppose (x_j) is a bounded net in M converging σ-strongly to 0. Since

$$px_j^* = pz_p x_j^* = z_p p x_j^* z_p$$

by cutting down by z_p we may assume p has central support 1. Choose a maximal family $\{p_i\}$ of mutually orthogonal projections in M with $p_i \precsim p$ for all i; then $\sum p_i = 1$ (III.1.1.12). There is a partial isometry v_i with $v_i^* v_i = p_i$ and $v_i v_i^* \leq p$, then $p_i = v_i^* p v_i$. It suffices to show that $\|px_j^* \xi\| \to 0$ whenever $\xi = p_i \xi$ for some i by I.3.1.2. For such i and ξ, we have

$$\|px_j^* \xi\| = \|px_j^* p_i \xi\| = \|px_j^* v_i^* p v_i \xi\| = \|(pv_i x_j p)^* pv_i \xi\|.$$

We have that $(pv_i x_j p)$ is a bounded net in pMp converging to 0 σ-strongly as $j \to \infty$, so by (ii) we have $(pv_i x_j p)^* \to 0$ σ-strongly and hence

$$\|(pv_i x_j p)^* pv_i \xi\| \to 0.$$

The next theorem will be useful in several contexts, including establishing the type of a tensor product of von Neumann algebras.

III.2.5.24 THEOREM. Let $N \subseteq M$ be von Neumann algebras, and θ a normal conditional expectation from M onto N. If N is Type III and p is a finite projection in M, then $\theta(p) = 0$.

PROOF: Suppose $\theta(p) \neq 0$, and let q be a nonzero projection in N with $\lambda q \leq \theta(p)$ for some $\lambda > 0$. Then there is a $y \in N$ with $y\theta(p) = q$. Let (x_j) be a bounded net in N converging σ-strongly to 0. Then $px_j^* \to 0$ σ-strongly by III.2.5.23, so $\theta(px_j^*) \to 0$ σ-strongly (III.2.2.2). But then

$$qx_j^* = y\theta(p)x_j^* = y\theta(px_j^*) \to 0$$

(σ-strongly), so q is finite by III.2.5.23, a contradiction.

III.2.5.25 COROLLARY. Let $N' \subseteq M$ be von Neumann algebras.

(i) If there is a normal conditional expectation $\theta : M \to N$, and M is semifinite, then N is semifinite.
(ii) If there is a a separating family $\{\theta_i : i \in I\}$ of normal conditional expectations from M to N (i.e. if $x \in M_+$, $x \neq 0$, then there is an i with $\theta_i(x) \neq 0$), and N is Type III, then M is Type III.

PROOF: (i): Suppose N is not semifinite, and let z be a central projection in N with Nz Type III. Then θ restricts to a normal conditional expectation from zMz to Nz, and zMz is semifinite. Thus, to obtain a contradiction, we may assume that N is Type III. Let $\{p_i\}$ be a family of mutually orthogonal finite projections in M with $\sum p_i = 1$. Then

$$1 = \theta(1) = \sum \theta(p_i) = 0,$$

a contradiction.
(ii) is immediate from the theorem.

One can also prove ([Tom59]; cf. [Str81, 10.21]):

III.2.5.26 THEOREM. Let $N \subseteq M$ be von Neumann algebras. If there is a normal conditional expectation $\theta : M \to N$, and M is Type I, then N is Type I.

See IV.2.2.3 for a related result.

Type of a Tensor Product

We can now establish the type of a tensor product of two von Neumann algebras. Throughout this subsection, M and N will be von Neumann algebras.

First note that tensor products respect direct sums, so to establish the type of $M \bar{\otimes} N$ we may assume that M and N are of pure type.

III.2.5.27 THEOREM. Let M and N be von Neumann algebras of pure type. Then

(i) $M\bar{\otimes}N$ is finite if and only if both M and N are finite.
(ii) $M\bar{\otimes}N$ is semifinite if and only if both M and N are semifinite.
(iii) $M\bar{\otimes}N$ is Type I if and only if both M and N are Type I.
(iv) $M\bar{\otimes}N$ is Type II if and only if both M and N are semifinite and one is Type II.
(v) $M\bar{\otimes}N$ is Type II$_1$ if and only if both M and N are finite and one is Type II$_1$.
(vi) $M\bar{\otimes}N$ is Type II$_\infty$ if and only if both M and N are semifinite, one is Type II, and not both are finite.
(vii) $M\bar{\otimes}N$ is Type III if and only if either M or N is Type III.

(There are, of course, redundancies in the statements of these parts.)

If $\{\phi_i : i \in I\}$ and $\{\psi_j : j \in J\}$ are separating families of normal tracial states on M and N respectively, then

$$\{\phi_i \otimes \psi_j : i \in I, j \in J\}$$

is a separating family of normal tracial states on $M\bar{\otimes}N$ (III.2.2.29, I.5.1.8). This shows that if M and N are both finite, then $M\bar{\otimes}N$ is finite. The converse direction of (i) follows from the fact that M and N embed in $M\bar{\otimes}N$, and a subalgebra of a finite algebra is finite.

For (ii), if p and q are finite projections in M and N respectively with central support 1, then $p \otimes q$ is a finite projection in $M\bar{\otimes}N$ (previous paragraph) with central support 1 (III.1.5.8), giving one direction. The converse direction of (ii) follows from (vii).

If M and N are Type I, and p and q are abelian projections in M and N with central support 1, then $p \otimes q$ is an abelian projection in $M\bar{\otimes}N$ with central support 1 (III.1.5.8), so $M\bar{\otimes}N$ is Type I. More specifically, it is easy to see that (Type I_m) $\bar{\otimes}$ (Type I_n) is Type I_{mn}. So to finish the proof of the theorem it suffices to prove the following two lemmas:

III.2.5.28 LEMMA. If M or N is Type III, then $M\bar{\otimes}N$ is Type III.

PROOF: This is almost an immediate corollary of III.2.5.25. Suppose N is Type III. As ϕ ranges over all normal states of M, the left slice maps L_ϕ (III.2.2.6) give a separating family of normal conditional expectations from $M\bar{\otimes}N$ to N: we have $s(L_\phi) = s(\phi) \otimes 1$, so

$$\bigvee_\phi s(L_\phi) = \bigvee_\phi [s(\phi) \otimes 1] = \left[\bigvee_\phi s(\phi)\right] \otimes 1 = 1 \otimes 1$$

(I.5.1.8).

III.2 Normal Linear Functionals and Spatial Theory   269

III.2.5.29 LEMMA. If M or N is continuous, then $M \bar{\otimes} N$ is continuous.

PROOF: If M or N is Type III, the result follows from III.2.5.28. Suppose M and N are semifinite, and p and q finite projections in M and N respectively with central support 1. Then $p \otimes q$ is a finite projection in $M \bar{\otimes} N$ with central support 1. It suffices to show that the finite von Neumann algebra $pMp \bar{\otimes} qNq$ is Type II$_1$ if M (and hence pMp) is Type II$_1$. If $pMp \bar{\otimes} qNq$ is not Type II$_1$, then there is a nonzero central projection $z \in pMp \bar{\otimes} qNq$ with $z(pMp \bar{\otimes} qNq)$ Type I$_n$ for some $n \in \mathbb{N}$. But $z(pMp \bar{\otimes} qNq)$ has a W*-subalgebra $z(pMp \otimes 1)$ which is isomorphic to a quotient of pMp by a weakly closed ideal and is therefore Type II$_1$, a contradiction since every W*-subalgebra of a Type I$_n$ von Neumann algebra is Type I.

III.2.6 Spatial Isomorphisms and Standard Forms

In this section, we elaborate on the essential uniqueness of the representation of a W*-algebra as a von Neumann algebra established in III.2.2.8.

III.2.6.1 DEFINITION. Let M be a von Neumann algebra, and ψ a faithful normal semifinite weight on M. The *ψ-standard form representation* of M is the GNS representation from ψ. A normal representation π of M is in *standard form* if it is (unitarily equivalent to) a ψ-standard form representation for some faithful normal semifinite weight ψ. A von Neumann algebra on $\mathcal{H}$ is in standard form if its identity representation is in standard form.

III.2.6.2 EXAMPLE. Let M be a type I factor, i.e. $M = \mathcal{L}(\mathcal{H})$ for some $\mathcal{H}$, and Tr the standard trace on M. The Tr-standard form representation of M is the representation as left multiplication operators on the ideal of Hilbert-Schmidt operators $\mathcal{L}^2(\mathcal{H})$. In this representation, M' is also isomorphic to M, acting by right multiplication. In particular, M and M' are the same "size."

The next two obvious observations are useful in decomposing and analyzing representations.

III.2.6.3 PROPOSITION. Let M be a von Neumann algebra on $\mathcal{H}$, and $\{z_i\}$ a set of mutually orthogonal central projections of M with $\sum z_i = 1$. Then M is in standard form on $\mathcal{H}$ if and only if Mz_i is in standard form on $z_i \mathcal{H}$ for all i.

III.2.6.4 PROPOSITION. Let M_1, M_2 be von Neumann algebras on $\mathcal{H}_1$ and $\mathcal{H}_2$ respectively. If M_1 and M_2 are in standard form, then $M_1 \bar{\otimes} M_2$ is in standard form on $\mathcal{H}_1 \otimes \mathcal{H}_2$.

The converse is false: if $\mathcal{H}_1$, $\mathcal{H}_2$ are separable, M_1 is Type III, and M_2 is arbitrary, then $M_1 \bar{\otimes} M_2$ is Type III (III.2.5.27) and hence in standard form (III.2.6.16).

III.2.6.5 PROPOSITION. Every von Neumann algebra has a standard form representation.

The existence of a standard form follows from the existence of a faithful normal semifinite weight on any von Neumann algebra (III.2.2.26).

III.2.6.6 THEOREM. Let M be a countably decomposable von Neumann algebra. Then the standard form representation of M is unique up to spatial isomorphism (i.e. any two standard form representations are unitarily equivalent). If M is in standard form, M' is also in standard form.

This theorem is also true without the hypothesis that M be countably decomposable, but the proof requires Modular Theory (III.4.5.7).

We first establish uniqueness of standard forms coming from faithful normal states, using III.2.3.3.

III.2.6.7 THEOREM. Let ϕ and ψ be faithful normal states on a von Neumann algebra M. Then π_ϕ and π_ψ are unitarily equivalent.

This is almost an immediate corollary of III.2.3.3. Let $\omega = \phi + \psi$; then $\phi \leq \omega$, so there is an $h \in M_+$ such that

$$\phi(x) = \omega(h^{1/2} x h^{1/2})$$

for all $x \in M$; since ϕ is faithful, the support projection of h is 1, so there is a sequence (x_n) in $C^*(h) \subseteq M$ such that $x_n h^{1/2} \nearrow 1$ in M. Form the GNS representations π_ϕ and π_ω on $\mathcal{H}_\phi$ and $\mathcal{H}_\omega$ with cyclic vectors ξ_ϕ and ξ_ω respectively. The map $x\xi_\phi \mapsto xh^{1/2}\xi_\omega$ extends to an isometry of $\mathcal{H}_\phi$ onto $\mathcal{X} = [Mh^{1/2}\xi_\omega]$ intertwining π_ϕ with $\pi_\omega|_\mathcal{X}$. But if $\eta \in \mathcal{H}_\omega$ (e.g. $\eta = \xi_\omega$), then $x_n h^{1/2} \eta \to \eta$ since $\pi_\phi(x_n h^{1/2}) \to 1$ strongly, so $\eta \in \mathcal{X}$, $\mathcal{X} = \mathcal{H}_\omega$ and π_ϕ is unitarily equivalent to π_ω. Similarly, π_ψ is unitarily equivalent to π_ω.

Note that if M is in ϕ-standard form from a state ϕ, then M' is obviously in standard form from the complementary state ϕ' defined by ξ_ϕ.

We can now finish the proof of III.2.6.6. If M is countably decomposable and ψ is a faithful normal semifinite weight on M, choose h as in the following lemma, and define

$$\phi(x) = \psi(h^{1/2} x h^{1/2})$$

for $x \in M_+$. Then ϕ defines a state on M, and exactly as in the previous proof π_ϕ is unitarily equivalent to π_ψ.

III.2.6.8 LEMMA. Let M be a countably decomposable von Neumann algebra, and ψ an unbounded normal semifinite weight on M. Then there is an $h \in \mathfrak{M}_\psi$, $0 \leq h \leq 1$, with $\psi(h) = 1$ and the support projection of h equal to 1.

PROOF: The unit ball of $(\mathfrak{M}_\psi)_+$ is σ-strongly dense in the unit ball of M_+ by the Kaplansky Density Theorem, and the unit ball of M is σ-strongly metrizable by III.2.2.27, so there is a sequence (h_n) of positive elements of $\mathfrak{M}_\psi$, of norm 1, converging σ-strongly to 1. Since ψ is unbounded, $\limsup \psi(h_n) = \infty$, so we may assume $\psi(h_n) \geq 1$ for all n. Set

$$h = \sum_{n=1}^{\infty} 2^{-n} \psi(h_n)^{-1} h_n.$$

(This h is in fact strictly positive in the norm-closure of $\mathfrak{M}_\psi$.)

III.2.6.9 COROLLARY. Let M be a countably decomposable von Neumann algebra on $\mathcal{H}$. Then M is in standard form if and only if M has a cyclic and separating vector. In particular, if M is in standard form, then M' is also countably decomposable.

Recall that if M is a von Neumann algebra which has a representation with a separating vector, or a faithful representation with a cyclic vector, then M is countably decomposable. A representation with a cyclic vector is "small" in the sense that any subrepresentation also has a cyclic vector (III.1.2.5(i)); a representation with a separating vector is "large" in the sense that every larger representation also has a separating vector. The standard form representation is the only one which simultaneously has both properties:

III.2.6.10 COROLLARY. Let M be a (necessarily countably decomposable) von Neumann algebra on $\mathcal{H}$. If M has both a cyclic vector and a separating vector, then M is in standard form.

PROOF: Let ξ be a cyclic vector and η a separating vector; let ρ be the identity representation of M, and σ the restriction of ρ to $[M\eta]$. For $x \in M$, let $\phi(x) = \langle x\xi, \xi \rangle + \langle x\eta, \eta \rangle$. Then, as in the proof of III.2.6.7, ρ and σ can be identified with subrepresentations of π_ϕ. Let $p, q \in \pi_\phi(M)'$ be the projections onto the subspace of ρ and σ respectively; then $q \leq p \leq 1$. But σ is in standard form, so $q \sim 1$ in $\pi_\phi(M)'$, and hence $p \sim 1$ in $\pi_\phi(M)'$ by the Schröder-Bernstein Theorem (III.1.1.9).

III.2.6.11 PROPOSITION. Let M be a countably decomposable von Neumann algebra on $\mathcal{H}$. Then there is a central projection z in M such that Mz has a cyclic vector on $z\mathcal{H}$ (i.e. z is cyclic under M), and $M(1-z)$ has a separating vector on $(1-z)\mathcal{H}$ (i.e. $1-z$ is cyclic under M').

PROOF: This is a corollary of III.2.2.11, comparing the identity representation of M with the standard form representation.

Type of the Commutant

If a von Neumann algebra M is in standard form, there is an intimate relation between the structure of M and M'. The most precise result is the deep theorem from modular theory (III.4.3.2) that M and M' are actually anti-isomorphic. However, we can by much more elementary arguments obtain less precise, but still useful information, such as that M and M' at least have the same type; to some extent this is still true if M is not in standard form.

III.2.6.12 THEOREM. Let M be a von Neumann algebra on $\mathcal{H}$. If M is Type I [*resp.* Type II, Type III], then M' is also Type I [*resp.* Type II, Type III]. If M is in standard form, and M is type I_n [*resp.* Type I_∞, Type II_1, Type II_∞], then M' is also Type I_n [*resp.* Type I_∞, Type II_1, Type II_∞].

The first statement for the Type I case has already been shown (III.1.5.20). Thus it suffices to prove the result for a Type I_n or II_1 von Neumann algebra in standard form; for, then, if M is any Type II_1 von Neumann algebra on a Hilbert space, then the identity representation is a reduction of an amplification of the standard form representation, and M' is thus Type II (it can have a nontrivial Type II_1 part and/or a nontrivial Type II_∞ part in general). If M is Type II_∞ on $\mathcal{H}$, and p is a finite projection in M with central support 1, then pMp is a Type II_1 von Neumann algebra on $p\mathcal{H}$, and $(pMp)' \cong M'$, hence M' is Type II. If M is in standard form, then M' cannot have a nonzero Type II_1 part by III.2.6.3, and hence is Type II_∞ (the II_∞ case also follows from III.2.6.4). Finally, if M is Type III, it follows that M' cannot have a Type I or Type II part and thus is Type III.

The result for general Type II_1 von Neumann algebras in standard form can be reduced to the countably decomposable case by III.2.6.3, so it suffices to prove:

III.2.6.13 PROPOSITION. Let τ be a faithful normal tracial state on a von Neumann algebra M in τ-standard form on $\mathcal{H}_\tau$. Then the complementary state τ' on M' defined by ξ_τ is also a (faithful normal) tracial state.

To prove this, note first that

$$\langle xy\xi_\tau, \xi_\tau \rangle = \tau(xy) = \tau(yx) = \langle yx\xi_\tau, \xi_\tau \rangle$$

for all $x, y \in M$ and

$$\|x\xi_\tau\|^2 = \langle x^*x\xi_\tau, \xi_\tau \rangle = \langle xx^*\xi_\tau, \xi_\tau \rangle = \|x^*\xi_\tau\|^2$$

for all $x \in M$. Now we show that $M'_{sa}\xi_\tau \subseteq (M_{sa}\xi_\tau)^-$. If $x = x^* \in M'$, there is a sequence (x_n) in M with $x_n\xi_\tau \to x\xi_\tau$. Since $(x_n\xi_\tau)$ is a Cauchy sequence, so is $(x_n^*\xi_\tau)$, i.e. $x_n^*\xi_\tau \to \eta$ for some η. If $y \in M$, then

$$\langle \eta, y\xi_\tau \rangle = \lim \langle x_n^*\xi_\tau, y\xi_\tau \rangle = \lim \langle \xi_\tau, x_n y\xi_\tau \rangle$$

$$= \lim\langle\xi_\tau, yx_n\xi_\tau\rangle = \langle\xi_\tau, yx\xi_\tau\rangle = \langle\xi_\tau, xy\xi_\tau\rangle = \langle x\xi_\tau, y\xi_\tau\rangle$$

and since ξ_τ is cyclic, it follows that $x_n^*\xi_\tau \to x\xi_\tau$, so $[(x_n + x_n^*)/2]\xi_\tau \to x\xi_\tau$.

Now suppose $x, y \in M'_{sa}$. Choose a sequence (x_n) in M_{sa} with $x_n\xi_\tau \to x\xi_\tau$. Then

$$\tau'(xy) = \langle xy\xi_\tau, \xi_\tau\rangle = \langle y\xi_\tau, x\xi_\tau\rangle = \lim\langle y\xi_\tau, x_n\xi_\tau\rangle$$

$$= \lim\langle x_n y\xi_\tau, \xi_\tau\rangle = \lim\langle yx_n\xi_\tau, \xi_\tau\rangle = \langle yx\xi_\tau, \xi_\tau\rangle = \tau'(yx)$$

so τ' is tracial.

III.2.6.14 Since the Type I_n factor case in standard form was done in III.2.6.2, by III.2.6.4 to finish off the I_n case of III.2.6.12 it suffices to show that a Type I_1 (i.e. commutative) von Neumann algebra Z is in standard form if and only if it is a masa. The countably decomposable case is III.1.5.19 and III.2.6.9, and the general case reduces immediately to the countably decomposable case (using III.2.6.3) since a commutative von Neumann algebra is locally countably decomposable (III.1.2.6).

Properly Infinite von Neumann Algebras

III.2.6.15 It follows from III.2.6.9, III.2.6.12, and III.2.2.9 that if M is properly infinite and countably decomposable, then M is in standard form if and only if M' is also properly infinite and countably decomposable. Thus, for von Neumann algebras on a separable Hilbert space, we get:

III.2.6.16 Corollary. Let M be a properly infinite von Neumann algebra on a separable Hilbert space. Then M is in standard form if and only if M' is also properly infinite. In particular, every Type III von Neumann algebra on a separable Hilbert space is in standard form.

The Coupling Function

III.2.6.17 If M is a finite von Neumann algebra, and $\mathcal{Z}(M) = \mathcal{Z}(M')$ is identified with $L^\infty(X, \mu)$ for some X, then there is a (locally) measurable function $\gamma(M, M')$ on X, taking values in $(0, \infty]$, which measures the deviation of the identity representation of M from the standard form representation. The function $\gamma(M, M')$ is called the *coupling function* of M and M'. (Of course, $\gamma(M, M')$ will only be defined locally a.e.)

If M' is properly infinite, define $\gamma(M, M')$ to be ∞ everywhere. (If M' is not locally countably decomposable, it will be necessary to instead use various infinite cardinals for the values of $\gamma(M, M')$.) If M' is finite, and π is the standard form representation of M, and the identity representation of M is equivalent to a subrepresentation of π corresponding to a projection $p \in \pi(M)'$, let $\gamma(M, M')$ be the function corresponding to $\theta(p)$, where θ is the center-valued trace on $\pi(M)'$. If π is equivalent to a subrepresentation of the identity representation of M corresponding to a projection $q \in M'$, let $\gamma(M, M')$ be the reciprocal of the function corresponding to $\theta'(q)$, where θ' is the center-valued trace on M'. (Since q has central support 1, this function will be nonzero locally a.e.)

In general, there are three central projections z_1, z_2, z_3 in M such that $z_1 + z_2 + z_3 = 1$, $z_1 M'$ and $z_2 M'$ are finite, $z_3 M'$ is properly infinite, the identity representation of $z_1 M$ is equivalent to a subrepresentation of the standard form representation, and the standard form representation of $z_2 M$ is equivalent to a subrepresentation of the identity representation. The coupling function can be defined on each piece as above.

III.2.6.18 The coupling function has the following properties:

(i) $\gamma(M, M')$ is finite locally a.e. (but not necessarily bounded) if and only if M' is finite.
(ii) If M' is finite, then $\gamma(M', M)$ is the reciprocal of $\gamma(M, M')$ locally a.e.
(iii) M is in standard form if and only if $\gamma(M, M') = 1$ locally a.e.
(iv) If M' is finite with center-valued trace θ', p is a projection in M' with central support 1, and $\theta'(p)$ is identified with a measurable function from X to $(0, 1]$, then $\gamma(pM, pM'p) = \theta'(p)\gamma(M, M')$ locally a.e.
(v) If π is a faithful normal representation of M, then π has an associated coupling function $\gamma(\pi(M), \pi(M)')$. Two faithful normal representations π and ρ are equivalent if and only if their coupling functions are equal locally a.e. The representation π is equivalent to a subrepresentation of ρ if and only if $\gamma(\pi(M), \pi(M)') \leq \gamma(\rho(M), \rho(M)')$ locally a.e.

III.2.6.19 If M is Type II_1, then every locally measurable function from X to $(0, \infty]$ occurs as the coupling function of a unique faithful normal representation of M. In the type I case, there are restrictions: on a Type I_n piece, the coupling function can only take the values $\{k/n : k \in \mathbb{N}\}$ and ∞.

III.2.6.20 If M is a factor, then X is reduced to a single point, and the coupling function is just a number called the *coupling constant*. If M is type II_1, the coupling constant can be any positive real number (or ∞ if M' is II_∞). If M is Type I_n, the constant is ∞ or k/n for some $k \in \mathbb{N}$.

Subfactors

III.2.6.21 Although it is rather off the subject of this volume, no modern treatment of von Neumann algebra theory would be complete without mention of the theory of subfactors.

If M is a factor, a *subfactor* (of M) is a factor N contained in M (with the same unit); the subfactor is usually denoted (N, M) or $(N \subseteq M)$. Then M is a (left or right) N-module in the obvious way, and it is not hard to show that M is a projective N-module. We say that N has *finite index* in M if M is finitely generated as a (left or right) N-module. If N is of finite index, a numerical index $[M:N]$ can be defined as the "dimension" of M as a finitely generated projective N-module. If M is Type II$_1$ in standard form from its trace, then $[M:N]$ is just the coupling constant $\gamma(N,N')$. Subfactors of the same index can have different "positions", so the index is far from a complete invariant.

The main goal of subfactor theory is to describe and, if possible, classify all subfactors of a fixed factor M up to the obvious notion of isomorphism. The theory is largely independent of the particular M chosen, and is perhaps most interesting if M is the (unique) hyperfinite II$_1$ factor $\mathcal{R}$ (III.3.1.4, III.3.4.3). The theory is extremely rich: for example, it includes as a (rather small) special case the classification of finite groups and, more generally, finite homogeneous spaces [if G is a finite or even countable group and H is a subgroup, there is a standard action α of G as outer automorphisms of $\mathcal{R}$, namely writing $\mathcal{R}$ as $\overline{\bigotimes}_{t \in G} \mathcal{R}$ and letting G permute the entries of the tensors; then the pair $(\mathcal{R} \bar{\rtimes}_\alpha H \subseteq \mathcal{R} \bar{\rtimes}_\alpha G)$ is a subfactor of index $[G:H]$, from which G/H can be recovered]. It is remarkable that the theory is so complex, with connections throughout a wide swath of mathematics and physics, and at the same time is not completely intractable.

The first remarkable discovery, by V. Jones [Jon83], was that the values taken by the index are precisely the numbers $\{4\cos^2(\pi/n) : n \geq 3\} \cup [4,\infty)$. Jones also discovered fascinating and unexpected connections with knot theory [Jon85] and, later, topological quantum field theory and statistical mechanics [Jon89], which revolutionized these subjects, and received the Fields Medal in 1990 for this work.

Subfactor theory has grown into a large industry with many applications. See e.g. [JS97] for a more complete treatment.

III.3 Examples and Constructions of Factors

So far, we have not seen any examples of non-type-I von Neumann algebras. In this section, we will give some of the most important constructions and examples of non-type-I factors. We will be most interested in factors on separable Hilbert spaces, but the constructions are quite general.

III.3.1 Infinite Tensor Products

Perhaps the simplest construction of non-type-I factors is via infinite tensor products.

III.3.1.1 Let $\{\mathcal{H}_i : i \in \Omega\}$ be a set of Hilbert spaces, and ξ_i a specified unit vector in $\mathcal{H}_i$. Then the *infinite tensor product of the $\mathcal{H}_i$ with respect to ξ_i*, denoted
$$\bigotimes_i (\mathcal{H}_i, \xi_i)$$

is constructed as follows. For each finite $F \subseteq \Omega$, let $\mathcal{H}_F = \odot_{i \in F} \mathcal{H}_i$ with its natural inner product (I.1.4.2). If $F \subseteq G$, identify $\mathcal{H}_F$ with the closed subspace

$$\mathcal{H}_F \otimes (\bigotimes_{i \in G \setminus F} \xi_i)$$

of $\mathcal{H}_G$. Then the inductive limit $\mathcal{H}_0$ has a natural structure as an inner product space; elements of $\mathcal{H}_0$ are linear combinations of elementary tensors $\otimes \eta_i$, where $\eta_i = \xi_i$ for all but finitely many i. Let $\bigotimes_i (\mathcal{H}_i, \xi_i)$ be the completion. $\bigotimes_i (\mathcal{H}_i, \xi_i)$ has a distinguished unit vector $\xi = \bigotimes \xi_i$. If each $\mathcal{H}_i$ is more than one dimensional, then $\bigotimes(\mathcal{H}_i, \xi_i)$ is separable if and only if each H_i is separable and Ω is countable.

III.3.1.2 If M_i is a unital C*-subalgebra of $\mathcal{L}(\mathcal{H}_i)$, then the infinite minimal C*-tensor product $\bigotimes_{\min} M_i$ acts naturally as a C*-algebra of operators on $\bigotimes_i (\mathcal{H}_i, \xi_i)$. If each M_i is a von Neumann algebra, we denote by

$$\overline{\bigotimes}_i (M_i, \xi_i)$$

the von Neumann algebra generated by the image, called the *infinite tensor product of the M_i with respect to ξ_i*. Although $\bigotimes_{\min} M_i$ is defined in a space-free manner, it turns out that the structure of $\overline{\bigotimes}_i (M_i, \xi_i)$ depends strongly on the choice of the ξ_i. There is a distinguished normal state on $\overline{\bigotimes}_i (M_i, \xi_i)$ coming from inner product with the distinguished unit vector ξ. The commutant of $\overline{\bigotimes}_i (M_i, \xi_i)$ is $\overline{\bigotimes}_i (M_i', \xi_i)$ (the general result requires the Commutation Theorem for Tensor Products (III.4.5.8), but we will be almost exclusively concerned with the case where each $\mathcal{H}_i$ is finite-dimensional, where the Commutation Theorem is elementary).

If $\{M_i\}$ is a collection of von Neumann algebras and ϕ_i is a normal state on M_i (usually assumed faithful), then we can form

$$\overline{\bigotimes}_i (M_i, \phi_i) = \overline{\bigotimes}_i (M_i, \xi_{\phi_i})$$

where $\xi_{\phi_i} \in \mathcal{H}_{\phi_i}$ comes from the GNS representation. The corresponding distinguished normal state on $\overline{\bigotimes}_i (M_i, \phi_i)$ is denoted $\bigotimes_i \phi_i$; the representation of $\overline{\bigotimes}_i (M_i, \phi_i)$ on $\bigotimes_i (\mathcal{H}_{\phi_i}, \xi_{\phi_i})$ is the GNS representation from $\bigotimes_i \phi_i$, and if each ϕ_i is faithful, then $\bigotimes_i \phi_i$ is faithful. If each ϕ_i is a trace, then ϕ is a trace.

III.3.1.3 PROPOSITION. If each M_i is a factor, then $M = \overline{\bigotimes}_i (M_i, \phi_i)$ is a factor.

PROOF: If F is a finite subset of Ω, write

$$M_F = \overline{\bigotimes}_{i \in F} M_i \quad \text{and} \quad N_F = \overline{\bigotimes}_{i \in \Omega \setminus F} (M_i, \xi_i).$$

Then $M = M_F \bar{\otimes} N_F$. Identify M_F with $M_F \otimes 1$, and let θ_F be the slice map (III.2.2.6) from M to M_F defined by $\bigotimes_{i \in \Omega \setminus F} \phi_i$; θ_F is a conditional expectation. If $\mathcal{H} = \bigotimes_i (\mathcal{H}_i, \xi_{\phi_i})$ and $\xi = \bigotimes_i \xi_{\phi_i}$, and $x \in M$ is identified with $x\xi \in \mathcal{H}$, θ_F is the orthogonal projection from $\mathcal{H}$ onto

$$\left(\bigotimes_{i \in F} \mathcal{H}_i\right) \otimes \left(\bigotimes_{i \in \Omega \setminus F} \xi_{\phi_i}\right).$$

Thus, for any $x \in M$, $\theta_F(x)\xi \to x\xi$ in $\mathcal{H}$ as $F \to \infty$.

Now suppose $x \in \mathcal{Z}(M)$. If $y \in M_F$, we have $y\theta_F(x) = \theta_F(yx) = \theta_F(xy) = \theta_F(x)y$; thus $\theta_F(x) \in \mathcal{Z}(M_F)$. Since each M_F is a factor (III.1.5.10), it follows that $\theta_F(x)\xi$ is a scalar multiple of ξ for all F. Since $\theta_F(x)\xi \to x\xi$, $x\xi$ is a multiple of ξ and hence $x \in \mathbb{C}1$ and M is a factor.

III.3.1.4 The simplest case is where each M_i ($i \in \mathbb{N}$) is $\mathbb{M}_2$ and each ϕ_i the (unique) tracial state. Then $\overline{\bigotimes}_i (M_i, \phi_i)$ is a II_1 factor, usually denoted $\mathcal{R}$ or $\mathcal{R}_1$. If A is the CAR algebra (II.8.2.2(iv)) and ϕ is the unique tracial state on A, then $\mathcal{R} \cong \pi_\phi(A)''$.

III.3.1.5 Similarly, a tensor product $\overline{\bigotimes}_i (M_i, \phi_i)$ as in III.3.1.4, but with an uncountable index set, yields a II_1 factor which cannot be represented on a separable Hilbert space.

III.3.1.6 A Type II_∞ factor can also be easily constructed: let $M = \mathcal{R} \bar{\otimes} \mathcal{L}(\mathcal{H})$ for a separable infinite-dimensional Hilbert space $\mathcal{H}$. This factor is usually denoted $\mathcal{R}_{0,1}$ (see III.3.1.13 for an explanation of the notation).

III.3.1.7 The next simplest case is the *Powers factors*. In a major advance, R. Powers [Pow67] in 1967 exhibited uncountably many mutually nonisomorphic type III factors on a separable Hilbert space; previously only three isomorphism classes were known.

Again let M_i ($i \in \mathbb{N}$) be $\mathbb{M}_2$. For $0 < \lambda < 1$ let ϕ_λ be the state on $\mathbb{M}_2$ defined by

$$\phi_\lambda((a_{ij})) = \alpha a_{11} + (1-\alpha)a_{22}$$

where $\alpha = \frac{\lambda}{1+\lambda}$ (so $\lambda = \frac{\alpha}{1-\alpha}$). Fix λ, and let $\phi_i = \phi_\lambda$ for each i, and $\mathcal{R}_\lambda = \overline{\bigotimes}_i (M_i, \phi_i)$. (The construction also works for $\lambda = 1$ and gives the above II_1 factor $\mathcal{R}_1$. The factors $\mathcal{R}_\lambda$ can also be constructed via the group measure space construction (III.3.2.14)).

III.3.1.8 THEOREM. *The $\mathcal{R}_\lambda$ ($0 < \lambda < 1$) form a pairwise nonisomorphic family of type III factors.*

III.3.1.9 The most efficient and elegant proof that the $\mathcal{R}_\lambda$ are mutually nonisomorphic and Type III uses modular theory, and will be given in III.4.6.8. Here we outline a direct argument, which leads to a generalization of the Powers factors and useful invariants. (See [Ped79, 6.5.15] for another proof that $\mathcal{R}_\lambda$ is Type III.)

First note that if ϕ is a state on $\mathbb{M}_n$, then there is a unique positive $a \in \mathbb{M}_n$ with $\text{Tr}(a) = 1$ and $\phi(x) = \text{Tr}(ax)$ for all x (cf. I.8.6.1). The eigenvalues $\alpha_1, \ldots, \alpha_n$ of a, listed in decreasing order with multiplicity, are called the *eigenvalue list* of ϕ and determine ϕ up to unitary equivalence. Since $\text{Tr}(a) = \sum \alpha_k = 1$, we have $0 \leq \alpha_k \leq 1$ for all k. The state ϕ is faithful if and only if all α_k are nonzero. The set $\{\alpha_i/\alpha_j : 1 \leq i, j \leq n\}$ is called the *ratio set* $r(\phi)$ of ϕ. The eigenvalue list of ϕ_λ is $(1-\alpha, \alpha)$, and the ratio set is $\{1, \lambda, \lambda^{-1}\}$. The ratio set of $\phi_1 \otimes \cdots \otimes \phi_k$ is

$$\{\lambda_1 \lambda_2 \cdots \lambda_k : \lambda_i \in r(\phi_i)\}.$$

Suppose $M = \overline{\bigotimes}(M_i, \phi_i)$ and $N = \overline{\bigotimes}(N_i, \psi_i)$, where $M_i = N_i = \mathbb{M}_2$, $\phi_i = \phi_{\lambda_1}$, $\psi_i = \phi_{\lambda_2}$ for all i, and $\lambda_1, \lambda_2 \in (0, 1]$. If $M \cong N$, since they are in standard form there is a spatial isomorphism σ from M onto N. Set $A_1 = M_1$ and $n_1 = 1$. There will be a $k = k_1$ such that $\sigma(A_1 \otimes 1)$ is approximately contained in $(\bigotimes_{j=1}^{k} N_j) \otimes 1$; suppose for simplicity the containment is exact. Then

$$\bigotimes_{j=1}^{k} N_j = B_1 \otimes B_2$$

where $B_1 = \sigma(A_1)$ and B_2 is the relative commutant. The state $\psi_1 \otimes \cdots \psi_k$ does not decompose into a tensor product state $\tilde{\psi}_1 \otimes \tilde{\psi}_2$ on $B_1 \otimes B_2$ in general; but up to similarity it does so approximately. Repeating the process, there is an $n = n_2$ such that $\sigma^{-1}(B_2)$ is (approximately) contained in $(\bigotimes_{i=n_1+1}^{n_2} M_i) \otimes 1$; write

$$\bigotimes_{i=n_1+1}^{n_2} M_i = A_2 \otimes A_3$$

with $A_2 = \sigma^{-1}(B_2)$. In this way, we rewrite M as $\overline{\bigotimes}(A_n, \tilde{\phi}_n)$ and N as $\overline{\bigotimes}(B_n, \tilde{\psi}_n)$. It then follows that the ratio sets of $\tilde{\phi}_n$ and $\tilde{\psi}_n$ must be "asymptotically" the same. But $r(\tilde{\phi}_n)$ is a set of powers of λ_1 and $r(\tilde{\psi}_n)$ powers of λ_2, so $\lambda_1 = \lambda_2$.

The fact that $\mathcal{R}_\lambda$ is Type III follows from III.3.1.12(iii).

III.3.1.10 A factor of the form $M = \overline{\bigotimes}_{i \in \mathbb{N}}(M_i, \phi_i)$, where the M_i are (finite-dimensional) matrix algebras and the ϕ_i faithful states, is called an *ITPFI factor* (ITPFI= "infinite tensor product of finite Type I.") A countably decomposable Type I_∞ factor in standard form is ITPFI (cf. III.3.1.12(i)), as is $\mathcal{R}_\lambda$ ($0 < \lambda \leq 1$). A tensor product (even countably infinite tensor product) of ITPFI factors is ITPFI; thus the II_∞ factor $\mathcal{R}_1 \bar{\otimes} \mathcal{L}(\mathcal{H})$ (in standard form) is ITPFI.

After work of H. Araki, D. Bures, C. Moore, and L. Pukanszky characterized the type of an ITPFI factor, Araki and E. J. Woods [AW69] made a systematic study of ITPFI factors. They found an isomorphism invariant $r_\infty(M)$ called the *asymptotic ratio set*, and an additional invariant $\rho(M)$. The invariant $r_\infty(M)$ is natural in light of the argument outlined above:

III.3.1.11 DEFINITION. Let $M = \overline{\bigotimes}_{i \in \mathbb{N}}(M_i, \phi_i)$ be an ITPFI factor. Then $r_\infty(M)$ is the set of all $x \geq 0$ such that, for all $k \in \mathbb{N}$ and $\epsilon > 0$, there is an n and a λ in the ratio set of $\phi_{k+1} \otimes \cdots \otimes \phi_{k+n}$ with $|x - \lambda| < \epsilon$.

It is easy to see that $r_\infty(M)$ is a closed subset of $[0, \infty)$, and that $r_\infty(M) \cap \mathbb{R}_+$ is a multiplicative subgroup of $\mathbb{R}_+$. Thus $r_\infty(M)$ is one of the following sets: $\{1\}$, $\{0, 1\}$, $\{0\} \cup \{\lambda^n : n \in \mathbb{Z}\}$ for some $0 < \lambda < 1$, or $[0, \infty)$. We have $r_\infty(\mathcal{R}_1) = \{1\}$ and $r_\infty(\mathcal{R}_\lambda) = \{0\} \cup \{\lambda^n : n \in \mathbb{Z}\}$ for $0 < \lambda < 1$.

It follows from the argument outlined in III.3.1.9 (cf. [AW69]) that if $M \cong N$, then $r_\infty(M) = r_\infty(N)$, i.e. $r_\infty(M)$ is an algebraic invariant of M and independent of the way M is written as an ITPFI factor. The technical details of the argument are substantial.

The main results about ITPFI factors are:

III.3.1.12 THEOREM. Let $M = \overline{\bigotimes}_{i \in \mathbb{N}}(M_i, \phi_i)$ be an ITPFI factor, with $M_i = \mathbb{M}_{n_i}$. Write $\alpha_k^{(i)}$ for the k'th element of the eigenvalue list of ϕ_i. Then

(i) ([Ara63], [Bur63]) M is Type I (I_∞) if and only if $\sum_i (1 - \alpha_1^{(i)}) < \infty$.
(ii) ([Bur63], [Puk56]) M is Type II_1 if and only if $\sum_{i,k} |(\alpha_k^{(i)})^{1/2} - n_i^{-1/2}|^2 < \infty$.
(iii) [Moo67] If there is a $\delta > 0$ with $\alpha_1^{(i)} \geq \delta$ for all i (in particular, if the n_i are bounded), then M is Type III if and only if

$$\sum_{i,k} \alpha_k^{(i)} \inf\{|\alpha_1^{(i)}/\alpha_k^{(i)} - 1|^2, C\} = \infty$$

for some (hence all) $C > 0$.

See III.3.2.14 for some comments on the proof.

III.3.1.13 THEOREM. [AW69] Let $M = \overline{\bigotimes}_{i \in \mathbb{N}}(M_i, \phi_i)$ and $N = \overline{\bigotimes}_{i \in \mathbb{N}}(N_i, \psi_i)$ be ITPFI factors. Then

(i) If $0 < \lambda < 1$, then $\lambda \in r_\infty(M)$ if and only if $M \cong M \bar{\otimes} \mathcal{R}_\lambda$. In particular, if there is a $\lambda \in r_\infty(M)$, $\lambda \neq 0, 1$, then M is Type III.
(ii) M is Type II_1 if and only if $r_\infty(M) = \{1\}$.
(iii) There is a Type III ITPFI factor $\mathcal{R}_\infty$ with $r_\infty(\mathcal{R}_\infty) = [0, \infty)$.
(iv) If $r_\infty(M) = \{0, 1\}$, then M could be Type I_∞, Type II_∞, or Type III. There are uncountably many mutually nonisomorphic Type III factors in this class.
(v) If M and N are Type II_∞, then $M \cong N$.
(vi) If $r_\infty(M) = r_\infty(N) \neq \{0, 1\}$, then $M \cong N$.

III.3.1.14 It follows from (iii) and (vi) that $\mathcal{R}_\infty$ is the unique ITPFI factor M with $r_\infty(M) = [0, \infty)$. This factor is called the ITPFI factor of Type III$_1$. (It might make sense to call this factor $\mathcal{R}_1$, but this name is normally used, with good reason, for the hyperfinite II$_1$ factor, since $r_\infty(\mathcal{R}_1) = \{1\}$. The asymptotic ratio set also explains the notation $\mathcal{R}_{0,1}$ for the unique ITPFI factor of Type II$_\infty$ (III.3.1.6), since $r_\infty(\mathcal{R}_{0,1}) = \{0, 1\}$.) The Powers factor $\mathcal{R}_\lambda$, which by (vi) is the unique ITPFI factor M with $r_\infty(M) = \{0\} \cup \{\lambda^n : n \in \mathbb{Z}\}$, is called the ITPFI factor of Type III$_\lambda$.

It is easy to see that if $0 < \lambda_1, \lambda_2 < 1$ and $\log \lambda_1 / \log \lambda_2$ is irrational, then $\mathcal{R}_{\lambda_1} \bar{\otimes} \mathcal{R}_{\lambda_2} \cong \mathcal{R}_\infty$. (If $\log \lambda_1 / \log \lambda_2$ is rational, then $\mathcal{R}_{\lambda_1} \bar{\otimes} \mathcal{R}_{\lambda_2} \cong \mathcal{R}_\lambda$, where $-\log \lambda$ is the largest number μ such that $-\log \lambda_1$ and $-\log \lambda_2$ are integer multiples of μ.) $\mathcal{R}_\infty$ absorbs all ITPFI factors: if M is any ITPFI factor, then $M \bar{\otimes} \mathcal{R}_\infty \cong \mathcal{R}_\infty$.

The factors occurring in quantum field theory are "generically" isomorphic to $\mathcal{R}_\infty$ [Ara64].

III.3.1.15 The Type III ITPFI factors with asymptotic ratio set $\{0, 1\}$ are called the ITPFI factors of Type III$_0$. An additional invariant ρ is useful here:

$$\rho(M) = \{\lambda \in [0, 1] : M \bar{\otimes} \mathcal{R}_\lambda \cong \mathcal{R}_\lambda\}$$

(let $\mathcal{R}_0$ be the I$_\infty$ factor). The set $\rho(M)$ is an algebraic invariant of M. We have:

$\rho(\mathcal{R}_0) = [0, 1)$
$\rho(\mathcal{R}_1) = (0, 1]$
$\rho(\mathcal{R}_{0,1}) = (0, 1)$
$\rho(\mathcal{R}_\lambda) = \{\lambda^{1/n} : n \in \mathbb{N}\}$ if $0 < \lambda < 1$
$\rho(\mathcal{R}_\infty) = \emptyset$

Of course, by III.3.1.13(vi), ρ gives no additional information in these cases.

If M is Type III$_0$, however, the set $\rho(M)$ can be very interesting. It turns out that

$$\{-2\pi/\log \lambda : \lambda \in \rho(M) \cap (0, 1)\}$$

is the positive part of a group $T(M)$ which is an invariant of M arising from modular theory (III.4.6.5). Uncountably many different such groups arise for ITPFI factors, including all countable subgroups of $\mathbb{R}$ [Con76] and many uncountable groups [Woo73]. See III.3.2.19 for a further discussion of the classification of ITPFI factors of Type III$_0$.

Historical comments on the theory of ITPFI factors and their connection with the classification program for injective factors can be found in [Woo82].

III.3.2 Crossed Products and the Group Measure Space Construction

The earliest construction of non-Type-I factors, first used by Murray and von Neumann, is the group measure space construction, a special case of the W*-crossed product.

Continuity of Group Actions

III.3.2.1 Let G be a locally compact group, N a von Neumann algebra on a Hilbert space $\mathcal{H}$, and $\alpha : G \to \mathrm{Aut}(N)$ a homomorphism. To construct a W*-analog of the C*-crossed product, the action α must be continuous for an appropriate topology on $\mathrm{Aut}(N)$, and if G is not discrete, there are some technical subtleties to consider.

The point-norm topology, used for the C*-crossed product, is too strong for the W* case; the most obvious candidate for the correct topology is pointwise weak, strong, σ-strong-*, etc., convergence (these topologies coincide on $\mathrm{Aut}(N)$ by I.3.2.9 and II.3.2.12). However, $\mathrm{Aut}(N)$ is not a topological group under this topology in general. It is sometimes useful to instead consider a stronger topology, under which $\mathrm{Aut}(N)$ is always a topological group. If $\alpha \in \mathrm{Aut}(N)$, α induces an isometry α^* on N^* by

$$\alpha^*(\phi) = \phi \circ \alpha.$$

This α^* leaves N_* invariant and thus induces an isometry α_* on N_*. Thus the topology of pointwise norm-convergence on N_* can be placed on $\mathrm{Aut}(N)$, i.e. $\alpha_i \to \alpha$ in this topology if $\phi \circ \alpha_i \to \phi \circ \alpha$ in norm for all $\phi \in N_*$. This topology is strictly stronger than the point-weak topology in general (e.g. for $L^\infty([0,1])$), although the topologies are the same for some infinite-dimensional von Neumann algebras (e.g. a von Neumann algebra with a faithful normal semifinite weight left invariant by all automorphisms, such as the trace on a factor of Type I or Type II$_1$). See [Haa75b] or [Str81, 2.23] for a description of these topologies and various others on the automorphism group of a von Neumann algebra.

Fortunately, for actions of locally compact groups, there is no ambiguity in the choice of the notion of continuity, due to the following result of W. Arveson [Arv74] (cf. [Str81, 13.5]):

III.3.2.2 THEOREM. Let G be a locally compact group, N a von Neumann algebra on a Hilbert space $\mathcal{H}$, and $\alpha : G \to \mathrm{Aut}(N)$ a homomorphism. Then the following are equivalent:

(i) α is continuous for the point-weak (point-strong, etc). topology.
(ii) α is continuous for the topology of pointwise norm-convergence on N_*.
(iii) The map $(g, \phi) \mapsto \phi \circ \alpha_g$ from $G \times N_*$ to N_* is norm-continuous.

A homomorphism α satisfying these conditions is called a *(W*-)continuous action* of G on N, and (N, G, α) is called a *W*-dynamical system*.

III.3.2.3 If α is a (W*-)continuous action of G on a von Neumann algebra N, then (N, G, α) is not in general a C*-dynamical system, i.e. α is not in general point-norm continuous (it is, of course, if G is discrete). But we do have the following result, which allows many results about W*-dynamical systems and W*-crossed products to be deduced from the corresponding C*-theory (although this is generally the reverse of the historical development, since in most cases the W* results were obtained first and the C*-theory developed later, largely in analogy with the W*-theory).

III.3.2.4 PROPOSITION. Let (N, G, α) be a W*-dynamical system. Set

$$N^c = \{x \in N \ : \ t \mapsto \alpha_t(x) \text{ is norm-continuous }\}.$$

Then N^c is a σ-weakly dense α-invariant C*-subalgebra of N, and (N^c, G, α) is a C*-dynamical system.

If $f \in L^1(G)$, and $x \in N$, set

$$\alpha_f(x) = \int_G f(t)\alpha_t(x)\,d\mu(t)$$

as in the C*-case; then it is easily seen that $\alpha_f(x)$ is well defined and in N^c, and x is in the closure of $\{\alpha_f(x) : f \in L^1(G)\}$ (cf. [Ped79, 7.5.1]; in fact, N^c is the norm closure of $\{\alpha_f(x) : f \in L^1(G), x \in N\}$.)

Spectral Analysis of Abelian Group Actions

There is an entire spectral theory for actions of locally compact abelian groups on von Neumann algebras (and, more generally, on dual Banach spaces), mostly due to W. Arveson. We will only give a few definitions; a more complete treatment can be found in [Tak03a, XI.1].

If α is an action of the locally compact abelian group G on a von Neumann algebra N, then for $f \in L^1(G)$ there is a map α_f from N to N defined by $\alpha_f(x) = \int f(-t)\alpha_t(x)\,dt$. If $x \in N$, set $I(x) = \{f \in L^1(G) : \alpha_f(x) = 0\}$, and $I(\alpha) = \cap_x I(x)$.

III.3.2.5 DEFINITION. The α-spectrum $Sp_\alpha(x)$ of $x \in N$ is

$$\{\gamma \in \hat{G} : \hat{f}(\gamma) = 0 \ \forall f \in I(x)\}.$$

The *Arveson spectrum* $Sp(\alpha)$ of α is

$$\{\gamma \in \hat{G} : \hat{f}(\gamma) = 0 \ \forall f \in I(\alpha)\}.$$

If p is a projection in the fixed-point algebra N^α, then α drops to an action α_p of G on pNp, and $Sp(\alpha_p) \subseteq Sp(\alpha)$. The set

$$\Gamma(\alpha) = \cap\{Sp(\alpha_p) : p \neq 0 \text{ a projection in } N^\alpha\}$$

is the *Connes spectrum* (or *essential spectrum*) of α.

If G is compact (so $\hat{G}$ is discrete), the subspaces

$$N^\gamma = \{x \in N : Sp_\alpha(x) = \{\gamma\}\}$$

for $\gamma \in \hat{G}$ span N σ-weakly, and decompose N into "spectral subspaces." There is an analogous "continuous spectral decomposition" if G is not compact.

W*-Crossed Products

III.3.2.6 Let α be a continuous action of a locally compact group G on a von Neumann algebra N. As in the case of reduced crossed products of C*-algebras, realize N and G on the Hilbert space

$$L^2(G, \mu, \mathcal{H}) \cong L^2(G, \mu) \otimes \mathcal{H}$$

(μ is left Haar measure on G) by

$$[\pi(x)f](s) = \alpha_s(x)f(s)$$

for $x \in N$ and

$$\lambda(t)f(s) = f(t^{-1}s)$$

for $t \in G$. Then (π, λ) is a covariant representation of (N, G, α). Let

$$N \bar{\rtimes}_\alpha G = \{\pi(N), \lambda(G)\}''$$

be the generated von Neumann algebra. $N \bar{\rtimes}_\alpha G$ is called the *W*-crossed product of N by G*.

It is easily seen using III.2.2.8 that up to isomorphism, $N \bar{\rtimes}_\alpha G$ depends only on N and α and not on the way N is represented on $\mathcal{H}$.

Strictly speaking, this crossed product should be called the reduced W*-crossed product; but there is no obvious W*-analog of the full C*-crossed product. The W*-crossed product is primarily used for amenable groups, usually abelian groups, where the full and reduced crossed products coincide.

Unlike the C*-case where multipliers must be considered, even if G is nondiscrete there are embeddings $\sigma : N \to N \bar{\rtimes}_\alpha G$ and $t \mapsto u_t \in \bar{\rtimes}_\alpha G$ for $t \in G$.

Note that $L^2(G, \mu, \mathcal{H})$ is separable if $\mathcal{H}$ is separable and G is second countable. Thus, if N has separable predual and G is second countable, then $N \bar{\rtimes}_\alpha G$ has separable predual.

III.3.2.7 $N \bar{\rtimes}_\alpha G$ contains a σ-weakly dense standard copy of the reduced C*-crossed product $N^c \rtimes_\alpha^r G$ (III.3.2.4); the containment is proper even if $N = N^c$, e.g. when G is discrete, unless G is a finite group.

Takesaki Duality

The von Neumann version of duality is due to M. Takesaki [Tak73], and closely resembles the C*-version (II.10.5). (As usual, the von Neumann version was obtained first.)

If α is an action of an abelian group G on a von Neumann algebra M, there is a dual action $\hat{\alpha}$ of $\hat{G}$ on $M \bar{\rtimes}_\alpha G$, defined as in the C*-case: $\hat{\alpha}_\gamma(\sigma(x)) = \sigma(x)$ for $x \in M$, and $\hat{\alpha}_\gamma(u_t) = \langle t, \gamma \rangle u_t$ for $t \in G$.

III.3.2.8 THEOREM. [TAKESAKI DUALITY] With notation as above, we have
$$(M\bar{\rtimes}_\alpha G)\bar{\rtimes}_{\hat\alpha} \hat G \cong M\bar\otimes\mathcal{L}(L^2(G)).$$

Under this isomorphism the second dual action $\hat{\hat\alpha}$ of G becomes the action $\alpha\otimes\lambda$ on $M\bar\otimes\mathcal{L}(L^2(G))$, where λ is the action on $\mathcal{L}(L^2(G))$ consisting of conjugating by the representation of left translation on $L^2(G)$.

For a complete discussion and proof of this theorem, see [Tak03a]. For a generalization to nonabelian groups via Hopf algebras, see II.10.8.19.

Dual Weights

III.3.2.9 If α is a W*-continuous action of a locally compact group G on a von Neumann algebra M, then the *-algebra $\mathcal{K}(G, M)$ of σ-strong-* continuous functions from G to M with compact support (with the usual convolution product and adjoint) sits naturally as a σ-weakly dense subalgebra of $M\bar{\rtimes}_\alpha G$. Define $\theta : \mathcal{K}(M, G) \to M$ by $\theta(f) = f(e_G)$. It turns out that θ extends to an operator-valued weight ("unbounded conditional expectation"), also called θ, from $M\bar{\rtimes}_\alpha G$ to M (regarded as a subalgebra). If G is abelian, then θ is just "averaging" via the dual action $\hat\alpha$; if G is discrete (so $\hat G$ is compact), then θ is a conditional expectation defined (at least on a dense subalgebra) as in II.6.10.4(v), using the dual action. In general, θ is unbounded, but faithful, normal, and "semifinite".

If ϕ is a weight on M, then $\phi \circ \theta$ is a weight on $M\bar{\rtimes}_\alpha G$, called the *dual weight* of ϕ, denoted $\hat\phi$. If ϕ is faithful, so is $\hat\phi$; if ϕ is normal and semifinite, so is $\hat\phi$.

The technical details are nontrivial, and require some modular theory; see [Tak03a], or [Str81] for an alternate exposition.

The Group Measure Space Construction

The case where the group G acts on a commutative von Neumann algebra $Z \cong L^\infty(X, \mathcal{B}, \mu)$ corresponds to an action of G on the measure space $(X, \mathcal{B}, \mu)$. We will be mostly interested in the case where $(X, \mathcal{B})$ is a standard Borel space and μ is a finite or σ-finite Radon measure; such an $(X, \mathcal{B}, \mu)$ will be called a *standard measure space*.

III.3.2.10 DEFINITION. An *automorphism* of a (locally finite) measure space $(X, \mathcal{B}, \mu)$ is an invertible transformation $T : X \to X$ with T and T^{-1} measurable, such that T preserves the measure class of μ, i.e. $\mu \circ T$ is equivalent to μ.

An *action* of G on $(X, \mathcal{B}, \mu)$ is a homomorphism α from G to the group of automorphisms of $(X, \mathcal{B}, \mu)$, which is continuous (as an action on $L^\infty(X, \mathcal{B}, \mu)$). Write α_g for the automorphism corresponding to $g \in G$. If T is an automorphism of $(X, \mathcal{B}, \mu)$, write α_T for the corresponding $\mathbb{Z}$-action.

The continuity can be expressed in many equivalent ways on the space level. One way is the requirement that $\mu(E \triangle \alpha_g(E)) \to 0$ as $g \to e$, for any $E \in \mathcal{B}$ with $\mu(E) < \infty$.

III.3.2.11 DEFINITION. An action α of G on $(X, \mathcal{B}, \mu)$ is *ergodic* if whenever $Y \subseteq X$ and Y is (globally) invariant under G, either $\mu(Y) = 0$ or $\mu(X \backslash Y) = 0$. The action α is *essentially free* if $\mu(F) = 0$, where

$$F = \{x \in X : \alpha_g(x) = x \text{ for some } e \neq g \in G\}.$$

III.3.2.12 THEOREM. [MvN43] Let α be an action of a group G on a measure space $(X, \mathcal{B}, \mu)$, and let $M = L^\infty(X, \mathcal{B}, \mu) \bar{\rtimes}_\alpha G$. If α is ergodic and essentially free, then M is a factor. Furthermore, in this case:

(i) M is finite if and only if there is a finite measure equivalent to μ and invariant under G.
(ii) M is semifinite if and only if there is a semifinite measure equivalent to μ and invariant under G.
(iii) M is Type I if and only if α is *essentially transitive*, i.e. there is an orbit Y with $\mu(X \backslash Y) = 0$.

Thus, if there is no semifinite invariant measure equivalent to μ, M is a Type III factor. In older references, an action with no semifinite invariant measure is called *non-measurable*, but this term has a different meaning today.

III.3.2.13 EXAMPLES. There are many examples with invariant measures. For example, let X be a locally compact group and G a dense subgroup with the discrete topology. G acts on X by (left) translation, and (left) Haar measure is invariant. If X is compact, the corresponding factor is Type II$_1$. A simple example is $\mathbb{Z}$ acting on $\mathbb{T}$ by an irrational rotation. If X is second countable and G is amenable, the resulting factor is approximately finite-dimensional (III.3.4.1); in particular, $L^\infty(\mathbb{T}, \mu) \bar{\rtimes}_\alpha \mathbb{Z}$ is isomorphic to $\mathcal{R}$.

III.3.2.14 Another class of examples is the ITPFI factors (III.3.1.10). Any ITPFI factor arises by the group measure space construction. Let X_i ($i \in \mathbb{N}$) be a finite space with n_i points, regarded as a group $G_i = \mathbb{Z}_{n_i}$, $X = \prod X_i$, $G = \bigoplus G_i$. G acts on X by translation. Let μ_i be a measure on X_i with point masses of weight $\alpha_k^{(i)}$ in decreasing order, $1 \leq k \leq n_i$, and $\mu = \prod \mu_i$. We must have all $\alpha_k^{(i)} > 0$ for translation by elements of G to be automorphisms. Then

$$M = L^\infty(X, \mu) \bar{\rtimes}_\alpha G \cong \overline{\bigotimes} (\mathbb{M}_{n_i}, \phi_i)$$

where ϕ_i is a state on $\mathbb{M}_{n_i}$ with eigenvalue list $(\alpha_k^{(i)})$ (III.3.1.9).

This construction sheds light on the proof of III.3.1.12. The measure μ is either purely atomic or purely nonatomic; it is purely atomic if and only if the condition of III.3.1.12(i) is satisfied. In this case, the atoms form an orbit whose complement has measure 0, and hence the action is essentially transitive and M is Type I (III.3.2.12(iii)); in the contrary case of nonatomic μ, all orbits have measure 0. In addition, the only nonzero probability measure on X invariant under G is Haar measure on X, with each $\alpha_k^{(i)} = n_i^{-1}$; it can be readily shown that this measure is equivalent to μ if and only if the condition of III.3.1.12(ii) is satisfied.

III.3.2.15 If G is discrete and the action α is essentially free, the von Neumann algebra $L^\infty(X,\mathcal{B},\mu)\bar{\rtimes}_\alpha G$ depends only weakly on the group action; the only really essential information is contained in the equivalence relation with orbits as equivalence classes. It is easily seen using left Hilbert algebras (III.4.4.1) that if β is an essentially free action of a discrete group H on $(X,\mathcal{B},\mu)$ with the same orbits, then

$$L^\infty(X,\mathcal{B},\mu)\bar{\rtimes}_\beta H \cong L^\infty(X,\mathcal{B},\mu)\bar{\rtimes}_\alpha G.$$

Since $L^\infty(X,\mathcal{B},\mu)\bar{\rtimes}_\alpha G$ also depends only on the measure class of μ, we conclude that

$$L^\infty(X,\mathcal{B},\mu)\bar{\rtimes}_\alpha G \cong L^\infty(Y,\mathcal{A},\nu)\bar{\rtimes}_\beta H$$

whenever $(X,\mathcal{B},\mu,G,\alpha)$ and $(Y,\mathcal{A},\nu,H,\beta)$ are isomorphic in the following sense:

III.3.2.16 DEFINITION. Let $(X,\mathcal{B},\mu,G,\alpha)$ and $(Y,\mathcal{A},\nu,H,\beta)$ be essentially free transformation groups, with G, H discrete. Then $(X,\mathcal{B},\mu,G,\alpha)$ and $(Y,\mathcal{A},\nu,H,\beta)$ are *orbit equivalent* if there is a Borel bijection $\theta : X \to Y$ which sends G-orbits to H-orbits a.e., with $\theta(\mu)$ equivalent to ν.

III.3.2.17 In fact, a group action is not necessary: a von Neumann algebra can be defined directly from a measured equivalence relation (again via left Hilbert algebras). If $(X,\mathcal{B},\mu)$ is a standard measure space and R is an equivalence relation on X with countable orbits, then R is a *measured equivalence relation* if R is a measurable subset of $X \times X$ (so the equivalence classes are measurable subsets of X), and the saturation of a set of measure 0 has measure 0. A von Neumann algebra denoted $L^\infty(R,\mu)$ can then be constructed; it depends on (R,μ) only up to isomorphism. Note that $L^\infty(R,\mu)$ is noncommutative in general despite the L^∞ notation: in fact, $L^\infty(R,\mu)$ is a factor if R is *ergodic* in the sense that if E is a saturated measurable set, then either $\mu(E) = 0$ or $\mu(X \setminus E) = 0$. If α is an essentially free action of a countable discrete group G and R_α is the corresponding equivalence relation, then

$$L^\infty(R_\alpha,\mu) \cong L^\infty(X,\mathcal{B},\mu)\bar{\rtimes}_\alpha G.$$

If T is an automorphism of $(X,\mathcal{B},\mu)$, write R_T for the corresponding equivalence relation. The definitive study of von Neumann algebras of measured equivalence relations was done by J. Feldman and C. Moore ([FM77a], [FM77b]). The construction can be further generalized to measured groupoids, which includes the case of group actions which are not essentially free, as well as other examples such as the von Neumann algebra of a foliation (cf. [Con94]).

Krieger Factors

III.3.2.18 If T is an ergodic automorphism of a standard measure space $(X, \mathcal{B}, \mu)$, then the corresponding factor

$$L^\infty(R_T, \mu) = L^\infty(X, \mathcal{B}, \mu) \bar{\rtimes}_{\alpha_T} \mathbb{Z}$$

is called a *Krieger factor*. W. Krieger [Kri76] made an extensive study of such factors.

EXAMPLE. Every ITPFI factor is a Krieger factor. If an ITPFI factor M corresponds to $(X, \mathcal{B}, \mu, G, \alpha)$ as in III.3.2.14, there is an "odometer transformation" T such that $(X, \mathcal{B}, \mu, G, \alpha)$ is isomorphic to $(X, \mathcal{B}, \mu, \mathbb{Z}, \alpha_T)$: T acts by "adding one" with appropriate carries. Thus $M \cong L^\infty(R_T, \mu)$ by III.3.2.15.

There are Krieger factors which are not ITPFI, but only ones of Type III_0. In fact, it turns out that every injective factor with separable predual is a Krieger factor (IV.2.7.4).

Krieger proved a converse to III.3.2.15 (along with III.3.2.15 itself), yielding:

III.3.2.19 THEOREM. Let T and S be automorphisms of standard measure spaces $(X, \mathcal{B}, \mu)$ and $(Y, \mathcal{A}, \nu)$ respectively. Then

$$L^\infty(R_T, \mu) \cong L^\infty(R_S, \nu)$$

if and only if $(X, \mathcal{B}, \mu, \mathbb{Z}, \alpha_T)$ and $(Y, \mathcal{A}, \nu, \mathbb{Z}, \alpha_S)$ are orbit equivalent (T and S are *weakly equivalent*).

III.3.2.20 Earlier, H. Dye [Dye59] proved that any two ergodic transformations of a standard measure space with a finite invariant measure are weakly equivalent. This result is very closely related to the uniqueness of the hyperfinite II_1 factor (III.3.4.3) (and follows from III.3.4.3 and III.3.2.19).

III.3.2.21 Krieger also generalized the Araki-Woods invariants $r_\infty(M)$ and $\rho(M)$ to this context. By III.3.2.19, these are expressible in terms of (R_T, μ). For example, if $\lambda > 0$, then $\lambda \in \rho(R_T, \mu)$ if and only if there is a measure ν equivalent to μ such that the Radon-Nikodym derivative $\left[\frac{d(\nu \circ T)}{d\nu}\right]$ has range contained in $\lambda^\mathbb{Z}$.

III.3.2.22 In light of Krieger's theorem, it is important to characterize which equivalence relations are isomorphic to ones arising from a $\mathbb{Z}$-action on a standard measure space and which hence give Krieger factors. This was definitively done in [CFW81]: an equivalence relation on a standard measure space, with countable equivalence classes, is isomorphic to one arising from a $\mathbb{Z}$-action if and only if it is amenable in the sense of R. Zimmer [Zim77]. See [CK77] and [OW80] for earlier partial results.

III.3.3 Regular Representations of Discrete Groups

A rather different instance of the crossed product construction, with $N = \mathbb{C}$ but using highly nonabelian groups, yields a very interesting supply of factors.

III.3.3.1 Let G be a locally compact group with left Haar measure μ, and λ the left regular representation of G on $L^2(G, \mu)$. We have that $\lambda = \pi_\psi$, the GNS representation from the weight ψ defined on $C_c(G) \subseteq L^1(G)$ by $\psi(f) = f(e_G)$ (called the *Plancherel weight*). Let

$$\mathfrak{L}(G) = \lambda(G)''.$$

$\mathfrak{L}(G)$ is called the *(left) von Neumann algebra* of G. $\mathfrak{L}(G)$ is in standard form, and $\mathfrak{L}(G)'$ can be identified with the von Neumann algebra $\mathfrak{R}(G)$ generated by the right regular representation ρ of G (this is elementary if G is discrete, but the nondiscrete case requires use of the basic theory of Hilbert algebras, and left Hilbert algebras for the nonunimodular case). $\mathfrak{L}(G)$ has separable predual if and only if G is second countable (in the discrete case, if and only if G is countable).

III.3.3.2 If G is unimodular, then ψ is a trace, so $\mathfrak{L}(G)$ is semifinite. If G is discrete, then ψ is a tracial state (provided μ is normalized), so $\mathfrak{L}(G)$ is finite.

III.3.3.3 Suppose G is discrete. If $\xi \in l^2(G)$ is the trace vector (ξ is the characteristic function of $\{e_G\}$), then identifying x with $x\xi$ we may identify elements of $\mathfrak{L}(G)$ with L^2-functions on G (i.e. square-summable sequences indexed by G). The multiplication in $\mathfrak{L}(G)$ corresponds to convolution in $l^2(G)$.

III.3.3.4 PROPOSITION. If G is discrete and $x \in \mathcal{Z}(\mathfrak{L}(G))$, the corresponding function on G is constant on conjugacy classes.

III.3.3.5 DEFINITION. A (discrete) group G is an *ICC group* if every conjugacy class in G, other than $\{e_G\}$, is infinite.

There is a large supply of interesting ICC groups. Here are some of the most elementary examples:

III.3 Examples and Constructions of Factors

III.3.3.6 EXAMPLES.

(i) A free group on more than one generator is an ICC group.
(ii) The group S_X of finite permutations of an infinite set X is a locally finite ICC group.
(iii) Many "$ax+b$" groups are ICC groups: for example,

$$\left\{ \begin{bmatrix} a & b \\ 0 & 1 \end{bmatrix} : a, b \in \mathbb{Q}, a > 0 \right\}.$$

III.3.3.7 PROPOSITION. If G is an ICC group, then $\mathfrak{L}(G)$ is a II$_1$ factor.

Indeed, by III.3.3.4 a central element of $\mathfrak{L}(G)$ is a square-summable sequence constant on conjugacy classes, hence is a multiple of ξ, so $\mathfrak{L}(G)$ is a factor. It is infinite-dimensional and finite, and hence II$_1$.

One can show that the $\mathfrak{L}(G)$ for countable ICC G comprise numerous isomorphism classes. Some of the most easily investigated properties used to distinguish these factors concern central sequences:

III.3.3.8 DEFINITION. Let M be a II$_1$ factor. A *central sequence* in M is a bounded sequence (x_n) in M such that $\|[x_n, y]\|_2 \to 0$ for all $y \in M$, where $[x, y] = xy - yx$ is the usual commutator.

The central sequences form a σ-weakly dense C*-subalgebra of the W*-algebra $l^\infty(M)$, called the *central sequence algebra* of M.

III.3.3.9 The factor $\mathcal{R}$ of III.3.1.4 has *Property* Γ: it has a central sequence consisting of unitaries of trace 0. In fact, it has a central sequence of self-adjoint unitaries of trace 0, or alternatively a central sequence of projections of trace 1/2 [let p_n be the elementary tensor whose n'th component is e_{11} and other components are 1.] It is easy to see that $\mathfrak{L}(S_\mathbb{N})$ (III.3.3.6(ii)) also has this property (in fact, $\mathfrak{L}(S_\mathbb{N}) \cong \mathcal{R}$ (III.3.4.3)).

III.3.3.10 PROPOSITION. Let $\mathbb{F}_n$ ($2 \leq n \leq \infty$) be the free group on n generators, and $\mathfrak{F}_n = \mathfrak{L}(\mathbb{F}_n)$. Then $\mathfrak{F}_n$ does not have Property Γ; hence $\mathfrak{F}_n$ is not isomorphic to $\mathcal{R}$.

III.3.3.11 Using variations and refinements of the notion of central sequence and the arguments in III.3.3.10, by the late 1960's nine isomorphism classes of II$_1$ factors with separable predual had been isolated ([MvN43], [Sch63], [Chi69], [Sak69], [DL69], [ZM69]). Then in 1968, making more delicate and sophisticated use of these ideas, D. McDuff [McD69] constructed an uncountable family G_i of countable ICC groups such that the $\mathfrak{L}(G_i)$ were mutually nonisomorphic.

Free Group Factors

III.3.3.12 The factors $\mathfrak{F}_n$ of III.3.3.10 are called the *free group factors*. One of the outstanding open questions of von Neumann algebra theory, which is a subject of much current research, is whether $\mathfrak{F}_n \cong \mathfrak{F}_m$ for $m \neq n$. The $\mathfrak{F}_n$ fit into a natural one-parameter family $\{\mathfrak{F}_t : t > 1\}$ of II_1 factors with separable predual, called the *interpolated free group factors* ([Dyk94], [Răd94]); it is known that

(i) If p is a projection in $\mathfrak{F}_t$ of trace λ, then $p\mathfrak{F}_t p \cong \mathfrak{F}_s$, where $s = \frac{t-1}{\lambda^2} + 1$.
(ii) Either all $\mathfrak{F}_t$ for $1 < t \leq \infty$ are isomorphic, or they are mutually nonisomorphic.

III.3.3.13 The idea from (i) is important, and can be systematized as follows. If N is an infinite semifinite factor with trace τ, and $\alpha \in \mathrm{Aut}(N)$, then $\tau \circ \alpha = \lambda \tau$ for some $\lambda > 0$, called the *modulus* of α. If N is type I, then λ is necessarily 1, so this concept is only interesting for II_∞ factors.

III.3.3.14 DEFINITION. Let M be a II_1 factor. The *fundamental group* of M is
$$F(M) = \{\lambda > 0 : \exists \alpha \in \mathrm{Aut}(M \bar{\otimes} \mathcal{L}(\mathcal{H})) \text{ with modulus } \lambda\}$$

It is easy to see that $F(M)$ is a multiplicative subgroup of $\mathbb{R}_+$, and that if $0 < \lambda < 1$ and p is a projection in M with trace λ, then $\lambda \in F(M)$ if and only if $pMp \cong M$. The group $F(M)$ is sometimes written as an additive subgroup of $\mathbb{R}$ by applying the logarithm function.

For "most" II_1 factors M, e.g. $M = \mathcal{R}$, $F(M) = \mathbb{R}_+$; the first example where this was shown to fail was constructed in [Con80]. Every countable subgroup of $\mathbb{R}_+$ occurs as the fundamental group of a II_1 factor on a separable Hilbert space [Pop04].

Property T

III.3.3.15 Factors with some unusual properties are obtained using ICC groups with Property T, introduced by D. Kazhdan [Kaž67]. An infinite (discrete) group G has *Property T* if the trivial representation is isolated in $\hat{G}$ in its natural topology. A Property T group is nonamenable, and has very rigid structure: for example, a Property T group cannot be written as a union of a strictly increasing sequence of subgroups (in particular, a countable Property T group is finitely generated). There is a corresponding notion of property T for nondiscrete groups. See [dlHV89] for a good treatment of the theory.

There are many examples of property T groups, For example, $SL(n, \mathbb{Z})$ has property T for $n \geq 3$. More generally, a lattice in any semisimple Lie group of real rank ≥ 2 has property T. For more examples, see e.g. [Kir94], [Gro87], and [CMS94]

III.3.3.16 If G is an ICC Property T group, $\mathcal{L}(G)$ is a *Property T factor*. A general definition of a Property T factor not necessarily arising from a group can be made using correspondences (cf. [Con94, V.B.ϵ]).

III.3.3.17 Property T factors have unusual rigidity properties. For example, if M is a property T factor, then:

(i) If (M_n) is an increasing sequence of von Neumann subalgebras of M with $M = [\cup M_n]''$, then $M_n = M$ for some n.
(ii) $\mathrm{Out}(M) = \mathrm{Aut}(M)/\mathrm{Inn}(M)$ is countable.
(iii) The fundamental group $F(M)$ (III.3.3.14) is countable.

Nondiscrete Groups

III.3.3.18 While the method of this section has primarily been used with discrete groups to construct II_1 factors, there exist nondiscrete second countable groups G (e.g. restricted direct products of p-adic $ax+b$ groups) such that the $\mathcal{L}(G)$ are factors of Type I_∞, II_∞, and III (cf. [Bla77b], [Bla83b]).

III.3.4 Uniqueness of the Hyperfinite II_1 Factor

One of the first significant classification results in the theory of operator algebras (approximately contemporaneous with the Gelfand-Naimark classification of commutative C*-algebras) was the proof of the uniqueness of the hyperfinite II_1 factor with separable predual by Murray and von Neumann [MvN43]. This uniqueness result was later greatly expanded by Connes.

III.3.4.1 DEFINITION. A von Neumann algebra M is *approximately finite dimensional*, or *hyperfinite*, if there is a directed collection $\{M_i\}$ of finite-dimensional *-subalgebras with $\cup M_i$ σ-weakly dense in M.

The M_i can be chosen to be unital subalgebras, hence W*-subalgebras. If M is approximately finite dimensional with separable predual, $\{M_i\}$ can be chosen to be an increasing sequence.

In some references, the term "hyperfinite" is reserved for II_1 factors, or sometimes for finite von Neumann algebras, as it seems terminologically inconsistent to have algebras which are hyperfinite but not finite.

III.3.4.2 EXAMPLES.

(i) It is easy to see that any Type I von Neumann algebra is approximately finite dimensional. If $\{M_i\}$ is a directed collection of Type I W*-subalgebras of M with dense union, then M is approximately finite dimensional.
(ii) An infinite tensor product of finite-dimensional (or Type I) von Neumann algebras is approximately finite dimensional. Thus the factor $\mathcal{R}$ (III.3.1.4) is a hyperfinite II_1 factor, and the ITPFI factors (III.3.1.10) are approximately finite-dimensional factors, mostly of Type III.

(iii) Let G be a locally finite group, a directed union of finite subgroups. If G acts freely and ergodically on a measure space X, then $L^\infty(X) \bar{\rtimes}_\alpha G$ is an approximately finite-dimensional factor by (i) and III.3.2.12.

(iv) If G is a locally finite ICC group, e.g. the group of finite permutations of a countable set, then $\mathfrak{L}(G)$ is a hyperfinite II$_1$ factor (III.3.3.7).

It turns out that the conclusions of (iii) and (iv) hold if "locally finite" is replaced by "amenable" (IV.2.2.15, IV.2.6.1).

The main result of this section is:

III.3.4.3 THEOREM. Let M be a hyperfinite II$_1$ factor with separable predual. Then M is isomorphic to $\mathcal{R}$.

III.3.4.4 COROLLARY. The II$_1$ factors of III.3.3.10 and III.3.3.11 are not hyperfinite.

III.3.4.5 We will now outline the proof of the theorem. Say that a II$_1$ factor M is of *CAR type* if M has an increasing sequence (A_n) of unital *-subalgebras, with weakly dense union, with $A_n \cong \mathbb{M}_{2^n}$, i.e. if M has a weakly dense unital C*-subalgebra isomorphic to the CAR algebra (II.8.2.2(iv)). It is clear that $\mathcal{R}$ is of CAR type, and that a factor of CAR type is hyperfinite.

III.3.4.6 PROPOSITION. Any II$_1$ factor of CAR type is isomorphic to $\mathcal{R}$.

PROOF: Let M be a II$_1$ factor of CAR type, with tracial state τ and weakly dense unital C*-subalgebra A isomorphic to the CAR algebra. Represent M in standard form using the GNS representation from τ. Then $\pi_\tau|_A$ can be identified with the GNS representation of A from the tracial state $\sigma = \tau|_A$. But σ is the unique tracial state of A, and $\pi_\sigma(A)'' \cong \mathcal{R}$.

III.3.4.7 The rest of the proof uses the "Hilbert-Schmidt" norm (2-norm) on a II$_1$ factor. If M is a II$_1$ factor with tracial state τ, for $x \in M$ define

$$\|x\|_2 = \|x\|_{2,\tau} = \tau(x^*x)^{1/2}$$

(III.2.2.16); $\|x\|_2$ is the norm of x regarded as a vector in $\mathcal{H}_\tau$. If (p_n) is a sequence of subprojections of p with $\tau(p_n) \to \tau(p)$, then $\|p - p_n\|_2 \to 0$. The 2-norm topology agrees with the σ-strong topology on the unit ball of M (cf. III.2.2.27).

A routine argument using the semiprojectivity of finite-dimensional matrix algebras (II.8.3.16(i)) to nest the subalgebras yields:

III.3.4.8 LEMMA. Let M be a II$_1$ factor with separable predual. Then M is of CAR type if and only if, for any $x_1, \ldots, x_k \in M$ and $\epsilon > 0$, there is an n and a unital C*-subalgebra B of M, isomorphic to $\mathbb{M}_{2^n}$, and elements $y_1, \ldots, y_k \in B$, with $\|x_i - y_i\|_2 < \epsilon$ for $1 \leq i \leq k$.

The proof of III.3.4.3 is then completed by applying the following theorem:

III.3.4.9 THEOREM. Let M be a II_1 factor, A a finite-dimensional C*-subalgebra of M, and $\epsilon > 0$. Then there is a unital C*-subalgebra B of M, isomorphic to $\mathbb{M}_{2^n}$ for some n, such that for every $a \in A$ there is a $b \in B$ with $\|a - b\|_2 < \epsilon\|a\|$.

To prove this theorem, first suppose

$$\{e_{ij} : 1 \leq i, j \leq m\}$$

is a set of matrix units in M (with $\sum e_{ii}$ not necessarily equal to 1), and $C = C^*(\{e_{ij}\})$. If p is a subprojection of e_{11} with $\tau(p)$ a dyadic rational number almost equal to $\tau(e_{11})$ (within ϵ/m^2 will do), and $f_{ij} = e_{i1}pe_{1j}$, then $\{f_{ij}\}$ is a set of matrix units for a (nonunital) matrix subalgebra D of M, and for each $c \in C$ there is a $d \in D$ with $\|c - d\|_2 < \epsilon\|c\|$.

Now let $C_1, \ldots, C_r$ be the central direct summands of A, and for each C_i choose p_i and D_i as above. Let q_i be the unit of D_i, and

$$q_{r+1} = 1 - \sum_{i=1}^{r} q_i$$

and $p_{r+1} = q_{r+1}$. There is an n such that $\tau(p_i) = 2^{-n}k_i$ for all i. Inside the II_1 factor $p_i M p_i$ there is a unital copy of $\mathbb{M}_{k_i}$; thus there is a matrix subalgebra B_i of M, with unit q_i, which contains D_i and in which the minimal projections have trace 2^{-n}. The minimal projections in the B_i are all equivalent in M, and hence with a bit of care a set of matrix units for all the B_i can be expanded to a set of matrix units for a unital copy of M_{2^n} in M.

III.4 Modular Theory

Modular theory is one of the most important and useful developments in the history of von Neumann algebras, giving a very precise and intimate connection between a von Neumann algebra and its commutant, along with a canonical one-parameter group of outer automorphisms. Modular theory was first developed by M. Tomita [Tom67a], and was refined and clarified and given a careful and complete exposition by M. Takesaki [Tak70] (some details have since been further simplified). The basic framework had previously been developed ([Nak50]; cf. [Dix69a]) in the much simpler case of semifinite von Neumann algebras. This theory made possible the great advances in the 1970's by Connes *et al.* on the classification of factors.

III.4.1 Notation and Basic Constructions

III.4.1.1 Let M be a von Neumann algebra in standard form (III.2.6) on a Hilbert space $\mathcal{H}$, i.e. the identity representation is the GNS representation from a faithful normal semifinite weight ϕ; recall that (at least in the countably decomposable case; cf. III.4.5.7) such a standard form is unique up to spatial isomorphism, and M' is also in standard form. The details are somewhat simpler in the case where M is countably decomposable, i.e. the weight can be taken to be a state, or alternatively, M has a cyclic and separating vector ξ (which is also cyclic and separating for M'), and we will concentrate on this case in the exposition.

III.4.1.2 We define two closed conjugate-linear densely defined operators S and F on $\mathcal{H}$. Let S_0 be the densely defined operator

$$x\xi = \eta \mapsto \eta^\sharp = x^*\xi$$

for $x \in M$, where ξ is the cyclic and separating vector associated with the faithful normal state ϕ, and F_0 the densely defined operator

$$y\xi = \eta \mapsto \eta^\flat = y^*\xi$$

for $y \in M'$. S_0 and F_0 are unbounded in general. If $x \in M$ and $y \in M'$, then

$$\langle S_0(x\xi), y\xi\rangle = \langle x^*\xi, y\xi\rangle = \langle y^*x^*\xi, \xi\rangle = \langle x^*y^*\xi, \xi\rangle = \langle y^*\xi, x\xi\rangle = \langle F_0(y\xi), x\xi\rangle$$

and thus $S_0^* \supseteq F_0$ and $F_0^* \supseteq S_0$, so S_0 and F_0 are closable. Let S be the closure of S_0, and $F = S^* = S_0^*$. (It turns out that F is the closure of F_0, and thus the situation is completely symmetric in M and M' – this is obvious after the fact using the main theorem, or see [Sak71, 2.8.2] for a direct proof. It is not necessary to use this for the development of the theory.) Since $S_0 = S_0^{-1}$, it follows that S and F are one-one with dense range, and $S = S^{-1}$, $F = F^{-1}$.

In the general case, define S_0 to be the operator whose domain is $\iota(\mathfrak{N}_\phi \cap \mathfrak{N}_\phi^*)$, where ι is the canonical embedding of $\mathfrak{N}_\phi$ into $\mathcal{H} = \mathcal{H}_\phi$, with $S_0(\iota(x)) = \iota(x^*)$. It must then be proved that S_0 is densely defined (a consequence of III.2.2.21) and closable, and that its closure S and adjoint $F = S^*$ have the right properties.

III.4.1.3 Set $\Delta = S^*S = FS$. Then Δ is an invertible densely defined (in general unbounded) positive operator with $\Delta^{-1} = SF$ (I.7.1.4). If

$$S = J\Delta^{1/2}$$

is the polar decomposition, then J is an invertible conjugate-linear isometry, and

$$S = S^{-1} = \Delta^{-1/2}J^{-1}$$

but

$$S = (SS^*)^{1/2}J = \Delta^{-1/2}J$$

so $J = J^{-1}$, i.e. J is an involution. We have $J\Delta J = \Delta^{-1}$ and

$$F = J\Delta^{-1/2} = \Delta^{1/2}J.$$

Since $S\xi = \xi$, we have $F\xi = \xi$, $J\xi = \xi$, $\Delta\xi = \xi$. Δ is called the *modular operator* for (M, ϕ), often denoted Δ_ϕ to emphasize its dependence on ϕ.

The situation is vastly simpler if M is semifinite and ϕ is a trace. In this case, $S = F = J$ and $\Delta = 1$. This is the case considered in [Dix69a].

III.4.1.4 EXAMPLE. A key motivating example comes from group representations. Let G be a locally compact nonunimodular group with left Haar measure μ, λ the left regular representation on $L^2(G,\mu)$, and $\mathcal{L}(G) = \lambda(G)''$ the left von Neumann algebra of G. Write λ also for the representation of $\mathcal{L}(G)$. Then $\lambda = \pi_\psi$ for the canonical Plancherel weight ψ (III.3.3.1). The modular operator Δ for this weight is exactly multiplication by the modular function Δ_G on $L^2(G,\mu)$.

III.4.2 Approach using Bounded Operators

III.4.2.1 There is an alternate approach using bounded operators, due to M. Rieffel and A. van Daele [RvD77] (cf. [Ped79, 8.13]). Let $\mathcal{H}_\mathbb{R}$ be $\mathcal{H}$ regarded as a real Hilbert space under $(\cdot,\cdot) = Re\langle\cdot,\cdot\rangle$. Let $\mathcal{X}$ and $\mathcal{Y}$ be the closures of the real subspaces $M_{sa}\xi$ and $iM_{sa}\xi$ respectively. Then $\mathcal{X}$ and $\mathcal{Y}$ are complementary subspaces of $\mathcal{H}_\mathbb{R}$, although they are not orthogonal unless ϕ is a trace (in fact, if $x,y \in M_{sa}$, then $x\xi \perp iy\xi$ if and only if $\phi(xy) = \phi(yx)$); however, $\mathcal{X}$ is orthogonal to $iM'_{sa}\xi$, from which it follows that $\mathcal{X} \cap \mathcal{Y} = \{0\}$. Let P and Q be the (real-linear) projections of $\mathcal{H}_\mathbb{R}$ onto $\mathcal{X}$ and $\mathcal{Y}$ respectively, and set $A = P + Q$, $C = P - Q$. Then the following are easy to verify:

(i) A is complex-linear, C is conjugate-linear, and both are "self-adjoint" (as operators on $\mathcal{H}$).
(ii) A and $2 - A$ are one-one with dense range, and $0 \leq A \leq 2$.
(iii) C^2 is one-one with dense range, $0 \leq C^2 \leq 4$; if $B = (C^2)^{1/2}$ and $C = JB$ is the polar decomposition, then J is an involution and $B = A^{1/2}(2-A)^{1/2}$ is one-one with dense range.
(iv) B commutes with P, Q, A, and J.
(v) $JP = (1-Q)J$, $JQ = (1-P)J$, and $JA = (2-A)J$.
(vi) If $\Delta = A^{-1}(2-A)$, then Δ is a (generally unbounded) densely defined, self-adjoint, positive invertible operator, $J\Delta J = \Delta^{-1}$, $M\xi \subseteq \mathcal{D}(\Delta^{1/2})$, and $J\Delta^{1/2}(x\xi) = x^*\xi$ for $x \in M$.
(vii) J and Δ agree with the J and Δ previously defined via S.

III.4.3 The Main Theorem

III.4.3.1 For $\alpha \in \mathbb{C}$, we can define the closed operator Δ^α as in I.7.4.7; in particular, for $t \in \mathbb{R}$, the operator Δ^{it} is unitary, and $t \mapsto \Delta^{it}$ is a strongly continuous one-parameter group of unitaries. Thus the automorphisms σ_t (often denoted σ_t^ϕ) defined by

$$\sigma_t(x) = \Delta^{it} x \Delta^{-it}$$

define a strongly continuous one-parameter group of automorphisms of $\mathcal{L}(\mathcal{H})$.

Because of the conjugate-linearity of J, from $J\Delta = \Delta^{-1}J$ we obtain $J\Delta^{it} = \Delta^{it}J$ for all $t \in \mathbb{R}$. Thus the automorphisms σ_t of $\mathcal{L}(\mathcal{H})$ commute with the conjugate-linear automorphism j of $\mathcal{L}(\mathcal{H})$ defined by $j(x) = JxJ$.

Here is the main theorem of modular theory.

III.4.3.2 THEOREM. [TOMITA-TAKESAKI] $j(M) = M'$ (so $j(M') = M$) and $\sigma_t(M) = M$, $\sigma_t(M') = M'$ for all $t \in \mathbb{R}$. If $x \in \mathcal{Z}(M)$, then $j(x) = x^*$ and $\sigma_t(x) = x$ for all t.

We give an argument (cf. [vD82]) that "explains" why the theorem is true. This argument as it stands is only valid in the relatively unusual case that S, and hence also F and Δ, is bounded (this implies that M is semifinite); but with considerable additional work it can be made rigorous in the general case ([Zsi75], [Kad78]; cf. [SZ79]).

If $x, y, z \in M$, then

$$SxSyz\xi = Sxz^*y^*\xi = yzx^*\xi = ySxz^*\xi = ySxSz\xi.$$

Since $\{z\xi : z \in M\}$ is dense in $\mathcal{H}$, it follows that $SxSy = ySxS$ for any $y \in M$, i.e. $SxS \in M'$; so $SMS \subseteq M'$. Similarly, $FM'F \subseteq M$, and taking adjoints we get $SM'S \subseteq M$, so $SMS = M'$ (and $SM'S = M$, $FMF = M'$, $FM'F = M$). Thus

$$\Delta M \Delta^{-1} = FSMSF = M.$$

Let $\mathcal{B}$ be the Banach algebra $\mathcal{L}(\mathcal{L}(\mathcal{H}))$. For $x \in \mathcal{L}(\mathcal{H})$ and for $\alpha \in \mathbb{C}$, define $\Phi_\alpha \in \mathcal{B}$ by $\Phi_\alpha(x) = \Delta^\alpha x \Delta^{-\alpha}$. If $\mathcal{A}$ is the closed subalgebra of $\mathcal{B}$ generated by Φ_1 and 1, then since Φ_1 leaves M invariant so does every element of $\mathcal{A}$. If $L, R \in \mathcal{B}$ are defined by $L(x) = \Delta x$, $R(x) = x\Delta^{-1}$, then $LR = RL = \Phi_1$, and $\sigma_\mathcal{B}(L) = \sigma_{\mathcal{L}(\mathcal{H})}(\Delta)$, $\sigma_\mathcal{B}(R) = \sigma_{\mathcal{L}(\mathcal{H})}(\Delta^{-1})$, so

$$\sigma_\mathcal{B}(\Phi_1) \subseteq \sigma_{\mathcal{L}(\mathcal{H})}(\Delta)\sigma_{\mathcal{L}(\mathcal{H})}(\Delta^{-1}) \subseteq (0, \infty)$$

by II.2.1.3. If f is the principal branch of the logarithm function, then f is analytic in a disk containing $\sigma_\mathcal{B}(\Phi_1)$, and $f(\Phi_1)$ can be defined by a power series in $\Phi_1 - \lambda 1$ for some $\lambda > 0$. Thus $\Psi = f(\Phi_1) \in \mathcal{A} \subseteq \mathcal{B}$, and $\Phi_\alpha = e^{\alpha \Psi} \in \mathcal{A}$ for all $\alpha \in \mathbb{C}$. (If $H = f(\Delta) \in \mathcal{L}(\mathcal{H})$, then $\Psi(x) = Hx - xH$ for $x \in \mathcal{L}(\mathcal{H})$.)

In particular, $\Phi_{it} = \sigma_t$ leaves M invariant. Also, $\Phi_{1/2}$ and $\Phi_{-1/2}$ leave M invariant, i.e. $\Delta^{-1/2} M \Delta^{1/2} = M$; thus

$$JMJ = S\Delta^{-1/2} M \Delta^{1/2} S = SMS = M'.$$

The last statement follows from the fact that if u is a unitary in $\mathcal{Z}(M)$, then $uS_0 u = S_0$, so $uSu = S$, $u^*Fu^* = F$, $u^*\Delta u = \Delta$, $uJu = J$.

III.4.4 Left Hilbert Algebras

The original approach of Tomita-Takesaki proceeded somewhat differently. It is convenient to abstract the situation.

III.4.4.1 A *left Hilbert algebra* (or *generalized Hilbert algebra*) is a complex algebra $\mathfrak{A}$ endowed with an inner product $\langle \cdot, \cdot \rangle$ and an (in general unbounded) involution $x \mapsto x^\sharp$ (conjugate-linear anti-automorphism of period 2), with the following properties (the topology on $\mathfrak{A}$ is the one coming from the inner product):

(i) For any $x, y, z \in \mathfrak{A}$, $\langle xy, z \rangle = \langle y, x^\sharp z \rangle$.
(ii) For each $x \in \mathfrak{A}$, the map $y \mapsto xy$ is bounded.
(iii) $\mathfrak{A}^2$ is dense in $\mathfrak{A}$.
(iv) The map $x \mapsto x^\sharp$ is closable as a map on the completion $\mathcal{H}$ of $\mathfrak{A}$.

If $\sharp$ is isometric, then $\mathfrak{A}$ is called a *Hilbert algebra*.

Hilbert algebras were first defined by H. Nakano in [Nak50] (a preliminary version appeared in [Amb49]). Left Hilbert algebras were introduced by Tomita, following the notion of quasi-Hilbert algebra of J. Dixmier [Dix69a].

In our previous context, M (or any weakly dense *-subalgebra) becomes a left Hilbert algebra with inner product $\langle x, y \rangle = \langle x\xi, y\xi \rangle$ and $x^\sharp = x^*$; in the general case, $\mathfrak{N}_\phi \cap \mathfrak{N}_\phi^*$ is a left Hilbert algebra with $\langle x, y \rangle = \phi(y^*x)$ and $x^\sharp = x^*$.

If S is the closure of $x \mapsto x^\sharp$, write $\mathcal{D}^\sharp$ for the domain of S and $x^\sharp$ for Sx if $x \in \mathcal{D}^\sharp$. If $F = S^*$, write $\mathcal{D}^\flat$ for the domain of F and $x^\flat$ for Fx if $x \in \mathcal{D}^\flat$. The map $x \mapsto x^\flat$ is also an (in general unbounded) involution, called the *adjoint involution*. The spaces $\mathcal{D}^\sharp$ and $\mathcal{D}^\flat$ are Hilbert spaces under the inner products

$$\langle x, y \rangle_\sharp = \langle x, y \rangle + \langle y^\sharp, x^\sharp \rangle \quad \text{and} \quad \langle x, y \rangle_\flat = \langle x, y \rangle + \langle y^\flat, x^\flat \rangle$$

respectively.

III.4.4.2 Each $x \in \mathfrak{A}$ defines a bounded operator $\pi(x)$ on $\mathcal{H}$ by $\pi(x)y = xy$. Since $\pi(x^\sharp) = \pi(x)^*$, π defines a nondegenerate *-representation of $\mathfrak{A}$ on $\mathcal{H}$. Let $\mathfrak{L}(\mathfrak{A}) = \pi(\mathfrak{A})''$ be the von Neumann algebra generated by $\pi(\mathfrak{A})$. $\mathfrak{L}(\mathfrak{A})$ is called the *left von Neumann algebra* of $\mathfrak{A}$.

III.4.4.3 Each $y \in \mathcal{H}$ defines an operator R_y with domain $\mathfrak{A}$ by $R_y x = \pi(x)y$. If $y \in \mathcal{D}^\flat$, a simple calculation shows $R_y^* \supseteq R_{y^\flat}$, so R_y has a closure $\pi'(y)$, and $\pi'(y)^* = \pi'(y^\flat)$. As in the proof of III.4.3.2, $\pi'(y)$ is permutable with $\pi(x)$ for all $x \in \mathfrak{A}$, i.e. $\pi'(y)$ is affiliated with $\mathfrak{L}(\mathfrak{A})'$.

Let $\mathfrak{A}'$ be the set of $y \in \mathcal{D}^\flat$ such that $\pi'(y)$ is bounded. If $x \in \mathcal{H}$ and $y \in \mathfrak{A}'$, define $xy = \pi'(y)x$. It is easy to see that $\mathfrak{A}'$ is closed under this multiplication and the involution $y \mapsto y^\flat$; and π' is a *-antihomomorphism of $\mathfrak{A}'$ into $\mathfrak{L}(\mathfrak{A})'$. A functional calculus argument shows that $\mathfrak{A}'$ and $(\mathfrak{A}')^2$ are dense in the Hilbert space $\mathcal{D}^\flat$, and therefore in $\mathcal{H}$. A simple calculation shows that if $y, z \in \mathfrak{A}'$ and $T \in \mathfrak{L}(\mathfrak{A})'$, then $\pi'(y)Tz \in \mathfrak{A}'$ and

$$\pi'(\pi'(y)Tz) = \pi'(y)T\pi'(z).$$

Then using an approximate unit for $\pi'(\mathfrak{A}')$ we obtain that $\pi'(\mathfrak{A}')'' = \mathfrak{L}(\mathfrak{A})'$.

Full Left Hilbert Algebras

III.4.4.4 Repeating the process, for $x \in \mathcal{H}$ define L_x with domain $\mathfrak{A}'$ by $L_x y = \pi'(y)x$. If $x \in \mathcal{D}^\sharp$, then $L_x^* \supseteq L_{x^\sharp}$; let $\pi(x)$ be the closure of L_x, and

$\mathfrak{A}''$ the set of $x \in \mathcal{D}^\sharp$ such that $\pi(x)$ is bounded. Then $\mathfrak{A} \subseteq \mathfrak{A}''$, and $\pi(x)$ agrees with the previously defined $\pi(x)$ for $x \in \mathfrak{A}$. $\mathfrak{A}''$ becomes a left Hilbert algebra with multiplication $xy = \pi(x)y$, and $\mathcal{L}(\mathfrak{A}'') = \mathcal{L}(\mathfrak{A})$. $(\mathfrak{A}'')' = \mathfrak{A}'$, so $(\mathfrak{A}'')'' = \mathfrak{A}''$. $\mathfrak{A}''$ is called the *full* (or *achieved*) *left Hilbert algebra of* $\mathfrak{A}$. Two left Hilbert algebras $\mathfrak{A}$ and $\mathfrak{B}$ are *equivalent* if $\mathfrak{A}''$ and $\mathfrak{B}''$ are isometrically *-isomorphic. Of course, $\mathfrak{A}$ is equivalent to $\mathfrak{A}''$ for any $\mathfrak{A}$.

It is easy to verify that a left Hilbert algebra arising from a faithful normal semifinite weight on a von Neumann algebra is full. Every full left Hilbert algebra arises in this manner, and thus there is a natural one-one correspondence between full left Hilbert algebras and von Neumann algebras with a specified faithful normal semifinite weight:

III.4.4.5 THEOREM. ([Com70]; cf. [Tak03a, VII.2.5]) Let $\mathfrak{A}$ be a full left Hilbert algebra. Then there is a unique faithful normal semifinite weight ϕ on $\mathcal{L}(\mathfrak{A})$ such that $\mathfrak{A} \cong \mathfrak{N}_\phi \cap \mathfrak{N}_\phi^*$ with its usual left Hilbert algebra structure.

The definition of ϕ is clear: if $a \in \mathcal{L}(\mathfrak{A})_+$, then $\phi(a) = \|x\|_\mathfrak{A}^2$ if $a = \pi(x)^*\pi(x)$ for some $x \in \mathfrak{A}$, and $\phi(a) = \infty$ otherwise.

The full left Hilbert algebra point of view for weights is a very useful one. For example, it gives a natural way to define the tensor product of two faithful normal semifinite weights, or the dual weight on a W*-crossed product, both of which would otherwise be obscure. For the tensor product, it suffices to observe that the algebraic tensor product of two left Hilbert algebras is a left Hilbert algebra.

Modular Hilbert Algebras

III.4.4.6 A left Hilbert algebra $\mathfrak{A}$ is a *modular Hilbert algebra*, or *Tomita algebra*, if it has a group $\{\delta_\alpha : \alpha \in \mathbb{C}\}$ (i.e. $\delta_{\alpha+\beta} = \delta_\alpha \circ \delta_\beta$) of automorphisms, called *modular automorphisms*, satisfying, for all $x, y \in \mathfrak{A}$:

(i) $(\delta_\alpha x)^\sharp = \delta_{-\bar\alpha}(x^\sharp)$ for all $\alpha \in \mathbb{C}$.
(ii) $\langle \delta_\alpha x, y \rangle = \langle x, \delta_{\bar\alpha} y \rangle$ for all $\alpha \in \mathbb{C}$.
(iii) $\langle \delta_1 x^\sharp, y^\sharp \rangle = \langle y, x \rangle$.
(iv) $\alpha \mapsto \langle \delta_\alpha x, y \rangle$ is an analytic function on $\mathbb{C}$.
(v) For all $t \in \mathbb{R}$, the set $(1 + \delta_t)\mathfrak{A}$ is dense in $\mathfrak{A}$.

Note that from (iii), if $x \in \mathfrak{A}$, then $x \in \mathcal{D}^\flat$ and $x^\flat = \delta_1(x^\sharp)$ $(= (\delta_{-1}x)^\sharp$ by (i)). In fact, axiom (iv) of III.4.4.1 follows from the other axioms for a modular Hilbert algebra; and $\mathfrak{A} \subseteq \mathfrak{A}'$.

A Hilbert algebra is a modular Hilbert algebra with $\delta_\alpha = id$ for all α.

For any $t \in \mathbb{R}$, the map

$$x \mapsto \delta_t(x^\sharp) = (\delta_{-t}x)^\sharp$$

is a (generally unbounded) involution on $\mathfrak{A}$. The involution

$$x \mapsto x^\natural = \delta_{1/2}(x^\sharp)$$

is isometric; let J be the closure of $x \mapsto x^\natural$ on the completion $\mathcal{H}$ of $\mathfrak{A}$. Then J is an involution on $\mathcal{H}$. Then, since $\mathfrak{A} \subseteq \mathfrak{A}'$, $\pi'(x)$ is defined for $x \in \mathfrak{A}$ and $\pi'(Jx) = J\pi(x)J$. Thus

$$J\mathfrak{L}(\mathfrak{A})J = \pi'(\mathfrak{A})'' \subseteq \mathfrak{L}(\mathfrak{A})'.$$

By (ii), δ_α is a closable densely defined operator on the completion $\mathcal{H}$ of $\mathfrak{A}$, and $\delta_\alpha^* \supseteq \delta_{\bar\alpha}$.

III.4.4.7 PROPOSITION. Let $\mathfrak{A}$ be a modular Hilbert algebra, with completion $\mathcal{H}$. Then there is a (generally unbounded) positive self-adjoint operator Δ on $\mathcal{H}$ such that Δ^α is the closure of δ_α for all $\alpha \in \mathbb{C}$.

The existence of Δ follows from Stone's Theorem (I.7.4.10): if U_t is the closure of δ_{it}, then $\{U_t\}$ is a strongly continuous one-parameter unitary group, and hence there is a self-adjoint operator H such that $U_t = e^{itH}$ for all t. Set $\Delta = e^H$. It is nontrivial but routine to check that Δ^α is the closure of δ_α for all α.

Δ is called the *modular operator* of $\mathfrak{A}$. As before, we have $J\Delta J = \Delta^{-1}$, $S = J\Delta^{1/2} = \Delta^{-1/2}J$, $F = \Delta^{1/2}J = J\Delta^{-1/2}$. J maps the Hilbert space $\mathcal{D}^\sharp$ isometrically onto $\mathcal{D}^\flat$. Since $J\mathfrak{A}^2 = \mathfrak{A}^2$ and $\mathfrak{A}^2$ is dense in $\mathcal{D}^\sharp$, $\mathfrak{A}^2$ is also dense in $\mathcal{D}^\flat$. Then by arguments as in III.4.4.3, one obtains:

III.4.4.8 THEOREM. If $\mathfrak{A}$ is a modular Hilbert algebra, then $\pi'(\mathfrak{A})$ generates $\mathfrak{L}(\mathfrak{A})'$ and $J\mathfrak{L}(\mathfrak{A})J = \mathfrak{L}(\mathfrak{A})'$. Also, the strongly continuous one-parameter group $\sigma_t = Ad\,\Delta^{it}$ of automorphisms of $\mathcal{L}(\mathcal{H})$ leaves $\mathfrak{L}(\mathfrak{A})$ and $\mathfrak{L}(\mathfrak{A})'$ invariant.

Thus, to prove III.4.3.2, it suffices to show that every von Neumann algebra M in standard form is isomorphic to $\mathfrak{L}(\mathfrak{A})$ for some modular Hilbert algebra $\mathfrak{A}$. Since $M \cong \mathfrak{L}(\mathfrak{A})$ for a left Hilbert algebra $\mathfrak{A}$, it suffices to prove:

III.4.4.9 THEOREM. Every left Hilbert algebra is equivalent to a modular Hilbert algebra.

Proving this theorem is, of course, the most difficult part of the theory. It is clear where the J and Δ come from for a left Hilbert algebra $\mathfrak{A}$: take the polar decomposition $S = J\Delta^{1/2}$ for the operator S. But a lot of work, including a considerable amount of complex analysis, is involved in finishing the argument. Even the specification of which elements of $\mathfrak{A}''$ are in the modular Hilbert algebra $\mathfrak{B}$ equivalent to $\mathfrak{A}$ is not obvious.

III.4.5 Corollaries of the Main Theorems

III.4.5.1 COROLLARY. Let M be a von Neumann algebra in standard form on a Hilbert space. Then M is linearly *-anti-isomorphic to M' under a map which is the identity on $\mathcal{Z}(M)$.

In fact, the map $x \mapsto j(x^*)$ is a linear anti-automorphism of $\mathcal{L}(\mathcal{H})$ carrying M onto M'.

There are examples of von Neumann algebras M (even II_1-factors with separable predual) such that M is not (linearly) *-isomorphic to M', i.e. M is not (linearly) *-anti-isomorphic to itself ([Con75a], [Con75b]).

Symmetric and Standard Forms

III.4.5.2 DEFINITION. Let M be a von Neumann algebra on $\mathcal{H}$. M is in *symmetric form* (sometimes called *hyperstandard form*) if there is an involution J of $\mathcal{H}$ such that $JMJ = M'$ and $JxJ = x^*$ for all $x \in \mathcal{Z}(M)$. A normal representation π is a *symmetric form representation* if $\pi(M)$ is in symmetric form.

III.4.5.3 PROPOSITION. Let M be a von Neumann algebra on $\mathcal{H}$, and z a central projection in M. Then M is in symmetric form on $\mathcal{H}$ if and only if zM and $(1-z)M$ are in symmetric form on $z\mathcal{H}$ and $(1-z)\mathcal{H}$ respectively.

III.4.5.4 THEOREM. Two symmetric form representations of a von Neumann algebra are unitarily equivalent.

The countably decomposable case follows immediately from III.2.6.6 and the following proposition. The general case can be derived from the countably decomposable case by an elementary but moderately involved argument (cf. [KR97b, 9.6.19-9.6.26]).

III.4.5.5 PROPOSITION. Let M be a countably decomposable von Neumann algebra in symmetric form on $\mathcal{H}$. Then M has a cyclic and separating vector and hence is in standard form.

PROOF: By III.2.6.11, there is a central projection z such that Mz has a cyclic vector on $z\mathcal{H}$ and $(1-z)M$ has a separating vector on $(1-z)\mathcal{H}$. But if ξ is a cyclic [*resp.* separating] vector, then $J\xi$ is a separating [*resp.* cyclic] vector. Apply III.2.6.10 and III.4.5.3.

An immediate consequence of the Tomita-Takesaki Theorem is:

III.4.5.6 COROLLARY. A standard form representation of a von Neumann algebra is in symmetric form.

Putting these results together, we obtain:

III.4.5.7 THEOREM. Let M be a von Neumann algebra. Then a representation of M is in symmetric form if and only if it is in standard form. M has such a representation, which is unique up to unitary equivalence.

PROOF: M has a standard form representation (III.2.6.5), which is symmetric (III.4.5.6); and any symmetric form representation is equivalent to this one by III.4.5.4.

The Commutation Theorem

The next theorem is the *commutation theorem for tensor products*:

III.4.5.8 THEOREM. Let M_1 and M_2 be von Neumann algebras on $\mathcal{H}_1$ and $\mathcal{H}_2$ respectively. Then
$$(M_1 \bar{\otimes} M_2)' = M_1' \bar{\otimes} M_2'$$
on $\mathcal{H}_1 \otimes \mathcal{H}_2$.

The modular Hilbert algebra formalism is especially suited to the proof of this theorem, which was one of Tomita's principal goals in developing the theory; in fact, the result was previously proved in the same way ([God51], [God54], [Seg53]) for semifinite von Neumann algebras, using Hilbert algebras (see also [Dix69a] for the origins of the general case). By routine arguments the result reduces to the case where M_1 and M_2 are in standard form, i.e. there are modular Hilbert algebras $\mathfrak{A}_1$ and $\mathfrak{A}_2$ with $M_i = \mathfrak{L}(\mathfrak{A}_i)$ ($i = 1, 2$). Then $\mathfrak{A} = \mathfrak{A}_1 \odot \mathfrak{A}_2$, with inner product
$$\langle x_1 \otimes x_2, y_1 \otimes y_2 \rangle = \langle x_1, y_1 \rangle_1 \langle x_2, y_2 \rangle_2,$$
product
$$(x_1 \otimes x_2)(y_1 \otimes y_2) = x_1 x_2 \otimes y_1 y_2,$$
and involution $(x \otimes y)^\sharp = x^\sharp \otimes y^\sharp$, is a left Hilbert algebra with $\mathfrak{L}(\mathfrak{A})$ naturally isomorphic to $\mathfrak{L}(\mathfrak{A}_1) \bar{\otimes} \mathfrak{L}(\mathfrak{A}_2)$. $\mathfrak{A}$ also has an obvious complex one-parameter group of tensor product automorphisms. Actually $\mathfrak{A}$ is a modular Hilbert algebra (axiom (v) is not obvious, but can be proved), but this is not really necessary for the proof; clearly $\mathfrak{A}$ and $\mathfrak{A}^2$ are contained in $\mathfrak{A}'$ and dense in $\mathcal{D}^\flat$, and thus $\pi'(\mathfrak{A})$ generates $\mathfrak{L}(\mathfrak{A})'$. But $\pi'(\mathfrak{A})''$ can be rather obviously identified with
$$(\pi'(\mathfrak{A}_1) \odot \pi'(\mathfrak{A}_2))'' = \mathfrak{L}(\mathfrak{A}_1)' \bar{\otimes} \mathfrak{L}(\mathfrak{A}_2)'.$$

Here is a useful variation of the Commutation Theorem. The proof uses the obvious fact that if M, N are von Neumann algebras on $\mathcal{H}$, then $(M \cup N)' = M' \cap N'$, and hence $(M \cap N)' = (M' \cup N')''$ (consider the commutants).

III.4.5.9 COROLLARY. Let M_i, N_i be von Neumann algebras on $\mathcal{H}_i$ ($i = 1, 2$). Then $(M_1 \bar{\otimes} M_2) \cap (N_1 \bar{\otimes} N_2) = (M_1 \cap N_1) \bar{\otimes} (M_2 \cap N_2)$.

PROOF:
$$[(M_1 \bar{\otimes} M_2) \cap (N_1 \bar{\otimes} N_2)]' = [(M_1 \bar{\otimes} M_2)' \cup (N_1 \bar{\otimes} N_2)']''$$
$$= [(M_1' \bar{\otimes} M_2') \cup (N_1' \bar{\otimes} N_2')]'' = [(M_1' \otimes I) \cup (N_1' \otimes I) \cup (I \otimes M_2') \cup (I \otimes N_2')]''$$
$$= ([[(M_1' \cup N_1')'' \otimes I] \cup [I \otimes (M_2' \cup N_2')'']])''$$
$$= (M_1 \cap N_1)' \bar{\otimes} (M_2 \cap N_2)' = [(M_1 \cap N_1) \bar{\otimes} (M_2 \cap N_2)]'.$$

On the other hand, III.4.5.9 implies III.4.5.8, since $(M_1 \bar{\otimes} M_2)'$ is clearly equal to $[M_1' \bar{\otimes} \mathcal{L}(\mathcal{H}_2)] \cap [\mathcal{L}(\mathcal{H}_1) \bar{\otimes} M_2']$.

The C*-algebra version of III.4.5.9 is false ([Hur79], [Kye84, 3.1]).

III.4.5.10 COROLLARY. Let M_i ($i = 1, 2$) be a von Neumann algebra on $\mathcal{H}_i$. Then $\mathcal{Z}(M_1 \bar{\otimes} M_2) = \mathcal{Z}(M_1) \bar{\otimes} \mathcal{Z}(M_2)$.

III.4.5.11 COROLLARY. Let M_i ($i = 1, 2$) be a von Neumann algebra on $\mathcal{H}_i$. If Z_i is a masa in M_i, then $Z_1 \bar{\otimes} Z_2$ is a masa in $M_1 \bar{\otimes} M_2$.

For the proof, note that if $Z \subseteq M$ are von Neumann algebras, then Z is a masa in M if and only if $Z = Z' \cap M$.

III.4.6 The Canonical Group of Outer Automorphisms and Connes' Invariants

Centralizer of a Weight

Using the modular automorphism group, a sensible definition of centralizer can be given for weights, and in particular for states:

III.4.6.1 DEFINITION. Let ϕ be a faithful normal semifinite weight on a von Neumann algebra M. The *centralizer* M^ϕ of ϕ is the centralizer (fixed-point algebra) M^{σ^ϕ}.

The centralizer M^ϕ of a weight ϕ is sometimes written M_ϕ, but this is easily confused with $\mathfrak{M}_\phi$.

M^ϕ is a von Neumann subalgebra of M which is, of course, invariant under σ^ϕ. There is an alternate characterization of M^ϕ, due to G. Pedersen and M. Takesaki [PT73] (cf. [Tak03a, VIII.2.6]):

III.4.6.2 THEOREM. Let ϕ be a faithful normal semifinite weight on a von Neumann algebra M, and $a \in M$. Then $a \in M^\phi$ if and only if

(i) a is a multiplier of $\mathfrak{M}_\phi$, i.e. $a\mathfrak{M}_\phi \subseteq \mathfrak{M}_\phi$ and $\mathfrak{M}_\phi a \subseteq \mathfrak{M}_\phi$.
(ii) $\phi(ax) = \phi(xa)$ for all $x \in \mathfrak{M}_\phi$.

The restriction of ϕ to M^ϕ is thus a trace; but it is not semifinite in general (cf. III.4.7.9).

If ϕ is a state, then $\mathfrak{M}_\phi = M$, and the condition in (ii) can be taken as the definition of M^ϕ.

III.4.6.3 If h is a positive operator affiliated with M^ϕ, then the formula

$$\psi(x) = \phi(hx) = \phi(xh) = \phi(h^{1/2} x h^{1/2})$$

defines a normal semifinite weight on M, which is faithful if h is invertible. In this case, we have

$$\sigma_t^\psi(x) = h^{it} \sigma_t^\phi(x) h^{-it}$$

for $x \in M$. This ψ is usually denoted ϕ_h.

A variation of this construction occurs if u is a partial isometry in M with $uu^* \in M^\phi$. Then $\psi(x) = \phi(uxu^*)$ defines a normal semifinite weight on M denoted ϕ_u. If ϕ is faithful, then $s(\phi_u) = u^*u$.

Both constructions have a common generalization using polar decomposition; we omit details since we will only need these two versions.

The Canonical Group of Outer Automorphisms

The modular automorphism group σ^ϕ depends on the choice of the weight ϕ. However, it is a remarkable and fundamental fact that this dependence is only up to inner automorphisms. The next theorem was proved by A. Connes [Con73b], using his "2×2 matrix trick."

THEOREM. Let M be a von Neumann algebra, and ϕ and ψ faithful normal semifinite weights on M with modular automorphism groups $\{\sigma_t^\phi\}$ and $\{\sigma_t^\psi\}$ respectively. Then the $\mathbb{R}$-actions σ^ϕ and σ^ψ are exterior equivalent (II.10.3.19). In particular, for each $t \in \mathbb{R}$, σ_t^ϕ and σ_t^ψ differ by an inner automorphism of M.

PROOF: Define a weight ω on $M_2(M)$ by

$$\omega\left(\begin{bmatrix} a & b \\ c & d \end{bmatrix}\right) = \phi(a) + \psi(d)$$

called the *balanced weight* of ϕ and ψ. The weight ω is obviously faithful and normal, and is also semifinite since $\mathfrak{N}_\omega = \begin{bmatrix} \mathfrak{N}_\phi & \mathfrak{N}_\psi \\ \mathfrak{N}_\phi & \mathfrak{N}_\psi \end{bmatrix}$. Also, it is easy to see that

$$\sigma_t^\omega(diag(x,y)) = diag(\sigma_t^\phi(x), \sigma_t^\psi(y))$$

for all $x, y \in M$, $t \in \mathbb{R}$. Apply II.10.3.20.

III.4.6.4 If $\mathrm{Out}(M) = \mathrm{Aut}(M)/\mathrm{Inn}(M)$ is the "outer automorphism group" of M, and $\delta(t)$ is the image of σ_t^ϕ in $\mathrm{Out}(M)$, then $\{\delta(t)\}$ is independent of the choice of ϕ, and is called the *canonical (one-parameter) group of outer automorphisms* of M. By [Con73b], $\delta(\mathbb{R})$ is contained in the center of $\mathrm{Out}(M)$.

Actually, any cross section of δ is a modular automorphism group, at least if M has separable predual: if σ is any action of $\mathbb{R}$ on M whose image in $\mathrm{Out}(M)$ is δ, there is a faithful normal semifinite weight ψ on M with $\sigma = \sigma^\psi$ (cf. III.4.7.5(ii)).

The invariant $T(M)$

III.4.6.5 Thus, if M is a von Neumann algebra, the set

$$T(M) = ker(\delta) = \{t \in \mathbb{R} : \sigma_t^\phi \text{ is an inner automorphism of } M\}$$

is an invariant of M independent of the choice of ϕ. This is one of Connes' two fundamental invariants.

If the spectrum of Δ_ϕ is contained in $\{0\} \cup \{\lambda^n : n \in \mathbb{Z}\}$, then $T(M)$ contains every t for which $\lambda^{it} = 1$, i.e.

$$T(M) \supseteq \left\{\frac{2\pi n}{\log \lambda} : n \in \mathbb{Z}\right\}.$$

Thus, if ϕ is a trace, so that $\Delta_\phi = 1$, $T(M) = \mathbb{R}$. The converse is also true if M has separable predual (but not in general):

III.4.6.6 THEOREM. If M is a von Neumann algebra with separable predual, then $T(M) = \mathbb{R}$ if and only if M is semifinite.

The converse uses the nontrivial fact, a generalization of Stone's theorem, that if α is an action of $\mathbb{R}$ on a von Neumann algebra N with separable predual, and α_t is inner for all t, then there is a strongly continuous one-parameter group $\{u_t\}$ in $\mathcal{U}(N)$ with $\alpha_t = Ad\, u_t$ for all t, i.e. there is a strongly continuous cross section for the map

$$\mathbb{R} \to \mathrm{Inn}(N) \cong \mathcal{U}(N)/\mathcal{U}(\mathcal{Z}(N)).$$

There is then an (in general unbounded) positive invertible h affiliated with N^α such that $u_t = h^{it}$ for all t. This result was obtained by Kadison [Kad65] (cf. [Bar54], [Han77]) in the factor case and R. Kallman [Kal71] and C. Moore [Moo76] in general.

If there is an invertible $h \geq 0$ affiliated with M^ϕ such that $u_t = h^{it}$ satisfies $\sigma_t^\phi = Ad\, u_t$, then $\psi(\cdot) = \phi(h\cdot)$ defines a faithful normal semifinite trace on M (some details need to be checked).

An interesting corollary of this result is:

III.4.6.7 COROLLARY. Every von Neumann algebra with separable predual and a nonzero Type III part has an outer automorphism.

In fact, if M is a Type III factor with separable predual, then $T(M)$ has measure 0 in $\mathbb{R}$ [KR97b, 14.4.16]. There is a countably decomposable Type III factor M (with nonseparable predual) with $T(M) = \mathbb{R}$ [KR97b, 14.4.20].

III.4.6.8 PROPOSITION. Let $\mathcal{R}_\lambda$ $(0 < \lambda < 1)$ be the Powers factor of type III_λ (III.3.1.7). Then

$$T(\mathcal{R}_\lambda) = \left\{\frac{2\pi n}{\log \lambda} : n \in \mathbb{Z}\right\}.$$

PROOF: The argument is similar to the proof of III.3.1.3. Write $(\mathcal{R}_\lambda, \phi) = \overline{\bigotimes}_{i \in \mathbb{N}}(\mathbb{M}_2, \phi_\lambda)$ as in III.3.1.7. Set

$$M^{(n)} = \left[\overline{\bigotimes}_{i=1}^{n}(\mathbb{M}_2, \phi_\lambda)\right] \otimes 1, \quad N^{(n)} = 1 \otimes \left[\overline{\bigotimes}_{i=n+1}^{\infty}(\mathbb{M}_2, \phi_\lambda)\right] = M^{(n)'} \cap \mathcal{R}_\lambda$$

and let θ_n be the unique ϕ-invariant conditional expectation from $\mathcal{R}_\lambda$ onto $M^{(n)}$. Suppose $t \in \mathbb{R}$, and $\sigma_t^\phi = Ad\, u$ for some $u \in \mathcal{U}(\mathcal{R}_\lambda)$. We have that $\sigma_t^{\phi_\lambda} = Ad\, v$ for some $v \in \mathcal{U}(\mathbb{M}_2)$, and

$$u = v \otimes \cdots \otimes v \otimes w$$

in the decomposition $\mathcal{R}_\lambda \cong M^{(n)} \bar{\otimes} N^{(n)}$, for some $w \in \mathcal{U}(N^{(n)})$; and hence

$$\theta_n(u) = v \otimes \cdots \otimes v \otimes \alpha 1$$

for some $\alpha \in \mathbb{C}$. We have that $\|\theta_n(u) - u\|_\phi \to 0$; if $0 < \epsilon < 1$, choose n so that $\|\theta_n(u) - u\|_\phi < \epsilon$. Comparing $\theta_n(u)$ and $\theta_{n+1}(u)$, we get that

$$\|v - \beta 1\|_{\phi_\lambda} < 2\epsilon/(1-\epsilon)$$

for some $\beta \in \mathbb{C}$. This implies that $v \in \mathbb{C}1$, $\sigma_t^{\phi_\lambda} = id$, $\lambda^{it} = 1$.

Once the machinery of modular theory is set up, this is probably the simplest argument to show that the Powers factors are Type III and mutually nonisomorphic.

Essentially the same argument shows:

III.4.6.9 PROPOSITION. Let $M = \overline{\otimes}_i(M_i, \phi_i)$ be an ITPFI factor, with eigenvalue list $\{\alpha_k^{(j)}\}$. Then

$$T(M) = \left\{ t \in \mathbb{R} : \sum_{j=1}^{\infty} \left(1 - \left|\sum_k (\alpha_k^{(j)})^{1+it}\right|\right) < \infty \right\}.$$

It follows easily for an ITPFI factor M, $T(M) = \{\frac{2\pi}{\log \lambda} : \lambda \in \rho(M)\}$ (III.3.1.15). More generally, if M is the von Neumann algebra of an equivalence relation $L^\infty(R, \mu)$ (III.3.2.17), then $T(M) = 2\pi/\log \rho(R)$, where $\rho(R)$ is Krieger's invariant (III.3.2.21). It follows from [Con73b] that every countable subgroup of $\mathbb{R}$, and also many uncountable subgroups, occur as $T(M)$ for an ITPFI factor M.

The invariant $S(M)$

III.4.6.10 The results above show that the spectrum of Δ_ϕ plays an important role. However, this spectrum is not independent of ϕ; for example, if ϕ_λ on $\mathbb{M}_2$ is as in III.3.1.7, then $\sigma(\Delta_{\phi_\lambda}) = \{1, \lambda, \lambda^{-1}\}$. The appropriate variation is to define, for a factor M,

$$S(M) = \bigcap_\phi \sigma(\Delta_\phi)$$

where ϕ runs over all faithful normal semifinite weights (states suffice in the countably decomposable case). This is Connes' second fundamental invariant. (In many cases, $S(M)$ can be realized by a single faithful normal semifinite weight; cf. III.4.8.6.)

There are several alternate descriptions of this invariant. Two of the most useful are:

III.4.6.11 THEOREM. Let M be a factor, and ϕ a faithful normal semifinite weight on M. Then

(i) $S(M) = \cap_p \sigma(\Delta_{\phi_p})$, where p runs over all nonzero projections in M^ϕ and ϕ_p is the weight on pMp given by restricting ϕ.
(ii) If $\hat{\mathbb{R}}$ is identified with the multiplicative group $\mathbb{R}_+$ by $\langle \lambda, t \rangle = \lambda^{it}$, then $S(M) \cap \mathbb{R}_+$ is the Connes spectrum (III.3.2.5) $\Gamma(\sigma^\phi)$.

An important corollary of (ii) is

III.4.6.12 COROLLARY. If M is a factor, then $S(M) \cap \mathbb{R}_+$ is a (multiplicative) subgroup.

Type Classification of Type III Factors

III.4.6.13 Since $S(M)$ is a closed subset of $[0, \infty)$ and $S(M) \cap \mathbb{R}_+$ is a multiplicative group, there are only the following possibilities for $S(M)$: $\{1\}$, $\{0, 1\}$, $\{0\} \cup \{\lambda^n : n \in \mathbb{Z}\}$ for some λ, $0 < \lambda < 1$, or $[0, \infty)$. M is semifinite if and only if $S(M) = \{1\}$.

If M is a Type III factor, then we say

M is *Type III_0* if $S(M) = \{0, 1\}$.
M is *Type III_λ* $(0 < \lambda < 1)$ if $S(M) = \{0\} \cup \{\lambda^n : n \in \mathbb{Z}\}$.
M is *Type III_1* if $S(M) = [0, \infty)$

III.4.6.14 It is easily seen from the results of this section that if M is a Type III ITPFI factor, then $S(M) = r_\infty(M)$, and hence these type definitions are consistent with the ones in III.3.1.14.

III.4.7 The KMS Condition and the Radon-Nikodym Theorem for Weights

There is an interesting and important connection between a weight ϕ and its modular automorphism group σ^ϕ.

III.4.7.1 Let M be a von Neumann algebra, σ an $\mathbb{R}$-action on M, ϕ a weight on M, and $\beta > 0$. Then ϕ satisfies the *KMS condition* or *modular condition* (or is a *KMS weight*) for σ at inverse temperature $-\beta$ if $\phi = \phi \circ \sigma_t$ for all t and, for all $x, y \in \mathfrak{N}_\phi$, there is a function f, bounded and continuous on the strip $\{z \in \mathbb{C} : 0 \leq Im(z) \leq \beta\}$ and analytic on $\{z \in \mathbb{C} : 0 < Im(z) < \beta\}$, such that, for $t \in \mathbb{R}$,

$$f(t) = \phi(\sigma_t(x)y), \quad f(t + i\beta) = \phi(y\sigma_t(x)).$$

A *KMS state* is a KMS weight which is a state.

This condition was first introduced (for states) by mathematical physicists R. Kubo [Kub57], P. Martin and J. Schwinger [MS59], and the connection with von Neumann algebras and modular theory discovered by R. Haag, N.

Hugenholtz, and M. Winnink [HHW67]. See the survey article [Kas76] for a discussion of KMS states and their connection with physics.

By rescaling the action σ to $\sigma'_t = \sigma_{-t/\beta}$, we may and will restrict to the case $\beta = 1$, i.e. "KMS weight" will mean "KMS weight at inverse temperature -1."

The fundamental observation of Takesaki [Tak70] and Winnink (extended to the case of weights) was:

III.4.7.2 THEOREM. Let ϕ be a faithful normal semifinite weight on a von Neumann algebra M. Then ϕ is a KMS weight for σ^ϕ, and σ^ϕ is the only continuous $\mathbb{R}$-action for which this is true.

If $\mathfrak{B}$ is a modular Hilbert algebra with $\mathfrak{L}(\mathfrak{B}) = M$ and modular operator Δ_ϕ, and $\eta, \zeta \in \mathfrak{B}$, then

$$f(z) = \langle \Delta_\phi^{-iz}\eta, \zeta^\sharp \rangle = \langle \zeta, \Delta_\phi^{1-i\bar{z}}\eta^\sharp \rangle$$

is entire and has the right values for η, ζ at $z = t$ and $z = t + i$. A limiting argument gives an f for arbitrary $x, y \in \mathfrak{N}_\phi$.

To prove the uniqueness of σ^ϕ, suppose ϕ is a KMS weight for a one-parameter automorphism group ρ. Then on the Hilbert space $\mathcal{H}_\phi$, there is a strongly continuous one-parameter unitary group $(U(t))$ such that $\rho_t = \mathrm{Ad}(U(t))$ since ϕ is invariant under ρ; by Stone's theorem, $U(t) = K^{it}$ for a self-adjoint operator K. If $H = e^K$, it must be shown that $H = \Delta$. This is a lengthy argument, which we omit. See [Tak03a, VIII.1.2] for details of both parts of the proof.

The Radon-Nikodym Theorem for Weights

The analog of III.2.3.3 for (normal semifinite) weights is false (see, for example, [Str81, 6.6-6.8]), so a correct version of the Radon-Nikodym theorem for weights requires considerable care. The definitive version comes out of the proof of III.4.6.3.

III.4.7.3 Given two faithful normal semifinite weights ϕ and ψ on a von Neumann algebra M, the proof of III.4.6.3 provides a strongly continuous path (u_t) of unitaries in M such that

$$\sigma_t^\psi = (\mathrm{Ad}\, u_t) \circ \sigma_t^\phi$$

for all t. This u_t is denoted $(\mathrm{D}\psi : \mathrm{D}\phi)_t$, and is called the *Radon-Nikodym derivative* of ψ with respect to ϕ at t. This Radon-Nikodym derivative satisfies the cocycle identity:

$$u_{t+s} = u_t \sigma_t^\phi(u_s) \ (s, t \in \mathbb{R}).$$

If $x \in M$, the element $u_t \sigma_t^\phi(x) = \sigma_t^\psi(x) u_t$ is denoted $\sigma_t^{\psi,\phi}(x)$. We have

$$\sigma^{\psi,\phi}_{t+s}(x) = \sigma^{\psi,\phi}_t(x)\sigma^{\psi,\phi}_s(x)$$

for $t, s \in \mathbb{R}$ (but note that

$$\sigma^{\psi,\phi}_t(xy) \neq \sigma^{\psi,\phi}_t(x)\sigma^{\psi,\phi}_t(y)$$

for $x, y \in M$ in general).

The definition of $(D\psi : D\phi)$ also works if ψ is not faithful. In this case, each u_t is an isometry with $u_t u_t^*$ equal to the support projection of ψ.

III.4.7.4 By applying the KMS condition to the balanced weight, we also have the *analytic extension property*: for each $x \in \mathfrak{N}_\phi \cap \mathfrak{N}_\psi^*$ and $y \in \mathfrak{N}_\psi \cap \mathfrak{N}_\phi^*$, there is an f, bounded and continuous on $\{z \in \mathbb{C} : 0 \leq Im(z) \leq 1\}$ and analytic on $\{z \in \mathbb{C} : 0 < Im(z) < 1\}$, such that, for all $t \in \mathbb{R}$,

$$f(t) = \psi(\sigma^{\psi,\phi}_t(x)y), \ f(t+i) = \phi(y\sigma^{\psi,\phi}_t(x))\ .$$

recovering ψ from ϕ

There is also a converse, obtained by extending u_t to an analytic function on the strip $\{z : 0 < Im(z) < 1\}$:

III.4.7.5 THEOREM. [CONNES-RADON-NIKODYM THEOREM FOR WEIGHTS] [Con73a] Let ϕ be a faithful normal semifinite weight on a von Neumann algebra M.

(i) If ψ is another faithful normal semifinite weight on M, then there is a unique strongly continuous path $(u_t) = (D\psi : D\phi)_t$ of unitaries in M, with the analytic extension property, with

$$\sigma^\psi_t(x) = u_t \sigma^\phi_t(x) u_t^* \text{ and } u_{t+s} = u_t \sigma^\phi_t(u_s)$$

for all $x \in M$, $t, s \in \mathbb{R}$.

(ii) Conversely, if (u_t) is a strongly continuous path of unitaries in M with $u_{t+s} = u_t \sigma^\phi_t(u_s)$ for all $t, s \in \mathbb{R}$, then there is a unique faithful normal semifinite weight ψ on M with $u_t = (D\psi : D\phi)_t$ for all t.

The Radon-Nikodym derivatives satisfy the expected rules: if ϕ, ψ, ω are faithful normal semifinite weights on M, and h is an invertible positive operator affiliated with M^ϕ (III.4.6.3), then

$$(D\omega : D\phi)_t = (D\omega : D\psi)_t (D\psi : D\phi)_t$$

$$(D\phi : D\psi)_t = (D\psi : D\phi)_t^{-1}$$

$$(D\phi_h : D\phi)_t = h^{it}\ .$$

The weights of the form ϕ_h for $h \geq 0$ affiliated with M^ϕ can also be characterized:

III.4.7.6 THEOREM. [PT73] Let ϕ and ψ be normal semifinite weights on a von Neumann algebra M. The following are equivalent:

(i) $\psi = \psi \circ \sigma_t^\phi$ for all t.
(ii) $(D\psi : D\phi)_t \in M^\phi$ for all t.
(iii) There is an $h \geq 0$ affiliated with M^ϕ with $\psi = \phi_h$.

If the conditions of this theorem are satisfied, then ψ *commutes with* ϕ. If ψ is also faithful and commutes with ϕ, then σ^ϕ and σ^ψ are commuting actions of $\mathbb{R}$ (the converse is false in general).

Weights and Conditional Expectations

There is a close connection between the modular automorphism group of a weight and the existence of conditional expectations onto subalgebras which preserve the weight:

III.4.7.7 THEOREM. Let M be a von Neumann algebra and N a von Neumann subalgebra, and let ϕ be a faithful normal semifinite weight on M such that $\psi = \phi|_N$ is semifinite. Then the following are equivalent:

(i) There is a conditional expectation $\theta : M \to N$ with $\phi = \phi \circ \theta$.
(ii) There is a faithful normal conditional expectation $\theta : M \to N$ with $\phi = \phi \circ \theta$.
(iii) N is invariant under σ^ϕ, i.e. $\sigma_t^\phi(N) = N$ for all t.

[Recall from II.6.10.10 that a conditional expectation as in (i), if it exists, is unique; (ii) asserts that such a conditional expectation is necessarily faithful and normal.]

Let $\mathcal{H}_\phi$ and $\mathcal{H}_\psi$ be the GNS Hilbert spaces and $\mathfrak{A} = \iota(\mathfrak{N}_\phi \cap \mathfrak{N}_\phi^*)$ and $\mathfrak{B} = \iota(\mathfrak{N}_\psi \cap \mathfrak{N}_\psi^*)$ the corresponding left Hilbert algebras, dense in $\mathcal{H}_\phi$ and $\mathcal{H}_\psi$ respectively. Regard $\mathcal{H}_\psi$ as a subspace of $\mathcal{H}_\phi$. Then by II.6.10.10 any θ as in (i) is necessarily induced by the orthogonal projection P from $\mathcal{H}_\phi$ onto $\mathcal{H}_\psi$. To prove (i) $\implies$ (iii), a simple computation shows that P is permutable with Δ_ϕ, and hence $\Delta_\phi^{it}\mathfrak{B} = \mathfrak{B}$ since $P\mathfrak{A} = \mathfrak{B}$. To show (iii) $\implies$ (ii), it must be shown that $P\mathfrak{A} = \mathfrak{B}$ and that $\pi_\psi(x) = P\pi_\phi(x)P$ for $x \in M$. The details can be found in [Tak03a, IX.4.2].

III.4.7.8 COROLLARY. Let M be a semifinite von Neumann algebra, and τ a faithful normal semifinite trace on M. Let N be von Neumann subalgebra of M. If $\tau|_N$ is semifinite, then there is a normal conditional expectation from M onto N. In particular, if M is finite and countably decomposable, and N is any von Neumann subalgebra of M, then there is a normal conditional expectation from M onto N.

Since M^ϕ is invariant under σ^ϕ, we obtain:

III.4.7.9 COROLLARY. ([Com71]; cf. [Str81, 10.9]) Let ϕ be a faithful normal semifinite weight on a von Neumann algebra M. The following are equivalent:

(i) ϕ is strictly semifinite (III.2.2.26).
(ii) $\phi|_{M^\phi}$ is semifinite.
(ii) There is a normal conditional expectation $\theta : M \to M^\phi$ such that $\phi = \phi \circ \theta$.

III.4.8 The Continuous and Discrete Decompositions of a von Neumann Algebra

The canonical group of outer automorphisms can be used to give a canonical description of a general properly infinite von Neumann algebra as a crossed product of a semifinite von Neumann algebra by a one-parameter group of automorphisms. This decomposition is closely related in spirit to (and was motivated by) the standard decomposition of a nonunimodular group as a semidirect product of a unimodular group by a subgroup of $\mathbb{R}_+$. In the case of a Type III$_\lambda$ factor, $0 \leq \lambda < 1$, the continuous decomposition can be replaced by a discrete decomposition (crossed product by $\mathbb{Z}$). The continuous and discrete decompositions allow many structural questions about Type III factors to be reduced to a study of semifinite von Neumann algebras and their automorphisms.

III.4.8.1 Let M be a von Neumann algebra, ϕ a faithful normal semifinite weight on M. Set $N = M \bar{\rtimes}_{\sigma^\phi} \mathbb{R}$. If ψ is another faithful normal semifinite weight, then σ^ψ is exterior equivalent to σ^ϕ, and hence up to a standard isomorphism N is independent of the choice of ϕ.

III.4.8.2 PROPOSITION. The dual weight $\hat{\phi}$ (III.3.2.9) is a faithful normal semifinite trace on N; thus N is semifinite.

III.4.8.3 If $\hat{\sigma}^\phi$ is the dual action of $\hat{\mathbb{R}} \cong \mathbb{R}$, then by Takesaki duality (III.3.2.8) $N \bar{\rtimes}_{\hat{\sigma}^\phi} \mathbb{R}$ is isomorphic to $M \bar{\otimes} \mathcal{L}(L^2(\mathbb{R}))$, and hence to M itself if M is properly infinite. This description of M as the crossed product of N by $\mathbb{R}$ is called the *continuous decomposition* of M.

If ψ is another faithful normal semifinite weight on M, and $M \bar{\rtimes}_{\sigma^\psi} \mathbb{R}$ is identified with N in the standard way, then $\hat{\sigma}^\psi$ is seen to be exterior equivalent to $\hat{\sigma}^\phi$, and hence the crossed product of N by $\mathbb{R}$ is canonical in a sense. However, although the identification of this crossed product with $M \bar{\otimes} \mathcal{L}(L^2(\mathbb{R}))$ by Takesaki duality is canonical, the isomorphism

$$M \bar{\otimes} \mathcal{L}(L^2(\mathbb{R})) \cong M$$

if M is properly infinite requires choices; so the isomorphism $N \bar{\rtimes}_{\hat{\sigma}^\phi} \mathbb{R} \cong M$ is not natural. This decomposition of M is nonetheless well defined up to isomorphism. In fact, the decomposition is unique up to conjugacy in a suitable sense [CT77] (cf. III.4.8.8). The dual action $\hat{\sigma}^\phi$ is usually denoted θ in this abstract setting to remove its notational dependency on ϕ.

III.4.8.4 PROPOSITION. *The automorphism $\hat{\sigma}_t^\phi$ scales the trace $\hat{\phi}$ on N by a factor of e^t.*

The Discrete Decomposition of a Type III Factor

If M is a Type III_λ factor, $0 \leq \lambda < 1$, then there is a closely related discrete decomposition. We first consider the case $0 < \lambda < 1$.

III.4.8.5 DEFINITION. If M is a factor of Type III_λ, $0 < \lambda < 1$, a *generalized trace* on M is a faithful normal semifinite weight ϕ such that $S(M) = \sigma(\Delta_\phi)$ and $\phi(1) = +\infty$.

III.4.8.6 THEOREM. *If M is a factor of Type III_λ, $0 < \lambda < 1$, then M has a generalized trace. If ϕ and ψ are generalized traces on M, then there is an inner automorphism α such that ψ is proportional to $\phi \circ \alpha$.*

III.4.8.7 THEOREM. *If M is a factor of Type III_λ, $0 < \lambda < 1$, with generalized trace ϕ, then the centralizer M^ϕ of ϕ is a Type II_∞ factor, and there is an automorphism α of M^ϕ, unique up to conjugacy, scaling the trace of M^ϕ by λ, such that $M \cong M^\phi \bar{\rtimes}_\alpha \mathbb{Z}$.* This decomposition is called the *discrete decomposition* of M.

The situation with Type III_0 factors is similar, but with an important difference:

III.4.8.8 THEOREM. *If M is a factor of Type III_0, then there is a von Neumann algebra N of Type II_∞, with diffuse center (no minimal central projections), and an automorphism α of N, such that $M \cong N \bar{\rtimes}_\alpha \mathbb{Z}$.*

There is also a uniqueness statement in this case, but it is more complicated to state since a decomposition can be "cut down" by a central projection (the uniqueness statement says that two decompositions have isomorphic cutdowns). The decomposition $M \cong N \bar{\rtimes}_\alpha \mathbb{Z}$ is called the *discrete decomposition* of M.

III.4.8.9 In the III_λ case ($0 < \lambda < 1$), there is a close connection between the discrete and continuous decompositions. In this case, if $t = -2\pi/\log \lambda$, then the action σ^ϕ of $\mathbb{R}$ is really an action ρ of $\mathbb{T} = \mathbb{R}/t\mathbb{Z}$. The discrete decomposition is then really $(M \bar{\rtimes}_\rho \mathbb{T}) \bar{\rtimes}_{\hat{\rho}} \mathbb{Z}$.

The connection between the continuous and discrete decompositions in the III_0 case is more complicated and less natural, made by building a "flow under a ceiling function." See [Tak03a] for details.

III.4.8.10 There is no discrete decomposition in the III_1 case in general; in fact, most III_1 factors cannot be written as the crossed product of a semifinite von Neumann algebra by a single automorphism.

III.4.8.1 The Flow of Weights.

III.4.8.1 Consider the continuous decomposition $M = N \bar{\rtimes}_\theta \mathbb{R}$. If
$$Z = \mathcal{Z}(N) \cong L^\infty(X, \mathcal{B}, \mu),$$
then the restriction of θ to Z gives a flow ($\mathbb{R}$-action) on $(X, \mathcal{B}, \mu)$ which is canonically associated with M, called the *flow of weights* on M.

There is an important equivalent description of the flow of weights on a von Neumann algebra which more clearly demonstrates the functoriality of the construction and explains the name. Our description is brief, and we limit discussion only to the case of separable predual; details can be found in [Tak03a] or [Str81].

III.4.8.2 A normal semifinite weight ϕ on a properly infinite von Neumann algebra M (with separable predual) is of *infinite multiplicity* if it is unitarily equivalent to $\phi \otimes \mathrm{Tr}$ on $M \cong M \bar{\otimes} \mathcal{L}(\mathcal{H})$. Equivalently, ϕ is of infinite multiplicity if and only if M^ϕ is properly infinite. Since $\mathrm{Tr} \cong \mathrm{Tr} \otimes \mathrm{Tr}$ on $\mathcal{L}(\mathcal{H}) \cong \mathcal{L}(\mathcal{H}) \bar{\otimes} \mathcal{L}(\mathcal{H})$, any weight of the form $\phi \otimes \mathrm{Tr}$ is of infinite multiplicity. Write $W_\infty(M)$ for the set of normal semifinite weights on M of infinite multiplicity.

III.4.8.3 Define an equivalence relation and a preorder on $W_\infty(M)$ by setting $\psi \precsim \phi$ if there is a partial isometry $u \in M$ with $u^*u = s(\psi)$, $uu^* \in M^\phi$, and $\psi = \phi_u$, i.e. $\psi(x) = \phi(uxu^*)$ for all $x \in M$. If $uu^* = s(\phi)$, then $\phi = \psi_{u^*}$ and also $\phi \precsim \psi$; say then $\phi \sim \psi$. In fact, $\phi \sim \psi$ if and only if $\phi \precsim \psi$ and $\psi \precsim \phi$. Denote the equivalence class of ϕ by $[\phi]$, and let $\mathfrak{B}$ be the set of equivalence classes. The preorder $\precsim$ drops to a partial order on $\mathfrak{B}$, also denoted $\precsim$. There is a natural addition on $\mathfrak{B}$:
$$[\phi] + [\psi] = [\phi' + \psi']$$
where $\phi' \sim \phi$, $\psi' \sim \psi$, and $s(\phi') \perp s(\psi')$. There is also a natural action of the multiplicative group $\mathbb{R}_+$ on $\mathfrak{B}$: $\lambda[\phi] = [\lambda\phi]$.

III.4.8.4 The set $\mathfrak{B}$ with its addition and partial order is order-isomorphic to the lattice of countably decomposable projections in a commutative von Neumann algebra A, and the action of $\mathbb{R}$ on $\mathfrak{B}$ comes from automorphisms of A. This group of automorphisms is, however, not continuous in general. If Z is the subalgebra on which the action is continuous (i.e. $x \in Z$ if $\lambda \mapsto \lambda x$ is σ-weakly continuous), then the induced action of $\mathbb{R}$ on Z is exactly the flow of weights. The action of $\mathbb{R}$ on all of A is called the *global flow of weights*; the flow of weights is sometimes called the *smooth flow of weights*.

Integrable and Dominant Weights

III.4.8.5 A (normal semifinite) weight ϕ for which $[\phi] \in Z$ is called an *integrable weight*. A weight ϕ is a *dominant weight* if it satisfies the equivalent conditions of the next theorem:

III.4.8.6 THEOREM. Let ϕ be a normal semifinite weight on a properly infinite von Neumann algebra M with separable predual. The following are equivalent:

(i) $s(\phi)$ has central support 1 and $\lambda\phi \sim \phi$ for all $\lambda > 0$.
(ii) ϕ is integrable and $\psi \precsim \phi$ for every integrable weight ψ.
(iii) ϕ is equivalent to the second dual weight (III.3.2.9) of a faithful normal semifinite weight.
(iv) Under the continuous decomposition $M \cong N \bar{\rtimes}_\theta \mathbb{R}$, ϕ is equivalent to the dual weight of a faithful normal semifinite trace on N.

Note that (iii) $\Longrightarrow$ (iv) comes from III.4.8.2.
From characterizations (ii) and (iii), we obtain:

III.4.8.7 COROLLARY. Let M be a properly infinite von Neumann algebra with separable predual. Then M has a faithful dominant weight, and any two dominant weights on M are equivalent.

III.4.8.8 If ϕ is a dominant weight, then M^ϕ is semifinite and isomorphic to the N of the continuous decomposition of M; thus $M \cong M^\phi \bar{\rtimes}_\theta \mathbb{R}$ for an action θ of $\mathbb{R}$ which can be explicitly described. This concrete realization of the continuous decomposition of M is used to prove the uniqueness of the continuous decomposition up to conjugacy (III.4.8.3).

III.5 Applications to Representation Theory of C*-Algebras

III.5.1 Decomposition Theory for Representations

The theory of von Neumann algebras allows a systematic study of the representation theory of C*-algebras, associating to a representation π of a C*-algebra A the von Neumann algebras $\pi(A)''$ and $\pi(A)'$. Note that the subrepresentations (invariant subspaces) of π correspond exactly to the projections in $\pi(A)'$. To avoid trivial complications which cause annoying verbosity in statements, we will assume throughout this section that all representations are *nondegenerate*.

Quasi-equivalence

III.5.1.1 Recall that two representations π and ρ of a C*-algebra A, on Hilbert spaces $\mathcal{H}_\pi$ and $\mathcal{H}_\rho$, are *(unitarily) equivalent*, denoted $\pi \approx \rho$, if there is a unitary $U \in \mathcal{L}(\mathcal{H}_\pi, \mathcal{H}_\rho)$ with $\rho(x) = U\pi(x)U^*$ for all $x \in A$.

There are some other useful related notions:

III.5.1.2 DEFINITION. Let π and ρ be two (nondegenerate) representations of a C*-algebra A, on Hilbert spaces $\mathcal{H}_\pi$ and $\mathcal{H}_\rho$.

(i) An *intertwiner* of π and ρ is an operator $T \in \mathcal{L}(\mathcal{H}_\pi, \mathcal{H}_\rho)$ with

$$T\pi(x) = \rho(x)T$$

for all $x \in A$. Denote by $Int(\pi, \rho)$ the set of intertwiners of π and ρ.

(ii) The representations π and ρ are *disjoint*, written $\pi \perp \rho$, if no nonzero subrepresentation of π is equivalent to a subrepresentation of ρ.

(iii) The representation π is *subordinate* to ρ, written $\pi \preceq \rho$, if no nonzero subrepresentation of π is disjoint from ρ (i.e. every nonzero subrepresentation of π contains a nonzero subrepresentation equivalent to a subrepresentation of ρ).

(iv) The representations π and ρ are *quasi-equivalent*, written $\pi \sim \rho$, if $\pi \preceq \rho$ and $\rho \preceq \pi$.

It is easily checked that $\perp$ is symmetric, $\preceq$ is reflexive and transitive, and $\sim$ is an equivalence relation. $Int(\pi, \rho)$ is a closed subspace of $\mathcal{L}(\mathcal{H}_\pi, \mathcal{H}_\rho)$, and

$$Int(\rho, \pi) = Int(\pi, \rho)^* = \{T^* : T \in Int(\pi, \rho)\}.$$

We have that $\pi \approx \rho$ if and only if there is a unitary in $Int(\pi, \rho)$. At the other extreme, it is easily seen that $\pi \perp \rho$ if, and only if, the only element of $Int(\pi, \rho)$ is 0 (if $T \in Int(\pi, \rho)$ has polar decomposition $T = U|T|$, then $|T| \in \pi(A)'$ and $U \in Int(\pi, \rho)$).

III.5.1.3 PROPOSITION. Let π be a representation of A on $\mathcal{H}$, and p, q projections in $\pi(A)'$, with central supports z_p, z_q. Let ρ, σ be the subrepresentations of π on $p\mathcal{H}$ and $q\mathcal{H}$ respectively. Then

(i) $Int(\rho, \sigma) = q\pi(A)'p$.
(ii) $\rho \perp \sigma$ if and only if $z_p \perp z_q$.
(iii) $\rho \preceq \sigma$ if and only if $z_p \leq z_q$.
(iv) $\rho \sim \sigma$ if and only if $z_p = z_q$.

The proof of (i) is a simple calculation, and the other parts follow immediately from (i) and III.1.1.4.

III.5.1.4 COROLLARY. If π is a factor representation of A, then any two nonzero subrepresentations of π are quasi-equivalent, and in particular any nonzero subrepresentation of π is quasi-equivalent to π. If ρ is another factor representation of A, then either $\pi \sim \rho$ or $\pi \perp \rho$.

A straightforward maximality argument then shows:

III.5.1.5 PROPOSITION. Let π, ρ be representations of a C*-algebra A, on $\mathcal{H}_\pi$ and $\mathcal{H}_\rho$. Then there are unique central projections p and q in $\pi(A)''$ and $\rho(A)''$ respectively, such that

(i) $\pi|_{p\mathcal{H}_\pi}$ and $\rho|_{q\mathcal{H}_\rho}$ are quasi-equivalent.
(ii) $\pi|_{(1-p)\mathcal{H}_\pi}$ is disjoint from ρ.
(iii) $\rho|_{(1-q)\mathcal{H}_\rho}$ is disjoint from π.

Combining this with III.2.2.8, we obtain:

III.5.1.6 PROPOSITION. Let π, ρ be (nondegenerate) representations of a C*-algebra A. Then the following are equivalent:

(i) π and ρ are quasi-equivalent.
(ii) Suitable amplifications of π and ρ are equivalent.
(iii) π and ρ have the same kernel J, and the identity map on A/J extends to an algebraic *-isomorphism from $\pi(A)''$ onto $\rho(A)''$.
(iv) π and ρ have the same kernel J, and the restriction of the σ-weak [σ-strong] topologies from $\pi(A)''$ and $\rho(A)''$ to A/J coincide.

Type of a Representation

III.5.1.7 A (nondegenerate) representation π of a C*-algebra A is said to be *(pure) Type I [resp. Type II, Type III, Type II_1, Type II_∞]* if $\pi(A)''$ is Type I [*resp.* Type II, Type III, Type II_1, Type II_∞]. By III.2.6.12, π is Type I [*resp.* Type II, Type III] if and only if $\pi(A)'$ is Type I [*resp.* Type II, Type III] (but this is not true for Type II_1 and Type II_∞ representations); and if π is Type I [*resp.* Type II, Type III, Type II_1, Type II_∞], so is every subrepresentation of π. The type of a representation is invariant under quasi-equivalence.

Every representation π can be uniquely decomposed as

$$\pi_I \oplus \pi_{II_1} \oplus \pi_{II_\infty} \oplus \pi_{III}$$

where the components are of pure type. Any factor representation is of pure type, of course.

Every C*-algebra has many Type I representations, e.g. direct sums of irreducible representations. Not every C*-algebra has Type II or Type III representations; see IV.1 for a discussion.

III.5.1.8 It would be possible to say that a representation π is finite [*resp.* semifinite, properly infinite, purely infinite] if $\pi(A)''$ is finite [*resp.* semifinite, etc.] However, it is customary, and more useful, to say that a representation π of A is finite [*resp.* properly infinite], or *of finite* [*resp. properly infinite*] *multiplicity*, if $\pi(A)'$ is finite [*resp.* properly infinite]. (For "semifinite" or "purely infinite" it makes no difference whether $\pi(A)'$ or $\pi(A)''$ is used.) This definition is more in line with standard terminology: π is of finite multiplicity if and only if π is not equivalent to a proper subrepresentation, if and only if it does not contain an infinite sequence of mutually orthogonal, equivalent nonzero subrepresentations. A representation π is of properly infinite multiplicity if and only if it is an infinite amplification of another representation; in fact, a representation of properly infinite multiplicity is equivalent to a countably infinite amplification of itself (III.1.3.5).

III.5.1.9 A representation π is *multiplicity-free* if any two subrepresentations of π on orthogonal subspaces are disjoint. Equivalently, π is multiplicity-free if $\pi(A)'$ is commutative. A multiplicity-free representation is Type I. Conversely, every Type I representation is quasi-equivalent to a multiplicity-free representation.

Similarly, we say that a representation π of A has *uniform multiplicity* n if $\pi(A)'$ is of Type I_n. Such a representation is an n-fold amplification of a multiplicity-free representation. Any Type I representation is canonically a direct sum of representations of uniform multiplicity (III.1.5.12).

A Type II or Type III representation cannot be reduced down to a multiple of a multiplicity-free representation; any subrepresentation can be written as a multiple of a smaller subrepresentation. Even a Type II representation of finite multiplicity is thus "infinite" in the sense that it is a direct sum of infinitely many mutually quasi-equivalent representations.

C*-algebras having only Type I representations are particularly well-behaved and important, and their structure will be analyzed in detail in IV.1.

Direct Integral Decompositions of Representations

From the results of III.1.6 we get a direct integral theory for representations of *separable* C*-algebras on *separable* Hilbert spaces. As a result, we get a canonical decomposition of a general representation as a direct integral of mutually disjoint factor representations, and a non-canonical decomposition as a direct integral of irreducible representations.

III.5.1.10 Let π be a (nondegenerate) representation of a C*-algebra A on a separable Hilbert space $\mathcal{H}$, and let Z be a commutative von Neumann subalgebra of $\pi(A)'$. Then there is a corresponding standard measure space (X, μ) and a direct integral decomposition

$$\mathcal{H} \cong \int_X^{\oplus} \mathcal{H}_x \, d\mu(x)$$

such that the operators in Z' are decomposable; in particular, for each $a \in A$, the operator $\pi(a)$ is decomposable and can be written

$$\pi(a) = \int_X^{\oplus} \pi(a)_x \, d\mu(x).$$

III.5.1.11 If, for each (or almost all) x, we write $\pi_x(a) = \pi(a)_x \in \mathcal{L}(\mathcal{H}_x)$, it turns out that, if A is separable, then π_x is a representation of A for almost all x, and we may sensibly write

$$\pi = \int_X^\oplus \pi_x \, d\mu(x)$$

as a "direct integral of representations."

III.5.1.12 If Z contains $\mathcal{Z}(\pi(A)'')$, then almost all the representations π_x are factor representations. If Z is a masa in $\pi(A)'$, then almost all of the π_x are irreducible.

III.5.1.13 If E is a measurable subset of X, then

$$\pi_E = \int_E^\oplus \pi_x \, d\mu(x)$$

is a subrepresentation of π. If Z is contained in $\mathcal{Z}(\pi(A)'')$, and E, F are disjoint subsets of X, then $\pi_E \perp \pi_F$.

The most natural choice for Z is $\mathcal{Z}(\pi(A)'')$. We summarize the properties of the corresponding direct integral decomposition, called the *central decomposition*:

III.5.1.14 THEOREM. Let π be a representation of a separable C*-algebra A on a separable Hilbert space $\mathcal{H}$. Then there is a canonical direct integral decomposition

$$\mathcal{H} = \int_X^\oplus \mathcal{H}_x \, d\mu(x), \quad \pi = \int_X^\oplus \pi_x \, d\mu(x)$$

such that

(i) π_x is a factor representation for almost all x.
(ii) $\pi_x \perp \pi_y$ for almost all (x, y); if E, F are disjoint measurable sets, then $\pi_E \perp \pi_F$.
(iii) Every subrepresentation ρ of π is of the form

$$\rho = \int_X^\oplus \rho_x \, d\mu(x)$$

where ρ_x is a subrepresentation of π_x.

III.5.1.15 The central decomposition of π respects the Type I, Type II$_1$, Type II$_\infty$, and Type III parts of π. In particular, a representation of pure type centrally decomposes into a direct integral of (a.e.) factor representations of the same type.

III.5.1.16 If π is multiplicity-free, then the π_x in the central decomposition are almost all irreducible, so π has a canonical decomposition as a direct integral of (a.e.) mutually inequivalent irreducible representations.

A general representation π can also be decomposed as a direct integral of (a.e.) irreducible representations, but the decomposition requires the choice of a masa in $\pi(A)'$ and is thus highly non-canonical, especially if π is not Type I. The representations π_x are also not necessarily mutually inequivalent, e.g. when π is a multiple of an irreducible representation. It turns out that if π has no Type I part, the π_x are (a.e.) mutually inequivalent (cf. IV.1.5.1(iii)$\Longrightarrow$(v)). However, in this case there is no natural relationship between the representation π and the irreducible representations π_x; for example, a different masa can yield a decomposition in which the irreducible representations appearing are (a.e.) inequivalent to any of the π_x.

III.5.1.17 There is an abstract (space-free) decomposition theory for states (cyclic representations) of C*-algebras, valid also in the nonseparable case; see [Sak71].

III.5.2 The Universal Representation and Second Dual

If A is any C*-algebra, then there is a "universal" representation ω of A, which contains every cyclic representation of A as a subrepresentation, and such that $\omega(A)''$ is isometrically isomorphic to the second dual A^{**}.

The Universal Representation

III.5.2.1 DEFINITION. Let A be a C*-algebra. The *universal representation* ω of A is

$$\bigoplus_{\phi \in A_+^*} \pi_\phi$$

where A_+^* is the set of positive linear functionals on A.

The universal representation of A is obviously faithful and nondegenerate.

III.5.2.2 The Hilbert space of the universal representation is highly nonseparable; in fact, its dimension is $Card(A^*)$. If π is any cyclic representation of A, then there are at least $2^{\aleph_0}$ mutually orthogonal subrepresentations of ω equivalent to π. The representation ω is not even quasi-equivalent to a representation on a separable Hilbert space except in the unusual case where A^* is separable, which implies that A is separable and Type I [Tom63].

III.5.2.3 If $A \neq \mathbb{C}$, then it can be shown that ω is equivalent to the subrepresentation

$$\bigoplus_{\phi \in \mathcal{S}(A)} \pi_\phi$$

where $\mathcal{S}(A)$ is the state space of A. [It is only necessary to show that each π_ϕ occurs $2^{\aleph_0}$ times in this subrepresentation; but if ϕ is a convex combination of ϕ_1 and ϕ_2, then π_{ϕ_1} and π_{ϕ_2} are equivalent to subrepresentations of π_ϕ (cf. the proof of III.2.6.7).]

III.5.2.4 Every cyclic representation of A is obviously equivalent to a subrepresentation of ω. Since every (nondegenerate) representation of A is a direct sum of cyclic representations, every (nondegenerate) representation of A is quasi-equivalent to a subrepresentation of ω. Any (nondegenerate) representation of A on a Hilbert space of dimension $\leq 2^{\aleph_0}$ is actually equivalent to a subrepresentation of ω.

The Second Dual

III.5.2.5 Let A be a C*-algebra, and set $M = \omega(A)''$. Then M is a von Neumann algebra containing $\omega(A) \cong A$ as a σ-weakly dense *-subalgebra. Any normal linear functional on M restricts to a bounded linear functional on A; thus we obtain a linear map $\rho : M_* \to A^*$.

III.5.2.6 PROPOSITION. The map ρ is an isometric isomorphism from M_* onto A^*.

PROOF: The Kaplansky Density Theorem (I.9.1.3) implies that ρ is isometric. Every state on A is a vector state for $\omega(A)$, hence extends to a normal state on M; thus ρ is surjective.

III.5.2.7 COROLLARY. The dual map $\rho^* : (M_*)^* = M \to A^{**}$ is an isometric isomorphism.

III.5.2.8 Thus there is a multiplication on A^{**}, extending the multiplication on A (naturally identified with a subspace of A^{**}), making A^{**} into a C*-algebra (which is automatically a W*-algebra by III.2.4.2). This multiplication can be alternately obtained as left or right Arens multiplication [Are51] on A^{**}.

III.5.2.9 If π is an arbitrary (nondegenerate) representation of A, then π is quasi-equivalent to a subrepresentation of ω, and hence $\pi(A)''$ is isomorphic to a direct summand of A^{**}. There is thus a normal *-homomorphism from A^{**} onto $\pi(A)''$, which extends the map

$$\pi : \omega(A) \cong A \to \pi(A)$$

i.e. the representation π extends to a normal representation of A^{**}. Since A is σ-weakly dense in A^{**}, the extension is unique and does not depend on how π is realized (up to quasi-equivalence) as a subrepresentation of ω.

III.5.2.10 If $\phi: A \to B$ is a bounded linear map between C*-algebras, then by general considerations $\phi^{**}: A^{**} \to B^{**}$ is a normal linear map of the same norm as ϕ. It is easily verified that

ϕ^{**} is injective if and only if ϕ is injective (in general, we have that $ker(\phi^{**}) = [ker(\phi)]^{**}$).
ϕ^{**} is surjective if and only if ϕ is surjective.
ϕ^{**} is positive if and only if ϕ is positive.
ϕ^{**} is completely positive if and only if ϕ is completely positive.
ϕ^{**} is a *-homomorphism if and only if ϕ is a *-homomorphism.
If $B \subseteq A$ (so $B^{**} \subseteq A^{**}$), ϕ^{**} is a conditional expectation if and only if ϕ is a conditional expectation.

III.5.2.11 If J is an ideal in A, then J^{**} is a weakly closed ideal in A^{**}, and is thus of the form pA^{**} for some central projection $p \in A^{**}$. If A^{**} is identified with $\omega(A)''$, then p is the projection onto the essential subspace of $\omega|_J$, and if (h_λ) is an increasing approximate unit for J, then p is the supremum (σ-strong limit) of (h_λ). We can identify the quotient

$$A^{**}/J^{**} \cong (1-p)A^{**}$$

with $(A/J)^{**}$, i.e. there is a natural isomorphism

$$A^{**} \cong J^{**} \oplus (A/J)^{**}.$$

III.5.2.12 If A is a C*-algebra, then $(M_n(A))^{**}$ is naturally isomorphic to $M_n(A^{**})$. However, it is not true that $(A \otimes B)^{**} \cong A^{**} \otimes B^{**}$ in general if A, B are infinite-dimensional. First of all, there are some rather subtle questions about which cross norms are appropriate to make proper sense of such an identification; these will be discussed in IV.2.3.1. And even if the right cross norms are taken, the left side of the equation is much larger than the right side in general.

Borel Functional Calculus Revisited

III.5.2.13 Let X be a compact Hausdorff space, and let $B(X)$ be the set of bounded Borel measurable functions on X. With pointwise operations and supremum norm, $B(X)$ becomes a commutative C*-algebra. Identifying $C(X)^*$ with the set of Radon measures on X by the Riesz Representation Theorem, each element f of $B(X)$ defines a bounded linear functional on $C(X)^*$ by $\mu \mapsto \int f \, d\mu$, and hence an element of $C(X)^{**}$. It is easily checked that this is an injective *-homomorphism from $B(X)$ to $C(X)^{**}$. The W*-algebra $l^\infty(X)$ of all bounded functions on X is a quotient (direct summand) of $C(X)^{**}$ [$l^\infty(X) \cong \pi(C(X))''$, where π is the sum of all irreducible representations of $C(X)$], and the composite embedding of $B(X)$ into $l^\infty(X)$ is the natural one. So the image of $B(X)$ in $C(X)^{**}$ is not all of $C(X)^{**}$ unless X is countable.

If π is a unital representation of $C(X)$ on a Hilbert space $\mathcal{H}$, then π extends to a representation of $B(X)$ on $\mathcal{H}$, and $\pi(B(X)) \subseteq \pi(C(X))''$. If $\mathcal{H}$ is separable, or more generally if π is quasi-equivalent to a representation with a cyclic vector, then the image of $B(X)$ in $\mathcal{L}(\mathcal{H})$ is all of $\pi(C(X))''$, since then $\pi(C(X))'' \cong L^\infty(X,\mu)$ for some Radon measure μ; but the image of $B(X)$ in $\mathcal{L}(\mathcal{H})$ is not all of $\pi(C(X))''$ in general. In particular, if T is a normal operator in $\mathcal{L}(\mathcal{H})$, and $A = C^*(T,I) \cong C(\sigma(T))$, then the identity representation of A extends to a *-homomorphism from $B(\sigma(T))$ to $\mathcal{L}(\mathcal{H})$ which coincides with the Borel functional calculus described in I.4.3.2. If B is a Borel subset of $\sigma(T)$, then the image of χ_B in $\mathcal{L}(\mathcal{H})$ is the spectral projection $E_B(T)$.

The Universal Normal Representation and Second Dual of a von Neumann Algebra

III.5.2.14 DEFINITION. Let M be a von Neumann algebra. The *universal normal representation* ν of M is

$$\bigoplus_{\phi \in (M_*)_+} \pi_\phi.$$

The universal normal representation of M is a direct summand of the universal representation of M, and is faithful and normal; hence $\nu(M)'' = \nu(M) \cong M$.

III.5.2.15 As with any representation, ν extends to a normal representation of M^{**}, i.e. a *-homomorphism from M^{**} onto $\nu(M)$. If $\nu(M)$ is identified with M, we get a normal *-homomorphism $\theta : M^{**} \to M$ which is the identity on M, i.e. a homomorphic normal conditional expectation. It is easily checked that this map is the dual of the natural inclusion of M_* into M^*.

Polar Decomposition in a C*-Algebra

Using representations, and in particular the universal representation, we can obtain a simple result about polar decomposition related to those of II.3.2.

III.5.2.16 PROPOSITION. Let A be a C*-algebra, $x \in A$, and $x = u|x|$ its polar decomposition in A^{**}. If $B = \overline{x^*Ax}$ is the hereditary C*-subalgebra of A generated by $|x|$, and $b \in B$, then $ub \in A$, and $(ub)^*(ub) = b^*b$.
PROOF: Suppose first that $b = p(|x|)$ for a polynomial p without constant term. Then $p(|x|) = |x|q(|x|)$ for a polynomial q, and $q(|x|) \in \tilde{A} \subseteq A^{**}$, so

$$up(|x|) = u|x|q(|x|) = xq(|x|) \in A.$$

Next suppose $f \in C_o(\sigma(|x|))$. Then f is the uniform limit on $\sigma(|x|)$ of a sequence (p_n) of polynomials without constant term, and so $uf(|x|) =$

$\lim_n up_n(|x|) \in A$. In particular, $u|x|^\alpha \in A$ for all $\alpha > 0$. Since $(|x|^\alpha)$ is an approximate unit for B, if $b \in B$ we have $ub = \lim_{\alpha \to 0} u|x|^\alpha b \in A$. Also, u^*u is a unit for x^*x, and hence also for $|x|^\alpha$, in A^{**}; so

$$(ub)^*(ub) = \lim_{\alpha \to 0} b^* u^* u |x|^\alpha b = \lim_{\alpha \to 0} b^* |x|^\alpha b = b^*b.$$

An alternate proof of II.3.2.1 can be easily given using I.5.2.4 and the argument of III.5.2.16. In fact, we obtain a generalization (II.3.2.1 is the special case $y = a^\alpha$, $z = a^{1/2-\alpha}$):

III.5.2.17 PROPOSITION. Let A be a C*-algebra, $x, y \in A$, $z \in \overline{Ay^*} = \overline{Ayy^*} \subseteq A$. If $x^*x \leq y^*z^*zy$, then there is a $u \in \overline{Ay^*}$, $\|u\| \leq \|z\|$, with $x = uy$.

PROOF: By I.5.2.4 there is a $w \in A^{**}$ with $x = wzy$ and $\|w\| \leq 1$. We show that $u = wz \in \overline{Ay^*} \subseteq A$. As in the proof of III.5.2.16 we have $wzyy^* = xy^* \in Ay^*$, so $wzp(yy^*) \in Ay^*$ for all polynomials p without constant term, and hence $wz(yy^*)^{1/n} \in \overline{Ay^*}$ for all n. But $z \in \overline{Ay^*}$, so $z(yy^*)^{1/n} \to z$, and hence $u(yy^*)^{1/n} \to u \in \overline{Ay^*}$.

III.5.2.18 We can also give a simple proof of II.3.2.3 using III.5.2.16. Let $x \in A$, and let $x = u(x^*x)^{1/2}$ be its polar decomposition in A^{**}. Then $y = u(x^*x)^{1/6} \in A$, and $y^*y = (x^*x)^{1/3}$, so

$$y(y^*y) = u(x^*x)^{1/6}(x^*x)^{1/3} = u(x^*x)^{1/2} = x.$$

For uniqueness, if $x = zz^*z$, then $x^*x = (z^*zz^*)(zz^*z) = (z^*z)^3$, so $z^*z = (x^*x)^{1/3}$, and similarly $zz^* = (xx^*)^{1/3}$. Thus, if $z = v(z^*z)^{1/2}$ is the polar decomposition, then the source and range projections of v are the same as for u, and

$$x = z(z^*z) = v(x^*x)^{1/6}(x^*x)^{1/3} = v(x^*x)^{1/2}$$

so by uniqueness of polar decomposition $v = u$, $z = y$.

III.5.2.19 There is nothing magical about using the universal representation in III.5.2.16; any faithful representation π of A can be used instead. But it is not transparent that the element ub of A is independent of the choice of π. It is not difficult to prove this directly, but it follows immediately from the universal representation construction: if $\bar{\pi}$ is the canonical extension of π to A^{**}, and $\pi(x) = v|\pi(x)|$ is the polar decomposition of $\pi(x)$ in $\pi(A)''$, then by uniqueness of polar decomposition we have $v = \bar{\pi}(u)$, so

$$v\pi(b) = \bar{\pi}(u)\pi(b) = \pi(ub).$$

IV

Further Structure

IV.1 Type I C*-Algebras

There is an important class of C*-algebras called the Type I C*-algebras, which can be characterized in several apparently different ways, which is precisely the class of C*-algebras with "tractable" representation theory and which therefore has traditionally been regarded as the class of "reasonable" C*-algebras whose structure can be "understood." Although in recent years the structure theory of C*-algebras has been largely divorced from representation theory and there is now a much larger class of C*-algebras whose structure can be "understood" in the modern sense, Type I C*-algebras are still very important; furthermore, the structure of Type I C*-algebras can be rather precisely described, forming a model for the structure and classification of more general classes of C*-algebras.

IV.1.1 First Definitions

A von Neumann algebra is Type I if it contains an abelian projection of central support 1 (III.1.4), or equivalently if every central summand (W*-quotient) contains a nonzero abelian projection. There are (at least) two reasonable C*-analogs:

IV.1.1.1 DEFINITION. Let A be a C*-algebra. An element $x \in A$ is *abelian* if the hereditary C*-subalgebra $[x^*Ax]^-$ is commutative.
A is *internally Type I* if every quotient of A contains a nonzero abelian element.
A is *bidual Type I* if A^{**} is a Type I von Neumann algebra, i.e. if $\pi(A)''$ is a Type I von Neumann algebra for every representation π of A.

These conditions, and several others, turn out to be equivalent, although (bidual Type I) $\Rightarrow$ (internally Type I) is a deep theorem (IV.1.5.7, IV.1.5.8); a C*-algebra satisfying these conditions is called a *Type I C*-algebra*. Both definitions in IV.1.1.1 have been used in the literature as the definition of a Type I C*-algebra; it will be convenient to give them separate provisional names.

IV.1.1.2 PROPOSITION. An internally type I C*-algebra is bidual Type I.

PROOF: Let A be internally type I, and π a non-Type-I representation of A. Passing to a subrepresentation corresponding to a central projection, we may assume $\pi(A)''$ has no Type I summand, i.e. contains no nonzero abelian projections. Let $J = ker(\pi)$. Then A/J is internally Type I, so replacing A by A/J we may assume π is faithful and identify A with $\pi(A)$. Let x be an abelian element of norm 1 in A, and $y = f_{1/2}(x^*x)$ (II.3.4.11). If p is the support projection for y in A'', then $z = f_{1/4}(x^*x)$ is a unit for p, and hence $pA''p$ is contained in the σ-weak closure of zAz, which is commutative. Thus p is a nonzero abelian projection in A'', a contradiction.

IV.1.1.3 Because of decomposition theory (III.1.6), a separable C*-algebra A is bidual Type I if and only if $\pi(A)''$ is a Type I factor, for every factor representation π of A, or, alternatively, if and only if every factor representation of A is a multiple of an irreducible representation.

This statement is also true if A is nonseparable, but is much harder to prove, requiring the machinery of the proof of Glimm's theorem (cf. IV.1.5.8).

IV.1.1.4 Every commutative C*-algebra is a Type I C*-algebra (by either definition), and it turns out that the Type I C*-algebras are built up from commutative C*-algebras by means of standard operations.

IV.1.1.5 Note that a Type I von Neumann algebra is not in general a Type I C*-algebra. For example, $\mathcal{L}(\mathcal{H})$ is not a Type I C*-algebra if $\mathcal{H}$ is infinite-dimensional (the Calkin algebra has no nonzero abelian elements). In fact, the only von Neumann algebras which are Type I C*-algebras are finite Type I von Neumann algebras of bounded degree, finite direct sums of matrix algebras over commutative von Neumann algebras. This slight ambiguity of terminology is unfortunate, but well established.

IV.1.1.6 DEFINITION. Let A be a C*-algebra.
A is *Type I_0* if the abelian elements of A generate A as a C*-algebra.
A is *antiliminal* if A contains no nonzero abelian elements, i.e. if every nonzero hereditary C*-subalgebra of A is noncommutative.

Any quotient of a Type I_0 C*-algebra is Type I_0. In particular, a Type I_0 C*-algebra is internally Type I. A C*-algebra is internally Type I if and only if it has no antiliminal quotients.

The next result follows immediately from II.6.1.9.

IV.1.1.7 PROPOSITION. Let A be a C*-algebra, $x \in A$. Then x is abelian if and only if rank $(\pi(x)) \leq 1$ for every $\pi \in \hat{A}$.

IV.1.1.8 It follows that x is abelian $\Leftrightarrow x^*x$ is abelian $\Leftrightarrow xx^*$ is abelian, and if x is abelian and $y \precsim x$ (II.3.4.3), then y is abelian. If A is a C*-algebra and $I_0(A)$ is the C*-subalgebra of A generated by the abelian elements of A, then it follows from II.3.2.7 that $I_0(A)$ is an ideal of A which is a Type I_0 C*-algebra. (In fact, $I_0(A)$ is the closed linear span of the abelian elements of A, since if $x, y \in A$ and x is abelian, then xy and yx are abelian.)

IV.1.1.9 EXAMPLES. Let $T = \mathbb{N} \cup \{\infty\}$ (or $T = [0, \infty]$). Define C*-subalgebras C_1, C_2, C_3 of $\mathbb{M}_2$ as follows:

$$C_1 = \left\{ \begin{bmatrix} \lambda & 0 \\ 0 & 0 \end{bmatrix} : \lambda \in \mathbb{C} \right\}$$

$$C_2 = \left\{ \begin{bmatrix} \lambda & 0 \\ 0 & \lambda \end{bmatrix} : \lambda \in \mathbb{C} \right\}$$

$$C_3 = \left\{ \begin{bmatrix} \lambda & 0 \\ 0 & \mu \end{bmatrix} : \lambda, \mu \in \mathbb{C} \right\}$$

Then, for $i = 1, 2, 3$, let

$$A_i = \{f \in C(T, \mathbb{M}_2) : f(\infty) \in C_i\}.$$

All three C*-algebras are internally Type I; A_1 and A_3 are Type I_0, but A_2 is not;

$$I_0(A_2) = \{f \in C(T, \mathbb{M}_2) : f(\infty) = 0\}.$$

IV.1.1.10 DEFINITION. Let A be a C*-algebra. A *subcomposition series* in A is a set (J_α) of closed ideals of A, indexed by (all) ordinals α, such that $J_\beta \subseteq J_\alpha$ if $\beta < \alpha$ and such that $J_\alpha = [\cup_{\beta < \alpha} J_\beta]^-$ if α is a limit ordinal. There is a γ such that $J_\alpha = J_\gamma$ for all $\alpha > \gamma$; J_γ is called the *limit* of the series. The series is called a *composition series* for A if its limit is A.

IV.1.1.11 PROPOSITION. If a C*-algebra A has a composition series (J_α) such that $J_{\alpha+1}/J_\alpha$ is internally type I for all α, then A is internally type I.
PROOF: Let I be an ideal in A, and β be the largest α such that $J_\alpha \subseteq I$. Then

$$(J_{\beta+1} + I)/I \cong J_{\beta+1}/(J_{\beta+1} \cap I)$$

is a nonzero ideal of A/I which is isomorphic to a quotient of $J_{\beta+1}/J_\beta$ and thus contains an abelian element.

IV.1.1.12 Set $I_0 = I_0(A)$, and inductively define $I_{\alpha+1}$ to be the inverse image in A of $I_0(A/I_\alpha)$. For limit α set $I_\alpha = [\cup_{\beta<\alpha} I_\beta]^-$. Then (I_α) is a subcomposition series in A with $I_{\alpha+1}/I_\alpha$ of Type I_0 for all α. Let $I(A)$ be the limit. Then $I(A)$ is an ideal of A which is an internally Type I C*-algebra, and $A/I(A)$ is antiliminal. In particular, A is internally Type I if and only if $I(A) = A$, i.e. A has a composition series in which successive quotients are Type I_0.

IV.1.2 Elementary C*-Algebras

IV.1.2.1 DEFINITION. A C*-algebra A is *elementary* if $A \cong \mathcal{K}(\mathcal{H})$ for some Hilbert space $\mathcal{H}$.

The separable elementary C*-algebras are the finite-dimensional matrix algebras and $\mathbb{K}$. Elementary C*-algebras are simple C*-algebras and have the most transparent structure of all C*-algebras. They are the "building blocks" for all Type I C*-algebras. An elementary C*-algebra is Type I_0.

An elementary C*-algebra has an obvious "identity representation," which is irreducible. It is often convenient to identify the C*-algebra with its image under this representation.

IV.1.2.2 PROPOSITION. Let A be an elementary C*-algebra. Then every nondegenerate representation of A is unitarily equivalent to a multiple of the identity representation. In particular, up to unitary equivalence A has only one irreducible representation, and every representation of A is a factor representation.

PROOF: The argument is very similar to that of III.1.5.3 (and the result follows immediately from III.1.5.3 if A is finite-dimensional). Let $A \cong \mathcal{K}(\mathcal{H})$ and let π be a nondegenerate representation of A on $\tilde{\mathcal{H}}$. Let $\{e_{ij} : i, j \in \Omega\}$ be a set of matrix units in A. Fix $j \in \Omega$ and ξ_j a unit vector in $e_{jj}\mathcal{H}$, and for each i let $\xi_i = e_{ij}\xi_j$. Then $\{\xi_i\}$ is an orthonormal basis for $\mathcal{H}$. Let $\mathcal{H}' = \pi(e_{jj})\tilde{\mathcal{H}}$, and $\{\eta_k : k \in \Lambda\}$ an orthonormal basis for $\mathcal{H}'$. For $i \in \Omega$ and $k \in \Lambda$, set $\zeta_{ik} = \pi(e_{ij})\eta_k$. Then $\{\zeta_{ik}\}$ is an orthonormal basis for $\tilde{H}$ by nondegeneracy of π, and $\zeta_{ik} \mapsto \xi_i \otimes \eta_k$ defines a unitary $U : \tilde{\mathcal{H}} \to \mathcal{H} \otimes \mathcal{H}'$ satisfying $U\pi(x)U^* = x \otimes 1$ (it is enough to check this for x a matrix unit, where it is obvious).

IV.1.2.3 Conversely, a separable C*-algebra with only one irreducible representation (up to unitary equivalence) must be elementary (IV.1.3.5, IV.1.5.1). It was a long-standing open question whether this is true also for nonseparable C*-algebras (the *Naimark problem*). Recently, C. Akemann and N. Weaver [AW04] constructed a nonelementary (non-Type I, unital, necessarily simple) nonseparable C*-algebra with only one irreducible representation, using an additional set-theoretic axiom (the "diamond principle"), at least showing that a positive answer in the nonseparable case cannot be proved in ZFC.

A C*-algebra for which every representation is a factor representation is simple and Type I, hence elementary.

IV.1.2.4 PROPOSITION. Let A be a C*-subalgebra of $\mathcal{K}(\mathcal{H})$ acting irreducibly on $\mathcal{H}$. Then $A = \mathcal{K}(\mathcal{H})$.

PROOF: If $\mathcal{H}$ is finite-dimensional, the result is obvious from the Bicommutant Theorem (I.9.1.1) (or just by some elementary linear algebra). For the general case, note that if $x \in \mathcal{K}(\mathcal{H})_+$, then $f_\epsilon(x)$ (II.3.4.11) is finite-rank for all $\epsilon > 0$. Thus the finite-rank operators in A are dense in A, and in particular A contains nonzero (finite-rank) projections. Let p be a minimal nonzero projection in A, and ξ a unit vector in $p\mathcal{H}$. Then pAp acts irreducibly on $p\mathcal{H}$ (II.6.1.9), so by the first remark $pAp = \mathcal{L}(p\mathcal{H})$, and thus $p\mathcal{H}$ is one-dimensional by minimality of p. Choose an orthonormal basis $\{\xi_i\}$ for $\mathcal{H}$ with $\xi_k = \xi$ for some fixed k, and let $\{e_{ij}\}$ be the corresponding matrix units in $\mathcal{K}(\mathcal{H})$ (so $e_{kk} = p$). For each i there is an $x \in A$ with $x\xi$ close to ξ_i; then xp is close to e_{ik}. It follows that A contains all the e_{ij}.

IV.1.2.5 COROLLARY. Let A be a C*-algebra, π an irreducible representation of A on $\mathcal{H}$. If $\pi(A) \cap \mathcal{K}(\mathcal{H}) \neq 0$, then $\pi(A) \supseteq \mathcal{K}(\mathcal{H})$.

PROOF: Let $J = \pi^{-1}(\mathcal{K}(\mathcal{H}))$. Then J is a closed ideal of A, and $\pi|_J$ is irreducible (II.6.1.6).

IV.1.2.6 COROLLARY. Let A be a simple C*-algebra. Then A is (internally) Type I if and only if A is elementary.

Indeed, if x is a nonzero abelian element of A and π a (necessarily faithful) irreducible representation of A on $\mathcal{H}$, then $\pi(x) \in \mathcal{K}(\mathcal{H})$ (IV.1.1.7). Hence $\pi(A) \subseteq \mathcal{K}(\mathcal{H})$ by simplicity, so $\pi(A) = \mathcal{K}(\mathcal{H})$.

IV.1.2.7 An argument similar to the proof of IV.1.2.4 shows that any C*-subalgebra of an elementary C*-algebra is a C*-direct sum of elementary C*-algebras.

IV.1.3 Liminal and Postliminal C*-Algebras

IV.1.3.1 DEFINITION. Let A be a C*-algebra, π an irreducible representation of A on a Hilbert space $\mathcal{H}$.
π is a *CCR* representation of A if $\pi(A) \subseteq \mathcal{K}(\mathcal{H})$ (and hence $\pi(A) = \mathcal{K}(\mathcal{H})$ by IV.1.2.5).
π is a *GCR* representation of A if $\pi(A) \cap \mathcal{K}(\mathcal{H}) \neq 0$ (so $\pi(A) \supseteq \mathcal{K}(\mathcal{H})$ by IV.1.2.5).
A C*-algebra A is *CCR* [*resp. GCR*] if every irreducible representation of A is CCR [resp. GCR].
A is *liminal* if it is CCR; A is *postliminal* if it has a composition series (J_α) such that $J_{\alpha+1}/J_\alpha$ is liminal for all α.
A (primitive) ideal J of A is *A-CCR* [*resp. A-GCR*] if J is the kernel of a CCR [resp. GCR] representation of A.

CCR stands for *completely continuous representations*, completely continuous being an old synonym for *compact*; *GCR* stands for *generalized CCR*. The terms CCR and GCR were introduced by Kaplansky and Glimm (as well as the term *NGCR*, synonymous with *antiliminal*); the original definition of GCR was different from the one given here (identical with our definition of *postliminal*). It is convenient for our exposition to have the different definitions, which turn out to be equivalent (IV.1.5.7).

The terms *liminal* [*postliminal, antiliminal*], sometimes written *liminary* [*postliminary, antiliminary*], are English "translations" of the French terms *liminaire*, meaning "prefatory," *postliminaire*, and *antiliminaire*, used in [Dix69b].

In some references such as [Sak71], GCR C*-algebras are called *smooth* (cf. IV.1.5.12).

IV.1.3.2 PROPOSITION. If a C*-algebra A is Type I_0, then A is liminal. If A is internally Type I, then A is postliminal.

PROOF: The first statement follows immediately from IV.1.1.7, and the second from IV.1.1.12.

A postliminal C*-algebra is internally Type I (IV.1.4.26), i.e. "internally Type I" is the same as "postliminal."

IV.1.3.3 EXAMPLES.

(i) Any elementary C*-algebra is liminal.
(ii) If every irreducible representation of A is finite-dimensional, then A is liminal. In particular, any commutative C*-algebra is liminal, and the examples of IV.1.1.9 are liminal. A_2 is liminal but not Type I_0.
(iii) $\tilde{\mathbb{K}}$ is postliminal but not liminal.

IV.1.3.4 If J is a primitive ideal in a C*-algebra A, then J is an A-CCR ideal if and only if A/J is elementary. In this case, J is a maximal ideal. An A-GCR ideal is an A-CCR ideal if and only if it is maximal. Thus every primitive ideal in a liminal C*-algebra is maximal, i.e. the primitive ideal space of a liminal C*-algebra is T_1. (By Example A_3 (IV.1.1.9, IV.1.3.3(ii)), it need not be Hausdorff.)

IV.1.3.5 PROPOSITION. Let J be a primitive ideal in a C*-algebra A. Then the following are equivalent:

(i) J is an A-GCR ideal.
(ii) There is an ideal K of A, containing J, such that K/J is elementary.
(iii) A/J is not antiliminal.

PROOF: (i) $\Rightarrow$ (ii) $\Rightarrow$ (iii) is trivial.
(iii) $\Rightarrow$ (i): Let I be a closed ideal of A containing J, such that I/J is liminal. Let π be an irreducible representation of A on $\mathcal{H}$ with kernel J. Then $\pi|_I$ is irreducible, and hence $\pi(I) \subseteq \mathcal{K}(\mathcal{H})$.

Combining IV.1.3.5 with II.6.1.6 and IV.1.2.2, we obtain:

IV.1.3.6 COROLLARY. If J is an A-GCR ideal in a C*-algebra A, then any two irreducible representations of A with kernel J are equivalent.

The converse is also true if A is separable (IV.1.5.1).

IV.1.3.7 COROLLARY. Let A be a C*-algebra. Then A is CCR [resp. GCR] if and only if every primitive ideal of A is an A-CCR [resp. A-GCR] ideal.

One direction is trivial, and the other follows from IV.1.3.5 and IV.1.2.2.

IV.1.3.8 PROPOSITION. Let A be a C*-algebra.

(i) If A is postliminal, then A is GCR.
(ii) If A is separable and GCR, then A is bidual Type I.

(The conditions are actually all equivalent, even in the nonseparable case, but (bidual Type I) $\Rightarrow$ (postliminal) is hard (IV.1.5.7), as is GCR $\Rightarrow$ (bidual Type I) in the nonseparable case (IV.1.5.8)).

PROOF: (i): Suppose (J_α) is a composition series for A with $J_{\alpha+1}/J_\alpha$ liminal for all α, and let π be an irreducible representation of A. Let β be the largest α such that $\pi|_{J_\alpha} = 0$. Then $\pi|_{J_{\beta+1}}$ is irreducible and defines an irreducible representation of $J_{\beta+1}/J_\beta$, so the image contains a nonzero compact operator.

(ii): Let π be a factor representation of A, and $J = \ker \pi$. Then J is primitive (II.6.1.11, II.6.5.15), so there is an ideal K of A containing J, with K/J elementary. Then $\pi|_K$ is a factor representation (II.6.1.6), hence is a multiple of an irreducible representation by IV.1.2.2 and therefore Type I. By II.6.1.6, $\pi(K)'' = \pi(A)''$.

IV.1.3.9 If A is any C*-algebra, let

$$J_0(A) = \{x \in A : \pi(x) \text{ is compact for all } \pi \in \hat{A}\}.$$

Then $J_0(A)$ is a closed ideal of A, and is the largest liminal ideal in A. Set $J_0 = J_0(A)$, and inductively define $J_{\alpha+1}$ to be the preimage of $J_0(A/J_\alpha)$ in A. Then (J_α) is a subcomposition series in A; let $J(A)$ be the limit. Then $J(A)$ is a closed ideal of A which is postliminal, and $A/J(A)$ is antiliminal. Because of IV.1.3.2 $I_0(A) \subseteq J_0(A)$ for all A, and hence $I(A) \subseteq J(A)$ for all A. Actually $I(A) = J(A)$ for all A by IV.1.4.26.

IV.1.3.10 If A is a liminal C*-algebra and B is a C*-subalgebra, then any irreducible representation of B can be extended to an irreducible representation of A, whose image consists of compact operators; thus B is liminal. If A is postliminal and $B \subseteq A$, and (J_α) is the liminal composition series for A, then $(B \cap J_\alpha)$ is a liminal composition series for B, so B is postliminal.

IV.1.4 Continuous Trace, Homogeneous, and Subhomogeneous C*-Algebras

Homogeneous and Subhomogeneous C*-Algebras

Some of the most important "building block" C*-algebras are the homogeneous and subhomogeneous C*-algebras. Their structure will be examined in more detail in IV.1.7.23.

IV.1.4.1 DEFINITION. Let $n \in \mathbb{N}$. A C*-algebra A is *n-homogeneous* if every irreducible representation of A is of dimension n. A is *n-subhomogeneous* if every irreducible representation of A has dimension $\leq n$. A is *homogeneous* [resp. *subhomogeneous*] if it is n-homogeneous [resp. n-subhomogeneous] for some n. A subhomogeneous C*-algebra is *locally homogeneous* if it is a (finite) direct sum of homogeneous C*-algebras.

There are also $\aleph_0$-homogeneous C*-algebras, which will be defined later (IV.1.7.12).

The standard example of an n-homogeneous C*-algebra is $C_0(T, \mathbb{M}_n)$, where T is locally compact. In fact, every homogeneous C*-algebra looks locally like one of these (IV.1.7.23).

The examples of IV.1.1.9 are 2-subhomogeneous. So a subhomogeneous C*-algebra need not have Hausdorff primitive ideal space. A closely related example is important:

IV.1.4.2 EXAMPLE. Let A be the free product $\mathbb{C} * \mathbb{C}$ (II.8.3.4), isomorphic to the universal C*-algebra generated by two projections p and q. We show that A is 2-subhomogeneous. The corner pAp is generated by the commuting positive elements p and pqp, and hence is commutative; so if π is any irreducible representation of A on a Hilbert space $\mathcal{H}$, II.6.1.9 implies that $\pi(p)\mathcal{H}$ is at most one-dimensional. Similarly, $dim(\pi(q)\mathcal{H}) \leq 1$. Since $\pi(p) \vee \pi(q) = I$, we have $dim(\mathcal{H}) \leq 2$.

There are three one-dimensional irreducible representations of A, $\pi_{0,p}$, $\pi_{0,q}$, and π_1, where

$$\pi_{0,p}(p) = 1, \ \pi_{0,p}(q) = 0, \ \pi_{0,q}(p) = 0, \ \pi_{0,q}(q) = 1, \ \pi_1(p) = \pi_1(q) = 1.$$

Suppose $dim(\mathcal{H}) = 2$. There is an orthonormal basis $\{\xi, \eta\}$ for $\mathcal{H}$ with respect to which $\pi(p) = \begin{bmatrix} 1 & 0 \\ 0 & 0 \end{bmatrix}$. Then

$$\pi(q) = \begin{bmatrix} t & \alpha\sqrt{t-t^2} \\ \bar{\alpha}\sqrt{t-t^2} & 1-t \end{bmatrix}$$

for some t and α, $0 < t < 1$, $|\alpha| = 1$; by scaling η we may make $\alpha = 1$. Thus π is unitarily equivalent to π_t for a unique t, $0 < t < 1$, where π_t is the irreducible representation of A on $\mathbb{C}^2$ with

$$\pi_t(p) = \begin{bmatrix} 1 & 0 \\ 0 & 0 \end{bmatrix}, \ \pi_t(q) = \begin{bmatrix} t & \sqrt{t-t^2} \\ \sqrt{t-t^2} & 1-t \end{bmatrix}.$$

Then $\pi_t(p)$ and $\pi_t(q)$ vary continuously in t in $\mathbb{M}_2$; as $t \to 0$, $\pi_t(p)$ and $\pi_t(q)$ approach $\begin{bmatrix} 1 & 0 \\ 0 & 0 \end{bmatrix}$ and $\begin{bmatrix} 0 & 0 \\ 0 & 1 \end{bmatrix}$ respectively, and as $t \to 1$, both approach $\begin{bmatrix} 1 & 0 \\ 0 & 0 \end{bmatrix}$. Thus there is a homomorphism ϕ from A to the C*-subalgebra

$$B = \left\{ f : [0,1] \to \mathbb{M}_2 \ : \ f(0) = \begin{bmatrix} \alpha & 0 \\ 0 & \beta \end{bmatrix}, f(1) = \begin{bmatrix} \gamma & 0 \\ 0 & 0 \end{bmatrix} (\alpha, \beta, \gamma \in \mathbb{C}) \right\}$$

of $C([0,1], \mathbb{M}_2)$. It is easily checked using the Stone-Weierstrass Theorem that ϕ is surjective, and every irreducible representation of A factors through ϕ, so ϕ is an isomorphism. Thus $\mathrm{Prim}(A) \cong [0,1]$ with the point at 0 "doubled".

Note that A is nonunital; $\tilde{A}$ is isomorphic to the C*-subalgebra of $C([0,1], \mathbb{M}_2)$ of functions for which $f(0)$ and $f(1)$ are diagonal.

Subalgebras of Homogeneous C*-Algebras

IV.1.4.3 PROPOSITION. A C*-algebra [*resp.* unital C*-algebra] is subhomogeneous if and only if it is isomorphic to a C*-subalgebra [*resp.* unital C*-subalgebra] of a unital homogeneous C*-algebra.

PROOF: If A is a C*-subalgebra of an n-homogeneous C*-algebra B, then every irreducible representation of A extends to an irreducible representation of B, hence has dimension $\leq n$. Conversely, suppose A is subhomogeneous. We may assume A is unital since $\tilde{A}$ is also subhomogeneous. Let $n_1, \ldots, n_r$ be the dimensions of the irreducible representations of A. Set

$$k = n_1 + \cdots + n_r$$

and let $\mathcal{H}$ be a k-dimensional Hilbert space, and identify $\mathcal{L}(\mathcal{H})$ with $\mathbb{M}_k$. Then for each $x \in A$ there is a unital *-homomorphism ϕ from A to $\mathbb{M}_k$ with $\phi(x) \neq 0$. Let T be the set of all unital *-homomorphisms from A to $\mathbb{M}_k$. Give T the topology of elementwise convergence. Then T can be identified with a closed subset of

$$\prod_{x \in B_1(A)} B_1(\mathbb{M}_k)$$

where $B_1(D)$ denotes the closed unit ball in D; thus T is compact and Hausdorff. For each $x \in A$ define $\hat{x} : T \to \mathbb{M}_k$ by $\hat{x}(\phi) = \phi(x)$. Then $x \mapsto \hat{x}$ is an injective unital *-homomorphism from A to $C(T, \mathbb{M}_k)$.

IV.1.4.4 The T constructed in the proof of IV.1.4.3 is rather complicated, although it is metrizable if A is separable. (See [Bla93] for a description of T as a stratified space in the homogeneous case.) In the separable case, as pointed out in [Phi01b], one can take a surjective continuous function from the Cantor set K to T, which induces an embedding of $C(T)$ into $C(K)$ and hence from $C(T, \mathbb{M}_k)$ into $C(K, \mathbb{M}_k)$. Thus every [unital] separable subhomogeneous C*-algebra is isomorphic to a [unital] C*-subalgebra of a matrix algebra over $C(K)$.

The k in the proof is not the smallest possible in general. We only need k to have the property that for each n_i, $k - n_i$ is either 0 or a sum of not necessarily distinct n_j's. For example, if A has irreducible representations of dimensions 1, 2, and 3, instead of $k = 6$ we may take $k = 3$. (In fact, if A has a one-dimensional representation, then we can always take $k = \max(n_i)$.) However, if (unital) A has all irreducible representations of dimensions 2 and 3 (e.g. $A = C(X, \mathbb{M}_2) \oplus C(Y, \mathbb{M}_3)$ for compact X, Y), then $k = 5$ is the best possible.

An alternate proof of IV.1.4.3 can be based on IV.1.4.6.

Algebraic Characterization of Subhomogeneity

IV.1.4.5 There is an algebraic characterization of n-subhomogeneous C*-algebras. For $r \in \mathbb{N}$, define a polynomial p_r in r noncommuting variables by
$$p_r(X_1, \ldots, X_r) = \sum_{\sigma \in S_r} \epsilon_\sigma X_{\sigma(1)} X_{\sigma(2)} \cdots X_{\sigma(r)}$$
where S_r is the symmetric group on r elements and $\epsilon_\sigma = \pm 1$ is the sign of σ. For each n there is an $r = r(n)$ (the smallest value of $r(n)$ is $2n$) such that, for any $x_1, \ldots, x_{r(n)} \in \mathbb{M}_n$, $p_{r(n)}(x_1, \ldots, x_{r(n)}) = 0$ (we say $\mathbb{M}_n$ satisfies the polynomial relation $p_{r(n)}$), but that $\mathbb{M}_{n+1}$ does not satisfy the relation $p_{r(n)}$, i.e. there are $x_1, \ldots, x_{r(n)} \in \mathbb{M}_{n+1}$ with $p_{r(n)}(x_1, \ldots, x_{r(n)}) \neq 0$ ([AL50], [Ros76]; cf. [Dix69b, 3.6.2]). Then the next result follows easily:

IV.1.4.6 PROPOSITION. Let A be a C*-algebra. The following are equivalent:

(i) A is n-subhomogeneous.
(ii) A satisfies the polynomial relation $p_{r(n)}$.
(iii) A^{**} satisfies the polynomial relation $p_{r(n)}$.
(iv) A^{**} is a direct sum of Type I_m von Neumann algebras for $m \leq n$.

In particular, a subhomogeneous C*-algebra is bidual Type I.

PROOF: If $\pi(A)$ satisfies $p_{r(n)}$ for every irreducible representation π of A, then A satisfies $p_{r(n)}$ (II.6.4.9), and thus (i) $\Longrightarrow$ (ii) by IV.1.4.5. For the proof of (ii) $\Longrightarrow$ (iii), if π is a representation of A and $\pi(A)$ satisfies the relation p_r, then so does $\pi(A)''$, by the Kaplansky Density Theorem and joint strong continuity of multiplication on bounded sets. (iii) $\Longrightarrow$ (iv) is III.1.5.14 (using IV.1.4.5), and (iv) $\Longrightarrow$ (i) is trivial.

By a refinement of this argument, it can be readily shown that a C*-algebra A is n-homogeneous if and only if A^{**} is a Type I_n von Neumann algebra.

IV.1.4.7 From this one sees that if A_i is n-subhomogeneous for each i, then $\prod A_i$ is n-subhomogeneous; and if (J_i) is a set of closed ideals in a C*-algebra A with $J = \cap J_i$, and A/J_i is n-subhomogeneous for each i, then A/J is n-subhomogeneous. Thus A has a smallest ideal J (often equal to A) such that A/J is n-subhomogeneous.

There is a useful reformulation. If A is a C*-algebra, let $_n\hat{A}$ be the set of equivalence classes of irreducible representations of A of dimension $\leq n$, and $\hat{A}_n$ the set of irreducible representations of dimension exactly n. Then $_n\hat{A}$ is closed in $\hat{A}$, and $\hat{A}_n = {_n\hat{A}} \setminus {_{n-1}\hat{A}}$ is relatively open in $_n\hat{A}$. (But note that $_n\hat{A} = \emptyset$ for "most" A and n.)

Continuous Trace C*-Algebras

IV.1.4.8 For any Hilbert space $\mathcal{H}$, we denote by Tr the usual trace on $\mathcal{L}(\mathcal{H})$ (i.e. the trace of a projection is its rank). If A is a C*-algebra and $x \in A_+$, let $\hat{x} : \hat{A} \to [0, \infty]$ be the function with $\hat{x}(\pi) = \text{Tr}(\pi(x))$. Then $\hat{x}$ drops to a well-defined function, also denoted $\hat{x}$, from $\text{Prim}(A)$ to $[0, \infty]$ (if $\pi(x) \neq 0$, then $\text{Tr}(\pi(x)) < \infty$ only if π is a GCR representation, in which case any $\rho \in \hat{A}$ with $ker(\rho) = ker(\pi)$ is equivalent to π (IV.1.3.6)). We obviously have $(x^*x)\check{} = (xx^*)\check{}$ for any $x \in A$.

IV.1.4.9 PROPOSITION. [Ped79, 4.4.9] For any A and $x \in A_+$,

$$\hat{x} = \sup(\sum_{i=1}^n \check{x}_i)$$

(II.6.5.6), where the supremum is taken over all finite subsets $\{x_1, \ldots, x_n\}$ of A_+ with $\sum_{i=1}^n x_i \leq x$. Thus $\hat{x}$ is lower semicontinuous.

IV.1.4.10 We say $x \in A_+$ *has continuous trace* if $\hat{x}$ is finite, bounded, and continuous on $\hat{A}$.

IV.1.4.11 PROPOSITION. Let A be a C*-algebra, and $\mathfrak{m}_+(A)$ the set of positive elements of A of continuous trace. Then $\mathfrak{m}_+(A)$ is a nonempty invariant hereditary cone in A_+, and thus the positive part of an ideal $\mathfrak{m}(A)$.

PROOF: It is clear that $\mathfrak{m}_+(A)$ is an invariant cone, and $0 \in \mathfrak{m}_+(A)$. If $x \in \mathfrak{m}_+(A)$ and $0 \leq y \leq x$, then $\hat{y}$ and $(x-y)\check{}$ are lower semicontinuous functions whose sum is the bounded continuous function $\hat{x}$, so $\hat{y}$ is bounded and continuous.

IV.1.4.12 DEFINITION. A C*-algebra A *has continuous trace* (or is a *continuous trace C*-algebra*) if $\mathfrak{m}(A)$ is dense in A.

IV.1.4.13 EXAMPLES. Any elementary C*-algebra has continuous trace. Of the examples of IV.1.1.9, A_1 has continuous trace but A_2 and A_3 do not.

IV.1.4.14 PROPOSITION. A locally homogeneous C*-algebra has continuous trace.

PROOF: This is obvious in the unital case, since if A is n-homogeneous, then $\hat{1}$ is the constant function n, and a (finite) direct sum of continuous trace C*-algebras has continuous trace. A slightly more complicated functional calculus argument is needed in the nonunital case.

IV.1.4.15 Conversely, a unital continuous trace C*-algebra is locally homogeneous (and it follows from IV.1.7.23 that a locally homogeneous C*-algebra A is unital if and only if $\hat{A}$ is compact). A subhomogeneous C*-algebra does not have continuous trace in general (IV.1.4.13).

It is obvious that a continuous trace C*-algebra is liminal, and hence $\hat{A} \cong Prim(A)$ is T_1. One can say more:

IV.1.4.16 PROPOSITION. If A is a continuous trace C*-algebra, then A is Type I_0 and $\hat{A}$ is Hausdorff.

PROOF: Let π, ρ be distinct points of $\hat{A}$. Since $ker(\pi) \not\subseteq ker(\rho)$, there is an $x \in A_+$ with $\|\pi(x)\| = 1$ and $\rho(x) = 0$. If $(y_n) \subseteq \mathfrak{m}_+(A)$ with $y_n \to x$, then, replacing x by a continuous function of $y_n x y_n$ for sufficiently large n (or by $f_\epsilon(x)$ (II.3.4.11), which is in $Ped(A) \subseteq \mathfrak{m}(A)$), we may assume $x \in \mathfrak{m}_+(A)$ and $\|x\| = 1$. Thus $\hat{x}$ is a continuous function separating π and ρ, so $\hat{A}$ is Hausdorff. The trace-class operator $\pi(x)$ has a nonzero eigenspace $\mathcal{H}_0$ with eigenvalue 1 (I.8.4.1); and $y = f(x)$ satisfies $\pi(y) = P_{\mathcal{H}_0}$ for a suitable continuous function f. Let P be a rank-one projection onto a subspace of $\mathcal{H}_0$; then by II.6.1.9 there is a $b \in [(yAy)^-]_+$ with $\pi(b) = P$, and replacing b by xbx we may assume that $b \in \mathfrak{m}_+(A)$. The continuous function $\check{b} - \check{b}^2$ is small ($\leq 1/8$) in a neighborhood U of π, and hence $\pi'(c)$ is a projection for all $\pi' \in U$, where $c = f_{1/4}(b)$. Since $\hat{c}$ is continuous and integer-valued on U, $\pi'(c)$ is a rank-one projection for all $\pi' \in U$. Let $z \in Z(M(A))_+$ with $\|\pi(z)\| = 1$ and $\sigma(z) = 0$ for all $\sigma \in \hat{A} \setminus U$ (II.6.5.10), and set $a = zc$. Then a is an abelian element of A not contained in $ker(\pi)$. Thus the ideal $I_0(A)$ is not contained in any primitive ideal of A, so $I_0(A) = A$, i.e. A is Type I_0.

Local Rank One Projections

The proof of IV.1.4.16 shows that a continuous trace C*-algebra satisfies the following condition (cf. [Fel61]):

IV.1.4.17 DEFINITION. A C*-algebra A satisfies *Fell's condition* if, for every $\pi \in \hat{A}$, there is a neighborhood U of π in $\hat{A}$ and an $x \in A_+$ such that $\rho(x)$ is a rank-one projection for all $\rho \in U$.

A C*-algebra satisfying Fell's condition must be GCR, and in fact CCR [if $ker(\rho) \subsetneq ker(\pi)$, and $x \in A_+$ with $\pi(x)$ a rank one projection, then $\rho(x)$ has infinite rank, and ρ is in any neighborhood of π]. Of the examples of IV.1.1.9, A_1 and A_3 satisfy Fell's condition, but A_2 does not.

A C*-algebra A of Type I_0 satisfies Fell's condition: if $\pi \in \hat{A}$, let x be a positive abelian element of A with $\pi(x) \neq 0$. If $0 < \epsilon < \|\pi(x)\|/2$, then $\rho(f_\epsilon(x))$ is a rank-one projection for all ρ in a neighborhood of π. The converse is unclear, but we have:

IV.1.4.18 PROPOSITION. Let A be a C*-algebra. The following are equivalent:

(i) A has continuous trace.
(ii) $\hat{A}$ is Hausdorff and A is Type I_0.
(iii) $\hat{A}$ is Hausdorff and A satisfies Fell's condition.

(i) $\iff$ (ii) is IV.1.4.16, (ii) $\implies$ (iii) is the argument above, and (iii) $\implies$ (i) by a partition of unity argument using II.6.5.10.

A C*-algebra with continuous trace thus has "local rank-one projections."

IV.1.4.19 COROLLARY. The class of continuous trace C*-algebras is closed under Morita equivalence.

PROOF: Morita equivalent C*-algebras have homeomorphic primitive ideal spaces (II.7.6.13), and it is easily seen that Fell's condition is preserved under Morita equivalence (cf. IV.1.7.8).

IV.1.4.20 DEFINITION. Let A be a C*-algebra. A *global rank-one projection* in A is a projection $p \in A$ with $\pi(p)$ rank-one for all $\pi \in \hat{A}$.

A C*-algebra with a full global rank-one projection must have continuous trace. In fact:

IV.1.4.21 PROPOSITION. If A is a C*-algebra with $T = \hat{A}$, and A has a full global rank-one projection, then T is compact and Hausdorff (and hence A has continuous trace), and A is Morita equivalent to $C(T)$.

T is compact by II.6.5.6, and if p is a full global rank-one projection in A, then Ap is a $A - C(T)$-imprimitivity bimodule.

IV.1.4.22 Thus if A is a continuous-trace C*-algebra with $T = \hat{A}$ compact, for A to be Morita equivalent to $C(T)$ it is sufficient for A to have a global rank-one projection. But this condition is not necessary: for example, let p be a trivial rank-one projection and q the Bott projection in $C(S^2, \mathbb{M}_2)$; then

$$A = \{f \in C([0,1], C(S^2, \mathbb{M}_2)) \mid f(0) \in \mathbb{C}p, f(1) \in \mathbb{C}q\}$$

is a continuous-trace algebra with $T = \hat{A} \cong [0,1] \times S^2$, and A is Morita equivalent to $C(T)$, but has no global rank-one projection. If A is separable and stable, then existence of a global rank-one projection is necessary for A

to be Morita equivalent to $C(T)$, since then $A \cong C(T) \otimes \mathbb{K}$ by the Brown-Green-Rieffel theorem (II.7.6.11).

If A is any continuous-trace C*-algebra (with $T = \hat{A}$ not necessarily compact), then A is "locally Morita equivalent" to $C_o(T)$ (IV.1.7.8), but there are global obstructions of a cohomological nature. Even if A is separable and stable, there is an obstruction in $H^3(T, \mathbb{Z})$ called the Dixmier-Douady invariant. This will be discussed in IV.1.7.10.

Homomorphisms

IV.1.4.23 Note that even unital *-homomorphisms between matrix algebras over commutative C*-algebras may relate in a complicated way to the underlying topological spaces. For example, let σ be the action of $\mathbb{Z}_2$ on $\mathbb{T}$ by $\sigma(z) = -z$. The embedding $C(\mathbb{T}) \to C(\mathbb{T}) \rtimes_\sigma \mathbb{Z}_2 \cong M_2(C(\mathbb{T}))$ is called a *twice-around embedding*, and is conjugate to the following homomorphism $\phi : C(\mathbb{T}) \to C(\mathbb{T}, M_2)$, where $\mathbb{T}$ is regarded as $[0,1]$ with 0 and 1 identified:

$$[\phi(f)](t) = \begin{bmatrix} \cos \frac{\pi t}{2} & -\sin \frac{\pi t}{2} \\ \sin \frac{\pi t}{2} & \cos \frac{\pi t}{2} \end{bmatrix} \begin{bmatrix} f\left(\frac{t}{2}\right) & 0 \\ 0 & f\left(\frac{t+1}{2}\right) \end{bmatrix} \begin{bmatrix} \cos \frac{\pi t}{2} & \sin \frac{\pi t}{2} \\ -\sin \frac{\pi t}{2} & \cos \frac{\pi t}{2} \end{bmatrix}$$

See [DNNP92] and [Bla93] for a description of a "classifying space" for homomorphisms between unital homogeneous C*-algebras.

IV.1.4.24 The homomorphisms from A to B (homogeneous, or more generally of continuous trace) which relate well to the topology of the underlying spaces $\hat{A}$ and $\hat{B}$ are the ones which "preserve rank." The cleanest way to precisely define this condition is to require that the image of any abelian element is abelian. If ϕ is such a homomorphism, and A, B, and ϕ are unital, then ϕ induces $\phi_* : \hat{B} \to \hat{A}$ just as in the commutative case. In the nonunital setting, ϕ is a proper continuous map from an open subset of $\hat{B}$ to $\hat{A}$.

Continuous Trace Ideals

We now have enough machinery to clean up a loose end from the previous section.

IV.1.4.25 PROPOSITION. A liminal C*-algebra is internally Type I.

PROOF: Since a quotient of a liminal C*-algebra is liminal, it suffices to show that if A is liminal, then A contains a nonzero abelian element. Let $x, y \in A_+$ with $y \neq 0$ and $xy = y$. Then if $\pi \in \hat{A}$, since $\pi(x)$ is compact it follows that $\pi(y)$ is finite-rank. Thus, if $B = [yAy]^-$ is the hereditary C*-subalgebra generated by y, then every irreducible representation of B is finite-dimensional. Since $(\hat{B} \setminus {}_n\hat{B})$ is a decreasing sequence of open sets in $\hat{B}$ with empty intersection, II.6.5.14 implies that ${}_n\hat{B}$ has nonempty interior for some n, and hence there is a closed ideal J of B such that

$$_{n-1}\hat{J} \neq {}_n\hat{J} = \hat{J}.$$

Then $\hat{J}_n$ is a nonempty open subset of $\hat{J}$, and hence there is an ideal I in J which is an n-homogeneous C*-algebra. I has continuous trace by IV.1.4.14 and therefore contains an abelian element by IV.1.4.16.

IV.1.4.26 COROLLARY. A C*-algebra is postliminal if and only if it is internally Type I.

PROOF: One direction is IV.1.3.2. The converse follows from IV.1.4.25 and IV.1.1.11.

The proof of IV.1.4.25, combined with II.5.1.4(iii) and the fact that the class of continuous-trace C*-algebras is closed under Morita equivalence (IV.1.4.19), gives:

IV.1.4.27 PROPOSITION. If A is a liminal C*-algebra, then A contains a nonzero ideal K which is a continuous-trace C*-algebra.

IV.1.4.28 COROLLARY. Every postliminal C*-algebra has a composition series (K_α) such that $K_{\alpha+1}/K_\alpha$ has continuous trace for all α.

Unlike the composition series I_α and J_α (IV.1.1.12, IV.1.3.9), the composition series (K_α) depends on choices in the construction and thus is non-canonical.

IV.1.4.29 There is a class of Type I C*-algebras called *generalized continuous trace C*-algebras* (abbreviated *GCT* or *GTC*) [Dix69b, 4.7.12], for which the subcomposition series (M_α) is a composition series, where $M_0(A) = \mathbf{m}(A)$ and $M_{\alpha+1}(A)$ is the preimage of $M_0(A/M_\alpha(A))$. A GCT C*-algebra is necessarily liminal (note, for example, that $M_0(\tilde{\mathbb{K}}) = 0$). Not every liminal C*-algebra is GCT [Dix61], but every C*-algebra with Hausdorff spectrum is GCT.

IV.1.5 Characterization of Type I C*-Algebras

The following fundamental theorem is a slight variation (cf. [BK01]) of a result of J. Glimm [Gli61], which was probably the deepest theorem in the subject of operator algebras when it was proved.

IV.1.5.1 THEOREM. Let J be a primitive ideal in a separable C*-algebra A. Then the following are equivalent:

(i) J is not an A-GCR ideal.
(ii) A/J is antiliminal.
(iii) J is the kernel of a non-type I factor representation of A.
(iv) There are two inequivalent irreducible representations of A with kernel J.

(v) There are uncountably many mutually inequivalent irreducible representations of A with kernel J.

PROOF: (i) $\Leftrightarrow$ (ii) is IV.1.3.5, (v) $\Rightarrow$ (iv) is trivial, and (iv) $\Rightarrow$ (i) is IV.1.3.6. (iii) $\Rightarrow$ (v)*(Outline)*: By replacing A by A/J, we may and will assume that $J = 0$, to simplify notation. If π is a faithful non-type-I factor representation of A on a separable Hilbert space and Z is a masa in $\pi(A)'$, the direct integral decomposition of π as

$$\int_X^\oplus \pi_x \, d\mu(x)$$

with respect to Z has almost all π_x faithful and irreducible. If, for a set E of nonzero measure, each π_x for $x \in E$ is equivalent to a fixed representation π_0, then

$$\int_E^\oplus \pi_x \, d\mu(x)$$

is a subrepresentation of π equivalent to a multiple of π_0, a contradiction. Thus, for each x, the set

$$E_x = \{y : \pi_y \sim \pi_x\}$$

has measure 0, so there must be uncountably many such sets.

IV.1.5.2 It remains to prove (ii) $\Rightarrow$ (iii). This is a complicated technical construction and will not even be outlined here. The rough idea is that in any hereditary C*-subalgebra of an antiliminal C*-algebra can be found a nonzero element x such that x^*x and xx^* are orthogonal (II.6.4.14). A delicate bisection process then leads to a sequence of carefully embedded approximate matrix units; as a consequence, one gets:

IV.1.5.3 THEOREM. Let A be an antiliminal C*-algebra. Then there is a C*-subalgebra B of A, and a closed ideal I of B, such that B/I is isomorphic to the CAR algebra (II.8.2.2(iv)).

IV.1.5.4 If A is separable, the C*-subalgebra B can be constructed in such a way that if τ is the tracial state on B/I, regarded as a state on B (so that $\pi_\tau(B)''$ is a II_1 factor), and τ is extended to a state ϕ on A, then $\pi_\phi(A)''$ is a Type II factor. The final conclusion of the construction is:

IV.1.5.5 THEOREM. Let A be a separable antiliminal C*-algebra, and (J_n) a decreasing sequence of essential ideals of A. Then there is a Type II factor representation π of A, such that $\pi|_{J_n}$ is nonzero for all n.

IV.1.5.6 To finish the proof of IV.1.5.1, note that if A is separable and primitive, then there is a decreasing sequence (J_n) of nonzero (not necessarily proper) ideals, which are automatically essential, such that every nonzero ideal of A contains J_n for some n; this is an immediate consequence of the fact that $\mathrm{Prim}(A)$ is a second countable T_0-space and 0 is a dense point.

We also obtain Glimm's original theorem:

IV.1.5.7 THEOREM. Let A be a separable C*-algebra. Then the following are equivalent:

(i) A is internally Type I.
(ii) A is bidual Type I.
(iii) A is postliminal.
(iv) A is GCR.
(v) Whenever π and ρ are irreducible representations of A with the same kernel, then π and ρ are equivalent (i.e. the map $\hat{A} \to \mathrm{Prim}(A)$ of II.6.5.13 is a bijection).

PROOF: (iii) $\Rightarrow$ (iv) $\Rightarrow$ (ii) by IV.1.3.8, (i) $\Rightarrow$ (iii) by IV.1.3.2, (iv) $\Rightarrow$ (v) by IV.1.3.5 and IV.1.2.2, and (v) $\implies$ (ii) is IV.1.5.1[(iii) $\implies$ (v)] (cf. IV.1.1.3). So we need only prove (ii) $\Rightarrow$ (i). If A is not internally Type I, let $B = A/I(A)$ (IV.1.1.12); then B is antiliminal. Applying IV.1.5.5 to B with $J_n = B$ for all n, we obtain a type II factor representation of B, and therefore of A.

IV.1.5.8 Note that separability is used in this proof only for (iv) $\implies$ (ii), (v) $\implies$ (ii), and (ii) $\implies$ (i); the implications (i) $\implies$ (iii) $\implies$ (iv) $\implies$ (v) are true in general (IV.1.3.2, IV.1.3.8, IV.1.3.6, IV.1.3.7). Also, by IV.1.4.26 and IV.1.1.2, (iii) $\implies$ (i) $\implies$ (ii) in general (these implications were not needed for the proof of IV.1.5.7). Using IV.1.5.3 and injectivity of the bidual of the CAR algebra, S. Sakai has shown that any antiliminal C*-algebra has a Type III factor representation, and hence (ii) $\implies$ (i) also in the nonseparable case. He then showed, using (i) $\implies$ (ii), that (iv) $\implies$ (ii) in general, and hence (i)–(iv) are equivalent even for a nonseparable A. (See [Sak71, 4.6] for a complete discussion.) However, (v) does not imply the other conditions in the nonseparable case (IV.1.2.3).

The same technique can be used to give the following version of O. Maréchal's refinement [Mar75] of Glimm's result:

IV.1.5.9 THEOREM. Let A be a separable unital primitive antiliminal C*-algebra. Then there is a unital sub-C*-algebra B of A and ideal J of B, such that:

(i) B/J is isomorphic to the CAR algebra D (write $\phi : B \to D$ for the quotient map)

(ii) For any cyclic representation π of D, there is a faithful cyclic representation ρ of A, and a projection

$$P \in \rho(B)'' \cap \rho(B)'$$

of central support 1 in $\rho(A)''$, such that the subrepresentation ρ_1 of $\rho|_B$ defined by P is equivalent to $\pi \circ \phi$ and $\rho_1(B)'' = P\rho(A)''P$.

IV.1.5.10 COROLLARY. Let A be a separable C*-algebra and J a non-A-GCR primitive ideal of A. If M is any properly infinite injective von Neumann algebra (in particular, any infinite injective factor) with separable predual, then there is a representation π of A with kernel J, such that $\pi(A)'' \cong M$.

IV.1.5.11 An important consequence of Glimm's Theorem (use IV.1.5.8 for the nonseparable case) is the nonobvious fact that a C*-subalgebra of a Type I C*-algebra is Type I (IV.1.3.10).

Separable Type I C*-algebras can also be characterized in terms of the Borel structure of the dual:

IV.1.5.12 THEOREM. [Gli61] Let A be a separable C*-algebra. Then the following are equivalent:

(i) A is Type I.
(ii) The Mackey Borel structure (II.6.5.16) on $\hat{A}$ is the Borel structure generated by the topology of $\hat{A}$.
(iii) The Mackey Borel structure on $\hat{A}$ is countably separated.
(iv) $\hat{A}$ with the Mackey Borel structure is a standard Borel space.

Because of this theorem, Type I C*-algebras are sometimes called "smooth."
Thus, if A is separable and Type I, $\hat{A}$ has a simple structure as a Borel space. On the other hand, if A is separable and not Type I, then $\hat{A}$ is not countably separated; this is the technical sense in which the representation theory of a (separable) non-Type-I C*-algebra "cannot be understood."

IV.1.6 Continuous Fields of C*-Algebras

The notion of a continuous field of C*-algebras is a topological analog of measurable fields of Hilbert spaces and von Neumann algebras (III.1.6).

IV.1.6.1 DEFINITION. Let T be a locally compact Hausdorff space. A *continuous field* of C*-algebras over T consists of the following data:

(i) a C*-algebra $A(t)$ for each $t \in T$.
(ii) a set Γ of sections $a : T \to \coprod_T A(t)$ satisfying:
   (1) $a(t) \in A(t)$ for all $a \in \Gamma$, $t \in T$.

(2) $[t \mapsto \|a(t)\|] \in C_o(T)$ for all $a \in \Gamma$.
(3) Γ is closed under scalar multiplication and under pointwise sum, product, and adjoint.
(4) For all $t_0 \in T$ and all $x \in A(t_0)$, there is an $a \in \Gamma$ with $a(t_0) = x$.
(5) Γ is closed under local uniform limits: if $b : T \to \coprod A(t)$ is a continuous section vanishing at infinity, i.e. $b(t) \in A(t)$ for all t and
$$[t \mapsto \|b(t)\|] \in C_o(T),$$
and for each $t_0 \in T$ and $\epsilon > 0$ there is a neighborhood U of t_0 and an $a \in \Gamma$ with $\|a(t) - b(t)\| < \epsilon$ for all $t \in U$, then $b \in \Gamma$.

We usually want to assume that $A(t) \neq \{0\}$ for all t. Such a continuous field will be called *full*.

IV.1.6.2 If $(T, \{A(t)\}, \Gamma)$ is a continuous field of C*-algebras, then Γ becomes a C*-algebra under the norm
$$\|a\| = \sup_t \|a(t)\| (= \max_t \|a(t)\|).$$

This is called the *C*-algebra of the continuous field*. Note that the continuous field contains more structure than simply the C*-algebra Γ – it also specifies how Γ is fibered over T.

IV.1.6.3 Isomorphism of continuous fields of C*-algebras over the same base space is defined in the obvious way: isomorphism of the C*-algebras of the continuous fields via a map respecting the fibers.

IV.1.6.4 It follows easily from the local uniform closure of Γ that Γ is closed under pointwise multiplication by $C_o(T)$ (or even by bounded continuous functions on T): if $a \in \Gamma$ and $f \in C_o(T)$, and $b(t) = f(t)a(t)$ for $t \in T$, then $b \in \Gamma$. Thus, if U is an open subset of X, the set of sections vanishing outside U forms a continuous field over U, called the *restriction* of Γ to U, denoted $\Gamma|_U$. The C*-algebra of this field can be naturally identified with a closed ideal of Γ, denoted Γ_U. Similarly, the restriction of Γ to a closed subset Z of T is a continuous field over Z, denoted $\Gamma|_Z$; the corresponding C*-algebra, denoted Γ^Z, can be identified with the quotient $\Gamma/\Gamma_{T\setminus Z}$.

If Γ is a nonfull field over T, then
$$U = \{t \in T : A(t) \neq \{0\}\}$$
is open in T, and Γ defines a full continuous field $\Gamma|_U$ over U which is isomorphic to Γ as a C*-algebra.

IV.1.6.5 The simplest example of a (full) continuous field over a space T is obtained by taking $A(t) = A$ for all t and $\Gamma = C_o(T, A)$, for some fixed C*-algebra A; such a field is called a *constant field*. A *trivial field* is one which is isomorphic to a constant field. A field over T is *locally trivial* if each $t \in T$ has a neighborhood U such that $\Gamma|_U$ is trivial.

IV.1.6.6 Theorem. Let A be a C*-algebra with $T = \mathrm{Prim}(A)$ Hausdorff. For $J \in \mathrm{Prim}(A)$, let $A(J) = A/J$. Let $\pi_J : A \to A/J$ be the quotient map. For $x \in A$, let $a_x(J) = \pi_J(x)$, and let $\Gamma = \{a_x : x \in A\}$. Then $(T, \{A(J)\}, \Gamma)$ is a full continuous field of simple C*-algebras, and the C*-algebra of the continuous field is naturally isomorphic to A.

Proof: The function $J \mapsto \|a_x(J)\|$ is just $\check{x}$, which is continuous by II.6.5.8. The local completeness follows from II.6.5.12 and a partition of unity argument. The isomorphism of Γ with A is obvious.

This result has a significant generalization. If A is a unital C*-algebra, with center Z, set $T = \mathrm{Prim}(Z) \cong \hat{Z}$. If $\pi \in \hat{A}$, then $\pi|_Z$ consists of scalars, and hence is a multiple of an irreducible representation of Z. Thus there is a well-defined surjective map ζ from $\hat{A}$ to T, which drops to a map from $\mathrm{Prim}(A)$ to T, also called ζ, defined by $\zeta(J) = J \cap Z$. It follows easily that ζ is continuous and open. If $t \in T$, let J_t be the ideal of A generated by t, and let $A(t) = A/J_t$, $\pi_t : A \to A/J_t$ the quotient map. Then

IV.1.6.7 Theorem. [DH68]

(i) The map ζ is the *complete regularization* of $\mathrm{Prim}(A)$, i.e. every continuous function from $\mathrm{Prim}(A)$ to a completely regular Hausdorff space factors through ζ.
(ii) For $x \in A$, let $a_x(t) = \pi_t(x)$, and let $\Gamma = \{a_x : x \in A\}$. Then $(T, \{A(t)\}, \Gamma)$ is a full continuous field of C*-algebras, and the C*-algebra of the continuous field is naturally isomorphic to A.

This result is also called the "Dauns-Hoffman Theorem," and is closely related to II.6.5.10. There is a version for nonunital C*-algebras. See [DG83] for a simplified proof.

Note that J_t is not a primitive ideal of A in general, so A_t is not simple or even prime in general.

IV.1.6.8 Examples.

(i) Let $T = \mathbb{N} \cup \{\infty\}$ (or $T = [0, \infty]$). For $t < \infty$, let
$$A_1(t) = A_2(t) = A_3(t) = \mathbb{M}_2.$$
Set $A_1(\infty) = A_2(\infty) = \mathbb{C}$, $A_3(\infty) = \mathbb{C}^2$. Identify $A_i(\infty)$ with the C*-subalgebra C_i of $\mathbb{M}_2$ (IV.1.1.9). Then, for $i = 1, 2, 3$, let
$$A_i = \Gamma_i = \{f \in C(T, \mathbb{M}_2) : f(\infty) \in A_i(\infty)\}.$$
If $U = T \setminus \{\infty\}$, then $\Gamma_i|_U$ is trivial for each i; but obviously none of the Γ_i are locally trivial at ∞.

(ii) For $i = 1, 2, 3$, let $B(t) = A(t) \otimes \mathbb{K}$, with $B_i(\infty)$ regarded as the obvious subalgebra of $\mathbb{M}_2 \otimes \mathbb{K} \cong M_2(\mathbb{K})$, e.g.

$$B_1(\infty) = \left\{ \begin{bmatrix} x & 0 \\ 0 & 0 \end{bmatrix} : x \in \mathbb{K} \right\},$$

Δ_i the obvious variation of Γ_i. Then Δ_3 is not locally trivial at infinity since $B_3(\infty) \cong \mathbb{K}^2 \not\cong B(t)$ for $t < \infty$. The field Δ_2 is also not locally trivial at ∞ even though $B_2(t) \cong \mathbb{K}$ for all t, since it does not satisfy Fell's condition at ∞. However, Δ_1 *is* a trivial continuous field (IV.1.7.22), as is not difficult to see directly.

(iii) Let $C(t) = \mathbb{C}$ for $t < \infty$, $C(\infty) = \mathbb{M}_2$. Regard $\mathbb{C}$ as $\mathbb{C}1 \subseteq \mathbb{M}_2$, and let

$$\Theta = \{f \in C(T, \mathbb{M}_2) : f(t) \in C(t) \text{ for all } t\}.$$

Then $(T, \{C(t)\}, \Theta)$ is not a continuous field since it violates IV.1.6.1(ii)(4) ($C(\infty)$ is "too large").

IV.1.6.9 A continuous field over T is frequently specified by giving a generating set Γ_0 of sections, satisfying

(1) $a(t) \in A(t)$ for all $a \in \Gamma_0$, $t \in T$
(2) $[t \mapsto \|a(t)\|]$ is continuous (but not necessarily vanishing at infinity) for all $a \in \Gamma_0$
(3) Γ_0 is closed under scalar multiplication and under pointwise sum, product, and adjoint
(4) For all $t_0 \in T$, $\{a(t_0) : a \in \Gamma_0\}$ is dense in $A(t_0)$.

The continuous field Γ then consists of all sections which vanish at infinity and which are local uniform limits of sections in Γ_0 (thus Γ_0 is not a subset of Γ in general if the sections in Γ_0 do not vanish at infinity).

IV.1.6.10 EXAMPLE. Let $T = \mathbb{T} \cong [0, 1]$ with 0 and 1 identified, and $A(\theta)$ the rotation algebra A_θ (II.8.3.3(i)), with generators $u(\theta)$, $v(\theta)$. Let Γ_0 be the *-algebra generated by the sections u and v. Then Γ_0 obviously satisfies (1), (3), and (4). Property (2) is not obvious, but may be proved as follows (cf. [Ell82]). On $L^2(\mathbb{T}^2)$, let $U(\theta)$ be defined for all θ by

$$[U(\theta)f](z, w) = zf(z, w)$$

and let $V(\theta)$ be defined by

$$[V(\theta)f](z, w) = wf(e^{2\pi i \theta}z, w).$$

Then $U(\theta)$ and $V(\theta)$ generate a C*-subalgebra of $\mathcal{L}(L^2(\mathbb{T}^2))$ isomorphic to $A(\theta)$; and if p is a polynomial in four noncommuting variables, then

$$\theta \mapsto p(U(\theta), U(\theta)^*, V(\theta), V(\theta)^*)$$

is strong-* continuous since $\theta \mapsto U(\theta)$ and $\theta \mapsto V(\theta)$ are strong-* continuous. Thus the norm of a section is lower semicontinuous. To show upper semicontinuity, suppose $\theta_n \to \theta$, and let

$$\pi : \prod_n A(\theta_n) \to (\prod_n A(\theta_n))/(\bigoplus_n A(\theta_n))$$

be the quotient map, and $\tilde{u} = \pi(\prod u(\theta_n))$ and $\tilde{v} = \pi(\prod v(\theta_n))$; then there is a homomorphism $\phi : A(\theta) \to C^*(\tilde{u}, \tilde{v})$ with $\phi(u(\theta)) = \tilde{u}$, $\phi(v(\theta)) = \tilde{v}$. Then, if p is as above, we have

$$\|p(u(\theta), u(\theta)^*, v(\theta), v(\theta)^*)\| \geq \|\phi(p(u(\theta), u(\theta)^*, v(\theta), v(\theta)^*))\|$$

$$= \|p(\tilde{u}, \tilde{u}^*, \tilde{v}, \tilde{v}^*)\| = \limsup_n \|p(u(\theta_n), u(\theta_n)^*, v(\theta_n), v(\theta_n)^*)\|.$$

Thus the rotation algebras form a continuous field over $\mathbb{T}$ via the generated field Γ. The C*-algebra of this field is the group C*-algebra $C^*(G)$ of the discrete Heisenberg group

$$G = \left\{ \begin{bmatrix} 1 & x & z \\ 0 & 1 & y \\ 0 & 0 & 1 \end{bmatrix} : x, y, z \in \mathbb{Z} \right\}$$

which has a presentation

$$\{u, v, w : uvu^{-1}v^{-1} = w, uw = wu, vw = wv\}$$

(u has $x = 1$, $y = z = 0$; v has $y = 1$, $x = z = 0$; and w has $z = 1$, $x = y = 0$).

IV.1.6.11 There is a theory of C*-bundles and Hilbert bundles (bundles of Hilbert spaces), developed primarily by J.M.G. Fell [Fel69] and M. Dupré [Dup74]; see [FD88a]–[FD88b] for a comprehensive treatment. The set of continuous sections of a C*-bundle is a continuous field of C*-algebras, and conversely. Hilbert bundles will be discussed in more detail in the next section.

IV.1.7 Structure of Continuous Trace C*-Algebras

In this section, we will give a classification up to Morita equivalence (and a partial description up to isomorphism) of the continuous trace C*-algebras over a base space T, in terms of a cohomological invariant in the (Čech) cohomology group $H^3(T, \mathbb{Z})$ called the *Dixmier-Douady invariant*. The theory works well only for paracompact spaces, so throughout this section we will for the most part assume the base space is paracompact. The case of separable continuous trace C*-algebras (where the base space is automatically paracompact) can be described more cleanly, and will be emphasized. There are many technicalities, particularly bundle- and sheaf-theoretic ones, which we will gloss over; a full treatment can be found in [RW98], to which we refer for definitions of sheaf-theoretic terms.

IV.1.7.1 Recall that a Hausdorff space T is *paracompact* if every open cover of T has a locally finite refinement. Every regular (Hausdorff) space which is metrizable or σ-compact is paracompact (see, for example, [Kel75] or [Mun75]), and every paracompact space is normal.

Sheaf Cohomology

Suppose T is paracompact.

IV.1.7.2 The Dixmier-Douady invariant actually is naturally an element of the sheaf cohomology group $H^2(T, \mathcal{S})$, where $\mathcal{S}$ is the sheaf of germs of continuous functions from T to $\mathbb{T}$. If $\mathcal{R}$ is the sheaf of germs of continuous functions from T to $\mathbb{R}$, and $\mathbf{Z}$ denotes the constant sheaf with stalk $\mathbb{Z}$, then there is a short exact sequence of sheaves

$$0 \longrightarrow \mathbf{Z} \xrightarrow{\iota} \mathcal{R} \xrightarrow{\exp} \mathcal{S} \longrightarrow 0$$

where exp denotes exponentiation. The sheaf $\mathcal{R}$ is fine, and hence $H^n(T, \mathcal{R}) = 0$ for all n. The long exact sequence of sheaf cohomology then yields a natural isomorphism $H^2(T, \mathcal{S}) \to H^3(T, \mathbf{Z}) \cong H^3(T, \mathbb{Z})$.

IV.1.7.3 There is another similar identification. Let $\mathcal{H}$ be a separable, infinite-dimensional Hilbert space, and $\mathcal{U}$ the sheaf of germs of continuous functions from T to $U(\mathcal{H})$ with the strong operator topology. Also, let $\mathcal{A}$ be the sheaf of germs of continuous functions from T to $\text{Aut}(\mathcal{K}(\mathcal{H}))$ with the point-norm topology. Since $\text{Aut}(\mathcal{K}(\mathcal{H})) \cong U(\mathcal{H})/\mathbb{T}$ as topological groups, we have an exact sequence of sheaves

$$0 \longrightarrow \mathcal{S} \longrightarrow \mathcal{U} \longrightarrow \mathcal{A} \longrightarrow 0$$

and since $U(\mathcal{H})$ is contractible (I.3.2.10), the sheaf $\mathcal{U}$ is soft. If the sheaves $\mathcal{U}$ and $\mathcal{A}$ were sheaves of abelian groups, one would then obtain from the long exact sequence an isomorphism $H^1(T, \mathcal{A}) \to H^2(T, \mathcal{S})$. In fact, even though the stalks are nonabelian and thus $H^1(T, \mathcal{A})$ does not have a natural group structure, and $H^2(T, \mathcal{U})$ is not even defined, such an isomorphism (bijection) from $H^1(T, \mathcal{A})$ to $H^2(T, \mathcal{S})$ can be constructed in this setting in the same manner as for sheaves of abelian groups.

Hilbert and C*-Bundles

IV.1.7.4 The connection between sheaf cohomology and homogeneous C*-algebras is made via Hilbert bundles (IV.1.6.11). For our purposes, a Hilbert bundle over a base space T (locally compact Hausdorff) is effectively the same thing as a Hilbert $C_o(T)$-module $\mathcal{E}$: $\mathcal{H}(t)$ is obtained from $\mathcal{E}$ using the inner product

$$\langle \xi, \eta \rangle_t = \langle \xi, \eta \rangle_{C_o(T)}(t),$$

dividing out by vectors of length 0 (it is not hard to see that this inner product space is already complete). Thus from the Stabilization Theorem we get that if $\mathcal{E}$ is countably generated (e.g. if each $\mathcal{H}(t)$ is separable and T is σ-compact) and full, then $\mathcal{E}^\infty$ is a trivial bundle, isomorphic to $\mathcal{H} \times T$.

IV.1.7.5 If $\mathcal{E}$ is a Hilbert $C_o(T)$-module, then there is a natural induced $C_o(T)$-module structure on $\mathcal{K}(\mathcal{E})$ (which, if $\mathcal{E}$ is full, coincides with the Dauns-Hoffman module structure when $\mathrm{Prim}(\mathcal{K}(\mathcal{E}))$ is identified with T). If A is a C*-algebra with a specified identification of $\mathrm{Prim}(A)$ with T, $\mathcal{E}$ is full, and ϕ is an isomorphism of A with $\mathcal{K}(\mathcal{E})$, then $\mathcal{E}$ is an $A - C_o(T)$-imprimitivity bimodule. The map ϕ induces a homeomorphism of $\mathrm{Prim}(A)$ with $\mathrm{Prim}(\mathcal{K}(\mathcal{E})) \cong T$; if this homeomorphism agrees with the specified identification of $\mathrm{Prim}(A)$ with T, then $\mathcal{E}$ is called an $A -_T C_o(T)$-*imprimitivity bimodule*, and A is said to be *Morita equivalent to* $C_o(T)$ *over* T.

IV.1.7.6 If $\mathcal{E} \cong (\mathcal{H}(t))$ is a Hilbert bundle over T, then there is a naturally associated C*-bundle whose fiber at T is $\mathcal{K}(\mathcal{H}(t))$, and the continuous sections form a continuous field of elementary C*-algebras over T.

IV.1.7.7 PROPOSITION. The C*-algebra A of a continuous field of elementary C*-algebras arising from a (full) Hilbert bundle over T satisfies Fell's condition (IV.1.4.17), i.e. A is a continuous trace C*-algebra with $\hat{A} \cong T$.

PROOF: Let $\mathcal{E}$ be an $A -_T C_o(T)$ imprimitivity bimodule, $t_0 \in T$, and p_0 a rank-one projection in $A(t_0)$. Let $\xi_0 \in \mathcal{H}(t_0)$ be a unit vector in the range of p_0, so that $p_0 = \Theta_{\xi_0,\xi_0}$. Let $\xi \in \mathcal{E}$ with $\xi(t_0) = \xi_0$; then $\langle \xi, \xi \rangle_{C_o(T)}$ is nonzero in a neighborhood of t_0, so multiplying ξ by an element $f \in C_o(T)$ with $f(t_0) = 1$ we may assume $\langle \xi, \xi \rangle_{C_o(T)}(t) = 1$ for all t in a neighborhood U of t_0. But then $_A\langle \xi, \xi \rangle(t) = \Theta_{\xi,\xi}(t)$ is a rank-one projection in $A(t)$ for all $t \in U$.

The converse is not true (globally) in general, but there is a local version:

IV.1.7.8 PROPOSITION. Let A be a continuous trace C*-algebra with $T = \hat{A}$. Then for each $t \in T$ there is a compact neighborhood Z of t such that A^Z is Morita equivalent to $C(Z)$ over Z.

To prove this, use Fell's condition to cover T with open sets U_i with compact closures Z_i such that A^{Z_i} has a full global rank-one projection p_i.

Conversely, if A is a C*-algebra with $\hat{A}$ Hausdorff, and A is locally Morita equivalent to commutative C*-algebras in this sense, then A has continuous trace.

The Dixmier-Douady Invariant

Suppose A is a continuous-trace C*-algebra with $T = \hat{A}$ paracompact. Choose a locally finite cover of T by open sets U_i with compact closures Z_i such that

A^{Z_i} has a full global rank-one projection p_i. Let $\mathcal{E}_i$ be the $A^{Z_i} -_T C_o(T)$-imprimitivity bimodule $A^{Z_i} p_i$, regarded as a Hilbert bundle over Z_i. If $U_{ij} = U_i \cap U_j$, then the identifications of $\mathcal{K}(\mathcal{E}_j|_{U_{ij}})$ and $\mathcal{K}(\mathcal{E}_i|_{U_{ij}})$ with $A_{U_{ij}}$ give an isometric isomorphism $\mathcal{E}_j|_{U_{ij}} \cong \mathcal{E}_i|_{U_{ij}}$, i.e. a unitary u_{ij} in $\mathcal{L}(\mathcal{E}_j|_{U_{ij}}, \mathcal{E}_i|_{U_{ij}})$. For any i,j,k, let

$$U_{ijk} = U_i \cap U_j \cap U_k.$$

Then, in $\mathcal{L}(\mathcal{E}_k|_{U_{ijk}})$, the unitary $u_{ik}u_{ij}u_{jk}$ commutes with the left action of $A_{U_{ijk}} \cong \mathcal{K}(\mathcal{E}_k|_{U_{ijk}})$, and hence it is easily seen that it is "central," multiplication by a continuous function ν_{ijk} from U_{ijk} to $\mathbb{T}$. A straightforward argument shows:

IV.1.7.9 PROPOSITION.

(i) The functions $\{\nu_{ijk}\}$ form an alternating 2-cocycle: for any i,j,k,l,

$$\nu_{ijk} = \nu_{jkl} \overline{\nu_{ikl}} \nu_{ijl}$$

on $U_{ijkl} = U_i \cap U_j \cap U_k \cap U_l$, and $\nu_{\sigma(i)\sigma(j)\sigma(k)} = \nu_{ijk}^{\epsilon_\sigma}$ on U_{ijk} for any permutation σ (ϵ_σ is the sign of σ).
(ii) The class of $\{\nu_{ijk}\}$ in $H^2(T,\mathcal{S})$ depends only on A and not on the choice of the $\mathcal{U}_i$ or the identifications.

IV.1.7.10 DEFINITION. The class of $\{\nu_{ijk}\}$ in $H^2(T,\mathcal{S}) \cong H^3(T,\mathbb{Z})$ is called the *Dixmier-Douady invariant* of A, denoted $\delta(A)$.

IV.1.7.11 THEOREM. Let A,B be continuous-trace C*-algebras with $T \cong \hat{A} \cong \hat{B}$ paracompact. Then A and B are Morita equivalent over T if and only if $\delta(A) = \delta(B)$.

IV.1.7.12 Every element of $H^3(T,\mathbb{Z})$ occurs as the Dixmier-Douady invariant of a continuous trace C*-algebra with spectrum T (if T is paracompact). For the isomorphism classes of locally trivial bundles over T with fiber $\mathbb{K}$ and structure group $\text{Aut}(\mathbb{K})$ are naturally parametrized by the elements of $H^1(T,\mathcal{A})$. Each such bundle is a continuous field of C*-algebras, and the C*-algebra of the continuous field is a continuous-trace C*-algebra called an $\aleph_0$-*homogeneous C*-algebra over* T. Once the set $H^1(T,\mathcal{A})$ is identified with $H^2(T,\mathcal{S}) \cong H^3(T,\mathbb{Z})$ as in IV.1.7.3, it is straightforward to verify that the Dixmier-Douady invariant of the corresponding $\aleph_0$-homogeneous C*-algebra is the element of $H^1(T,\mathcal{A})$ corresponding to the bundle.

Functoriality

IV.1.7.13 Since $H^3(\cdot,\mathbb{Z})$ is (contravariantly) functorial, one would expect the Dixmier-Douady invariant to be (covariantly) functorial. There is, however, a serious complication: a homomorphism from A to B does not induce a continuous function from $\hat{B}$ to $\hat{A}$ in general, even in the unital case, since local rank need not be preserved (IV.1.4.23). Even if the homomorphism preserves rank in the sense of IV.1.4.24, it does not induce a continuous function from $\hat{B}$ to $\hat{A}$ if A and B are nonunital (there is a proper continuous map from an open subset of $\hat{B}$ to $\hat{A}$ in general if A and B have continuous trace). One cannot just avoid the problem by adding a unit since the unitization of a continuous-trace C*-algebra rarely has continuous trace.

IV.1.7.14 PROPOSITION. Let A and B be continuous-trace C*-algebras with $\hat{A}$, $\hat{B}$ paracompact, and $\phi : A \to B$ a *-homomorphism such that ϕ induces a proper continuous map $\phi^* : \hat{B} \to \hat{A}$, and hence a homomorphism

$$\phi_* : H^3(\hat{A}, \mathbb{Z}) \to H^3(\hat{B}, \mathbb{Z}).$$

Then $\phi_*(\delta(A)) = \delta(B)$.

This condition will always hold if $\hat{A}$ and $\hat{B}$ are compact and ϕ is rank-preserving (i.e. sends abelian elements to abelian elements).

We can then rephrase IV.1.7.11 as:

IV.1.7.15 THEOREM. Let A, B be separable continuous-trace C*-algebras. Then A and B are Morita equivalent if and only if there is a homeomorphism $\phi : \hat{A} \to \hat{B}$ with $\phi_*(\delta(A)) = \delta(B)$.

IV.1.7.16 If A and B are $\aleph_0$-homogeneous C*-algebras with $\hat{A} \cong \hat{B} \cong T$, then A and B are isomorphic over T if and only if $\delta(A) = \delta(B)$. But it can happen that $A \cong B$ even if $\delta(A) \neq \delta(B)$: in fact, $A \cong B$ if and only if there is a homeomorphism $\phi : T \to T$ with $\phi_*(\delta(A)) = \delta(B)$.

There is an interesting consequence. Suppose T is a closed orientable 3-manifold. Then $H^3(T, \mathbb{Z}) \cong \mathbb{Z}$. If α is a generator, then a homeomorphism ϕ of T sends α to $-\alpha$ if and only if ϕ is orientation-reversing. Thus, for example, if T is S^3 or $\mathbb{T}^3$, the $\aleph_0$-homogeneous C*-algebras over T with Dixmier-Douady invariants $\pm\alpha$ are isomorphic. But there is a closed orientable 3-manifold M (a suitable lens space) with no orientation-reversing homeomorphism ([Jac80], [Bon83], [Lu88]); the $\aleph_0$-homogeneous C*-algebras over M with Dixmier-Douady invariants $\pm\alpha$ are not isomorphic.

If A is a continuous-trace C*-algebra over T with Dixmier-Douady invariant δ, then the *opposite algebra* A^{op}, obtained from A by reversing multiplication, is also a continuous-trace C*-algebra over T, and it is easily verified that the Dixmier-Douady invariant of A^{op} is $-\delta$. So if M and α are as above, and A is the $\aleph_0$-homogeneous C*-algebra over M with Dixmier-Douady invariant α, then A is not isomorphic to A^{op}. This is perhaps the simplest example of a C*-algebra which is not anti-isomorphic to itself. (See [Phi01a] for other interesting examples along this line.) It is an important open question whether there exists a separable simple nuclear C*-algebra not anti-isomorphic to itself. (There are II_1 factors with this property, and a separable simple nonnuclear example can be constructed from these by the method of II.8.5; cf. [Phi04].)

Separable Continuous-Trace C*-Algebras

If A is a separable continuous-trace C*-algebra, then $\hat{A}$ is second countable and thus paracompact. Combining IV.1.7.15 with the Brown-Green-Rieffel theorem (II.7.6.11), we obtain:

IV.1.7.17 THEOREM. Let A, B be separable continuous-trace C*-algebras. Then A and B are stably isomorphic if and only if there is a homeomorphism $\phi : \hat{A} \to \hat{B}$ with $\phi_*(\delta(A)) = \delta(B)$.

IV.1.7.18 If A is a separable continuous-trace C*-algebra with $T = \hat{A}$, let $\{U_i\}$ be an open cover as in IV.1.7.8. The local imprimitivity bimodules (Hilbert bundles) $\mathcal{E}_i$ are countably generated, with separable fibers, and by II.7.6.12 $\mathcal{E}_i^\infty$ is trivial. Thus the corresponding open cover for $A \otimes \mathbb{K}$ represents $A \otimes \mathbb{K}$ as a locally trivial bundle with fiber $\mathbb{K}$, i.e. $A \otimes \mathbb{K}$ is an $\aleph_0$-homogeneous C*-algebra. Thus we obtain:

IV.1.7.19 THEOREM. The stable separable continuous-trace C*-algebras are precisely the separable $\aleph_0$-homogeneous C*-algebras.

IV.1.7.20 Of course, if A is a separable stable continuous-trace C*-algebra (or, more generally, a separable stable liminal C*-algebra), then the quotient of A by any primitive ideal must be isomorphic to $\mathbb{K}$, i.e. the corresponding bundle of elementary C*-algebras has fiber $\mathbb{K}$ everywhere. The converse (that if A is a separable continuous-trace C*-algebra and each fiber is $\mathbb{K}$, then A is stable) is not quite true, since IV.1.7.19 gives an additional necessary condition: the bundle must be locally trivial. This is not automatic: for example, if T is a countably infinite product of copies of S^2, then there is a bundle over T with fiber $\mathbb{K}$, satisfying Fell's condition, which is not locally trivial at any point [DD63]. However, we have:

IV.1.7.21 THEOREM. [DD63], [Dix69b, 10.8.8] Let T be a second countable locally compact Hausdorff space of finite (covering) dimension. Then every bundle over T, with fiber $\mathbb{K}$, satisfying Fell's condition, is locally trivial.

IV.1.7.22 COROLLARY. Let A be a separable continuous-trace C*-algebra with $\hat{A}$ finite-dimensional. Then the following are equivalent:

(i) A has no finite-dimensional irreducible representations.
(ii) A is $\aleph_0$-homogeneous.
(iii) A is stable.

Finite Homogeneous C*-Algebras

A finite homogeneous C*-algebra has continuous trace (IV.1.4.14), and thus has a Dixmier-Douady invariant (if the base space is paracompact). This situation is special:

IV.1.7.23 THEOREM. Let A be an n-homogeneous C*-algebra over a paracompact space T. Then

(i) The corresponding bundle of elementary C*-algebras ($\mathbb{M}_n$-bundle) is locally trivial.
(ii) The Dixmier-Douady invariant $\delta(A)$ is an n-torsion element of $H^3(T, \mathbb{Z})$.

IV.1.7.24 Conversely, if δ is an n-torsion element of $H^3(T, \mathbb{Z})$, then there is a k-homogeneous C*-algebra A over T with $\delta(A) = \delta$, for some k (which may be much larger than n) [Gro68, 1.7].

IV.1.7.25 If A is a continuous-trace C*-algebra over T (assumed compact), and p is a full projection in A, then pAp is a unital continuous-trace C*-algebra over T, hence (locally) homogeneous. Thus $\delta(A)$ must be a torsion element. So we obtain:

IV.1.7.26 COROLLARY. Let A be a continuous-trace C*-algebra whose base space T is connected and compact, and such that $\delta(A)$ is not a torsion element of $H^3(T, \mathbb{Z})$. Then A contains no nonzero projections.

If A is any continuous-trace C*-algebra whose base space T is connected and noncompact, then A can contain no nonzero projections by II.6.5.6.

Structure of Type I C*-Algebras

IV.1.7.27 If A is a separable stable continuous-trace C*-algebra, then by IV.1.7.19 A is built up from algebras of the form $C_o(T, \mathbb{K})$ by extensions, finitely many if $\hat{A}$ is compact but transfinitely (although countably) many in general.

Since every Type I C*-algebra has a composition series in which the successive quotients have continuous trace (IV.1.4.28), the same can be said for any separable stable Type I C*-algebra. Thus every separable Type I C*-algebra can be built up out of commutative C*-algebras by successive applications of the following operations:

Stabilization (tensoring with $\mathbb{K}$).
Cutting down to a corner.
Taking extensions, perhaps transfinitely (but countably) many.

IV.2 Classification of Injective Factors

One of the greatest achievements so far in the subject of operator algebras is the complete classification of injective factors with separable predual: up to isomorphism, the examples constructed in III.3.2.18 are the only ones. This technical *tour de force* was mostly accomplished by A. Connes [Con76] (his

Fields medal was primarily for this work), building on important previous work by several authors, with the final case settled by U. Haagerup [Haa87]. Some of Connes' arguments have been simplified, but the overall results remain very deep.

There are close connections and analogies with the theory of nuclear C*-algebras (IV.3). The two theories developed somewhat in parallel, and ideas from each were influential in the other.

It is far beyond the scope of this volume to give a complete treatment of the theory of injective factors. We will only outline the main results, mostly without proof. For a comprehensive treatment, see the volume by Takesaki [Tak03a].

The results fall naturally into two parts:

(i) Proof that several natural classes of factors (injective, semidiscrete, approximately finite dimensional, amenable) coincide;
(ii) Explicit description of all injective factors.

Because of the importance of this class of algebras, the fact that it can be characterized in several different ways has led to discussion of what name should be generically used for it. Some leading experts (e.g. in [Con94]) have advocated using "amenable," with considerable merit. A similar discussion has taken place in the C*-algebra setting, concerning the equally important class characterized by various conditions equivalent to nuclearity (cf. IV.3), where the term "amenable" has been proposed with equal merit. There are two drawbacks to adopting "amenable" as a generic term in either or both cases, however:

(1) The same ambiguity which occurs with the term "Type I" would arise, since an amenable von Neumann algebra is not in general an amenable C*-algebra. It thus seems preferable to use separate terms in the two situations.
(2) Other terms have come to be commonly used, particularly in the C*-algebra case where the term "nuclear" is nearly universal. Terminology in the von Neumann setting is less firmly established, but the term "injective" is the most common. The next most common term is probably "hyperfinite," but this term has its own drawbacks (III.3.4.1).

We have chosen to use the terms "injective" and "nuclear" generically for these reasons. Of course, in discussing the equivalence of the characterizing conditions, the various names will be used and separately defined.

The classes of injective von Neumann algebras and nuclear C*-algebras are the most important classes of operator algebras, at least in the present state of the subject.

IV.2.1 Injective C*-Algebras

IV.2.1.1 DEFINITION. A C*-algebra M is an *injective C*-algebra* if, whenever A is a C*-algebra, B a C*-subalgebra of A, and $\phi : B \to M$ a completely positive contraction, then ϕ extends to a completely positive contraction $\psi : A \to M$ (i.e. M is an injective object in the category of C*-algebras and completely positive contractions).

IV.2.1.2 EXAMPLES.

(i) For any $\mathcal{H}$, $\mathcal{L}(\mathcal{H})$ is an injective C*-algebra by the Arveson extension theorem (II.6.9.12). In particular, $\mathbb{M}_n$ is injective (more generally, any finite-dimensional C*-algebra is injective).
(ii) Any product (in particular, a finite direct sum) of injective C*-algebras is injective.
(iii) If N is a C*-subalgebra of M, M is injective, and there is a conditional expectation θ (projection of norm 1; cf. II.6.10.2) from M onto N, then N is injective [if $B \subseteq A$ and $\phi : B \to N \subseteq M$, extend ϕ to $\psi : A \to M$; then $\theta \circ \psi : A \to N$ is an extension of ϕ.] In particular, if M is injective and p is a projection in M, then pMp is injective.

IV.2.1.3 There is a "converse" to (iii): if N is a C*-subalgebra of a C*-algebra M, and N is injective, then the identity map from N to N extends to a completely positive idempotent contraction from M to N; thus there is a conditional expectation from M onto N.

Combining this with (iii), we get:

IV.2.1.4 PROPOSITION. Let M be a C*-algebra. Then the following are equivalent:

(i) M is injective.
(ii) There is a faithful representation π of M on a Hilbert space $\mathcal{H}$, such that there is a conditional expectation from $\mathcal{L}(\mathcal{H})$ onto $\pi(M)$.
(iii) For every faithful representation π of M on a Hilbert space $\mathcal{H}$, there is a conditional expectation from $\mathcal{L}(\mathcal{H})$ onto $\pi(M)$.
(iv) Whenever M is a C*-subalgebra of a C*-algebra A, there is a conditional expectation from A to M.

IV.2.1.5 COROLLARY. A matrix algebra over an injective C*-algebra is injective.

PROOF: If $M \subseteq \mathcal{L}(\mathcal{H})$ and $\theta : \mathcal{L}(\mathcal{H}) \to M$ is a conditional expectation, so is

$$\theta \otimes id : \mathcal{L}(\mathcal{H}) \otimes \mathbb{M}_n \cong \mathcal{L}(\mathcal{H}^n) \to M \otimes \mathbb{M}_n \cong \mathbb{M}_n(M).$$

IV.2.1.6 By an argument similar to IV.2.1.3, if X is an operator system which is an injective object in the category of operator systems and completely positive contractions (the existence of an identity in these operator systems is not essential, i.e. one can work in the category of closed positively generated *-subspaces of C*-algebras), embed X into a C*-algebra A and extend the identity map on X to an idempotent completely positive contraction from A onto X. Then X is completely order isomorphic to an (injective) C*-algebra by II.6.10.11. Conversely, if M is an injective C*-algebra, and $Y \subseteq X$ are operator systems and $\phi : Y \to M$ is a completely positive contraction, embed M into $\mathcal{L}(\mathcal{H})$, extend ϕ to $\omega : X \to \mathcal{L}(\mathcal{H})$ by the Arveson Extension Theorem, and compose with a conditional expectation from $\mathcal{L}(\mathcal{H})$ to M to get an extension of ϕ to $\psi : X \to M$. Thus the injective C*-algebras are also precisely the injective objects in the larger category of operator systems and completely positive contractions.

IV.2.1.7 PROPOSITION. An injective C*-algebra is an AW*-algebra. An injective C*-algebra which can be faithfully represented on a separable Hilbert space, whose center is a von Neumann algebra (W*-algebra), is a von Neumann algebra.

PROOF: Let M be an injective C*-algebra, faithfully represented on a Hilbert space $\mathcal{H}$, and let θ be a conditional expectation from $\mathcal{L}(\mathcal{H})$ onto M. Let S be a subset of M. The right annihilator of S in $\mathcal{L}(\mathcal{H})$ is $p\mathcal{L}(\mathcal{H})$ for a projection p. We have $x\theta(p) = \theta(xp) = 0$ for all $x \in S$, and if $y \in M$ is in the right annihilator of S, we have $py = y$, so

$$y = \theta(y) = \theta(py) = \theta(p)y.$$

Thus, taking $y = \theta(p)$, $\theta(p)$ is a projection in M; and the right annihilator of S in M is $\theta(p)M$. So M is an AW*-algebra by III.1.8.2. The last statement follows from III.1.8.6.

In particular, every injective C*-algebra is unital (this can be seen more easily: if A is nonunital, there cannot be a conditional expectation from $\tilde{A}$ to A by II.6.10.5).

IV.2.1.8 There are injective C*-algebras which are not W*-algebras. For example, every commutative AW*-algebra is injective (see IV.2.2.10 for the W*-case). In fact, a commutative AW*-algebra is injective in the larger category of Banach spaces and linear contractions ([Nac50], [Goo49], [Has58]; the converse is also true [Kel52], [Has58]. See also [Lac74, §11].) There are even wild injective AW*-factors [Ham86] (but only on nonseparable Hilbert spaces).

IV.2.2 Injective von Neumann Algebras

From now on, we consider only injective von Neumann algebras.

Normal Conditional Expectations

IV.2.2.1 If M is an injective von Neumann algebra on $\mathcal{H}$, it is not true in general that there is a *normal* conditional expectation from $\mathcal{L}(\mathcal{H})$ onto M, as the next result shows. This fact creates considerable technical difficulties in the theory of injective von Neumann algebras.

A von Neumann algebra M is *purely atomic* if M contains a set $\{p_i\}$ of minimal projections with $\sum p_i = 1$. Since a minimal projection is abelian, a purely atomic von Neumann algebra is Type I. But many Type I von Neumann algebras are not purely atomic: for example, $L^\infty(X, \mu)$ is purely atomic if and only if μ is purely atomic. In fact, a von Neumann algebra is purely atomic if and only if it is a product of Type I factors. It is easily seen that a von Neumann algebra is purely atomic if and only if it has a purely atomic masa.

IV.2.2.2 THEOREM. [Tom59] Let M be a von Neumann algebra on $\mathcal{H}$. Then there is a normal conditional expectation from $\mathcal{L}(\mathcal{H})$ onto M if and only if M is purely atomic.

PROOF: If M is purely atomic, then $\mathcal{Z}(M)$ is purely atomic. If $\{z_i\}$ is a family of minimal projections in $\mathcal{Z}(M)$ with $\sum_i z_i = 1$, for $x \in \mathcal{L}(\mathcal{H})$ define $\omega(x) = \sum_i z_i x z_i$; then ω is a normal conditional expectation from $\mathcal{L}(\mathcal{H})$ onto

$$\prod z_i \mathcal{L}(\mathcal{H}) z_i = \prod \mathcal{L}(z_i \mathcal{H}).$$

Since Mz_i is a Type I factor on $z_i \mathcal{H}$, there is a normal conditional expectation θ_i from $\mathcal{L}(z_i \mathcal{H})$ onto Mz_i; then $(\prod \theta_i) \circ \omega$ is a normal conditional expectation from $\mathcal{L}(\mathcal{H})$ onto M.

Conversely, suppose first that A is a commutative von Neumann algebra on $\mathcal{H}$ which is not purely atomic. If $\{p_i\}$ is a maximal family of mutually orthogonal minimal projections in A, set $q = 1 - \sum p_i$. If there is a normal conditional expectation $\theta: \mathcal{L}(\mathcal{H}) \to A$, then θ restricts to a normal conditional expectation from

$$q\mathcal{L}(\mathcal{H})q = \mathcal{L}(q\mathcal{H})$$

to qA. Thus, to obtain a contradiction, we may assume A has no minimal projections. If γ is a character (pure state) of A, and p is a nonzero projection in A, then $p = q + r$ for nonzero projections q, r. Since $\gamma(q), \gamma(r)$, and $\gamma(p) = \gamma(q) + \gamma(r)$ are each either 0 or 1, either $\gamma(q)$ or $\gamma(r)$ (or both) is 0, i.e. p dominates a nonzero projection p' with $\gamma(p') = 0$. Let $\{q_i\}$ be a maximal set of mutually orthogonal nonzero projections in A with $\gamma(q_i) = 0$ for all i; then $\sum q_i = 1$. If ψ is a normal positive linear functional on A with $\psi \leq \gamma$, then

$$\|\psi\| = \psi(1) = \sum \psi(q_i) \leq \sum \gamma(q_i) = 0.$$

Thus γ is a singular state on A, and hence $\gamma \circ \theta$ is a singular state on $\mathcal{L}(\mathcal{H})$. So if p is a rank-one projection in $\mathcal{L}(\mathcal{H})$, then $\gamma(\theta(p)) = 0$ (III.2.1.12). This is true for all pure states γ of A, so $\theta(p) = 0$. If $\{p_i\}$ is a maximal family of mutually orthogonal minimal projections in $\mathcal{L}(\mathcal{H})$, then

$$1 = \theta(1) = \sum \theta(p_i) = 0,$$

a contradiction. Thus there is no normal conditional expectation from $\mathcal{L}(\mathcal{H})$ onto A.

Now suppose M is a general von Neumann algebra on $\mathcal{H}$, and suppose there is a normal conditional expectation $\theta : \mathcal{L}(\mathcal{H}) \to M$. Then M is semifinite by III.2.5.25. If q is a finite projection in M of central support 1, then as above θ induces a normal conditional expectation from $\mathcal{L}(q\mathcal{H})$ onto qMq. If M is not purely atomic, neither is qMq. Thus, to obtain a contradiction, we may assume M is finite. If M is not purely atomic, then M has a masa A which is not purely atomic. There is a normal conditional expectation $\omega : M \to A$ by III.4.7.8. But then $\omega \circ \theta$ is a normal conditional expectation from $\mathcal{L}(\mathcal{H})$ to A, contradicting the previous paragraph.

An almost identical argument shows more generally:

IV.2.2.3 THEOREM. Let $N \subseteq M$ be von Neumann algebras. If there is a normal conditional expectation from M to N, and M is purely atomic, then N is purely atomic.

Thus, if (X, μ) is a measure space, then there is always a conditional expectation from $\mathcal{L}(L^2(X))$ onto $L^\infty(X)$, but there is a normal conditional expectation (if and) only if μ is purely atomic. If M is not Type I, there is never a normal conditional expectation from $\mathcal{L}(\mathcal{H})$ onto M. See III.4.7.7–III.4.7.8 for conditions under which normal conditional expectations exist on more general von Neumann algebras.

A simple consequence of IV.2.2.2 is useful:

IV.2.2.4 COROLLARY. Let $N \subseteq M$ be von Neumann algebras, with N finite-dimensional. Then there is a normal conditional expectation from M onto N.

This also follows easily from IV.3.2.16, or from III.1.5.3 and III.2.2.6.

Amplifications and Commutants

We now show that the class of injective von Neumann algebras is closed under amplifications and commutants.

IV.2.2.5 PROPOSITION. Let $N \subseteq M$ be von Neumann algebras on $\mathcal{H}$, θ a conditional expectation from M onto N, and let $\mathcal{H}'$ be another Hilbert space. Then there is a conditional expectation Θ from $M \bar{\otimes} \mathcal{L}(\mathcal{H}')$ onto $N \bar{\otimes} \mathcal{L}(\mathcal{H}')$ such that

$$\Theta(x \otimes y) = \theta(x) \otimes y$$

for all $x \in M$, $y \in \mathcal{L}(\mathcal{H}')$.

PROOF: The result is immediate from III.2.2.5 if θ is normal, but θ is not necessarily normal (and there need not be any normal conditional expectation from M onto N). We thus argue as follows. Let $\{e_{ij}|i,j \in \Omega\}$ be a set of matrix units in $\mathcal{L}(\mathcal{H}')$, and $\mathcal{F}$ the set of finite subsets of Ω. If $F \in \mathcal{F}$, let $p_F = \sum_{i \in F} e_{ii}$. Then $p_F \mathcal{L}(\mathcal{H}') p_F$ is a finite-dimensional matrix algebra, and there is a conditional expectation ϕ_F from

$$M \bar{\otimes} p_F \mathcal{L}(\mathcal{H}') p_F = M \odot p_F \mathcal{L}(\mathcal{H}') p_F$$

onto

$$N \bar{\otimes} p_F \mathcal{L}(\mathcal{H}') p_F = N \odot p_F \mathcal{L}(\mathcal{H}') p_F$$

defined by

$$\phi_F\left(\sum x_n \otimes y_n\right) = \sum \theta(x_n) \otimes y_n.$$

From this we get a conditional expectation Θ_F from $M \bar{\otimes} \mathcal{L}(\mathcal{H}')$ to $N \bar{\otimes} p_F \mathcal{L}(\mathcal{H}') p_F$ defined by

$$\Theta_F(x) = \phi_F((1 \otimes p_F) x (1 \otimes p_F)).$$

Now let ω be a Banach limit (I.3.2.4) on $\mathcal{F}$ in $\mathcal{L}(\mathcal{H} \otimes \mathcal{H}')$, and for each $x \in M \bar{\otimes} \mathcal{L}(\mathcal{H}')$ set $\Theta(x) = \lim_\omega \Theta_F(x)$. Then Θ is a map from $M \bar{\otimes} \mathcal{L}(\mathcal{H}')$ into $N \bar{\otimes} \mathcal{L}(\mathcal{H}')$ with $\|\Theta\| \leq 1$; and if $y \in N \bar{\otimes} \mathcal{L}(\mathcal{H}')$, then

$$(1 \otimes p_F) y (1 \otimes p_F) \in N \bar{\otimes} p_F \mathcal{L}(\mathcal{H}') p_F$$

so

$$\Theta_F(y) = (1 \otimes p_F) y (1 \otimes p_F)$$

for all F, and hence

$$\Theta(y) = \lim_\omega ((1 \otimes p_F) y (1 \otimes p_F)) = y$$

since $(1 \otimes p_F) y (1 \otimes p_F) \to y$ weakly (in fact strongly). Thus Θ is a conditional expectation by II.6.10.2. (It is easy to prove directly that Θ is completely positive.)

IV.2.2.6 COROLLARY. Let M be an injective von Neumann algebra on $\mathcal{H}$, and let $\mathcal{H}'$ be another Hilbert space. Then $M \bar{\otimes} \mathcal{L}(\mathcal{H}')$ is injective.

IV.2.2.7 THEOREM. Let M be a von Neumann algebra on $\mathcal{H}$. Then M is injective if and only if M' is injective.

PROOF: By III.2.2.8, IV.2.2.6, and the last sentence of IV.2.1.2(iii), if π and ρ are faithful normal representations of M, then $\pi(M)'$ is injective if and only if $\rho(M)'$ is injective. But if M is in standard form, it is obvious from Modular Theory that M is injective if and only if M' is injective.

IV.2.2.8 PROPOSITION.

(i) Let M be a von Neumann algebra on $\mathcal{H}$. If there is an increasing net (M_i) of injective von Neumann subalgebras of M, with $\cup M_i$ σ-weakly dense in M, then M is injective.
(ii) If (M_i) is a decreasing net of injective von Neumann algebras on $\mathcal{H}$, then $M = \cap M_i$ is injective.

The second part follows easily from the fact that the set of unital completely positive maps from $\mathcal{L}(\mathcal{H})$ to $\mathcal{L}(\mathcal{H})$ is compact in the point-σ-weak topology: if θ_i is a conditional expectation from $\mathcal{L}(\mathcal{H})$ onto M_i, then any point-σ-weak limit of the θ_i is a conditional expectation from $\mathcal{L}(\mathcal{H})$ onto M. The first part cannot be quite proved the same way because the conditional expectations are not necessarily normal; but (i) follows from (ii) and IV.2.2.7 since $(\cup M_i)' = \cap M_i'$.

IV.2.2.9 COROLLARY. Every approximately finite dimensional (III.3.4.1) von Neumann algebra is injective.

Since every Type I von Neumann algebra is approximately finite-dimensional, we obtain:

IV.2.2.10 COROLLARY. Every Type I von Neumann algebra (in particular, every commutative von Neumann algebra) is injective.

The next result is an immediate corollary of III.4.7.8:

IV.2.2.11 COROLLARY. If M is an injective countably decomposable finite von Neumann algebra, and N is any von Neumann subalgebra of M, then N is injective.

Since the hyperfinite II_1 factor $\mathcal{R}$ is injective, any von Neumann subalgebra of $\mathcal{R}$ is also injective.

By arguments similar to those in IV.2.2.8, one can show:

IV.2.2.12 PROPOSITION. Let M be a von Neumann algebra on a separable Hilbert space, with central decomposition $\int_X^\oplus M_x \, d\mu(x)$. Then M is injective $\iff M_x$ is injective for almost all x.

Thus the study of injective von Neumann algebras can be effectively reduced to studying injective factors.

Nuclearity and Injectivity

There is a close relationship between nuclearity and injectivity, the most elementary part being the next result, which gives many examples of injective von Neumann algebras. The argument in the proof is due to E. C. Lance [Lan73].

IV.2.2.13 THEOREM. Let A be a quasinuclear C*-algebra (II.9.6.4). If π is any representation of A, then $\pi(A)''$ is injective. In particular, A^{**} is injective.

PROOF: Let $\pi : A \to \mathcal{L}(\mathcal{H})$ be a representation. Then π extends to a representation ρ of $A \otimes_{\max} \pi(A)'$ on $\mathcal{H}$. We have

$$A \otimes_{\max} \pi(A)' \subseteq A \otimes_{\max} \mathcal{L}(\mathcal{H})$$

since A is quasinuclear, so ρ extends to a representation σ of $\tilde{A} \otimes_{\max} \mathcal{L}(\mathcal{H})$ on a larger Hilbert space $\mathcal{H}'$. If P is the projection from $\mathcal{H}'$ onto $\mathcal{H}$, and $x \in \mathcal{L}(\mathcal{H})$, set $\theta(x) = P\sigma(1 \otimes x)|_{\mathcal{H}}$. Then $\theta(x) \in \mathcal{L}(\mathcal{H})$, and since P and $\sigma(1 \otimes x)$ commute with $\sigma(a \otimes 1)$ for $a \in A$, it follows that $\theta(x) \in \pi(A)'$; and $\theta : \mathcal{L}(\mathcal{H}) \to \pi(A)'$ is a conditional expectation. Apply IV.2.2.7.

The converse of this result also holds (IV.3.1.12); in fact, if A^{**} is injective, then A is nuclear (so a quasinuclear C*-algebra is nuclear.) This is, however, a deep result.

Injectivity and Amenable Group Actions

The next result, a close analog of II.6.10.4(v), provides additional examples of injective von Neumann algebras.

IV.2.2.14 PROPOSITION. Let α be an action of a locally compact group G on a von Neumann algebra M. If G is amenable, then there is a conditional expectation from M onto $M^G (= M^\alpha)$.

PROOF: Let m be an invariant mean on $C_b(G)$. If $x \in M$, $\phi \in M_*$, set

$$\theta_x(\phi) = m(g \mapsto \phi \circ \alpha_g(x)).$$

Then $\theta_x \in (M_*)^*$, hence corresponds to an element $\theta(x) \in M$. It is easy to check that $\theta(x) \in M^G$, and $\theta : M \to M^G$ is a projection of norm 1.

IV.2.2.15 COROLLARY. Let π be a representation of an amenable locally compact group G. Then $\pi(G)'$ and $\pi(G)''$ are injective. In particular, $\mathfrak{L}(G)$ is injective.

IV.2.2.16 COROLLARY. Let α be an action of a locally compact group G on a von Neumann algebra N. If N is injective and G is amenable, then $M = N \bar{\rtimes}_\alpha G$ is injective.

PROOF: Let M (and hence N) act on $\mathcal{H}$. Inside M is a canonical group $\mathcal{G}$ of unitaries, isomorphic to G, normalizing N and inducing α. $\mathcal{G}$ also normalizes N', hence defines an action of G on N'; and $M' = (N')^G$. The conditional expectation of IV.2.2.14 from N' to $(N')^G$ can be composed with a conditional expectation from $\mathcal{L}(\mathcal{H})$ onto N' to get a conditional expectation from $\mathcal{L}(\mathcal{H})$ onto M'.

IV.2.2.17 Combining IV.2.2.16 with Takesaki Duality (III.3.2.8) and either the continuous or discrete decomposition (III.4.8), if M is a von Neumann algebra of Type III and $N \cong N_0 \bar\otimes \mathcal{L}(\mathcal{H})$ is the associated von Neumann algebra of Type II$_\infty$, where N_0 is Type II$_1$, then M is injective if and only if N_0 is injective. This fact is used to reduce the proof that injective implies approximately finite dimensional to the II$_1$ case (and, using IV.2.2.12, to the II$_1$ factor case).

IV.2.2.18 It is not true in general that if an amenable group G acts on a C*-algebra A, that there is a conditional expectation from A onto A^G, even if $G = \mathbb{Z}$ and A is commutative (an example similar to the one in II.6.10.5 can be constructed).

Property P

IV.2.2.19 Injectivity for von Neumann algebras was first discussed in [HT67] (cf. [Tom57]), where it was called *Property E*. An apparently stronger property, related to the Dixmier property (III.2.5.16), had been previously considered by J. T. Schwartz [Sch63] (cf. [Sak71, 4.4.14-4.4.21]):

IV.2.2.20 DEFINITION. A von Neumann algebra M on a Hilbert space $\mathcal{H}$ has *Property P* if, for every $T \in \mathcal{L}(\mathcal{H})$, the σ-weakly closed convex hull $C(T)$ of
$$\{uTu^* : u \in \mathcal{U}(M)\}$$
contains an element of M'.

It is easily proved using III.2.2.8 that Property P is independent of the way M is represented on $\mathcal{H}$, and hence is an algebraic invariant of M. So by modular theory M has Property P if and only if M' does.

Schwartz proved that if M has Property P, then M' is injective, using a Banach limit type argument (one must choose in a linear fashion one element from $C(T) \cap M'$ for each T), and that if G is a discrete group, then $\mathfrak{L}(G)$ has property P if and only if G is amenable.

It is easily seen using II.6.10.4(iv) and the σ-weak compactness of the unit ball of $\mathcal{L}(\mathcal{H})$ that an approximately finite dimensional von Neumann algebra has Property P. It is not obvious that an injective von Neumann algebra has Property P; however, since an injective von Neumann algebra is approximately finite dimensional, it turns out that Property P is equivalent to injectivity.

Hypertraces

IV.2.2.21 DEFINITION. If M is a von Neumann algebra on a Hilbert space $\mathcal{H}$, a *hypertrace* for M is a state ϕ on $\mathcal{L}(\mathcal{H})$ with the property that $\phi(xy) = \phi(yx)$ for all $x \in M$, $y \in \mathcal{L}(\mathcal{H})$.

If ϕ is a hypertrace for M, the restriction of ϕ to M is a tracial state on M. Thus if M is a factor, it must be finite.

A hypertrace for M is not a normal state on $\mathcal{L}(\mathcal{H})$ in general, although the restriction to M is normal if M is a factor. In fact, if M is a von Neumann algebra with a normal hypertrace whose restriction to M is faithful, then M must be purely atomic and hence Type I.

If M is a finite injective von Neumann algebra, then M has a hypertrace: if $\theta : \mathcal{L}(\mathcal{H}) \to M$ is a conditional expectation and τ a tracial state on M, then $\tau \circ \theta$ is a hypertrace for M. The converse also turns out to be true for factors [Con76], i.e.

IV.2.2.22 THEOREM. A II_1 factor M has a hypertrace if and only if it is injective.

More generally, a von Neumann algebra M with a family of hypertraces whose restrictions to M are a faithful normal family must be injective (and finite).

IV.2.3 Normal Cross Norms.

If one or both factors in a tensor product are von Neumann algebras, there are interesting norms besides the maximal and minimal ones:

IV.2.3.1 DEFINITION. Let A and B be C*-algebras.

(i) If A is a von Neumann algebra, and $x = \sum_{k=1}^{n} a_k \otimes b_k \in A \odot B$, set

$$\|x\|_{\mathrm{lnor}} = \sup\{\|\sum_{k=1}^{n} \pi(a_k)\rho(b_k)\|\}$$

where π, ρ range over commuting representations of A and B with π normal.

(ii) If B is a von Neumann algebra, and $x = \sum_{k=1}^{n} a_k \otimes b_k \in A \odot B$, set

$$\|x\|_{\mathrm{rnor}} = \sup\{\|\sum_{k=1}^{n} \pi(a_k)\rho(b_k)\|\}$$

where π, ρ range over commuting representations of A and B with ρ normal.

(iii) If A and B are von Neumann algebras, and $x = \sum_{k=1}^{n} a_k \otimes b_k \in A \odot B$, set

$$\|x\|_{\mathrm{bin}} = \sup\{\|\sum_{k=1}^{n} \pi(a_k)\rho(b_k)\|\}$$

where π, ρ range over commuting normal representations of A and B.

Write $A \otimes_{\text{lnor}} B$, $A \otimes_{\text{rnor}} B$, $A \otimes_{\text{bin}} B$ for the completions with respect to these norms (when defined); these are called the *left normal*, *right normal*, and *binormal* tensor product of A and B respectively.

Caution: although the notation does not reflect it, as with $\|\cdot\|_{\max}$, the norms $\|\cdot\|_{\text{lnor}}$, etc., depend on the choice of the C*-algebras A and B, i.e. if $A \subseteq A_1$, $B \subseteq B_1$, then the restriction of $\|\cdot\|_{\text{lnor}}$ on $A_1 \odot B_1$ to $A \odot B$ is not $\|\cdot\|_{\text{lnor}}$ on $A \odot B$ in general, and similarly for $\|\cdot\|_{\text{rnor}}$, $\|\cdot\|_{\text{bin}}$.

Note that if A and B are C*-algebras, the restriction of $\|\cdot\|_{\text{rnor}}$ on $A \odot B^{**}$ to $A \odot B$ is $\|\cdot\|_{\max}$ on $A \odot B$ (and hence the restriction of $\|\cdot\|_{\max}$ on $A \odot B^{**}$ to $A \odot B$ is $\|\cdot\|_{\max}$ on $A \odot B$); and if A is a von Neumann algebra, then the restriction of $\|\cdot\|_{\text{bin}}$ on $A \odot B^{**}$ to $A \odot B$ is $\|\cdot\|_{\text{lnor}}$ on $A \odot B$.

The next result is an immediate corollary of III.1.5.3.

IV.2.3.2 PROPOSITION. If B is any C*-algebra, then $\|\cdot\|_{\text{lnor}} = \|\cdot\|_{\min}$ on $\mathcal{L}(\mathcal{H}) \odot B$ (and $\|\cdot\|_{\text{rnor}} = \|\cdot\|_{\min}$ on $B \odot \mathcal{L}(\mathcal{H})$). If M is a von Neumann algebra, then $\|\cdot\|_{\text{bin}} = \|\cdot\|_{\min}$ on $M \odot \mathcal{L}(\mathcal{H})$ (or $\mathcal{L}(\mathcal{H}) \odot M$).

There are normal versions of II.9.7.2 and II.9.7.3 (cf. III.2.2.4):

IV.2.3.3 LEMMA. Under the hypotheses of II.9.7.2, if A and/or B is a von Neumann algebra and ϕ and/or ψ is normal, then π and/or ρ can be chosen to be normal.

IV.2.3.4 COROLLARY. Let A_1, A_2, B_1, B_2 be C*-algebras and $\phi : A_1 \to A_2$ and $\psi : B_1 \to B_2$ completely positive contractions.

(i) If A_1, A_2 are von Neumann algebras and ϕ is normal, then the map $\phi \otimes \psi$ from $A_1 \odot B_1$ to $A_2 \odot B_2$ given by

$$(\phi \otimes \psi)\left(\sum x_i \otimes y_i\right) = \sum \phi(x_i) \otimes \psi(y_i)$$

extends to a completely positive contraction, also denoted $\phi \otimes \psi$, from $A_1 \otimes_{\text{lnor}} B_1$ to $A_2 \otimes_{\text{lnor}} B_2$.

(ii) If A_1, A_2, B_1, B_2 are von Neumann algebras and ϕ and ψ are normal, then the map $\phi \otimes \psi$ from $A_1 \odot B_1$ to $A_2 \odot B_2$ given by

$$(\phi \otimes \psi)\left(\sum x_i \otimes y_i\right) = \sum \phi(x_i) \otimes \psi(y_i)$$

extends to a completely positive contraction, also denoted $\phi \otimes \psi$, from $A_1 \otimes_{\text{bin}} B_1$ to $A_2 \otimes_{\text{bin}} B_2$.

IV.2.3.5 Note that if M is a von Neumann algebra on $\mathcal{H}$, then there is a *-homomorphism from $M \otimes_{\max} M'$ onto $C^*(M, M')$, which is injective on $M \odot M'$ if (and only if) M is a factor [MvN43], and hence an induced C*-seminorm $\|\cdot\|_\sigma$ on $M \odot M'$ (which is a norm if M is a factor). $C^*(M, M') \cong M \otimes_\sigma M'$. From III.2.2.8 it follows easily that:

IV.2.3.6 PROPOSITION. If M is a factor, then $\|\cdot\|_\sigma = \|\cdot\|_{\text{bin}}$ on $M \odot M'$.

The proof of (ii) $\Longrightarrow$ (i) in the next proposition is a slight variation of Lance's argument in the proof of IV.2.2.13.

IV.2.3.7 PROPOSITION. Let M be a von Neumann algebra on a Hilbert space $\mathcal{H}$. Then the following are equivalent:

(i) M is injective.
(ii) The restriction of $\|\cdot\|_{\max}$ on $M \odot \mathcal{L}(\mathcal{H})$ to $M \odot M'$ dominates $\|\cdot\|_\sigma$.
(iii) The restriction of $\|\cdot\|_{\text{lnor}}$ on $M \odot \mathcal{L}(\mathcal{H})$ to $M \odot M'$ dominates $\|\cdot\|_{\text{lnor}}$.

PROOF: (i) $\Longrightarrow$ (iii): Let $\theta : \mathcal{L}(\mathcal{H}) \to M'$ be a conditional expectation. Then there is a conditional expectation $id \otimes \theta$ from $M \otimes_{\text{lnor}} \mathcal{L}(\mathcal{H})$ onto $M \otimes_{\text{lnor}} M'$ (IV.2.3.4) which is the identity on $M \odot M'$. Since $id \otimes \theta$ is a contraction, the result follows.
(iii) $\Longrightarrow$ (ii) is trivial since $\|\cdot\|_\sigma \leq \|\cdot\|_{\text{lnor}}$ on $M \odot M'$.
(ii) $\Longrightarrow$ (i): Let $\|\cdot\|_\mu$ be the restriction of $\|\cdot\|_{\max}$ on $M \odot \mathcal{L}(\mathcal{H})$ to $M \odot M'$, i.e. $M \otimes_\mu M'$ is the closure of $M \odot M'$ in $M \otimes_{\max} \mathcal{L}(\mathcal{H})$. Then

$$\sum x_k \otimes y_k \mapsto \sum x_k y_k$$

yields a representation of $M \odot M'$ on $\mathcal{H}$, which extends to a representation π of $M \otimes_\mu M'$ by assumption. Then π extends to a representation ρ of $M \otimes_{\max} \mathcal{L}(\mathcal{H})$ on a larger Hilbert space $\mathcal{H}'$. Let P be the projection from $\mathcal{H}'$ onto $\mathcal{H}$, and define $\theta : \mathcal{L}(\mathcal{H}) \to \mathcal{L}(\mathcal{H})$ by $\theta(x) = P\rho(1 \otimes x)|_\mathcal{H}$. Then $\theta|_{M'}$ is the identity; and $\theta(\mathcal{L}(\mathcal{H}))$ is contained in M' since, for $x \in M$, $\rho(x \otimes 1)$ commutes with P and $\rho(1 \otimes \mathcal{L}(\mathcal{H}))$, and $P\rho(x \otimes 1)|_\mathcal{H} = x$. Thus θ is a conditional expectation from $\mathcal{L}(\mathcal{H})$ onto M'. The result then follows from IV.2.2.7.

IV.2.4 Semidiscrete Factors

IV.2.4.1 One could also consider the following natural conditions on a von Neumann algebra M in addition to conditions (i)–(iii) of IV.2.3.7:

(iv) If $B \subseteq B_1$ are C*-algebras, then the restriction of $\|\cdot\|_{\text{lnor}}$ on $M \odot B_1$ to $M \odot B$ is $\|\cdot\|_{\text{lnor}}$ on $M \odot B$.
(v) If $N \subseteq N_1$ are von Neumann algebras, then the restriction of $\|\cdot\|_{\text{bin}}$ on $M \odot N_1$ to $M \odot N$ is $\|\cdot\|_{\text{bin}}$ on $M \odot N$.

But by IV.2.3.2, condition (v) (with $N_1 = \mathcal{L}(\mathcal{H})$) implies (hence is equivalent to)

(vi) If N is any von Neumann algebra, then $\|\cdot\|_{\text{bin}} = \|\cdot\|_{\min}$ on $M \odot N$.

Applying (vi) to B^{**}, we obtain

(vii) If B is any C*-algebra, then $\|\cdot\|_{\text{lnor}} = \|\cdot\|_{\min}$ on $M \odot B$.

Conversely, (vii) $\implies$ (vi); also, (vii) implies (iv), which in turn implies (iii), and thus (v) $\iff$ (vi) $\iff$ (vii) $\implies$ (iv) $\implies$ (iii) $\iff$ (ii) $\iff$ (i).

Condition (v) would follow easily from (i) as in the proof of (i) $\implies$ (ii) *if* there were a normal conditional expectation from $\mathcal{L}(\mathcal{H})$ onto M'. But such a conditional expectation does not exist in general (IV.2.2.1).

It turns out that all of these conditions (i)–(vii) are in fact equivalent (IV.2.4.9).

Since, by IV.2.3.2, $\mathcal{L}(\mathcal{H})$ and, more generally, any Type I (discrete) von Neumann algebra satisfies (v)–(vii), it is natural to make the following definition:

IV.2.4.2 DEFINITION. Let M be a von Neumann algebra. M is *semidiscrete* if the natural map $M \odot M' \to C^*(M, M')$ extends to a *-homomorphism from $M \otimes_{\min} M'$ onto $C^*(M, M')$ (i.e. $\|\cdot\|_\sigma \leq \|\cdot\|_{\min}$ on $M \odot M'$).

If M is a factor, then M is semidiscrete if and only if $\|\cdot\|_{\text{bin}} = \|\cdot\|_{\min}$ on $M \odot M'$ by IV.2.3.6. Obviously, M is semidiscrete if and only if M' is semidiscrete.

Semidiscrete von Neumann algebras were first studied by Effros and Lance in [EL77]; the original definition was different but equivalent (IV.2.4.4(ii)). We have chosen the above form of the definition because it fits our exposition better and also better justifies the name "semidiscrete."

IV.2.4.3 DEFINITION. If X and Y are matrix ordered spaces (II.6.9.20) and $\phi : X \to Y$ is a completely positive contraction, then ϕ *approximately factors through matrix algebras* (in a specified topology) if there are n_i and completely positive contractions

$$\alpha_i : X \to \mathbb{M}_{n_i}, \ \beta_i : \mathbb{M}_{n_i} \to Y$$

such that $\beta_i \circ \alpha_i \to \phi$ in the specified topology.

IV.2.4.4 THEOREM. Let M be a von Neumann algebra on $\mathcal{H}$. The following are equivalent:

(i) M is semidiscrete.
(ii) The identity map on M approximately factors through matrix algebras in the point-σ-weak topology.
(iii) The identity map on M_* approximately factors through matrix algebras in the point-norm topology.
(iv) The identity map on M is a point-σ-weak limit of completely positive finite-rank maps.
(iv') The identity map on M_* is a point-norm limit of completely positive finite-rank maps.
(v)–(vii) Conditions (v)–(vii) of IV.2.4.1.

PROOF: (Outline) (iii) $\Longrightarrow$ (ii) by dualization (using $\mathbb{M}_n^* \cong \mathbb{M}_n$), and (ii) $\Longrightarrow$ (iii) by a simple convexity argument. Similarly, (iv) $\Longleftrightarrow$ (iv'). (iii) $\Longrightarrow$ (i) by an application of II.9.4.8, and (i) $\Longrightarrow$ (ii), although more complicated, is also based on II.9.4.8. The implications (vi) $\Longrightarrow$ (i) $\Longrightarrow$ (vii) are obvious ((i) $\Longrightarrow$ (vii) uses observation IV.2.4.5(i)).

IV.2.4.5 The following are simple corollaries of IV.2.4.4:

(i) The property of being semidiscrete is an algebraic property of M and independent of the way M is represented as a von Neumann algebra.
(ii) A semidiscrete von Neumann algebra is injective.
(iii) If $N \subseteq M$ are von Neumann algebras, M is semidiscrete, and there is a normal conditional expectation of M onto N, then N is semidiscrete. In particular, if M is semidiscrete and p is a projection in M, then pMp is semidiscrete.
(iv) $M \bar{\otimes} N$ is semidiscrete if and only if both M and N are semidiscrete.
(v) $\prod M_i$ is semidiscrete if and only if each M_i is semidiscrete.

IV.2.4.6 Generalizing (v), if M is a von Neumann algebra on a separable Hilbert space, with central decomposition $\int_X^\oplus M_x \, d\mu(x)$, then M is semidiscrete $\iff M_x$ is semidiscrete for almost all x. A direct proof is possible, but messy: cf. [CE76b]. This result was originally needed (along with additional arguments to handle the nonseparable case [CE77b]) to extend Connes' equivalence of injectivity and semidiscreteness for factors to general von Neumann algebras, but there is now a direct proof of this equivalence (IV.2.4.9).

IV.2.4.7 It is not obvious that approximately finite dimensional von Neumann algebras are semidiscrete. An argument which almost shows this, but not quite, is to let (M_i) be an increasing net of finite-dimensional unital subalgebras of M with σ-weakly dense union; then there is a normal conditional expectation $\theta_i : M \to M_i$ (IV.2.2.4), which is a unital finite-rank completely positive map from M to M. But it is not necessarily true that $\theta_i \to id$ in the point-σ-weak topology: if θ is a point-σ-weak limit of the θ_i, then θ is the identity on the σ-weakly dense set $\cup M_i$; but θ is not normal in general.

This argument can be made to work for M finite:

IV.2.4.8 PROPOSITION. *The hyperfinite factor $\mathcal{R}$ is semidiscrete.*
PROOF: This follows almost immediately from II.6.10.7: if $\mathcal{R} = [\cup M_k]^-$ with M_k a unital matrix subalgebra, and τ is the tracial state on $\mathcal{R}$, then there is a unique conditional expectation θ_k from $\mathcal{R}$ onto M_k commuting with τ, given by orthogonal projection from $L^2(\mathcal{R}, \tau)$ onto the finite-dimensional subspace $L^2(M_k, \tau)$ [$\mathcal{R} \cong M_k \bar{\otimes} (M_k' \cap \mathcal{R})$ respecting τ.] As maps from $\mathcal{R}$ to $\mathcal{R}$, $\theta_k \to id_\mathcal{R}$ in the point-2-norm topology and hence in the point-σ-weak topology.

It follows that any approximately finite-dimensional von Neumann algebra with separable predual is semidiscrete. The result reduces to the factor case by IV.2.4.6. The finite factor case is IV.2.4.8, and the semifinite case follows from IV.2.4.5(iv). The result for the Type III case can then be obtained indirectly using the discrete or continuous decomposition (III.4.8). This result is subsumed in the next very important theorem, whose proof has the same general outline (cf. [Tak03b, XV.3.1]):

IV.2.4.9 THEOREM. ([Was77], [Con79]) Every injective von Neumann algebra is semidiscrete. Thus a von Neumann algebra is semidiscrete if and only if it is injective, and conditions (i)–(vii) of IV.2.3.7 and IV.2.4.1 are equivalent.

This result, while not easy, is not nearly as difficult as the equivalence of approximate finite dimensionality with injectivity and semidiscreteness.

IV.2.5 Amenable von Neumann Algebras

Amenability for von Neumann algebras is defined in analogy with the notion of C*-amenability, which is discussed in detail in IV.3.3. If M is a von Neumann algebra and $\mathcal{X}^*$ is a dual Banach M-module, then $\mathcal{X}^*$ is said to be *normal* if $a \mapsto a\phi$ and $a \mapsto \phi a$ are continuous maps from M with the σ-weak topology to $\mathcal{X}^*$ with the weak-* topology, for every $\phi \in \mathcal{X}^*$.

IV.2.5.1 DEFINITION. A von Neumann algebra M is *amenable* if every (bounded) derivation from M into a dual normal Banach M-module is inner.

The next result gives a connection with amenable groups and helps motivate the term "amenable" for von Neumann algebras, and also gives half of the equivalence between amenability and the other equivalent conditions of this section. This result is due to B. Johnson, R. Kadison, and J. Ringrose [JKR72] (cf. [KR97b, 12.4.38]).

IV.2.5.2 THEOREM. Let M be a von Neumann algebra. If there is a subgroup G of $\mathcal{U}(M)$ which is amenable as a discrete group, with $G'' = M$, then M is amenable.

For the proof, suppose $\mathcal{X}$ is a Banach M-module with $\mathcal{X}^*$ normal, and $\delta : M \to \mathcal{X}^*$ a derivation. For $x \in \mathcal{X}$, define $f_x \in l^\infty(G)$ by

$$f_x(u) = [u^*\delta(u)](x)$$

and $\phi \in \mathcal{X}^*$ by $\phi(x) = m(f_x)$, where m is an invariant mean on G. Then $\delta(a) = \phi a - a\phi$ for all $a \in M$. (There are some nontrivial details to be checked.)

IV.2.5.3 COROLLARY. An approximately finite dimensional von Neumann algebra with separable predual is amenable.

PROOF: If A is a finite-dimensional C*-algebra with a fixed set of matrix units, then the group of unitaries in A which are linear combinations of matrix units with coefficients 0, 1, or -1, form a finite group which generates A. If A is contained in a larger finite-dimensional C*-algebra B, then a set of matrix units for B can be found for which each of the matrix units for A is a sum of matrix units for B; thus the generating finite group for A can be expanded to a finite generating group for B. By repeating the process, any approximately finite-dimensional von Neumann algebra with separable predual is generated by a locally finite group of unitaries, which is amenable.

Using the argument of the proof and the deep fact that an amenable von Neumann algebra is approximately finite dimensional, the converse of IV.2.5.2 is true: every amenable von Neumann algebra (with separable predual) is generated by an amenable group of unitaries. (See [dlH79] and [Pat92] for related results.)

The "converse" to IV.2.5.3, that an amenable von Neumann algebra is injective, was proved by Connes [Con78] using a virtual diagonal argument (cf. IV.3.3.2). This argument was simplified and placed in a more natural context in [Eff88]. We outline a more elementary argument due to J. Bunce and W. Paschke [BP78].

IV.2.5.4 THEOREM. An amenable von Neumann algebra is injective.

To prove this, let M be an amenable von Neumann algebra on $\mathcal{H}$. We construct a bounded idempotent map $\psi : \mathcal{L}(\mathcal{H}) \to M'$ such that $\psi(axb) = a\psi(x)b$ for all $a, b \in M'$, $x \in \mathcal{L}(\mathcal{H})$. The map ψ is not quite a conditional expectation since it may not have norm 1 (it is not necessarily positive), but a conditional expectation can be made from ψ by a polar decomposition argument.

Let $\mathcal{X}$ be the projective tensor product $\mathcal{L}(\mathcal{H}) \hat{\otimes} \mathcal{L}(\mathcal{H})_*$ (II.9.2.4). $\mathcal{X}$ is a Banach $\mathcal{L}(\mathcal{H})$-module, and hence a Banach M-module, via $a(x \otimes \phi) = x \otimes a\phi$, $(x \otimes \phi)a = x \otimes \phi a$. The dual $\mathcal{X}^*$ is isometrically isomorphic to $\mathcal{L}(\mathcal{L}(\mathcal{H}))$ via

$$T \in \mathcal{L}(\mathcal{L}(\mathcal{H})) \leftrightarrow (x \otimes \phi \mapsto \phi(T(x)))$$

and the M-actions become $[aT](x) = T(x)a$, $[Ta](x) = aT(x)$. Using the characterization of III.2.1.4(i), it is straightforward to show that $\mathcal{X}^*$ is a normal $\mathcal{L}(\mathcal{H})$-module (hence a dual normal M-module).

Let

$$\mathcal{Y}^* = \{T \in \mathcal{L}(\mathcal{L}(\mathcal{H})) : T(xc) = T(x)c, T(cx) = cT(x),$$
$$T(c) = 0 \text{ for all } x \in \mathcal{L}(\mathcal{H}), c \in M'\}.$$

Then $\mathcal{Y}^*$ is a weak-* closed submodule of $\mathcal{X}^*$, and is the dual of $\mathcal{Y} = \mathcal{X}/\mathcal{Z}$, where $\mathcal{Z} \subseteq \mathcal{X}$ is the submodule spanned by

$$\{xc \otimes \phi - x \otimes c\phi, cx \otimes \phi - x \otimes \phi c, c \otimes \phi : c \in M', \phi \in \mathcal{L}(\mathcal{H})_*\}.$$

The derivation $\delta : M \to \mathcal{X}^*$ defined by $\delta(a)T = aT - Ta$, maps M into $\mathcal{Y}^*$, which is a dual normal M-module; hence there is an $S \in \mathcal{Y}^*$ with $\delta(a) = aS - Sa$. Set $\psi = I - S$.

IV.2.6 Approximate Finite Dimensionality

The deepest result in the theory of injective von Neumann algebras is:

IV.2.6.1 THEOREM. Every injective von Neumann algebra with separable predual is approximately finite dimensional.

Combining this with IV.2.2.9, IV.2.4.5(ii), IV.2.4.9, IV.2.5.3, and IV.2.5.4, we obtain:

IV.2.6.2 COROLLARY. Let M be a von Neumann algebra with separable predual. The following are equivalent:

(i) M is approximately finite dimensional.
(ii) M is injective.
(iii) M is semidiscrete.
(iv) M is an amenable von Neumann algebra.

IV.2.6.3 Theorem IV.2.6.1 was proved by Connes in [Con76]. The proof was enormously complicated, and contained many important ideas. Connes' proof concentrated on the II_1 case, deducing the properly infinite case from it by standard arguments. U. Haagerup [Haa85] gave a much simpler argument, including a relatively easy direct proof of the properly infinite case. Then S. Popa [Pop86] gave an alternate simplified argument for the II_1 case, which is still the most difficult of the cases. These arguments are given in detail in [Tak03b].

IV.2.7 Invariants and the Classification of Injective Factors

The first classification consequence of the theorem of the last section is the uniqueness of the injective II_1 factor, using III.3.4.3:

IV.2.7.1 COROLLARY. Every injective II_1 factor with separable predual is isomorphic to $\mathcal{R}$ (III.3.1.4).

IV.2.7.2 One of the main long-standing open questions of the subject, which motivated much of the early work on classification, was whether the approximately finite dimensional II_∞ factor is unique, i.e. whether $\mathcal{R}_{0,1} = \mathcal{R} \bar{\otimes} \mathcal{L}(\mathcal{H})$ (III.3.1.6) is the only one. If M is an injective II_∞ factor with separable predual, and p is a nonzero finite projection in M, then

$$M \cong pMp \bar{\otimes} \mathcal{L}(\mathcal{H}).$$

Since pMp is an injective II_1 factor, the uniqueness follows from the previous result.

IV.2.7.3 COROLLARY. Every injective II$_\infty$ factor with separable predual is isomorphic to $\mathcal{R}_{0,1}$.

Classification of injective Type III factors requires a detailed study of the automorphism group of $\mathcal{R}$ and of $\mathcal{R}_{0,1}$. Connes carried out much of the necessary analysis as part of his proof of IV.2.6.1. U. Haagerup [Haa87] finished the III$_1$ case. The intricate details are found in [Tak03b]; we only state the deceptively simple conclusion.

IV.2.7.4 THEOREM. Every injective factor with separable predual is a Krieger factor (III.3.2.18). For $0 < \lambda \leq 1$, there is a unique injective Type III$_\lambda$ factor with separable predual, the Powers factor $\mathcal{R}_\lambda$ (III.3.1.7) if $0 < \lambda < 1$ and the factor $\mathcal{R}_\infty$ (III.3.1.13(iv)-(vi)) if $\lambda = 1$. There are uncountably many injective Type III$_0$ factors with separable predual, classified by ergodic flows (III.3.2.19).

IV.3 Nuclear and Exact C*-Algebras

In this section, we will give a detailed description of nuclear and exact C*-algebras, and several important alternate characterizations of each class. Roughly speaking, the nuclear C*-algebras turn out to be the C*-algebras which are "approximately finite dimensional" in an appropriate order-theoretic sense; there is a close analogy, and an intimate connection, with the theory of injective von Neumann algebras (IV.2). See the introduction to IV.2 for a discussion of terminology for the class of nuclear C*-algebras.

IV.3.1 Nuclear C*-Algebras

Recall (II.9.4) that a C*-algebra A is *nuclear* if $\|\cdot\|_{\max} = \|\cdot\|_{\min}$ on $A \odot B$ for every C*-algebra B. The class of nuclear C*-algebras contains all commutative C*-algebras (II.9.4.4) and finite-dimensional C*-algebras (II.9.4.2), and is closed under inductive limits (the case of injective connecting maps is elementary, but the general case is rather delicate (cf. IV.3.1.13)) and (minimal) tensor products; in fact:

IV.3.1.1 PROPOSITION. Let A and B be C*-algebras. Then $A \otimes_{\min} B$ is nuclear if and only if both A and B are nuclear.

For the proof, note that if B is nonnuclear and C is a C*-algebra for which $\|\cdot\|_{\max} \neq \|\cdot\|_{\min}$ on $B \odot C$, the natural homomorphism

$$(A \otimes_{\min} B) \otimes_{\max} C \to A \otimes_{\min} (B \otimes_{\max} C) \to A \otimes_{\min} (B \otimes_{\min} C) \cong (A \otimes_{\min} B) \otimes_{\min} C$$

has nontrivial kernel, so $A \otimes_{\min} B$ is nonnuclear. Conversely, if A and B are nuclear and C is any C*-algebra, then

$$(A \otimes_{\min} B) \otimes_{\max} C \cong (A \otimes_{\max} B) \otimes_{\max} C \cong A \otimes_{\max} (B \otimes_{\max} C)$$

$$\cong A \otimes_{\max} (B \otimes_{\min} C) \cong A \otimes_{\min} (B \otimes_{\min} C) \cong (A \otimes_{\min} B) \otimes_{\min} C.$$

As a corollary, since $\mathbb{K}$ is nuclear, the class of nuclear C*-algebras is closed under stable isomorphism.

IV.3.1.2 A closed ideal in a nuclear C*-algebra is nuclear (II.9.6.3); in fact, a hereditary C*-subalgebra of a nuclear C*-algebra is nuclear (in the separable case, every hereditary C*-subalgebra of A is stably isomorphic to the closed ideal of A it generates; see IV.3.1.14 for the general case). The class of nuclear C*-algebras is closed under extensions:

IV.3.1.3 PROPOSITION. Let A be a C*-algebra and J a closed ideal in A. If J and A/J are nuclear, then A is nuclear.

PROOF: Let B be a C*-algebra. Consider the following diagram:

$$\begin{array}{ccccccccc} 0 & \longrightarrow & J \otimes_{\max} B & \longrightarrow & A \otimes_{\max} B & \longrightarrow & (A/J) \otimes_{\max} B & \longrightarrow & 0 \\ & & \downarrow \pi_J & & \downarrow \pi_A & & \downarrow \pi_{A/J} & & \\ 0 & \longrightarrow & J \otimes_{\min} B & \longrightarrow & A \otimes_{\min} B & \longrightarrow & (A/J) \otimes_{\min} B & \longrightarrow & 0 \end{array}$$

where π_C ($C = J, A, A/J$) denotes the quotient map from $C \otimes_{\max} B$ to $C \otimes_{\min} B$. The diagram commutes, and the top row is exact (cf. II.9.6.6). Also,

$$(A \odot B)/(J \odot B) \cong (A/J) \odot B,$$

and the quotient of $\|\cdot\|_{\min}$ on $A \odot B$ defines a C*-norm on $(A/J) \odot B$ which is the minimal cross norm on $(A/J) \odot B$ because A/J is nuclear. Thus the bottom row is also exact. Since π_J and $\pi_{A/J}$ are isomorphisms, π_A is also an isomorphism by the Five Lemma.

It follows that every Type I C*-algebra is nuclear, since every stable Type I C*-algebra is built up from commutative C*-algebras by stabilization, extensions, and inductive limits (IV.1.7.27). (A more direct proof that Type I C*-algebras are nuclear is to note that if A is Type I, then A^{**} is semidiscrete (cf. IV.3.1.12)).

It is also true that any quotient of a nuclear C*-algebra is nuclear, but this is a surprisingly delicate result (IV.3.1.13). However, a C*-subalgebra of a nuclear C*-algebra is not nuclear in general (IV.3.5.7).

Factorization Through Matrix Algebras

Nuclearity turns out to be characterized by the existence of completely positive finite-rank approximations of the identity (cf. II.9.4.9).

IV.3.1.4 DEFINITION. If A and B are C*-algebras, a completely positive contraction $\phi : A \to B$ is *nuclear* if ϕ approximately factors through matrix algebras in the point-norm topology (IV.2.4.3), i.e. if, for any $x_1, \ldots, x_k \in A$ and $\epsilon > 0$, there is an n and completely positive contractions

$$\alpha : A \to \mathbb{M}_n, \ \beta : \mathbb{M}_n \to A$$

such that $\|x_j - \beta \circ \alpha(x_j)\| < \epsilon$ for $1 \leq j \leq k$.

The Main Theorem

The principal theorem of the theory of nuclear C*-algebras is a combination of several deep results:

IV.3.1.5 THEOREM. Let A be a C*-algebra. The following are equivalent:
  (i) A is nuclear.
  (ii) The identity map on A is nuclear.
  (iii) The identity map on A is a point-norm limit of completely positive finite-rank contractions.
  (iv) A^{**} is injective.
  (v) A is C*-amenable (IV.3.3).

IV.3.1.5 shows that the nuclear C*-algebras form an extremely natural and well-behaved class; further evidence comes from the results in IV.3.5 about group C*-algebras and crossed products. Nuclear C*-algebras also have additional nice structure properties; perhaps the most important of these is the Choi-Effros Lifting Theorem (IV.3.2.4).

We now outline the proof of the equivalence of IV.3.1.5(i)–(iv). The equivalence of (v) with the other conditions will be discussed in IV.3.3.

We first describe the equivalence between IV.3.1.5(i), (ii), and (iii). Recall that the equivalence between (i), (vii), and (viii) of the next lemma was II.9.4.9.

IV.3.1.6 LEMMA. Let A be a C*-algebra. The following are equivalent:
  (i) A is nuclear.
  (ii) The identity map on A is nuclear.
  (iii) The identity map on A is a point-norm limit of completely positive finite-rank contractions.
  (iv) The identity map on A^* approximately factors through matrix algebras in the point-weak-* topology (IV.2.4.3).
  (v) The identity map on A^* is a point-weak-* limit of completely positive finite-rank contractions.

(vi) For any C*-algebra B, any completely positive contraction from A to B^* approximately factors through matrix algebras in the point-weak-* topology.
(vii) For any C*-algebra B, any completely positive contraction from A to B^* is a point-weak-* limit of completely positive finite-rank contractions.
(viii) For any C*-algebra B, any completely positive contraction from B to A^* is a point-weak-* limit of completely positive finite-rank contractions.

For the proof, note first that (ii) $\implies$ (iii), (iv) $\implies$ (v), and (vi) $\implies$ (vii) are trivial, (v) $\implies$ (viii) is virtually trivial, and (ii) $\implies$ (iv) and (iii) $\implies$ (v) are simple dualizations using the fact that $\mathbb{M}_n^* \cong \mathbb{M}_n$ as matrix-ordered spaces. (And (i) $\iff$ (vii) $\iff$ (viii) is II.9.4.9.)

The implication (i) $\implies$ (vi) [in fact, (i) $\iff$ (vi) $\iff$ (vii)] was shown in [CE78] by showing that every map of the form T_ϕ as in II.9.3.7 exactly factors through a matrix algebra. The argument is nontrivial, but not difficult.

The remaining and most difficult implication is (i) $\implies$ (ii). This was proved independently in [CE78] and [Kir77a]; both proofs are straightforward in broad outline, but with subtle technicalities. The idea of the proof (cf. [Lan82]) of [CE78], once (i) $\implies$ (vi) was established, is to consider a (faithful) representation π of A on $\mathcal{H}$ with a cyclic vector ξ of norm 1. Then, identifying A with $\pi(A)$, there is a completely positive contraction $\phi : A'' \to (A')^*$ given by $[\phi(x)](y) = \langle xy\xi, \xi \rangle$, which is injective since ξ is separating for A'. Composing with the embedding of A into A'' gives a completely positive contraction $\psi : A \to (A')^*$, which approximately factors through matrix algebras in the weak-* topology by (vi). A perturbation argument (to make the range of β contained in the range of ϕ) shows that the embedding of A into A'' approximately factors through matrix algebras in the point-σ-weak topology. The same is true for direct sums of cyclic representations, hence for the universal representation, so the embedding of A into A^{**} approximately factors through matrix algebras in the point-σ-weak topology. An application of the Kaplansky Density Theorem (to make the range of β contained in A) and a convexity argument (to obtain point-norm approximation) then show that the identity map on A is nuclear.

IV.3.1.7 Conditions (ii)–(v) of IV.3.1.6 are collectively called the *completely positive approximation property* or *CPAP* (this term has been used in the literature for various of the conditions, although logically it should most naturally mean (iii)). A C*-algebra with the CPAP has the *metric approximation property* of Banach space theory (see e.g. [LT77]); but there exist C*-algebras with the metric approximation property (e.g. $C^*(\mathbb{F}_2)$ [Haa79]) which do not have the CPAP. (There are also C*-algebras which fail to have the metric approximation property, e.g. $\mathcal{L}(\mathcal{H})$ [Sza81]; a separable example can be constructed from this as in II.8.5.)

A simple but interesting consequence of the CPAP is the fact that nuclear C*-algebras are "approximately injective:"

IV.3.1.8 PROPOSITION. Let D be a nuclear C*-algebra, A a C*-algebra, B a C*-subalgebra of A, and $\phi : B \to D$ a completely positive contraction. Then there is a net (ψ_i) of completely positive contractions from A to D such that $\psi_i(x) \to \phi(x)$ for all $x \in B$ (i.e. $\psi_i|_B \to \phi$ in the point-norm topology). The ψ_i can be chosen to be of finite rank. In particular, if $D \subseteq A$, there is a net (ψ_i) of (finite-rank) completely positive contractions from A to D such that $\psi_i(x) \to x$ for all $x \in D$.

PROOF: Choose completely positive contractions

$$\alpha_i : D \to \mathbb{M}_{k_i},\ \beta_i : \mathbb{M}_{k_i} \to D$$

such that $\beta_i \circ \alpha_i \to id_D$ in the point-norm topology. Using injectivity of $\mathbb{M}_{k_i}$, extend $\alpha_i \circ \phi : B \to \mathbb{M}_{k_i}$ to $\gamma_i : A \to \mathbb{M}_{k_i}$, and set $\psi_i = \beta_i \circ \gamma_i$. The last statement is the case $B = D$, $\phi = id_D$.

Using the tensor product definition of nuclearity, it is not obvious how to show that nuclearity is an (SI) property (II.8.5.1), since it is not clear that there are countably many "test algebras" to tensor with to check the nuclearity of a separable C*-algebra. But using the CPAP characterization of nuclearity, the proof is routine:

IV.3.1.9 PROPOSITION. Nuclearity is an (SI) property.

PROOF: Part (ii) of the definition (II.8.5.1) is satisfied by II.9.6.5. For part (i), let A be a nuclear C*-algebra and B a separable C*-subalgebra. Let $\{x_1, x_2, \dots\}$ be a countable dense set in B. For each $k \in \mathbb{N}$, let $\psi_{1,k} : A \to A$ be a finite-rank cp-contraction such that

$$\|x_i - \psi_{1,k}(x_i)\| \leq \frac{1}{k}$$

for $1 \leq i \leq k$. If $\{y_1, \dots, y_m\} \subseteq B$ and $\epsilon > 0$, there is a k such that

$$\|y_i - \psi_{1,k}(y_i)\| < \epsilon$$

for $1 \leq i \leq m$. Let $\tilde{B}_1$ be the C*-subalgebra of A generated by B and $\cup_k \psi_{1,k}(A)$. Since $\psi_{1,k}(A)$ is finite-dimensional for all k, $\tilde{B}_1$ is separable. Repeat the construction with $\tilde{B}_1$ in place of B to obtain $\psi_{2,k} : A \to A$, and let $\tilde{B}_2$ be the C*-subalgebra of A generated by $\tilde{B}_1$ and $\cup_k \psi_{2,k}(A)$. Iterate the construction to get an increasing sequence $(\tilde{B}_j)$ and maps $\psi_{j,k} : A \to A$. Set $\tilde{B} = [\cup \tilde{B}_j]^-$. Then $\tilde{B}$ is separable. If $\{z_1, \dots, z_m\} \subseteq \tilde{B}$ and $\epsilon > 0$, choose j and $\{y_1, \dots, y_m\} \subseteq \tilde{B}_{j-1}$ with

$$\|y_i - z_i\| < \epsilon/3$$

for $1 \leq i \leq m$. Then there is a k such that

$$\|y_i - \psi_{j,k}(y_i)\| < \epsilon/3$$

for $1 \leq i \leq m$. Since the range of $\psi_{j,k}$ is contained in $\tilde{B}_j$, $\psi_{j,k}(\tilde{B}) \subseteq \tilde{B}$. Thus, if $\phi = \psi_{j,k}|_{\tilde{B}}$, then ϕ is a finite-rank cp-contraction from $\tilde{B}$ to $\tilde{B}$, and $\|z_i - \phi(z_i)\| < \epsilon$ for $1 \leq i \leq m$. Thus $\tilde{B}$ is nuclear by IV.3.1.5(iii).

IV.3.1.10 COROLLARY. Every nuclear C*-algebra is an inductive limit of a system (with injective connecting maps) of separable nuclear C*-algebras.

IV.3.1.11 If A is a nuclear C*-algebra, it is interesting to consider the conditions under which the maps α and β approximating the identity map through a matrix algebra can be chosen to be almost multiplicative (on finite sets). It is not hard to show that if the map β can be chosen almost multiplicative (in a suitable sense), then A must be an AF algebra. It is much more interesting to examine when the α can be chosen almost multiplicative. This cannot always be done, since there are infinite nuclear C*-algebras such as the Toeplitz algebra (or the Cuntz algebras). We will return to this question in V.4.3.9.

Nuclearity and the Second Dual

We now turn to the equivalence of IV.3.1.5(i) and (iv). This crucial consequence of IV.2.4.9 for C*-algebras was obtained by Choi and Effros [CE77b] (some additional argument was needed for the nonseparable case):

IV.3.1.12 THEOREM. Let A be a C*-algebra. The following are equivalent:

(i) A is nuclear.
(ii) A is quasinuclear (II.9.6.4).
(iii) A^{**} is semidiscrete (IV.2.4.2).
(iv) A^{**} is injective.

We have that (i) $\Longrightarrow$ (ii) is trivial, (ii) $\Longrightarrow$ (iv) is IV.2.2.13, and (iv) $\Longrightarrow$ (iii) is IV.2.4.9. The implication (iii) $\Longrightarrow$ (i) follows from comparing IV.2.4.4(iii) for A^{**} with IV.3.1.6(iv).

IV.3.1.13 COROLLARY. A quotient of a nuclear C*-algebra is nuclear.
PROOF: If J is a closed ideal in a C*-algebra A, then
$$A^{**} \cong J^{**} \oplus (A/J)^{**}$$
(III.5.2.11). If A is nuclear, then A^{**} is injective, so $(A/J)^{**}$ is also injective.

A similar argument also shows that an ideal in a nuclear C*-algebra is nuclear, and that an extension of nuclear C*-algebras is nuclear, but there are elementary proofs of these facts (II.9.6.3, IV.3.1.3). (Interestingly, the result about ideals was an early observation, but the extension result was first proved using IV.3.1.12.) However, there is no known proof of IV.3.1.13 which does not use IV.2.4.9.

Another corollary concerns hereditary C*-subalgebras:

IV.3.1.14 COROLLARY. A hereditary C*-subalgebra of a nuclear C*-algebra is nuclear.

For the proof, note that if B is a hereditary C*-subalgebra of a C*-algebra A, then $B^{**} \cong pA^{**}p$ for a projection p (p is the supremum in A^{**} of any approximate unit for B).

The Bootstrap Category

The most important unresolved general structural question about the class of nuclear C*-algebras is whether the class can be described constructively as the smallest class of C*-algebras closed under certain standard operations. Many different sets of constructions can be considered, but it is particularly important to consider sets under which K-theory is nicely behaved. A minimal set of possibilities is:

IV.3.1.15 DEFINITION. Let $\mathfrak{N}$ be the smallest class of separable nuclear C*-algebras with the following properties:

(i) $\mathfrak{N}$ contains $\mathbb{C}$.
(ii) $\mathfrak{N}$ is closed under stable isomorphism.
(iii) $\mathfrak{N}$ is closed under inductive limits.
(iv) $\mathfrak{N}$ is closed under crossed products by $\mathbb{Z}$.
(v) If $0 \to J \to A \to A/J \to 0$ is an exact sequence, and two of $J, A, A/J$ are in $\mathfrak{N}$, so is the third.

The class $\mathfrak{N}$ is called the *small bootstrap class* (or *category*); property (v) is called the *two-out-of-three property*.

IV.3.1.16 QUESTION. Is $\mathfrak{N}$ the class of all separable nuclear C*-algebras?

The class $\mathfrak{N}$ contains all known separable nuclear C*-algebras, including all inductive limits of Type I C*-algebras and Cuntz-Krieger algebras. (It is possible that (ii) is redundant; even without (ii) the class is closed under stabilization, and if (vi) below is added, it is also closed under taking matrix algebras.) However, it does not seem to be directly provable that $\mathfrak{N}$ is closed under homotopy equivalence, or even that $A \oplus B \in \mathfrak{N}$ implies $A, B \in \mathfrak{N}$. To remedy the last defect, it is harmless (from the K-theory standpoint) to add

(vi) If $0 \to J \to A \to A/J \to 0$ is a split exact sequence, and $A \in \mathfrak{N}$, so are J and A/J.

A potentially larger class is obtained if it is required to be closed under some type of homotopy equivalence. The weakest assumption is that the class contain all contractible (separable nuclear) C*-algebras. In the presence of (vi), this implies that the class is closed under homotopy equivalence (II.5.5.8). Even (apparently) stronger assumptions, that the class be closed under shape equivalence or KK-equivalence, give the *large bootstrap class* $\mathcal{N}$, which will be discussed in more detail in V.1.5.4.

IV.3.2 Completely Positive Liftings

One of the most useful consequences of nuclearity is the Choi-Effros lifting theorem for completely positive maps into quotients. This result has important applications in K-theory.

IV.3.2.1 Let A and B be C*-algebras, and J a closed ideal of B. A completely positive contraction (*cp*-contraction) $\phi : A \to B/J$ is *liftable* (to B) if there is a *cp*-contraction $\psi : A \to B$ such that $\phi = \pi \circ \psi$, where $\pi : B \to B/J$ is the quotient map. ψ is called a *lifting* of ϕ.

A has the *lifting property* if ϕ is liftable for every B, J, and ϕ.

IV.3.2.2 An important special case comes when $A = B/J$ and ϕ is the identity map. A lifting ψ is then a completely positive cross section for π, so the extension $0 \to J \to B \to B/J \to 0$ is semisplit (II.8.4.23). Semisplit extensions are well behaved for many purposes related to K-theory (cf. II.8.4.24).

IV.3.2.3 If A, B, and ϕ are unital and ϕ is liftable, then ϕ has a unital lift: if ψ is any lift and ω a state on B/J, then ψ' is a unital lift, where

$$\psi'(x) = \psi(x) + \omega(\phi(x))(1_B - \psi(1_A)).$$

A related observation is that if a *cp*-contraction $\phi : A \to B/J$ lifts to a completely positive *map* $\psi : A \to B$, then it is liftable (i.e. lifts to a completely positive *contraction*). This is easily seen if A is unital (the only case we will use): if f is the continuous function on $[0, \infty)$ with $f(t) = 1$ for $0 \le t \le 1$ and $f(t) = t^{-1/2}$ for $t > 1$, then

$$\psi'(x) = f(\psi(1))\psi(x)f(\psi(1))$$

defines a completely positive contractive lift of ϕ. The case of nonunital A is trickier, and can be found in [CS86].

IV.3.2.4 THEOREM.[CHOI-EFFROS LIFTING] Let A and B be C*-algebras, and J a closed ideal of B. If A is separable and $\phi : A \to B/J$ is a nuclear completely positive contraction, then ϕ is liftable to B. In particular, every separable nuclear C*-algebra has the lifting property.

Before giving the proof, we describe some consequences and variations.

IV.3.2.5 COROLLARY. If $0 \to J \to B \to B/J \to 0$ is an extension of C*-algebras, and B/J is separable and nuclear, then the extension is semisplit.

Using IV.3.1.13, it suffices that B is nuclear (and B/J separable). But see IV.3.2.12.

The results can be generalized easily to the case where A is replaced by a separable operator system. There is also a version in the exact case ([EH85]; cf. [Was94, 6.10]):

IV.3.2.6 THEOREM. Let B be a C*-algebra, J a nuclear closed ideal of B. Suppose that for every C*-algebra A, the sequence

$$0 \longrightarrow A \otimes_{\min} J \longrightarrow A \otimes_{\min} B \longrightarrow A \otimes_{\min} B/J \longrightarrow 0$$

is exact. If X is a separable operator system and $\phi : X \to B/J$ is a completely positive contraction, then ϕ is liftable to B.

IV.3.2.7 COROLLARY. Let B be an exact C*-algebra, J a nuclear closed ideal in B. If X is a separable operator system, and $\phi : X \to B/J$ is a completely positive contraction, then ϕ is liftable to B.

For the proof, combine IV.3.2.6 with IV.3.4.14 and IV.3.4.18.

IV.3.2.8 COROLLARY. If $0 \to J \to B \to B/J \to 0$ is an extension of C*-algebras, B is exact, J is nuclear, and B/J is separable, then the extension is semisplit.

IV.3.2.9 The separability hypotheses in these results cannot be removed: the exact sequence
$$0 \longrightarrow c_o \longrightarrow l^\infty \longrightarrow l^\infty/c_o \longrightarrow 0$$
of commutative C*-algebras is not semisplit since there is not even a closed subspace of l^∞ complementary to c_o [LT77, 2.a.7]. However, there is an approximate result valid in the nonseparable case:

IV.3.2.10 COROLLARY. If $0 \to J \to B \to B/J \to 0$ is an extension of C*-algebras, and B/J is nuclear, then the extension is "approximately semisplit" in the sense that if $\{x_1, \ldots, x_n\} \subseteq B/J$ and $\epsilon > 0$, there is a cp-contraction $\sigma : B/J \to B$ such that $\|x_j - \pi \circ \sigma(x_j)\| < \epsilon$ for $1 \leq j \leq n$, where $\pi : B \to B/J$ is the quotient map. The map σ can be chosen of finite rank.

PROOF: Let $C = C^*(\{x_1, \ldots, x_n\}) \subseteq B/J$, and $\bar{B}$ a separable nuclear C*-subalgebra of B/J containing C (IV.3.1.9). If $D = \pi^{-1}(\bar{B})$, then $J \subseteq D$ and $D/J \cong \bar{B}$. Let $\rho : D \to D/J$ be the quotient map (i.e. $\rho = \pi|_D$), and let $\tau : D/J \to D$ be a cp-splitting (IV.3.2.5). Since D/J is nuclear, there are cp-contractions $\alpha : D/J \to \mathbb{M}_m$ and $\beta : \mathbb{M}_m \to D/J$ such that $\|x_j - \beta \circ \alpha(x_j)\| < \epsilon$ for $1 \leq j \leq n$. Extend α to a cp-contraction $\gamma : B/J \to \mathbb{M}_m$ (II.6.9.12), and let $\sigma = \tau \circ \beta \circ \gamma$.

IV.3.2.11 Actually, for the exact sequence
$$0 \longrightarrow c_o \longrightarrow l^\infty \longrightarrow l^\infty/c_o \longrightarrow 0$$
one can do better: by a similar argument using injectivity of l^∞, if $\{x_1, x_2, \ldots\}$ is a sequence in l^∞/c_o, there is a cp-contraction $\sigma : l^\infty/c_o \to l^\infty$ such that $\pi(\sigma(x_j)) = x_j$ for all j, i.e. the exact sequence is "countably semisplit."

IV.3.2.12 Note that the hypotheses of IV.3.2.5 involve nuclearity of B/J, not of B, and thus the result is useless for proving IV.3.1.13. If the hypothesis were that B is nuclear, then the nuclearity of B/J would follow easily in the separable case by taking a cp-splitting σ for $\pi : B \to B/J$ and considering $\pi \circ \phi_i \circ \sigma$, where (ϕ_i) is a net of finite-rank cp-contractions from B to B converging to the identity in the point-norm topology. The nonseparable case would then follow from IV.3.1.10. Such an argument is possible using IV.3.2.8, but the proof of IV.3.2.8 via IV.3.4.14 even in the case where B is nuclear uses IV.2.4.9 (in IV.3.4.8).

Proof of the Lifting Theorem

Our proof of IV.3.2.4, following [Arv77] (cf. [Was94]), is based on the next two facts.

IV.3.2.13 PROPOSITION. Let $\phi_1, \phi_2 : A \to B/J$ be liftable *cp*-contractions, $x_1, \ldots, x_n \in A$, and $\epsilon > 0$. If ψ_1 is a lifting of ϕ_1, then there is a lifting ψ_2 of ϕ_2 such that

$$\|\psi_2(x_k) - \psi_1(x_k)\| \leq \|\phi_2(x_k) - \phi_1(x_k)\| + \epsilon$$

for all $1 \leq k \leq n$. If A, B, ϕ_1, ϕ_2, and ψ_1 are unital, we may choose ψ_2 to be unital.

PROOF: Let ψ be any [unital] lifting of ϕ_2, and let (h_λ) be an approximate unit for J which is quasicentral for B (II.4.3.1). By II.8.1.5 we have that, for all $x \in A$,

$$\lim_{\lambda \to \infty} \|(1 - h_\lambda)^{1/2}(\psi(x) - \psi_1(x))(1 - h_\lambda)^{1/2}\| = \|\phi_2(x) - \phi_1(x)\|.$$

By II.8.1.5 we also have that, as $\lambda \to \infty$,

$$\|\psi_1(x) - h_\lambda^{1/2}\psi_1(x)h_\lambda^{1/2} - (1-h_\lambda)^{1/2}\psi_1(x)(1-h_\lambda)^{1/2}\|$$
$$\leq \|\psi_1(x)h_\lambda - h_\lambda^{1/2}\psi_1(x)h_\lambda^{1/2}\| + \|\psi_1(x)(1-h_\lambda)$$
$$- (1-h_\lambda)^{1/2}\psi_1(x)(1-h_\lambda)^{1/2}\| \to 0.$$

Thus we can take $\psi_2(x) = (1-h_\lambda)^{1/2}\psi(x)(1-h_\lambda)^{1/2} + h_\lambda^{1/2}\psi_1(x)h_\lambda^{1/2}$ for sufficiently large λ.

IV.3.2.14 PROPOSITION. Let A and B be C*-algebras, with A separable, and J a closed ideal in B. Then the set of liftable *cp*-contractions from A to B/J is closed in the point-norm topology.

PROOF: Let Φ be the set of liftable *cp*-contractions from A to B/J, and ϕ a point-norm limit point of Φ. Let (x_n) be a dense sequence in the unit ball of A. For each n choose $\phi_n \in \Phi$ with

$$\|\phi_n(x_k) - \phi(x_k)\| < 2^{-n-2}$$

for $1 \leq k \leq n$; then

$$\|\phi_{n+1}(x_k) - \phi_n(x_k)\| < 2^{-n-1}$$

for all n and all $k \leq n$. Let ψ_1 be any lifting of ϕ_1, and inductively using IV.3.2.13 choose a lifting ψ_n of ϕ_n such that

$$\|\psi_{n+1}(x_k) - \psi_n(x_k)\| < 2^{-n}$$

for all n and all $k \leq n$. The sequence (ψ_n) converges in the point-norm topology to a lift ψ for ϕ.

IV.3.2.15 To prove IV.3.2.4, if ϕ is nuclear it can be approximated in the point-norm topology by maps of the form $\beta \circ \alpha$, where $\alpha : A \to \mathbb{M}_n$ and $\beta : \mathbb{M}_n \to B/J$ are cp-contractions. If β is always liftable to B, so is $\beta \circ \alpha$, and thus ϕ is liftable by IV.3.2.14. So the proof can be reduced to the case where $A = \mathbb{M}_n$.

IV.3.2.16 LEMMA. Let B be a C*-algebra and $\phi : \mathbb{M}_n \to B$ a linear map. Then ϕ is completely positive if and only if the element $\sum_{i,j} \phi(e_{ij}) \otimes e_{ij}$ is positive in $B \otimes \mathbb{M}_n$.

PROOF: The necessity is obvious, since $\sum_{i,j} e_{ij} \otimes e_{ij}$ is positive in $\mathbb{M}_n \otimes \mathbb{M}_n$ (it is a multiple of a projection). Conversely, we show that $\phi \otimes id : \mathbb{M}_n \otimes \mathbb{M}_m \to B \otimes \mathbb{M}_m$ is positive for any m. If $x_1, \ldots, x_n \in \mathbb{M}_m$, then $\sum_{k,l} e_{kl} \otimes x_k^* x_l \geq 0$ in $\mathbb{M}_n \otimes \mathbb{M}_m$ (cf. II.6.9.8), so, in $B \otimes \mathbb{M}_n \otimes \mathbb{M}_n \otimes \mathbb{M}_m$,

$$\sum_{i,j,k,l} \phi(e_{ij}) \otimes e_{ij} \otimes e_{kl} \otimes x_k^* x_l = (\sum_{i,j} \phi(e_{ij}) \otimes e_{ij}) \otimes (\sum_{k,l} e_{kl} \otimes x_k^* x_l) \geq 0$$

$$0 \leq \sum_{r,s} (\sum_t 1 \otimes e_{rt} \otimes e_{st} \otimes 1)(\sum_{i,j,k,l} \phi(e_{ij}) \otimes e_{ij} \otimes e_{kl} \otimes x_k^* x_l)(\sum_t 1 \otimes e_{rt} \otimes e_{st} \otimes 1)^*$$

$$= \sum_{r,s} (\sum_{i,j} \phi(e_{ij}) \otimes e_{rr} \otimes e_{ss} \otimes x_i^* x_j) = \sum_{i,j} \phi(e_{ij}) \otimes 1 \otimes 1 \otimes x_i^* x_j$$

So $\sum_{i,j} \phi(e_{ij}) \otimes x_i^* x_j \geq 0$ in $B \otimes \mathbb{M}_m$. If $x = \sum_{i,j} e_{ij} \otimes x_{ij}$ is an arbitrary element of $\mathbb{M}_n \otimes \mathbb{M}_m$, then

$$(\phi \otimes id)(x^* x) = \sum_k (\sum_{i,j} \phi(e_{ij}) \otimes x_{ki}^* x_{kj}) \geq 0.$$

IV.3.2.17 Thus, if $\phi : \mathbb{M}_n \to B/J$ is a cp-contraction and $y_{ij} = \phi(e_{ij})$, then the element $\sum_{i,j} y_{ij} \otimes e_{ij}$ is positive in $(B/J) \otimes \mathbb{M}_n \cong (B \otimes \mathbb{M}_n)/(J \otimes \mathbb{M}_n)$, so there is a positive preimage $\sum_{i,j} x_{ij} \otimes e_{ij}$ in $B \otimes \mathbb{M}_n$. The map ψ defined by $\psi(e_{ij}) = x_{ij}$ is completely positive by IV.3.2.16. Thus by IV.3.2.3, ϕ is liftable. This completes the proof of IV.3.2.4.

IV.3.3 Amenability for C*-Algebras

Recall (II.5.5.17(iii)) that a C*-algebra A is *amenable* if every (bounded) derivation from A to a dual Banach A-module is inner. The same definition can be made for general Banach algebras.

Fundamental work on amenability for Banach algebras was done by B. Johnson, individually [Joh72] and in collaboration with R. Kadison and J. Ringrose [JKR72], who had begun the cohomology theory of Banach algebras [KR71a], [KR71b]. Johnson [Joh72] proved the following theorem, justifying the term "amenable" in the Banach algebra context. Many of the algebraic ideas concerning amenability previously arose in the study of Hochschild homology [Hoc45].

IV.3.3.1 THEOREM. If G is a locally compact group, then the Banach algebra $L^1(G)$ is amenable if and only if G is amenable.

A C*-algebra version of this theorem will be discussed in IV.3.5.2 and IV.3.5.5.

Virtual Diagonals

IV.3.3.2 An alternate characterization of amenability uses the notion of a virtual diagonal. If $\{e_{ij}\}$ is a set of matrix units in $\mathbb{M}_n$, the element

$$d = \sum_{i=1}^n e_{ii} \otimes e_{ii}$$

of $\mathbb{M}_n \otimes \mathbb{M}_n$ has the properties that $ad = da$ when $\mathbb{M}_n \otimes \mathbb{M}_n$ is regarded as an $\mathbb{M}_n$-bimodule in the natural way, and $\pi(d) = 1$, where $\pi : \mathbb{M}_n \otimes \mathbb{M}_n \to \mathbb{M}_n$ is multiplication $\pi(a \otimes b) = ab$. If $\delta : \mathbb{M}_n \to \mathcal{X}$ is a derivation, where $\mathcal{X}$ is any $\mathbb{M}_n$-bimodule, set

$$x = \sum_{i=1}^n e_{ii} \delta(e_{ii}).$$

Then a simple calculation shows that $\delta(a) = ax - xa$ for all $a \in \mathbb{M}_n$.

If A is a C*-algebra (or Banach algebra), the same argument can be made to work if there is just a $d \in A \hat{\otimes} A$ (II.9.2.4), or even in $(A \hat{\otimes} A)^{**}$, satisfying $ad = da$ and $a\pi^{**}(d) = \pi^{**}(d)a = a$ for all $a \in A$ (regarding $(A \hat{\otimes} A)^{**}$ and A^{**} as dual Banach A-modules in the obvious way), provided in the second case that $\mathcal{X}$ is a dual Banach A-module. Such a d is called a *virtual diagonal* for A. Thus, if A has a virtual diagonal, it is amenable. The converse is also true [Joh72]:

IV.3.3.3 THEOREM. A Banach algebra A is amenable if and only if it has a virtual diagonal.

To obtain a virtual diagonal for an amenable A, let (h_λ) be a bounded approximate unit for A (for an amenable Banach algebra such an approximate unit must be proved to exist). Then $(h_\lambda \otimes h_\lambda)$ converges weak-* to some $\phi \in (A \hat{\otimes} A)^{**}$. The derivation $\delta(a) = a\phi - \phi a$ maps A into the dual Banach A-module $ker(\pi^{**})$; by amenability there is a $\psi \in ker(\pi^{**})$ such that $\delta(a) = a\psi - \psi a$. Then $d = \phi - \psi$ is a virtual diagonal.

Permanence Properties

The first permanence property is virtually trivial, since if A is a Banach algebra and J a closed ideal, any Banach A/J-module can be regarded as a Banach A-module, and similarly with derivations:

IV.3.3.4 PROPOSITION. Any quotient of an amenable Banach algebra is amenable.

Contrast this result with the depth of the corresponding statement for nuclearity (IV.3.1.13).

IV.3.3.5 PROPOSITION. Let A be an amenable Banach algebra, and J a closed ideal of A. Then J is an amenable Banach algebra if and only if it has a bounded approximate unit. In particular, if A is an amenable C*-algebra, so is J.

In the C*-case, a dual Banach J-module extends to a Banach $M(J)$-module, and hence to a Banach A-module; similarly, a derivation on J to a dual Banach J-module extends to a derivation of $M(J)$ and hence A.

IV.3.3.6 PROPOSITION. Let $0 \to J \to A \to A/J \to 0$ be an exact sequence of Banach algebras. If J and A/J are amenable, so is A.

If $\delta : A \to \mathcal{X}^*$ is a derivation, $\delta|_J$ is inner, so an inner derivation can be subtracted from δ to obtain $\delta|_J = 0$. If $\mathcal{X}_J$ is the closed subspace of $\mathcal{X}$ spanned by $J\mathcal{X} + \mathcal{X}J$, then δ can be regarded as a derivation from A/J to the dual Banach A/J-module $(\mathcal{X}/\mathcal{X}_J)^*$.

The next result is easily proved using virtual diagonals:

IV.3.3.7 PROPOSITION. Let A and B be Banach algebras. Then $A \hat{\otimes} B$ is amenable if and only if A and B are amenable. If A and B are C*-algebras, then $A \otimes_{\max} B$ is amenable if and only if A and B are amenable.

IV.3.3.8 THEOREM. An inductive limit of amenable C*-algebras is amenable.

This almost has a trivial proof: if $A = \varinjlim A_i$, $\mathcal{X}^*$ is a dual Banach A-module, and $\delta : A \to \mathcal{X}^*$ is a derivation, choose $\phi_i \in \mathcal{X}^*$ such that $\delta(a) = a\phi_i - \phi_i a$ for all $a \in A_i$. If (ϕ_i) is bounded, any weak-* limit point ϕ satisfies $\delta(a) = a\phi - \phi a$ for all $a \in A$. But it is nontrivial to prove that the ϕ_i can be chosen bounded (in fact, ϕ_i can be chosen with $\|\phi_i\| \leq \|\delta\|$ [Haa83]).

Since $\mathbb{M}_n$ is amenable by the argument of IV.3.3.2, $\mathbb{K}$ is amenable. Combining this with IV.3.3.7, we obtain:

IV.3.3.9 COROLLARY. The class of amenable C*-algebras is closed under stable isomorphism.

Amenability of Type I C*-Algebras

Although the next two results will be subsumed in IV.3.3.15, it is worth giving an elementary argument for them:

IV.3.3.10 PROPOSITION. Every commutative C*-algebra is amenable.

This is an immediate consequence of IV.3.3.4, IV.3.3.5, and IV.3.5.5, taking G to be $\mathcal{U}(A)$ with the discrete topology.

Combining IV.3.3.10 with IV.3.3.9 and IV.3.3.6, by the usual bootstrapping we obtain:

IV.3.3.11 COROLLARY. Every Type I C*-algebra is amenable.

Amenability vs. Nuclearity

Both directions of the equivalence of amenability and nuclearity are nontrivial and require some heavy machinery. The easier direction, and the earlier result [Con78], is that amenability implies nuclearity. The first preliminary result is nearly trivial, using the σ-weak density of A in A^{**}:

IV.3.3.12 PROPOSITION. If A is an amenable C*-algebra, then A^{**} is an amenable von Neumann algebra.

IV.3.3.13 THEOREM. Every amenable C*-algebra is nuclear.

This is an immediate consequence of IV.3.3.12 and IV.2.5.4, combined with IV.3.1.12.

The converse proved elusive, and was finally proved by U. Haagerup in [Haa83]:

IV.3.3.14 THEOREM. Every nuclear C*-algebra is amenable.

There is a well-known fallacious argument for this result (cf. [BP80]). If A is a C*-algebra and $\mathcal{X}$ a Banach A-module, then $\mathcal{X}^*$ can be made into a Banach A^{**}-module as follows [Joh72]. If $a \in A$, $x \in \mathcal{X}$, $\phi \in \mathcal{X}^*$, define $a\phi, \phi a \in \mathcal{X}^*$ by

$$(a\phi)(x) = \phi(xa), \ (\phi a)(x) = \phi(ax)$$

and then define $(x, \phi), [\phi, x] \in A^*$ by

$$(x, \phi)(a) = \phi(ax), \ [\phi, x](a) = \phi(xa).$$

If $b \in A^{**}$, $\phi \in \mathcal{X}^*$, define $b\phi, \phi b \in \mathcal{X}^*$ by

$$(b\phi)(x) = \langle [\phi, x], b \rangle, \ (\phi b)(x) = \langle (x, \phi), b \rangle$$

where $\langle \cdot, \cdot \rangle$ is the pairing between A^* and A^{**} It can be checked that this action makes $\mathcal{X}^*$ into a normal Banach A^{**}-module.

The dual action of A^{**} on X^{**} thus makes X^{**} a dual Banach A^{**}-module. There is an alternate description of a Banach A^{**}-module structure on X^{**}, obtained by applying the construction of the previous paragraph to the Banach A-module $\mathcal{X}^*$; this makes X^{**} into a normal Banach A^{**}-module.

Assuming that the two module structures always coincide (it is easy to check that they coincide on $A \subseteq A^{**}$), at least if A is nuclear, making $\mathcal{X}^{**}$ a dual normal Banach A^{**}-module, we could show that a nuclear C*-algebra A is amenable as follows. Suppose $\mathcal{X} = \mathcal{Y}^*$ is a dual Banach A-module, and $\delta : A \to \mathcal{X}$ is a derivation. Then $\delta^{**} : A^{**} \to \mathcal{X}^{**}$ is a derivation, and A^{**} is injective and hence W*-amenable, so there is a $z \in \mathcal{X}^{**} = \mathcal{Y}^{***}$ with $\delta^{**}(b) = bz - zb$ for all $b \in A^{**}$. The inclusion $j : \mathcal{Y} \hookrightarrow \mathcal{Y}^{**}$ is an A-module map, so the projection $j^* : Y^{***} \to \mathcal{Y}^*$ is also an A-module map, and hence if $x = j^*(z) \in \mathcal{Y}^* = \mathcal{X}$, we have $\delta(a) = ax - xa$ for all $a \in A$.

Unfortunately, as pointed out in [BP80], the two module structures do *not* coincide in general, even for $A = \mathbb{K}$, and there is no natural way to make $\mathcal{X}^{**}$ into a dual normal Banach A^{**}-module.

The argument of Haagerup proceeds differently, showing the existence of a virtual diagonal in a nuclear C*-algebra using his generalization of the Grothendieck-Pisier inequality.

IV.3.3.15 COROLLARY. A C*-algebra is nuclear if and only if it is amenable.

Weak and Strong Amenability

There are two variations of amenability worth mentioning:

IV.3.3.16 DEFINITION. A C*-algebra (or Banach algebra) A is *weakly amenable* if every (bounded) derivation from A to A^* is inner.
A unital C*-algebra A is *strongly amenable* if, whenever $\mathcal{X}$ is a unital Banach A-module and $\delta : A \to \mathcal{X}^*$ is a (bounded) derivation, there is an x in the weak-* closed convex hull of

$$\{\delta(u)u^* : u \in \mathcal{U}(A)\}$$

with $\delta(a) = xa - ax$ for all $a \in A$. A general C*-algebra A is strongly amenable if $\tilde{A}$ is strongly amenable.

Obviously A strongly amenable $\Longrightarrow$ A amenable $\Longrightarrow$ A weakly amenable. Neither implication can be reversed:

IV.3.3.17 THEOREM. [Haa83] Every C*-algebra is weakly amenable.

Thus a C*-algebra such as $C^*(\mathbb{F}_2)$ is weakly amenable but not amenable.

IV.3.3.18 THEOREM. [Bun72] Let A be a unital C*-algebra. Then A is strongly amenable if and only if, whenever $\mathcal{X}$ is a Banach A-module and $C \subseteq \mathcal{X}^*$ is a weak-* compact convex subset such that $uCu^* \subseteq C$ for all $u \in \mathcal{U}(A)$, there is a $c \in C$ with $ucu^* = c$ for all $u \in \mathcal{U}(A)$.

Applying this result to the state space $\mathcal{S}(A) \subseteq A^*$, we obtain:

IV.3.3.19 COROLLARY. If A is a strongly amenable unital C*-algebra, then A has a tracial state.

Thus an algebra such as O_n (II.8.3.3(ii)) is amenable but not strongly amenable. Note the strong similarity, and close connection, between strong amenability and the Dixmier property (III.2.5.16).

There is no analog of IV.3.3.19 for nonunital C*-algebras. In fact:

IV.3.3.20 THEOREM. [Haa83] Let A be an amenable C*-algebra. Then $A \otimes \mathbb{K}$ is strongly amenable.

IV.3.3.21 THEOREM. [Ros77] If α is an action of a discrete amenable group G on a strongly amenable C*-algebra A, then $A \rtimes_\alpha G$ is strongly amenable. In particular, $C^*(G)$ is strongly amenable.

Finally, one might wonder why only dual Banach modules are used in the theory of amenability. Dual modules are natural in this theory, and are nice because of the presence of the weak-* topology. But they are also necessary as shown in [Hel93, VII.1.75] (cf. [Run02, 4.1.6]), using a virtual diagonal argument (IV.3.3.2):

IV.3.3.22 THEOREM. Let A be a C*-algebra. Suppose every (bounded) derivation from A into a Banach A-module is inner. Then A is finite-dimensional.

The converse is true: every derivation from a finite-dimensional C*-algebra A into an A-(bi)module is inner (IV.3.3.2). The generalization of IV.3.3.22 to general Banach algebras appears to still be open.

IV.3.4 Exactness and Subnuclearity

Recall (II.9.6.6) that a C*-algebra A is *exact* if tensoring a short exact sequence with A preserves exactness for the minimal cross norm. There is an alternate characterization in terms of slice maps (II.9.7.1):

IV.3.4.1 PROPOSITION. Let A and B be C*-algebras, J a closed ideal of B, and $\pi : B \to B/J$ the quotient map. Then the kernel K_J of $id \otimes \pi : A \otimes_{\min} B \to A \otimes_{\min} B/J$ is

$$S_J = \{x \in A \otimes_{\min} B \mid R_\phi(x^*x) \in J \ \forall \phi \in \mathcal{S}(A)\}$$

where $R_\phi : A \otimes_{\min} B \to B$ is the right slice map $a \otimes b \to \phi(a)b$. In particular, if $x \in S_J \cap (A \odot B)$, then $x \in A \odot J$.

PROOF: Regard states on B/J as states on B vanishing on J. The representations $\pi_\phi \otimes \pi_\psi$, as ϕ and ψ range over $\mathcal{S}(A)$ and $\mathcal{S}(B/J)$ respectively, separate the points of $A \otimes_{\min} B/J$. If $x \in S_J$, then

$$(\phi \otimes \psi)(x^*x) = \psi(R_\phi(x^*x)) = 0$$

for all $\phi \in \mathcal{S}(A)$, $\psi \in \mathcal{S}(B/J)$. If $(\pi_\phi \otimes \pi_\psi)(x^*x) \neq 0$ for some such ϕ, ψ, then by I.2.6.16 there are vector states $\phi' \in \mathcal{S}(A)$, $\psi' \in \mathcal{S}(B/J)$ with $(\phi' \otimes \psi')(x^*x) \neq 0$, a contradiction; so $x^*x \in K_J$, $x \in K_J$. Conversely, if $x \notin S_J$, there is a state ϕ on A with $R_\phi(x^*x) \notin J$, and for this ϕ there is a state ψ on B, vanishing on J, with $\psi(R_\phi(x^*x)) \neq 0$. But

$$\psi(R_\phi(x^*x)) = (\phi \otimes \psi)(x^*x)$$

so $(\pi_\phi \otimes \pi_\psi)(x^*x) \neq 0$, $(\pi_\phi \otimes \pi_\psi)(x) \neq 0$.

IV.3.4.2 COROLLARY. Let A be a C*-algebra. Then A is exact if and only if, for every C*-algebra B and closed ideal J of B,

$$A \otimes_{\min} J = \{x \in A \otimes_{\min} B \mid R_\phi(x^*x) \in J \; \forall \phi \in \mathcal{S}(A)\} \;.$$

The next corollary is an important property of exactness, proved independently by E. Kirchberg (cf. [Kir95b]) and S. Wassermann (cf. [Was76]), which is in contrast with the fact that a C*-subalgebra of a nuclear C*-algebra is not nuclear in general:

IV.3.4.3 THEOREM. A C*-subalgebra of an exact C*-algebra is exact. In particular, every C*-subalgebra of a nuclear C*-algebra is exact.

PROOF: Let A be an exact C*-algebra, and C a C*-subalgebra of A. If B is a C*-algebra with closed ideal J, suppose $x \in C \otimes_{\min} B$ and $R_\phi(x^*x) \in J$ for all $\phi \in \mathcal{S}(C)$. Then, if $\psi \in \mathcal{S}(A)$, and $\phi = \psi|_C$, then $R_\psi(x^*x) = R_\phi(x^*x) \in J$, and thus $x \in A \otimes_{\min} J$ since A is exact. Thus

$$x \in (A \otimes_{\min} J) \cap (C \otimes_{\min} B) = C \otimes_{\min} J$$

(II.9.6.9).

IV.3.4.4 PROPOSITION. An inductive limit (with *injective* connecting maps) of exact C*-algebras is exact.

This result is also true if the connecting maps are not injective, but this requires the deep fact IV.3.4.19.

PROOF: Let $A = [\cup A_i]^-$, with each A_i exact, and let $0 \to J \to B \to B/J \to 0$ be an exact sequence. Then $(A \otimes_{\min} J) \cap (A_i \otimes_{\min} B) = A_i \otimes_{\min} J$ for all i (II.9.6.9), so

$$(A \otimes_{\min} B)/(A \otimes_{\min} J) = [\cup (A_i \otimes_{\min} B)/(A_i \otimes_{\min} J)]^-$$
$$= [\cup (A_i \otimes_{\min} (B/J))]^- = A \otimes_{\min} (B/J) \;.$$

Combining the last two results, we obtain:

IV.3.4.5 COROLLARY. A C*-algebra A is exact if and only if every separable C*-subalgebra of A is exact.

Property C

We next discuss a property introduced by R. Archbold and C. Batty [AB80], which implies exactness and turns out to be equivalent, at least in the separable case (IV.3.4.18).

IV.3.4.6 Let A and B be C*-algebras, and ω the universal representation of $A \otimes_{\min} B$. Then ω defines commuting representations ω_A and ω_B of A and B, and hence commuting normal representations of A^{**} and B^{**}, and so a representation ρ of $A^{**} \odot B^{**}$, which is easily seen to be faithful. Since $\rho(A^{**} \odot B^{**}) \subseteq \omega(A \otimes_{\min} B)'' \cong (A \otimes_{\min} B)^{**}$, we get an embedding

$$\iota_{A,B} : A^{**} \odot B^{**} \to (A \otimes_{\min} B)^{**}$$

and hence a C*-norm on $A^{**} \odot B^{**}$, denoted $\|\cdot\|_C$. We have $\|\cdot\|_C \leq \|\cdot\|_{\mathrm{bin}}$.

IV.3.4.7 DEFINITION. A C*-algebra A has *Property C* if, for every C*-algebra B, $\|\cdot\|_C = \|\cdot\|_{\min}$ on $A^{**} \odot B^{**}$.

IV.3.4.8 PROPOSITION. Every nuclear C*-algebra has Property C.

This is an immediate consequence of IV.2.4.4(vi), since A^{**} is semidiscrete if A is nuclear (IV.2.2.13, IV.2.4.9).

The C-norm has good permanence properties:

IV.3.4.9 PROPOSITION. Let A and B be C*-algebras, D a C*-subalgebra of B, and J a closed ideal in B.
(i) The restriction of the C-norm on $A^{**} \odot B^{**}$ to $A^{**} \odot D^{**}$ is the C-norm on $A^{**} \odot D^{**}$.
(ii) The embedding of $A^{**} \odot (B/J)^{**}$ into $(A^{**} \otimes_C B^{**})/(A^{**} \otimes_C J^{**})$ gives an isometry of $A^{**} \otimes_C (B/J)^{**}$ onto $(A^{**} \otimes_C B^{**})/(A^{**} \otimes_C J^{**})$.

PROOF: (i): The diagram

$$\begin{array}{ccc} A^{**} \odot D^{**} & \longrightarrow & A^{**} \odot B^{**} \\ \iota_{A,D} \downarrow & & \downarrow \iota_{A,B} \\ (A \otimes_{\min} D)^{**} & \longrightarrow & (A \otimes_{\min} B)^{**} \end{array}$$

is easily seen to be commutative, and the sides and bottom are isometries.
(ii): $J^{**} = pB^{**}$ for a central projection $p \in B^{**}$. If $\pi : B \to B/J$ is the quotient map, the kernel of $(id \otimes \pi)^{**} : (A \otimes_{\min} B)^{**} \to (A \otimes_{\min} (B/J))^{**}$ is

$q(A \otimes_{\min} B)^{**}$ for a central projection $q \in (A \otimes_{\min} B)^{**}$. It is straightforward to check that $q = \iota_{A,B}(1 \otimes p)$, so that the homomorphism $x \to (1-q)x$ from $A^{**} \odot B^{**}$ to $(A \otimes_{\min} B)^{**}$ extends to an isomorphism ψ from $(A \otimes_{\min} (B/J))^{**}$ onto $(1-q)(A \otimes_{\min} B)^{**}$ (although the kernel of $(id \otimes \pi)^{**}$ cannot be identified with $q(A \otimes_{\min} B)^{**}$ in general). It is also straightforward to check that the diagram

$$\begin{array}{ccc} A^{**} \odot (B/J)^{**} & \xrightarrow{0 \oplus id} & (A^{**} \odot pB^{**}) \oplus (A^{**} \odot (1-p)B^{**}) \cong A^{**} \odot B^{**} \\ \iota_{A,B/J} \downarrow & & \downarrow \iota_{A,B} \\ (A \otimes_{\min} (B/J))^{**} & \xrightarrow{0 \oplus \psi} & q(A \otimes_{\min} B)^{**} \oplus (1-q)(A \otimes_{\min} B)^{**} \cong (A \otimes_{\min} B)^{**} \end{array}$$

is commutative. The result is then obvious.

IV.3.4.10 COROLLARY. *If A has Property C, then every C*-subalgebra or quotient of A has Property C. In particular, every subquotient of a nuclear C*-algebra has Property C.*

IV.3.4.11 COROLLARY. *Let A and B be C*-algebras, and J be a closed ideal of B. If $\|\cdot\|_C = \|\cdot\|_{\min}$ on $A^{**} \odot B^{**}$, then the sequence*

$$0 \to A \otimes_{\min} J \to A \otimes_{\min} B \to A \otimes_{\min} (B/J) \to 0$$

is exact.

IV.3.4.12 COROLLARY. *A C*-algebra with Property C is exact.*

Property C also leads to an interesting variation of the notion of exactness:

IV.3.4.13 DEFINITION. *Let B be a C*-algebra, J a closed ideal of B. (B, J) is a* coexact pair *if, for every C*-algebra A, the sequence*

$$0 \to A \otimes_{\min} J \to A \otimes_{\min} B \to A \otimes_{\min} (B/J) \to 0$$

is exact.
Perhaps a better term would be "exact pair" in retrospect, in light of the next result and IV.3.4.18.

IV.3.4.14 PROPOSITION. *Let B be a C*-algebra, J a closed ideal of B. Then (B, J) is a coexact pair under either of the following conditions:*

(i) *B has Property C.*
(ii) *B/J is nuclear.*

Part (i) is an immediate corollary of IV.3.4.11, and (ii) is obvious.

IV.3 Nuclear and Exact C*-Algebras   387

Exactness and Nuclear Embeddability

If A is a nuclear C*-algebra, then the identity map on A is nuclear (IV.3.1.4). It is interesting to consider the more general property of a C*-algebra A that there is a nuclear embedding of A into some larger C*-algebra D; any subnuclear C*-algebra clearly has this property.

IV.3.4.15 PROPOSITION. Let A be a C*-algebra. The following are equivalent:

(i) There is a C*-algebra D and a nuclear embedding (nuclear injective *-homomorphism) from A to D.
(ii) There is a nuclear embedding of A into $\mathcal{L}(\mathcal{H})$ for some Hilbert space $\mathcal{H}$.
(iii) There is a C*-algebra D and a nuclear complete order embedding from A to D.

A C*-algebra A satisfying these properties is *nuclearly embeddable*.

IV.3.4.16 EXAMPLES.

(i) Any nuclear C*-algebra is nuclearly embeddable.
(ii) Any C*-subalgebra of a nuclearly embeddable C*-algebra is nuclearly embeddable. In particular, every subnuclear C*-algebra is nuclearly embeddable.

IV.3.4.17 THEOREM. A C*-algebra is nuclearly embeddable if and only if it is exact.

One direction of this theorem is relatively easy, that a nuclearly embeddable C*-algebra is exact; we give the argument. Suppose A is nuclearly embeddable, $\iota : A \to D$ is a nuclear embedding, and $0 \to J \to B \to B/J \to 0$ is an exact sequence of C*-algebras. By IV.3.4.2, it suffices to show that the set $S_J \subseteq A \otimes B$ is just $A \otimes J$. Let $x \in S_J$ and $\epsilon > 0$, and let $\alpha : A \to \mathbb{M}_n$, $\beta : \mathbb{M}_n \to D$ be cp-contractions such that

$$\|[(\beta \circ \alpha) \otimes id](x^*x) - [\iota \otimes id](x^*x)\| < \epsilon$$

(it is easily seen that such α, β exist by approximating x^*x by an element of $A \odot B$). If $\phi \in \mathcal{S}(\mathbb{M}_n)$, then $\phi \circ \alpha \in \mathcal{S}(A)$, so

$$R_\phi((\alpha \otimes id)(x^*x)) = R_{\phi \circ \alpha}(x^*x) \in J.$$

Thus

$$(\alpha \otimes id)(x^*x) \in \{0 \leq y \in \mathbb{M}_n \otimes B : R_\phi(y) \in J \,\forall \phi \in \mathcal{S}(\mathbb{M}_n)\}.$$

Since $\mathbb{M}_n \otimes B = \mathbb{M}_n \odot B$, $(\alpha \otimes id)(x^*x) \in \mathbb{M}_n \otimes J$ by IV.3.4.1. Thus $[(\beta \circ \alpha) \otimes id](x^*x) \in D \otimes J$; since ϵ is arbitrary, $(\iota \otimes id)(x^*x) \in D \otimes J$. So

$$(\iota \otimes id)(x^*x) \in (\iota(A) \otimes B) \cap (D \otimes J) = \iota(A) \otimes J$$

(II.9.6.9), $x \in A \otimes J$.

The converse, that an exact C*-algebra is nuclearly embeddable, is a technically difficult argument due to Kirchberg ([Kir95b, 4.1]; cf. [Was94, Theorem 7.3]).

Characterization of Separable Exact C*-Algebras

The deepest results about exact C*-algebras are due to Kirchberg, and are summarized (along with IV.3.4.17) in the next theorem:

IV.3.4.18 THEOREM. Let A be a separable C*-algebra. Then the following are equivalent:

(i) A is exact.
(ii) A is subnuclear.
(iii) A embeds in the Cuntz algebra O_2.
(iv) A is a subquotient of the CAR algebra B.
(v) There is a complete order embedding of A into the CAR algebra B.
(vi) A has Property C.

A is nuclear if and only if there is a complete order embedding $\phi : A \to B$ such that there is an idempotent cp-projection from B onto $\phi(A)$.

(iii) $\Longrightarrow$ (ii) is trivial, and (ii) $\Longrightarrow$ (vi) by IV.3.4.10. (vi) $\Longrightarrow$ (i) is IV.3.4.12. (i) $\Longrightarrow$ (v) was proved (along with the last assertion of the theorem) in [Kir95b, 1.4] (cf. [Was94, 9.1]); the argument is complicated and we do not outline it here. (v) $\Longrightarrow$ (iv) by II.6.10.12, and (iv) $\Longrightarrow$ (vi) follows from IV.3.4.10 since the CAR algebra is nuclear.

(i) $\Longrightarrow$ (iii) is a difficult argument which will not be described here; see [KP00].

Note that (iii) $\Longrightarrow$ (iv) by the result [Bla85a] that O_2 is a subquotient of the CAR algebra; however, this is not needed for the proof of the theorem.

IV.3.4.19 COROLLARY. A quotient of an exact C*-algebra is exact.

This is actually only a corollary of IV.3.4.18 in the separable case (property (iv) is clearly preserved under quotients); the general case can be easily obtained from the separable case and IV.3.4.5.

The fact that the class of exact C*-algebras is closed under quotients, like the corresponding property for nuclearity, is a deep result.

Kirchberg ([Kir93], [Kir95b]) has also proved:

IV.3.4.20 THEOREM. An extension of exact C*-algebras is not exact in general; however, a semisplit extension (II.8.4.23) of exact C*-algebras is exact.

See [Kir95a] or [Was94] for a more complete description of the theory of exact C*-algebras.

Primitive Ideal Space of a Tensor Product

Throughout this subsection, as is our usual convention, "$\otimes$" will mean "$\otimes_{\min}$" unless otherwise specified.

If A and B are C*-algebras, we wish to describe the primitive ideal space $\mathrm{Prim}(A \otimes B)$ in terms of $\mathrm{Prim}(A)$ and $\mathrm{Prim}(B)$. In the commutative case, if $A = C_o(X)$, $B = C_o(Y)$, we have $A \otimes B \cong C_o(X \times Y)$, so

$$\mathrm{Prim}(A \otimes B) \cong X \times Y \cong \mathrm{Prim}(A) \times \mathrm{Prim}(B)$$

(II.9.4.4); in fact, in this case we have $\widehat{A \otimes B} \cong \hat{A} \times \hat{B}$. We examine to what extent these relations hold in general.

The results of this subsection are essentially due to J. Tomiyama [Tom67b, Theorem 5] (cf. [Bla77a, 3.3]) and E. Kirchberg (cf. [BK04, §2.8]).

IV.3.4.21 If A and B are arbitrary C*-algebras, there is an injective map

$$\Pi : \hat{A} \times \hat{B} \to \widehat{A \otimes B}$$

given by $\Pi(\rho, \sigma) = \rho \otimes \sigma$. This is a continuous map relative to the natural topologies (II.6.5.13), and drops to a well-defined map, also denoted Π, from $\mathrm{Prim}(A) \times \mathrm{Prim}(B)$ to $\mathrm{Prim}(A \otimes B)$; this Π is injective, continuous (it preserves containment in the appropriate sense), and its range is dense in $\mathrm{Prim}(A \otimes B)$ since the intersection of the kernels of the representations $\{\rho \otimes \sigma : \rho \in \hat{A}, \sigma \in \hat{B}\}$ is 0.

IV.3.4.22 If $I = \ker(\rho)$, $J = \ker(\sigma)$, then $\Pi(I, J)$ is the kernel of the natural map from $A \otimes B$ to $(A/I) \otimes (B/J)$. Thus $\Pi(I, J)$ contains $I \otimes B + A \otimes J$. In general,

$$(A \otimes B)/(I \otimes B + A \otimes J) \cong (A/I) \otimes_\gamma (B/J)$$

for some cross norm γ which is not the minimal cross norm in general.

IV.3.4.23 PROPOSITION. Under any of the following conditions, we have $\gamma = \min$, so $\Pi(I, J) = I \otimes B + A \otimes J$:

(i) Either A/I or B/J is nuclear.
(ii) (A, I) is a coexact pair (IV.3.4.13) and A/I is exact.
(iii) (B, J) is a coexact pair and B/J is exact.
(iv) Either A or B is separable and exact.

PROOF: (i) is obvious. For (ii), under the hypotheses we have exact sequences

$$0 \to I \otimes B \to A \otimes B \to (A/I) \otimes B \to 0$$

$$0 \to (A/I) \otimes J \to (A/I) \otimes B \to (A/I) \otimes (B/J) \to 0$$

so it follows that the kernel of the composite map $A \otimes B \to (A/I) \otimes B \to (A/I) \otimes (B/J)$ is exactly $I \otimes B + A \otimes J$. (iii) is similar, and (iv) follows from IV.3.4.14, IV.3.4.18, and IV.3.4.10.

IV.3.4.24 There is a way of generating an "inverse" for Π, at least in good cases. If π is an irreducible representation of $A \otimes B$, then π_A and π_B are factor representations of A and B respectively (II.9.2.1). If $I = \ker(\pi_A)$, $J = \ker(\pi_B)$, then I and J are closed prime ideals (II.6.1.11) and hence primitive if A and B are separable (II.6.5.15) or if either A or B is Type I (since then both π_A and π_B are Type I, and it follows easily from III.1.5.5 that $\pi \cong \rho \otimes \sigma$, where ρ and σ are irreducible representations quasi-equivalent to π_A and π_B respectively). The ideals I and J clearly depend only on $\ker(\pi)$, so we obtain a well-defined map
$$\Delta : \mathrm{Prim}(A \otimes B) \to \mathrm{Prim}(A) \times \mathrm{Prim}(B)$$
at least if A and B are separable. In fact, if $K \in \mathrm{Prim}(A \otimes B)$, then $\Delta(K) = (I, J)$, where I and J are the kernels of the composite maps
$$A \to A \otimes 1 \subseteq A \otimes \tilde{B} \to (A \otimes \tilde{B})/K$$
$$B \to 1 \otimes B \subseteq \tilde{A} \otimes B \to (\tilde{A} \otimes B)/K.$$
It is easy to check that Δ is continuous, and that $\Delta \circ \Pi$ is the identity on $\mathrm{Prim}(A) \times \mathrm{Prim}(B)$.

If $K \in \mathrm{Prim}(A \otimes B)$, and $\Delta(K) = (I, J)$, then $I \otimes B + A \otimes J \subseteq K$. Also, the natural map from $(A/I) \odot (B/J)$ to $(A \otimes B)/K$ is injective (II.9.5.2), hence $K \subseteq \Pi(I, J)$. Thus, under any of the conditions in IV.3.4.23, we have
$$K = I \otimes B + A \otimes J = \Pi(I, J)$$
and therefore $\Pi \circ \Delta(K) = K$. The conclusion is:

IV.3.4.25 THEOREM. If A and B are separable C*-algebras and either A or B is exact, then Π is a homeomorphism from $\mathrm{Prim}(A) \times \mathrm{Prim}(B)$ to $\mathrm{Prim}(A \otimes_{\min} B)$, with inverse Δ.

Note that the result also holds even if A and B are nonseparable if either A or B is Type I (or if both are simple by II.9.5.3); in fact, it holds in general in the nonseparable case (if A or B is exact) if primitive ideals are replaced by closed prime ideals [BK04].

IV.3.4.26 The conclusion of IV.3.4.25 does not hold in full generality, even for separable C*-algebras. Let $A = B = C^*(\mathbb{F}_2)$. Then the left and right regular representations of $\mathbb{F}_2$ are factor representations (III.3.3.7) and give an irreducible representation π of $A \otimes_{\min} B$ (II.9.4.6(iii)), whose kernel is not in the image of $\Pi(\mathrm{Prim}(A) \times \mathrm{Prim}(B))$.

IV.3.4.27 What about $\widehat{A \otimes B}$? If $\pi \in \widehat{A \otimes B}$, and π_A or π_B is Type I, then as observed above, the other is also Type I and $\pi \cong \rho \otimes \sigma$ for ρ, σ irreducible representations quasi-equivalent to π_A and π_B respectively. Thus, if A or B is Type I (no separability necessary), the map $\Pi : \hat{A} \times \hat{B} \to \widehat{A \otimes B}$ is surjective, and it is easily verified to be a homeomorphism.

Surjectivity of Π does not hold in general. For example, if M is an injective factor on a separable $\mathcal{H}$, which is not Type I, and C and D are σ-weakly dense C*-subalgebras of M and M' respectively, then since M is semidiscrete the identity representations of C and D define an irreducible representation of $C \otimes D$ which does not decompose as a tensor product, i.e. is not in $\Pi(\hat{C} \times \hat{D})$.

If M is Type III, and A and B are separable C*-algebras which are not Type I, then by IV.1.5.10 there are representations ρ of A and σ of B on $\mathcal{H}$, such that $\rho(A)'' = M$, $\sigma(B)'' = M'$. ρ and σ define an irreducible representation of $A \otimes B$ which is not in the range of Π, so $\Pi : \hat{A} \times \hat{B} \to \widehat{A \otimes B}$ is not surjective if A and B are separable and not Type I.

IV.3.5 Group C*-Algebras and Crossed Products

In this section, we discuss the relation between nuclearity and exactness for group C*-algebras and crossed products by a locally compact group G and structural properties of G, particularly amenability. An expanded version of this discussion can be found in [Bla04a].

The first result has an easy and elementary proof:

IV.3.5.1 PROPOSITION. Let A and B be C*-algebras, G a locally compact group, and α an action of G on A. Write β for the action $\alpha \otimes id$ on $A \otimes_{\max} B$ or $A \otimes_{\min} B$. Then there are natural isomorphisms

$$\phi_{\max} : (A \rtimes_\alpha G) \otimes_{\max} B \to (A \otimes_{\max} B) \rtimes_\beta G$$

$$\phi_{\min} : (A \rtimes_\alpha^r G) \otimes_{\min} B \to (A \otimes_{\min} B) \rtimes_\beta^r G$$

making the following diagram commute, where π is the quotient map from the full to the reduced crossed product:

$$\begin{array}{ccc}
(A \rtimes_\alpha G) \otimes_{\max} B & \xrightarrow{\phi_{\max}} & (A \otimes_{\max} B) \rtimes_\beta G \\
\pi \otimes id \downarrow & & \downarrow \pi \\
(A \rtimes_\alpha^r G) \otimes_{\min} B & \xrightarrow{\phi_{\min}} & (A \otimes_{\min} B) \rtimes_\beta^r G
\end{array}$$

PROOF: To obtain $\phi_{\max}$, both $(A \rtimes_\alpha G) \otimes_{\max} B$ and $(A \otimes_{\max} B) \rtimes_\beta G$ are isomorphic to the universal C*-algebra generated by

$$\{\pi(f)\rho(a)\sigma(b) : f \in L^1(G), a \in A, b \in B\}$$

where π, ρ, and σ are representations of G, A, and B respectively on the same Hilbert space, with π and ρ α-covariant and σ commuting with π and ρ. The proof for $\phi_{\min}$ is similar, using the obvious natural isomorphism between $L^2(G, \mathcal{H}_1) \otimes \mathcal{H}_2$ and $L^2(G, \mathcal{H}_1 \otimes \mathcal{H}_2)$ for Hilbert spaces $\mathcal{H}_1$, $\mathcal{H}_2$.

Combining this result with II.10.3.14, we obtain:

IV.3.5.2 COROLLARY. Let A be a nuclear C*-algebra, G an amenable locally compact group, and α an action of G on A. Then $A \rtimes_\alpha G$ is nuclear. In particular, if G is amenable, then $C^*(G)$ is nuclear.

The last statement was first observed by A. Guichardet [Gui69], using essentially the argument of IV.3.5.1.

IV.3.5.3 IV.3.5.2 implies that many standard C*-algebras are nuclear, for example the noncommutative tori (II.8.3.3(i), II.10.4.12(i)) and the Cuntz and Cuntz-Krieger algebras (II.8.3.3(ii)–(iii)).

IV.3.5.4 The converse of IV.3.5.2 is not valid in general for crossed products: a (full) crossed product by a nonamenable group can be an amenable C*-algebra (e.g. II.10.4.3). Even the converse of the result about group C*-algebras can fail if G is not discrete: there are nonamenable Lie groups (e.g. $SL_2(\mathbb{R})$) whose group C*-algebra is Type I [HC57]. In fact, the group C*-algebra of any connected group is amenable (nuclear) [Con76, 6.9(c)].

The converse of the group C*-algebra part of IV.3.5.2 for discrete G is true, but more difficult; however, it avoids the heaviest machinery of IV.2.6.2. The next result was proved in [Joh72] and [Bun76]:

IV.3.5.5 THEOREM. Let G be a discrete group. Then the following are equivalent:

(i) $C^*(G)$ is an amenable C*-algebra.
(ii) $C_r^*(G)$ is an amenable C*-algebra.
(iii) G is an amenable group.

PROOF: (i) $\Longrightarrow$ (ii) is IV.3.3.4, and (iii) $\Longrightarrow$ (i) is almost an immediate consequence of IV.3.3.1, since a Banach $C^*(G)$-module can be regarded as a Banach $L^1(G)$-module.

(ii) $\Longrightarrow$ (iii): Let τ be the Plancherel trace on $C_r^*(G)$, i.e. $\tau(f) = \langle \lambda(f)\chi_e, \chi_e \rangle$. Regard $\mathcal{Y} = C_r^*(G)$ as a C*-subalgebra of $\mathcal{X} = \mathcal{L}(l^2(G))$, and extend τ to $\mathcal{L}(l^2(G))$ by the same formula. $\mathcal{X}$ is a Banach $C_r^*(G)$-module and $\tau \in \mathcal{X}^*$. Define a derivation δ from $C_r^*(G)$ to $\mathcal{X}^*$ by

$$\delta(f) = \lambda(f)\tau - \tau\lambda(f).$$

Then, for any f, $\delta(f) \in \mathcal{Y}^\perp \cong (\mathcal{X}/\mathcal{Y})^*$, so δ is a derivation from $C_r^*(G)$ to $(\mathcal{X}/\mathcal{Y})^*$. Since $C_r^*(G)$ is amenable, there is a $\psi \in \mathcal{Y}^\perp$ with $\delta(f) = \lambda(f)\psi - \psi\lambda(f)$. If $\phi = \tau - \psi$, then $\lambda(f)\phi = \phi\lambda(f)$ for all $f \in C_r^*(G)$. If $h \in C_b(G)$, set $\mu(h) = \phi(M_h)$, where M_h is the multiplication operator of h; then μ is a right invariant mean on G.

This argument can be modified to show that if G is discrete and $C_r^*(G)$ is nuclear, then G is amenable, without using the deep result that a nuclear C*-algebra is amenable: if $C_r^*(G)$ is nuclear, then $\mathfrak{L}(G)$ (III.3.3.1) is injective and thus has a hypertrace (IV.2.2.21). The last part of the proof then shows that G is amenable.

Putting this together with IV.3.5.2 and IV.3.1.13, we obtain:

IV.3.5.6 THEOREM. Let G be a discrete group. Then the following are equivalent:

(i) $C^*(G)$ is a nuclear C*-algebra.
(ii) $C_r^*(G)$ is a nuclear C*-algebra.
(iii) G is an amenable group.

IV.3.5.5 was proved before either direction of the equivalence of nuclearity and amenability was known. Then after Connes' work on injective factors, IV.3.5.6 was proved, providing evidence for the equivalence of nuclearity and amenability.

Nonnuclear Subalgebras of Nuclear C*-Algebras

IV.3.5.7 From the preceding results, we can obtain an explicit example of a nonnuclear C*-subalgebra of a nuclear C*-algebra, due to Choi [Cho79]. Let $G = \mathbb{Z}_2 * \mathbb{Z}_3$ with generators u and v, $u^2 = 1$, $v^3 = 1$; then G contains a copy of $\mathbb{F}_2$, and is thus not amenable, so $C_r^*(G)$ is not nuclear. Let $C_r^*(G)$ act on $l^2(G)$ via λ. There is a subset S of G such that $S \cap (uS) = S \cap (vS) = \emptyset$ and $S \cup (uS) = S \cup (vS) \cup (v^{-1}S) = G$ (e.g. S is the set inductively defined as the set of reduced words xy in $\{u, v, v^{-1}\}$ of length $n \geq 1$ with x of length 1, y of length $n-1$, $x \neq v^{-1}$, $y \notin S$, i.e. $S = \{u, v, uv^{-1}, uvu, uv^{-1}u, vuv, \dots\}$). Let p be the projection of $l^2(G)$ onto $\mathcal{X} = \text{span}(S)$, and let A be the C*-subalgebra of $\mathcal{L}(l^2(G))$ generated by $\lambda(C_r^*(G))$ and p. Then $A \cong \mathcal{O}_2$, which is nuclear. See [Cho79, 2.6] for details.

Exactness in Group C*-Algebras

IV.3.5.8 In analogy with the C*-theory, there is a recent theory of exact locally compact groups, initiated by Kirchberg and S. Wassermann [KW99]. If α is an action of a locally compact group G on a C*-algebra A, and J is a (closed two-sided) ideal of A which is globally invariant under α, so that α can be restricted to J, then there is also an induced action of G on A/J, also denoted α, and quotient maps $A \rtimes_\alpha G \to (A/J) \rtimes_\alpha G$ and $A \rtimes_\alpha^r G \to (A/J) \rtimes_\alpha^r G$. It is routine to show that $J \rtimes_\alpha G$ and $J \rtimes_\alpha^r G$ sit naturally as ideals in $A \rtimes_\alpha G$ and $A \rtimes_\alpha^r G$ respectively, and that the following sequence is exact:

$$0 \to J \rtimes_\alpha G \to A \rtimes_\alpha G \to (A/J) \rtimes_\alpha G \to 0.$$

However, the reduced sequence

$$0 \to J \rtimes_\alpha^r G \to A \rtimes_\alpha^r G \to (A/J) \rtimes_\alpha^r G \to 0$$

is not obviously exact in general. The locally compact group G is *exact* if the reduced sequence is exact for every A, J, α. Every amenable group is exact.

There is a condition introduced by N. Higson, J. Roe, and G. Yu which turns out to be closely related:

IV.3.5.9 DEFINITION. A locally compact group is *amenable at infinity* if it has an amenable action on a compact space.

Since the trivial action of an amenable group on a point is amenable, amenable groups are amenable at infinity.

IV.3.5.10 THEOREM. Let G be a locally compact group. Consider the following conditions:

(i) G is amenable at infinity.
(ii) G is exact.
(iii) Whenever α is an action of G on an exact C*-algebra A, $A \rtimes_\alpha^r G$ is exact.
(iv) $C_r^*(G)$ is an exact C*-algebra.
(v) There is a uniform embedding of G into a Hilbert space.

Then (i) $\Longrightarrow$ (ii) $\Longrightarrow$ (iii) $\Longrightarrow$ (iv) and (i) $\Longrightarrow$ (v); if G is discrete, then (i)–(iv) are all equivalent.

Parts of this theorem were proved in [KW99], [AD02], [Oza00], [HR00], and [Yu00].

IV.3.5.11 It is unknown whether the conditions (i)–(iv) are equivalent for nondiscrete groups. They are equivalent for almost connected groups; in fact, they are always satisfied for almost connected groups. It is also unknown whether (v) is equivalent to (i)–(iv) for discrete groups.

IV.3.5.12 An example was recently given by Ozawa [Oza00], using an unpublished construction of M. Gromov [Gro00], of a discrete group G not satisfying the conditions of IV.3.5.10.

IV.3.5.13 The conditions of IV.3.5.10 are closely related to the *Baum-Connes Conjecture*. The Baum-Connes Conjecture is discussed in detail in [BCH94], [Hig98], and [Val02] (and a number of other references), and we will not even state it precisely; roughly, it says that if G is a (discrete) group, then the K-theory of $C_r^*(G)$, and more generally of certain G-spaces, can be described geometrically in a natural way. The Baum-Connes Conjecture for a group G implies the Strong Novikov Conjecture on homotopy invariance of higher signatures.

V
K-Theory and Finiteness

In this chapter, we give a discussion of the basics of K-theory for C*-algebras, as well as properties related to finiteness in C*-algebras, including the important notion of quasidiagonality, which may be regarded as a type of strong finiteness. Another strong notion of finiteness, stable rank one, is included in a general discussion of stable rank. K-Theory is a vast subject (other parts of the theory are described more comprehensively in [CST04]), and we concentrate only on those aspects directly related to the structure and classification of C*-algebras, most notably ordered K-theory for finite C*-algebras; thus the topics treated in this chapter have a natural unity from this point of view.

V.1 K-Theory for C*-Algebras

One of the most profound developments in the subject of operator algebras over the last thirty years has been the incorporation of ideas and techniques from algebraic topology, to the extent that the whole subject is often thought of as "noncommutative topology." K-Theory and its sophisticated generalizations form the core of this aspect of the subject, and the language of K-theory pervades almost all parts of the modern theory of operator algebras.

We will only give an overview of K-theory and its generalizations, with very few proofs and even only modest explanation of the origin of the ideas and the connections with topology. A much more complete treatment can be found in [Bla98], [WO93], or [RLL00], and [Con94] contains an excellent description of the uses of K-theory beyond those described here. The treatment here is largely excerpted from [Bla98].

Topological K-theory is the study of vector bundles by algebraic means. The first ideas are due to A. Grothendieck [Gro58], and K-theory as a branch of algebraic topology was first developed by M. Atiyah and F. Hirzebruch [AH59]. The notions of K-theory were translated into algebraic language, leading to algebraic K-theory (the theory of projective modules) and a version of K-theory suitable for Banach algebras. Good references are [Ati67] and

[Kar78] for topological K-theory and [Ros94] and [Sri91] for algebraic K-theory; these references give a complete historical account of the development of the subject.

In this section, we describe the Banach algebra version of K-theory, specializing to C*-algebras.

V.1.1 K_0-Theory

The goal of K_0-theory is to associate to a C*-algebra (or pre-C*-algebra) A an abelian group $K_0(A)$ whose elements are "formal differences of equivalence classes of projections over A." Most of the theory works equally well for an arbitrary ring.

Equivalence of Projections

V.1.1.1 Recall (II.3.3.3) that projections p and q are equivalent in A, written $p \sim q$, if there is a partial isometry $u \in A$ with $u^*u = p$, $uu^* = q$. There are two other natural notions of equivalence: p and q are *unitarily equivalent* (in A), written $p \sim_u q$, if there is a unitary $v \in \tilde{A}$ with $vpv^* = q$, and p and q are *homotopic* (in A), written $p \sim_h q$, if there is a norm-continuous path (p_t) ($0 \le t \le 1$) of projections in A with $p_0 = p$, $p_1 = q$.

$$p \sim_u q \Longrightarrow p \sim q$$

(set $u = vp$), and

$$p \sim_h q \Longrightarrow p \sim_u q$$

by II.3.3.4. The converse implications do not hold in general, but they are true "stably":

V.1.1.2 PROPOSITION. If $p \sim q$ in A, then $diag(p,0) \sim_u diag(q,0)$ in $M_2(A)$.
PROOF: If $u \in A$ is a partial isometry from p to q, then

$$v = \begin{bmatrix} 1-q & q \\ q & 1-q \end{bmatrix} \begin{bmatrix} 1-p & u^* \\ u & 1-q \end{bmatrix}$$

is a unitary in $\widetilde{M_2(A)}$ with $v \cdot diag(p,0) \cdot v^* = diag(q,0)$.

V.1.1.3 PROPOSITION. If $p \sim_u q$ in A, then $diag(p,0) \sim_h diag(q,0)$ in $M_2(A)$.
PROOF: If $vpv^* = q$, for $0 \le t \le 1$ let

$$w_t = \begin{bmatrix} v & 0 \\ 0 & 1 \end{bmatrix} \begin{bmatrix} \cos(\frac{\pi}{2}t)1 & -\sin(\frac{\pi}{2}t)1 \\ \sin(\frac{\pi}{2}t)1 & \cos(\frac{\pi}{2}t)1 \end{bmatrix} \begin{bmatrix} v^* & 0 \\ 0 & 1 \end{bmatrix} \begin{bmatrix} \cos(\frac{\pi}{2}t)1 & \sin(\frac{\pi}{2}t)1 \\ -\sin(\frac{\pi}{2}t)1 & \cos(\frac{\pi}{2}t)1 \end{bmatrix}$$

in $\widetilde{M_2(A)}$, and set $p_t = w_t \cdot diag(p,0) \cdot w_t^*$. Then $p_0 = diag(p,0)$ and $p_1 = diag(q,0)$.

The Semigroup $V(A)$

Recall (II.6.6.9) that $M_\infty(A)$ is the algebra of all infinite matrices over A with only finitely many nonzero entries. Whenever it is convenient, we will identify $M_n(A)$ with its image in the upper left-hand corner of $M_{n+k}(A)$ or $M_\infty(A)$. The algebra $M_\infty(A)$ has a natural norm, and the completion is the stable algebra $A \otimes \mathbb{K}$ of A (II.6.6.11).

V.1.1.4 DEFINITION. $V_0(A)$ is the set of equivalence classes of projections in A. We set $V(A) = V_0(M_\infty(A))$.

There is a binary operation (orthogonal addition) on $V(A)$: if $[p], [q] \in V(A)$, choose $p' \in [p]$ and $q' \in [q]$ with $p' \perp q'$ (this is always possible by "moving down the diagonal," as in the proof of V.1.1.3), and define

$$[p] + [q] = [p' + q'].$$

This operation is well defined, and makes $V(A)$ into an abelian semigroup with identity $[0]$.

V.1.1.5 If A is unital, $V(A)$ can also be described as the set of isomorphism classes of finitely generated projective right A-modules (one could equally well use left modules instead). The binary operation on $V(A)$ corresponds to direct sum of modules. If $A = C(X)$, then isomorphism classes of finitely generated projective modules over A are in natural one-one correspondence with isomorphism classes of (complex) vector bundles over X, the module associated to a bundle being the set of continuous sections (II.3.3.4 is used to show that every projective module comes from a bundle).

Because of V.1.1.2 and V.1.1.3, one obtains exactly the same semigroup starting with $\sim_u$ or $\sim_h$ instead of $\sim$ as the equivalence, since the three notions coincide on $M_\infty(A)$.

$V(A)$ depends on A only up to stable isomorphism: if $M_\infty(A) \cong M_\infty(B)$, or more generally if $A \otimes \mathbb{K} \cong B \otimes \mathbb{K}$ (V.1.1.10), then $V(A) \cong V(B)$. In particular, $V(M_n(A)) \cong V(A)$.

If A is separable, then $V(A)$ is countable (this follows easily from II.3.3.4).

V.1.1.6 EXAMPLES.

(i) $V(\mathbb{C}) \cong V(\mathbb{M}_n) \cong V(\mathbb{K}) \cong \mathbb{N} \cup \{0\}$.
(ii) If $A = \mathcal{L}(\mathcal{H})$ with $\mathcal{H}$ separable and infinite-dimensional, then $V(A) \cong \{0\} \cup \mathbb{N} \cup \{\infty\}$. If A is a II$_1$ factor, then

$$V(A) \cong \mathbb{R}_+ \cup \{0\}.$$

If A is a countably decomposable II$_\infty$ factor, then

$$V(A) \cong \{0\} \cup \mathbb{R}_+ \cup \{\infty\}.$$

If A is a countably decomposable type III factor, then $V(A) = \{0, \infty\}$. In each case the operation is the ordinary one with $\infty + x = \infty$ for all x. (If A is not countably decomposable, then $V(A)$ will have other infinite cardinals.)

(iii) Let $A = C(S^2)$. There is an extremely important projection in $M_2(A)$, called the *Bott projection*, which can be defined by identifying S^2 with $\mathbb{C}P^1$ and identifying a point of $\mathbb{C}P^1$ (a one-dimensional subspace of $\mathbb{C}^2$) with the projection onto the subspace (a projection in $\mathbb{M}_2$). In this way a projection in
$$C(\mathbb{C}P^1, \mathbb{M}_2) \cong M_2(C(S^2))$$
is obtained. The corresponding line bundle on S^2 is nontrivial. There are countably many isomorphism classes of line bundles on S^2, defined by "clutching" over the equator, naturally parametrized by $\pi_1(\mathcal{U}_1(\mathbb{C})) \cong \mathbb{Z}$. Since every vector bundle on S^2 is a sum of line bundles, $V(A) \cong \mathbb{N} \times \mathbb{Z}$, with $[1_A] = (1,0)$, and $(1,1)$ the class of the Bott projection.

Example (iii) shows the necessity of considering projections in matrix algebras over A, since unexpected projections sometimes appear which have nothing to do with projections in A. This can even happen in simple unital C*-algebras [Bla81, 4.11] (but not in a factor). It turns out to be necessary for K-theory to take such projections into account. Consideration of matrix algebras is more natural if one associates projections with projective modules.

Example (ii) shows that the semigroup $V(A)$ can fail to have cancellation. Even $V(C(X))$ fails to have cancellation for many compact differentiable manifolds X (e.g. S^5 or $\mathbb{T}^5$; cf. [Hus66]).

Properties of $V(A)$

V.1.1.7 *Functoriality:* If $\phi : A \to B$ is a *-homomorphism, then ϕ induces a map $\phi_* : V_0(A) \to V_0(B)$. Then ϕ extends to a homomorphism from $M_\infty(A)$ to $M_\infty(B)$, which induces a semigroup homomorphism, also denoted ϕ_*, from $V(A)$ to $V(B)$. So V is a covariant functor from the category of C*-algebras to the category of abelian semigroups.

V.1.1.8 *Homotopy Invariance:* If $\phi, \psi : A \to B$ are homotopic (II.5.5.6), then $\phi(p) \sim_h \psi(p)$ for any projection $p \in M_\infty(A)$, and hence $\phi_* = \psi_*$, i.e. V is a *homotopy-invariant functor*.

V.1.1.9 *Direct Sums:* If $A = A_1 \oplus A_2$, then
$$M_\infty(A) \cong M_\infty(A_1) \oplus M_\infty(A_2)$$
and equivalence is coordinatewise; hence
$$V(A) \cong V(A_1) \oplus V(A_2).$$

V.1.1.10 *Inductive Limits:* Suppose $A = \lim_\rightarrow (A_i, \phi_{ij})$ (II.8.2.1). It is easy to see that if A_0 is the algebraic direct limit of (A_i, ϕ_{ij}), then $V(A_0)$ is the algebraic direct limit of $(V(A_i), \phi_{ij*})$. However, since A is the completion of A_0, it is not obvious that the natural map from $V(A_0)$ to $V(A)$ is either injective or surjective. In fact, if B is a dense *-subalgebra of a C*-algebra A, then the map from $V(B)$ into $V(A)$ is neither injective nor surjective in general. But it is true in the inductive limit case, i.e. $V(A)$ is the algebraic direct limit of $(V(A_i), \phi_{ij*})$. This is an immediate consequence of II.3.3.4 and the next result, which follows from a functional calculus argument.

V.1.1.11 PROPOSITION. Let A be the inductive limit of (A_i, ϕ_{ij}), p a projection in A, and $\epsilon > 0$. Then for sufficiently large i there is a projection $p_0 \in A_i$ with $\|p - \phi_i(p_0)\| < \epsilon$. If $p, q \in A$ with $p \sim_u q$, then there are projections p_0 and q_0 in A_i for sufficiently large i with $\|p - \phi_i(p_0)\| < \epsilon$, $\|q - \phi(q_0)\| < \epsilon$, and $p_0 \sim_u q_0$ in A_i.

V.1.1.12 EXAMPLE. Let A be the CAR algebra (II.8.2.2(iv)). Then $V(A) \cong \mathbb{D}_+$, the nonnegative dyadic rational numbers.

The Grothendieck Group

V.1.1.13 If H is an abelian semigroup, then there is a universal enveloping abelian group $G(H)$ called the *Grothendieck group* of H. $G(H)$ can be constructed in a number of ways. For example, $G(H)$ may be defined to be the quotient of $H \times H$ under the equivalence relation $(x_1, y_1) \sim (x_2, y_2)$ if and only if there is a z with

$$x_1 + y_2 + z = x_2 + y_1 + z.$$

$G(H)$ may be thought of as the group of (equivalence classes of) formal differences of elements of H, thinking of (x, y) as $x - y$. The prototype example of this construction is the construction of $\mathbb{Z}$ from $\mathbb{N}$. $G(H)$ may also be defined by generators and relations, with generators $\{\langle x \rangle : x \in H\}$ and relations $\{\langle x \rangle + \langle y \rangle = \langle x + y \rangle : x, y \in H\}$.

V.1.1.14 There is a canonical homomorphism from H into $G(H)$ which sends x to $[(x + x, x)]$. This homomorphism is injective if and only if H has cancellation. $G(H)$ has the universal property that any homomorphism from H into an abelian group factors through $G(H)$. G gives a covariant functor from abelian semigroups to abelian groups.

Definition of $K_0(A)$

We might be tempted to define $K_0(A)$ to be the Grothendieck group of $V(A)$; but it turns out that this is not the proper definition for A nonunital. We begin with the unital case.

V.1.1.15 DEFINITION. If A is a unital C*-algebra, $K_0(A)$ is the Grothendieck group of $V(A)$.

K_0 is a covariant functor from unital C*-algebras (or even rings) to abelian groups satisfying the properties of V.1.1.7-V.1.1.10. Elements of $K_0(A)$ may be pictured as formal differences $[p] - [q]$, where $[p_1] - [q_1] = [p_2] - [q_2]$ if there are orthogonal projections p'_i, q'_i, r in $M_\infty(A)$ with $p'_i \sim p_i$, $q'_i \sim q_i$, and

$$p'_1 + q'_2 + r \sim p'_2 + q'_1 + r.$$

This is the *Standard Picture* of K_0 for unital A.

V.1.1.16 EXAMPLES.

(i) $K_0(\mathbb{C}) \cong K_0(\mathbb{M}_n) \cong \mathbb{Z}$.
(ii) If A is a II$_1$ factor, then $K_0(A) \cong \mathbb{R}$. If A is an infinite factor, then $K_0(A) = 0$.
(iii) $K_0(C(S^1)) \cong \mathbb{Z}$; $K_0(C(S^2)) \cong \mathbb{Z}^2$.
(iv) Let A be the CAR algebra (II.8.2.2(iv)). Then $K_0(A) \cong \mathbb{D}$, the dyadic rational numbers. Similarly, if B is the UHF algebra of type 3^∞, then $K_0(A)$ is the "triadic" rationals, rationals whose denominator is a power of 3. (Thus in particular A and B are not isomorphic or even stably isomorphic.) More generally, the K_0-groups of the UHF algebras are precisely the dense subgroups of $\mathbb{Q}$ containing $\mathbb{Z}$. These are in one-one correspondence with the *generalized integers* (or *supernatural numbers*), formal products

$$q = 2^{m_2} 3^{m_3} 5^{m_5} \cdots$$

where an infinite number of primes and infinite exponents are allowed. The subgroup of $\mathbb{Q}$ corresponding to q is the group, denoted $\mathbb{Z}_{(q)}$, of all rational numbers whose denominators "divide" q. If $A = \varinjlim(A_n, \phi_{mn})$, where A_n is a matrix algebra with $A_1 = \mathbb{C}$ and $\phi_{n,n+1}$ is a (unital) embedding of multiplicity k_n, then $K_0(A) = \mathbb{Z}_{(q)}$, where $q = \prod k_n$. It turns out that two UHF algebras are isomorphic if (and only if) their generalized integers are the same. This classification is due to J. Glimm [Gli60], and is a special case of the classification of AF algebras (V.2.4.19).

If A is nonunital, there is a natural *-homomorphism $\mu : A^\dagger = \tilde{A} \to \mathbb{C}$ with kernel A, and hence there is an induced homomorphism

$$\mu_* : K_0(A^\dagger) \to K_0(\mathbb{C}) \cong \mathbb{Z}.$$

V.1.1.17 DEFINITION. If A is nonunital, then $K_0(A) = \ker \mu_*$.

This definition is consistent with the previous one if A is unital; then $A^\dagger \cong A \oplus \mathbb{C}$, and μ is projection onto the second coordinate. Thus $K_0(A)$ may be viewed as the set of formal differences $[p] - [q]$, where $p, q \in M_\infty(A^\dagger)$ with $p \sim q \mod M_\infty(A)$, with the usual notion of equivalence of formal differences in $K_0(A^\dagger)$. In fact, any element of $K_0(A)$ may be written $[p] - [q_n]$, where

$$q_n = diag(1,\ldots,1,0,\ldots)$$

(with n ones on the diagonal) and $p \equiv q_n \mod M_\infty(A)$: if n is large enough, $q \leq q_n$, and

$$[p] - [q] = [p' + (q_n - q)] - [q_n]$$

where $p' \sim p$ and $p' \perp q_n$. This is the *Standard Picture* of $K_0(A)$ for general A.

K_0 is a covariant functor from C*-algebras (general rings) to abelian groups which has the properties of V.1.1.7-V.1.1.10.

V.1.1.18 For any A, there is a homomorphism from $V(A^\dagger)$ to $K_0(A)$, given by $[p] \to [p] - [q_n]$, where $\mu(p)$ is a rank n projection in $M_\infty(\mathbb{C})$. Composing this with the canonical map from $V(A)$ to $V(A^\dagger)$ yields a homomorphism from $V(A)$ to $K_0(A)$. If A is unital, or more generally *stably unital* ($A \otimes \mathbb{K}$ has an approximate unit of projections), then the image of $V(A)$ generates $K_0(A)$, but this is not true for general A.

V.1.1.19 EXAMPLES.
(i) $\mathbb{K}$ is stably unital, so $K_0(\mathbb{K}) \cong \mathbb{Z}$ since $V(\mathbb{K}) \cong \mathbb{N} \cup \{0\}$.
(ii) $K_0(C_o(\mathbb{R}^2)) \cong \mathbb{Z}$ because of V.1.1.16(iii); but $V(C_o(\mathbb{R}^2)) = 0$.

Exactness

V.1.1.20 THEOREM. If J is a (closed two-sided) ideal in A, then the sequence

$$K_0(J) \xrightarrow{\iota_*} K_0(A) \xrightarrow{\pi_*} K_0(A/J)$$

is exact in the middle, i.e. $\ker(\pi_*) = \operatorname{im}(\iota_*)$.

The proof is an easy calculation.

This theorem is one of the most important reasons for defining K_0 the way we did. If we instead had defined $K_0(A)$ to be the Grothendieck group of $V(A)$ in general, the result would fail for the exact sequence $0 \to C_o(\mathbb{R}^2) \to C(S^2) \to \mathbb{C} \to 0$. Thus the more complicated definition of K_0 is necessary to make the desired exact sequences work.

V.1.1.21 It is important to realize that an exact sequence

$$0 \longrightarrow J \xrightarrow{\iota} A \xrightarrow{\pi} A/J \longrightarrow 0$$

does *not* yield an exact sequence

$$0 \longrightarrow K_0(J) \xrightarrow{\iota_*} K_0(A) \xrightarrow{\pi_*} K_0(A/J) \longrightarrow 0$$

in general, i.e. ι_* is not always injective (for example, $A = \mathcal{L}(\mathcal{H})$, $J = \mathcal{K}(\mathcal{H})$) and π_* not always surjective (for example, $A = C([0,1])$, $J = C_o((0,1))$). The problem with π_* is that projections in a quotient do not in general lift to projections. The exact sequence of V.1.1.20 can be expanded to a larger exact sequence, but K_1 and Bott periodicity are needed.

V.1.2 K_1-Theory and Exact Sequences

We now associate to a C*-algebra A another group $K_1(A)$, which is a stabilized version of the group $\mathcal{U}(A)/\mathcal{U}(A)_o$ of connected components of the unitary group of A. (This construction can only be done for Banach algebras, although there is a related construction in algebraic K-theory for general rings.) There are intimate connections between K_0 and K_1, using the notion of suspension.

Definition of $K_1(A)$

V.1.2.1 Let A be a C*-algebra (or pre-C*-algebra). Define

$$\mathcal{U}_n(A) = \{x \in \mathcal{U}(M_n(A^\dagger)) : x \equiv 1_n \mod M_n(A)\}$$

(if A is unital, then $\mathcal{U}_n(A)$ is just isomorphic to $\mathcal{U}(M_n(A))$). Similarly, we can define

$$\mathrm{GL}_n(A) = \{x \in \mathrm{GL}_n(A^\dagger) : x \equiv 1_n \mod M_n(A)\}.$$

$\mathcal{U}_n(A)$ [resp. $\mathrm{GL}_n(A)$] is a closed normal subgroup of $\mathcal{U}_n(A^\dagger)$ [resp. $\mathrm{GL}_n(A^\dagger)$]. We embed $\mathcal{U}_n(A)$ into $\mathcal{U}_{n+1}(A)$ (and $\mathrm{GL}_n(A)$ into $\mathrm{GL}_{n+1}(A)$) by $x \to diag(x, 1)$. [This embedding is the "exponential" of the embedding of $M_n(A)$ into $M_{n+1}(A)$ considered in V.1.1. This is the appropriate analog, since the connection between K_0 and K_1 is given by exponentiation.]

Let $\mathcal{U}_\infty(A) = \varinjlim \mathcal{U}_n(A)$ and $\mathrm{GL}_\infty(A) = \varinjlim \mathrm{GL}_n(A)$. These are topological groups with the inductive limit topology or the norm topology in $M_\infty(A^\dagger)$. $\mathcal{U}_\infty(A)$ and $\mathrm{GL}_\infty(A)$ can be thought of as the group of unitary or invertible infinite matrices which have diagonal elements in $1_{A^\dagger} + A$, off-diagonal elements in A, and only finitely many entries different from 0 or $1_{A^\dagger}$. We will identify elements of $\mathcal{U}_n(A)$ or $\mathrm{GL}_n(A)$ with their images in $\mathcal{U}_\infty(A)$ or $\mathrm{GL}_\infty(A)$.

The connected component of the identity $\mathcal{U}_n(A)_o$ is a path-connected open subgroup of $\mathcal{U}_n(A)$ (and similarly for $\mathrm{GL}_n(A)_o$). The embedding of $\mathcal{U}_n(A)$ into $\mathcal{U}_{n+1}(A)$ maps $\mathcal{U}_n(A)_o$ into $\mathcal{U}_{n+1}(A)_o$, and $\mathcal{U}_\infty(A)_o = \varinjlim \mathcal{U}_n(A)_o$. Similarly, $\mathrm{GL}_\infty(A)_o = \varinjlim \mathrm{GL}_n(A)_o$. There is a deformation retraction of $\mathrm{GL}_n(A)$ onto $\mathcal{U}_n(A)$ given by polar decomposition, and hence $\mathrm{GL}_n(A)/\mathrm{GL}_n(A)_o$ is naturally isomorphic to $\mathcal{U}_n(A)/\mathcal{U}_n(A)_o$.

V.1.2.2 DEFINITION.

$$K_1(A) = \mathcal{U}_\infty(A)/\mathcal{U}_\infty(A)_o = \varinjlim [\mathcal{U}_n(A)/\mathcal{U}_n(A)_o]$$

or alternatively

$$K_1(A) = \mathrm{GL}_\infty(A)/\mathrm{GL}_\infty(A)_o = \varinjlim [\mathrm{GL}_n(A)/\mathrm{GL}_n(A)_o].$$

This is related to, but not the same as, the group K_1^{alg} of algebraic K-theory: $K_1^{alg}(A)$ is the quotient of $\mathrm{GL}_\infty(A)$ by its commutator subgroup. See [Kar78, II.6.13] for the relationship.

It is easily seen that $K_1(A)$ is also isomorphic to $\mathcal{U}_1(A \otimes \mathbb{K})/\mathcal{U}_1(A \otimes \mathbb{K})_o$.

$K_1(A)$ is countable if A is separable, since nearby invertible elements are in the same component.

V.1.2.3 EXAMPLES.

(i) $K_1(\mathbb{C}) = 0$, and more generally K_1 of any von Neumann algebra (or AW*-algebra) is 0 (the unitary group of a von Neumann algebra is connected by spectral theory). K_1 of any AF algebra is also 0, since the path component of any unitary is open and thus contains a unitary with finite spectrum.
(ii) If $A = C(S^1)$, then $\mathcal{U}_1(A)/\mathcal{U}_1(A)_o \cong \mathbb{Z}$ by sending a function to its winding number around 0. The map from $\mathcal{U}_1(A)/\mathcal{U}_1(A)_o$ to $K_1(A)$ is an isomorphism in this case, so $K_1(C(S^1)) \cong \mathbb{Z}$. We also have $K_1(C_o(\mathbb{R})) \cong \mathbb{Z}$ by the same argument.
(iii) Let $\mathcal{Q}$ be the Calkin algebra (I.8.2.1). Then $K_1(\mathcal{Q}) \cong \mathbb{Z}$, with the isomorphism being given by Fredholm index.

V.1.2.4 The map from $\mathcal{U}_n(A)/\mathcal{U}_n(A)_o$ to $\mathcal{U}_{n+1}(A)/\mathcal{U}_{n+1}(A)_o$ need not be an isomorphism, and hence the map from $\mathcal{U}_n(A)/\mathcal{U}_n(A)_o$ to $K_1(A)$ need not be an isomorphism. For example, let $A = C(S^3)$. Then $\mathcal{U}_1(A)/\mathcal{U}_1(A)_o$ is trivial since every map from S^3 to S^1 is homotopic to a constant; but the homeomorphism $S^3 \cong SU(2)$ gives a unitary in $C(S^3, \mathbb{M}_2) \cong M_2(A)$ which is not in the connected component of the identity.

The group $\mathcal{U}_n(A)$ is almost never abelian. Even the group $\mathcal{U}_n(A)/\mathcal{U}_n(A)_o$ need not be abelian in general (even if A is commutative). However, we have the following. If $u \in \mathrm{GL}_n(A)$, we write $[u]$ for its image in $K_1(A)$.

V.1.2.5 PROPOSITION. $K_1(A)$ is an abelian group; in fact,

$$[u][v] = [diag(u,v)].$$

PROOF: If $u, v \in \mathcal{U}_n(A)$, then by an argument similar to the proof of V.1.1.3, we have that $diag(uv, 1)$, $diag(vu, 1)$, and $diag(u, v)$ are all in the same connected component of $\mathcal{U}_{2n}(A)$.

V.1.2.6 The theory of stable rank (V.3.1) asserts that the map from the group $\mathcal{U}_n(A)/\mathcal{U}_n(A)_o$ to $K_1(A)$ is an isomorphism for sufficiently large n if $sr(A)$ is finite, and gives some information on the smallest n for which this is true. The question of when this map is injective or surjective is a typical question of *nonstable K-theory*; see V.3.1.26 for more details.

V.1.2.7 If $\phi : A \to B$, then ϕ extends uniquely to a unital map from $A^\dagger$ to $B^\dagger$, and hence defines a homomorphism $\phi_* : K_1(A) \to K_1(B)$. Also, if $A = \varinjlim(A_i, \phi_{ij})$, then it is easy to see that

$$K_1(A) \cong \varinjlim(K_1(A_i), \phi_{ij*}).$$

So K_1 is a functor from C*-algebras to abelian groups which commutes with inductive limits.

Suspensions

The connection with K_0 uses the fundamental notion of suspension (II.5.5.10). Since $SA \cong C_o(\mathbb{R}) \otimes A$ (II.9.4.4), we have

$$S(M_n(A)) \cong M_n(SA) \quad \text{and} \quad S(A \otimes \mathbb{K}) \cong SA \otimes \mathbb{K}$$

and the map $\phi : A \to B$ induces $S\phi : SA \to SB$ (suspension is *functorial*). We have that $(SA)^\dagger$ is isomorphic to

$$\{f : [0,1] \to A^\dagger \mid f \text{ continuous}, f(0) = f(1) = \lambda 1, f(t) = \lambda 1 + x_t \text{ for } x_t \in A\}$$
$$\cong \{f : S^1 \to A \mid f \text{ continuous}, f(z) = \lambda 1 + x_z, x_z \in A, x_1 = 0\}.$$

V.1.2.8 There is a map $\theta_A : K_1(A) \to K_0(SA)$ defined as follows. Let $v \in \mathcal{U}_n(A)$. Take a path w_t from 1_{2n} to $diag(v, v^*)$ in $\mathcal{U}_{2n}(A)$ as in the proof of V.1.1.3. Set $p_t = w_t q_n w_t^*$. Then $p = (p_t)$ is a projection in $M_{2n}((SA)^\dagger)$. Set $\theta_A([v]) = [p] - [q_n]$ (where q_n also denotes the corresponding element of $M_\infty((SA)^\dagger)$, i.e. the constant function q_n).

V.1.2.9 THEOREM. $\theta_A : K_1(A) \to K_0(SA)$ is an isomorphism. Furthermore, the isomorphism is natural, i.e. θ gives an invertible natural transformation from K_1 to $K_0 \circ S$.

The proof is quite straightforward and elementary, but rather long.

Long Exact Sequence

V.1.2.10 COROLLARY. If

$$0 \to J \xrightarrow{\iota} A \xrightarrow{\pi} A/J \to 0$$

is an exact sequence of C*-algebras, then the induced sequence

$$K_1(J) \xrightarrow{\iota_*} K_1(A) \xrightarrow{\pi_*} K_1(A/J)$$

is exact in the middle.

V.1.2.11 Just as with K_0, we cannot make the sequence exact at the ends by adding 0's. For example, if $A = C([0,1])$, $J = C_o((0,1))$, then the unitary $u(t) = e^{2\pi i t}$ in $J^\dagger$ gives a nontrivial element of $K_1(J)$ which becomes trivial in $K_1(A)$, so the map from $K_1(J)$ to $K_1(A)$ is not injective. Similarly, if $A = C(\bar{D})$, $J = C_o(D)$, where D is the open unit disk, then $A/J \cong C(S^1)$. $K_1(A)$ is trivial, but $K_1(A/J) \cong \mathbb{Z}$, so the map from $K_1(A)$ to $K_1(A/J)$ is not surjective.

Instead, we can define a connecting map $\partial : K_1(A/J) \to K_0(A)$ which makes a long exact sequence

$$K_1(J) \xrightarrow{\iota_*} K_1(A) \xrightarrow{\pi_*} K_1(A/J) \xrightarrow{\partial} K_0(J) \xrightarrow{\iota_*} K_0(A) \xrightarrow{\pi_*} K_0(A/J).$$

V.1.2.12 DEFINITION. Let $u \in \mathcal{U}_n(A/J)$, and let $w \in \mathcal{U}_{2n}(A)$ be a lift of $diag(u, u^{-1})$ (II.1.6.8, V.1.2.5). Define

$$\partial([u]) = [wq_n w^{-1}] - [q_n] \in K_0(J)$$

(we have $\partial([u]) \in K_0(J)$ because $diag(u, u^{-1})$ commutes with q_n, so $wq_n w^* \in M_{2n}(J^\dagger)$, and its image mod $M_{2n}(J)$ is q_n).

The map ∂ is well defined (a straightforward calculation), and is obviously a homomorphism since both operations are diagonal sum.

V.1.2.13 The map ∂ is called the *index map*. The reason is the following. Suppose A is a unital C*-algebra and u is a unitary in $M_n(A/J)$. If u lifts to a partial isometry $v \in M_n(A)$, then $diag(u, u^*)$ lifts to the unitary

$$w = \begin{bmatrix} v & 1 - vv^* \\ 1 - v^*v & v^* \end{bmatrix}$$

so

$$\partial([u]) = [wq_n w^{-1}] - [q_n] = [diag(vv^*, 1 - v^*v)] - [q_n] = [1 - v^*v] - [1 - vv^*].$$

In the special case $A = \mathcal{L}(\mathcal{H})$, $J = \mathcal{K}(\mathcal{H})$, and $K_0(\mathcal{K}(\mathcal{H}))$ is identified with $\mathbb{Z}$ in the standard way, the map ∂ is exactly the map which sends a unitary in the Calkin algebra to its Fredholm index.

Unitaries in a quotient do not lift to partial isometries in general, so the definition of ∂ must be stated in the more complicated way given above.

V.1.2.14 PROPOSITION. ∂ makes the sequence exact at $K_1(A/J)$ and at $K_0(J)$.

The proof is a simple calculation.

V.1.2.15 We can also define higher K-groups by

$$K_2(A) = K_1(SA) = K_0(S^2 A), \ldots, K_n(A) = K_0(S^n A).$$

We then have connecting maps from $K_{n+1}(A/J)$ to $K_n(J)$ for each n by suspension, and an infinite long exact sequence

$$\cdots \xrightarrow{\partial} K_n(J) \xrightarrow{\iota_*} K_n(A) \xrightarrow{\pi_*} K_n(A/J) \xrightarrow{\partial} K_{n-1}(J) \xrightarrow{\iota_*} \cdots \xrightarrow{\pi_*} K_0(A/J).$$

V.1.2.16 PROPOSITION. If
$$0 \longrightarrow J \longrightarrow A \longrightarrow A/J \longrightarrow 0$$
is a split exact sequence of C*-algebras, then
$$0 \longrightarrow K_n(J) \longrightarrow K_n(A) \longrightarrow K_n(A/J) \longrightarrow 0$$
is a split exact sequence for all n. Thus
$$K_n(A) \cong K_n(J) \oplus K_n(A/J)$$
for all n.

PROOF: All the connecting maps are 0 since everything in $K_n(A/J)$ lifts.

V.1.2.17 COROLLARY. $K_1(A) \cong K_1(A^\dagger)$ for all A. More precisely, the inclusion of A into $A^\dagger$ induces an isomorphism.

PROOF: Use V.1.2.16 plus the fact that $K_1(\mathbb{C}) = 0$.

Bott Periodicity and the Six-Term Exact Sequence

In fact, $K_0(A)$ is naturally isomorphic to $K_1(SA)$, and hence to $K_2(A)$. As a consequence, the long exact sequence of V.1.2.15 becomes a cyclic 6-term exact sequence.

V.1.2.18 We have a split exact sequence
$$0 \longrightarrow SA \longrightarrow \Omega A \longrightarrow A \longrightarrow 0$$
where $\Omega A = C(S^1, A)$, which induces a split exact sequence
$$0 \longrightarrow K_1(SA) \longrightarrow K_1(\Omega A) \longrightarrow K_1(A) \longrightarrow 0$$
so $K_1(SA) = \ker \eta_*$, where $\eta : \Omega A \to A$ is evaluation at 1. This will be our *standard picture* of $K_1(SA)$. So $K_1(SA)$ may be viewed as the group of homotopy equivalence classes of loops in $\mathcal{U}_\infty(A)$ with base point 1. The group operation is pointwise multiplication, but may alternately be taken as the ordinary concatenation multiplication of loops [Spa66, 1.6.10], i.e. $K_1(SA) \cong \pi_1(\mathcal{U}_\infty(A))$.

If p is a projection in $M_n(A^\dagger)$, write
$$f_p(z) = zp + (1-p) \in \mathcal{U}_n(\Omega(A^\dagger)) \cong C(S^1, \mathcal{U}_n(A^\dagger)).$$

If $p_1 \equiv p_2 \mod M_n(A)$, then $f_{p_1} f_{p_2}^* \in \mathcal{U}_n(\Omega A)$, taking the value 1 at $z = 1$. If $p_1 \sim_h p_2$, then f_{p_1} is homotopic to f_{p_2} as elements of $\mathcal{U}_n(\Omega(A^\dagger))$ taking the value 1 at 1, i.e. as loops in $\mathcal{U}_n(A^\dagger)$ with base point 1.

V.1.2.19 DEFINITION. The homomorphism $\beta_A : K_0(A) \to K_1(SA)$ defined by
$$\beta_A([e] - [q_n]) = [f_e f_{q_n}^{-1}]$$
is called the *Bott map* for A.

β_A is well defined, and the Bott map construction is clearly functorial (natural).

V.1.2.20 THEOREM. [BOTT PERIODICITY] β_A is an isomorphism.

This theorem is probably the central result of K-theory. There are several essentially different proofs known; all require nontrivial arguments and calculations. Perhaps the most elegant proof is an argument of Cuntz [Cun81b], using a "Toeplitz extension." The first proof of the theorem in this form was given by Atiyah [Ati68].

V.1.2.21 Combining V.1.2.20 with V.1.2.9, $\theta_{SA} \circ \beta_A$ gives an isomorphism between $K_0(A)$ and $K_0(S^2A)$ which is natural in A. In the case $A = \mathbb{C}$, this map provides an isomorphism from $K_0(\mathbb{C})$ to $K_0(C_o(\mathbb{R}^2))$. This map can be described as the map which sends $[1]$ to $[p] - [q_1]$, where p is the Bott projection in $M_2(C_o(\mathbb{R}^2)^\dagger) \cong M_2(C(S^2))$.

Combining V.1.2.20 with V.1.2.15, we obtain the most fundamental exact sequence in K-theory:

V.1.2.22 COROLLARY. [STANDARD SIX-TERM EXACT SEQUENCE] Let $0 \longrightarrow J \xrightarrow{\iota} A \xrightarrow{\pi} A/J \longrightarrow 0$ be an exact sequence of C*-algebras. Then the following six-term cyclic sequence is exact:

$$\begin{array}{ccccc} K_0(J) & \xrightarrow{\iota_*} & K_0(A) & \xrightarrow{\pi_*} & K_0(A/J) \\ \partial \uparrow & & & & \downarrow \partial \\ K_1(A/J) & \xleftarrow{\pi_*} & K_1(A) & \xleftarrow{\iota_*} & K_1(J) \end{array}$$

The map $\partial : K_0(A/J) \to K_1(J)$ is the composition of the suspended index map $\partial : K_2(A/J) \to K_1(J)$ with the Bott map.

V.1.2.23 The connecting map $\partial : K_0(A/J) \to K_1(J)$ is called the *exponential map*. An explicit formula for this map is given by
$$\partial([p] - [q_n]) = [\exp(2\pi i x)]$$
where p is a projection in $M_\infty((A/J)^\dagger)$ with $p \equiv q_n \mod M_\infty(A/J)$ and $x \in M_\infty(A^\dagger)_+$ with $\pi(x) = p$. The derivation of this formula is an easy exercise.

If p lifts to a projection in $M_\infty(A^\dagger)$, then $\partial([p] - [q_n]) = 0$. The exponential map is the obstruction to (stably) lifting projections from quotients, just as the index map is the obstruction to (stably) lifting unitaries.

V.1.3 Further Topics

K-Theory of Crossed Products

It is not easy to determine the K-theory of a crossed product. However, there are two fundamental results which allow computation of the K-theory of crossed products by $\mathbb{Z}$ or $\mathbb{R}$:

V.1.3.1 THEOREM. [PIMSNER–VOICULESCU EXACT SEQUENCE] Let A be a C*-algebra and $\alpha \in \operatorname{Aut}(A)$. Then there is a cyclic six-term exact sequence

$$\begin{array}{ccccc}
K_0(A) & \xrightarrow{1-\alpha_*} & K_0(A) & \xrightarrow{\iota_*} & K_0(A \rtimes_\alpha \mathbb{Z}) \\
\partial \uparrow & & & & \downarrow \partial \\
K_1(A \rtimes_\alpha \mathbb{Z}) & \xleftarrow{\iota_*} & K_1(A) & \xleftarrow{1-\alpha_*} & K_1(A)
\end{array}$$

where $\iota : A \to A \otimes_\alpha \mathbb{Z}$ is the inclusion and $\alpha_* : K_n(A) \to K_n(A)$ is the induced automorphism.

V.1.3.2 THEOREM. [CONNES' THOM ISOMORPHISM] If $\alpha : \mathbb{R} \to \operatorname{Aut}(A)$, then $K_i(A \rtimes_\alpha \mathbb{R}) \cong K_{1-i}(A)$ ($i = 0, 1$).

Connes' Thom Isomorphism [Con81] is a generalization of Bott Periodicity (the case of trivial action), and is an analog (though not a generalization) of the ordinary Thom isomorphism, which says that if E is a K-oriented n-dimensional real vector bundle over X, and E is itself regarded as a locally compact Hausdorff space, then $K_i(C_o(E)) \cong K_{i+n \bmod 2}(C(X))$. The result is a bit surprising at first glance, since it says that the K-theory of a crossed product by $\mathbb{R}$ is independent of the action. An intuitive argument for this fact is that any action of $\mathbb{R}$ can be continuously deformed to a trivial action, and K-theory is insensitive to continuous deformations. This rough argument can be used as the basis of a proof, using KK-theory ([FS81]; see also [ENN93]). The Pimsner-Voiculescu (P-V) exact sequence can then be obtained rather easily from the Thom isomorphism via a mapping torus construction (it is just the six-term exact sequence of a suitable extension).

V.1.3.3 The P-V exact sequence [PV80], which predates Connes' result, shows that the K-theory of a crossed product by $\mathbb{Z}$ is not independent of the action; but (roughly speaking) it depends only on the induced action on the K-theory of A. The P-V exact sequence is not in general enough to completely determine the K-theory of the crossed product, except in special cases (e.g. when the K-groups of A are free abelian groups); but it is nonetheless a powerful tool in determining the possibilities.

V.1.3.4 EXAMPLE. Let O_n be the Cuntz algebra (II.8.3.3). Write

$$O_n \otimes \mathbb{K} = B \rtimes_\sigma \mathbb{Z}$$

as in II.10.3.15(iv). Since $K_0(B)$ is $\mathbb{Z}_{(n^\infty)}$, $K_1(B) = 0$, and $\alpha_* : K_0(B) \to K_0(B)$ is multiplication by n, we have

$$K_1(O_n) = \ker \alpha_* = 0$$

$$K_0(O_n) \cong \mathbb{Z}_{(n^\infty)}/(1-n)\mathbb{Z}_{(n^\infty)} \cong \mathbb{Z}_{n-1}.$$

(Cuntz's original argument computing the K-theory of O_n was more elementary and did not use the P-V sequence.) In fact, $[1_{O_n}]$ is a generator of $K_0(O_n)$; if $s_1, \ldots, s_n$ are the generators of O_n and $p_k = s_k s_k^*$, then

$$[1] = [p_k] = [p_1] + \cdots + [p_n] = n[1]$$

so $(n-1)[1] = 0$. A similar argument using the crossed product description of $O_A \otimes \mathbb{K}$ for an $n \times n$ matrix A shows that $K_1(O_A) \cong \ker(I - A) \subseteq \mathbb{Z}^n$ and $K_0(O_A) = \mathbb{Z}^n/(I - A)\mathbb{Z}^n$.

V.1.3.5 It is harder to describe the K-theory of crossed products by $\mathbb{Z}_n$ or $\mathbb{T}$. There are six-term exact sequences [Bla98, 10.6,10.7.1], but they give much less information than the P-V sequence. Computing the K-theory of crossed products by most finite groups is harder yet! The P-V sequence generalizes to reduced crossed products by free groups.

K-Theory as Cohomology

If X is a locally compact Hausdorff space, define $K^{-n}(X) = K_n(C_o(X))$ for $n = 0, 1$ (or $n \in \mathbb{N}$). $K^*(X)$ is called the *complex K-theory of X (with compact supports)*, and is sometimes written $K_\mathbb{C}^*(X)$ or $KU^*(X)$ (the U in KU stands for "unitary.") Because of the contravariant relationship between X and $C_o(X)$, K^n is a contravariant functor, which is homotopy-invariant. Relative K-groups can be defined by $K^n(X, Y) = K_n(C_o(X \setminus Y))$.

V.1.3.6 THEOREM. Complex K-theory is an extraordinary cohomology theory, that is, it is a sequence of homotopy-invariant contravariant functors from compact spaces and compact pairs to abelian groups, with a long exact sequence, and satisfying the excision and continuity axioms (but not the dimension axiom).

V.1.3.7 THEOREM. [CHERN CHARACTER] Let X be compact. Then there are isomorphisms

$$\chi^0 : K^0(X) \otimes \mathbb{Q} \to \bigoplus_{n \text{ even}} H^n(X; \mathbb{Q})$$

$$\chi^1 : K^{-1}(X) \otimes \mathbb{Q} \to \bigoplus_{n \text{ odd}} H^n(X; \mathbb{Q})$$

where $H^n(X; \mathbb{Q})$ denotes the n-th ordinary (Alexander or Čech) cohomology group of X with coefficients in $\mathbb{Q}$.

So, at least rationally, $K^0(X)$ is just the direct sum of the even cohomology groups of X, and $K^{-1}(X)$ the sum of the odd ones. (The result is not always true if the groups are not rationalized.)

See [Kar78] for a discussion of topological K-theory and proofs of these theorems.

Additional Observations

V.1.3.8 It is sometimes useful to unify K_0 and K_1 by regarding K-theory as a $\mathbb{Z}_2$-graded theory, writing $K_*(A) = K_0(A) \oplus K_1(A)$ with the K_0 and K_1 the even and odd parts respectively. A *-homomorphism $\phi : A \to B$ gives a homomorphism $\phi_* : K_*(A) \to K_*(B)$ of degree 0; the six-term exact sequence V.1.2.22 becomes a triangular three-term exact sequence with $\partial : K_*(A/J) \to K_*(J)$ a homomorphism of degree 1.

V.1.3.9 There is an alternate way of viewing the groups $K_0(A)$ and $K_1(A)$ using the outer multiplier algebra of A, which helps to motivate KK-theory (V.1.4.4): $K_i(A)$ is isomorphic to K_{1-i} of the stable outer multiplier algebra $M(A \otimes \mathbb{K})/(A \otimes \mathbb{K})$ of A. The proof uses the six-term exact sequence and the fact that the K-theory of the stable multiplier algebra $M(A \otimes \mathbb{K})$ is trivial.

V.1.3.10 Here are some considerations which help to motivate E-theory (V.1.4.7). If A and B are C*-algebras, denote by $[A, B]$ the set of homotopy classes of *-homomorphisms from A to B (II.5.5.6).

(i) For any A and B, there is a notion of "orthogonal direct sum" on $[A, B \otimes \mathbb{K}]$ making it into an abelian semigroup [use an isomorphism $M_2(\mathbb{K}) \cong \mathbb{K}$].
(ii) A homomorphism from $\mathbb{C}$ into a C*-algebra B is just a choice of projection in B (the image of 1). So for any B, we may identify $[\mathbb{C}, B \otimes \mathbb{K}]$ with $V(B)$.
(iii) A homomorphism ϕ from $S = C_o(\mathbb{R}) \cong C_o((0,1))$ into a C*-algebra B is just a choice of unitary in $B^\dagger$ which is 1 mod B (the image of $f(t) = e^{2\pi i t}$ under the extension of ϕ to $\tilde{\phi} : C(\mathbb{T}) \to B^\dagger$). Thus $[S, B \otimes \mathbb{K}]$ may be identified with $K_1(B)$.
(iv) Using Bott periodicity, $[S, SB \otimes \mathbb{K}]$ may be identified with $K_0(B)$. If B is unital, this identification agrees with the suspension of the identification induced by the identification of (ii).

V.1.4 Bivariant Theories

K-Theory is just one aspect of a large class of noncommutative homology/ cohomology theories on C*-algebras and more general topological algebras. The most general theories are bivariant, i.e. bifunctors which are contravariant in one variable and covariant in the other. These theories are discussed in detail in [CST04], so we give only a very brief description here.

Extension Theory

V.1.4.1 The first bivariant theory is the theory of extensions of C*-algebras, discussed in II.8.4. Given (separable) C*-algebras A and B, a semigroup $Ext(A, B)$ is defined. For fixed B, $Ext(\cdot, B)$ is a contravariant functor from C*-algebras to abelian semigroups, and for fixed A, $Ext(A, \cdot)$ is a covariant functor. The group $Ext^{-1}(A, B)$ (which coincides with $Ext(A, B)$ in many cases, e.g. when A is nuclear) is of particular importance, and is a homotopy-invariant bifunctor with six-term exact sequences and Bott Periodicity in each variable separately (under mild hypotheses).

V.1.4.2 L. Brown, R. Douglas, and P. Fillmore [BDF77] showed that if X is compact, then

$$Ext(C(X)) := Ext(C(X), \mathbb{C}) \cong K_1(X),$$

the first K-homology group of X (K-homology is the homology theory which is dual to complex K-theory). This result attracted great attention from topologists, and may fairly be regarded as the beginning of noncommutative topology as a discipline (although work such as Murray-von Neumann dimension theory, Fredholm index theory, and the Atiyah–Karoubi approach to topological K-theory can in retrospect be considered to be pioneering work in noncommutative topology).

V.1.4.3 It turns out, and is not hard to see, that we have $Ext(\mathbb{C}, B) \cong K_1(B)$ for any (separable) B, and hence by Bott Periodicity

$$K_0(B) \cong Ext(C_o(\mathbb{R}), B)$$

for any (separable) B. Thus the bivariant Ext theory includes K-theory for separable C*-algebras, with a dimension shift.

In light of V.1.4.2, it makes sense to define

$$K^1(A) = Ext(A) := Ext(A, \mathbb{C})$$

for a general separable (nuclear) C*-algebra, and by Bott Periodicity to let $K^0(A) = Ext(A, C_o(\mathbb{R}))$. These groups are sometimes called the "K-homology" groups of A, but this terminology is misleading since K^0 and K^1 are actually contravariant functors from C*-algebras to abelian groups.

KK-Theory

V.1.4.4 A very important advance in the development of operator K-theory is the KK-theory of Kasparov [Kas80b]. Given a pair of C*-algebras (A, B), with A separable and B containing a strictly positive element, we define an abelian group $KK(A, B)$. There are two standard ways of viewing the elements of $KK(A, B)$, as Fredholm modules or as quasihomomorphisms; each is useful in certain applications. The quasihomomorphism approach (due to J. Cuntz) is perhaps more intuitive. A quasihomomorphism from A to B is a pair $(\phi, \overline{\phi})$ of homomorphisms from A to $M(B \otimes \mathbb{K})$ which agree modulo $B \otimes \mathbb{K}$. Then $KK(A, B)$ is the set of equivalence classes of quasihomomorphisms from A to B, under a suitable notion of homotopy; the group operation, similar to that on Ext, is essentially orthogonal direct sum. KK is a homotopy-invariant bifunctor from pairs of C*-algebras to abelian groups, contravariant in the first variable and covariant in the second. $KK(\mathbb{C}, B) \cong K_0(B)$.

V.1.4.5 The central tool in the theory is the *intersection product* (or *Kasparov product*), a method of combining an element of $KK(A, B)$ with one in $KK(B, C)$ to yield an element of $KK(A, C)$. In the quasihomomorphism picture, the intersection product gives a way of "composing" quasihomomorphisms; if one of the quasihomomorphisms is an actual homomorphism, then the intersection product is really just composition, defined in a straightforward way. The technical details of establishing the properties of the intersection product are formidable.

V.1.4.6 Armed with the intersection product, it is fairly easy to prove all the important properties of the KK-groups. First, $KK(SA, B) \cong KK(A, SB)$; call this $KK^1(A, B)$. Then we have Bott periodicity

$$KK^1(SA, B) \cong KK^1(A, SB) \cong KK(A, B) \cong KK(SA, SB).$$

Moreover $KK^1(\mathbb{C}, B) \cong K_1(B)$; more generally, if A is nuclear, then

$$KK^1(A, B) \cong Ext(A, B).$$

There are cyclic six-term exact sequences in each variable separately under some mild restrictions. We have $KK(\mathbb{C}, B) \cong K_0(B)$, $KK^1(\mathbb{C}, B) \cong K_1(B)$, $KK(A, \mathbb{C}) \cong K^0(A)$, $KK^1(A, \mathbb{C}) \cong K^1(A)$ for any (separable) A, B.

E-Theory

V.1.4.7 There are some situations where the six-term exact sequences do not hold in KK-theory, however. This shortcoming and some important potential applications led A. Connes and N. Higson [CH90] to develop E-theory, which may be regarded as a "variant" of KK-theory in which the six-term exact sequences hold in complete generality (for separable C*-algebras). The basic objects of study in E-theory are *asymptotic morphisms*, paths of maps indexed by $[1,\infty)$ which become asymptotically *-linear and multiplicative. $E(A,B)$ is the set of homotopy classes of asymptotic morphisms from A to $B \otimes \mathbb{K}$. As with KK, there is a natural way of "composing" asymptotic morphisms (well defined up to homotopy), giving a product in E-theory analogous to the intersection product.

For any separable A, B, there is a natural map from $KK(A,B)$ to $E(A,B)$, which is an isomorphism if A is nuclear.

Equivariant and General Bivariant K-Theory

V.1.4.8 These bivariant theories can be abstracted and axiomatized to apply to more general classes of topological algebras and classes of extensions which will lead to six-term exact sequences.

All the theories can be adapted to incorporate actions of second countable locally compact topological groups, giving equivariant theories.

V.1.4.9 Operator K-theory has led both to spectacular advances within the subject of operator algebras and to deep applications to problems in geometry and topology. K-Theory, including the bivariant theories, has become a standard tool in the subject of operator algebras. Perhaps the greatest achievement of K-theory within operator algebras has been to give some insight into the previously mysterious (and still rather mysterious) internal structure of crossed products and free products of C*-algebras. The two most notable applications to geometry and topology so far have been the various generalizations of the Atiyah–Singer Index Theorem due to Connes, Skandalis, Kasparov, Moscovici, Miscenko, Fomenko, and Teleman, and the work on the homotopy invariance of higher signatures of manifolds by Kasparov and Miscenko and vanishing of "higher $\hat{A}$-genera" of manifolds of positive scalar curvature by Rosenberg (parallel to work of Gromov and Lawson). It would appear certain that the work so far has only scratched the surface of the possibilities. Until recently it would have been difficult to conceive of theorems in differential topology whose proofs require the use of operator algebras in an essential way.

V.1.5 Axiomatic K-Theory and the Universal Coefficient Theorem

Work of J. Cuntz, N. Higson, J. Rosenberg, and C. Schochet shows that K-theory can be characterized (at least for suitably nice C*-algebras) by a simple set of axioms analogous to the Steenrod axioms of cohomology, and has established generalizations for KK-theory of the Universal Coefficient Theorem and Künneth Theorem of topology, which provide powerful tools for the calculation of the KK-groups of many C*-algebras (essentially reducing the problem to calculation of K-groups).

KK as a Category

V.1.5.1 The KK-groups can be regarded as the morphisms of a category. Let **KK** be the category whose objects are separable C*-algebras, and for which the set of morphisms from A to B is $KK(A, B)$. Composition of morphisms is via the intersection product.

One can similarly form a category **E** with the same objects, with $E(A, B)$ the morphisms from A to B.

KK-Equivalence and the Bootstrap Class

V.1.5.2 An isomorphism in the category **KK** is called a *KK-equivalence*. KK-Equivalence is a fairly weak equivalence relation on (separable) C*-algebras, with the following properties:

(i) If A is a separable C*-algebra, then A is KK-equivalent to $A \otimes \mathbb{K}$. Thus stably isomorphic C*-algebras are KK-equivalent.
(ii) Homotopy-equivalent C*-algebras are KK-equivalent.
(iii) If $0 \to J \to A \to A/J \to 0$ is a split exact sequence of C*-algebras, then A is KK-equivalent to $J \oplus A/J$.

One can also define E-equivalence, using isomorphisms in **E**. KK-equivalence implies E-equivalence.

V.1.5.3 If A and B are KK-equivalent, then for any D we have

$$KK(A, D) \cong KK(B, D) \quad \text{and} \quad KK(D, A) \cong KK(D, B).$$

In particular,

$$K_0(A) \cong KK(\mathbb{C}, A) \cong KK(\mathbb{C}, B) \cong K_0(B)$$

and similarly $K_1(A) \cong K_1(B)$.

It turns out that for suitably nice A and B, the converse is true (V.1.5.13).

V.1.5.4 The *(large) bootstrap class* is the smallest class $\mathcal{N}$ of separable nuclear C*-algebras with properties (i), (iii), and (v) of IV.3.1.15, i.e. $\mathbb{C} \in \mathcal{N}$, $\mathcal{N}$ is closed under countable inductive limits, and $\mathcal{N}$ has the two-out-of-three property, and in addition $\mathcal{N}$ is closed under KK-equivalence. Then $\mathcal{N}$ also satisfies (ii) ($\mathcal{N}$ is closed under stable isomorphism) and (iv) (a Toeplitz extension yields that $\mathcal{N}$ is closed under crossed products by $\mathbb{Z}$), and in addition $\mathcal{N}$ satisfies (vi) of IV.3.1.15 and $\mathcal{N}$ is closed under homotopy equivalence.

$\mathcal{N}$ contains the small bootstrap class $\mathfrak{N}$ of IV.3.1.15, and in particular contains all commutative C*-algebras. It is an important unsolved problem whether $\mathcal{N}$ is the class of all separable nuclear C*-algebras; it certainly contains all separable nuclear C*-algebras for which there is hope of a classification via K-theoretic invariants.

The Universal Coefficient Theorem and Künneth Theorems

We now state several results which allow computation of KK-groups of "nice" C*-algebras. Among other things, KK-equivalence is completely determined by isomorphism of the K_0 and K_1 groups for C*-algebras in the class $\mathcal{N}$ defined in V.1.5.4, and consequently that any C*-algebra in $\mathcal{N}$ is KK-equivalent to a commutative C*-algebra. The results of this section are due to J. Rosenberg and C. Schochet ([RS87]; [Sch82]), based on earlier work by L. Brown [Bro84]. Cuntz had also previously obtained the Universal Coefficient sequence for the O_A.

All C*-algebras in this subsection will be *separable*. We will consider K-theory and KK-theory to be $\mathbb{Z}_2$-graded theories in order to simplify notation: $KK^*(A,B)$ will denote $KK(A,B) \oplus KK^1(A,B)$, and similarly

$$K_*(A) = K_0(A) \oplus K_1(A), \ K^*(A) = K^0(A) \oplus K^1(A).$$

If G and H are graded abelian groups, then $\operatorname{Hom}(G,H)$ is also graded by degree-preserving/degree-reversing maps. A tensor product of abelian groups is, of course, the ordinary tensor product over $\mathbb{Z}$. A tensor product of graded groups has a natural and obvious grading.

V.1.5.5 For any separable C*-algebras A and B, there are maps

$$\alpha : K_*(A) \otimes K_*(B) \to K_*(A \otimes B)$$

$$\beta : K^*(A) \otimes K_*(B) \to KK^*(A,B)$$

$$\gamma : KK^*(A,B) \to \operatorname{Hom}(K_*(A), K_*(B))$$

which are natural in A and B.

These maps are defined using the intersection product. α comes from the pairing $KK^*(\mathbb{C},A) \times KK^*(\mathbb{C},B) \to KK^*(\mathbb{C}, A \otimes B)$. More specifically, α is induced by the four pairings

$$KK(\mathbb{C},A) \times KK(\mathbb{C},B) \to KK(\mathbb{C}, A \otimes B)$$

$$KK(C_o(\mathbb{R}),A) \times KK(\mathbb{C},B) \to KK(C_o(\mathbb{R}), A \otimes B)$$

$$KK(\mathbb{C},A) \times KK(C_o(\mathbb{R}),B) \to KK(C_o(\mathbb{R}), A \otimes B)$$

$$KK(C_o(\mathbb{R}),A) \times KK(C_o(\mathbb{R}),B) \to KK(C_o(\mathbb{R}^2), A \otimes B) \cong KK(\mathbb{C}, A \otimes B)$$

The map β comes from the pairing $KK^*(A,\mathbb{C}) \times KK^*(\mathbb{C},B) \to KK^*(A,B)$. Finally, γ is the adjoint of the pairing $KK^*(\mathbb{C},A) \times KK^*(A,B) \to KK^*(\mathbb{C},B)$.

When $A = B$, $KK^*(A,A)$ is a graded ring, and γ is a ring-homomorphism (if the intersection product is written composition-style).

The surjectivity of β measures to what extent a general KK-element factors through $\mathbb{C}$ or $C_o(\mathbb{R})$ (or through $C(S^1)$, which is KK-equivalent to $\mathbb{C} \oplus C_o(\mathbb{R})$).

V.1.5.6 It is not necessary to use KK-theory to define α, β, or γ. α has a straightforward K-theoretic definition (which actually defines a map from $K_*(A) \otimes K_*(B)$ to $K_*(A \otimes_{\max} B)$), and for β everything may be rephrased in terms of the pairing between K-theory and "K-homology". γ has a nice interpretation in terms of extensions, which could be taken as an alternate definition. If $\tau \in KK^1(A, B)$ is represented by the extension

$$0 \longrightarrow B \longrightarrow D \longrightarrow A \longrightarrow 0$$

then $\gamma(\tau)$ is given by the connecting maps in the associated six-term exact sequence of K-theory.

V.1.5.7 One might hope that α, β, and γ would be isomorphisms, but they cannot be in general for (essentially) homological algebra reasons. If the sequences are modified in the appropriate way to incorporate the homological algebra obstructions, sequences are obtained which are valid at least for C*-algebras in $\mathcal{N}$.

The additional ingredient is easiest to describe in the case of γ. If $\gamma(\tau) = 0$ for an extension τ, then the six-term K-theory sequence degenerates into two short exact sequences of the form

$$0 \longrightarrow K_i(B) \longrightarrow K_i(D) \longrightarrow K_i(A) \longrightarrow 0$$

and thus determines an element $\kappa(\tau) \in \mathrm{Ext}_{\mathbb{Z}}^1(K_*(A), K_*(B))$ [this is the $\mathrm{Ext}_{\mathbb{Z}}^1$-group of homological algebra, the derived functor of the Hom functor, *not* the Ext-group of II.8.4.18.] Note that κ reverses degree. The maps γ and κ are generalizations of the Adams d and e operations in topological K-theory.

The obstruction for α and β is an element of $\mathrm{Tor}_1^{\mathbb{Z}}$, the derived functor of the tensor product functor. It is natural to expect a $\mathrm{Tor}_1^{\mathbb{Z}}$ obstruction, since $\mathrm{Tor}_1^{\mathbb{Z}}$ measures the deviation from exactness of the tensor product functor on groups.

The statements of the theorems are as follows.

V.1.5.8 THEOREM. [UNIVERSAL COEFFICIENT THEOREM (UCT)] [RS87] Let A and B be separable C*-algebras, with $A \in \mathcal{N}$. Then there is a short exact sequence

$$0 \longrightarrow \mathrm{Ext}_{\mathbb{Z}}^1(K_*(A), K_*(B)) \overset{\delta}{\longrightarrow} KK^*(A, B) \overset{\gamma}{\longrightarrow} \mathrm{Hom}(K_*(A), K_*(B)) \longrightarrow 0$$

The map γ has degree 0 and δ has degree 1. The sequence is natural in each variable, and splits unnaturally. So if $K_*(A)$ is free or $K_*(B)$ is divisible, then γ is an isomorphism.

V.1.5.9 THEOREM. [KÜNNETH THEOREM (KT)] [RS87] Let A and B be separable C*-algebras, with $A \in \mathcal{N}$, and suppose $K_*(A)$ or $K_*(B)$ is finitely generated. Then there is a short exact sequence

$$0 \longrightarrow K^*(A) \otimes K_*(B) \xrightarrow{\beta} KK^*(A,B) \xrightarrow{\rho} \operatorname{Tor}_1^{\mathbb{Z}}(K^*(A), K_*(B)) \longrightarrow 0.$$

The map β has degree 0 and ρ has degree 1. The sequence is natural in each variable, and splits unnaturally. So if $K^*(A)$ or $K_*(B)$ is torsion-free, β is an isomorphism.

V.1.5.10 THEOREM. [KÜNNETH THEOREM FOR TENSOR PRODUCTS (KTP)] [Sch82] Let A and B be C*-algebras, with $A \in \mathcal{N}$. Then there is a short exact sequence

$$0 \longrightarrow K_*(A) \otimes K_*(B) \xrightarrow{\alpha} K_*(A \otimes B) \xrightarrow{\sigma} \operatorname{Tor}_1^{\mathbb{Z}}(K_*(A), K_*(B)) \longrightarrow 0.$$

The map α has degree 0 and σ has degree 1. The sequence is natural in each variable, and splits unnaturally. So if $K_*(A)$ or $K_*(B)$ is torsion-free, α is an isomorphism.

The names given to these theorems reflect the fact that they are analogs of the ordinary Universal Coefficient Theorem and Künneth Theorem of algebraic topology. The UCT can also be regarded as a theorem about K-theory with coefficients.

The KT and especially the KTP can be stated and proved without reference to KK-theory. The proofs, however, are similar to and require some of the same machinery as the proof of the UCT, so it is most efficient to consider all three together.

V.1.5.11 The strategy of proof for all three theorems is the same. One first proves the theorems by bootstrap methods for arbitrary A, with fixed B of a form making α, β, γ isomorphisms. Then the general results are deduced by "abstract nonsense", using an appropriate exact sequence giving a resolution of a general B into ones of the special form.

V.1.5.12 Let $\mathcal{N}'$ be the class of all separable nuclear C*-algebras A such that the exact sequence in the statement of the UCT holds for every separable C*-algebra B. The UCT then says that $\mathcal{N} \subseteq \mathcal{N}'$. A straightforward application of the Five Lemma yields:

V.1.5.13 PROPOSITION. Let $A, B \in \mathcal{N}'$. If $x \in KK^*(A,B)$ and $\gamma(x) \in \operatorname{Hom}(K_*(A), K_*(B))$ is an isomorphism, then x is a KK-equivalence. Thus, if $K_i(A) \cong K_i(B)$ for $i = 0, 1$, then A and B are KK-equivalent.

V.1.5.14 Given countable abelian groups G_0 and G_1, there is a standard construction (using mapping cones) of a second countable locally compact Hausdorff space X of dimension ≤ 3 such that $K_i(C_o(X)) \cong G_i$ ($i = 0, 1$) (cf. [Bla98, 23.10.3]). Since every separable commutative C*-algebra is in $\mathcal{N}$, we obtain:

V.1.5.15 COROLLARY. Let A be a separable nuclear C*-algebra. The following are equivalent:

(i) $A \in \mathcal{N}$.
(ii) $A \in \mathcal{N}'$.
(iii) A is KK-equivalent to a commutative C*-algebra.

V.1.5.16 There are many (separable) nonnuclear C*-algebras which are in the class $\mathcal{H}$ of A such that the UCT sequence holds for any separable B. For example, any separable C*-algebra KK-equivalent to an element of $\mathcal{N}$ is in $\mathcal{H}$; in particular, every (separable) contractible C*-algebra (e.g. CA for any A) is in $\mathcal{H}$. There are separable C*-algebras A which are not in $\mathcal{H}$ because $KK^*(A, \cdot)$ does not have exact sequences for arbitrary extensions (it does if A is nuclear).

However, nonexistence of exact sequences is not the root cause of the failure of the UCT in general. An analog of the UCT exact sequence can be formulated using $E(A, B)$ in place of $KK(A, B)$. There are C*-algebras A, B for which both the E-UCT and the KK-UCT fail to hold, e.g. $A = B = C_r^*(G)$ for Ozawa's example of a countable discrete group G not satisfying the Baum-Connes Conjecture [Oza00].

V.2 Finiteness

In this section, we discuss properties related to finiteness in C*-algebras, including an order structure on the K-theory of a (stably) finite C*-algebra which plays a crucial role in the classification program.

V.2.1 Finite and Properly Infinite Unital C*-Algebras

Recall the definitions of finite, infinite, and properly infinite projections and unital C*-algebras from III.1.3.1. These properties behave somewhat differently for general C*-algebras than for von Neumann algebras. Throughout this subsection, all C*-algebras will be unital unless stated otherwise.

V.2.1.1 A commutative C*-algebra is obviously finite (cf. V.2.1.13). A Type I C*-algebra need not be finite: the Toeplitz algebra (II.8.3.2(v)) is a counterexample. A properly infinite C*-algebra contains a copy of the non-Type-I C*-algebra O_∞ (III.1.3.3, IV.1.2.6); since a C*-subalgebra of a Type I C*-algebra is Type I (IV.1.5.11), a Type I C*-algebra cannot be properly infinite.

V.2.1.2 A quotient of an infinite C*-algebra need not be infinite: $C(S^1)$ (and also $\mathbb{C}$) is a quotient of the Toeplitz algebra. Any quotient of a properly infinite C*-algebra is properly infinite.

On the other hand, a quotient of a finite C*-algebra need not be finite; in fact, every C*-algebra is a quotient of a finite C*-algebra. This is most easily seen using cones (II.5.5.10): note that $C_1(A)$ is finite for all A. In fact, $C_1(A)$ is *projectionless* (contains no nontrivial projections). If A is unital, then A is a quotient of $C_1(A)$ via evaluation at 1.

V.2.1.3 DEFINITION. A (unital) C*-algebra A is *residually finite* if every quotient of A is finite.

V.2.1.4 There is a related condition. We say a (unital) C*-algebra A has the *unitary extreme property* if every extreme point in the unit ball of A is unitary (cf. II.3.2.17). A C*-algebra with the unitary extreme property is obviously finite. A residually finite C*-algebra has the unitary extreme property: if x is an extreme point in the unit ball of A, and $1 - xx^*$ is nonzero, then $1 - xx^*$ is not in the closed ideal J generated by $1 - x^*x$ by II.3.2.17, and hence the image of x in A/J is a nonunitary isometry. In particular, a finite simple C*-algebra has the unitary extreme property.

There are C*-algebras with the unitary extreme property which are not residually finite: for example, if A is any unital C*-algebra, then the unital cone $C_1(A)$ has the unitary extreme property. There are also finite C*-algebras which do not have the unitary extreme property, e.g. $C^*(S \oplus S^*)$, where S is the unilateral shift.

It turns out that the most useful and important notion of finiteness involves matrix algebras:

V.2.1.5 DEFINITION. A (unital) C*-algebra A is *stably finite* if $M_n(A)$ is finite for all n. A is *residually stably finite* if every quotient of A is stably finite.

V.2.1.6 EXAMPLE. Not every finite C*-algebra is stably finite. The first explicit example, due to N. Clarke [Cla86], is the Toeplitz algebra on the 3-sphere S^3 [Cob74]. Let $H^2(S^3) \subseteq L^2(S^3)$ be the "Hardy space" of functions analytic on the open unit ball of $\mathbb{C}^2$ with square-integrable boundary values. If $f \in C(S^3)$, the compression of the multiplication operator M_f (I.2.4.3(i)) on $L^2(S^3)$, to $H^2(S^3)$, is called the *Toeplitz operator* with symbol f, denoted T_f. Let $A = \mathcal{T}(S^3)$ be the C*-subalgebra of $\mathcal{L}(H^2(S^3))$ generated by $\{T_f : f \in C(S^3)\}$. Then A contains $\mathcal{K}(H^2(S^3))$, and $A/\mathcal{K} \cong C(S^3)$. Since $\pi^1(S^3) \cong \pi_3(S^1) = 0$, the unitary group of $C(S^3)$ is connected. Every isometry in A has unitary image in $C(S^3)$, hence has index 0, i.e. A contains no nonunitary isometries. But $M_2(A)$ does contain a nonunitary isometry; in fact, the image of the multiplication operator with symbol

$$f(z,w) = \begin{bmatrix} z & -\bar{w} \\ w & \bar{z} \end{bmatrix}$$

is a Fredholm operator of nonzero index.

This example works because, although $K_1(C(S^3))$ is nontrivial, the nontrivial elements first appear in the 2×2 matrix algebra.

By considering the Toeplitz algebra on the unit ball of $\mathbb{C}^n$, one can make the nonunitary isometries first appear in the $n \times n$ matrix algebra.

Note that this C*-algebra is Type I. $M_2(A)$ is thus not properly infinite (V.2.1.1); in fact, if v is a nonunitary isometry in $M_2(A)$, then $1 - vv^*$ is contained in a proper ideal.

V.2.1.7 In an important advance, M. Rørdam [Rør03] recently constructed an example of a finite simple unital nuclear C*-algebra C*-algebra A such that $M_2(A)$ is infinite (and hence properly infinite by V.2.3.1). Rørdam previously gave an example of a finite (non-simple) unital C*-algebra A such that $M_2(A)$ is properly infinite [Rør98].

V.2.1.8 PROPOSITION. Let (A_i, ϕ_{ij}) be an inductive system of unital C*-algebras (with the ϕ_{ij} unital), and $A = \varinjlim(A_i, \phi_{ij})$. If each A_i is [stably] finite, then A is [stably] finite.

PROOF: Let u be an isometry in A, and $x \in A_i$ with $\|x\| = 1$ and $\|\phi_i(x) - u\|$ small. Then $\phi_i(x^*x)$ is close to 1_A, and by increasing i we may assume that x^*x is close to 1_{A_i}. Thus, if $v = x(x^*x)^{-1/2}$, then v is an isometry in A_i and $\phi_i(v)$ is close to u. Since A_i is finite, $vv^* = 1_{A_i}$, and thus uu^* is close to 1 and hence invertible. Thus u is invertible and therefore unitary.

As a corollary, several types of C*-algebras important in the classification program are stably finite:

V.2.1.9 DEFINITION. A (not necessarily unital) C*-algebra A is *approximately homogeneous*, or an *AH algebra*, if it is isomorphic to an inductive limit of locally homogeneous C*-algebras (in the sense of IV.1.4.1). A is *approximately subhomogeneous*, or an *ASH algebra*, if it is isomorphic to an inductive limit of subhomogeneous C*-algebras.

AH and ASH algebras are usually required to be separable.

Every AH algebra is an ASH algebra. Every ASH algebra is finite, and since a matrix algebra or quotient of a locally homogeneous [subhomogeneous] C*-algebra is locally homogeneous [subhomogeneous], a matrix algebra or quotient of an AH [ASH] algebra is AH [ASH]. Thus an ASH algebra is residually stably finite. If A is ASH, so is $\tilde{A}$. If A is AH, then $\tilde{A}$ is not necessarily AH, but it is if A has an approximate unit of projections (e.g. if A is an inductive limit of unital locally homogeneous C*-algebras).

These acronym names are in analogy with the AF algebras (II.8.2.2(iv)). Other similar names for classes of interesting C*-algebras are the AI-algebras and A$\mathbb{T}$-algebras, inductive limits of direct sums of matrix algebras over $C([0,1])$ and $C(\mathbb{T})$ respectively; these classes are sometimes (rather misleadingly) called "interval algebras" and "circle algebras" respectively.

Another interesting class of stably finite C*-algebras is:

V.2.1.10 DEFINITION. A (not necessarily unital) C*-algebra A is *residually finite-dimensional* if it has a separating family of finite-dimensional representations (i.e. a separating family of finite-dimensional quotients).
[Note that the meaning of "residually" in this definition is not the same as in V.2.1.3 or V.2.1.5.]

Any C*-subalgebra of a residually finite-dimensional C*-algebra is residually finite-dimensional. If A is residually finite-dimensional, so is $\tilde{A}$. A matrix algebra over a residually finite-dimensional C*-algebra is residually finite-dimensional.

Any C*-algebra, all of whose irreducible representations are finite-dimensional (e.g. any subhomogeneous C*-algebra), is residually finite-dimensional. A more interesting example is:

V.2.1.11 THEOREM. [Cho80] If F is any free group, then $C^*(F)$ is residually finite-dimensional.

V.2.1.12 COROLLARY. Every C*-algebra is a quotient of a residually finite-dimensional C*-algebra.

In fact, every separable unital C*-algebra is a quotient of $C^*(F)$ for a free group F.

V.2.1.13 PROPOSITION. A unital residually finite-dimensional C*-algebra is stably finite.
PROOF: Let A be residually finite-dimensional. Since a matrix algebra over A is also residually finite-dimensional, it suffices to show that A is finite. If u is a nonunitary isometry in A, let J be a closed ideal in A with A/J finite-dimensional and $\pi(1 - uu^*) \neq 0$, where $\pi : A \to A/J$ is the quotient map. Then $\pi(u)$ is a nonunitary isometry in A/J, a contradiction.

In particular, every subhomogeneous C*-algebra (e.g. every commutative C*-algebra) is stably finite.

V.2.1.14 A C*-algebra A with a separating family of tracial states (or, more generally, normalized quasitraces (II.6.8.15)) must be finite: if $x \in A$ with $x^*x = 1_A$, then $\tau(1_A - xx^*) = 0$ for every normalized quasitrace τ. Since $M_n(A)$ also has a separating family of normalized quasitraces, A must be stably finite.

Even a residually stably finite C*-algebra need not have a separating family of normalized quasitraces: $\tilde{\mathbb{K}}$ is a counterexample. But there is a very important partial converse to the previous paragraph:

V.2.1.15 THEOREM. ([Han81], [BH82]) Let A be a stably finite unital C*-algebra. Then A has a normalized quasitrace.

Combining this result with II.6.8.17, we obtain:

V.2.1.16 COROLLARY. Let A be a stably finite unital exact C*-algebra. Then A has a tracial state.

V.2.1.17 The proof of V.2.1.15 is not difficult given the proper framework (the comparison theory of II.3.4), although there are some technical details to be handled. The proof has a strong K-theoretic flavor.

The first step, done in [Cun78] in the simple case and [Han81] in general, is to construct an ordered group $K_0^*(A)$ whose states (V.2.4.21) are exactly the dimension functions on A. $K_0^*(A)$ is constructed in a manner very similar to $K_0(A)$ except that one starts with all (positive) elements of $M_\infty(A)$, using the relation $\precsim$ of II.3.4.3. Since a dimension function on A measures the "size" of the "support projections" of the elements of A, the correspondence between dimension functions and states of $K_0^*(A)$ is natural.

Begin with a (not necessarily unital) C*-algebra A (or a somewhat more general *-subalgebra such as a hereditary *-subalgebra of a C*-algebra). The relations $\precsim$ and $\approx$ make sense on $M_\infty(A)_+$ (II.6.6.9); the set $W(A)$ of equivalence classes of elements of $M_\infty(A)_+$ has a binary operation ("orthogonal addition"), well defined by II.3.4.7(ii), making it an abelian semigroup with identity [0]. There is a natural partial order on $W(A)$ induced by $\precsim$. The Grothendieck group (V.1.1.13) of formal differences from $W(A)$ becomes a preordered abelian group with the induced order; this preordered group is denoted $K_0^*(A)$ in [Cun78] and $D(A)$ in other references. The dimension functions taking finite values on A are in natural one-one correspondence with the order-preserving homomorphisms from $K_0^*(A)$ to $\mathbb{R}$; if A is unital, the normalized dimension functions are the homomorphisms taking the value 1 on [1]. In most reasonable cases, e.g. when A is σ-unital or simple, $K_0^*(A)$ will contain an order unit. If $K_0^*(A)$ is nontrivial and contains an order unit, then the Goodearl-Handelman theory (V.2.4.24) insures a nonzero order-preserving homomorphism and thus a nonzero finite dimension function on A.

It is immediate from the construction that if $a \in M_\infty(A)_+$, then $[a] = [0]$ in $K_0^*(A)$ if and only if there is a $b \in M_\infty(A)_+$, $b \perp a$, with $a + b \precsim b$. It follows easily that if A is unital, then $K_0^*(A) = \{0\}$ if and only if $[1_A] = [0]$, which occurs if and only if $1_A + b \precsim b$ for some b which has n ones on the diagonal and zeroes elsewhere, for some n; this will be the case if and only if $M_{n+1}(A)$ is infinite. Thus, if A is unital, there is a normalized dimension function on A if and only if A is stably finite.

V.2.1.18 The final step is the transition from a dimension function to a quasitrace. This was first done in [Han81] for AW*-algebras using a ring-theoretic argument, and the general C*-algebra case was reduced to the AW*-case in [BH82]. A quicker argument was given in [BR92]: if D is a dimension function on a C*-algebra A (or hereditary *-subalgebra), define $D'(a) = \sup_{\epsilon > 0} D(f_\epsilon(a))$; then it is easily checked that D' is a dimension function, $D' \leq D$, and D' is lower semicontinuous by II.3.4.15. Thus D' corresponds to a quasitrace by II.6.8.14. $D'(p) = D(p)$ for any projection p, and in particular, if A is unital, D' is normalized if D is normalized. (If A is nonunital, it can happen that $D' = 0$ even if $D \neq 0$, but only if D vanishes on $Ped(A)$.)

This completes the (outlined) proof of V.2.1.15.

V.2.1.19 As a byproduct of the construction used in the proof, it is easy to see that if B is a full hereditary C*-subalgebra of a C*-algebra A contained in $Ped(A)$, then $K_0^*(B)$ is naturally isomorphic to $K_0^*(Ped(A))$; thus there is a finite nonzero dimension function on B if and only if there is one on $Ped(A)$. (But $K_0^*(A)$ can be quite different: $K_0^*(\mathcal{C}(\mathcal{H})) \cong \mathbb{Z}$, but $K_0^*(\mathcal{K}(\mathcal{H})) = \{0\}$. The K_0^*-group is much more sensitive to completion than the K-groups.)

V.2.2 Nonunital C*-Algebras

It is not straightforward to rephrase the definitions of finite, infinite, and properly infinite C*-algebras in the nonunital case in an interesting and useful way, and indeed there seems to be no completely satisfactory way to proceed except in the stable simple case, where most reasonable finiteness conditions coincide. In this subsection, we will discuss some of the alternatives; some are of interest also in the unital case.

Finite Algebras

V.2.2.1 The most commonly used definition of a finite C*-algebra, which is reasonably satisfactory, is to declare A [stably] finite if $\tilde{A}$ is [stably] finite. This definition is fairly broad: for example, the cone $C(A)$ is stably finite by this definition, for any A. Thus the nonunital version of V.2.1.15 is false: a stably finite C*-algebra under this definition (e.g. $C(O_2)$) need not have a nondegenerate quasitrace (one taking values other than 0 and ∞) at all.

V.2.2.2 An even broader definition of finiteness for a C*-algebra A is to simply require that A contain no infinite projections. A C*-algebra which is finite in the sense of V.2.2.1 has this property, but the converse is not true (V.2.2.14(i))(although it is true for simple C*-algebras (V.2.3.6)), and this broader notion does not seem to be too useful. The problem is that general C*-algebras have a dearth of projections, and unexpected projections and partial isometries can appear from standard constructions, even adding a unit.

V.2.2.3 A far more restrictive notion of finiteness is to require the existence of a separating family of normalized quasitraces (or equivalently in the separable case, a faithful normalized quasitrace). This notion of finiteness excludes such examples as $\mathbb{K}$ (which is arguably only "semifinite", not finite). One could relax this condition to just require a separating family of densely defined (quasi)traces; this notion of finiteness would include $\mathbb{K}$ (but not $\tilde{\mathbb{K}}$).

A closely related condition is the existence of nonnormal hyponormal elements:

V.2.2.4 DEFINITION. An element x in a C*-algebra A is *hyponormal* if $xx^* \leq x^*x$.

Note that the hyponormality of an element is independent of the containing C*-algebra. For a full discussion of hyponormality and related topics, see [Con91].

V.2.2.5 A strong finiteness condition on a C*-algebra A is the requirement that every hyponormal element in A is normal. A C*-algebra with a separating family of tracial states has this property; it is unknown whether the same is true for a separating family of normalized quasitraces, even for simple unital C*-algebras. The converse is false: $\mathbb{K}$ does not contain a nonnormal hyponormal element ([Con91], [Hal67, Problem 206]).

It is known [CP79] that a C*-algebra A has a separating family of tracial states if and only if, whenever $x_1, \ldots, x_n \in A$ and $\sum x_i x_i^* \leq \sum x_i^* x_i$, we have $\sum x_i x_i^* = \sum x_i^* x_i$.

V.2.2.6 If $\tilde{A}$ is infinite, then by multiplying an isometry by a suitable scalar $\tilde{A}$ contains a nonunitary isometry of the form $1 + x$, where $x \in A$. From $(1+x)^*(1+x) = 1$ we obtain $x + x^* = -x^*x$.

$$(1+x)(1+x)^* = 1 + x + x^* + xx^* = 1 + xx^* - x^*x$$

is a nontrivial projection, and thus $x^*x - xx^*$ is a nonzero projection in A. So x is a nonnormal hyponormal element in A (and actually has a much stronger property). Thus a C*-algebra containing no nonnormal hyponormal element is finite in the sense of V.2.2.1.

V.2.2.7 The property of containing no nonnormal hyponormal element is much stronger than finiteness (in the sense of V.2.2.1), however: for example, $C(O_2)$ is finite but does contain a nonnormal hyponormal. In fact, by [Spi88], the suspension of the Toeplitz algebra can be embedded in an AF algebra, so even an AF algebra can contain a nonnormal hyponormal. Even $C(O_2)$ can be embedded in an AF algebra; in fact, the cone over any separable exact C*-algebra is AF-embeddable [Oza03].

A nonnormal hyponormal element is a generalization of a partial isometry implementing an equivalence between an infinite projection and a proper subprojection. A more restrictive generalization of such partial isometries is a scaling element [BC82]:

V.2.2.8 DEFINITION. Let x be an element of a C*-algebra A. Then x is a *scaling element* if $xx^* \neq x^*x$ and $xx^* \ll x^*x$, i.e. $(x^*x)(xx^*) = xx^*$.

The condition $xx^* \neq x^*x$ simply rules out the case where $x^*x = xx^*$ is a projection; if x is a scaling element, then

$$\|x^*x - xx^*\| \geq 1/2.$$

It is easily seen that the condition $(x^*x)(xx^*) = xx^*$ is equivalent to $(x^*x)x = x$. The property of being a scaling element is independent of the containing C*-algebra. If x is a scaling element in A, then $xx^* \in A_+^c$ (II.5.2.4), so $x \in Ped(A)$.

Simple functional calculus arguments show the following:

V.2.2.9 PROPOSITION. Let A be a C*-algebra. Then

(i) If A contains a scaling element, then it contains a scaling element x and $a, b \geq 0$ such that $\|x\| = \|a\| = \|b\| = 1$ and $a \ll x^*x$, $a \perp xx^*$, $b \ll a$.
(ii) If A contains an element y such that $\|y\| = 1$, $\|y^*y - yy^*\| > 3/4$ and $\|(y^*y)(yy^*) - yy^*\| < 1/4$, then A contains a scaling element.

V.2.2.10 COROLLARY. Let $A = \lim_\to A_i$. If A contains a scaling element, then some A_i contains a scaling element.

V.2.2.11 The existence of a scaling element in a C*-algebra is a useful notion of infiniteness, but this notion is really only interesting in the nonunital case: if A is a unital C*-algebra containing a scaling element, then A contains an infinite projection. In fact, if x is a scaling element in A as in V.2.2.9, then an easy calculation shows that $u = x + (1 - x^*x)^{1/2}$ is a nonunitary isometry.

So if A is finite (in the sense of V.2.2.1), then A does not contain a scaling element. The converse is false (V.2.2.14(i)).

V.2.2.12 A C*-algebra containing a scaling element at least contains a nonzero projection: if x is as in V.2.2.11, and $u = x + (1 - x^*x)^{1/2} \in \tilde{A}$, then $1 - uu^*$ is a nonzero projection in A. In fact, we have:

V.2.2.13 PROPOSITION. Let x be a scaling element of norm one in a C*-algebra A. Then there is a sequence (p_n) of mutually equivalent, mutually orthogonal nonzero projections in A with $p_n \ll x^*x$ for all n.

PROOF: Note that $y = x(xx^*)^{1/2}$ is also a scaling element: $y^*y = xx^*$, and

$$(xx^*)x^2 = x(x^*x)x = x^2$$

since $(x^*x)x = x$. Let

$$u = y + (1 - y^*y)^{1/2} \in \tilde{A}$$

and $p_n = u^n(1 - uu^*)u^{*n}$. Since $u \in \widetilde{C^*(y)}$ and $1 - uu^* \in C^*(y)$, $p_n \in C^*(y)$ and x^*x is a unit for p_n.

V.2.2.14 EXAMPLES.

(i) Let S be the unilateral shift (I.2.4.3(ii)), and $A = C^*(S-I)$. Then $\tilde{A} \cong C^*(A, I) = C^*(S)$; thus $\tilde{A}$ is not finite. However, A contains no scaling element. For A contains $\mathbb{K}$, and $A/\mathbb{K} \cong C_o(\mathbb{R})$. If $\pi : A \to C_o(\mathbb{R})$ is the quotient map, and x is a scaling element in A, then $\pi(x^*x) = \pi(xx^*)$ is a projection in $C_o(\mathbb{R})$, and thus $\pi(x) = 0$, $x \in \mathbb{K}$. But $\mathbb{K}$ contains no scaling elements.

(ii) Let B be the universal C*-algebra generated by a scaling element of norm 1. Then B contains three important essential closed ideals I (generated by the element $x^*x - (x^*x)^2$), J (generated by $x^*x - xx^*$), and K (generated by $x - x^*x$). We have $I \subseteq J \subseteq K$, $I \cong C_o(\mathbb{R}) \otimes \mathbb{K}$, $B/I \cong C^*(S)$, $J/I \cong \mathbb{K}$, $K/I \cong C^*(S-I)$, $B/K \cong \mathbb{C}$. The extension $0 \to I \to K \to C^*(S-I) \to 0$ is split: if $u = x + (1 - x^*x)^{1/2}$, then u is an isometry in $\tilde{B}$ and $S - I \mapsto y = u - 1$ is a cross section. The extension $0 \to K \to B \to \mathbb{C} \to 0$ is not split. Every projection in B is thus in K, and hence actually in J; it follows easily that B has no infinite projections. (B does have nonzero projections (V.2.2.12).) Thus B is finite in the sense of V.2.2.2, but not in the sense of V.2.2.1 (V.2.2.11). See [Kat04a] for details.

Note that both these examples are Type I C*-algebras.

We summarize most of the various notions in a theorem:

V.2.2.15 THEOREM. Let A be a C*-algebra. Then the following are successively more restrictive notions of finiteness for A:

(i) A contains no infinite projections.
(ii) A contains no scaling element.
(iii) $\tilde{A}$ is finite.
(iv) A contains no nonnormal hyponormal element.
(v) A has a separating family of tracial states.

All the conditions are distinct.

Properly Infinite Algebras

It is not obvious how to extend the definition of properly infinite C*-algebras to the nonunital case. We can make the following definition for elements (cf. [KR00]):

V.2.2.16 DEFINITION. Let A be a C*-algebra, $a \in A_+$. Then a is an *infinite element of* A if $diag(a, b) \precsim diag(a, 0)$ in $M_2(A)$ (II.3.4.3) for some $b \in A_+$, $b \neq 0$, and a is a *properly infinite element of* A if $diag(a, a) \precsim diag(a, 0)$ in $M_2(A)$.

An infinite [resp. properly infinite] projection is an infinite [resp. properly infinite] element.

V.2.2.17 One could then say that a C*-algebra A is *properly infinite* if it has a properly infinite strictly positive element. This would extend the unital definition, but exclude some natural candidates which are not σ-unital. A more inclusive definition would require just that the properly infinite positive elements in A generate A as an ideal, i.e. that every quotient of A contains a nonzero properly infinite positive element. However, it is unclear that this definition is the same as the usual one in the unital case. An intermediate possibility is to require A to have a full properly infinite positive element; but both objections would arise with this definition.

None of these definitions seems satisfactory, since the property of containing a properly infinite element is quite weak. For example, even $\mathbb{K}$ contains a properly infinite element; in fact, any infinite-rank positive element is properly infinite.

V.2.2.18 PROPOSITION. If A is a C*-algebra and a is properly infinite, then for any $b \in (J_a)_+$, $b \precsim a$.

To prove this, note that by induction

$$diag(a, a, \ldots, a) \precsim diag(a, 0, \ldots, 0)$$

in $M_n(A)$ for all n, and apply II.5.2.12.

If A is a C*-algebra and $a \in A_+$, set

$$I(a) = \{x \in A : diag(a, |x|) \precsim diag(a, 0) \text{ in } M_2(A)\}.$$

V.2.2.19 PROPOSITION. [KR00] $I(a)$ is a closed ideal of A contained in $J_a = \overline{span(AaA)}$; $I(a) = \{0\}$ if and only if a is finite, and $I(a) = J_a$ if and only if a is properly infinite. The image of a in $A/I(a)$ is finite.

V.2.2.20 COROLLARY. If A is a C*-algebra, $a \in A_+$, then a is properly infinite if and only if the image of a in any quotient of A is either 0 or infinite.

Purely Infinite Algebras

V.2.2.21 DEFINITION. Let A be a C*-algebra. Then A is *purely infinite* if A has no one-dimensional quotients and, whenever $a, b \in A_+$, $b \in J_a = \overline{span(AaA)}$ (the closed ideal of A generated by a), for every $\epsilon > 0$ there is an $x \in A$ with $\|b - x^*ax\| < \epsilon$.

In other words, if b can be approximated by a sum $\sum x_k^* a x_k$ (II.5.2.13), it can be approximated by a single such term.

Purely infinite C*-algebras were first defined in the simple unital case ([Cun77a], [Cun81a]; cf. V.2.3.3). The definition in V.2.2.21 is from [KR00].

V.2.2.22 PROPOSITION. Every quotient of a purely infinite C*-algebra is purely infinite.

PROOF: Let A be purely infinite, J a closed ideal, and $\pi : A \to A/J$ the quotient map. Suppose $c, d \in (A/J)_+$, with d in the closed ideal of A/J generated by c. Then $d = \sum_{k=1}^{\infty} z_k^* c z_k$ for some $z_k \in A/J$ (II.5.2.13). Let $y_k \in A$ with $z_k = \pi(y_k)$ for each k. If $\epsilon > 0$, fix r with

$$\|d - \sum_{k=1}^{r} z_k^* c z_k\| < \epsilon/2.$$

If $a \in A_+$ with $\pi(a) = c$, then $b = \sum_{k=1}^{r} y_k^* a y_k$ is in the ideal of A generated by a, so there is an $x \in A$ with $\|b - x^* a x\| < \epsilon/2$. Then

$$\|d - \pi(x)^* c \pi(x)\| \leq \|d - \pi(b)\| + \|\pi(b) - \pi(x^* a x)\| < \epsilon.$$

V.2.2.23 PROPOSITION. Every hereditary C*-subalgebra of a purely infinite C*-algebra is purely infinite.

PROOF: Let B be a hereditary C*-subalgebra of a purely infinite C*-algebra A. Let $a, b \in B_+$, $\|b\| = 1$, b in the closed ideal of B generated by a, and $\epsilon > 0$. Then $b^{1/2}$ is in the closed ideal of A generated by a^2, so there is an $x \in A$ with $\|b^{1/2} - x^* a^2 x\| < \epsilon$. Then

$$\|b - (b^{1/4} x^* a^{1/2}) a (a^{1/2} x b^{1/4})\| < \epsilon$$

and $a^{1/2} x b^{1/4} \in B$.

It remains to show that B cannot have a one-dimensional quotient. But a quotient of B is a hereditary C*-subalgebra of a quotient of A (II.5.1.6), so it suffices to show that B cannot be one-dimensional, i.e. that A cannot contain a projection p with pAp one-dimensional. If there were such a p, then pAp could not be an ideal (direct summand) in A, so there is a $y \in A$ with $y^* p y \notin pAp$. Then $p + y^* p y$ can be approximated by $x^* p x$ for some $x \in A$, and $p x x^* p = \lambda p$ for some $\lambda > 0$, so $u = \lambda^{-1/2} p x$ is a partial isometry with $u u^* = p$, and $p \lneq u^* u$. Then $u p u^*$ is a proper subprojection of p, a contradiction.

V.2.2.24 PROPOSITION.

(i) Every nonzero projection in a purely infinite C*-algebra is infinite.
(ii) Every full hereditary C*-subalgebra of a purely infinite unital C*-algebra contains a (necessarily infinite) projection equivalent to the identity.

PROOF: The proofs are nearly identical. For (i), by V.2.2.23 it suffices to show that if A is purely infinite and unital, then 1 is an infinite projection. Let a be a noninvertible full positive element in A (II.6.4.15). For (ii), let B

be a full hereditary C*-subalgebra of A, and a a full positive element in B (II.5.3.11). In either case, there is an $x \in A$ with $\|1 - x^*ax\| < 1$, so $r = x^*ax$ is invertible. Set $u = a^{1/2}xr^{-1/2}$; then $u^*u = 1$, so uu^* is a projection which is in the hereditary C*-subalgebra generated by a.

A purely infinite C*-algebra, even an ideal in a unital purely infinite C*-algebra, need not contain any nonzero projections (V.2.2.31).

V.2.2.25 It can be shown easily using II.3.4.15 and II.3.4.17 that suitable limits of properly infinite elements are properly infinite. From this we obtain:

V.2.2.26 PROPOSITION. An inductive limit of purely infinite C*-algebras is purely infinite.

It follows from V.2.2.19 that we have:

V.2.2.27 THEOREM. A C*-algebra A is purely infinite if and only if every nonzero positive element of A is properly infinite.

V.2.2.28 THEOREM. [KR00] A C*-algebra stably isomorphic to a purely infinite C*-algebra is purely infinite. In particular, if A is purely infinite, so is $A \otimes \mathbb{K}$ and $M_n(A)$ for all n.

The last statement is proved first, and follows easily from V.2.2.27; the rest then follows from V.2.2.23 and V.2.2.26.

V.2.2.29 PROPOSITION. A purely infinite C*-algebra has no nondegenerate lower semicontinuous trace.

V.2.2.30 THEOREM. [KR02] Let O_∞ be the Cuntz algebra (II.8.3.3(ii)). If A is any C*-algebra, then $A \otimes O_\infty$ is purely infinite.

This result follows easily from the fact (cf. [KP00]) that $O_\infty \cong \bigotimes_1^\infty O_\infty$. In fact, if B is any C*-algebra with $B \cong B \otimes O_\infty$ (e.g. any purely infinite simple nuclear C*-algebra), and A is any C*-algebra, then $A \otimes_{\min} B$ is purely infinite.

V.2.2.31 EXAMPLES. O_∞ itself is purely infinite, as are the other Cuntz algebras [Cun77a]. $C([0,1]) \otimes O_\infty$ is purely infinite, as is its ideal $C_o((0,1)) \otimes O_\infty$ which contains no nonzero projections.

V.2.2.32 It is not true in general that if A is purely infinite, then $A \cong A \otimes O_\infty$, even if A is separable and simple. This is an open question for separable nuclear C*-algebras; several variations of the definition of a purely infinite C*-algebra were considered in [KR02] in an attempt to answer this question.

V.2.3 Finiteness in Simple C*-Algebras

The finiteness situation is considerably simpler for simple C*-algebras.

Simple Unital C*-Algebras

The first observation is that if a simple unital C*-algebra is infinite, it is properly infinite. Compare the next result to III.1.3.2.

V.2.3.1 PROPOSITION. Let A be a simple unital C*-algebra. Then the following are equivalent:

(i) A is infinite.
(ii) A is properly infinite.
(iii) There is a sequence of mutually orthogonal, mutually equivalent nonzero projections in A.
(iv) There is a sequence of mutually orthogonal projections in A, each equivalent to 1_A, i.e. A contains a unital copy of O_∞ (II.8.3.3(ii)).
(v) A contains a left [or right] invertible element which is not invertible.

PROOF: (i) $\Longrightarrow$ (v), (ii) $\Longrightarrow$ (i), and (iv) $\Longrightarrow$ (ii) are trivial, and (i) $\Longrightarrow$ (iii) is III.1.3.2. (Also, (ii) $\Longrightarrow$ (iv) is III.1.3.3, although this is not needed for this proof.)

(v) $\Longrightarrow$ (i): If x is left invertible, say $yx = 1$, then $1 = x^*y^*yx \le \|y\|^2 x^*x$, so x^*x is invertible, and $u = x(x^*x)^{-1/2}$ is an isometry. We have that

$$uu^* = x(x^*x)^{-1}x^* \le \|(x^*x)^{-1}\|xx^*$$

so if x is not invertible (i.e. xx^* is not invertible), then uu^* is not invertible. The argument if x is right invertible is similar (or just note that if x is right invertible, then x^* is left invertible).

(iii) $\Longrightarrow$ (iv): Let (p_n) be a sequence of mutually orthogonal nonzero equivalent projections in A. Let $p = p_1$, and u_n a partial isometry with $u_n^*u_n = p$ and $u_nu_n^* = p_n$ for each n; so $p = u_n^*p_nu_n$. Then, since A is algebraically simple, there are elements x_i with

$$1 = \sum_{i=1}^n x_i^* p x_i = \sum_{i=1}^n x_i^* u_i^* p_i u_i x_i$$

for some n (II.5.4.3). If $q = \sum_{i=1}^n p_i$ and $v = \sum_{i=1}^n u_i x_i$, then $1 = v^*qv$, so qv is an isometry with range projection $qvv^*q \le q$. Similarly, there is a projection equivalent to 1 under $\sum_{i=kn+1}^{(k+1)n} p_i$ for every k.

V.2.3.2 COROLLARY. Let A be a simple unital C*-algebra, and p an infinite projection in A. Then p is properly infinite and contains a subprojection equivalent to 1_A.

PROOF: Apply V.2.3.1 to pAp to conclude that p is properly infinite and contains a sequence of mutually orthogonal, mutually equivalent subprojections. Then exactly as in the proof of (iii) $\Longrightarrow$ (iv) we obtain a subprojection of p equivalent to 1_A (and, in fact, an entire sequence of mutually orthogonal subprojections each equivalent to 1_A).

Purely Infinite Simple C*-Algebras

Purely infinite simple unital C*-algebras have several equivalent characterizations. Examples of purely infinite simple unital C*-algebras are (countably decomposable) type III factors, the Calkin algebra, the Cuntz algebras (II.8.3.3(ii)), and the simple Cuntz-Krieger algebras (II.8.3.3(iii)).

See V.2.3.13 for the nonunital case; a nonunital purely infinite simple C*-algebra has many infinite projections, and a nonunital σ-unital purely infinite simple C*-algebra is isomorphic to the stabilization of a simple unital C*-algebra (V.3.2.14).

V.2.3.3 PROPOSITION. If A is a simple unital C*-algebra, $A \neq \mathbb{C}$, then the following are equivalent:

(i) A is purely infinite.
(ii) $A \not\cong \mathbb{C}$ and, for every nonzero $a \in A$, there are $x, y \in A$ with $xay = 1$.
(iii) Every nonzero hereditary C*-subalgebra of A contains an infinite projection.
(iv) Every nonzero hereditary C*-subalgebra of A contains a projection equivalent to 1.

The equivalence (i) $\Longleftrightarrow$ (iii) holds also for nonunital purely infinite simple C*-algebras (V.2.3.12).

PROOF: (i) $\Longrightarrow$ (ii): Let $a \in A$. Then 1 is in the ideal generated by a^*a, so there is a $z \in A$ with $\|1 - z^*a^*az\| < 1$. Then $r = z^*a^*az$ is invertible, so $r^{-1}z^*a^*az = 1$.
(ii) $\Longrightarrow$ (i): Let $a, b \in A_+$ with $a \neq 0$, and let $x, y \in A$ with $xay = 1$. Then $1 = y^*ax^*xay \leq \|x\|^2 y^*a^2 y$, so $r = y^*a^2 y$ is invertible, and we have

$$b = (b^{1/2}r^{-1/2}y^*a^{1/2})a(a^{1/2}yr^{-1/2}b^{1/2}).$$

(i) $\Longrightarrow$ (iii) and (i) $\Longrightarrow$ (iv) are V.2.2.24; (iii) $\Longrightarrow$ (iv) is V.2.3.2.
(iv) $\Longrightarrow$ (ii): Let a be a nonzero element of A, and B the hereditary C*-subalgebra generated by $f_\epsilon(a^*a)$ (II.3.4.11) for ϵ small enough that $f_\epsilon(a^*a) \neq 0$. Let u be an isometry in A with $p = uu^* \in B$. Then $a^*a \geq (\epsilon/2)p$, so $z = u^*a^*au$ is invertible, and $z^{-1}u^*a^*au = 1$.

V.2.3.4 PROPOSITION. A matrix algebra or stable algebra over a simple unital purely infinite C*-algebra A is purely infinite.

This result is of course a special case of V.2.2.28, but has a simple direct proof: 1_A is properly infinite and hence contains a sequence of mutually orthogonal subprojections equivalent to 1_A. Thus $M_n(A)$ for any n and $A \otimes \mathbb{K}$ are isomorphic to hereditary C*-subalgebras of A. Apply V.2.2.23.

V.2.3.5 An infinite simple unital C*-algebra need not be purely infinite: if A is Rørdam's example (V.2.1.7), then $M_2(A)$ is infinite but not purely infinite since it contains a finite projection.

It is conceivable that there is a simple unital C*-algebra in which every nonzero projection is infinite, but in which there is a nonzero projectionless hereditary C*-subalgebra. Such a C*-algebra would not be purely infinite.

Nonunital and Stable Simple C*-Algebras

The next theorem, the main result of this section, shows that most of the notions of finiteness coincide for stable simple C*-algebras.

V.2.3.6 THEOREM. [BC82] Let A be a simple C*-algebra. Then the following are equivalent:

(i) A contains a scaling element.
(ii) A contains an infinite projection.
(iii) $\tilde{A}$ is infinite (contains an infinite projection).

These conditions imply the following, which are equivalent:

(iv) There is no finite nonzero dimension function on $Ped(A)$.
(v) There is no nondegenerate quasitrace (one taking values other than 0 and ∞) on A.

These imply:

(vi) A is algebraically simple.

If A is stable, then (i)–(vi) are all equivalent.
If A is exact, (iv) and (v) are also equivalent to

(vii) There is no nondegenerate trace on A.

Note that most of this theorem follows immediately from V.2.1.15 and V.2.1.19 (and II.6.8.17 for (vii)) if A is assumed to contain a nonzero projection; but it is crucial to prove the result without this assumption.

The implications (ii) $\implies$ (i), (ii) $\implies$ (iii), and (iv) $\implies$ (v) $\implies$ (vii) are trivial, (i) $\implies$ (iii) is V.2.2.11, (v) $\implies$ (iv) is V.2.1.18 (a nonzero dimension function on $Ped(A)$ is automatically faithful since $Ped(A)$ is algebraically simple), and (vii) $\implies$ (v) for exact A is II.6.8.17. Also, (ii) $\implies$ (v) is obvious, since if D is a (necessarily faithful) finite nonzero dimension function on $Ped(A)$, and $p \sim q \lneq p$, then $D(q) = D(p) = D(q) + D(p-q)$, a contradiction.

V.2.3.7 The proof of (iii) $\implies$ (ii) is very similar to the proof of (iii) $\implies$ (iv) of V.2.3.1. Let $1+x$ be a nonunitary isometry in $\tilde{A}$. Then $(f_\epsilon(x^*x))$ is a right approximate unit for x, and is also a left approximate unit since $xx^* \le x^*x$ (V.2.2.6); thus $f_\epsilon(x^*x)xf_\epsilon(x^*x) \to x$ as $\epsilon \to 0$. Let $y = f_\epsilon(x^*x)xf_\epsilon(x^*x)$ for small enough ϵ that $\|y-x\| < 1/6$; then $\|y\| \le 2$, so $a = (1+y^*)(1+y)$ is invertible in $\tilde{A}$ since

$$\|a - 1\| = \|(1+y^*)(1+y) - (1+x^*)(1+x)\| < 1$$

but $(1+y)(1+y^*)$ is not invertible (II.3.2.19). Set $z = (1+y)a^{-1/2}-1$. Then $z \in C^*(y)$, so $b = f_{\epsilon/2}(x^*x)$ is a unit for $C^*(z)$, and $1+z$ is a nonunitary isometry in $\tilde{A}$. Thus $C^*(z) \cong C^*(S-I)$, where S is the unilateral shift, and hence there is a sequence (p_n) of mutually orthogonal, mutually equivalent nonzero projections in $C^*(z)$. Let $q_m = \sum_{n=1}^m p_n$ for each m. Since $b \in Ped(A)_+$, there is an m and elements y_k with $b = \sum_{k=1}^m y_k^* p_k y_k$. Thus, if $u = \sum_{k=1}^m p_k y_k q_{m+1}$, then $u^*u = q_{m+1}$ and $uu^* \le q_m$, so q_{m+1} is an infinite projection.

V.2.3.8 We next show (vi) $\implies$ (i) if A is stable. Let $a, b \in A_+$ with $\|a\| = \|b\| = 1$ and $b \ll a \ne b$. Set

$$c = \sum_{k=1}^\infty 2^{-k} a \otimes e_{kk} \in A \otimes \mathbb{K}.$$

Then $c \in Ped(A \otimes \mathbb{K}) = A \otimes \mathbb{K}$, so by an argument identical to the proof of V.2.3.1[(iii) $\implies$ (iv)] there is an n and a $y \in A \otimes \mathbb{K}$ with $c = y^*(\sum_{k=1}^n b \otimes e_{kk})y$. Let

$$z = \sum_{k=1}^n 2^n b \otimes e_{kk}$$

and set

$$x = (\sum_{k=1}^n b \otimes e_{kk})^{1/2} y z^{1/2}.$$

Then $x^*x = \sum_{k=1}^n a^2 \otimes e_{kk}$ and $xx^* \le M(\sum_{k=1}^n b \otimes e_{kk})$ for some M, so x is a scaling element.

V.2.3.9 Now we show (i) $\implies$ (vi). Let $x, a \in A$ be as in V.2.2.9(i), and let $c \in A_+$. Write $c = \sum_{k=1}^\infty c_k$ with $c_k \in Ped(A)_+$ and $\|c_k\| \le 2^{-k}$ for $k > 1$ (II.5.2.6). Set $a_n = x^n a x^{*n}$; then $\|a_n\| = 1$ and $a_n \ll x^*x$ for all n, $a_n \perp a_m$ for $n \ne m$, $a_m = x^{m-n} a_n x^{*m-n}$ for $m > n$, and $a_m = x^{*n-m} a_n x^{n-m}$ for $m < n$. As in the proof of V.2.3.1, there are pairwise disjoint finite subsets $F_k \subseteq \mathbb{N}$ and elements $y_k \in A$ with $c_k = y_k^*(\sum_{j \in F_k} a_j) y_k$. Set

$$z = \sum_{k=1}^\infty y_k (\sum_{j \in F_k} a_j)^{1/2}.$$

Then $zz^* = c$ and $z^*z \ll x^*x$, so $z^*z \in A_+^c$ and thus $c = zz^* \in Ped(A)$.

V.2.3.10 Next we prove (iv) $\Longrightarrow$ (i) if A is stable. We only outline the argument. Let $a, b \in Ped(A)_+$ with $\|a\| = \|b\| = 1$ and $ab = b \neq a$. Since $K_0^*(Ped(A)) = \{0\}$ (V.2.1.17), we have $a + c \precsim c$ for some $c \in Ped(A)_+$, $c \perp a$. In $A \otimes \mathbb{K}$, set $d = c \otimes e_{11}$, $a_k = a \otimes e_{kk}$, $b_k = b \otimes e_{kk}$. Then, by iterating $c + a \precsim c$, we get that

$$\sum_{k=1}^{n} a_k \precsim d + \sum_{k=1}^{n} a_k \precsim d$$

for all n. As in the proof of V.2.3.1, there is an n and $y \in A$ with $d = y^*(\sum_{k=1}^{n} b_k)y$, and since $\sum_{k=1}^{n} a_k \precsim d$, there is a $z \in A$ with $z^* dz$ close to $\sum_{i=1}^{n} a_k$. Then, if

$$x = (\sum_{k=1}^{n} b_k)^{1/2} yz,$$

then x^*x is close to $\sum_{k=1}^{n} a_k$ and $xx^* \leq M \sum_{k=1}^{n} b_k$ for some M; thus x is an approximate scaling element and $C^*(x) \subseteq A$ contains a scaling element by V.2.2.9(ii).

V.2.3.11 Finally, we prove (iv) $\Longrightarrow$ (vi) for general simple A. If (iv) holds, then there is no dimension function on $Ped(A \otimes \mathbb{K})$, and thus $A \otimes \mathbb{K}$ contains a scaling element and is therefore algebraically simple, by the last two paragraphs. But then A is also algebraically simple by II.5.4.2.

This completes the proof of V.2.3.6.

V.2.3.12 COROLLARY. *Let A be a purely infinite simple C*-algebra. Then A contains an infinite projection.*

PROOF: There is no dimension function on $Ped(A \otimes \mathbb{K})$, so $A \otimes \mathbb{K}$ contains an infinite projection p, and is purely infinite by V.2.2.28. If $x \in A$ is nonzero, then

$$\overline{x^*Ax} \otimes \mathbb{K} \cong p(A \otimes \mathbb{K})p \otimes \mathbb{K}$$

(II.7.6.11). Every hereditary C*-subalgebra of $p(A \otimes \mathbb{K})p \otimes \mathbb{K}$ contains an infinite projection by V.2.2.24.

V.2.3.13 COROLLARY. *A simple C*-algebra A is purely infinite if and only if every hereditary C*-subalgebra of A contains a nonzero infinite projection.*

V.2.4 Ordered K-Theory

V.2.4.1 DEFINITION. A *preordered group* is an abelian group G with a translation-invariant preorder (transitive relation) $\leq$. $G_+ = \{x \in G : 0 \leq x\}$ is the *positive cone* of G. If $x, y \in G$, $x \leq y$ if and only if $y - x \in G_+$.

V.2.4.2 If A is a C*-algebra, the semigroup $V(A)$ really contains the essential information about the projections of $M_\infty(A)$; however, semigroups (particularly ones without cancellation) can be nasty algebraic objects, and for technical reasons it is necessary to pass to the group $K_0(A)$ in order to apply techniques from topology and homological algebra to the study of projections over A. But it is desirable to keep the original semigroup in the picture as much as possible. One way to do this is to put a (pre)ordering on $K_0(A)$ by taking the image $K_0(A)_+$ of $V(A)$ in $K_0(A)$ to be the positive cone. (Even at this point we may lose information, since the map from $V(A)$ into $K_0(A)$ will be injective only if $V(A)$ has cancellation.) Just as the elements of $K_0(A)$ determine the (stable) equivalence of projections in $M_\infty(A)$, the ordering will determine the (stable) comparability of projections.

The order structure on K_0 seems to have played only a minimal role in topological K-theory; the order on $K_0(C(X))$ is usually either rather trivial or else badly behaved. But the ordering is crucial in many of the applications of K-theory to C*-algebras, particularly to the classification problem.

V.2.4.3 We also define the *scale* $\Sigma(A)$ to be the image of $\text{Proj}(A)$ in $K_0(A)$. If A is unital, the scale is simply the elements of $K_0(A)_+$ which are $\leq [1_A]$, so the scale can be described by simply specifying $[1_A]$, and by slight abuse of terminology we will frequently do so. We will mostly be concerned with the unital case. The triple $(K_0(A), K_0(A)_+, \Sigma(A))$ is called the *scaled preordered K_0-group* of A.

The preordered group $(K_0(A), K_0(A)_+)$ depends on A only up to stable isomorphism; but the scale gives a finer invariant, which can be used to distinguish between algebras in the same stable isomorphism class (V.2.4.16).

If $\phi : A \to B$ is a homomorphism, then $\phi_* : K_0(A) \to K_0(B)$ is a homomorphism of scaled preordered groups, i.e. $\phi_*(K_0(A)_+) \subseteq K_0(B)_+$ and $\phi_*(\Sigma(A)) \subseteq \Sigma(B)$.

Ordered Groups

V.2.4.4 DEFINITION. An *ordered group* (G, G_+) is an abelian group G with a distinguished subsemigroup G_+ containing the identity 0, called the positive cone of G, having these properties:

(1) $G_+ - G_+ = G$.
(2) $G_+ \cap (-G_+) = \{0\}$.

G_+ induces a translation-invariant partial ordering on G by $y \leq x$ if $x - y \in G_+$. By $y < x$ we will mean that $y \leq x$ and $y \neq x$.

An element $u \in G_+$ is called an *order unit* if for any $x \in G$ there is an $n > 0$ with $x \leq nu$ (in other words, the order ideal [hereditary subgroup] generated by u is all of G). A triple (G, G_+, u) consisting of an ordered group (G, G_+) with a fixed order unit u is called a *scaled ordered group*. We say G is a *simple* ordered group if G has no proper order ideals, i.e. if every nonzero positive element is an order unit.

V.2.4.5 EXAMPLES.

(i) On $\mathbb{Z}^n$ or $\mathbb{R}^n$, there are two standard orderings: the *ordinary ordering*, with positive cone
$$\{(x_1,\ldots,x_n) \mid x_1,\ldots,x_n \geq 0\}$$
and the *strict ordering*, with positive cone
$$\{0\} \cup \{(x_1,\ldots,x_n) \mid x_1,\ldots,x_n > 0\}.$$
These orderings coincide for $n=1$.

(ii) More generally, if X is a set and G is a positively generated additive group of real-valued functions on X, then G can be given the ordinary ordering with
$$G_+ = \{f : f \geq 0 \text{ everywhere }\}$$
or the strict ordering with
$$G_+ = \{0\} \cup \{f : f > 0 \text{ everywhere }\}.$$
The positive cone in the strict ordering is sometimes denoted G_{++}. Even more generally, if ρ is a homomorphism from G into the additive group of real-valued functions on X whose image is positively generated, then G may be given the *strict ordering from ρ*, with
$$G_+ = \{0\} \cup \{a \in G : \rho(a) > 0 \text{ everywhere }\}.$$
There is no analog of the ordinary ordering if ρ is not injective.

(iii) If ρ is a homomorphism from G into an ordered group H whose range is positively generated, then G can be given the *strict ordering from ρ* by taking
$$G_+ = \{0\} \cup \{x : \rho(x) > 0\}.$$

A group G with the strict ordering in the sense of (ii), or in the sense of (iii) with H simple, is a simple ordered group.

V.2.4.6 The set $K_0(A)_+$ does not satisfy (1) of V.2.4.4 in general. For example, if $A = C_o(\mathbb{R}^2)$, then $K_0(A) \cong \mathbb{Z}$ and $K_0(A)_+ = 0$. However, if A is (stably) unital, then $K_0(A)_+$ does satisfy condition (1).

From now on in this section, we will assume A is *unital*. Almost everything carries through (with appropriate technical modifications) to the stably unital case.

V.2.4.7 EXAMPLE. The set $K_0(A)_+$ does not satisfy (2) of V.2.4.4 in general. Let O_n be the Cuntz algebra (II.8.3.3, V.1.3.4); then
$$K_0(O_n)_+ = \Sigma(O_n) = K_0(O_n) \cong \mathbb{Z}_{n-1}.$$
Actually, we have $K_0(A)_+ = \Sigma(A) = K_0(A)$ whenever A is a properly infinite C*-algebra (e.g. a simple unital C*-algebra containing a nonunitary isometry (V.2.3.1)).

V.2.4.8 PROPOSITION. If A is stably finite, then $(K_0(A), K_0(A)_+)$ is an ordered group.

V.2.4.9 EXAMPLES.

(i) The ordering on $K_0(\mathbb{C})$ and $K_0(\mathbb{M}_n)$ is the ordinary ordering on $\mathbb{Z}$.

$$\Sigma(\mathbb{M}_n) = \{0, \ldots, n\}.$$

(ii) If A is a II_1 factor, then the ordering on $K_0(A)$ is the ordinary ordering on $\mathbb{R}$, and $\Sigma(A) = [0,1]$.

(iii) If $A = \mathbb{C}^2$, then $K_0(A) \cong \mathbb{Z}^2$ with the ordinary ordering.

$$\Sigma(A) = \{(0,0),\ (0,1),\ (1,0),\ (1,1)\}.$$

(iv) $K_0(C(S^2))$ is $\mathbb{Z}^2$ with the strict ordering from the first coordinate, i.e.

$$K_0(C(S^2))_+ = \{(0,0)\} \cup \{(m,n) \mid m > 0\}$$

$$\Sigma(C(S^2)) = \{(0,0), (1,0)\}.$$

Examples (iii) and (iv) show that $K_0(A)$ and $K_0(B)$ can be isomorphic as groups without being isomorphic as ordered groups. Thus the order structure can be used to distinguish between algebras.

$K_0(A)$ is frequently a simple ordered group. If $p \in M_n(A)$ is a projection, identify p with $diag(p, 0)$ in $M_{n+m}(A)$ for any m, and for $k \in \mathbb{N}$ write $k \cdot p$ for $diag(p, \ldots, p)$ in $M_k(M_n(A)) \cong M_{kn}(A)$ (or in $M_{kn+m}(A)$).

V.2.4.10 PROPOSITION. Let A be a C*-algebra, and p and q projections in $M_n(A)$, with q full (II.5.3.10). Then $p \precsim k \cdot q$ for some $k \in \mathbb{N}$.

PROOF: [Cun77b] Since p is in $Ped(M_n(A))$, which is the ideal of $M_n(A)$ generated algebraically by q, there are elements $x_1, \ldots, x_k \in M_n(A)$ with $p = \sum_{i=1}^{k} x_i^* q x_i$ (II.5.2.10). We may assume $x_i = q x_i p$ for all i. Let

$$X \in M_k(M_n(A)) \cong M_{kn}(A)$$

be the matrix with $x_1, \ldots, x_k$ in the first column and zeroes elsewhere, $P = diag(p, 0, \ldots, 0)$, $Q = diag(q, q, \ldots, q)$. Then $X^* Q X = P$, so $U = QX$ is a partial isometry with $U^* U = P$. UU^* is a projection, and Q is a unit for UU^*, so $UU^* \leq Q$.

This result can also be deduced from II.7.6.11.

V.2.4.11 COROLLARY. If A is stably finite, and every nonzero projection in $M_\infty(A)$ is full, then $K_0(A)$ is a simple ordered group.

V.2.4.12 COROLLARY. If A is a stably finite C*-algebra and Prim(A) contains no nontrivial compact open subsets, then $K_0(A)$ is a simple ordered group. So if Prim(A) is Hausdorff and connected, $K_0(A)$ is simple. In particular, if A is simple or if $A = C(X)$, X connected, then $K_0(A)$ is simple.

Cancellation

V.2.4.13 The semigroup $V(A)$ does not have cancellation in general. For example, if u is a nonunitary isometry in A and $p = uu^*$, then $[p] + [0] = [p] = [1_A] = [p] + [1_A - p]$, and $[1_A - p] \neq [0]$. More interestingly, cancellation in $V(A)$ can fail even if A is stably finite: for example, $V(C(S^5))$ does not have cancellation [Hus66]. We say a C*-algebra A has *cancellation* if $V(\tilde{A})$ has cancellation (if A is nonunital, this is strictly stronger than requiring just that $V(A)$ has cancellation). Since $V(A)$ includes projections from all matrix algebras over A, cancellation for A implies cancellation for $M_n(A)$ and for $A \otimes \mathbb{K}$. A von Neumann algebra has cancellation if and only if it is finite (III.1.3.8).

A has cancellation if, whenever p, q, r are projections in $M_n(\tilde{A})$ for some n, with $p \perp r$, $q \perp r$, and $(p + r) \sim (q + r)$, then $p \sim q$. We say A has *strict cancellation* if, for any such p, q, r, $(p+r) \prec (q+r)$ implies $p \prec q$. Cancellation implies strict cancellation (cf. III.1.3.8); the converse is unclear.

V.2.4.14 PROPOSITION. let A be a unital C*-algebra. The following are equivalent:

(i) A has cancellation.
(ii) If p, q are projections in $M_n(A)$ for some n, and $p \sim q$, then $(1 - p) \sim (1 - q)$.
(iii) If p, q are projections in $M_n(A)$ for some n, and $p \sim q$, then $p \sim_u q$.

PROOF: (iii) $\Longrightarrow$ (ii) and (i) $\Longrightarrow$ (ii) are trivial. (ii) $\Longrightarrow$ (i): suppose p, q, r are projections in $M_n(A)$ with $p \perp r$, $q \perp r$, and $(p + r) \sim (q + r)$. By (ii), $(1 - p - r) \sim (1 - q - r)$, and hence

$$(1 - p) = (1 - p - r) + r \sim (1 - q - r) + r = (1 - q)$$

so again by (ii) $p \sim q$. (ii) $\Longrightarrow$ (iii): if $u^*u = p$, $uu^* = q$, then there is a partial isometry $v \in M_n(A)$ with $v^*v = 1 - p$, $vv^* = 1 - q$, and $w = u + v$ is a unitary with $wpw^* = q$.

Perforation

One difficulty which can occur in ordered groups is perforation.

V.2.4.15 DEFINITION. An ordered group (G, G_+) is *unperforated* if $nx \geq 0$ for some $n > 0$ implies $x \geq 0$; G is *weakly unperforated* if $nx > 0$ for some $n > 0$ implies $x > 0$.

An unperforated group must be torsion-free. A weakly unperforated group can have torsion: for example, $\mathbb{Z} \oplus \mathbb{Z}_2$ with strict ordering from the first coordinate. If (G, G_+) is weakly unperforated, H is the torsion subgroup of G, and $\pi : G \to G/H$ the quotient map, then $(G/H, \pi(G_+))$ is an unperforated ordered group. Hence a weakly unperforated group is "unperforated up to torsion." Conversely, if (K, K_+) is a (weakly) unperforated ordered group and $\rho : G \to K$ is a homomorphism with positively generated image, then G is weakly unperforated if given the strict ordering from ρ. A weakly unperforated group is unperforated if and only if it is torsion-free.

V.2.4.16 EXAMPLES.

(i) Let $G = \mathbb{Z}$, $G_+ = \{0\} \cup \{n : n \geq 2\}$. Then (G, G_+) is not weakly unperforated.
(ii) $K_0(C(\mathbb{R}P^2)) \cong \mathbb{Z} \oplus \mathbb{Z}_2$ with strict ordering from the first coordinate [Kar78, IV.6.47]. $K_0(C(\mathbb{T}^4)) \cong \mathbb{Z}^8$ is perforated, where $\mathbb{T}^4$ is the 4-torus [Bla98, 6.10.2]. There are stably finite simple unital C*-algebras A with torsion in $K_0(A)$. There are even stably finite simple C*-algebras whose K_0 is not weakly unperforated [Vil98].
(iii) Perforation in K_0 can be eliminated by "rationalizing": if R is the (unique) UHF algebra with $K_0(R) = \mathbb{Q}$, then for any A we have

$$K_0(A \otimes R) \cong K_0(A) \otimes \mathbb{Q}$$

with

$$K_0(A \otimes R)_+ = K_0(A)_+ \otimes \mathbb{Q}_+,$$

and $K_0(A \otimes R)$ is unperforated.

Classification of Stably Isomorphic C*-Algebras

V.2.4.17 Suppose that A is a unital C*-algebra with cancellation. Then the scale $\Sigma(A)$ is a hereditary subset of $K_0(A)_+$; in fact, $\Sigma(A)$ is the closed interval

$$[0, [1_A]] = \{x \in K_0(A)_+ \mid x \leq [1_A]\}.$$

Although $\Sigma(A)$ does not always generate $K_0(A)$ as a group, $[1_A]$ is always an order unit, so the order ideal generated by $\Sigma(A)$ is $K_0(A)$. If B is a unital C*-algebra stably isomorphic to A, then $K_0(B)$ is order-isomorphic to $K_0(A)$, and the image of $\Sigma(B)$ in $K_0(A)$ will be an interval $[0, u]$ for some order unit u. Conversely, if u is an order unit in $K_0(A)$, then there is a unital C*-algebra B stably isomorphic to A with $\Sigma(B) = [0, u]$: let $u = [p]$ for some projection $p \in M_n(A)$, and take $B = pM_n(A)p$. So one can nearly classify all unital

C*-algebras stably isomorphic to A by the order units in $K_0(A)$. The correspondence is, however, not one-to-one in general: the algebras corresponding to u and v may be isomorphic if there is an order-automorphism of $K_0(A)$ taking u to v. (But the existence of such an order-automorphism does not guarantee that the algebras are isomorphic; not every order-automorphism of $K_0(A)$ is induced from an isomorphism on the algebra level in general.)

V.2.4.18 One can extend the above classification to certain nonunital C*-algebras as follows. If $u_1, u_2, \ldots$ is an increasing sequence of elements of $K_0(A)_+$, then one can find an increasing sequence of projections $p_1, p_2, \ldots$ in $A \otimes \mathbb{K}$ with $[p_n] = u_n$. If $\{u_1, u_2, \ldots\}$ generates $K_0(A)_+$ as an order ideal, then the C*-algebra

$$B = \left(\bigcup p_n (A \otimes \mathbb{K}) p_n \right)^-$$

is stably isomorphic to A and corresponds naturally to the interval $\bigcup [0, u_n] \subseteq K_0(A)$. A hereditary subset Σ of $K_0(A)_+$ is of this form if and only if it generates G as an order ideal and is countably generated and upward directed, i.e. if $x, y \in \Sigma$ there is a $z \in \Sigma$ with $x \leq z$ and $y \leq z$. The C*-algebra corresponding to Σ by the above construction depends up to isomorphism only on Σ and not on the choice of the u_n or p_n. Conversely, every C*-algebra which is stably isomorphic to A and which has an approximate unit consisting of a sequence of projections is obtained in this way. So if A is separable with real rank zero (V.3.2.7), one obtains a complete classification (modulo the possible identifications through order-automorphisms).

Classification of AF Algebras

This procedure works especially well for AF algebras (II.8.2.2(iv)), and leads to a complete classification via ordered K-theory. This classification, due to G. Elliott, was one of the first explicit appearances of K-theory in operator algebras. It is the prototype of the more advanced classification results described in [Rør02b], which also includes a full treatment of the AF case. See also [Eff81] and [Bla98].

V.2.4.19 THEOREM. [Ell76] Let A be an AF algebra. Then the scaled ordered K_0-group $(K_0(A), K_0(A)_+, \Sigma(A))$ (called the *dimension group* of A) is a complete isomorphism invariant for A among AF algebras: if B is an AF algebra and

$$(K_0(B), K_0(B)_+, \Sigma(B)) \cong (K_0(A), K_0(A)_+, \Sigma(A)),$$

then $B \cong A$.

The scaled ordered groups which can occur as dimension groups can also be characterized abstractly, due to the Effros-Handelman-Shen Theorem:

V.2.4.20 THEOREM. [EHS80] Let (G, G_+, Σ) be a scaled ordered group. Then (G, G_+, Σ) is the dimension group of an AF algebra if and only if it has the following properties:

(i) (G, G_+) is countable and unperforated (in particular, G is torsion-free).
(ii) (G, G_+) has the *Riesz Interpolation Property*: if x_1, x_2, y_1, y_2 are in G and $x_i \leq y_j$ for all i, j, then there is a $z \in G$ with $x_i \leq z \leq y_j$ for all i, j.
(iii) Σ is an upward directed hereditary subset of G_+ which generates G.

States on Ordered Groups

The order structure on an ordered group is at least partially (and in good cases completely) determined by the states, which in the K_0 case are closely related to the tracial states on the algebra.

V.2.4.21 DEFINITION. A *state* on a scaled ordered group (G, G_+, u) is an order-preserving homomorphism f from G to $\mathbb{R}$ with $f(u) = 1$.

The set $S(G, G_+, u)$ (or just denoted $S(G)$ when there is no confusion) of all states on (G, G_+, u) is a compact convex set in the topology of pointwise convergence. $S(G)$ is called the *state space* of G.

We now develop some properties of $S(G)$ due to K. Goodearl and D. Handelman [GH76], including a Hahn–Banach type existence theorem.

V.2.4.22 LEMMA. Let (G, G_+, u) be a scaled ordered group. Let H be a subgroup of G containing u, and f a state on $(H, H \cap G_+, u)$ (we do not assume H is positively generated). Let $t \in G_+$, and

$$p = \sup\{f(x)/m \mid x \in H, m > 0, x \leq mt\}$$

$$q = \inf\{f(y)/n \mid y \in H, n > 0, nt \leq y\}.$$

Then:

(i) $0 \leq p \leq q < \infty$.
(ii) If g is a state on $(H + \mathbb{Z}t, u)$ which extends f, then $p \leq g(t) \leq q$.
(iii) If $p \leq r \leq q$, then there is a unique state g on $(H + \mathbb{Z}t, u)$ which extends f with $g(t) = r$.

The proof is a simple calculation. The next theorem then follows from a Zorn's Lemma argument:

V.2.4.23 THEOREM. Let (G, G_+, u) be a scaled ordered group, and let H be a subgroup of G containing u. If f is any state on $(H, H \cap G_+, u)$, then f extends to a state on (G, G_+, u).

V.2.4.24 COROLLARY. Let (G, G_+, u) be a scaled ordered group, and let $t \in G_+$. Let $f_*(t) = p, f^*(t) = q$ defined as in V.2.4.22 with $H = \mathbb{Z}u$. Then:

(i) $0 \leq f_*(t) \leq f^*(t) < \infty$.
(ii) If f is any state on G, then $f_*(t) \leq f(t) \leq f^*(t)$.
(iii) If $f_*(t) \leq r \leq f^*(t)$, then there is a state g on G with $g(t) = r$.

$f_*(t)$ and $f^*(t)$ can be more elegantly described as

$$f_*(t) = \sup\{n/m \mid nu \leq mt\}$$

$$f^*(t) = \inf\{n/m \mid mt \leq nu\}.$$

V.2.4.25 THEOREM. Let (G, G_+, u) be a simple weakly unperforated scaled ordered group. Then G has the strict ordering from its states, i.e. $G_+ = \{0\} \cup \{x \mid f(x) > 0 \text{ for all } f \in S(G)\}$.

PROOF: If $x > 0$, then x is an order unit, so $u \leq mx$ for some $m > 0$. Then $0 < 1/m \leq f_*(x)$, so $f(x) > 0$ for all $f \in S(G)$. Conversely, suppose $f(x) > 0$ for all $f \in S(G)$. By compactness, we have

$$f_*(x) = \inf_{f \in S(G)} f(x) > 0$$

so there are positive integers n and m with $0 < nu \leq mx$, and therefore by weak unperforation $x > 0$. (Note that this implication does not require G to be simple.)

There is an alternate way to view V.2.4.25. If $x \in G$, then x induces a continuous affine function $\hat{x}$ on $S(G)$ by $\hat{x}(f) = f(x)$. Thus there is a homomorphism ρ from G to $\text{Aff}(S(G))$, the group of all continuous real-valued affine functions on $S(G)$. V.2.4.25 then says that G has the strict ordering from ρ, in the sense of V.2.4.5(ii).

V.2.4.26 It is easy to identify at least some (in fact, all) of the states on $(K_0(A), K_0(A)_+, [1_A])$. If τ is a tracial state on A, or more generally a quasitrace on A (II.6.8.15), then τ induces a state on $K_0(A)$ in an obvious way. Let $T(A)$ and $QT(A)$ denote respectively the tracial states and normalized quasitraces on A. We have $T(A) \subseteq QT(A)$; it is a very important and difficult question whether $T(A) = QT(A)$ for all A (II.6.8.16). It is at least true that $QT(A)$, like $T(A)$, is always a Choquet simplex, which is metrizable if A is separable [BH82, II.4.4]. For the ordering on $K_0(A)$ the set $QT(A)$ is the more natural and important set to consider, due to the fact that the quasitraces on A are in one-one correspondence with the lower semicontinuous dimension functions on A (II.6.8.14).

V.2.4.27 The correspondence described above gives a continuous affine map

$$\chi : QT(A) \to S = S(K_0(A)).$$

This map is not injective in general. For example, if $A = C(S^1)$, then $QT(A) = T(A)$ is the state space of A, while S is a singleton. Also, if $A = C^*(\mathbb{Z}_2 * \mathbb{Z}_2)$, then S is a square and not even a simplex, so χ is not injective (there are similar simple examples [ET94]). It is injective if A has real rank zero (V.3.2.19). Under some mild additional hypotheses, a precise description can then be given of the ordered group $K_0(A)$.

The map χ is always surjective [BR92]. In other words, if A is any C*-algebra, then every state on $K_0(A)$ comes from a quasitrace on A; if A is nuclear (or exact), then every state on $K_0(A)$ comes from a tracial state. Thus there are enough tracial states on A to completely determine the order structure on $K_0(A)$ (up to perforation).

So if A is a stably finite unital C*-algebra with real rank zero, then χ is a bijection, hence a homeomorphism.

K_0 of a Crossed Product

V.2.4.28 It is a highly nontrivial matter to determine the order structure on $K_0(A \rtimes_\alpha G)$, even if $G = \mathbb{Z}$. Even if the action is trivial, so that $A \rtimes_\alpha \mathbb{Z} \cong C(S^1, A)$, there is no known way to calculate the order structure. For example, if $A = C(\mathbb{T}^3)$, then $K_0(A) \cong \mathbb{Z}^4$ with strict ordering from the first coordinate; but $K_0(C(\mathbb{T}^4)) \cong \mathbb{Z}^8$ is perforated.

One can, however, obtain quite a bit of partial information: it is (at least theoretically) possible to calculate the range of any state on $K_0(A \rtimes_\alpha \mathbb{Z})$ which comes from a tracial state on the crossed product (by [BR92], every state on K_0 arises this way, at least if A is exact). In good cases this calculation can be done rather easily and gives complete information on the order structure.

Using these results, one can fairly easily give examples of simple unital C*-algebras with no nontrivial projections. The first such examples were given in [Bla80] and [Bla81].

See [Bla98, 10.10] for details and references.

Order Structure on $K_*(A)$

For some purposes in classification theory, the ordering on $K_0(A)$ is not enough; one needs a finer ordering on the $\mathbb{Z}_2$-graded group $K_*(A) = K_0(A) \oplus K_1(A)$.

V.2.4.29 DEFINITION. Let A be a unital C*-algebra. Set

$$K_*(A)_+ = \{([p], [u]) : p \in M_\infty(A), u \in \mathcal{U}(pM_\infty(A)p)\}$$

and $\Sigma_*(A) = \{([p], [u]) \in K_*(A)_+ : p \in A\}$. $(K_*(A), K_*(A)_+, \Sigma_*(A))$ is called the *scaled (pre)ordered K_*-group* of A.

V.2.4.30 In many cases, the ordering on $K_*(A)$ just comes from the ordering on the K_0 part. For example, if A is properly infinite, then $K_*(A)_+ = \Sigma_*(A) = K_*(A)$. And if A is simple with $\mathrm{sr}(A) = 1$, then $K_*(A)$ has the strict ordering from the K_0 part, i.e.

$$K_*(A)_+ = \{(0,0)\} \cup \{(x,y) : 0 < x \text{ in } K_0(A)\}$$
$$\Sigma_*(A) = \{(0,0),(1,0)\} \cup \{(x,y) : 0 < x < 1 \text{ in } K_0(A)\}.$$

However, this is not always the case: for example, if $A = C(S^3)$, then $K_0(A) \cong K_1(A) \cong \mathbb{Z}$, so $K_*(A) \cong \mathbb{Z}^2$, and

$$K_*(A)_+ = \{(0,0),(1,0)\} \cup \{(m,n) : m \geq 2\}.$$

There are examples of stably finite simple C*-algebras where the same phenomenon happens [Vil02].

V.2.4.31 There is a clean alternate description of the ordering on $K_*(A)$. From the split exact sequence

$$0 \longrightarrow SA \longrightarrow C(S^1) \otimes A \longrightarrow A \longrightarrow 0$$

we obtain that

$$K_*(A) \cong K_0(C(S^1) \otimes A)$$

and under this isomorphism $K_*(A)_+$ and $\Sigma_*(A)$ just become $K_0(C(S^1) \otimes A)_+$ and $\Sigma(C(S^1) \otimes A)$ respectively.

V.3 Stable Rank and Real Rank

An algebraic theory of stable rank in rings was developed by H. Bass [Bas64], primarily to handle cancellation problems in algebraic K-theory. M. Rieffel adapted the theory to C*-algebras (and more general topological algebras) [Rie83]. This theory was formally modeled on dimension theory for compact Hausdorff spaces, but it was quickly realized that stable rank does not resemble a dimension theory very closely in the noncommutative case. The theory has nonetheless proved interesting and useful, particularly with regard to nonstable K-theory questions. The principal reason that stable rank does not behave like a dimension theory in the noncommutative case, and also the reason it gives nonstable K-theory information, is its behavior under forming matrix algebras ([Vn71], [Rie83, 6.1]): the stable rank $\mathrm{sr}(M_n(A))$ is roughly $\mathrm{sr}(A)$ divided by n (see V.3.1.16 for the precise formula).

L. Brown and G. Pedersen [BP91] developed an analogous theory of real rank for C*-algebras. The base case of this theory, the case of real rank zero, is the most important existence of projections property in the theory of C*-algebras.

In this section, we briefly develop the general theories, with particular emphasis on the base cases (stable rank one, real rank zero), which are both the simplest and the most important aspects of the theories. Much of this exposition (which largely follows [Rie83]) first appeared in [Bla04c].

V.3.1 Stable Rank

Both stable rank and real rank are motivated by the following fact from topology ([HW41, VI.1], [Pea75, 3.3.2]; cf. [Kat50] and [GJ76] for related results).

V.3.1.1 THEOREM. Let X be a compact metrizable space. Then all dimension theories coincide on X, and $dim(X)$ can be characterized as the smallest n with the following property: any continuous function $f : X \to \mathbb{R}^{n+1}$ can be uniformly approximated arbitrarily closely by $g : X \to \mathbb{R}^{n+1}$ such that $g(X)$ does not contain the origin in $\mathbb{R}^{n+1}$.

V.3.1.2 DEFINITION. Let A be a unital C*-algebra. Let $Lg_n(A)$ be the set of $(x_1, \ldots, x_n) \in A^n$ such that there exists $(y_1, \ldots, y_n) \in A^n$ with $\sum_{i=1}^n y_i x_i = 1$.

The *stable rank* of A, denoted $sr(A)$, is the smallest n such that $Lg_n(A)$ is dense in A^n. If there is no such n, set $sr(A) = \infty$.

If A is nonunital, then $sr(A)$ is defined to be $sr(\tilde{A})$.

V.3.1.3 If A is unital, the elements of $Lg_n(A)$ are the n-tuples which generate A as a left A-module. It is easily seen that $(x_1, \ldots, x_n) \in Lg_n(A)$ if and only if $\sum_{i=1}^n x_i^* x_i$ is invertible. In particular, if $A = C(X)$, then $(f_1, \ldots, f_n) \in Lg_n(A)$ if and only if for every $t \in X$ there is an i such that $f_i(t) \neq 0$. Thus, since $(f_1, \ldots, f_n) \in A^n$ can be regarded as a continuous function from X to $\mathbb{C}^n \cong \mathbb{R}^{2n}$, we obtain from V.3.1.1 that

$$sr(C(X)) = \left\lfloor \frac{dim(X)}{2} \right\rfloor + 1$$

where $\lfloor \cdot \rfloor$ denotes "integer part of."

V.3.1.4 The number $sr(A)$ defined in V.3.1.2 is properly called the *left topological stable rank* of A, denoted $ltsr(A)$ in [Rie83]; $Rg_n(A)$ and $rtsr(A)$ can be defined analogously. Because of the involution, there is an obvious correspondence between $Lg_n(A)$ and $Rg_n(A)$, and $ltsr(A) = rtsr(A)$; this number is called $tsr(A)$ in [Rie83] to distinguish it from the Bass stable rank $Bsr(A)$. The topological stable rank of a C*-algebra was shown to coincide with the Bass stable rank in [HV84]; thus we may use the term "stable rank" and the notation $sr(A)$ unambiguously.

Stable Rank One

We now examine the case of stable rank one more carefully. A has stable rank one if and only if the left invertible elements of $\tilde{A}$ are dense in $\tilde{A}$.

V K-Theory and Finiteness

V.3.1.5 PROPOSITION. Let A be a unital C*-algebra of stable rank one. Then every left or right invertible element of A is invertible (and, in particular, the invertible elements of A are dense in A).

PROOF: Because of the involution, it suffices to show that left invertible elements are invertible. Let x be left invertible with left inverse r. If y is a left invertible element with $\|y-r\| < \|x\|^{-1}$, then $\|yx-1\| < 1$, so yx is invertible and hence y is right invertible and thus invertible; therefore x is also invertible.

Thus a C*-algebra of stable rank one must be finite. By V.3.1.16, such a C*-algebra must even be stably finite. Until fairly recently, no stably finite simple C*-algebra was known to have stable rank greater than 1. Examples have been constructed by Villadsen [Vil99].

The Generalized Matrix Picture

V.3.1.6 Elements of A^n may be regarded as $n \times 1$ matrices over A, and $Lg_n(A)$ becomes the set of left invertible $n \times 1$ matrices, so stable rank can be defined in terms of density of left invertible matrices.

V.3.1.7 If p and q are projections in a C*-algebra A, it is useful to think of the subspace pAq of A as a space of "nonsquare matrices" with "p rows" and "q columns". If r and s are other projections orthogonal to p and q respectively, then a "$(p+r) \times (q+s)$ matrix" (an element of $(p+r)A(q+s)$) may be symbolically written as a 2×2 "block matrix": write $x \in (p+r)A(q+s)$ as $\begin{bmatrix} pxq & pxs \\ rxq & rxs \end{bmatrix}$. The algebraic operations in these sets (as subsets of A) can be calculated by formal matrix algebra.

It is convenient to generalize the notion of left invertibility:

V.3.1.8 DEFINITION. Let A be a C*-algebra, p and q projections in A. An element $x \in pAq$ is *left invertible* (in pAq or with respect to (p,q)) if there is $y \in qAp$ with $yx = q$. We write $Lg_{(p,q)}(A)$ for the set of left invertible elements of pAq.

It is easily seen that $x \in pAq$ is left invertible in pAq if and only if x^*x is invertible in qAq. (Thus left invertibility with respect to (p,q) really depends only on the q.)

If $p \sim p'$, $q \sim q'$, then there is an obvious isometric isomorphism from pAq onto $p'Aq'$ sending $Lg_{(p,q)}(A)$ onto $Lg_{(p',q')}(A)$, given by $x \mapsto uxv$, where $u^*u = p$, $uu^* = p'$, $vv^* = q$, $v^*v = q'$.

V.3.1.9 Recall (II.3.4.3) that $q \precsim p$ if there is a $u \in A$ with $u^*u = q$ and $uu^* \leq p$. We will write $n \cdot q \precsim p$ if there are mutually orthogonal subprojections $p_1, \ldots, p_n$ of p, each equivalent to q.

V.3.1.10 PROPOSITION. Let A be a C*-algebra, p, q projections in A.
(i) There exists a left invertible element in pAq if and only if $q \precsim p$.
(ii) If $sr(qAq) = n < \infty$ and $n \cdot q \precsim p$, then the left invertible elements of pAq are dense in pAq.
(iii) If p is equivalent to a proper subprojection of q, then the left invertible elements of pAq are not dense in pAq.

PROOF: (i): if $u^*u = q$ and $uu^* \leq p$, then $u \in pAq$ and is left invertible. Conversely, if $x \in pAq$ is left invertible, with $y \in qAp$ with $yx = q$, then
$$q = x^*y^*yx \leq \|y\|^2 x^*x$$
so x^*x is invertible in qAq. If $r \in qAq$ with $x^*xr = q$, set $u = xr^{1/2}$; then $u^*u = q$ and uu^* is a projection in pAp.

(ii): if $p_1, \ldots, p_n$ are mutually orthogonal subprojections of p, and u_i satisfies $u_i^* u_i = p_i$ and $u_i u_i^* = q$ for $1 \leq i \leq n$, set $r = p - \sum p_i$. If $x \in pAq$, write $x_i = p_i x$ for $1 \leq i \leq n$ and $x_{n+1} = rx$. Set $y_i = u_i x_i \in qAq$ for $1 \leq i \leq n$. By assumption, $(y_1, \ldots, y_n)$ can be approximated by $(z_1, \ldots, z_n) \in Lg_n(qAq)$, i.e. with $\sum_{i=1}^n z_i^* z_i$ invertible in qAq. Set $w_i = u_i^* z_i$ for $1 \leq i \leq n$, and
$$w = \sum_{i=1}^n w_i + x_{n+1}.$$
Then w closely approximates x, and
$$w^*w = \sum_{i=1}^n z_i^* z_i + x_{n+1}^* x_{n+1}$$
is invertible in qAq, so w is left invertible in pAq.

(iii): We may assume $p \lneq q$. Suppose $x = pxq \in pAq$ is left invertible and approximates p closely enough that $xp = pxp$ is invertible in pAp. Let $y = qyp \in qAp$ with $yx = q$. If $r = q - p$, then
$$[ryp][pxp] = ryxp = (q-p)p = 0$$
and since pxp is invertible in pAp, $ryp = 0$. But
$$[ryp][pxr] = ryxr = rqr = r \neq 0,$$
a contradiction.

Part (iii) is a version of the fact (II.3.2.19) that a proper isometry in a (unital) C*-algebra cannot be a limit of invertible elements. See [Rør88] for more detailed results along this line.

V.3.1.11 COROLLARY. ([Rie83, 3.1,6.5], [Rob80]) Let A be a properly infinite C*-algebra. Then $sr(A) = \infty$.

This follows from V.3.1.10(ii)–(iii), since if A is properly infinite, $n \cdot 1$ is equivalent to a proper subprojection of 1 for every n.

V.3.1.12 Although a C*-algebra of stable rank 1 must be stably finite, an infinite C*-algebra need not have infinite stable rank: if T is the Toeplitz algebra (II.8.3.2(v)), then it is not difficult to show that $sr(T) = 2$. This also follows from V.3.1.5 and V.3.1.21.

Expanding and Contracting Matrices

The next two propositions can be used to establish the behavior of stable rank for matrix algebras and full corners. Note that the elements of pAp act on pAq by left multiplication, and left multiplication by an invertible element sends $Lg_{(p,q)}(A)$ onto itself.

Recall that an element in a unital ring of the form $1 + x$, x nilpotent, is called *unipotent*. A unipotent element is invertible.

V.3.1.13 PROPOSITION. (cf. [Rie83, 3.4]) Let A be a C*-algebra, p, q, r projections in A with $p \perp r$, $q \perp r$. If $Lg_{(p+r,q+r)}(A)$ is dense in $(p+r)A(q+r)$, then $Lg_{(p,q)}(A)$ is dense in pAq.

PROOF: Let $x \in pAq$. Let $0 < \epsilon < 1$, and approximate $x + r$ within ϵ by an element $y \in Lg_{(p+r,q+r)}(A)$. Then $\|ryq\| < \epsilon$, $\|pyr\| < \epsilon$, and $\|r - ryr\| < \epsilon$. Thus there is an $a \in rAr$ with $\|a\| < (1-\epsilon)^{-1}$ and $a(ryr) = r$. We then have

$$(p + r - pyra)y(q + r - aryq) = pyq - (pyr)a(ryq) + ryr.$$

Also, $p + r - pyra$ and $q + r - aryq$ are unipotent and hence invertible in $(p+r)A(p+r)$ and $(q+r)A(q+r)$ respectively, and y is left invertible in $(p+r)A(q+r)$, so $pyq - (pyr)a(ryq) + ryr$ is left invertible in $(p+r)A(q+r)$, and hence $pyq - (pyr)a(ryq)$ is left invertible in pAq. But

$$\|x - [pyq - (pyr)a(ryq)]\| \leq \|x - pyq\| + \|pyr\|\|a\|\|ryq\|$$
$$< \epsilon + \epsilon^2(1-\epsilon)^{-1} = \epsilon(1-\epsilon)^{-1}.$$

V.3.1.14 PROPOSITION. Let A be a C*-algebra, p, q, r projections in A with $p \perp r$, $q \perp r$, $r \precsim n \cdot q$ for some n. If $Lg_{(p,q)}(A)$ is dense in pAq, then $Lg_{(p+r,q+r)}(A)$ is dense in $(p+r)A(q+r)$.

PROOF: We may assume $q \leq p$ by V.3.1.10(i). It suffices to prove the result for $r \sim q$, for then the case where $r \sim (2^n - 1) \cdot q$ follows by induction, and V.3.1.13 then gives the case $r \precsim n \cdot q$. Since $s = p - q + r \sim p$, $Lg_{(s,q)}(A)$ and $Lg_{(s,r)}(A)$ are dense in sAq and sAr respectively. If $x \in (p+r)A(q+r)$ and $\epsilon > 0$, then x can be approximated within $\epsilon/2$ by $y \in (p+r)A(q+r)$ such that $a = syq \in Lg_{(s,q)}(A)$. We will show there is an invertible $z \in (p+r)A(p+r)$ such that $zyq = q$. This will suffice to prove the statement, since by hypothesis there will be $w \in Lg_{(s,r)}(A)$ approximating $szyr$ within $\epsilon/2\|z^{-1}\|$, and then $q + w + qzyr$ will be an element of $Lg_{(p+r,q+r)}(A)$ approximating

$$q + szyr + qzyr = zyq + (p+r)zyr = zy$$

within $\epsilon/2\|z^{-1}\|$, so

$$z^{-1}(q + w + qzyr) \in Lg_{(p+r,q+r)}(A)$$

approximates y within $\epsilon/2$ and therefore approximates x within ϵ.

To find z, first note that if $b \in qAs$ with $ba = q$, then $u = p+r+(q-qyq)b$ is unipotent and hence invertible in $(p+r)A(p+r)$; if $v = p+r-syq$, then v is unipotent in $(p+r)A(p+r)$ and $vuyq = q$. Set $z = vu$.

The proofs of V.3.1.13 and V.3.1.14 are easier to follow and understand if elements are written symbolically as matrices as in V.3.1.7.

Note that V.3.1.14 is not true in general if r is "too large" compared to q: let $A = \mathcal{L}(\mathcal{H})$, $p = q$ a finite-rank projection, and $r = 1 - p$.

V.3.1.15 COROLLARY. Let A be a unital C*-algebra and $m \geq 0$. If the left invertible $(m+1) \times 1$ matrices are dense in the $(m+1) \times 1$ matrices over A, then the left invertible $(m+k) \times k$ matrices are dense in the $(m+k) \times k$ matrices over A for all k. Conversely, if the left invertible $(m+k) \times k$ matrices are dense for some k, then the left invertible $(m+1) \times 1$ matrices over A are dense in A^{m+1}.

PROOF: Apply the previous two propositions to $M_{m+k}(A)$ with p, q, and r diagonal projections of rank $m+1$, 1, and $k-1$ respectively.

Matrix Algebras and Stable Algebras

V.3.1.16 COROLLARY. [Rie83, 6.1] Let A be a C*-algebra. Then, for any n, we have

$$sr(M_n(A)) = \left\lceil \frac{sr(A) - 1}{n} \right\rceil + 1$$

where $\lceil \cdot \rceil$ denotes "least integer $\geq$." In particular, $sr(M_n(A)) = 1$ [resp. ∞] if and only if $sr(A) = 1$ [resp. ∞].

Indeed, the $r \times 1$ matrices over $M_n(A)$ can be identified with the $nr \times n$ matrices over A. Thus, if $sr(A) = m + 1$, $Lg_r(M_n(A))$ is dense in $(M_n(A))^r$ if and only if $m + n \leq nr$. (In the nonunital case, one must also show that $sr(M_n(\tilde{A})) = sr(\widetilde{M_n(A)})$, which follows from the sharper form of V.3.1.21.)

Thus, if $1 < sr(A) < \infty$, then $sr(M_n(A)) = 2$ for sufficiently large n. A related result holds for stable algebras:

V.3.1.17 PROPOSITION. [Rie83, 3.6–6.4] Let A be a C*-algebra. If $sr(A) = 1$, then $sr(A \otimes \mathbb{K}) = 1$; if $1 < sr(A) \leq \infty$, then $sr(A \otimes \mathbb{K}) = 2$.

Stable Rank of Full Corners

Another corollary of V.3.1.14 is:

V.3.1.18 COROLLARY. Let A be a unital C*-algebra and p a full projection in A. Then $sr(pAp) \geq sr(A)$. If $sr(A) < \infty$, then $sr(pAp) < \infty$; and $sr(pAp) = 1$ if and only if $sr(A) = 1$.

PROOF: We have that $1 \precsim n \cdot p$ for some n (V.2.4.10), and hence $1 - p \precsim n \cdot p$. If $m = sr(pAp) < \infty$, then $Lg_{(m \cdot p, p)}(A)$ is dense in $(pAp)^m$, and hence by V.3.1.14 $Lg_{(m \cdot p + 1 - p, p + 1 - p)}(A)$ is dense in the corresponding column space. The same is then true for $Lg_{(m \cdot p + m \cdot (1-p), 1)}(A) \cong Lg_m(A)$ in A^m, i.e. $sr(A) \leq m$. On the other hand, 1 is equivalent to a full projection in $M_n(pAp)$, so

$$sr(A) \geq sr(M_n(pAp)) = \left\lceil \frac{sr(pAp) - 1}{n} \right\rceil + 1.$$

In particular, if $sr(pAp) = \infty$, then $sr(A) = \infty$. Also, if $sr(pAp) = 1$, then $sr(A) \leq sr(pAp) = 1$, and conversely if $sr(A) = 1$, then $sr(M_n(pAp)) \leq sr(A) = 1$, and hence $sr(pAp) = 1$ by V.3.1.16.

Simple counterexamples show that the hypothesis that p be full is necessary: for example, $A = \mathcal{B}(\mathcal{H})$ for infinite-dimensional $\mathcal{H}$, p a finite-rank projection; or $A = \mathbb{C} \oplus O_2$, $p = (1, 0)$.

What about the nonunital case? It is very plausible that if p is a full projection in A, then $sr(pAp) \geq sr(A)$ even if A is nonunital; however, this does not seem to follow from V.3.1.18 except in a special case:

V.3.1.19 COROLLARY. Let A be a C*-algebra with an approximate unit (q_i) of projections, and let p be a full projection in A. Then $sr(pAp) \geq sr(A)$.

PROOF: We have $sr(A) \leq \liminf sr(q_i A q_i)$, and there is an i_0 such that $p \succsim q_i$ for all $i \geq i_0$, so $sr(pAp) \geq sr(q_i A q_i)$ for $i \geq i_0$.

V.3.1.20 It is less clear whether a general full corner in a nonunital C*-algebra A (II.7.6.5(iii)) also satisfies the inequality. As pointed out by N. Elhage Hassan [EH93], the inequality fails for general full hereditary C*-subalgebras: if A is a purely infinite simple unital C*-algebra, e.g. O_2, and B is a nonunital hereditary C*-subalgebra, then B is stable, so $2 = sr(B) < sr(A) = \infty$.

Other Properties of Stable Rank

V.3.1.21 THEOREM. [Rie83, 4.3–4.4–4.12] Let $0 \to J \to A \to B \to 0$ be an exact sequence of C*-algebras. Then

$$\max\{sr(J), sr(B)\} \leq sr(A) \leq \max\{sr(J), sr(B) + 1\}.$$

A more precise statement can be made using the notion of connected stable rank [Rie83, 4.11].

The next proposition is obvious. The inequality cannot be replaced by an equality: there are obvious commutative counterexamples, and also some quite nonobvious noncommutative ones (cf. [BBEK92]).

V.3.1.22 PROPOSITION. Let $A = \varinjlim(A_i, \phi_{ij})$. Then
$$sr(A) \leq \liminf_i sr(A_i).$$

V.3.1.23 THEOREM. [Rie83, 7.1] Let A be a C*-algebra and $\alpha \in \mathrm{Aut}(A)$. Then $sr(A \rtimes_\alpha \mathbb{Z}) \leq sr(A) + 1$.

Nonstable K-Theory

One of the important applications of stable rank is to cancellation and related questions. Cancellation questions are part of what is known as *nonstable K-theory*, which is concerned with relating the K-theory data of A (which is "stable" data) to the actual structure of A.

V.3.1.24 PROPOSITION. Let A be a C*-algebra. If $sr(A) = 1$, then A has cancellation.

OUTLINE OF PROOF: We may assume A is unital. Let p and q be projections in A, with $p \sim q$, and let u be a partial isometry with $u^*u = p$, $uu^* = q$. Approximate u closely by an invertible element $x \in A$. Then $x^*x \approx p$, so $\sigma(x^*x)$ consists of a small piece σ_0 near 0 and a small piece σ_1 near 1. Let $f(t) = t^{-1/2}$ on σ_0 and $f(t) = 0$ on σ_1, $g(t) = 0$ on σ_0 and $g(t) = t^{-1/2}$ on σ_1, and let $v = xf(x^*x)$, $w = xg(x^*x)$. Then v and w are partial isometries with complementary support and range projections, $v + w$ is the unitary in the polar decomposition of x, and $w \approx x \approx u$, so $p' = w^*w \approx p$, $q' = ww^* \approx q$. By II.3.3.4, $p' \sim p$, $q' \sim q$, $(1-p') \sim (1-p)$, $(1-q') \sim (1-q)$. Since $v^*v = 1-p'$, $vv^* = 1-q'$, we have $(1-p) \sim (1-p') \sim (1-q') \sim (1-q)$.

The exact general relationship between stable rank and cancellation is not known. There are some significant generalizations of V.3.1.24, however; see, for example, [Rie83] and [Bla83a]. Cancellation results are typically of the form: if $p \oplus r \sim q \oplus r$, and p and q are "large enough" or r is "small enough," then $p \sim q$. See [Hus66, Chapter 8] for some results of this sort for vector bundles. We cite one example related to stable rank:

V.3.1.25 THEOREM. ([Rie83, 10.13], [War80]) Let p, q, r be projections in a unital C*-algebra A, and $n \geq sr(rAr)$. If $p \oplus n \cdot r \oplus r \sim q \oplus r$, then $p \oplus n \cdot r \sim q$.

Another nonstable K-theory question is: if A is a unital C*-algebra, for which n (if any) is the natural map from $\mathcal{U}_n(A)/\mathcal{U}_n(A)_o$ to $\mathcal{U}_{n+1}(A)/\mathcal{U}_{n+1}(A)_o$ (or to $K_1(A)$) injective? surjective? The best result for C*-algebras is the following [Rie87, 2.10]; the best results valid for general rings are weaker ([Bas74, 2.4], [Vas69]; cf. [Rie83, §10]).

V.3.1.26 THEOREM. Let A be a unital C*-algebra and $n \geq sr(A)$. Then the map from $\mathcal{U}_n(A)/\mathcal{U}_n(A)_o$ to $\mathcal{U}_{n+1}(A)/\mathcal{U}_{n+1}(A)_o$ is an isomorphism. Hence the map from $\mathcal{U}_n(A)/\mathcal{U}_n(A)_o$ to $K_1(A)$ is an isomorphism.

The condition is the best possible in general, but the conclusion also holds under other circumstances, for example:

V.3.1.27 PROPOSITION. Let A be a purely infinite simple unital C*-algebra. Then the map from $\mathcal{U}_1(A)/\mathcal{U}_1(A)_o$ (or $\mathcal{U}_n(A)/\mathcal{U}_n(A)_o$ for any n) to $K_1(A)$ is an isomorphism.

The proof basically consists of noting that A has a hereditary C*-subalgebra B isomorphic to $A \otimes \mathbb{K}$, with each unitary in A homotopic (via functional calculus) to a unitary in $A \otimes e_{11} \subseteq B$.

V.3.2 Real Rank

The theory of real rank for C*-algebras, developed by Brown and Pedersen [BP91], formally resembles the theory of stable rank, but there are important differences under the surface.

V.3.2.1 DEFINITION. Let A be a unital C*-algebra. Let $slg_n(A)$ be the set of $(x_1, \ldots, x_n)$ in $(A_{sa})^n$ such that $\sum_{i=1}^n x_i^2$ is invertible.

The *real rank* of A, denoted $rr(A)$, is the smallest n such that $slg_{n+1}(A)$ is dense in $(A_{sa})^{n+1}$. If there is no such n, set $rr(A) = \infty$.

If A is nonunital, then $rr(A)$ is defined to be $rr(\tilde{A})$.

V.3.2.2 It follows easily from V.3.1.1 that $rr(C(X)) = dim(X)$ for any compact Hausdorff space X.

V.3.2.3 Note that there is a difference in numbering convention between stable rank and real rank: real rank n corresponds (very roughly) to stable rank $n+1$ (or, perhaps more accurately, stable rank $[n/2]+1$; cf. V.3.2.4). In particular, "real rank zero" is the analog of "stable rank one." This difference in convention is somewhat unfortunate, but well established. Real rank is more closely related to dimension than stable rank, and the convention in real rank was chosen so that $rr(C(X)) = dim(X)$; on the other hand, the convention in stable rank was chosen to be consistent with the numbering in Bass stable rank, which is natural from an algebraic standpoint.

There is one weak relation between stable rank and real rank:

V.3.2.4 PROPOSITION. If A is a C*-algebra, then $rr(A) \leq 2sr(A) - 1$.

For the proof, note that the natural map from A^n to $(A_{sa})^{2n}$ by taking real and imaginary parts sends $Lg_n(A)$ into $slg_{2n}(A)$ by II.3.1.9(ii).

This inequality is the best possible in general, but equality often fails; if $A = C(X)$, then equality holds if and only if $dim(X)$ is odd. In fact, there is no general inequality in the opposite direction: there are C*-algebras with real rank zero and stable rank infinity (V.3.1.11, V.3.2.12).

V.3.2.5 The theory of real rank is not as well developed as the theory of stable rank, and analogs of V.3.1.16 and V.3.1.23 are not known in general. In fact, not much is known about the theory of real rank for C*-algebras of real rank greater than zero. One of the few results is a partial analog of V.3.1.16:

V.3.2.6 THEOREM. [BE91] If X is a compact Hausdorff space, then

$$rr(M_n(C(X))) = \left\lceil \frac{dim(X)}{2n-1} \right\rceil .$$

Real Rank Zero

V.3.2.7 On the other hand, the real rank zero property is one of the most significant properties that a C*-algebra can have. Many experts are of the opinion that C*-algebras of real rank zero are the most appropriate noncommutative analogs of zero-dimensional topological spaces.

A unital C*-algebra A has real rank zero if the invertible self-adjoint elements of A are dense in A_{sa}. This turns out to be a strong existence of projections property, equivalent to several other important properties:

V.3.2.8 DEFINITION. Let A be a C*-algebra.

(i) A has the *(HP) property* if every hereditary C*-subalgebra of A has an idempotent approximate unit (approximate unit of projections).
(ii) A has the *(FS) property* if the self-adjoint elements of A of finite spectrum are dense in A_{sa}.

It follows from II.5.3.12 that if A is a C*-algebra with (HP), then any positively generated hereditary *-subalgebra of A has an idempotent approximate unit.

The idempotent approximate unit in a hereditary C*-subalgebra of a C*-algebra with (HP) is not assumed to be increasing. It can be shown that if A is separable with (HP), then every hereditary C*-subalgebra of A has an increasing sequential idempotent approximate unit.

V.3.2.9 THEOREM. Let A be a C*-algebra. Then the following are equivalent:

(i) A has (HP).
(ii) The well-supported (II.3.2.8) self-adjoint elements of A are dense in A_{sa}.
(iii) A has real rank zero.
(iv) A has (FS).

454   V K-Theory and Finiteness

OUTLINE OF PROOF: (i) $\implies$ (ii): if x is self-adjoint, $y = |x|^{1/2}$, and p is a projection in yAy which is an approximate unit for y and hence for x, then pxp is a well-supported self-adjoint element (II.4.2.6) closely approximating x.
(ii) $\implies$ (iii): if x is well-supported, then $x + \lambda 1$ is invertible for all sufficiently small nonzero λ.
(iii) $\implies$ (iv): suppose $x = x^*$ is given. We may assume $0 \leq x \leq 1$. Let $\{\lambda_1, \lambda_2, \ldots\}$ be the rationals in $[0, 1]$. Set $x_1 = x$. For each n, let y_n be a well-supported self-adjoint element closely approximating $x_n - \lambda_n 1$, and set $x_{n+1} = y_n + \lambda_n 1$. Then x_n approximates x and its spectrum has gaps around $\lambda_1, \ldots, \lambda_n$; the approximate to x with finite spectrum can then be made from x_n by functional calculus.
(iv) $\implies$ (i) is the trickiest part; see [Ped80].

V.3.2.10 It is obvious that (FS) is preserved in inductive limits, and it is not hard to show that real rank zero passes to matrix algebras. (HP) obviously passes to hereditary C*-subalgebras. Thus all the properties are preserved under stable isomorphism (and even under Morita equivalence).

There are other interesting existence of projections properties for C*-algebras; for example, the defining property of AW*-algebras, that the right annihilator of any subset is generated by a projection, is a strong existence of projections property. See [Bla94] for a complete discussion.

V.3.2.11 PROPOSITION. If A has (HP) and p and q are projections of A, then the well-supported elements of qAp are dense in qAp.
PROOF: Let $x \in qAp$. Let r be a projection in $\cup_{\epsilon > 0} f_\epsilon(|x|) A f_\epsilon(|x|)$ which is almost a unit for $|x| = (x^*x)^{1/2}$, and set $y = xr$. Then $y \in qAr \subseteq qAp$, and since $r \leq n(x^*x)$ for some sufficiently large n,

$$y^*y = rx^*xr \geq (1/n)r$$

so y^*y is invertible in rAr. Thus y is well-supported, and closely approximates x.

In particular, if $rr(A) = 0$, then the well-supported elements of A are dense in A. The converse is not true in general: for example, if A is unital and $sr(A) = 1$, the invertible elements (which are well-supported) are dense in A, but A need not have real rank 0.

A C*-algebra with real rank zero need not have stable rank 1. For example, any von Neumann algebra has real rank zero. Here is another interesting class of examples:

V.3.2.12 PROPOSITION. [Zha90] Every purely infinite simple C*-algebra has real rank 0.
PROOF: Let A be a purely infinite simple C*-algebra. We may assume A is unital (V.2.3.12, V.3.2.10). Let $x = x^* \in A$, and $0 < \epsilon \ll 1$. Set $y = x f_\epsilon(x^2)$;

then y approximates x and there is an $h \in A_+$, $h \neq 0$, $h \perp y$. Let s be a projection in $\overline{hAh}$ which is equivalent to 1_A, and set $p = 1_A - s$. Then $y = pyp$ and $p \precsim s$; let v be a partial isometry in A with $v^*v = p$, $vv^* = q \leq s$, and set $r = s - q$. Then $p + q + r = 1_A$, and

$$z = y + \epsilon(v + v^* + r)$$

is invertible, with

$$z^{-1} = -\epsilon^{-2} vyv^* + \epsilon^{-1}(v + v^* + r),$$

and z approximates x. [Using symbolic matrix notation (V.3.1.7) with respect to $\{p, q, r\}$, $z = \begin{bmatrix} y & \epsilon & 0 \\ \epsilon & 0 & 0 \\ 0 & 0 & \epsilon \end{bmatrix}$.]

V.3.2.13 COROLLARY. Let A be a simple C*-algebra. Then A is purely infinite if and only if A has real rank zero and every nonzero projection in A is infinite.

PROOF: One direction follows immediately from V.3.2.12 and V.2.2.24. Conversely, if A has real rank zero, then A has (HP) and every nonzero hereditary C*-subalgebra contains a nonzero projection. Then, if every nonzero projection in A is infinite, A is purely infinite by V.2.3.13.

V.3.2.14 PROPOSITION. A σ-unital purely infinite simple nonunital C*-algebra is stable.

PROOF: Suppose A is separable, simple, nonunital, and purely infinite. Then A has a strictly increasing approximate unit (p_n) of projections, and $p_{n+1} - p_n$ is infinite for all n. It is routine using V.2.3.2 to construct a strictly increasing sequence (q_n) of projections such that $p_n \leq q_n \leq p_{n+1}$ and $(q_{n+1} - q_n) \sim q_1$ for all n. The sequence (q_n) is then also an approximate unit for A, and the isomorphism of A with $q_1 A q_1 \otimes \mathbb{K}$ is then obvious.

In the presence of real rank 0, there is a "converse" to V.3.1.24 [BH82]:

V.3.2.15 PROPOSITION. Let A be a C*-algebra. If $rr(A) = 0$ and A has cancellation, then $sr(A) = 1$.

OUTLINE OF PROOF: We may assume A is unital. Let $x \in A$, and let y be a well-supported element closely approximating x. Let p and q be the source and range projections of y. Then $p \sim q$, so $(1 - p) \sim (1 - q)$ by cancellation. Let v be a partial isometry in A with $v^*v = 1 - p$, $vv^* = 1 - q$. Then $y + \epsilon v$ is invertible for any $\epsilon \neq 0$, and closely approximates x if ϵ is small.

There is no known example of a stably finite C*-algebra with real rank zero which does not have cancellation (stable rank 1).

V.3.2.16 If A has real rank zero and $a, b \in A_+$ with $b \ll a$ and $\|a\| = \|b\| = 1$, it would be very helpful to have an interpolating projection p with $b \leq p \leq a$ (hence $b \ll p \ll a$). This cannot be done in general (e.g. in $C(K)$, where K is the Cantor set); but if there is a little "room" between a and b it can be:

V.3.2.17 PROPOSITION. Let A be a C*-algebra with real rank zero, $a, b \in A_+$. If there is a $c \in A_+$ with $b \ll c \ll a$, then there is a projection $p \in A$ with $b \ll p \ll a$.

PROOF: We may assume A is unital since $\tilde{A}$ also has real rank zero and a $p \in \tilde{A}$ satisfying the conclusion is automatically in A. We have $b \ll f_{1/2}(c) \ll f_{1/4}(c) \ll a$. Changing notation, we have

$$b \ll c \ll d \ll a$$

for some $c, d \in A_+$. Replacing b by $g(b)$, etc., where $g(t) = \min(t, 1)$, we may assume a, b, c, d have norm 1; thus

$$1 - a \ll 1 - d \ll 1 - c \ll 1 - b.$$

Using (HP), let r be a projection in $\overline{(1-c)A(1-c)}$ which is almost a unit for $1 - c$, and let $q = 1 - r$. Since $b \perp 1 - c$, $qb = b$. Then d is almost a unit for q, so qdq is approximately equal to q and thus invertible in qAq. So $x = d^{1/2}q$ is well-supported. Let $x = u|x|$ be the polar decomposition; $u^*u = q$ and $p = uu^* \in \overline{dAd}$ and thus $p \ll a$. Since

$$d^{1/2}b = bd^{1/2} = b = qb = bq$$

we have $xb = bx = b$ and thus $pb = b$.

V.3.2.18 L. Brown and G. Pedersen [BP95] considered a variation of stable rank 1, extremal richness, which is particularly interesting in the case of real rank zero. A unital C*-algebra A is *extremally rich* if the set A_q of quasi-invertible elements (II.3.2.21) of A is dense in A. If $sr(A) = 1$, then A is extremally rich, but extremal richness is strictly weaker than stable rank 1 in general. For example, a purely infinite simple C*-algebra is extremally rich.

Ordered K_0-Group of a Real Rank Zero C*-Algebra

If A is a stably finite unital C*-algebra with real rank zero, then the map χ of V.2.4.27 is a bijection, hence a homeomorphism.

V.3.2.19 COROLLARY. [BH82, III] If A is a stably finite simple unital C*-algebra with real rank zero, such that $K_0(A)$ is weakly unperforated, then the state space of $K_0(A)$ is the simplex $QT(A)$, and $K_0(A)$ has the strict ordering induced from $\rho : K_0(A) \to \text{Aff}(QT(A))$, i.e. $[p] < [q]$ if and only if $\tau(p) < \tau(q)$

for every quasitrace τ. If A has cancellation (i.e. stable rank 1), and p and q are projections in $M_\infty(A)$ with $\tau(p) < \tau(q)$ for all τ, then $p \precsim q$.

It is not true that $\tau(p) \leq \tau(q)$ for all τ implies $p \precsim q$, even in a simple unital AF algebra.

V.3.2.20 THEOREM. [BH82, III] Let A be simple, unital, stably finite, with real rank zero and cancellation, and with $K_0(A)$ weakly unperforated. Then the range of $\rho : K_0(A) \to \mathrm{Aff}(QT(A))$ is uniformly dense.

V.4 Quasidiagonality

It has become increasingly recognized in recent years that quasidiagonality is a key notion in the structure and classification of C*-algebras, even though it is still somewhat mysterious and far from completely understood. Quasidiagonality is a sort of approximate finite-dimensionality, and thus has close connections with finiteness and nuclearity, although the exact relationships between these concepts are not yet known.

The theory works best for separable C*-algebras and separable Hilbert spaces, so we will concentrate on this case whenever convenient.

For a more complete discussion of quasidiagonality, see [Bro04].

V.4.1 Quasidiagonal Sets of Operators

V.4.1.1 DEFINITION. Let $\mathcal{H}$ be a Hilbert space.
A *block system* in $\mathcal{L}(\mathcal{H})$ is a set $\{Q_i\}$ of mutually orthogonal finite-rank projections on $\mathcal{H}$ with $\sum_i Q_i = I$.
An operator $T \in \mathcal{L}(\mathcal{H})$ is *block diagonal* with respect to the block system $\{Q_i\}$ if $T = \sum_i Q_i T Q_i$, i.e. $Q_i T Q_j = 0$ for all $i \neq j$. If each Q_i is one-dimensional, T is *diagonal*. An operator T is *block-diagonalizable* [resp. *diagonalizable*] if it is block diagonal [resp. diagonal] with respect to some block system.
If $\mathcal{H}$ is separable, an operator $T \in \mathcal{L}(\mathcal{H})$ is *quasidiagonal* if it is block-diagonalizable up to compacts, i.e. it is a compact perturbation of a block-diagonalizable operator.

V.4.1.2 EXAMPLES.

(i) Any compact operator is quasidiagonal. In particular, if $\mathcal{H}$ is finite-dimensional, every operator on $\mathcal{H}$ is quasidiagonal (in fact, block-diagonalizable).

(ii) An operator $T \in \mathcal{L}(\mathcal{H})$ is diagonalizable if and only if there is an orthonormal basis for $\mathcal{H}$ consisting of eigenvectors for T. Any diagonalizable operator is normal. Conversely, by the Weyl-von Neumann-Berg Theorem [Ber71], any normal operator on a separable Hilbert space is a compact perturbation of a diagonalizable operator, hence is quasidiagonal.

(iii) A block-diagonalizable semi-Fredholm operator clearly is Fredholm of index 0, and hence a quasidiagonal semi-Fredholm operator on a separable Hilbert space is Fredholm of index 0. In particular, a nonunitary isometry is not quasidiagonal. (See V.4.1.11 for the nonseparable case.)

Separable Quasidiagonal Sets and C*-Algebras of Operators

V.4.1.3 More generally, if $\mathcal{H}$ is separable and $\mathcal{S} \subseteq \mathcal{L}(\mathcal{H})$ is norm-separable, then $\mathcal{S}$ is a *quasidiagonal set of operators* if the operators in $\mathcal{S}$ are simultaneously block-diagonalizable up to compacts, i.e. if there is a block system $\{Q_i\}$ such that each $S \in \mathcal{S}$ is a compact perturbation of an operator which is block diagonal with respect to $\{Q_i\}$. (See V.4.1.6 for the general definition of a quasidiagonal set of operators.)

If $\mathcal{S}$ is a quasidiagonal set of operators, then so is $C^*(\mathcal{S}) + \mathcal{K}(\mathcal{H}) + \mathbb{C}I$. Thus quasidiagonality is fundamentally a property of concrete C*-algebras of operators.

If $\mathcal{H}$ is finite-dimensional, then $\mathcal{L}(\mathcal{H})$ (or any subset) is a quasidiagonal set of operators.

V.4.1.4 Any subset of a quasidiagonal set of operators is quasidiagonal. In particular, if $\mathcal{S}$ is a quasidiagonal set of operators, then any $S \in \mathcal{S}$ is quasidiagonal. The converse is false: if S is the unilateral shift and S_1, S_2 are its real and imaginary parts, then S_1 and S_2 are quasidiagonal (V.4.1.2(ii)), but $\{S_1, S_2\}$ is not quasidiagonal (V.4.1.2(iii)).

The proof of the next proposition is a straightforward exercise (II.3.3.5 is needed to prove (iv) $\Longrightarrow$ (iii)). Recall that the commutator $[S, T]$ is $ST - TS$ for $S, T \in \mathcal{L}(\mathcal{H})$.

V.4.1.5 PROPOSITION. Let $\mathcal{H}$ be a separable Hilbert space and $\mathcal{S} \subseteq \mathcal{L}(\mathcal{H})$ be norm-separable. Then the following are equivalent:

(i) $\mathcal{S}$ is a quasidiagonal set of operators.
(ii) There is an increasing sequence (P_n) of finite-rank projections on $\mathcal{H}$, with $\bigvee_n P_n = I$ (i.e. $P_n \to I$ strongly), such that $\lim_n \|[P_n, S]\| = 0$ for all $S \in \mathcal{S}$.
(iii) For every finite-rank projection $Q \in \mathcal{L}(\mathcal{H})$, $S_i, \ldots, S_n \in \mathcal{S}$, and $\epsilon > 0$, there is a finite-rank projection $P \in \mathcal{L}(\mathcal{H})$ with $Q < P$ and $\|[P, S_i]\| < \epsilon$ for $1 \leq i \leq n$.
(iv) For every finite-rank projection $Q \in \mathcal{L}(\mathcal{H})$, $S_i, \ldots, S_n \in \mathcal{S}$, and $\epsilon > 0$, there is a finite-rank projection $P \in \mathcal{L}(\mathcal{H})$ with $\|P\xi - \xi\| \leq \epsilon\|\xi\|$ for all $\xi \in Q\mathcal{H}$ and $\|[P, S_i]\| < \epsilon$ for $1 \leq i \leq n$.

General Quasidiagonal Sets of Operators

V.4.1.6 If $\mathcal{H}$ or $\mathcal{S}$ is not separable, then V.4.1.5(ii)–(iv) are still equivalent (with "sequence" replaced by "net" in (ii)), but these conditions do not imply that the operators in $\mathcal{S}$ can be simultaneously block-diagonalized up to compacts. For example, if $\mathcal{S} = \{S\}$, where S is an uncountable amplification of a compact operator which is not block-diagonalizable (e.g. a weighted shift), then $\mathcal{S}$ satisfies (ii)–(iv) but cannot be block-diagonalized up to compacts. See [Bro04] for an example of a nonseparable $\mathcal{S}$ on a separable $\mathcal{H}$, satisfying (ii)–(iv), which cannot be simultaneously block-diagonalized up to compacts.

It is the consensus of experts that conditions (ii)–(iv) should be taken as the definition of a general quasidiagonal operator or quasidiagonal set of operators on a general Hilbert space, and we will do so. Thus, a set $\mathcal{S}$ of operators is quasidiagonal if and only if each finite subset of $\mathcal{S}$ is quasidiagonal.

V.4.1.7 COROLLARY. Let A be a concrete C*-algebra of operators on $\mathcal{H}$, containing $\mathcal{K}(\mathcal{H})$. Then A is a quasidiagonal C*-algebra of operators if and only if there is an approximate unit for $\mathcal{K}(\mathcal{H})$, consisting of projections, which is quasicentral for A.

From V.4.1.5(iii) we obtain:

V.4.1.8 COROLLARY. Let (A_i) be a nested family of quasidiagonal C*-algebras of operators on $\mathcal{H}$. Then $[\cup A_i]^-$ is a quasidiagonal C*-algebra of operators.

V.4.1.9 PROPOSITION. Let A be a quasidiagonal C*-algebra of operators on $\mathcal{H}$. Then, for any n, $M_n(A)$ is a quasidiagonal C*-algebra of operators on $\mathcal{H}^n$, and $A \otimes \mathbb{K}$ is a quasidiagonal C*-algebra of operators on $\mathcal{H}^\infty$. More generally, if A_i is a quasidiagonal C*-algebra of operators on $\mathcal{H}_i$ ($i = 1, 2$), then $A_1 \otimes_{\min} A_2$ is a quasidiagonal C*-algebra of operators on $\mathcal{H}_1 \otimes \mathcal{H}_2$.

The proof is very simple and straightforward.

Combining V.4.1.9 with V.4.1.2(iii), we obtain:

V.4.1.10 COROLLARY. Let A be a quasidiagonal C*-algebra of operators. Then $\tilde{A}$ is stably finite.

Actually, V.4.1.2(iii) only gives this result on a separable Hilbert space. For the general case, use:

V.4.1.11 PROPOSITION. A quasidiagonal isometry on any Hilbert space is unitary.

OUTLINE OF PROOF: Let T be a quasidiagonal isometry on $\mathcal{H}$, ξ a unit vector in $\mathcal{H}$, and P a finite-rank projection on $\mathcal{H}$ almost commuting with T, such that $P\xi = \xi$. Then PTP is almost an isometry on the finite-dimensional space $P\mathcal{H}$, hence invertible; so there is an $\eta \in P\mathcal{H}$ with

$$\xi = PTP\eta \approx TPP\eta = T\eta.$$

Thus ξ is in the closure of the range of T.

V.4.2 Quasidiagonal C*-Algebras

V.4.2.1 DEFINITION. Let A be an (abstract) C*-algebra.
A representation π of A is *quasidiagonal* if $\pi(A)$ is a quasidiagonal C*-algebra of operators.
A is *quasidiagonal* if it has a faithful quasidiagonal representation.
A is *inner quasidiagonal* if it has a separating family of quasidiagonal irreducible representations.
A is *strongly quasidiagonal* if every representation of A is quasidiagonal.

Any strongly quasidiagonal C*-algebra is inner quasidiagonal (and also quasidiagonal).

V.4.2.2 It is easily seen that any sum of quasidiagonal representations is quasidiagonal. In particular, any amplification of a quasidiagonal representation is quasidiagonal. Thus an inner quasidiagonal C*-algebra is quasidiagonal. It also follows that every quasidiagonal C*-algebra A has a faithful quasidiagonal representation π on a Hilbert space $\mathcal{H}$ such that $\pi(A) \cap \mathcal{K}(\mathcal{H}) = \{0\}$.

It can be easily shown using V.4.1.5(iii) that a separable quasidiagonal C*-algebra has a faithful quasidiagonal representation on a separable Hilbert space; in fact, any faithful quasidiagonal representation has a faithful quasidiagonal subrepresentation on a separable subspace.

The next theorem is an immediate corollary of Voiculescu's Weyl-von Neumann Theorem (II.8.4.29).

V.4.2.3 THEOREM. Let A be a separable quasidiagonal C*-algebra. Then any faithful representation π of A on a separable Hilbert space $\mathcal{H}$ such that

$$\pi(A) \cap \mathcal{K}(\mathcal{H}) = \{0\}$$

is quasidiagonal.

V.4.2.4 EXAMPLES.

(i) Any quasidiagonal concrete C*-algebra of operators is a quasidiagonal (abstract) C*-algebra.
(ii) Every residually finite-dimensional C*-algebra (V.2.1.10) is inner quasidiagonal. In particular, every commutative C*-algebra is inner quasidiagonal, and $C^*(F)$ is inner quasidiagonal for any free group F (V.2.1.11). A residually finite-dimensional C*-algebra is not necessarily strongly quasidiagonal: every unital C*-algebra is a quotient of $C^*(F)$ for a free group F on suitably many generators.
(iii) Let S be the unilateral shift, and $A = C^*(S \oplus S^*)$. Since $S \oplus S^*$ is a compact perturbation of a unitary, A is quasidiagonal. But A has an ideal isomorphic to $\mathbb{K} \oplus \mathbb{K}$, and has precisely two irreducible representations which are nonzero on this ideal (sending $S \oplus S^*$ to S and S^* respectively); neither of these is quasidiagonal by V.4.1.2(iii), and so A is not inner quasidiagonal.
(iv) Let A be as in (iii), and $\pi : A \to \mathcal{L}(\mathcal{H})$ an infinite amplification of the identity representation of A. Then π is quasidiagonal. Set $B = \pi(A) + \mathcal{K}(\mathcal{H})$; then B is a quasidiagonal C*-algebra of operators on $\mathcal{H}$, and the identity representation of B is irreducible, so B is inner quasidiagonal. But A is a quotient of B, and A has non-quasidiagonal representations, so B is not strongly quasidiagonal.

Examples (iii) and (iv) are Type I.

The next two results are immediate corollaries of V.4.1.9 and V.4.1.10.

V.4.2.5 PROPOSITION. If A and B are quasidiagonal [*resp.* inner quasidiagonal] C*-algebras, then $A \otimes_{\min} B$ is quasidiagonal [*resp.* inner quasidiagonal].

It is unclear whether the same result is true for strong quasidiagonality unless A and B are separable and one is exact, in which case it can be routinely proved using IV.3.4.25 and II.8.4.29. The situation with $A \otimes_{\max} B$ is also unclear in general.

V.4.2.6 PROPOSITION. If A is a quasidiagonal C*-algebra, then $\tilde{A}$ is stably finite.

V.4.2.7 PROPOSITION. A unital quasidiagonal C*-algebra has a tracial state.

PROOF: Let π be a unital quasidiagonal representation (not necessarily faithful) of a unital C*-algebra A on $\mathcal{H}$. Let (P_i) be a net of finite-rank projections in $\mathcal{L}(\mathcal{H})$ for $\pi(A)$ as in V.4.1.5(ii). For each i let τ_i be the normalized trace on the matrix algebra $\mathcal{L}(P_i\mathcal{H})$, and for $x \in A$ define $\phi_i(x) = \tau_i(P_i\pi(x)P_i)$. Then ϕ_i is a state on A, and for $x, y \in A$,

$$\phi_i(xy) = \tau_i(P_i\pi(x)\pi(y)P_i) \approx \tau_i(P_i\pi(x)P_i\pi(y)P_i)$$

$$= \tau_i(P_i\pi(y)P_i\pi(x)P_i) \approx \tau_i(P_i\pi(y)\pi(x)P_i) = \phi_i(yx)$$

for i large, so any limit of (ϕ_i) is a tracial state of A.

See V.4.3.8 for a generalization.

There is in important alternate characterization of quasidiagonal C*-algebras, due to D. Voiculescu [Voi91, Theorem 1]:

V.4.2.8 THEOREM. A C*-algebra A is quasidiagonal if and only if, for every $x_1, \ldots, x_n \in A$ and $\epsilon > 0$, there is a representation π of A on a Hilbert space $\mathcal{H}$ and a finite-rank projection $P \in \mathcal{L}(\mathcal{H})$ with $\|P\pi(x_j)P\| > \|x_j\| - \epsilon$ and $\|[P, \pi(x_j)]\| < \epsilon$ for $1 \leq j \leq n$.

V.4.2.9 COROLLARY. A C*-algebra A is quasidiagonal if and only if every finitely generated C*-subalgebra is quasidiagonal.

Homotopy Invariance of Quasidiagonality

An important consequence of V.4.2.8 is the fact that quasidiagonality is preserved under homotopy equivalence [Voi91, Theorem 5]:

V.4.2.10 THEOREM. Let A and B be C*-algebras. If A is quasidiagonal and A homotopically dominates B (II.5.5.8), then B is quasidiagonal.

V.4.2.11 COROLLARY. Any subcontractible C*-algebra (II.5.5.8) is quasidiagonal. In particular, if A is any C*-algebra, then CA and SA are quasidiagonal.

Indeed, a contractible C*-algebra is homotopy equivalent to the 0 C*-algebra.

Thus even such an "infinite" C*-algebra as SO_2 is quasidiagonal. So a nonunital quasidiagonal C*-algebra need not have a nonzero trace (cf. V.4.2.7).

Quasidiagonality vs. Nuclearity

V.4.2.12 Quasidiagonality is more of a finiteness condition than an amenability or nuclearity property (however, see V.4.2.13 and V.4.2.14). There are nuclear C*-algebras which are not finite and hence not quasidiagonal, for example the Toeplitz algebra or the Cuntz algebras. On the other hand, there are many quasidiagonal C*-algebras which are not nuclear, or even exact: for example, $C^*(\mathbb{F}_2)$ is residually finite dimensional and hence quasidiagonal (or just apply V.4.2.11). A quotient of a quasidiagonal C*-algebra is not quasidiagonal in general: in fact, every C*-algebra is a quotient of a quasidiagonal

C*-algebra (this can be seen by applying V.4.2.11, or by noting that every unital C*-algebra is a quotient of the full C*-algebra of a free group). Even strong quasidiagonality does not imply nuclearity: there is a separable simple unital C*-algebra which is quasidiagonal but not nuclear, in fact not subnuclear (exact) [Dad00].

However, there is no known example of a stably finite nuclear C*-algebra which is not quasidiagonal. We will discuss this matter in more detail in V.4.3.13.

Quasidiagonal Group C*-Algebras

Quasidiagonality does seem to have some flavor of the Følner condition, and using this idea J. Rosenberg proved the following result [Had87, Appendix]:

V.4.2.13 THEOREM. Let G be a discrete group. If $C_r^*(G)$ is quasidiagonal, then G is amenable.

(The statement of this result in [Had87, A1] is more restrictive, but Rosenberg's argument works also with an infinite amplification of the left regular representation to give the result stated here.)

PROOF: [Was94, Prop. 4.2] Let λ be the left regular representation of G on $l^2(G)$, π the representation of $l^\infty(G) = C_b(G)$ on $l^2(G)$ by multiplication operators, and $(P_i)_{i \in \Omega}$ a net of finite-rank projections in $\mathcal{L}(l^2(G))$ with $\lim_i P_i = 1$ strongly and $\lim_i [P_i, \lambda(g)] = 0$ for all $g \in G$. Let ω be a Banach Limit (I.3.2.3) on Ω, and τ_i the normalized trace on the matrix algebra $\mathcal{L}(P_i l^2(G))$. For $\phi \in l^\infty(G)$, set $m(\phi) = \lim_\omega \tau_i(P_i \pi(\phi) P_i)$. Since, for fixed $g \in G$, $\phi \in l^\infty(G)$,

$$\tau_i(P_i \pi(g \cdot \phi) P_i) = \tau_i(P_i \lambda(g) \pi(\phi) \lambda(g)^* P_i) \approx \tau_i(P_i \lambda(g) P_i \pi(\phi) \lambda(g)^* P_i)$$

$$= \tau_i(P_i \pi(\phi) \lambda(g)^* P_i \lambda(g) P_i) \approx \tau_i(P_i \pi(\phi) \lambda(g) \lambda(g)^* P_i) = \tau_i(P_i \pi(\phi) P_i)$$

for i large, $m(g \cdot \phi) = m(\phi)$. Thus m is a left invariant mean on G.

This argument is almost identical to the one in V.4.2.7.

Combining this with IV.3.5.6, it follows that if G is discrete and $C_r^*(G)$ is quasidiagonal, then it is nuclear!

The case of $\mathbb{F}_2$ shows that $C_r^*(G)$ cannot be replaced by $C^*(G)$ in V.4.2.13. The converse of V.4.2.13 is open.

Space-Free Characterization of Quasidiagonality

The next theorem, due to Voiculescu, gives an alternate space-free characterization of quasidiagonality which is reminiscent of the definition of nuclearity, and variants of which characterize NF and strong NF algebras (V.4.3.13 and V.4.3.19).

V.4.2.14 THEOREM. [Voi91, Theorem 1] Let A be a C*-algebra. Then A is quasidiagonal if and only if, for every $x_1, \ldots, x_n \in A$ and $\epsilon > 0$, there is a finite-dimensional C*-algebra B and a completely positive contraction $\alpha : A \to B$ such that $\|\alpha(x_i)\| \geq \|x_i\| - \epsilon$ and

$$\|\alpha(x_i x_j) - \alpha(x_i)\alpha(x_j)\| < \epsilon$$

for $1 \leq i, j \leq n$.

Quasidiagonal Extensions by $\mathbb{K}$

V.4.2.15 A (separable) C*-algebra A has an essential quasidiagonal extension by $\mathbb{K}$ if there is a quasidiagonal C*-algebra of operators $B \subseteq \mathcal{L}(\mathcal{H})$ for separable $\mathcal{H}$, with $\mathbb{K} \subseteq B$, and $B/\mathbb{K} \cong A$. A is a quasidiagonal C*-algebra if and only if it has a split essential quasidiagonal extension by $\mathbb{K}$. The property of having an essential quasidiagonal extension by $\mathbb{K}$ is strictly weaker than quasidiagonality in general: $C_r^*(\mathbb{F}_2)$ is not quasidiagonal (V.4.2.13), but does have an essential quasidiagonal extension by $\mathbb{K}$ [Was91]. The two notions coincide for nuclear C*-algebras [DHS89].

V.4.2.16 PROPOSITION. [BK97, 3.1.3] Let A be a separable C*-algebra. Then A has an essential quasidiagonal extension by $\mathbb{K}$ if and only if A can be embedded in $(\prod \mathbb{M}_{k_n})/(\bigoplus \mathbb{M}_{k_n})$ for some sequence $\langle k_n \rangle$.

In fact, if A has a quasidiagonal extension by $\mathbb{K}$ and $B \subseteq \mathcal{L}(\mathcal{H})$ is as in V.4.2.15, let (P_n) be an increasing sequence of finite-rank projections as in V.4.1.5(ii); set $Q_n = P_n - P_{n-1}$ ($Q_1 = P_1$), $k_n = rank(Q_n)$; then $Q_n \mathcal{L}(\mathcal{H}) Q_n \cong \mathbb{M}_{k_n}$. To the element $x \in B$ associate the sequence $(Q_n x Q_n) \in \prod \mathbb{M}_{k_n}$. Modulo $\bigoplus \mathbb{M}_{k_n}$ this set of sequences is isomorphic to A. Conversely, if

$$A \subseteq [(\prod \mathbb{M}_{k_n})/(\bigoplus \mathbb{M}_{k_n})]$$

let $\mathcal{H} = \bigoplus \mathbb{C}^{k_n}$, and for each $x \in (\prod \mathbb{M}_{k_n})/(\bigoplus \mathbb{M}_{k_n})$, let $\rho(x) = (x_n)$, where (x_n) is a sequence in $\prod \mathbb{M}_{k_n}$ projecting to x, and $\prod \mathbb{M}_{k_n}$ is regarded as

$$\prod (\mathcal{L}(\mathbb{C}^{k_n})) \subseteq \mathcal{L}(\mathcal{H}).$$

For fixed x, any two choices of (x_n) differ by an element of $\bigoplus \mathbb{M}_{k_n} \subseteq \mathbb{K} = \mathcal{K}(\mathcal{H})$. Then the C*-subalgebra B of $\mathcal{L}(\mathcal{H})$ generated by $\mathbb{K}$ and all the $\rho(x)$ for $x \in A$ is a quasidiagonal extension of A by $\mathbb{K}$.

V.4.3 Generalized Inductive Limits

In this subsection, we summarize some of the results of [BK97], [BK01], and [BK]; cf. [BK00], [Bla04a].

One of the principles coming out of the work on classification of C*-algebras is that in considering inductive systems of C*-algebras, asymptotic behavior is all that matters; exact good behavior at each step is not necessary. (This is also very much in the spirit of the E-theory of Connes and Higson (V.1.4.7), and of the characterization of quasidiagonality in V.4.2.14.) In the classification work this principle is primarily applied to intertwinings of inductive systems. However, it is also possible, and interesting, to relax the requirements on the connecting maps themselves, requiring them only to be asymptotically additive, *-preserving, and multiplicative. We therefore consider *generalized inductive systems* of C*-algebras, where the connecting maps only asymptotically preserve the structure of the algebras.

The algebras thus obtained from finite-dimensional C*-algebras have very close connections with quasidiagonal C*-algebras.

We will primarily restrict attention to separable C*-algebras, although the basic definitions can be made in general.

V.4.3.1 DEFINITION. A *generalized inductive system of C*-algebras* is a sequence (A_n) of C*-algebras, with coherent maps $\phi_{m,n} : A_m \to A_n$ for $m < n$, such that for all k and all $x, y \in A_k$, $\lambda \in \mathbb{C}$, and all $\epsilon > 0$, there is an M such that, for all $M \leq m < n$,

(i) $\|\phi_{m,n}(\phi_{k,m}(x) + \phi_{k,m}(y)) - (\phi_{k,n}(x) + \phi_{k,n}(y))\| < \epsilon$
(ii) $\|\phi_{m,n}(\lambda\phi_{k,m}(x)) - \lambda\phi_{k,n}(x)\| < \epsilon$
(iii) $\|\phi_{m,n}(\phi_{k,m}(x)^*) - \phi_{k,n}(x)^*\| < \epsilon$
(iv) $\|\phi_{m,n}(\phi_{k,m}(x)\phi_{k,m}(y)) - \phi_{k,n}(x)\phi_{k,n}(y)\| < \epsilon$
(v) $\sup_r \|\phi_{k,r}(x)\| < \infty$.

A system satisfying (i) [resp. (iv)] is called *asymptotically additive* [resp. *asymptotically multiplicative*]. A generalized inductive system in which all $\phi_{m,n}$ are linear is called a *linear generalized inductive system*; if all the $\phi_{m,n}$ also preserve adjoints, the system is called **-linear*. A system is *contractive* if all the connecting maps are contractions. Of course, any ordinary inductive system (II.8.2.1) is a (contractive, *-linear) generalized inductive system.

At least if all the A_n are finite-dimensional, there is no loss of generality in assuming that all the connecting maps are *-linear (V.4.3.5).

V.4.3.2 Suppose $(A_n, \phi_{m,n})$ is a generalized inductive system of C*-algebras. The inductive limit of the system can be defined abstractly, but it is more useful to give a more concrete description of the construction of the inductive limit as follows. Let $\prod A_n$ be the full C*-direct product of the A_n, i.e. the set of bounded sequences $\langle x_n \rangle$, with $x_n \in A_n$, with pointwise operations and sup norm; and let $\bigoplus A_n$ be the C*-direct sum, the set of sequences converging to zero in norm. Then $\prod A_n$ is a C*-algebra and $\bigoplus A_n$ is a closed two-sided ideal; let π be the quotient map from $\prod A_n$ to $(\prod A_n)/(\bigoplus A_n)$. Each element x of A_m naturally defines an element

$$\phi_m(x) = \pi(\langle \phi_{m,n}(x) \rangle)$$

of $(\prod A_n)/(\bigoplus A_n)$. The closure of the set of all such elements (for all m) is a C*-subalgebra of $(\prod A_n)/(\bigoplus A_n)$, which we may call $\lim_{\to}(A_n, \phi_{m,n})$. Thus a C*-algebra which is an inductive limit of a generalized inductive system $(A_n, \phi_{m,n})$ can be embedded in $(\prod A_n)/(\bigoplus A_n)$.

V.4.3.3 PROPOSITION. Let $(A_n, \phi_{m,n})$ be a generalized inductive system, with $A = \lim_{\to}(A_n, \phi_{m,n})$. If each A_n is commutative [resp. finite, stably finite], then A is commutative [resp. finite, stably finite].

In fact, it is a simple argument to show that $(\prod A_n)/(\bigoplus A_n)$ is commutative [resp. finite, stably finite] if all but finitely many of the A_n are commutative [resp. finite, stably finite].

We will primarily consider three classes of generalized inductive limits:

V.4.3.4 DEFINITION. A separable C*-algebra A is an *MF algebra* if it can be written as $\lim_{\to}(A_n, \phi_{m,n})$ for a generalized inductive system with the A_n finite-dimensional. If the connecting maps $\phi_{m,n}$ can be chosen to be completely positive (*-linear) contractions, then A is an *NF algebra*, and A is a *strong NF algebra* if the $\phi_{m,n}$ can be chosen to be complete order embeddings (completely positive and completely isometric *-linear maps).

A generalized inductive system $(A_n, \phi_{m,n})$ in which each A_n is finite-dimensional and each $\phi_{m,n}$ is a completely positive contraction [resp. a complete order embedding] is called an *NF system* [resp. a *strong NF system*].

If $(A_n, \phi_{m,n})$ is an NF system [resp. strong NF system] and

$$A = \lim_{\to}(A_n, \phi_{m,n}),$$

then the natural map $\phi_n : A_n \to A$ is a completely positive contraction [resp. complete order embedding].

An MF algebra is separable. It is obvious that (Strong NF) $\implies$ (NF) $\implies$ (MF), and from V.4.3.3 it follows that every MF algebra is stably finite. Every AF algebra is a strong NF algebra, but the class of strong NF algebras is much larger than just the AF algebras.

The study of these classes of C*-algebras, particularly of the strong NF algebras, may be regarded as "noncommutative piecewise-linear (PL) topology" (cf. V.4.3.41).

MF Algebras

We now give a number of characterizations of MF algebras, which show that they are a natural and important class to consider.

V.4.3.5 THEOREM. [BK97, 3.2.2] Let A be a separable C*-algebra. Then the following are equivalent:

(i) A is an MF algebra.
(ii) A is isomorphic to $\lim_{\to}(A_n, \phi_{m,n})$ for a *-linear generalized inductive system of finite-dimensional C*-algebras.
(iii) A can be embedded as a C*-subalgebra of $(\prod \mathbb{M}_{k_n})/(\bigoplus \mathbb{M}_{k_n})$ for some sequence $\langle k_n \rangle$.
(iv) A has an essential quasidiagonal extension by the compact operators $\mathbb{K}$.
(v) There is a continuous field of C*-algebras $\langle B(t) \rangle$ over $\mathbb{N} \cup \{\infty\}$ with $B(\infty) \cong A$ and $B(n)$ finite-dimensional for each $n < \infty$.
(vi) There is a continuous field of C*-algebras $\langle B(t) \rangle$ over $\mathbb{N} \cup \{\infty\}$ with $B(\infty) \cong A$ and $B(n) \cong \mathbb{M}_{k_n}$ for each $n < \infty$, for some sequence $\langle k_n \rangle$.

The key implication is (iii) $\Longrightarrow$ (ii); the rest of the implications are trivial or routine ((iii) $\Longleftrightarrow$ (iv) is V.4.2.16).

There is thus an intimate connection between MF algebras and quasidiagonality, and also with continuous fields ((vi) is the inspiration for the name "MF algebra," which stands for "matricial field" or "M. Fell," who was the first to consider such algebras and propose that they are interesting objects for study).

Here are some simple consequences of the theorem; most are not obvious from the definition of MF algebras, and some may even be a little surprising.

V.4.3.6 COROLLARY.

(i) Every C*-subalgebra of an MF algebra is MF.
(ii) Every separable residually finite-dimensional C*-algebra (V.2.1.10) is an MF algebra.
(iii) If A is any separable C*-algebra, then CA and SA are MF algebras.
(iv) Every separable C*-algebra is a quotient of an MF algebra [A is a quotient of CA].
(v) If A and B are MF algebras, then $A \otimes_\alpha B$ is an MF algebra for some cross norm α. If one of them is nuclear, then $A \otimes B = A \otimes_{\min} B$ is an MF algebra.
(vi) If A is MF, then every separable C*-algebra Morita equivalent to A is MF.
(vii) Let $\langle B(t) \rangle$ be a continuous field of separable C*-algebras over $\mathbb{N} \cup \{\infty\}$. If each $B(t)$ for $t \in \mathbb{N}$ is an MF algebra, then $B(\infty)$ is an MF algebra.
(viii) Let $\langle A_n \rangle$ be a sequence of MF algebras. Then any separable C*-subalgebra of $(\prod A_n)/(\bigoplus A_n)$ is an MF algebra.
(ix) Let $(A_n, \phi_{m,n})$ be a generalized inductive system of C*-algebras, with $A = \lim_{\to}(A_n, \phi_{m,n})$. If each A_n is an MF algebra, then A is an MF algebra.

V.4.3.7 Not every MF algebra is residually finite-dimensional - many AF algebras are not residually finite-dimensional.

Every MF algebra is stably finite (V.4.3.3). So there are even type I C*-algebras (e.g. the Toeplitz algebra) which are not MF. It is unknown whether every stably finite type I C*-algebra is MF. Every residually stably finite type I C*-algebra is MF [Spi88].

There are apparently no known examples of a stably finite separable C*-algebra which is not MF, but $C_r^*(G)$ for a property (T) group G which is not residually finite dimensional (such groups have been constructed by Gromov [Gro00]) is a good candidate.

An MF algebra can fail to be strongly quasidiagonal, e.g. $C^*(S \oplus S^*)$. An MF algebra can even fail to be (weakly) quasidiagonal [Was91, Prop. 5].

V.4.3.8 PROPOSITION. A unital MF algebra has a tracial state.

PROOF: It suffices to note that $(\prod \mathbb{M}_{k_n})/(\bigoplus \mathbb{M}_{k_n})$ has a tracial state. Let τ_n be the tracial state on $\prod \mathbb{M}_{k_n}$ obtained by applying the normalized trace on $\mathbb{M}_{k_n}$ to the n'th coordinate. Then any weak-* limit of (τ_n) is a tracial state on $\prod \mathbb{M}_{k_n}$ which vanishes on $\bigoplus \mathbb{M}_{k_n}$.

NF Algebras

V.4.3.9 We now describe the structure of NF algebras. Of course, every NF algebra is an MF algebra. The key observation in developing the further properties of NF algebras is that NF algebras are also nuclear. Since a *strong* NF algebra has an increasing sequence of finite-dimensional subspaces completely order isomorphic to (finite-dimensional) C*-algebras, with dense union, it follows easily from the injectivity of finite-dimensional C*-algebras that every strong NF algebra is nuclear (and, in fact, the identity map can be approximated in the point-norm topology by *idempotent* finite-rank completely positive contractions). Nuclearity of NF algebras then follows from the next simple construction.

V.4.3.10 PROPOSITION. Every NF algebra is a quotient of a strong NF algebra.

PROOF: Let $(A_n, \phi_{m,n})$ be an NF system for an NF algebra A, and for each n let $B_n = A_1 \oplus \cdots \oplus A_n$, $\psi_{n,n+1} : B_n \to B_{n+1}$ defined by

$$\psi_{n,n+1}(x_1, \ldots, x_n) = (x_1, \ldots, x_n, \phi_{n,n+1}(x_n)).$$

Then $(B_n, \psi_{m,n})$ is a strong NF system whose inductive limit is a strong NF algebra having A as a quotient in an evident way.

See V.4.3.27 for a refinement of this result.

V.4.3.11 More generally, it can be shown by a very similar argument that if $(A_n, \phi_{m,n})$ is a generalized inductive system with each A_n nuclear and each $\phi_{m,n}$ a completely positive contraction, then $\lim_{\to} (A_n, \phi_{m,n})$ is nuclear.

V.4.3.12 So an NF algebra $A = \lim_\to (A_n, \phi_{m,n})$ is a separable nuclear C*-algebra which can be embedded in $(\prod A_n)/(\bigoplus A_n)$. The crucial consequence of nuclearity is that there is a completely positive contractive lifting of the embedding to $\sigma : A \to \prod A_n$ (IV.3.2.4). Composing σ with the quotient map to A_n yields a sequence $\gamma_n : A \to A_n$ of completely positive, asymptotically multiplicative contractions such that $\phi_n \circ \gamma_n$ converges to id_A in the point-norm topology. This argument applies more generally to any nuclear C*-subalgebra of $(\prod \mathbb{M}_{k_n})/(\bigoplus \mathbb{M}_{k_n})$ for a sequence k_n, i.e. for any nuclear MF algebra.

In fact, we have the following characterizations of NF algebras:

V.4.3.13 THEOREM. [BK97, 5.2.2] Let A be a separable C*-algebra. The following are equivalent:

(i) A is an NF algebra.
(ii) A is a nuclear MF algebra.
(iii) A is nuclear and can be embedded as a C*-subalgebra of $(\prod \mathbb{M}_{k_n})/(\bigoplus \mathbb{M}_{k_n})$ for a sequence k_n.
(iv) A is nuclear and quasidiagonal.
(v) The identity map on A can be approximated in the point-norm topology by completely positive approximately multiplicative contractions through finite-dimensional C*-algebras, i.e. given $x_1, \ldots, x_n \in A$ and $\epsilon > 0$, there is a finite-dimensional C*-algebra B and completely positive contractions $\alpha : A \to B$ and $\beta : B \to A$ such that $\|x_i - \beta \circ \alpha(x_i)\| < \epsilon$ and
$$\|\alpha(x_i x_j) - \alpha(x_i)\alpha(x_j)\| < \epsilon$$
for all i, j.

Compare (v) to V.4.2.14.

(ii) $\iff$ (iii) $\iff$ (iv) follows from V.4.3.5 and [DHS89]. (i) $\implies$ (ii) and (iii) $\implies$ (v) are described above. The remaining implication (v) $\implies$ (i) takes the most work and requires some perturbation results for completely positive maps, but is fairly straightforward; it is similar in spirit to the proof that an "AF algebra in the local sense" is an AF algebra (i.e. that finite-dimensional C*-subalgebras can be nested).

V.4.3.14 So a separable C*-algebra is an NF algebra if and only if the identity map on A can be approximately factored by completely positive almost multiplicative contractions through matrix algebras. Thus the NF algebras form a very natural class of nuclear C*-algebras, the ones in which not only the complete order structure but also the multiplication can be approximately modeled in finite-dimensional C*-algebras.

Here are some consequences of the theorem, similar to the results of V.4.3.6.

V.4.3.15 COROLLARY.

(i) A nuclear C*-subalgebra of an NF algebra is NF (although a general C*-subalgebra of an NF algebra need not be NF because it is not necessarily nuclear).
(ii) Every separable nuclear residually finite-dimensional C*-algebra (in particular, every subhomogeneous C*-algebra) is an NF algebra.
(iii) If A is any separable nuclear C*-algebra, then CA and SA are NF algebras.
(iv) Every separable nuclear C*-algebra is a quotient of an NF algebra.
(v) Let $(A_n, \phi_{m,n})$ be a generalized inductive system of C*-algebras, with each $\phi_{m,n}$ a completely positive contraction (e.g. an ordinary inductive system). If each A_n is an NF algebra, then A is an NF algebra. In particular, every approximately subhomogeneous (ASH) C*-algebra (V.2.1.9) is NF.
(vi) The class of NF algebras is closed under stable isomorphism (Morita equivalence).

V.4.3.16 Not every NF algebra is ASH, since an ASH algebra is residually stably finite and an NF algebra is not necessarily residually stably finite by (iv). $C^*(S \oplus S^*)$ and the examples of [Bro84], [BD96], and [DL94] are residually stably finite and NF, but not ASH; they are also not strongly quasidiagonal C*-algebras.

V.4.3.17 An NF algebra must of course be stably finite. There is no known example of a stably finite separable nuclear C*-algebra which is not NF. The name NF stands for "nuclear MF," but might also hopefully mean "nuclear (stably) finite."

V.4.3.18 Using the same types of perturbation theorems for completely positive maps needed for the proof of V.4.3.13, one can show that any two NF systems for an NF algebra are asymptotically equivalent in a certain natural but technical sense [BK97, 2.4.1], via completely positive contractions [BK97, 5.3.9].

Strong NF Algebras

The difference between the classes of NF and strong NF algebras is a subtle one; the fact that the classes are indeed different was not discovered until [BK01], and the seemingly definitive description of the essential difference not given until [BK].

The property of strong NF algebras described in V.4.3.9 turns out to be a characterization (V.4.3.19(iv)). As in the case of MF and NF algebras, there are several equivalent descriptions of strong NF algebras:

V.4.3.19 THEOREM. ([BK97, 6.1.1], [BK]) Let A be a separable C*-algebra. The following are equivalent:

(i) A is a strong NF algebra.
(ii) There is an increasing sequence $\langle S_n \rangle$ of finite-dimensional *-subspaces of A, each completely order isomorphic to a (finite-dimensional) C*-algebra, with dense union.
(iii) Given $x_1, \ldots, x_n \in A$ and $\epsilon > 0$, there is a finite-dimensional C*-algebra B, a complete order embedding ϕ of B into A, and elements $b_1, \ldots, b_n \in B$ with $\|x_i - \phi(b_i)\| < \epsilon$ for $1 \leq i \leq n$.
(iv) The identity map on A can be approximated in the point-norm topology by idempotent completely positive finite-rank contractions from A to A, i.e. given $x_1, \ldots, x_n \in A$ and $\epsilon > 0$, there is an idempotent completely positive finite-rank contraction $\omega : A \to A$ with $\|x_i - \omega(x_i)\| < \epsilon$ for $1 \leq i \leq n$.
(v) The identity map on A can be approximated in the point-norm topology by completely positive approximately multiplicative retractive contractions through finite-dimensional C*-algebras, i.e. given $x_1, \ldots, x_n \in A$ and $\epsilon > 0$, there is a finite-dimensional C*-algebra B and completely positive contractions $\alpha : A \to B$ and $\beta : B \to A$ with $\alpha \circ \beta = id_B$ (β is then automatically a complete order embedding), such that $\|x_i - \beta \circ \alpha(x_i)\| < \epsilon$ and $\|\alpha(x_i x_j) - \alpha(x_i)\alpha(x_j)\| < \epsilon$ for all i, j.
(vi) Same as (v) with the "approximately multiplicative" condition on α deleted.
(vii) Given $x_1, \ldots, x_n \in A$ and $\epsilon > 0$, there is a finite-dimensional C*-algebra B and completely positive contractions $\alpha : A \to B$ and $\beta : B \to A$ with β a complete order embedding, such that $\|x_i - \beta \circ \alpha(x_i)\| < \epsilon$ and α is approximately multiplicative, i.e. $\|\alpha(x_i x_j) - \alpha(x_i)\alpha(x_j)\| < \epsilon$ for all i, j.
(viii) Same as (vii) with the "approximately multiplicative" condition on α deleted.
(ix) A is nuclear and inner quasidiagonal.

For the proof, (i) $\Longrightarrow$ (ii) is obvious, and (ii) $\Longrightarrow$ (iv) was described in Section V.4.3.8. (v) $\Longrightarrow$ (ii) is similar to the proof of (v) $\Longrightarrow$ (i) in V.4.3.13, but is more subtle and requires a more sophisticated perturbation result for completely positive maps (V.4.3.42). The implication (ii) $\Longrightarrow$ (i) and the equivalence of (iii)–(viii) are easy consequences of injectivity of finite-dimensional C*-algebras and II.6.10.11. The equivalence of (ix) with the other conditions, proved in [BK], is a difficult technical argument using ultraproducts of Hilbert spaces.

V.4.3.20 Condition (iii) could be taken as the definition of a "strong NF algebra in the local sense." One then has that a strong NF algebra in the local sense is a strong NF algebra. Similarly, one could define an AF [resp. AH, ASH] algebra in the local sense to be a C*-algebra A with the property that, for any $x_1, \ldots, x_n \in A$ and $\epsilon > 0$, there is a finite-dimensional [resp. locally homogeneous, subhomogeneous] C*-algebra B, elements $b_1, \cdots, b_n \in B$, and a *-homomorphism $\phi : B \to A$ with $\|x_i - \phi(b_i)\| < \epsilon$ for $1 \le i \le n$. A (separable) AF algebra in the local sense is an AF algebra (II.8.3.24), but a (separable) AH algebra in the local sense is not necessarily an AH algebra [DE99]. It is not known in general whether a (separable) ASH algebra in the local sense is an ASH algebra.

Some consequences of V.4.3.19:

V.4.3.21 COROLLARY. Every separable nuclear residually finite-dimensional C*-algebra (in particular, every subhomogeneous C*-algebra) is a strong NF algebra.

V.4.3.22 COROLLARY. Let $(A_n, \phi_{m,n})$ be a generalized inductive system of C*-algebras, with each $\phi_{m,n}$ a complete order embedding (e.g. an ordinary inductive system with injective connecting maps). If each A_n is a strong NF algebra, then A is a strong NF algebra. In particular, every ASH C*-algebra is strong NF.

However, even an ordinary inductive limit of strong NF algebras need not be strong NF if the connecting maps are not injective [BK01, 5.16].

Combining V.4.3.10 and V.4.3.15(iv), we obtain:

V.4.3.23 COROLLARY. Every separable nuclear C*-algebra is a quotient of a strong NF algebra.

V.4.3.24 Not every strong NF algebra is ASH, since an ASH algebra is residually stably finite and an NF algebra is not necessarily residually stably finite by V.4.3.23.

The examples of [Bro84], [BD96], and [DL94] are residually stably finite and NF, but not strong NF by V.4.3.19(ix). $C^*(S \oplus S^*)$ is also NF but not strong NF for the same reason. For a rather different example, it follows easily from V.4.3.19(ix) that if A is separable and nuclear, then CA and SA are strong NF if and only if A is; thus, for example, SO_2 is NF but not strong NF.

It is somewhat easier to prove the following result than the equivalence of V.4.3.19(i) and (ix):

V.4.3.25 THEOREM. [BK01, 5.4] If A is separable, nuclear, and prime, then A is a strong NF algebra if and only if some (hence every) faithful irreducible representation of A is quasidiagonal.

V.4.3.26 COROLLARY. Every antiliminal prime NF algebra is a strong NF algebra. Every simple NF algebra is a strong NF algebra.

V.4.3.27 COROLLARY. Let A be any NF algebra, and let B be a split essential extension of A by $\mathbb{K}$. Then B is a strong NF algebra. So A can be embedded as a C*-subalgebra of a strong NF algebra B with a retraction (homomorphic conditional expectation) from B onto A.

V.4.3.28 Thus, in particular, a nuclear C*-subalgebra of a strong NF algebra need not be strong NF. However, a *hereditary* C*-subalgebra of a strong NF algebra is strong NF ([BK97, 6.1.7]; this can be proved more easily using V.4.3.19(ix)). It is also easily seen from V.4.3.19(ix) that the class of strong NF algebras is closed under stable isomorphism (Morita equivalence). We also have:

V.4.3.29 PROPOSITION. [BK01, 5.17] A tensor product of strong NF algebras is strong NF.

Since an NF algebra can be embedded in a strong NF algebra, using V.4.3.15(i) we obtain:

V.4.3.30 COROLLARY. A tensor product of NF algebras is NF.

Combining V.4.3.25 with Voiculescu's Weyl-von Neumann Theorem (II.8.4.29), we obtain:

V.4.3.31 COROLLARY. Let A be a separable nuclear C*-algebra. The following are equivalent:

(i) Every quotient of A is a strong NF algebra.
(ii) Every primitive quotient of A is a strong NF algebra.
(iii) Every irreducible representation of A is quasidiagonal.
(iv) A is strongly quasidiagonal.

V.4.3.32 We also have a stronger version of the existence of the cross sections discussed in V.4.3.12. If $(A_n, \phi_{m,n})$ is a strong NF system for a strong NF algebra A, then by injectivity of finite-dimensional C*-algebras there are idempotent completely positive contractions

$$\gamma_{n+1,n} : A_{n+1} \to \phi_{n,n+1}(A_n)$$

for each n, which may be thought of as going from A_{n+1} to A_n. These induce maps $\gamma_{n,m} : A_n \to A_m$ for $n > m$, and thus maps $\gamma_n : A \to A_n$, which are completely positive contractions and coherent in the obvious sense, and for which the maps $\phi_{m,n}$ and ϕ_n are cross sections. The $\gamma_{n,m}$ and γ_n are also asymptotically multiplicative.

RF Algebras

V.4.3.33 DEFINITION. A strong NF algebra is an *RF algebra* if it has a strong NF system $(A_n, \phi_{m,n})$ such that the $\gamma_{n,m}$ of V.4.3.32 (and hence also the γ_n) can be chosen exactly multiplicative.

An RF algebra is obviously nuclear and residually finite-dimensional. The converse is true:

V.4.3.34 THEOREM. [BK] A separable C*-algebra is an RF algebra if and only if it is nuclear and residually finite-dimensional.

An interesting consequence of the above results is that every strong NF algebra can be written as an *ordinary* inductive limit of RF algebras:

V.4.3.35 THEOREM. [BK97, 6.1.6] Let A be a strong NF algebra. Then there is an increasing sequence (C_k) of C*-subalgebras of A, with dense union, such that each C_k is an RF algebra.

PROOF: Let $(A_n, \phi_{m,n})$ be a strong NF system for A. Fix k. For $n \geq k$, inductively define $C_{k,n}$ by taking $C_{k,k} = A_k$ and $C_{n,k}$ the C*-subalgebra of A_n generated by $\phi_{n-1,n}(C_{k,n-1})$ for $n > k$. Then

$$C_k = [\cup_{n>k} \phi_n(C_{k,n})]^-$$

is a C*-subalgebra of A which is a strong NF algebra. For $k < n$, the map $\gamma_{n,n-1}|_{C_{k,n}}$ is a homomorphism from $C_{k,n}$ onto $C_{k,n-1}$ by II.6.10.12(ii); so by composition, if $k \leq m < n$, $\gamma_{n,m}|_{C_{k,n}}$ is a homomorphism from $C_{k,n}$ onto $C_{k,m}$. Thus, by letting $n \to \infty$, for $k \leq m$, the map $\gamma_m|_{C_k}$ is a homomorphism from C_k onto $C_{k,m}$. So C_k is residually finite-dimensional. $\phi_k(A_k) \subseteq C_k \subseteq C_{k+1}$ for all k, so $\cup C_k$ is dense in A.

Note that $C_{k,n}$ is strictly larger than $D_{k,n} = C^*(\phi_{k,n}(A_k))$ in general, so C_k is strictly larger than $D_k = C^*(\phi_k(A_k))$. D_k is also residually finite-dimensional (since it is a C*-subalgebra of C_k), but it is not obvious that it is nuclear. If it is, it is an RF algebra by V.4.3.34. It appears that D_k should be a generalized inductive limit of the $D_{k,n}$, where the connecting map from $D_{k,n}$ to $D_{k,n+1}$ is $\phi_{n,n+1}|_{D_{k,n}}$ followed by a conditional expectation of $C_{k,n+1}$ onto $D_{k,n+1}$.

V.4.3.36 One might hope that the conclusion of V.4.3.35 could be strengthened to replace "residually finite-dimensional" by "subhomogeneous" (i.e. to show that every strong NF algebra is approximately subhomogeneous). But this cannot be true. Since a quotient of an ASH algebra is ASH, a necessary condition for a C*-algebra to be ASH is that it be residually strong NF, or equivalently strongly quasidiagonal (V.4.3.31). Even then, there is an enormous gap between RF algebras and subhomogeneous C*-algebras. Nonetheless, V.4.3.35 gives the beginnings of a connection between this theory and the part of Elliott's classification program concerned with inductive limits of subhomogeneous building blocks.

It is difficult to compute structural invariants of a (strong) NF algebra from a (strong) NF system. In particular, the ideal structure, trace space, and K-theory are all difficult to compute. But there is a nice combinatorial way to compute the Čech cohomology of a compact metrizable space X in terms of a strong NF system for $C(X)$ as described in V.4.3.41, which suggests that there could be a combinatorial approach to the invariants and, perhaps, classification of (strong) NF algebras.

Commutative C*-Algebras and PL Topology

By V.4.3.21, every (separable) commutative C*-algebra is a strong NF algebra. In this subsection, we give a direct proof of the essential part of this fact, and describe the connection with PL topology.

We will show that commutative C*-algebras satisfy V.4.3.19(iii); the rest of the proof that commutative C*-algebras are strong NF consists of applying a perturbation result to show that the images of the finite-dimensional commutative C*-algebras can be chosen nested (the perturbation results needed are much simpler in the commutative case).

For simplicity, we will work only with unital C*-algebras; the general result follows easily from the unital case.

V.4.3.37 DEFINITION. Let X be a compact Hausdorff space. A *weak triangulation* of X is a continuous function ϕ from X to the underlying space of a simplicial complex K, such that every face of K contains an interior point in the image (in particular, every vertex of K lies in the image).

A *triangulation* is a weak triangulation in which ϕ is a homeomorphism.

V.4.3.38 PROPOSITION. Let X be a compact Hausdorff space. Then there is a natural one-one correspondence between the following sets:

(i) Weak triangulations of X
(ii) Effective partitions of unity on X
(iii) Unital complete order embeddings of finite-dimensional commutative C*-algebras into $C(X)$.

(A partition of unity $\{f_1, \ldots, f_n\}$ is *effective* if $\max f_i(x) = 1$ for all i, i.e. if there are points $x_1, \ldots, x_n \in X$ such that $f_j(x_i) = \delta_{ij}$.)

The correspondences go like this: if $\phi : X \to K$ is a weak triangulation, and $(e_1, \ldots, e_n)$ are the vertices of K (in a fixed order), for $x \in X$ write $\phi(x) = \sum_{i=1}^n \lambda_i e_i$, and set $f_i(x) = \lambda_i$. Then $\{f_1, \ldots, f_n\}$ is an effective partition of unity on X. The corresponding complete order embedding ψ of $\mathbb{C}^n$ into $C(X)$ sends $(\lambda_1, \ldots, \lambda_n)$ to $\sum_{i=1}^n \lambda_i f_i$. There is a corresponding idempotent completely positive contraction γ from $C(X)$ onto the image: send g to $\sum_{i=1}^n g(x_i) f_i$.

Conversely, if $\{f_1, \ldots, f_n\}$ is an effective partition of unity of X, then there are x_i such that $f_j(x_i) = \delta_{ij}$. Let

$$U_i = \{x \in X : f_i(x) > 0\}$$

and let K be a simplicial complex with vertices $e_1, \ldots, e_n$, such that if $I = \{i_1, \ldots, i_r\}$ is a subset of $\{1, \ldots, n\}$, then there is a face in K with vertices $e_{i_1}, \ldots, e_{i_r}$ if and only if $U_{i_1} \cap \cdots \cap U_{i_r} \neq \emptyset$. Define a map $\phi : X \to K$ by

$$\phi(x) = \sum_{i=1}^{n} f_i(x) e_i.$$

Then ϕ is a weak triangulation.

Similarly, if $\psi : \mathbb{C}^n \to C(X)$ is a unital complete order embedding, let v_i be the i'th standard basis vector in $\mathbb{C}^n$, and set $f_i = \phi(v_i)$. Because $\|f_i\| = 1$ for all i, $\{f_1, \ldots, f_n\}$ is an effective partition of unity of X.

V.4.3.39 By this correspondence, a strong NF system for $C(X)$ can be described as consisting of:

(i) A sequence $\langle K_i \rangle$ of simplicial complexes with underlying spaces X_i
(ii) An inverse system (X_i, ϕ_{ij}), where the $\phi_{ij} : X_j \to X_i$ $(i < j)$ are piecewise-linear weak triangulations
(iii) A homeomorphism (identification) of X with $\varprojlim (X_i, \phi_{ij})$ such that the natural maps $\phi_i : X \to X_i$ are weak triangulations, which become "sufficiently fine" for large i in the sense that any $g \in C(X)$ can be uniformly approximated arbitrarily closely by a function of the form $f \circ \phi_i$, where f is a piecewise-linear function from X_i to $\mathbb{C}$, for some sufficiently large i.

It is well known to topologists that every compact metrizable space can be so written; proofs (which are essentially equivalent to the proof outlined here) can be found in several topology books.

The next result is just a restatement of II.6.9.3(v).

V.4.3.40 PROPOSITION. Let X be a compact Hausdorff space, $g_1, \ldots, g_m \in C(X)$, and $\epsilon > 0$. Then there is an n, a unital complete order embedding ψ of $\mathbb{C}^n$ into $C(X)$, and $w_1, \ldots, w_m \in \mathbb{C}^n$, such that $\|g_k - \psi(w_k)\| < \epsilon$ for $1 \leq k \leq m$.

V.4.3.41 To complete the argument that separable commutative C*-algebras are strong NF, choose a sequence ψ_n of complete order embeddings of $\mathbb{C}^{k_n}$ into $C(X)$ such that finite subsets of a dense sequence in $C(X)$ can be approximated more and more closely as $n \to \infty$, with retractions $\gamma_n : C(X) \to \mathbb{C}^{k_n}$ satisfying $\gamma_n \circ \psi_n = id_{\mathbb{C}^{k_n}}$. By passing to a subsequence converging rapidly enough, we may insure that for each n,

$$\|\psi_{n+1} \circ \gamma_{n+1} \circ \psi_n - \psi_n\| < 2^{-n}.$$

Then for each n the sequence

$$(\psi_n, \psi_{n+1} \circ \gamma_{n+1} \circ \psi_n, \psi_{n+2} \circ \gamma_{n+2} \circ \psi_{n+1} \circ \gamma_{n+1} \circ \psi_n, \ldots)$$

converges in the point-norm topology to a completely positive map $\theta_n : \mathbb{C}^{k_n} \to C(X)$, with $\cup_n \theta_n(\mathbb{C}^{k_n})$ dense in $C(X)$.

However, θ_n will not be a complete order embedding in general. But if the approximations are done closely enough, $\gamma_{n+1} \circ \psi_n : \mathbb{C}^{k_n} \to \mathbb{C}^{k_{n+1}}$ will be "almost" a complete order embedding, and we need to know that it can be perturbed slightly to an actual complete order embedding. This perturbation result requires a theorem (V.4.3.42) in general, but is almost obvious in the commutative case: one needs only to perturb $\gamma_{n+1} \circ \psi_n$ to a map ω such that $\|\omega(v_i)\| = 1$ for all i (where v_i is the i'th standard basis vector in $\mathbb{C}^{k_n}$). If $\|\gamma_{n+1} \circ \psi_n(v_i)\| \geq 1 - \epsilon$ for all i, for some ϵ, $0 < \epsilon < 1/2$, for each i choose a coordinate m_i such that the m_i'th coordinate of $\gamma_{n+1} \circ \psi_n(v_i)$ is at least $1 - \epsilon$. The m_i are necessarily distinct. Let $\omega(v_i)$ be the vector with m_i'th coordinate 1, m_j'th coordinate 0 for $j \neq i$, and m'th coordinate the same as the m'th coordinate of $\gamma_{n+1} \circ \psi_n(v_i)$ for m not one of the m_j. Extend ω by linearity; then ω is a complete order embedding satisfying

$$\|\omega(v_i) - \gamma_{n+1} \circ \psi_n(v_i)\| < \epsilon$$

for all i. So if the subsequence is chosen so that $\|\gamma_{n+1} \circ \psi_n(v_i)\| \geq 1 - (2^{-n}/k_n)$ and ω_n is the approximant, then the sequence

$$(\psi_n, \psi_{n+1} \circ \omega_n, \psi_{n+2} \circ \omega_{n+1} \circ \psi_n, \ldots)$$

converges in the point-norm topology to a complete order embedding $\phi_n : \mathbb{C}^{k_n} \to C(X)$, satisfying $\phi_n = \phi_{n+1} \circ \omega_n$ for all n, with $\cup_n \phi_n(\mathbb{C}^{k_n})$ dense in $C(X)$. These give the desired coherent family.

Here is the perturbation result needed in general:

V.4.3.42 THEOREM. [BK97, 4.2.7] Let $d > 0$ and $\epsilon > 0$. Then there is a $\delta > 0$ such that, whenever A is a finite-dimensional C*-algebra with $dim(A) \leq d$ and matrix units $\{e_{ij}^r : 1 \leq i, j \leq k_r\}$, B is a finite-dimensional C*-algebra (or just a C*-algebra of real rank zero), and ϕ is a completely positive contraction from A to B with

$$\|\phi(e_{12}^r)\phi(e_{23}^r) \cdots \phi(e_{k_r-1,k_r}^r)\| > 1 - \delta$$

for all r, then there is a complete order embedding ψ from A to B with $\|\psi - \phi\| < \epsilon$.

V.4.3.43 Thus a strong NF system for $C(X)$ may be regarded as a piecewise-linear structure on X, and the piecewise-linear connecting maps can be described by combinatorial data. So it is reasonable to think of a strong NF system for a strong NF algebra A as giving a "piecewise-linear" structure on A, or a "combinatorial description" of A.

References

[AB80] R. J. Archbold and C. J. K. Batty. C^*-tensor norms and slice maps. *J. London Math. Soc. (2)*, 22(1):127–138, 1980.

[AD02] Claire Anantharaman-Delaroche. Amenability and exactness for dynamical systems and their C^*-algebras. *Trans. Amer. Math. Soc.*, 354(10):4153–4178 (electronic), 2002.

[AEPT76] Charles A. Akemann, George A. Elliott, Gert K. Pedersen, and Jun Tomiyama. Derivations and multipliers of C^*-algebras. *Amer. J. Math.*, 98(3):679–708, 1976.

[AH59] M. F. Atiyah and F. Hirzebruch. Riemann-Roch theorems for differentiable manifolds. *Bull. Amer. Math. Soc.*, 65:276–281, 1959.

[Ake81] Charles A. Akemann. Operator algebras associated with Fuchsian groups. *Houston J. Math.*, 7(3):295–301, 1981.

[AL50] A. S. Amitsur and J. Levitzki. Minimal identities for algebras. *Proc. Amer. Math. Soc.*, 1:449–463, 1950.

[Amb49] W. Ambrose. The L_2-system of a unimodular group. I. *Trans. Amer. Math. Soc.*, 65:27–48, 1949.

[And78] Joel Anderson. A C^*-algebra $\mathcal{A}$ for which $\mathrm{Ext}(\mathcal{A})$ is not a group. *Ann. of Math. (2)*, 107(3):455–458, 1978.

[AP77] Charles A. Akemann and Gert K. Pedersen. Ideal perturbations of elements in C^*-algebras. *Math. Scand.*, 41(1):117–139, 1977.

[APT73] Charles A. Akemann, Gert K. Pedersen, and Jun Tomiyama. Multipliers of C^*-algebras. *J. Functional Analysis*, 13:277–301, 1973.

[Ara63] Huzihiro Araki. A lattice of von Neumann algebras associated with the quantum theory of a free Bose field. *J. Mathematical Phys.*, 4:1343–1362, 1963.

[Ara64] Huzihiro Araki. Type of von Neumann algebra associated with free field. *Progr. Theoret. Phys.*, 32:956–965, 1964.

[Are51] Richard Arens. The adjoint of a bilinear operation. *Proc. Amer. Math. Soc.*, 2:839–848, 1951.

[Arv74] William Arveson. On groups of automorphisms of operator algebras. *J. Functional Analysis*, 15:217–243, 1974.

[Arv76] William Arveson. *An invitation to C^*-algebras*. Springer-Verlag, New York, 1976.

[Arv77]   William Arveson. Notes on extensions of C^*-algebras. *Duke Math. J.*, 44(2):329–355, 1977.
[AS03]   Erik M. Alfsen and Frederic W. Shultz. *Geometry of state spaces of operator algebras*. Mathematics: Theory & Applications. Birkhäuser Boston Inc., Boston, MA, 2003.
[Ati67]   M. F. Atiyah. *K-theory*. W. A. Benjamin, New York and Amsterdam, 1967. Notes by D. W. Anderson. Second edition: Addison-Wesley, Reading, MA, 1989.
[Ati68]   M. F. Atiyah. Bott periodicity and the index of elliptic operators. *Quart. J. Math. Oxford Ser. (2)*, 19:113–140, 1968.
[Atk53]   F. V. Atkinson. On relatively regular operators. *Acta Sci. Math. Szeged*, 15:38–56, 1953.
[Avi82]   Daniel Avitzour. Free products of C^*-algebras. *Trans. Amer. Math. Soc.*, 271(2):423–435, 1982.
[AW69]   Huzihiro Araki and E. J. Woods. A classification of factors. *Publ. Res. Inst. Math. Sci. Ser. A*, 4:51–130, 1968/1969.
[AW04]   Charles Akemann and Nik Weaver. Consistency of a counterexample to Naimark's problem. *Proc. Natl. Acad. Sci. USA*, 101(20):7522–7525 (electronic), 2004.
[Ban32]   Stefan Banach. *Théorie des opérations linéaires*. Warsaw, 1932.
[Bar54]   V. Bargmann. On unitary ray representations of continuous groups. *Ann. of Math. (2)*, 59:1–46, 1954.
[Bas64]   H. Bass. K-theory and stable algebra. *Inst. Hautes Études Sci. Publ. Math.*, (22):5–60, 1964.
[Bas74]   Hyman Bass. *Introduction to some methods of algebraic K-theory*. American Mathematical Society, Providence, R.I., 1974.
[BBEK92]   Bruce Blackadar, Ola Bratteli, George A. Elliott, and Alexander Kumjian. Reduction of real rank in inductive limits of C^*-algebras. *Math. Ann.*, 292(1):111–126, 1992.
[BC82]   Bruce E. Blackadar and Joachim Cuntz. The structure of stable algebraically simple C^*-algebras. *Amer. J. Math.*, 104(4):813–822, 1982.
[BCH94]   Paul Baum, Alain Connes, and Nigel Higson. Classifying space for proper actions and K-theory of group C^*-algebras. In C^*-*algebras: 1943–1993 (San Antonio, TX, 1993)*, volume 167 of *Contemp. Math.*, pages 240–291. Amer. Math. Soc., Providence, RI, 1994.
[BD96]   Lawrence G. Brown and Marius Dadarlat. Extensions of C^*-algebras and quasidiagonality. *J. London Math. Soc. (2)*, 53(3):582–600, 1996.
[BDF73]   L. G. Brown, R. G. Douglas, and P. A. Fillmore. Unitary equivalence modulo the compact operators and extensions of C^*-algebras. In *Proceedings of a Conference on Operator Theory (Dalhousie Univ., Halifax, N.S., 1973)*, pages 58–128. Lecture Notes in Math., Vol. 345, Berlin, 1973. Springer.
[BDF77]   L. G. Brown, R. G. Douglas, and P. A. Fillmore. Extensions of C^*-algebras and K-homology. *Ann. of Math. (2)*, 105(2):265–324, 1977.
[BE91]   Edwin J. Beggs and David E. Evans. The real rank of algebras of matrix valued functions. *Internat. J. Math.*, 2(2):131–138, 1991.
[Béd91]   Erik Bédos. Discrete groups and simple C^*-algebras. *Math. Proc. Cambridge Philos. Soc.*, 109(3):521–537, 1991.
[Ber71]   I. David Berg. An extension of the Weyl-von Neumann theorem to normal operators. *Trans. Amer. Math. Soc.*, 160:365–371, 1971.

[Ber72] Sterling K. Berberian. *Baer ∗-rings*. Number 195 in Grundlehren der mathematischen Wissenschaften. Springer, New York, 1972.

[BGR77] Lawrence G. Brown, Philip Green, and Marc A. Rieffel. Stable isomorphism and strong Morita equivalence of C^*-algebras. *Pacific J. Math.*, 71(2):349–363, 1977.

[BH82] Bruce Blackadar and David Handelman. Dimension functions and traces on C^*-algebras. *J. Funct. Anal.*, 45(3):297–340, 1982.

[BK] Bruce Blackadar and Eberhard Kirchberg. Irreducible representations of inner quasidiagonal C^*-algebras. to appear.

[BK97] Bruce Blackadar and Eberhard Kirchberg. Generalized inductive limits of finite-dimensional C^*-algebras. *Math. Ann.*, 307(3):343–380, 1997.

[BK00] Bruce Blackadar and Eberhard Kirchberg. Generalized inductive limits and quasidiagonality. In *C^*-algebras (Münster, 1999)*, pages 23–41. Springer, Berlin, 2000.

[BK01] Bruce Blackadar and Eberhard Kirchberg. Inner quasidiagonality and strong NF algebras. *Pacific J. Math.*, 198(2):307–329, 2001.

[BK04] Etienne Blanchard and Eberhard Kirchberg. Non-simple purely infinite C^*-algebras: the Hausdorff case. *J. Funct. Anal.*, 207(2):461–513, 2004.

[Bla77a] Bruce E. Blackadar. Infinite tensor products of C^*-algebras. *Pacific J. Math.*, 72(2):313–334, 1977.

[Bla77b] Bruce E. Blackadar. The regular representation of restricted direct product groups. *J. Functional Analysis*, 25(3):267–274, 1977.

[Bla78] Bruce E. Blackadar. Weak expectations and nuclear C^*-algebras. *Indiana Univ. Math. J.*, 27(6):1021–1026, 1978.

[Bla80] Bruce E. Blackadar. A simple C^*-algebra with no nontrivial projections. *Proc. Amer. Math. Soc.*, 78(4):504–508, 1980.

[Bla81] Bruce E. Blackadar. A simple unital projectionless C^*-algebra. *J. Operator Theory*, 5(1):63–71, 1981.

[Bla83a] Bruce Blackadar. A stable cancellation theorem for simple C^*-algebras. *Proc. London Math. Soc. (3)*, 47(2):303–305, 1983. Appendix to: "The cancellation theorem for projective modules over irrational rotation C^*-algebras" [Proc. London Math. Soc. (3) **47** (1983), no. 2, 285–302] by M. A. Rieffel.

[Bla83b] Bruce E. Blackadar. The regular representation of local affine motion groups. *Pacific J. Math.*, 108(2):265–274, 1983.

[Bla85a] Bruce Blackadar. Nonnuclear subalgebras of C^*-algebras. *J. Operator Theory*, 14(2):347–350, 1985.

[Bla85b] Bruce Blackadar. Shape theory for C^*-algebras. *Math. Scand.*, 56(2):249–275, 1985.

[Bla93] Bruce Blackadar. Matricial and ultramatricial topology. In *Operator algebras, mathematical physics, and low-dimensional topology (Istanbul, 1991)*, volume 5 of *Res. Notes Math.*, pages 11–38. A K Peters, Wellesley, MA, 1993.

[Bla94] Bruce Blackadar. Projections in C^*-algebras. In *C^*-algebras: 1943–1993 (San Antonio, TX, 1993)*, volume 167 of *Contemp. Math.*, pages 130–149. Amer. Math. Soc., Providence, RI, 1994.

[Bla98] Bruce Blackadar. *K-theory for operator algebras*, volume 5 of *Mathematical Sciences Research Institute Publications*. Cambridge University Press, Cambridge, second edition, 1998.

[Bla04a]   Bruce Blackadar. The algebraization of dynamics: amenability, nuclearity, quasidiagonality, and approximate finite dimensionality. In *Operator algebras, quantization, and noncommutative geometry*, volume 365 of *Contemp. Math.*, pages 51–83. Amer. Math. Soc., Providence, RI, 2004.

[Bla04b]   Bruce Blackadar. Semiprojectivity in simple C^*-algebras. In *Operator algebras and applications*, volume 38 of *Adv. Stud. Pure Math.*, pages 1–17. Math. Soc. Japan, Tokyo, 2004.

[Bla04c]   Bruce Blackadar. The stable rank of full corners in C^*-algebras. *Proc. Amer. Math. Soc.*, 132(10):2945–2950 (electronic), 2004.

[BM85]   Robert J. Blattner and Susan Montgomery. A duality theorem for Hopf module algebras. *J. Algebra*, 95(1):153–172, 1985.

[BN88]   Florin Boca and Viorel Niţică. Combinatorial properties of groups and simple C^*-algebras with a unique trace. *J. Operator Theory*, 20(1):183–196, 1988.

[Bon83]   Francis Bonahon. Difféotopies des espaces lenticulaires. *Topology*, 22(3):305–314, 1983.

[BP78]   John W. Bunce and William L. Paschke. Quasi-expectations and injective operator algebras. In *C^*-algebras and applications to physics (Proc. Second Japan-USA Sem., Los Angeles, Calif., 1977)*, volume 650 of *Lecture Notes in Math.*, pages 123–125. Springer, Berlin, 1978.

[BP80]   John W. Bunce and William L. Paschke. Derivations on a C^*-algebra and its double dual. *J. Funct. Anal.*, 37(2):235–247, 1980.

[BP91]   Lawrence G. Brown and Gert K. Pedersen. C^*-algebras of real rank zero. *J. Funct. Anal.*, 99(1):131–149, 1991.

[BP95]   Lawrence G. Brown and Gert K. Pedersen. On the geometry of the unit ball of a C^*-algebra. *J. Reine Angew. Math.*, 469:113–147, 1995.

[BR87]   Ola Bratteli and Derek W. Robinson. *Operator algebras and quantum statistical mechanics. 1*. Texts and Monographs in Physics. Springer-Verlag, New York, second edition, 1987. C^*- and W^*-algebras, symmetry groups, decomposition of states.

[BR92]   Bruce Blackadar and Mikael Rørdam. Extending states on preordered semigroups and the existence of quasitraces on C^*-algebras. *J. Algebra*, 152(1):240–247, 1992.

[Bra86]   Ola Bratteli. *Derivations, dissipations and group actions on C^*-algebras*, volume 1229 of *Lecture Notes in Mathematics*. Springer-Verlag, Berlin, 1986.

[Bro77]   Lawrence G. Brown. Stable isomorphism of hereditary subalgebras of C^*-algebras. *Pacific J. Math.*, 71(2):335–348, 1977.

[Bro84]   L. G. Brown. The universal coefficient theorem for Ext and quasidiagonality. In *Operator algebras and group representations, Vol. I (Neptun, 1980)*, volume 17 of *Monogr. Stud. Math.*, pages 60–64. Pitman, Boston, MA, 1984.

[Bro04]   Nathanial P. Brown. On quasidiagonal C^*-algebras. In *Operator algebras and applications*, volume 38 of *Adv. Stud. Pure Math.*, pages 19–64. Math. Soc. Japan, Tokyo, 2004.

[BS27]   S. Banach and H. Steinhaus. Sur le principe de condesation des singularités. *Fund. Math.*, 9:50–61, 1927.

[BS93]   Saad Baaj and Georges Skandalis. Unitaires multiplicatifs et dualité pour les produits croisés de C^*-algèbres. *Ann. Sci. École Norm. Sup. (4)*, 26(4):425–488, 1993.

[Bun72]   John Bunce. Characterizations of amenable and strongly amenable C^*-algebras. *Pacific J. Math.*, 43:563–572, 1972.

[Bun76]   John W. Bunce. Finite operators and amenable C^*-algebras. *Proc. Amer. Math. Soc.*, 56:145–151, 1976.

[Bur63]   D. J. C. Bures. Certain factors constructed as infinite tensor products. *Compositio Math.*, 15:169–191 (1963), 1963.

[Bus68]   Robert C. Busby. Double centralizers and extensions of C^*-algebras. *Trans. Amer. Math. Soc.*, 132:79–99, 1968.

[BW76]   Jochen Brüning and Wolfgang Willgerodt. Eine Verallgemeinerung eines Satzes von N. Kuiper. *Math. Ann.*, 220(1):47–58, 1976.

[CDS98]   Alain Connes, Michael R. Douglas, and Albert Schwarz. Noncommutative geometry and matrix theory: compactification on tori. *J. High Energy Phys.*, (2):Paper 3, 35 pp. (electronic), 1998.

[CE76a]   Man Duen Choi and Edward G. Effros. The completely positive lifting problem for C^*-algebras. *Ann. of Math. (2)*, 104(3):585–609, 1976.

[CE76b]   Man Duen Choi and Edward G. Effros. Separable nuclear C^*-algebras and injectivity. *Duke Math. J.*, 43(2):309–322, 1976.

[CE77a]   Man Duen Choi and Edward G. Effros. Injectivity and operator spaces. *J. Functional Analysis*, 24(2):156–209, 1977.

[CE77b]   Man Duen Choi and Edward G. Effros. Nuclear C^*-algebras and injectivity: the general case. *Indiana Univ. Math. J.*, 26(3):443–446, 1977.

[CE78]   Man Duen Choi and Edward G. Effros. Nuclear C^*-algebras and the approximation property. *Amer. J. Math.*, 100(1):61–79, 1978.

[CE81]   Joachim Cuntz and David E. Evans. Some remarks on the C^*-algebras associated with certain topological Markov chains. *Math. Scand.*, 48(2):235–240, 1981.

[CFW81]   A. Connes, J. Feldman, and B. Weiss. An amenable equivalence relation is generated by a single transformation. *Ergodic Theory Dynamical Systems*, 1(4):431–450 (1982), 1981.

[CH87]   Joachim Cuntz and Nigel Higson. Kuiper's theorem for Hilbert modules. In Palle E. T. Jorgensen and Paul S. Muhly, editors, *Operator algebras and mathematical physics*, number 62 in Contemp. Math., pages 429–435. Amer. Math. Soc., Providence, 1987.

[CH90]   Alain Connes and Nigel Higson. Déformations, morphismes asymptotiques et K-théorie bivariante. *C. R. Acad. Sci. Paris Sér. I Math.*, 311(2):101–106, 1990.

[Chi69]   Wai-mee Ching. Non-isomorphic non-hyperfinite factors. *Canad. J. Math.*, 21:1293–1308, 1969.

[Cho74]   Man Duen Choi. A Schwarz inequality for positive linear maps on C^*-algebras. *Illinois J. Math.*, 18:565–574, 1974.

[Cho79]   Man Duen Choi. A simple C^*-algebra generated by two finite-order unitaries. *Canad. J. Math.*, 31(4):867–880, 1979.

[Cho80]   Man Duen Choi. The full C^*-algebra of the free group on two generators. *Pacific J. Math.*, 87(1):41–48, 1980.

[Chr81]   Erik Christensen. On nonselfadjoint representations of C^*-algebras. *Amer. J. Math.*, 103(5):817–833, 1981.

[CK77]   Alain Connes and Wolfgang Krieger. Measure space automorphisms, the normalizers of their full groups, and approximate finiteness. *J. Functional Analysis*, 24(4):336–352, 1977.

[CK80] Joachim Cuntz and Wolfgang Krieger. A class of C^*-algebras and topological Markov chains. *Invent. Math.*, 56(3):251–268, 1980.

[Cla86] N. P. Clarke. A finite but not stably finite C^*-algebra. *Proc. Amer. Math. Soc.*, 96(1):85–88, 1986.

[CM84] M. Cohen and S. Montgomery. Group-graded rings, smash products, and group actions. *Trans. Amer. Math. Soc.*, 282(1):237–258, 1984.

[CMS94] D. I. Cartwright, W. Młotkowski, and T. Steger. Property (T) and $\tilde{A}_2$ groups. *Ann. Inst. Fourier (Grenoble)*, 44(1):213–248, 1994.

[Cob74] L. A. Coburn. Singular integral operators and Toeplitz operators on odd spheres. *Indiana Univ. Math. J.*, 23:433–439, 1973/74.

[Coh59] Paul J. Cohen. Factorization in group algebras. *Duke Math. J*, 26:199–205, 1959.

[Com68] François Combes. Poids sur une C^*-algèbre. *J. Math. Pures Appl. (9)*, 47:57–100, 1968.

[Com70] François Combes. Poids associé à une algèbre hilbertienne généralisée. *C. R. Acad. Sci. Paris Sér. A-B*, 270:A33–A36, 1970.

[Com71] François Combes. Poids et espérances conditionnelles dans les algèbres de von Neumann. *Bull. Soc. Math. France*, 99:73–112, 1971.

[Con73a] Alain Connes. Sur le théorème de Radon-Nikodym pour les poids normaux fidèles semi-finis. *Bull. Sci. Math. (2)*, 97:253–258 (1974), 1973.

[Con73b] Alain Connes. Une classification des facteurs de type III. *Ann. Sci. École Norm. Sup. (4)*, 6:133–252, 1973.

[Con75a] A. Connes. A factor not anti-isomorphic to itself. *Bull. London Math. Soc.*, 7:171–174, 1975.

[Con75b] Alain Connes. Sur la classification des facteurs de type II. *C. R. Acad. Sci. Paris Sér. A-B*, 281(1):Aii, A13–A15, 1975.

[Con76] A. Connes. Classification of injective factors. Cases II_1, II_∞, III_λ, $\lambda \neq 1$. *Ann. of Math. (2)*, 104(1):73–115, 1976.

[Con78] A. Connes. On the cohomology of operator algebras. *J. Functional Analysis*, 28(2):248–253, 1978.

[Con79] A. Connes. On the equivalence between injectivity and semidiscreteness for operator algebras. In *Algèbres d'opérateurs et leurs applications en physique mathématique (Proc. Colloq., Marseille, 1977)*, volume 274 of *Colloq. Internat. CNRS*, pages 107–112. CNRS, Paris, 1979.

[Con80] A. Connes. A factor of type II_1 with countable fundamental group. *J. Operator Theory*, 4(1):151–153, 1980.

[Con81] A. Connes. An analogue of the Thom isomorphism for crossed products of a C^*-algebra by an action of **R**. *Adv. in Math.*, 39(1):31–55, 1981.

[Con91] John B. Conway. *The theory of subnormal operators*, volume 36 of *Mathematical Surveys and Monographs*. American Mathematical Society, Providence, RI, 1991.

[Con94] Alain Connes. *Noncommutative geometry*. Academic Press Inc., San Diego, CA, 1994.

[Con01] Corneliu Constantinescu. C^*-*algebras. Vol. 1–5*, volume 58–62 of *North-Holland Mathematical Library*. North-Holland Publishing Co., Amsterdam, 2001.

[CP79] Joachim Cuntz and Gert Kjaergård Pedersen. Equivalence and traces on C^*-algebras. *J. Funct. Anal.*, 33(2):135–164, 1979.

[CS86] J. Cuntz and G. Skandalis. Mapping cones and exact sequences in KK-theory. *J. Operator Theory*, 15(1):163–180, 1986.

[CST04]   Joachim Cuntz, Georges Skandalis, and Boris Tsygan. *Cyclic homology in non-commutative geometry*, volume 121 of *Encyclopaedia of Mathematical Sciences*. Springer-Verlag, Berlin, 2004.

[CT77]   Alain Connes and Masamichi Takesaki. The flow of weights on factors of type III. *Tôhoku Math. J. (2)*, 29(4):473–575, 1977.

[Cun77a]   Joachim Cuntz. Simple C^*-algebras generated by isometries. *Comm. Math. Phys.*, 57(2):173–185, 1977.

[Cun77b]   Joachim Cuntz. The structure of multiplication and addition in simple C^*-algebras. *Math. Scand.*, 40(2):215–233, 1977.

[Cun78]   Joachim Cuntz. Dimension functions on simple C^*-algebras. *Math. Ann.*, 233(2):145–153, 1978.

[Cun81a]   Joachim Cuntz. K-theory for certain C^*-algebras. *Ann. of Math. (2)*, 113(1):181–197, 1981.

[Cun81b]   Joachim Cuntz. K-theory for certain C^*-algebras. II. *J. Operator Theory*, 5(1):101–108, 1981.

[Dad00]   Marius Dadarlat. Nonnuclear subalgebras of AF algebras. *Amer. J. Math.*, 122(3):581–597, 2000.

[Dav96]   Kenneth R. Davidson. C^*-*algebras by example*, volume 6 of *Fields Institute Monographs*. American Mathematical Society, Providence, RI, 1996.

[DB86]   Robert S. Doran and Victor A. Belfi. *Characterizations of C^*-algebras*, volume 101 of *Monographs and Textbooks in Pure and Applied Mathematics*. Marcel Dekker Inc., New York, 1986.

[DD63]   Jacques Dixmier and Adrien Douady. Champs continus d'espaces hilbertiens et de C^*-algèbres. *Bull. Soc. Math. France*, 91:227–284, 1963.

[DE99]   Marius Dădărlat and Søren Eilers. Approximate homogeneity is not a local property. *J. Reine Angew. Math.*, 507:1–13, 1999.

[DG83]   Maurice J. Dupré and R. M. Gillette. *Banach bundles, Banach modules and automorphisms of C^*-algebras*, volume 92 of *Research Notes in Mathematics*. Pitman (Advanced Publishing Program), Boston, MA, 1983.

[DH68]   John Dauns and Karl Heinrich Hofmann. *Representation of rings by sections*. Memoirs of the American Mathematical Society, No. 83. American Mathematical Society, Providence, R.I., 1968.

[DHS89]   K. R. Davidson, D. A. Herrero, and N. Salinas. Quasidiagonal operators, approximation, and C^*-algebras. *Indiana Univ. Math. J.*, 38(4):973–998, 1989.

[Dix49]   J. Dixmier. Les anneaux d'opérateurs de classe finie. *Ann. Sci. École Norm. Sup. (3)*, 66:209–261, 1949.

[Dix51]   J. Dixmier. Sur certains espaces considérés par M. H. Stone. *Summa Brasil. Math.*, 2:151–182, 1951.

[Dix61]   J. Dixmier. Points séparés dans le spectre d'une C^*-algèbre. *Acta Sci. Math. Szeged*, 22:115–128, 1961.

[Dix69a]   Jacques Dixmier. *Les algèbres d'opérateurs dans l'espace hilbertien (algèbres de von Neumann)*. Les Grands Classiques Gauthier-Villars. [Gauthier-Villars Great Classics]. Éditions Jacques Gabay, Paris, 1969. Translated as *Von Neumann algebras*, North-Holland, Amsterdam, 1981. First Edition 1957.

[Dix69b]   Jacques Dixmier. *Les C^*-algèbres et leurs représentations*. Number 29 in Cahiers Scientifiques. Gauthier-Villars, Paris, 2nd edition, 1969. Reprinted by Éditions Jacques Gabay, Paris, 1996. Translated as C^*-algebras, North-Holland, Amsterdam, 1977. First Edition 1964.

[DL69]   J. Dixmier and E. C. Lance. Deux nouveaux facteurs de type II_1. *Invent. Math.*, 7:226–234, 1969.

[DL94]   M. Dădărlat and A. Loring. Extensions of certain real rank zero C^*-algebras. *Ann. Inst. Fourier (Grenoble)*, 44(3):907–925, 1994.

[dlH79]   P. de la Harpe. Moyennabilité du groupe unitaire et propriété P de Schwartz des algèbres de von Neumann. In *Algèbres d'opérateurs (Sém., Les Plans-sur-Bex, 1978)*, volume 725 of *Lecture Notes in Math.*, pages 220–227. Springer, Berlin, 1979.

[dlH85]   Pierre de la Harpe. Reduced C^*-algebras of discrete groups which are simple with a unique trace. In *Operator algebras and their connections with topology and ergodic theory (Buşteni, 1983)*, volume 1132 of *Lecture Notes in Math.*, pages 230–253. Springer, Berlin, 1985.

[dlHV89]   Pierre de la Harpe and Alain Valette. La propriété (T) de Kazhdan pour les groupes localement compacts (avec un appendice de Marc Burger). *Astérisque*, (175):158, 1989.

[DNNP92]   Marius Dădărlat, Gabriel Nagy, András Némethi, and Cornel Pasnicu. Reduction of topological stable rank in inductive limits of C^*-algebras. *Pacific J. Math.*, 153(2):267–276, 1992.

[Dri87]   V. G. Drinfel'd. Quantum groups. In *Proceedings of the International Congress of Mathematicians, Vol. 1, 2 (Berkeley, Calif., 1986)*, pages 798–820, Providence, RI, 1987. Amer. Math. Soc.

[DS88a]   Nelson Dunford and Jacob T. Schwartz. *Linear operators. Part I*. Wiley Classics Library. John Wiley & Sons Inc., New York, 1988.

[DS88b]   Nelson Dunford and Jacob T. Schwartz. *Linear operators. Part II*. Wiley Classics Library. John Wiley & Sons Inc., New York, 1988.

[Dup74]   Maurice J. Dupré. Classifying Hilbert bundles. *J. Functional Analysis*, 15:244–278, 1974.

[Dye59]   H. A. Dye. On groups of measure preserving transformation. I. *Amer. J. Math.*, 81:119–159, 1959.

[Dyk94]   Ken Dykema. Interpolated free group factors. *Pacific J. Math.*, 163(1):123–135, 1994.

[Ech93]   Siegfried Echterhoff. Regularizations of twisted covariant systems and crossed products with continuous trace. *J. Funct. Anal.*, 116(2):277–313, 1993.

[Eff81]   Edward G. Effros. *Dimensions and C^*-algebras*. Number 46 in CBMS Regional Conference Series in Mathematics. Amer. Math. Soc., Providence, 1981.

[Eff88]   Edward G. Effros. Amenability and virtual diagonals for von Neumann algebras. *J. Funct. Anal.*, 78(1):137–153, 1988.

[EH85]   Edward G. Effros and Uffe Haagerup. Lifting problems and local reflexivity for C^*-algebras. *Duke Math. J.*, 52(1):103–128, 1985.

[EH93]   Nawfal Elhage Hassan. Sur les rangs stables des C^*-algèbres. *C. R. Acad. Sci. Paris Sér. I Math.*, 317(8):773–776, 1993.

[EHS80]   Edward G. Effros, David E. Handelman, and Chao Liang Shen. Dimension groups and their affine representations. *Amer. J. Math.*, 102(2):385–407, 1980.

[EK86] E. G. Effros and J. Kaminker. Homotopy continuity and shape theory for C^*-algebras. In *Geometric methods in operator algebras (Kyoto, 1983)*, volume 123 of *Pitman Res. Notes Math. Ser.*, pages 152–180. Longman Sci. Tech., Harlow, 1986.

[EL77] Edward G. Effros and E. Christopher Lance. Tensor products of operator algebras. *Adv. Math.*, 25(1):1–34, 1977.

[Ell76] George A. Elliott. On the classification of inductive limits of sequences of semisimple finite-dimensional algebras. *J. Algebra*, 38(1):29–44, 1976.

[Ell77] George A. Elliott. Some C^*-algebras with outer derivations. III. *Ann. Math. (2)*, 106(1):121–143, 1977.

[Ell82] George A. Elliott. Gaps in the spectrum of an almost periodic Schrödinger operator. *C. R. Math. Rep. Acad. Sci. Canada*, 4(5):255–259, 1982.

[ELP99] Søren Eilers, Terry A. Loring, and Gert K. Pedersen. Morphisms of extensions of C^*-algebras: pushing forward the Busby invariant. *Adv. Math.*, 147(1):74–109, 1999.

[Enf73] Per Enflo. A counterexample to the approximation problem in Banach spaces. *Acta Math.*, 130:309–317, 1973.

[ENN93] George A. Elliott, Toshikazu Natsume, and Ryszard Nest. The Heisenberg group and K-theory. *K-Theory*, 7(5):409–428, 1993.

[Eno77] Michel Enock. Produit croisé d'une algèbre de von Neumann par une algèbre de Kac. *J. Functional Analysis*, 26(1):16–47, 1977.

[Eno03] Michel Enock. Quantum groupoids and pseudo-multiplicative unitaries. In *Locally compact quantum groups and groupoids (Strasbourg, 2002)*, volume 2 of *IRMA Lect. Math. Theor. Phys.*, pages 17–47. de Gruyter, Berlin, 2003.

[EO74] George A. Elliott and Dorte Olesen. A simple proof of the Dauns-Hofmann theorem. *Math. Scand.*, 34:231–234, 1974.

[ER94] Edward G. Effros and Zhong-Jin Ruan. Discrete quantum groups. I. The Haar measure. *Internat. J. Math.*, 5(5):681–723, 1994.

[ER00] Edward G. Effros and Zhong-Jin Ruan. *Operator spaces*, volume 23 of *London Mathematical Society Monographs. New Series*. The Clarendon Press Oxford University Press, New York, 2000.

[Ern67] John Ernest. Hopf-von Neumann algebras. In *Functional Analysis (Proc. Conf., Irvine, Calif., 1966)*, pages pp 195–215. Academic Press, London, 1967.

[ES75] Michel Enock and Jean-Marie Schwartz. *Une dualité dans les algèbres de von Neumann*. Société Mathématique de France, Paris, 1975.

[ES80] Michel Enock and Jean-Marie Schwartz. Produit croisé d'une algèbre de von Neumann par une algèbre de Kac. II. *Publ. Res. Inst. Math. Sci.*, 16(1):189–232, 1980.

[ES92] Michel Enock and Jean-Marie Schwartz. *Kac algebras and duality of locally compact groups*. Springer-Verlag, Berlin, 1992.

[ET94] G. A. Elliott and K. Thomsen. The state space of the K_0-group of a simple separable C^*-algebra. *Geom. Funct. Anal.*, 4(5):522–538, 1994.

[EV93] Michel Enock and Jean-Michel Vallin. C^*-algèbres de Kac et algèbres de Kac. *Proc. London Math. Soc. (3)*, 66(3):619–650, 1993.

[Eym64] Pierre Eymard. L'algèbre de Fourier d'un groupe localement compact. *Bull. Soc. Math. France*, 92:181–236, 1964.

[FD88a] J. M. G. Fell and R. S. Doran. *Representations of *-algebras, locally compact groups, and Banach *-algebraic bundles. Vol. 1*, volume 125 of *Pure and Applied Mathematics*. Academic Press Inc., Boston, MA, 1988.

[FD88b] J. M. G. Fell and R. S. Doran. *Representations of *-algebras, locally compact groups, and Banach *-algebraic bundles. Vol. 2*, volume 126 of *Pure and Applied Mathematics*. Academic Press Inc., Boston, MA, 1988.

[Fel61] J. M. G. Fell. The structure of algebras of operator fields. *Acta Math.*, 106:233–280, 1961.

[Fel69] J. M. G. Fell. *An extension of Mackey's method to Banach * algebraic bundles*. Memoirs of the American Mathematical Society, No. 90. American Mathematical Society, Providence, R.I., 1969.

[Fil96] Peter A. Fillmore. *A user's guide to operator algebras*. Canadian Mathematical Society Series of Monographs and Advanced Texts. John Wiley & Sons Inc., New York, 1996.

[FM77a] Jacob Feldman and Calvin C. Moore. Ergodic equivalence relations, cohomology, and von Neumann algebras. I. *Trans. Amer. Math. Soc.*, 234(2):289–324, 1977.

[FM77b] Jacob Feldman and Calvin C. Moore. Ergodic equivalence relations, cohomology, and von Neumann algebras. II. *Trans. Amer. Math. Soc.*, 234(2):325–359, 1977.

[FS81] Thierry Fack and Georges Skandalis. Connes' analogue of the Thom isomorphism for the Kasparov groups. *Invent. Math.*, 64(1):7–14, 1981.

[Fug50] Bent Fuglede. A commutativity theorem for normal operators. *Proc. Nat. Acad. Sci. U. S. A.*, 36:35–40, 1950.

[Fuk52] Masanori Fukamiya. On a theorem of Gelfand and Neumark and the B^*-algebra. *Kumamoto J. Sci. Ser. A.*, 1(1):17–22, 1952.

[Gar65] L. Terrell Gardner. On isomorphisms of C^*-algebras. *Amer. J. Math.*, 87:384–396, 1965.

[Gar84] L. Terrell Gardner. An elementary proof of the Russo-Dye theorem. *Proc. Amer. Math. Soc.*, 90(1):171, 1984.

[Gel41] I. Gelfand. Normierte Ringe. *Rec. Math. [Mat. Sbornik] N. S.*, 9 (51):3–24, 1941.

[GH76] K. R. Goodearl and D. Handelman. Rank functions and K_O of regular rings. *J. Pure Appl. Algebra*, 7(2):195–216, 1976.

[GJ76] Leonard Gillman and Meyer Jerison. *Rings of continuous functions*. Springer-Verlag, New York, 1976. Reprint of the 1960 edition, Graduate Texts in Mathematics, No. 43.

[GKn69] I. C. Gohberg and M. G. Kreĭn. *Introduction to the theory of linear nonselfadjoint operators*. Translated from the Russian by A. Feinstein. Translations of Mathematical Monographs, Vol. 18. American Mathematical Society, Providence, R.I., 1969.

[Gli60] James G. Glimm. On a certain class of operator algebras. *Trans. Amer. Math. Soc.*, 95:318–340, 1960.

[Gli61] James Glimm. Type I C^*-algebras. *Ann. of Math. (2)*, 73:572–612, 1961.

[GN43] I. Gelfand and M. Neumark. On the imbedding of normed rings into the ring of operators in Hilbert space. *Rec. Math. [Mat. Sbornik] N.S.*, 12(54):197–213, 1943.

[God51] Roger Godement. Mémoire sur la théorie des caractères dans les groupes localement compacts unimodulaires. *J. Math. Pures Appl. (9)*, 30:1–110, 1951.

[God54] Roger Godement. Théorie des caractères. I. Algèbres unitaires. *Ann. of Math. (2)*, 59:47–62, 1954.

[Goo49] Dwight Benjamin Goodner. *Projections in Normed Linear Spaces*. Abstract of a Thesis, University of Illinois, 1949.

[Goo82] K. R. Goodearl. *Notes on real and complex C^*-algebras*, volume 5 of *Shiva Mathematics Series*. Shiva Publishing Ltd., Nantwich, 1982.

[Goo91] K. R. Goodearl. *von Neumann regular rings*. Robert E. Krieger Publishing Co. Inc., Malabar, FL, second edition, 1991.

[GR79] Elliot C. Gootman and Jonathan Rosenberg. The structure of crossed product C^*-algebras: a proof of the generalized Effros-Hahn conjecture. *Invent. Math.*, 52(3):283–298, 1979.

[Gre69] Frederick P. Greenleaf. *Invariant means on topological groups and their applications*. Van Nostrand Mathematical Studies, No. 16. Van Nostrand Reinhold Co., New York, 1969.

[Gre78] Philip Green. The local structure of twisted covariance algebras. *Acta Math.*, 140(3-4):191–250, 1978.

[Gro55] Alexandre Grothendieck. Produits tensoriels topologiques et espaces nucléaires. *Mem. Amer. Math. Soc.*, 1955(16):140, 1955.

[Gro58] Alexander Grothendieck. La théorie des classes de Chern. *Bull. Soc. Math. France*, 86:137–154, 1958.

[Gro68] Alexander Grothendieck. Le groupe de Brauer. I. Algèbres d'Azumaya et interprétations diverses. In *Dix Exposés sur la Cohomologie des Schémas*, pages 46–66. North-Holland, Amsterdam, 1968.

[Gro87] M. Gromov. Hyperbolic groups. In *Essays in group theory*, volume 8 of *Math. Sci. Res. Inst. Publ.*, pages 75–263. Springer, New York, 1987.

[Gro00] Misha Gromov. Spaces and questions. *Geom. Funct. Anal.*, (Special Volume, Part I):118–161, 2000. GAFA 2000 (Tel Aviv, 1999).

[Gui69] Alain Guichardet. Tensor products of C^*-algebras. Aarhus University Lecture Notes Series no. 12, 1969.

[Haa75a] Uffe Haagerup. Normal weights on W^*-algebras. *J. Functional Analysis*, 19:302–317, 1975.

[Haa75b] Uffe Haagerup. The standard form of von Neumann algebras. *Math. Scand.*, 37(2):271–283, 1975.

[Haa83] U. Haagerup. All nuclear C^*-algebras are amenable. *Invent. Math.*, 74(2):305–319, 1983.

[Haa85] Uffe Haagerup. A new proof of the equivalence of injectivity and hyperfiniteness for factors on a separable Hilbert space. *J. Funct. Anal.*, 62(2):160–201, 1985.

[Haa87] Uffe Haagerup. Connes' bicentralizer problem and uniqueness of the injective factor of type III_1. *Acta Math.*, 158(1-2):95–148, 1987.

[Haa90] Uffe Haagerup. On convex combinations of unitary operators in C^*-algebras. In *Mappings of operator algebras (Philadelphia, PA, 1988)*, volume 84 of *Progr. Math.*, pages 1–13. Birkhäuser Boston, Boston, MA, 1990.

[Haa92] U. Haagerup. Quasitraces on exact C^*-algebras are traces. Preprint, 1992.

[Haa79]   Uffe Haagerup. An example of a nonnuclear C^*-algebra, which has the metric approximation property. *Invent. Math.*, 50(3):279–293, 1978/79.

[Had87]   Don Hadwin. Strongly quasidiagonal C^*-algebras. *J. Operator Theory*, 18(1):3–18, 1987. With an appendix by Jonathan Rosenberg.

[Hal67]   Paul R. Halmos. *A Hilbert space problem book*. Van Nostrand, Princeton and Toronto, 1967. Second edition, revised and enlarged: Graduate Texts in Mathematics **19**, Springer, New York, 1982.

[Ham86]   M. Hamana. Injective envelopes and regular monotone completions of C^*-algebras. In *Geometric methods in operator algebras (Kyoto, 1983)*, volume 123 of *Pitman Res. Notes Math. Ser.*, pages 211–219. Longman Sci. Tech., Harlow, 1986.

[Han77]   Frank Hansen. Inner one-parameter groups acting on a factor. *Math. Scand.*, 41(1):113–116, 1977.

[Han81]   David Handelman. Homomorphisms of C^* algebras to finite AW^* algebras. *Michigan Math. J.*, 28(2):229–240, 1981.

[Har72]   Lawrence A. Harris. Banach algebras with involution and Möbius transformations. *J. Functional Analysis*, 11:1–16, 1972.

[Has58]   Morisuke Hasumi. The extension property of complex Banach spaces. *Tôhoku Math. J. (2)*, 10:135–142, 1958.

[HC57]   Harish-Chandra. Representations of semisimple Lie groups. In *Proceedings of the International Congress of Mathematicians, Amsterdam, 1954, Vol. 1*, pages 299–304. Erven P. Noordhoff N.V., Groningen, 1957.

[Hel93]   A. Ya. Helemskii. *Banach and locally convex algebras*. Oxford Science Publications. The Clarendon Press Oxford University Press, New York, 1993. Translated from the Russian by A. West.

[Hew64]   Edwin Hewitt. The ranges of certain convolution operators. *Math. Scand.*, 15:147–155, 1964.

[HHW67]   R. Haag, N. M. Hugenholtz, and M. Winnink. On the equilibrium states in quantum statistical mechanics. *Comm. Math. Phys.*, 5:215–236, 1967.

[Hig98]   Nigel Higson. The Baum-Connes conjecture. In *Proceedings of the International Congress of Mathematicians, Vol. II (Berlin, 1998)*, number Extra Vol. II, pages 637–646 (electronic), 1998.

[Hil43]   Einar Hille. Review of "On the embedding of normed rings into the ring of operators in Hilbert space," by I. Gelfand and M. Neumark. *Math. Reviews*, MR0009426 (5,147d), 1943.

[Hoc45]   G. Hochschild. On the cohomology groups of an associative algebra. *Ann. of Math. (2)*, 46:58–67, 1945.

[HP74]   Einar Hille and Ralph S. Phillips. *Functional analysis and semi-groups*. American Mathematical Society, Providence, R. I., 1974.

[HR63]   Edwin Hewitt and Kenneth A. Ross. *Abstract harmonic analysis. Vol. I: Structure of topological groups. Integration theory, group representations*. Die Grundlehren der mathematischen Wissenschaften, Bd. 115. Academic Press Inc., Publishers, New York, 1963.

[HR70]   Edwin Hewitt and Kenneth A. Ross. *Abstract harmonic analysis. Vol. II: Structure and analysis for compact groups. Analysis on locally compact Abelian groups*. Die Grundlehren der mathematischen Wissenschaften, Band 152. Springer-Verlag, New York, 1970.

[HR89]   Roger E. Howe and Jonathan Rosenberg. The unitary representation theory of $GL(n)$ of an infinite discrete field. *Israel J. Math.*, 67(1):67–81, 1989.

[HR93] Uffe Haagerup and Mikael Rørdam. C^*-algebras of unitary rank two. *J. Operator Theory*, 30(1):161–171, 1993.

[HR00] Nigel Higson and John Roe. Amenable group actions and the Novikov conjecture. *J. Reine Angew. Math.*, 519:143–153, 2000.

[HT67] Jôsuke Hakeda and Jun Tomiyama. On some extension properties of von Neumann algebras. *Tôhoku Math. J. (2)*, 19:315–323, 1967.

[HT99] U. Haagerup and S. Thorbjørnsen. Random matrices and K-theory for exact C^*-algebras. *Doc. Math.*, 4:341–450 (electronic), 1999.

[Hur79] Tadasi Huruya. An intersection result for tensor products of C^*-algebras. *Proc. Amer. Math. Soc.*, 75(1):186–187, 1979.

[Hus66] Dale Husemoller. *Fibre bundles*. McGraw-Hill, New York, 1966. Second edition, Graduate Texts in Mathematics, **20**, Springer, New York, 1975; third edition, 1994.

[HV84] Richard H. Herman and Leonid N. Vaserstein. The stable range of C^*-algebras. *Invent. Math.*, 77(3):553–555, 1984.

[HW41] Witold Hurewicz and Henry Wallman. *Dimension Theory*. Princeton Mathematical Series, v. 4. Princeton University Press, Princeton, N. J., 1941.

[HZ84] Uffe Haagerup and László Zsidó. Sur la propriété de Dixmier pour les C^*-algèbres. *C. R. Acad. Sci. Paris Sér. I Math.*, 298(8):173–176, 1984.

[Iór80] Valéria B. de Magalh aes Iório. Hopf C^*-algebras and locally compact groups. *Pacific J. Math.*, 87(1):75–96, 1980.

[Jac80] William Jaco. *Lectures on three-manifold topology*, volume 43 of *CBMS Regional Conference Series in Mathematics*. American Mathematical Society, Providence, R.I., 1980.

[JKR72] B. E. Johnson, R. V. Kadison, and J. R. Ringrose. Cohomology of operator algebras. III. Reduction to normal cohomology. *Bull. Soc. Math. France*, 100:73–96, 1972.

[Joh72] Barry Edward Johnson. *Cohomology in Banach algebras*. Number 127 in Memoirs of the Amer. Math. Soc. Amer. Math. Soc., Providence, 1972.

[Jon83] V. F. R. Jones. Index for subfactors. *Invent. Math.*, 72(1):1–25, 1983.

[Jon85] Vaughan F. R. Jones. A polynomial invariant for knots via von Neumann algebras. *Bull. Amer. Math. Soc. (N.S.)*, 12(1):103–111, 1985.

[Jon89] V. F. R. Jones. On knot invariants related to some statistical mechanical models. *Pacific J. Math.*, 137(2):311–334, 1989.

[Jon90] V. F. R. Jones. Notes on subfactors and statistical mechanics. *Internat. J. Modern Phys. A*, 5(3):441–460, 1990.

[JS97] V. Jones and V. S. Sunder. *Introduction to subfactors*, volume 234 of *London Mathematical Society Lecture Note Series*. Cambridge University Press, Cambridge, 1997.

[Kac63] G. I. Kac. Ring groups and the duality principle. *Trudy Moskov. Mat. Obšč.*, 12:259–301, 1963.

[Kac65] G. I. Kac. Annular groups and the principle of duality. II. *Trudy Moskov. Mat. Obšč.*, 13:84–113, 1965.

[Kad51] Richard V. Kadison. Isometries of operator algebras. *Ann. Of Math. (2)*, 54:325–338, 1951.

[Kad55] Richard V. Kadison. On the additivity of the trace in finite factors. *Proc. Nat. Acad. Sci. U. S. A.*, 41:385–387, 1955.

[Kad57] Richard V. Kadison. Irreducible operator algebras. *Proc. Nat. Acad. Sci. U.S.A.*, 43:273–276, 1957.

[Kad61] Richard V. Kadison. The trace in finite operator algebras. *Proc. Amer. Math. Soc.*, 12:973–977, 1961.

[Kad65] Richard V. Kadison. Transformations of states in operator theory and dynamics. *Topology*, 3(suppl. 2):177–198, 1965.

[Kad78] Richard V. Kadison. Similarity of operator algebras. *Acta Math.*, 141(3-4):147–163, 1978.

[Kal71] Robert R. Kallman. Groups of inner automorphisms of von Neumann algebras. *J. Functional Analysis*, 7:43–60, 1971.

[Kap51a] Irving Kaplansky. Projections in Banach algebras. *Ann. of Math. (2)*, 53:235–249, 1951.

[Kap51b] Irving Kaplansky. A theorem on rings of operators. *Pacific J. Math.*, 1:227–232, 1951.

[Kap53] Irving Kaplansky. Modules over operator algebras. *Amer. J. Math.*, 75:839–858, 1953.

[Kap68] Irving Kaplansky. *Rings of operators*. W. A. Benjamin, Inc., New York-Amsterdam, 1968.

[Kar78] Max Karoubi. *K-theory: An introduction*, volume 226 of *Grundlehren der Mathematischen Wissenschaften*. Springer, Berlin, 1978.

[Kas76] Daniel Kastler. Equilibrium states of matter and operator algebras. In *Symposia Mathematica, Vol. XX (Convegno sulle Algebre C^* e loro Applicazioni in Fisica Teorica, Convegno Sulla Teoria degli Operatori Indice e Teoria K, INDAM, Rome, 1975)*, pages 49–107. Academic Press, London, 1976.

[Kas80a] G. G. Kasparov. Hilbert C^*-modules: theorems of Stinespring and Voiculescu. *J. Operator Theory*, 4(1):133–150, 1980.

[Kas80b] G. G. Kasparov. The operator K-functor and extensions of C^*-algebras. *Izv. Akad. Nauk SSSR Ser. Mat.*, 44(3):571–636, 719, 1980. In Russian; translation in *Math. USSR Izvestija* **16** (1981), 513-572.

[Kas95] Christian Kassel. *Quantum groups*, volume 155 of *Graduate Texts in Mathematics*. Springer-Verlag, New York, 1995.

[Kat50] Miroslav Katětov. On rings of continuous functions and the dimension of compact spaces. *Časopis Pěst. Mat. Fys.*, 75:1–16, 1950.

[Kat04a] T. Katsura. C^*-algebras generated by scaling elements. Preprint, 2004.

[Kat04b] Takeshi Katsura. On C^*-algebras associated with C^*-correspondences. *J. Funct. Anal.*, 217(2):366–401, 2004.

[Kaž67] D. A. Každan. On the connection of the dual space of a group with the structure of its closed subgroups. *Funkcional. Anal. i Priložen.*, 1:71–74, 1967.

[Kel52] J. L. Kelley. Banach spaces with the extension property. *Trans. Amer. Math. Soc.*, 72:323–326, 1952.

[Kel75] John L. Kelley. *General topology*. Springer-Verlag, New York, 1975.

[Kir77a] Eberhard Kirchberg. C^*-nuclearity implies CPAP. *Math. Nachr.*, 76:203–212, 1977.

[Kir77b] Eberhard Kirchberg. Representations of coinvolutive Hopf-W^*-algebras and non-Abelian duality. *Bull. Acad. Polon. Sci. Sér. Sci. Math. Astronom. Phys.*, 25(2):117–122, 1977.

[Kir93] Eberhard Kirchberg. On nonsemisplit extensions, tensor products and exactness of group C^*-algebras. *Invent. Math.*, 112(3):449–489, 1993.

[Kir94]   Eberhard Kirchberg. Discrete groups with Kazhdan's property T and factorization property are residually finite. *Math. Ann.*, 299(3):551–563, 1994.

[Kir95a]   Eberhard Kirchberg. Exact C^*-algebras, tensor products, and the classification of purely infinite algebras. In *Proceedings of the International Congress of Mathematicians, Vol. 1, 2 (Zürich, 1994)*, pages 943–954, Basel, 1995. Birkhäuser.

[Kir95b]   Eberhard Kirchberg. On subalgebras of the CAR-algebra. *J. Funct. Anal.*, 129(1):35–63, 1995.

[Kir97]   Eberhard Kirchberg. On the existence of traces on exact stably projectionless simple C^*-algebras. In *Operator algebras and their applications (Waterloo, ON, 1994/1995)*, volume 13 of *Fields Inst. Commun.*, pages 171–172. Amer. Math. Soc., Providence, RI, 1997.

[Kis80]   Akitaka Kishimoto. Simple crossed products of C^*-algebras by locally compact abelian groups. *Yokohama Math. J.*, 28(1-2):69–85, 1980.

[KM46]   Shizuo Kakutani and George W. Mackey. Ring and lattice characterization of complex Hilbert space. *Bull. Amer. Math. Soc.*, 52:727–733, 1946.

[KP85]   Richard V. Kadison and Gert K. Pedersen. Means and convex combinations of unitary operators. *Math. Scand.*, 57(2):249–266, 1985.

[KP00]   Eberhard Kirchberg and N. Christopher Phillips. Embedding of exact C^*-algebras in the Cuntz algebra $\mathcal{O}_2$. *J. Reine Angew. Math.*, 525:17–53, 2000.

[KPR98]   Alex Kumjian, David Pask, and Iain Raeburn. Cuntz-Krieger algebras of directed graphs. *Pacific J. Math.*, 184(1):161–174, 1998.

[KR71a]   Richard V. Kadison and John R. Ringrose. Cohomology of operator algebras. I. Type I von Neumann algebras. *Acta Math.*, 126:227–243, 1971.

[KR71b]   Richard V. Kadison and John R. Ringrose. Cohomology of operator algebras. II. Extended cobounding and the hyperfinite case. *Ark. Mat.*, 9:55–63, 1971.

[KR97a]   Richard V. Kadison and John R. Ringrose. *Fundamentals of the theory of operator algebras. Vol. I*, volume 15 of *Graduate Studies in Mathematics*. American Mathematical Society, Providence, RI, 1997. Elementary theory, Reprint of the 1983 original.

[KR97b]   Richard V. Kadison and John R. Ringrose. *Fundamentals of the theory of operator algebras. Vol. II*, volume 16 of *Graduate Studies in Mathematics*. American Mathematical Society, Providence, RI, 1997. Advanced theory, Corrected reprint of the 1986 original.

[KR00]   Eberhard Kirchberg and Mikael Rørdam. Non-simple purely infinite C^*-algebras. *Amer. J. Math.*, 122(3):637–666, 2000.

[KR02]   Eberhard Kirchberg and Mikael Rørdam. Infinite non-simple C^*-algebras: absorbing the Cuntz algebras $\mathcal{O}_\infty$. *Adv. Math.*, 167(2):195–264, 2002.

[Kre49]   M. Kreĭn. A principle of duality for bicompact groups and quadratic block algebras. *Doklady Akad. Nauk SSSR (N.S.)*, 69:725–728, 1949.

[Kri76]   Wolfgang Krieger. On ergodic flows and the isomorphism of factors. *Math. Ann.*, 223(1):19–70, 1976.

[KS97]   Anatoli Klimyk and Konrad Schmüdgen. *Quantum groups and their representations.* Texts and Monographs in Physics. Springer-Verlag, Berlin, 1997.

[KT99]   Johan Kustermans and Lars Tuset. A survey of C^*-algebraic quantum groups. I. *Irish Math. Soc. Bull.*, (43):8–63, 1999.

[KT00]   Johan Kustermans and Lars Tuset. A survey of C^*-algebraic quantum groups. II. *Irish Math. Soc. Bull.*, (44):6–54, 2000.

[Kub57]   Ryogo Kubo. Statistical-mechanical theory of irreversible processes. I. General theory and simple applications to magnetic and conduction problems. *J. Phys. Soc. Japan*, 12:570–586, 1957.

[Kui65]   Nicolaas H. Kuiper. The homotopy type of the unitary group of Hilbert space. *Topology*, 3:19–30, 1965.

[KV53]   J. L. Kelley and R. L. Vaught. The positive cone in Banach algebras. *Trans. Amer. Math. Soc.*, 74:44–55, 1953.

[KV00]   Johan Kustermans and Stefaan Vaes. Locally compact quantum groups. *Ann. Sci. École Norm. Sup. (4)*, 33(6):837–934, 2000.

[KW99]   Eberhard Kirchberg and Simon Wassermann. Exact groups and continuous bundles of C^*-algebras. *Math. Ann.*, 315(2):169–203, 1999.

[Kye84]   Seung-Hyeok Kye. *On the Fubini products of C^*- algebras.* Seoul National University College of Natural Sciences, Seoul, 1984.

[Lac74]   H. Elton Lacey. *The isometric theory of classical Banach spaces.* Springer-Verlag, New York, 1974.

[Lan73]   Christopher Lance. On nuclear C^*-algebras. *J. Functional Analysis*, 12:157–176, 1973.

[Lan77]   Magnus B. Landstad. Duality for dual covariance algebras. *Comm. Math. Phys.*, 52(2):191–202, 1977.

[Lan79]   Magnus B. Landstad. Duality theory for covariant systems. *Trans. Amer. Math. Soc.*, 248(2):223–267, 1979.

[Lan82]   C. Lance. Tensor products and nuclear c^*-algebras. In R. V. Kadison, editor, *Operator Algebras and Applications, 1*, number 38 in Proc. Sympos. Pure Math., pages 379–399. Amer. Math. Soc., Providence, 1982.

[Lan95]   E. C. Lance. *Hilbert C^*-modules*, volume 210 of *London Mathematical Society Lecture Note Series*. Cambridge University Press, Cambridge, 1995.

[Loo52]   L. H. Loomis. Note on a theorem of Mackey. *Duke Math. J.*, 19:641–645, 1952.

[Lor97]   Terry A. Loring. *Lifting solutions to perturbing problems in C^*-algebras*, volume 8 of *Fields Institute Monographs*. American Mathematical Society, Providence, RI, 1997.

[LT77]   Joram Lindenstrauss and Lior Tzafriri. *Classical Banach spaces. I.* Springer-Verlag, Berlin, 1977.

[Lu88]   Ning Lu. Simple proofs of two theorems on the lens spaces. *Stud. Cerc. Mat.*, 40(4):297–303, 1988.

[Lus93]   George Lusztig. *Introduction to quantum groups*, volume 110 of *Progress in Mathematics*. Birkhäuser Boston Inc., Boston, MA, 1993.

[Mac49]   George W. Mackey. A theorem of Stone and von Neumann. *Duke Math. J.*, 16:313–326, 1949.

[MT05]   V.M. Manuilov and E.V. Troitsky. *Hilbert C^*-modules*, volume 226 of *Translations of Mathematical Monographs*. American Mathematical Society, Providence, RI, 2005.

[Mar75]   Odile Maréchal. Une remarque sur un théorème de Glimm. *Bull. Sci. Math. (2)*, 99(1):41–44, 1975.
[McD69]   Dusa McDuff. Uncountably many II_1 factors. *Ann. of Math. (2)*, 90:372–377, 1969.
[Miš79]   A. S. Miščenko. Banach algebras, pseudodifferential operators and their applications to K-theory. *Uspekhi Mat. Nauk*, 34(6(210)):67–79, 1979.
[MN94]   Tetsuya Masuda and Yoshiomi Nakagami. A von Neumann algebra framework for the duality of the quantum groups. *Publ. Res. Inst. Math. Sci.*, 30(5):799–850, 1994.
[MNW03]   T. Masuda, Y. Nakagami, and S. L. Woronowicz. A C^*-algebraic framework for quantum groups. *Internat. J. Math.*, 14(9):903–1001, 2003.
[Moo67]   Calvin C. Moore. Invariant measures on product spaces. In *Proc. Fifth Berkeley Sympos. Math. Statist. and Probability (Berkeley, Calif., 1965/66), Vol. II: Contributions to Probability Theory, Part 2*, pages 447–459. Univ. California Press, Berkeley, Calif., 1967.
[Moo76]   Calvin C. Moore. Group extensions and cohomology for locally compact groups. IV. *Trans. Amer. Math. Soc.*, 221(1):35–58, 1976.
[Mor58]   Kiiti Morita. Duality for modules and its applications to the theory of rings with minimum condition. *Sci. Rep. Tokyo Kyoiku Daigaku Sect. A*, 6:83–142, 1958.
[MP84]   J. A. Mingo and W. J. Phillips. Equivariant triviality theorems for Hilbert C^*-modules. *Proc. Amer. Math. Soc.*, 91(2):225–230, 1984.
[MS59]   Paul C. Martin and Julian Schwinger. Theory of many-particle systems. I. *Phys. Rev. (2)*, 115:1342–1373, 1959.
[Mun75]   James R. Munkres. *Topology: a first course*. Prentice-Hall Inc., Englewood Cliffs, N.J., 1975.
[Mur90]   Gerard J. Murphy. C^*-*algebras and operator theory*. Academic Press Inc., Boston, MA, 1990.
[MVN36]   F. J. Murray and J. Von Neumann. On rings of operators. *Ann. of Math. (2)*, 37(1):116–229, 1936.
[MvN37]   F. J. Murray and J. von Neumann. On rings of operators. II. *Trans. Amer. Math. Soc.*, 41(2):208–248, 1937.
[MvN43]   F. J. Murray and J. von Neumann. On rings of operators. IV. *Ann. of Math. (2)*, 44:716–808, 1943.
[Nac50]   Leopoldo Nachbin. A theorem of the Hahn-Banach type for linear transformations. *Trans. Amer. Math. Soc.*, 68:28–46, 1950.
[Nak50]   Hidegorô Nakano. Hilbert algebras. *Tôhoku Math. J. (2)*, 2:4–23, 1950.
[Nak77]   Yoshiomi Nakagami. Dual action on a von Neumann algebra and Takesaki's duality for a locally compact group. *Publ. Res. Inst. Math. Sci.*, 12(3):727–775, 1976/77.
[Neu28]   John von Neumann. Die zerlegung eines intervalles in abzählbar viele kongruente teilmengen. *Fund. Math.*, 11:230–238, 1928.
[Neu29]   John von Neumann. Zur allgemeinen theorie des masses. *Fund. Math.*, 13:73–116, 1929.
[Ole74]   Dorte Olesen. Derivations of AW^*-algebras are inner. *Pacific J. Math.*, 53:555–561, 1974.
[OW80]   Donald S. Ornstein and Benjamin Weiss. Ergodic theory of amenable group actions. I. The Rohlin lemma. *Bull. Amer. Math. Soc. (N.S.)*, 2(1):161–164, 1980.

[Oza00] Narutaka Ozawa. Amenable actions and exactness for discrete groups. *C. R. Acad. Sci. Paris Sér. I Math.*, 330(8):691–695, 2000.

[Oza03] N. Ozawa. Homotopy invariance of AF-embeddability. *Geom. Funct. Anal.*, 13(1):216–222, 2003.

[Pac94] Judith Packer. Transformation group C^*-algebras: a selective survey. In *C^*-algebras: 1943–1993 (San Antonio, TX, 1993)*, volume 167 of *Contemp. Math.*, pages 182–217. Amer. Math. Soc., Providence, RI, 1994.

[Pas73] William L. Paschke. Inner product modules over B^*-algebras. *Trans. Amer. Math. Soc.*, 182:443–468, 1973.

[Pat88] Alan L. T. Paterson. *Amenability*, volume 29 of *Mathematical Surveys and Monographs*. American Mathematical Society, Providence, RI, 1988.

[Pat92] Alan L. T. Paterson. Nuclear C^*-algebras have amenable unitary groups. *Proc. Amer. Math. Soc.*, 114(3):719–721, 1992.

[Pau84] Vern I. Paulsen. Every completely polynomially bounded operator is similar to a contraction. *J. Funct. Anal.*, 55(1):1–17, 1984.

[Pau02] Vern Paulsen. *Completely bounded maps and operator algebras*, volume 78 of *Cambridge Studies in Advanced Mathematics*. Cambridge University Press, Cambridge, 2002.

[Pea75] A. R. Pears. *Dimension theory of general spaces*. Cambridge University Press, Cambridge, England, 1975.

[Ped69] Gert Kjaergård Pedersen. Measure theory for C^* algebras. III, IV. *Math. Scand. 25 (1969)*, 71-93; ibid., 25:121–127, 1969.

[Ped72] Gert K. Pedersen. Operator algebras with weakly closed abelian subalgebras. *Bull. London Math. Soc.*, 4:171–175, 1972.

[Ped79] Gert K. Pedersen. *C^*-algebras and their automorphism groups*. Number 14 in London Mathematical Society Monographs. Academic Press, London, 1979.

[Ped80] Gert K. Pedersen. The linear span of projections in simple C^*-algebras. *J. Operator Theory*, 4(2):289–296, 1980.

[Ped89a] Gert K. Pedersen. *Analysis now*, volume 118 of *Graduate Texts in Mathematics*. Springer-Verlag, New York, 1989.

[Ped89b] Gert K. Pedersen. Three quavers on unitary elements in C^*-algebras. *Pacific J. Math.*, 137(1):169–179, 1989.

[Ped98] Gert K. Pedersen. Factorization in C^*-algebras. *Exposition. Math.*, 16(2):145–156, 1998.

[Ped99] Gert K. Pedersen. Pullback and pushout constructions in C^*-algebra theory. *J. Funct. Anal.*, 167(2):243–344, 1999.

[Phi87] N. Christopher Phillips. *Equivariant K-theory and freeness of group actions on C^*-algebras*, volume 1274 of *Lecture Notes in Mathematics*. Springer-Verlag, Berlin, 1987.

[Phi89] N. Christopher Phillips. *Equivariant K-theory for proper actions*. Number 178 in Pitman Research Notes in Mathematics Series. Longman, Harlow, UK, 1989.

[Phi01a] N. Christopher Phillips. Continuous-trace C^*-algebras not isomorphic to their opposite algebras. *Internat. J. Math.*, 12(3):263–275, 2001.

[Phi01b] N. Christopher Phillips. Recursive subhomogeneous C^*-algebras. Preprint, 2001.

[Phi04]   N. Christopher Phillips. A simple separable C^*-algebra not isomorphic to its opposite algebra. *Proc. Amer. Math. Soc.*, 132(10):2997–3005 (electronic), 2004.

[Pie84]   Jean-Paul Pier. *Amenable locally compact groups*. Pure and Applied Mathematics. John Wiley & Sons Inc., New York, 1984. A Wiley-Interscience Publication.

[Pim97]   Michael V. Pimsner. A class of C^*-algebras generalizing both Cuntz-Krieger algebras and crossed products by **Z**. In *Free probability theory (Waterloo, ON, 1995)*, volume 12 of *Fields Inst. Commun.*, pages 189–212. Amer. Math. Soc., Providence, RI, 1997.

[Pop86]   Sorin Popa. A short proof of "injectivity implies hyperfiniteness" for finite von Neumann algebras. *J. Operator Theory*, 16(2):261–272, 1986.

[Pop04]   Sorin Popa. On the fundamental group of type II_1 factors. *Proc. Natl. Acad. Sci. USA*, 101(3):723–726 (electronic), 2004.

[Pow67]   Robert T. Powers. Representations of uniformly hyperfinite algebras and their associated von Neumann rings. *Ann. of Math. (2)*, 86:138–171, 1967.

[Pow75]   Robert T. Powers. Simplicity of the C^*-algebra associated with the free group on two generators. *Duke Math. J.*, 42:151–156, 1975.

[PR88]   Ian F. Putnam and Mikael Rørdam. The maximum unitary rank of some C^*-algebras. *Math. Scand.*, 63(2):297–304, 1988.

[PS79]   William L. Paschke and Norberto Salinas. C^*-algebras associated with free products of groups. *Pacific J. Math.*, 82(1):211–221, 1979.

[PT73]   Gert K. Pedersen and Masamichi Takesaki. The Radon-Nikodym theorem for von Neumann algebras. *Acta Math.*, 130:53–87, 1973.

[Puk56]   L. Pukánszky. Some examples of factors. *Publ. Math. Debrecen*, 4:135–156, 1956.

[Put51]   C. R. Putnam. On normal operators in Hilbert space. *Amer. J. Math.*, 73:357–362, 1951.

[PV80]   M. Pimsner and D. Voiculescu. Exact sequences for K-groups and Ext-groups of certain cross-product C^*-algebras. *J. Operator Theory*, 4(1):93–118, 1980.

[PW90]   P. Podleś and S. L. Woronowicz. Quantum deformation of Lorentz group. *Comm. Math. Phys.*, 130(2):381–431, 1990.

[Răd94]   Florin Rădulescu. Random matrices, amalgamated free products and subfactors of the von Neumann algebra of a free group, of noninteger index. *Invent. Math.*, 115(2):347–389, 1994.

[RD66]   B. Russo and H. A. Dye. A note on unitary operators in C^*-algebras. *Duke Math. J.*, 33:413–416, 1966.

[Ren80]   Jean Renault. *A groupoid approach to C^*-algebras*, volume 793 of *Lecture Notes in Mathematics*. Springer, Berlin, 1980.

[Rie13]   F. Riesz. *Les systèmes d'équations linéaires à une infinité d'inconnues*. Gauthier-Villars, Paris, 1913.

[Rie72]   Marc A. Rieffel. On the uniqueness of the Heisenberg commutation relations. *Duke Math. J.*, 39:745–752, 1972.

[Rie74]   Marc A. Rieffel. Induced representations of C^*-algebras. *Advances in Math.*, 13:176–257, 1974.

[Rie81]   Marc A. Rieffel. C^*-algebras associated with irrational rotations. *Pacific J. Math.*, 93(2):415–429, 1981.

References

[Rie83]   Marc A. Rieffel. Dimension and stable rank in the K-theory of C^*-algebras. *Proc. London Math. Soc. (3)*, 46(2):301–333, 1983.

[Rie87]   Marc A. Rieffel. The homotopy groups of the unitary groups of non-commutative tori. *J. Operator Theory*, 17(2):237–254, 1987.

[Rie90]   Marc A. Rieffel. Proper actions of groups on C^*-algebras. In *Mappings of operator algebras (Philadelphia, PA, 1988)*, volume 84 of *Progr. Math.*, pages 141–182. Birkhäuser Boston, Boston, MA, 1990.

[RLL00] M. Rørdam, F. Larsen, and N. Laustsen. *An introduction to K-theory for C^*-algebras*, volume 49 of *London Mathematical Society Student Texts*. Cambridge University Press, Cambridge, 2000.

[RN67]   Czesław Ryll-Nardzewski. On fixed points of semigroups of endomorphisms of linear spaces. In *Proc. Fifth Berkeley Sympos. Math. Statist. and Probability (Berkeley, Calif., 1965/66), Vol. II: Contributions to Probability Theory, Part 1*, pages 55–61. Univ. California Press, Berkeley, Calif., 1967.

[Rob76] John E. Roberts. Cross products of von Neumann algebras by group duals. In *Symposia Mathematica, Vol. XX (Convegno sulle Algebre C^* e loro Applicazioni in Fisica Teorica, Convegno sulla Teoria degli Operatori Indice e Teoria K, INDAm, Rome, 1974)*, pages 335–363. Academic Press, London, 1976.

[Rob80] A. Guyan Robertson. Stable range in C^*-algebras. *Math. Proc. Cambridge Philos. Soc.*, 87(3):413–418, 1980.

[Rør88] Mikael Rørdam. Advances in the theory of unitary rank and regular approximation. *Ann. of Math. (2)*, 128(1):153–172, 1988.

[Rør92] Mikael Rørdam. On the structure of simple C^*-algebras tensored with a UHF-algebra. II. *J. Funct. Anal.*, 107(2):255–269, 1992.

[Rør98] Mikael Rørdam. On sums of finite projections. In *Operator algebras and operator theory (Shanghai, 1997)*, volume 228 of *Contemp. Math.*, pages 327–340. Amer. Math. Soc., Providence, RI, 1998.

[Rør02a] M. Rørdam. Classification of nuclear, simple C^*-algebras. In *Classification of nuclear C^*-algebras. Entropy in operator algebras*, volume 126 of *Encyclopaedia Math. Sci.*, pages 1–145. Springer, Berlin, 2002.

[Rør02b] M. Rørdam. Classification of nuclear, simple C^*-algebras. In *Classification of nuclear C^*-algebras. Entropy in operator algebras*, volume 126 of *Encyclopaedia Math. Sci.*, pages 1–145. Springer, Berlin, 2002.

[Rør03] Mikael Rørdam. A simple C^*-algebra with a finite and an infinite projection. *Acta Math.*, 191(1):109–142, 2003.

[Ros58] M. Rosenblum. On a theorem of Fuglede and Putnam. *J. London Math. Soc.*, 33:376–377, 1958.

[Ros76] Shmuel Rosset. A new proof of the Amitsur-Levitski identity. *Israel J. Math.*, 23(2):187–188, 1976.

[Ros77] Jonathan Rosenberg. Amenability of crossed products of C^*-algebras. *Comm. Math. Phys.*, 57(2):187–191, 1977.

[Ros79] Jonathan Rosenberg. Appendix to: "Crossed products of UHF algebras by product type actions" [Duke Math. J. **46** (1979), no. 1, 1–23; MR 82a:46063 above] by O. Bratteli. *Duke Math. J.*, 46(1):25–26, 1979.

[Ros82] Jonathan Rosenberg. Homological invariants of extensions of C^*-algebras. In *Operator algebras and applications, Part 1 (Kingston, Ont., 1980)*, volume 38 of *Proc. Sympos. Pure Math.*, pages 35–75. Amer. Math. Soc., Providence, RI, 1982.

[Ros94]   Jonathan Rosenberg. *Algebraic K-theory and its applications*, volume 147 of *Graduate Texts in Mathematics*. Springer-Verlag, New York, 1994.

[Ros04]   Jonathan Rosenberg. A selective history of the Stone-von Neumann theorem. In *Operator algebras, quantization, and noncommutative geometry*, volume 365 of *Contemp. Math.*, pages 331–353. Amer. Math. Soc., Providence, RI, 2004.

[RS87]   Jonathan Rosenberg and Claude Schochet. The Künneth theorem and the universal coefficient theorem for Kasparov's generalized K-functor. *Duke Math. J.*, 55(2):431–474, 1987.

[RSN55]   Frigyes Riesz and Béla Sz.-Nagy. *Functional analysis*. Frederick Ungar Publishing Co., New York, 1955.

[Rua88]   Zhong-Jin Ruan. Subspaces of C^*-algebras. *J. Funct. Anal.*, 76(1):217–230, 1988.

[Run02]   Volker Runde. *Lectures on amenability*, volume 1774 of *Lecture Notes in Mathematics*. Springer-Verlag, Berlin, 2002.

[RvD77]   Marc A. Rieffel and Alfons van Daele. A bounded operator approach to Tomita-Takesaki theory. *Pacific J. Math.*, 69(1):187–221, 1977.

[RW93]   Iain Raeburn and Dana P. Williams. Topological invariants associated with the spectrum of crossed product C^*-algebras. *J. Funct. Anal.*, 116(2):245–276, 1993.

[RW98]   Iain Raeburn and Dana P. Williams. *Morita equivalence and continuous-trace C^*-algebras*, volume 60 of *Mathematical Surveys and Monographs*. American Mathematical Society, Providence, RI, 1998.

[Sai79]   Kazuyuki Saitô. AW^*-algebras with monotone convergence property and examples by Takenouchi and Dyer. *Tôhoku Math. J. (2)*, 31(1):31–40, 1979.

[Sak56]   Shôichirô Sakai. A characterization of W^*-algebras. *Pacific J. Math.*, 6:763–773, 1956.

[Sak58]   Shôichirô Sakai. On linear functionals of W^*-algebras. *Proc. Japan Acad.*, 34:571–574, 1958.

[Sak60]   Shôichirô Sakai. On a conjecture of Kaplansky. *Tôhoku Math. J. (2)*, 12:31–33, 1960.

[Sak65]   Shôichirô Sakai. A Radon-Nikodým theorem in W^*-algebras. *Bull. Amer. Math. Soc.*, 71:149–151, 1965.

[Sak66]   Shôichirô Sakai. Derivations of W^*-algebras. *Ann. of Math. (2)*, 83:273–279, 1966.

[Sak68]   Shôichirô Sakai. Derivations of simple C^*-algebras. *J. Functional Analysis*, 2:202–206, 1968.

[Sak69]   Shôichirô Sakai. Asymptotically abelian II_1-factors. *Publ. Res. Inst. Math. Sci. Ser. A*, 4:299–307, 1968/1969.

[Sak71]   Shôichirô Sakai. *C^*-algebras and W^*-algebras*. Classics in Mathematics. Springer-Verlag, Berlin, 1971. 1998 Reprint of 1971 Original.

[Sak91]   Shôichirô Sakai. *Operator algebras in dynamical systems*, volume 41 of *Encyclopedia of Mathematics and its Applications*. Cambridge University Press, Cambridge, 1991. The theory of unbounded derivations in C^*-algebras.

[Saw68]   Parfeny P. Saworotnow. A generalized Hilbert space. *Duke Math. J.*, 35:191–197, 1968.

[Sch52] J. A. Schatz. Review of "On a theorem of Gelfand and Neumark and the B*-algebra," by M. Fukamiya. *Math. Reviews*, MR0054176 (14,884a), 1952.
[Sch63] J. Schwartz. Two finite, non-hyperfinite, non-isomorphic factors. *Comm. Pure Appl. Math.*, 16:19–26, 1963.
[Sch82] Claude Schochet. Topological methods for C^*-algebras. II. Geometry resolutions and the Künneth formula. *Pacific J. Math.*, 98(2):443–458, 1982.
[Sch93] Herbert Schröder. *K-theory for real C^*-algebras and applications*, volume 290 of *Pitman Research Notes in Mathematics Series*. Longman Scientific & Technical, Harlow, 1993.
[Seg47] I. E. Segal. Irreducible representations of operator algebras. *Bull. Amer. Math. Soc.*, 53:73–88, 1947.
[Seg50] I. E. Segal. An extension of Plancherel's formula to separable unimodular groups. *Ann. of Math. (2)*, 52:272–292, 1950.
[Seg53] I. E. Segal. A non-commutative extension of abstract integration. *Ann. of Math. (2)*, 57:401–457, 1953.
[Spa66] Edwin H. Spanier. *Algebraic topology*. McGraw-Hill, New York, 1966. Corrected reprint, Springer, New York, 1981.
[Spi88] John S. Spielberg. Embedding C^*-algebra extensions into AF algebras. *J. Funct. Anal.*, 81(2):325–344, 1988.
[Spi01] John Spielberg. Semiprojectivity for certain purely infinite c*-algebras. Preprint, 2001.
[Sri91] V. Srinivas. *Algebraic K-theory*, volume 90 of *Progress in Mathematics*. Birkhäuser Boston Inc., Boston, MA, 1991.
[Sti55] W. Forrest Stinespring. Positive functions on C^*-algebras. *Proc. Amer. Math. Soc.*, 6:211–216, 1955.
[Sti59] W. Forrest Stinespring. Integration theorems for gages and duality for unimodular groups. *Trans. Amer. Math. Soc.*, 90:15–56, 1959.
[Sto30] Marshall H. Stone. Linear transformations in Hilbert space III. operational methods and group theory. *Proc. Nat. Acad. Sci*, 16:172–175, 1930.
[Sto32] M. H. Stone. On one-parameter unitary groups in Hilbert space. *Ann. of Math. (2)*, 33(3):643–648, 1932.
[Str81] Şerban Strătilă. *Modular theory in operator algebras*. Editura Academiei Republicii Socialiste România, Bucharest, 1981. Translated from the Romanian by the author.
[SVZ76] Serban Strătilă, Dan Voiculescu, and László Zsidó. Sur les produits croisés. *C. R. Acad. Sci. Paris Sér. A-B*, 282(15):Ai, A779–A782, 1976.
[SVZ77] Şerban Strătilă, Dan Voiculescu, and László Zsidó. On crossed products. II. *Rev. Roumaine Math. Pures Appl.*, 22(1):83–117, 1977.
[Swe69] Moss E. Sweedler. *Hopf algebras*. Mathematics Lecture Note Series. W. A. Benjamin, Inc., New York, 1969.
[SZ79] Şerban Strătilă and László Zsidó. *Lectures on von Neumann algebras*. Editura Academiei, Bucharest, 1979. Revision of the 1975 original, Translated from the Romanian by Silviu Teleman.
[Sza81] Andrzej Szankowski. $B(\mathcal{H})$ does not have the approximation property. *Acta Math.*, 147(1-2):89–108, 1981.
[Tak64] Masamichi Takesaki. On the cross-norm of the direct product of C^*-algebras. *Tôhoku Math. J. (2)*, 16:111–122, 1964.

[Tak70]	M. Takesaki. *Tomita's theory of modular Hilbert algebras and its applications*. Lecture Notes in Mathematics, Vol. 128. Springer-Verlag, Berlin, 1970.
[Tak72]	Masamichi Takesaki. Duality and von Neumann algebras. In *Lectures on operator algebras; Tulane Univ. Ring and Operator Theory Year, 1970–1971, Vol. II; (dedicated to the memory of David M. Topping)*, pages 665–786. Lecture Notes in Math., Vol. 247. Springer, Berlin, 1972.
[Tak73]	Masamichi Takesaki. Duality for crossed products and the structure of von Neumann algebras of type III. *Acta Math.*, 131:249–310, 1973.
[Tak75]	Hiroshi Takai. On a duality for crossed products of C^*-algebras. *J. Functional Analysis*, 19:25–39, 1975.
[Tak78]	Osamu Takenouchi. A non-W^*, AW^*-factor. In C^*-*algebras and applications to physics (Proc. Second Japan-USA Sem., Los Angeles, Calif., 1977)*, volume 650 of *Lecture Notes in Math.*, pages 135–139. Springer, Berlin, 1978.
[Tak02]	M. Takesaki. *Theory of operator algebras. I*, volume 124 of *Encyclopaedia of Mathematical Sciences*. Springer-Verlag, Berlin, 2002. Reprint of the first (1979) edition, Operator Algebras and Non-commutative Geometry, 5.
[Tak03a]	M. Takesaki. *Theory of operator algebras. II*, volume 125 of *Encyclopaedia of Mathematical Sciences*. Springer-Verlag, Berlin, 2003. Operator Algebras and Non-commutative Geometry, 6.
[Tak03b]	M. Takesaki. *Theory of operator algebras. III*, volume 127 of *Encyclopaedia of Mathematical Sciences*. Springer-Verlag, Berlin, 2003. Operator Algebras and Non-commutative Geometry, 8.
[Tan39]	T. Tannaka. Über den dualitätssatz der nichtkommutativen topologischen gruppen. *Tôhoku Math. J.*, 45:1–12, 1939.
[Tat67]	Nobuhiko Tatsuuma. A duality theorem for locally compact groups. *J. Math. Kyoto Univ.*, 6:187–293, 1967.
[Tom57]	Jun Tomiyama. On the projection of norm one in W^*-algebras. *Proc. Japan Acad.*, 33:608–612, 1957.
[Tom59]	Jun Tomiyama. On the projection of norm one in W^*-algebras. III. *Tôhoku Math. J. (2)*, 11:125–129, 1959.
[Tom63]	Jun Tomiyama. A characterization of C^*-algebras whose conjugate spaces are separable. *Tôhoku Math. J. (2)*, 15:96–102, 1963.
[Tom67a]	Minoru Tomita. On canonical forms of von Neumann algebras. In *Fifth Functional Analysis Sympos. (Tôhoku Univ., Sendai, 1967) (Japanese)*, pages 101–102. Math. Inst., Tôhoku Univ., Sendai, 1967.
[Tom67b]	Jun Tomiyama. Applications of Fubini type theorem to the tensor products of C^*-algebras. *Tôhoku Math. J. (2)*, 19:213–226, 1967.
[Tom82]	Jun Tomiyama. On the difference of n-positivity and complete positivity in C^*-algebras. *J. Funct. Anal.*, 49(1):1–9, 1982.
[Tur52]	Takasi Turumaru. On the direct-product of operator algebras. I. *Tôhoku Math. J. (2)*, 4:242–251, 1952.
[Val85]	Jean-Michel Vallin. C^*-algèbres de Hopf et C^*-algèbres de Kac. *Proc. London Math. Soc. (3)*, 50(1):131–174, 1985.
[Val02]	Alain Valette. *Introduction to the Baum-Connes conjecture*. Lectures in Mathematics ETH Zürich. Birkhäuser Verlag, Basel, 2002.
[Vas69]	L. N. Vaserstein. On the stabilization of the general linear group over a ring. *Math. USSR-Sb.*, 8:383–400, 1969.

[vD82]   Alfons van Daele. Celebration of Tomita's theorem. In *Operator algebras and applications, Part 2 (Kingston, Ont., 1980)*, volume 38 of *Proc. Sympos. Pure Math.*, pages 1–4. Amer. Math. Soc., Providence, R.I., 1982.

[VD96]   A. Van Daele. Discrete quantum groups. *J. Algebra*, 180(2):431–444, 1996.

[VD98]   A. Van Daele. An algebraic framework for group duality. *Adv. Math.*, 140(2):323–366, 1998.

[VDZ00]   Alfons Van Daele and Yinhuo Zhang. The duality theorem for coactions of multiplier Hopf algebras. In *Interactions between ring theory and representations of algebras (Murcia)*, volume 210 of *Lecture Notes in Pure and Appl. Math.*, pages 413–422. Dekker, New York, 2000.

[Vil98]   Jesper Villadsen. Simple C^*-algebras with perforation. *J. Funct. Anal.*, 154(1):110–116, 1998.

[Vil99]   Jesper Villadsen. On the stable rank of simple C^*-algebras. *J. Amer. Math. Soc.*, 12(4):1091–1102, 1999.

[Vil02]   Jesper Villadsen. Comparison of projections and unitary elements in simple C^*-algebras. *J. Reine Angew. Math.*, 549:23–45, 2002.

[VK74]   L. I. Vaĭnerman and G. I. Kac. Nonunimodular ring groups and Hopf-von Neumann algebras. *Mat. Sb. (N.S.)*, 94(136):194–225, 335, 1974.

[vN30a]   John von Neumann. Allgemeine Eigenwerttheorie Hermitescher Funktionaloperatoren. *Math. Ann.*, 102:49–131, 1930.

[vN30b]   John von Neumann. Zur Algebra der Funktionaloperatoren und Theorie der normalen Operatoren. *Math. Ann.*, 102:370–427, 1930.

[vN31]   John von Neumann. Die Eindeutigkeit der Schrödingerschen Operatoren. *Math. Ann.*, 104:570–578, 1931.

[Vn71]   L. N. Vaseršteĭ n. The stable range of rings and the dimension of topological spaces. *Funkcional. Anal. i Priložen.*, 5(2):17–27, 1971.

[Voi76]   Dan Voiculescu. A non-commutative Weyl-von Neumann theorem. *Rev. Roumaine Math. Pures Appl.*, 21(1):97–113, 1976.

[Voi91]   Dan Voiculescu. A note on quasi-diagonal C^*-algebras and homotopy. *Duke Math. J.*, 62(2):267–271, 1991.

[Voi00]   Dan Voiculescu. Lectures on free probability theory. In *Lectures on probability theory and statistics (Saint-Flour, 1998)*, volume 1738 of *Lecture Notes in Math.*, pages 279–349. Springer, Berlin, 2000.

[VVD01]   Stefaan Vaes and Alfons Van Daele. Hopf C^*-algebras. *Proc. London Math. Soc. (3)*, 82(2):337–384, 2001.

[Wag93]   Stan Wagon. *The Banach-Tarski paradox*. Cambridge University Press, Cambridge, 1993. With a foreword by Jan Mycielski, Corrected reprint of the 1985 original.

[Wal03]   Martin E. Walter. Algebraic structures determined by 3 by 3 matrix geometry. *Proc. Amer. Math. Soc.*, 131(7):2129–2131 (electronic), 2003.

[War80]   R. B. Warfield, Jr. Cancellation of modules and groups and stable range of endomorphism rings. *Pacific J. Math.*, 91(2):457–485, 1980.

[Was76]   Simon Wassermann. The slice map problem for C^*-algebras. *Proc. London Math. Soc. (3)*, 32(3):537–559, 1976.

[Was77]   Simon Wassermann. Injective W^*-algebras. *Math. Proc. Cambridge Philos. Soc.*, 82(1):39–47, 1977.

[Was90]   Simon Wassermann. Tensor products of free-group C^*-algebras. *Bull. London Math. Soc.*, 22(4):375–380, 1990.

[Was91] Simon Wassermann. A separable quasidiagonal C^*-algebra with a non-quasidiagonal quotient by the compact operators. *Math. Proc. Cambridge Philos. Soc.*, 110(1):143–145, 1991.

[Was94] Simon Wassermann. *Exact C^*-algebras and related topics*, volume 19 of *Lecture Notes Series*. Seoul National University Research Institute of Mathematics Global Analysis Research Center, Seoul, 1994.

[Wea03] Nik Weaver. A prime C^*-algebra that is not primitive. *J. Funct. Anal.*, 203(2):356–361, 2003.

[Wen88] Hans Wenzl. Hecke algebras of type A_n and subfactors. *Invent. Math.*, 92(2):349–383, 1988.

[Wit84] G. Wittstock. Extension of completely bounded C^*-module homomorphisms. In *Operator algebras and group representations, Vol. II (Neptun, 1980)*, volume 18 of *Monogr. Stud. Math.*, pages 238–250. Pitman, Boston, MA, 1984.

[WO93] N. E. Wegge-Olsen. *K-theory and C^*-algebras*. Oxford Science Publications. The Clarendon Press Oxford University Press, New York, 1993.

[Woo73] E. J. Woods. The classification of factors is not smooth. *Canad. J. Math.*, 25:96–102, 1973.

[Woo82] E. J. Woods. ITPFI factors—a survey. In *Operator algebras and applications, Part 2 (Kingston, Ont., 1980)*, volume 38 of *Proc. Sympos. Pure Math.*, pages 25–41. Amer. Math. Soc., Providence, R.I., 1982.

[Wor87] S. L. Woronowicz. Twisted SU(2) group. An example of a noncommutative differential calculus. *Publ. Res. Inst. Math. Sci.*, 23(1):117–181, 1987.

[Wor96] S. L. Woronowicz. From multiplicative unitaries to quantum groups. *Internat. J. Math.*, 7(1):127–149, 1996.

[Wri76] J. D. Maitland Wright. Wild AW^*-factors and Kaplansky-Rickart algebras. *J. London Math. Soc. (2)*, 13(1):83–89, 1976.

[Wri80] J. D. Maitland Wright. On some problems of Kaplansky in the theory of rings of operators. *Math. Z.*, 172(2):131–141, 1980.

[Wri76] J. D. Maitland Wright. On AW^*-algebras of finite type. *J. London Math. Soc. (2)*, 12(4):431–439, 1975/76.

[Yea71] F. J. Yeadon. A new proof of the existence of a trace in a finite von Neumann algebra. *Bull. Amer. Math. Soc.*, 77:257–260, 1971.

[Yu00] Guoliang Yu. The coarse Baum-Connes conjecture for spaces which admit a uniform embedding into Hilbert space. *Invent. Math.*, 139(1):201–240, 2000.

[Zha90] Shuang Zhang. A property of purely infinite simple C^*-algebras. *Proc. Amer. Math. Soc.*, 109(3):717–720, 1990.

[Zim77] Robert J. Zimmer. Hyperfinite factors and amenable ergodic actions. *Invent. Math.*, 41(1):23–31, 1977.

[ZM69] G. Zeller-Meier. Deux nouveaux facteurs de type II$_1$. *Invent. Math.*, 7:235–242, 1969.

[Zsi75] László Zsidó. A proof of Tomita's fundamental theorem in the theory of standard von Neumann algebras. *Rev. Roumaine Math. Pures Appl.*, 20(5):609–619, 1975.

[Zsi78] László Zsidó. On the equality of two weights. *Rev. Roumaine Math. Pures Appl.*, 23(4):631–646, 1978.

Index

$(A \odot B)^*_+$ 182
$(A \odot B)^*_{++}$ 183
$(A \odot B)^d$ 182
$A \rtimes_\alpha G$ 201
$A \rtimes^r_\alpha G$ 203
$A \otimes \mathbb{K}$ 117
A_o^{-1} 56
A^G 134
$A^\dagger$ 53
A^{**} 319
A^{-1} 56
A^{op} 348
A_+ 63
A_+^c 86
A_+^f 86
A_ω 156
A_θ 160
A_q 72
A_{sa} 57
$C(A)$ 97
CA 97
$CP(X, Y)$ 131
$C_o(X)$ 52
$C^*(A, G, \alpha)$ 201
$C^*(G)$ 159, 198
$C^*(T)$ 48
$C^*(\mathcal{G}|\mathcal{R})$ 158
$C_r^*(A, G, \alpha)$ 203
$C_r^*(G)$ 159, 198
$C_1(A)$ 97
$E(A, B)$ 413
$EN(X)$ 175
$Ext(A, B)$ 172

$Ext(X)$ 174
$I(A)$ 325
$I_0(A)$ 325
I_a 88
I_a^o 88
$Int(\pi, \rho)$ 314
$J(A)$ 329
$J_0(A)$ 329
J_a 88
K-homology   174
$KK(A, B)$ 412
$K_0(A)$ 400
$K_1(A)$ 402
$K_1^{alg}(A)$ 403
$L(C)$ 89
L_ψ 190
$Lg_n(A)$ 445
$Lg_{(p,q)}(A)$ 446
$M(A)$ 144
$M(G)$ 194
M^ϕ 302
M_* 244
$M_\infty(A)$ 117
$M_n(A)$ 116
$N \bar{\rtimes}_\alpha G$ 283
N_ϕ 119
O_A 160
O_∞ 160
O_n 160
$Q(B)$ 168
$QT(A)$ 124
R_ϕ 190
$S(A)$ 97

Index

$S(G)$ 441
$S(M)$ 305
SA 97
$T(M)$ 303
T_+ 19
T_- 19
$Tr(T)$ 41
$V(A)$ 397
$VN(T)$ 48
$V_0(A)$ 397
$W(T)$ 11
$\text{Aff}(S(G))$ 442
$\text{Aut}(A)$ 95
$\mathcal{L}(\mathcal{E})$ 142
$\mathcal{L}(\mathcal{E}, \mathcal{F})$ 142
$\mathcal{L}(\mathcal{X})$ 5
$\mathcal{L}(\mathcal{X}, \mathcal{Y})$ 5
$\mathbf{Ext}(A, B)$ 171
$\mathbb{C}G$ 193
Δ_G 193
Δ_ϕ 294
$\mathcal{E}^\infty$ 139
$\text{GL}_1(A)$ 56
$\text{GL}_1(A)_\circ$ 56
$\text{GL}_\infty(A)$ 402
$\text{GL}_n(A)$ 402
$\Gamma(\alpha)$ 282
$\text{Hom}(A, B)$ 95
$\text{In}(A)$ 97
$\text{Inn}(A)$ 98
$\text{Irr}_\mathcal{H}(A)$ 115
$\mathbb{K}$ 52
Λ_I 80
$\mathbb{M}_n$ 52
$\text{Prim}(A)$ 59, 111
$\text{Prim}^J(A)$ 112
$\text{Prim}_J(A)$ 112
$\text{Proj}(\mathcal{H})$ 20
$\Sigma(A)$ 435
$\Theta_{\xi, \eta}$ 142
α-hereditary 93
$\mathbf{KK}$ 414
$\bar\otimes$ 233
$\mathbb{F}_n$ 159
$\bigoplus_{i \in \Omega} A_i$ 155
$\mathcal{C}(\mathcal{X}, \mathcal{Y})$ 36
$\mathcal{H}^\infty$ 4
$\mathcal{H}^n$ 4
$\mathcal{H}_B$ 139

$\mathcal{H}_\phi$ 107
$\mathcal{K}(\mathcal{E})$ 142
$\mathcal{K}(\mathcal{X}, \mathcal{Y})$ 36
$\mathcal{K}(\mathcal{H})$ 36
$\mathcal{M}(\mathfrak{N})$ 92
$\mathcal{N}$ 374, 414
$\mathcal{N}(T)$ 7
$\mathcal{P}(A)$ 105
$\mathcal{R}$ 277
$\mathcal{R}(T)$ 7
$\mathcal{R}_1$ 277
$\mathcal{R}_\lambda$ 277
$\mathcal{R}_{0,1}$ 277
$\mathcal{S}(A)$ 104
$\mathcal{T}$ 160
$\mathcal{T}(A)$ 123
$\mathcal{U}(A)$ 58
$\mathcal{U}(A)_\circ$ 58
$\mathcal{U}(\mathcal{H})$ 16
$\mathcal{U}_\infty(A)$ 402
$\mathcal{U}_n(A)$ 402
$\check{x}$ 113
$\delta(A)$ 347
ℓ_ϵ 78
$\mathfrak{F}_n$ 289
$\mathfrak{L}(G)$ 288
$\mathfrak{L}(\mathfrak{A})$ 297
$\mathfrak{M}_\phi$ 119
$\mathfrak{N}$ 374
$\mathfrak{N}_\phi$ 119
$\mathbf{m}(A)$ 333
$*A_i$ 162
$\gamma(M, M')$ 273
$\hat{A}_n$ 333
$\hat{A}$ 59, 115
$\hat{G}$ 196
$\hat{f}$ 197
$\hat{x}$ 333
$\hat\otimes$ 181
$\leq_a$ 183
$\ll$ 76
$\mu_n(T)$ 41
$\odot$ 180
$\otimes$ 184
$\otimes_A$ 148
$\otimes_{\max}$ 181
$\otimes_{\min}$ 180
$Ped(A)$ 86
$\perp$ 67

Index   507

$\perp\perp$ 67
$\pi \preceq \rho$ 314
π_ϕ 107
$\prod_{i\in\Omega} A_i$ 155
$\rho(M)$ 280, 287
$\precsim$ 73, 76
σ-strong operator topology   14, 44
σ-strong topology
    on a Hilbert module   144
σ-strong-* operator topology   14, 44
σ-unital C*-algebra   81
σ-weak operator topology   14, 43
σ-weak topology
    on a Hilbert module   144
$\sigma(T)$ 10
$\sigma_A(x)$ 54
$\sigma_N(T)$ 41
$\sigma_e(T)$ 37
$\tilde{A}$ 53
$\varinjlim A_i$ 156
$\overrightarrow{\xi_\phi}$ 107
$\|\cdot\|_C$ 385
$\|\cdot\|_J$ 112
$\|\cdot\|_{\mathrm{bin}}$ 360
$\|\cdot\|_{\mathrm{lnor}}$ 360
$\|\cdot\|_{\max}$ 181
$\|\cdot\|_{\min}$ 180
$\|\cdot\|_{\mathrm{rnor}}$ 360
$b \approx a$ 76
$b \ll a$ 76
$b \gtrapprox a$ 77
$b \precapprox a$ 77
$b \preceq a$ 77
$b \precsim a$ 76
$b \sim a$ 77
$b \simeq a$ 77
c 73
$c(T)$ 33
c_0 43
e^x 56
$exp(A)$ 56
$f * g$ 194
f_ϵ 77
g_ϵ 78
$index(T)$ 37
l^1 43
l^2 2
l^∞ 43
$p \sim q$ 73

$r(T)$ 11
$r(x)$ 54
$r_\infty(M)$ 279, 287
$rr(A)$ 452
$s(\phi)$ 254
$slg_n(A)$ 452
$sr(A)$ 445
$w(T)$ 11
x_+ 63
x_- 63
$\mathcal{C}(\mathcal{H})$ 43
$\mathcal{L}^1(\mathcal{H})$ 42
$\mathcal{L}^2(\mathcal{H})$ 42
$\mathcal{L}^p(\mathcal{H})$ 45
$\mathcal{S}'$ 10
$\mathcal{S}''$ 10
$_n\hat{A}$ 333
$|T|$ 19
$|x|$ 65
E-theory   412
KK-equivalence   414
KK-theory   412
K-homology   411

abelian
    element   323
    projection   227
action
    of a Hopf algebra   216
adjoint
    conjugate-linear operator   7
    Hilbert module operator   141
    operator   7
    unbounded operator   29
adjointable
    operator on a Hilbert module   141
    unbounded operator   29
AF algebra   157
    local sense   167
    nonseparable   167
AH algebra   420
amalgamated free product   162
amenable
    at infinity   394
    Banach algebra   378
    C*-algebra   100, 378
    group   195
    von Neumann algebra   365
amplification

of a Hilbert space 4
of a representation 101, 250
operator 9
annihilator
  of an ideal 95
  right 243
ANR 162
antiliminal C*-algebra 324
approximate unit 79
  almost idempotent 79
  continuous 79
  idempotent 79
  quasicentral 82
  sequential 79
approximate unitary equivalence 97
approximately finite dimensional 291
approximately homogeneous C*-algebra 420
approximately subhomogeneous C*-algebra 420
AR 162
Arveson Extension Theorem 129
Arveson spectrum 282
ASH algebra 420
asymptotic morphism 413
asymptotic ratio set 279
asymptotic unitary equivalence 97
Atkinson's theorem 38
automorphism 95
  approximately inner 97
  inner 97
  of a C*-algebra 95
  of a measure space 284
AW*-algebra 243

B*-algebra 52
Baire space 115
Banach *-algebra 51
Banach algebra 51
Banach limit 15
Banach module 99
  dual Banach A-module 99
  right 90
Baum-Connes Conjecture 394
Berg's theorem 176
bicommutant 10
Bicommutant Theorem 47
bidual Type I 323
block-diagonalizable operator 457

bootstrap class 414
  large 374, 414
  small 374
Bott map 407
Bott periodicity 407, 412
Bott projection 398
bounded operator 5
Brown-Douglas-Fillmore Theorem 175
Brown-Green-Rieffel Theorem 153
Busby invariant 168

C*-algebra 51
  $\aleph_0$-homogeneous 347
  σ-unital 81
  abstract 51
  algebraically simple 93
  amenable 100, 378
  antiliminal 324
  approximately homogeneous 420
  approximately subhomogeneous 420
  bidual Type I 323
  CCR 327
  concrete 47
  continuous trace 333
  contractible 96
  elementary 326
  exact 189, 383
  extremally rich 456
  finite 227
  GCR 327
  GCT 337
  generalized continuous trace 337
  group 159, 198
  GTC 337
  homogeneous 330
  infinite 227
  injective 129, 352
  inner quasidiagonal 460
  internally Type I 323
  liminal 327
  locally homogeneous 330
  NGCR 327
  nuclear 184, 368
  nuclearly embeddable 387
  postliminal 327
  prime 94
  projectionless 419
  projective 163
  properly infinite 227

purely infinite   160, 427
quasidiagonal   460
quasinuclear   188
reduced group   159, 198
residually finite   419
residually finite-dimensional   421
residually stably finite   419
seminuclear   188
semiprojective   162
simple   93
smooth   328, 340
stable   118
stably finite   419
stably unital   401
strongly amenable   382
strongly quasidiagonal   460
subcontractible   96
subhomogeneous   330
transformation group   205
Type I_0 324
Type I   323
universal   158
weakly amenable   382
C*-axiom   51
C*-bundle   344
C*-dynamical system   199
Calkin algebra   37
cancellation   438
    strict   438
canonical group of outer automorphisms   303
CAR algebra   157
carrier
    central   223
    of a normal weight   254
Cayley transform   33
CBS inequality   1
    for Hilbert modules   138
    for states   104
CCR
    C*-algebra   327
    representation   327
central carrier   223
central decomposition
    of a representation   317
    of a von Neumann algebra   239
central sequence   289
centralizer
    of a weight   302

characteristic list   41
characteristic number   41
Chern character   410
Choi-Effros Lifting Theorem   375
closable operator   29
closed
    operator   29
Closed Graph Theorem   6
coisometry   57
commutant   10
Commutation Theorem   301
compact operator   36
complete contraction   131
complete order embedding   125, 130
complete order isomorphism   130
completely bounded map   131
completely continuous   36
completely hereditary   93
completely positive approximation
    property   371
completely positive map   125
composition series   325
compression   125
conditional expectation   132
cone   64
cone over a C*-algebra   97
Connes spectrum   282
Connes' Thom Isomorphism   408
continuous decomposition   310
continuous field of C*-algebras   340
continuous trace
    C*-algebra   333
    element   333
contractible C*-algebra   96
convolution   194
core (for unbounded operators)   28
corner   147
correspondence   148
coupling constant   274
coupling function   273
covariant representation   200
covariant system   199
CPAP   371
cross norm   187
crossed product
    of C*-algebras   201
        reduced   203
    of von Neumann algebras   283
Cuntz algebra   160

Cuntz-Krieger algebra 161
cyclic vector 107, 226

Dauns-Hoffman Theorem 114
densely defined operator 27
derivation 99
  *-derivation 99
  antihermitian 99
  closed 99
  inner 99
diagonalizable operator 457
dimension function 123
dimension group 440
direct integral
  of Hilbert spaces 238
  of representations 316–318
  of von Neumann algebras 239
direct product of C*-algebras 154
direct sum
  of C*-algebras 154, 155
  of Hilbert modules 139
  of Hilbert spaces 4
discrete decomposition 311
discrete Heisenberg group 344
Dixmier property 264–265
  strong 264
Dixmier trace 47
Dixmier-Douady invariant 347
dominant weight 312
double centralizer 144
dual Banach A-module 99
dual group 196
dual weight 284

Effros-Handelman-Shen Theorem 440
eigenvalue list
  of a compact operator 41
  of a state 278
element with closed range 69
elementary C*-algebra 326
equivalence
  of extensions 170–171
  of projections 73
  of representations 101
essential spectrum 37
essential subspace 101
essentially normal operator 175
exact C*-algebra 189, 383
exact group 393

exact sequence 167
exponential map (of K-theory) 407
extension of C*-algebras 167
  absorbing 174
  essential 168
  semisplit 173
  split 172
  trivial 172
extremally rich C*-algebra 456

factor 48
  approximately finite dimensional 291
  free group 290
  hyperfinite 291
  injective 352
  ITPFI 278
  Krieger 287
  Powers 277
  Property T 291
  semidiscrete 363
  Type I 232
  Type I_∞ 236
  Type I_n 235
  Type II 232
  Type II_1 232
  Type II_∞ 232
  Type III 232
  Type III_0 280, 306
  Type III_1 280, 306
  Type III_λ 280, 306
  wild 244
factor representation 102
Fell's condition 334
finite ascent 38
finite descent 38
flow of weights 312
Fourier-Plancherel transform 197
Fredholm Alternative 39
Fredholm index 37
Fredholm operator 37
free group factor 290
free product 162
  amalgamated 162
  reduced 162
  unital 162
FS Property 453
Fuglede's Theorem 26, 110
  approximate 155

full element 91
full group C*-algebra 198
full hereditary C*-subalgebra 91
functional calculus
  Borel 19, 25
  continuous
    bounded operators 18–19
    for C*-algebras 61–62
  holomorphic 55–56
  unbounded self-adjoint operators 34–35
fundamental group 290
Følner condition 195

GCR
  C*-algebra 327
  representation 327
Gelfand transform 60
Gelfand-Mazur Theorem 55
Gelfand-Naimark Theorem 60, 109
Generalized Comparability 224
generalized continuous trace 337
generalized inductive limit 466
generalized inductive system 465
generalized integers 400
generalized trace 311
Glimm's Theorem 339
global rank-one projection 335
GNS construction
  for positive linear functionals 107
  for weights 119–120
GNS representation 107
Grothendieck group 399
group
  amenable 195
  amenable at infinity 394
  discrete Heisenberg 344
  exact 393
  Grothendieck 399
  ICC 288
  locally compact 193
  ordered 435
  Pontrjagin dual 196
  preordered 434
  Property T 290
  quantum 218
  unimodular 193
group action
  ergodic 285
  essentially free 285
  essentially transitive 285
  on a C*-algebra 199
  on a measure space 284
  on a topological space 192
  on a von Neumann algebra 281
  orbit equivalence 286
group actions
  cocycle conjugate 204
  conjugate 204
  exterior equivalent 205
  outer conjugate 204
group algebra 193
group C*-algebra 159, 198
group measure space construction 284–288

Haar measure 193
Heisenberg commutation relations 206
hereditary C*-subalgebra 75
hereditary subalgebra
  full 91
Hilbert algebra 297
  generalized 296
  left 296
  modular 298
Hilbert bundle 344, 345
Hilbert module 138
  full 141
Hilbert space 2
  real 5
Hilbert-Schmidt operator 42
homogeneous C*-algebra 330
  $\aleph_0$-homogeneous 347
  n-homogeneous 330
homomorphism 95
homotopy equivalent C*-algebras 96
homotopy of homomorphisms 96
HP property 453
hyperfinite 291
hypertrace 359
hyponormal 424

ICC group 288
ideal 53
  A-CCR 327
  A-GCR 327
  essential 95
  left 89

Pedersen   86
primitive   111
strongly invariant   86
weakly closed   225
weighted   93
imaginary part
   of an element of a C*-algebra   57
imprimitivity bimodule   150
$A -_T C_o(T)$-imprimitivity bimodule   346
index map   405
inductive limit   156
   generalized   466
inductive system
   generalized   465
infinite element   426
injective C*-algebra   129, 352
inner product   1
   B-valued   137
inner quasidiagonal C*-algebra   460
integrable weight   312
internally Type I   323
intersection product   412
intertwiner   314
involution
   Banach algebra   51
   Hilbert space   8
   unbounded   296
irrational rotation algebra   160, 208
isometry
   in a C*-algebra   57
   operator   8
ITPFI factor   278

Künneth Theorem   416
Künneth Theorem for Tensor Products   417
Kac algebra   215
Kadison Transitivity Theorem   103
Kadison's inequality   129
Kaplansky Density Theorem   48
Kasparov product   412
KK-theory   412
KMS condition   306
Krieger factor   287

large bootstrap class   374, 414
left centralizer   145
left Hilbert algebra   296

full   298
lifting property   375
liminal C*-algebra   327
linear functional
   normal   245
   positive   104
linking algebra   152
local AF algebra   167

Macaev ideal   46
Mackey Borel structure   115
mapping cone   170
masa   221
matrix ordered space   131
matrix units   117, 159, 232
   approximate   166
   of type D 160
measurable field
   of bounded operators   238
   of Hilbert spaces   238
measure algebra   194
metric approximation property   371
MF algebra   466
modular condition   306
modular function   193
modular Hilbert algebra   298
modular operator   294, 299
Modular Theory   293–313
modulus   290
Morita equivalence   152
multiple
   of a representation   101
multiplication operator   8
multiplier algebra   144
Murray-von Neumann equivalence   73

NF algebra   466
   strong   466
NF system   466
noncommutative torus   160, 213
nondegenerate
   homomorphism   146
   representation   101
nonstable K-theory   403
normal
   element of a C*-algebra   57
   linear functional   245
   map   248
   operator   8

weight   255
normal Stinespring theorem   249
nuclear C*-algebra   184, 368
nuclearly embeddable   387
numerical range   11

Open Mapping Theorem   6
operator
  block-diagonalizable   457
  bounded   5
  bounded below   10
  compact   36
  diagonalizable   457
  essentially normal   175
  Fredholm   37
  isometry   8
  multiplication   8
  normal   8
  positive   11
  projection   8
  quasidiagonal   457, 459
  self-adjoint   8
  semi-Fredholm   40
  unitary   8
operator algebra   47
operator space   106
operator system   106
opposite algebra   348
order embedding   125, 130
order unit   435
ordered group   435
  simple   435
  unperforated   439
  weakly unperforated   439
ordinary ordering   436
orthogonality
  in a C*-algebra   67
  in Hilbert space   2
orthomodular law   20
orthonormal basis   3
outer multiplier algebra   168

paracompact space   345
paradoxical decomposition   195
partial isometry
  in a C*-algebra   57
  operator   22
partially defined operator   27
partially liftable homomorphism   163

Pedersen ideal   86
perforation   439
permutable operators   28
Pimsner–Voiculescu exact sequence   408
Plancherel Theorem   197
Plancherel weight   288
point-norm topology   95
polar decomposition
  for an operator   21–23
  for an unbounded operator   35
  for normal linear functionals   257
  in a C*-algebra   67–69, 321
polarization   1, 65
polynomial relation   332
Pontrjagin dual group   196
positive
  element of a C*-algebra   63
  linear functional   104
  map   125
  operator   11
  unbounded operator   32
positively generated *-subalgebra   92
postliminal C*-algebra   327
Powers factor   277
pre-Hilbert module   137
pre-inner product   1
predual   244
preordered group   434
prime C*-algebra   94
primitive ideal   111
product
  of C*-algebras   155
projection
  abelian   227
  continuous   231
  countably decomposable   225
  cyclic   226
  discrete   231
  final   22, 63
  finite   227
  full   91
  global rank-one   335
  in a C*-algebra   57
  infinite   227
  initial   22, 63
  monic   261
  operator   8, 20
  orthogonal   67

properly infinite 227
purely infinite 231
range 22, 63
semifinite 231
source 22, 63
subordinate 73
very orthogonal 67
projection-valued measure 24
projective C*-algebra 163
projective tensor product 181
proper
  group action 207
properly infinite element 426
Property Γ 289
Property C 385
Property E 359
Property FS 453
Property HP 453
Property P 359
Property T 290
pullback 169
purely infinite C*-algebra 160, 427

quantum group 218
  finite 215
quasi-equivalence 314
quasi-invertible element 72
quasidiagonal
  C*-algebra 460
  operator 457, 459
  representation 460
  set of operators 458, 459
quasitrace 124

Radon-Nikodym derivative 307
Radon-Nikodym Theorem
  for normal positive linear functionals 258
  for positive linear functionals 108
  for weights 308
real part
  of an element of a C*-algebra 57
real rank 452
real rank zero 453
reduced crossed product 203
reduced group C*-algebra 159, 198
reduction of a representation 250
regular representation 159, 198
representation 101

amplification 250
CCR 327
covariant 200
disjoint 314
factor 102
faithful 101
finite multiplicity 315
GCR 327
GNS 107
irreducible 101
multiplicity-free 316
nondegenerate 101
of a group 197
quasi-equivalence 314
quasidiagonal 460
reduction 250
standard form 269
subordinate 314
symmetric form 300
type 315
universal 318
universal normal 321
residually finite-dimensional C*-algebra 421
resolvent 55
RF algebra 474
Riemann-Lebesgue lemma 197
Riesz Decomposition Property 68
Riesz Interpolation Property 441
Riesz-Fréchet theorem 3
right Banach module 90
rotation algebra 160
Russo-Dye Theorem 70

scale 435
scaled ordered group 435
scaled preordered K_0-group 435
scaling element 424
Schatten ideal 45
Schröder-Bernstein Theorem 223
self-adjoint
  element of a C*-algebra 57
  operator 8
  unbounded operator 30
semi-Fredholm operator 40
semiprojective C*-algebra 162
semisplit extension 173
separably inheritable property 176
separating vector 226

sesquilinear form 6
shape theory 162
SI property 176
simple C*-algebra 93
simple ordered group 435
six-term exact sequence 407
slice map
   for C*-algebras 190
   normal 250
small bootstrap class 374
space
   paracompact 345
spectral projection 24
spectral radius 11, 54
Spectral Theorem
   normal operators 25
   self-adjoint operators 24
   unbounded self-adjoint operators 33
spectrum
   of a commutative Banach algebra 60
   of an element of a Banach algebra 54
   of an operator 10
   of an unbounded operator 30
square root
   for positive operators 19
   in a C*-algebra 64
Stabilization Theorem 141
stable algebra 118
stable C*-algebra 118
stable rank 445
stable relations 166
stably finite C*-algebra 419
stably isomorphic C*-algebras 118
stably unital 401
standard form 269
standard measure space 284
state 104
   center-valued 260
   completely additive 245
   normal 245
   of a scaled ordered group 441
   pure 105
   singular 248
   tracial 121
   vector 104
state space 104, 441
Stinespring's Theorem 127

for Hilbert modules 150
for normal tensor products 361
for tensor products 190
normal 249
Stone's Theorem 35
Stone-von Neumann Theorem 206
strict cancellation 438
strict homomorphism 147
strict mapping 149
strict ordering 436
strict topology 144, 146
strictly positive element 81
strong NF algebra 466
strong NF system 466
strong operator topology 13
σ-strong operator topology 14, 44
strong-* operator topology 14
σ-strong-* operator topology 14, 44
strongly amenable C*-algebra 382
strongly quasidiagonal C*-algebra 460
subcomposition series 325
subcontractible C*-algebra 96
subfactor 274
subhomogeneous C*-algebra 330
subordinate 73
   representation 314
subprojection 73
subrepresentation 101
sum
   of C*-algebras 155
   of representations 101
support projection
   central 223
   in a C*-algebra 69
   of a normal weight 254
   operator 21
suspension of a C*-algebra 97
symmetric form 300
symmetric operator 30
symmetrically normed ideal 45

Takesaki duality 283
tensor product
   binormal 361
   infinite
      of C*-algebras 191
      of Hilbert spaces 275
      of von Neumann algebras 275–276
   normal 361

of C*-algebras 179–191
  maximal 181
  minimal 180
  spatial 180
of completely positive maps 190
of Hilbert modules 148
of Hilbert spaces 4
of homomorphisms 187
of normal completely positive maps 361
of normal maps 250
of normal semifinite weights 257
of von Neumann algebras 233
projective 181
Toeplitz algebra 160
Tomita algebra 298
Tomita-Takesaki Theorem 296
trace
  center-valued 260
  generalized 311
  of an operator 41
  on a C*-algebra 121
  semifinite 122
trace-class operator 42
transformation group C*-algebra 205
triangulation 475
  weak 475
twice-around embedding 336
two-out-of-three property 374
Type I
  C*-algebra 323
  representation 315
  von Neumann algebra 231

UCT 416
UHF algebra 157
ultraproduct 156
Uniform Boundedness Theorem 6
unilateral shift 9
unimodular group 193
unipotent 448
unitarily invariant cone 86
unitary
  in a C*-algebra 57
  operator 8
unitary extreme property 419
unitization 53
universal C*-algebra 158
Universal Coefficient Theorem 416

universal representation 318
unperforated group 439

very orthogonal 67
virtual diagonal 379
Voiculescu's Weyl-von Neumann Theorem 174
von Neumann algebra 47
  amenable 365
  approximately finite dimensional 291
  continuous 231
  continuous decomposition 310
  countably decomposable 225
  discrete 231
  finite 227
  hyperfinite 291
  infinite 227
  injective 352
  locally countably decomposable 226
  properly infinite 227
  purely atomic 354
  purely infinite 231
  semidiscrete 363
  semifinite 231
  Type I 231
  Type I_∞ 236
  Type I_n 235
  Type II 232
  Type II_1 232
  Type II_∞ 232
  Type III 231
von Neumann regular element 69

W*-algebra 243
W*-crossed product 283
W*-dynamical system 281
weak operator topology 13
  σ-weak operator topology 14, 43
weak triangulation 475
weakly amenable C*-algebra 382
weakly unperforated group 439
weight 118
  centralizer 302
  dominant 312
  faithful 118
  integrable 312
  normal 255
  Plancherel 288

semifinite 120
strictly semifinite 255
weighted ideal 93

well-supported element 69
Weyl-von Neumann Theorem 176
   Voiculescu's Noncommutative 174